Lecture Notes in Networks and Systems

Volume 5589

The series "Lecture Notes in Networks and Systems" publishes the latest developments in Networks and Systems—quickly, informally and with high quality. Original research reported in proceedings and post-proceedings represents the core of LNNS.

Volumes published in LNNS embrace all aspects and subfields of, as well as new challenges in, Networks and Systems.

The series contains proceedings and edited volumes in systems and networks, spanning the areas of Cyber-Physical Systems, Autonomous Systems, Sensor Networks, Control Systems, Energy Systems, Automotive Systems, Biological Systems, Vehicular Networking and Connected Vehicles, Aerospace Systems, Automation, Manufacturing, Smart Grids, Nonlinear Systems, Power Systems, Robotics, Social Systems, Economic Systems and other. Of particular value to both the contributors and the readership are the short publication timeframe and the world-wide distribution and exposure which enable both a wide and rapid dissemination of research output.

The series covers the theory, applications, and perspectives on the state of the art and future developments relevant to systems and networks, decision making, control, complex processes and related areas, as embedded in the fields of interdisciplinary and applied sciences, engineering, computer science, physics, economics, social, and life sciences, as well as the paradigms and methodologies behind them.

Indexed by SCOPUS, EI Compendex, INSPEC, WTI Frankfurt eG, zbMATH, SCImago.

All books published in the series are submitted for consideration in Web of Science.

For proposals from Asia please contact Aninda Bose (aninda.bose@springer.com).

Harish Sharma · Antorweep Chakravorty ·
Shahid Hussain · Rajani Kumari
Editors

Artificial Intelligence: Theory and Applications

Proceedings of AITA 2024, Volume 2

Springer

Editors
Harish Sharma
Department of Computer Science
and Engineering
Rajasthan Technical University
Kota, Rajasthan, India

Antorweep Chakravorty
Department of Computer Science
and Electrical Engineering
University of Stavanger
Stavanger, Norway

Shahid Hussain
Biomedical Robotics, Information,
Technology (IT) and Systems
University of Canberra
Bruce, ACT, Australia

Rajani Kumari
Off-Campus Centre of ICFAI Foundation
for Higher Education (IFHE) University
IBS Bangalore
Bengaluru, Karnataka, India

ISSN 2367-3370 ISSN 2367-3389 (electronic)
Lecture Notes in Networks and Systems
ISBN 978-981-96-1686-2 ISBN 978-981-96-1687-9 (eBook)
https://doi.org/10.1007/978-981-96-1687-9

This Springer imprint is published by the registered company Springer Nature Singapore Pte Ltd.
The registered company address is: 152 Beach Road, #21-01/04 Gateway East, Singapore 189721, Singapore

AITA 2024 Organizing Committee

General Chairs

Antorweep Chakravorty, University of Stavanger, Norway
Rajani Kumari, IBS, Bangalore, Off-Campus Center of IFHE University
Harish Sharma, Rajasthan Technical University, Kota, India
Shahid Hussain, Biomedical Robotics, University of Canberra, Australia

Organizing Chairs

Dr. Manisha Singh, IBS, Bangalore, Off-Campus Center of IFHE University
Rajani Kumari, IBS, Bangalore, Off-Campus Center of IFHE University

Program Chairs

P. K. Chandra Shekhar, IBS, Bangalore, Off-Campus Center of IFHE University
Shweta Puneet, IBS, Bangalore, Off-Campus Center of IFHE University
Leena Sidenur, IBS, Bangalore, Off-Campus Center of IFHE University
Raghavendra Rao, IBS, Bangalore, Off-Campus Center of IFHE University
Sandeep Kumar, CHRIST (Deemed to be University), Kengeri Campus, Bengaluru
Susheela Girisaballa, IBS, Bangalore, Off-Campus Center of IFHE University

Publicity Chairs

Aruna Bala, IBS, Bangalore, Off-Campus Center of IFHE University
S. Bharathi Gopal, IBS, Bangalore, Off-Campus Center of IFHE University
Harisankar Muralidharan, IBS, Bangalore, Off-Campus Center of IFHE University
B. Surjyabrat, IBS, Bangalore, Off-Campus Center of IFHE University
R. Srinivasan, IBS, Bangalore, Off-Campus Center of IFHE University

Technical Program Committee

Abdullah Umar Nasib, BRAC University, Bangladesh
Malaviya National Institute of Technology Jaipur
Md. Golam Rashed, University of Rajshahi, Bangladesh
Md. Tofael Ahmed, Comilla University, Bangladesh
Monali Prajapati, Gujarat Technological University, Gujarat, India
Monica Bhutani, Guru Gobind Singh Indraprastha University, New Delhi; Indian Institute of Technology, New Delhi
Nahid Fatima, Prince Sultan University Riyadh
Suhaib Ahmed, Model Institute of Engineering and Technology, Jammu
Suman Nehra, Bikaner Technical Education, India
Sunil, Jamia Millia Islamia (A Central University), New Delhi
Vivekananda Jayaram, Florida International University, USA
Kalpana Rai, Sagar Institute of Research and Technology, Bhopal
Ehab Saleh, Leibniz Supercomputing Centre, Germany
El Bachir Lahmer, Faculty of Sciences Oujda, Morocco
Eriona Çela, University of New York Tirana
Farhana Sharmin, Founder and CEO, Global Emerging Technology Network, USA
Fátima Rodrigues, ISEP, Instituto Politécnico do Porto, Portugal
Fernaz Narin Nur, Notre Damn University, Bangladesh
Kavita Choudhary, Rajasthan Technical University, India
Keerthana Madhavamoorthy, Thiagarajar College of Engineering, India
Keyur Mahant, CHARUSAT University, India
Kishore Kumar Perumalsamy, Adobe USA
Koushik Kumar Ganeeb, Salesforce Inc., USA
Krishna Kumar Mohbey, Central University of Rajasthan, Rajasthan, India
Kristy Tiampo, University of Colorado Boulder, USA
R. Krithiga, VIT Chennai, India
Kumar Rahul, NIFTEM, USA
Kyryl Korobchynskyi, National Aerospace University, Kharkov, Ukraine
Lakshmanan Sethu Sankaranarayanan, Google USA
Lav Kumar, Salesforce Inc., USA
Lei Kou, TU Dresden, Germany

Loong Chuen Lee, Universiti Kebangsaan Malaysia
Luiz Guerreiro Lopes, University of Madeira, Portugal
Madhu Kirola, Uttaranchal University, Dehradun, India
Manish Kumar, SRMIST Tiruchirappalli, India
Manjit Chakraborty, Amazon USA
Mayukh Maitra, Walmart, USA
Md. Afroz, Al Faisal University, Prince Sultan College of Business, Jeddah
M. Meharunnisa, Ethiraj College for Women, Tamil Nadu, India
Mhamad Bakro, Engineering and Technology Faculty, Cordoba Private University, Syria
Michael Moses, Christ University, Bengaluru, India
Mijanur Rahaman Seikh, Kazi Nazrul University, West Bengal, India
Milind Shah, SVIT Vasad, India
Mitali Nayak, ITER, Siksha 'O' Anusandhan, India
Mitul Kumar Ahirwal, Maulana Azad National Institute of Technology Bhopal, India
Mohd Naseem, Baba Ghulam Shah Badshah University, Rajouri, Jammu and Kashmir, India
Mohit Kumar Saini, Dev Bhomi Uttranchal University, India
Mohit Tiwari, Bharati Vidyapeeth's College of Engineering, Delhi
Moti Lal Rinawa, Government Engineering College Jhalawar, India
Dasari Anantha Reddy, KL University, Hyderabad, India
Mragank Kumar Yadav, Apple Inc., USA
Mrutyunjaya S. Yalawar, CMR Engineering College, Hyderabad
Mukesh Saraswat, Jaypee Institute of Information Technology, Noida
Nagarajan Madhavan, Birla Institute of Technology and Science, Pilani
Namrata Nagpal, Amity University Uttar Pradesh, India
Naveen Kumar, University of Delhi, India
Nazmun Nessa Moon, Daffodil International University, Bangladesh
Neeta Maitre, Cummins College of Engineering for Women, India
T. N. Nisha, Symbiosis Centre for Information Technology, India
Nistala Murthy, Andhra University, India
Nitin Srinivasan, Google USA
Omar H. Alhazmi, Taibah University, Saudi Arabia
Omprakash Netula, SR University, India
Parth Santpurkar, Netflix, USA
Pavan Kumar, VIT Bhopal University, India
Pavan N. Kunchur, GIT, Belgaum, India
Pavlo Maruschak, Ternopil National Ivan Puluj Technical University, Ukraine
Pawan Mandal, Indian Institute of Technology Mandi, India
Philip Moore, Lanzhou University, China
Pradeep Chintale, SEI Investment Company, USA
Prakash Murugesan, Verizon, USA
Praveen Kumar Sridhar, Citizen, USA
Preetham Vemasani, Uber, USA
Preeti Malik, Graphic Era University, Dehradun

Prema Kumar Veerapaneni, Mphasis Corporation
V. V. Subrahmanyam, Indira Gandhi National Open University, India
Puneet Gangrade, Fordham University, New York, USA
Punidha Angusamy, KPR Institute of Engineering and Technology, India
Purva Joshi, University of Pisa, Italy
R. Harish, Deputy Director, IBS Bangalore
Radhika Kotecha, K. J. Somiaya Institute of Engineering and Information Technology, Mumbai, India
Raghavendra Rao, IBS, Bangalore, India
G. N. Raghavendra Rao, ICFAI Foundation for Higher Education, India
M. Raja, CMR Institute of Technology, Bengaluru, Karnataka, India
Rajesh Kolli, Kansas Board of Public Utilities, USA
Rama Krushna Rath, Indian Institute of Information Technology (IIIT) Sri City, India
Rambabu Nalagandla, IEEE Senior Member, USA
Rashmi Benni, KLE Technological University, India
Ravi Kumar Maddila, Malaviya National Institute of Technology Jaipur
Rita Yi Man Li, Hong Kong Shue Yan University, Hong Kong
Sangeeta Mangesh Karyakarte, Dr. A. P. J. Abdul Kalam Technical University, Lucknow, India
Satya Srinivas Maddipati, Sasi Institute of Technology and Engineering, India
Satyasai Jagannath Nanda, Malaviya National Institute of Technology Jaipur
Saydul Murad, Universiti Malaysia Pahang, Malaysia
Seemant Tiwari, Southern Taiwan University of Science and Technology, Tainan City, Taiwan
Shahid Hussain, Biomedical Robotics, University of Canberra
Shantanu Kumar, Senior Software Engineer, Amazon
Sheikh Imroza Manzoor, Central University of Jammu, India
Shilpa Mehta, Presidency University, Bengaluru, India
Shruti Singh, Washington State University, USA
Siddheshwari Mishra, NITTTR, Chandigarh, India
P. Sivakumar, PSG College of Technology, India
Soukaina Merzouk, Hassan II University of Casablanca, Faculty of Sciences Ben M'sik, Morocco
Sujatha Kokatnoor, CHRIST (Deemed to be University), India
Sumit Bhatnagar, JP Morgan Chase, USA
Sunita Kumawat, Amity University Haryana, Gurgaon
Surabhi Sinha, University of Southern California, USA
Swathi Narayanan, Vellore Institute of Technology, India
Syed Mohd Mahmood, GD Goenka University, India
N. L. Taranath, Adichunchanagiri Institute of Technology, India
Thangavel Murugan, United Arab Emirates University, UAE
Usha Kosarkar, GHRIET, India
Venkata Ashok K. Gorantla, Georgia Institute of Technology, USA
Vincent Nyangaresi, Tom Mboya University College, Kenya
Vipin C. Nair, Amity Global Business School, Bengaluru, India

Preface

This book contains outstanding research papers as the proceedings of the 2nd International Conference on Artificial Intelligence: Theory and Applications (AITA 2024). AITA 2024 has been organized by ICFAI Business School (IBS) Bangalore, India, and is technically sponsored by the Soft Computing Research Society, India. The conference serves as a platform for exchanging ideas and research findings from academia and industry, aimed at addressing the challenges and advancements in artificial intelligence and its applications to solve complex real-world problems. This book will help in strengthening congenial networking between academia and industry. We have tried our best to enrich the quality of the AITA 2024 through the stringent and careful peer-review process. This book presents novel contributions to artificial intelligence and serves as reference material for advanced research.

From autonomous robotics and financial fraud detection to advancements in health care with AI-driven diagnostics, such as glaucoma classification and smart asthma management, AITA 2024 showcases a wide range of innovative solutions. The conference also features groundbreaking work in energy sustainability, agricultural technologies, and AI applications in real-world challenges like intrusion detection and cybersecurity.

Through a rigorous peer-review process, AITA 2024 received 793 submissions from across the globe, with only 109 high-quality papers being selected for presentation. This book presents the second volume, featuring 53 papers contributing novel insights into artificial intelligence and its diverse applications. We hope that this book will serve as a valuable reference for advanced research and foster meaningful collaboration between academia and industry.

Kota, India — Harish Sharma
Stavanger, Norway — Antorweep Chakravorty
Canberra, Australia — Shahid Hussain
Bengaluru, India — Rajani Kumari

Contents

About the Editors

Dr. Harish Sharma is Associate Professor at Rajasthan Technical University, Kota, in Department of Computer Science and Engineering. He has worked at Vardhaman Mahaveer Open University Kota and Government Engineering College Jhalawar. He received his B.Tech. and M.Tech. degree in computer engineering from Government Engineering College, Kota, and Rajasthan Technical University, Kota, in 2003 and 2009, respectively. He obtained his Ph.D. from ABV—Indian Institute of Information Technology and Management, Gwalior, India. He is Secretary and one of the Founder Members of Soft Computing Research Society of India. He is Life Time Member of Cryptology Research Society of India, ISI, Kolkata. He is Associate Editor of *International Journal of Swarm Intelligence (IJSI)* published by Inderscience. He has also edited special issues of the many reputed journals like *Memetic Computing*, *Journal of Experimental and Theoretical Artificial Intelligence*, *Evolutionary Intelligence*, etc. His primary area of interest is nature-inspired optimization techniques. He has contributed in more than 105 papers published in various international journals and conferences.

Dr. Antorweep Chakravorty is Associate Professor at the University of Stavanger. His current research and development work is in the field of applied blockchains, big data, large-scale machine learning, and data privacy. He has an interest in real-world problems, especially development of privacy-enabled data-driven services in smart energy, health care, and smart city domains. Antorweep completed his Ph.D. in 2015 with a thesis on Privacy Preserving Big Data Analytics at the University of Stavanger, Norway. Along with having a background in applied research in data-driven solutions, he is also involved in mentoring, teaching, and supervision.

Dr. Shahid Hussain is working at University of Canberra as Associate Professor of biomedical robotics. Prior to that he has worked as Lecturer at University of Wollongong, Australia. Dr. Hussain has obtained his Ph.D. in mechanical engineering from the University of Auckland, New Zealand, in 2013. His research interests include assistive and rehabilitation robotics, compliant actuation of robots, robot mechanism design and optimization, nonlinear dynamics and control of robotic systems, human–robot interaction, biomechanical modeling, engineering education, and micro electro-mechanical systems (MEMS). Dr. Hussain has published more than 65 papers in the prestigious journals of the field.

Dr. Rajani Kumari is currently Assistant Professor at IBS, Bangalore, Off-Campus Centre of ICFAI Foundation for Higher Education (IFHE) University, India. Previously she was Assistant Professor at IIIM, Jaipur, and St. Xavier's College Jaipur, CHRIST University. She received the Ph.D. degree in computer science in 2015 and the MCA and BCA from University of Rajasthan in 2010 and 2006, respectively. She has published more than forty research papers in various international journals/ conferences and participated in many national and international conferences and workshops. She edited some special issue in Taylor & Francis and Inderscience journals including *Journal of Information and Optimization Sciences (JIOS)* and *International Journal of Intelligent Information and Database Systems (IJIIDS).* Her research interests include nature-inspired algorithms, swarm intelligence, soft computing, and computational intelligence.

Construction and Application of Knowledge Graphs in Requirement Engineering Process

Preeti Malik, Lata Nautiyal, and Varsha Mittal

Abstract Software requirements play a very important role in the development and management process of software systems. Development and analysis of requirement specification is a crucial and time-consuming task if it is done manually. Software quality is determined by the accuracy, consistency, and completeness of the requirement specification document. Between business people, requirements researchers, and software developers, there are also frequent discrepancies and ambiguities in the realizing requirements during the requirements elicitation process. Knowledge graphs, which describe information as a semantic graph, have raised a great deal of controversy in both the academic and professional worlds. Their ability to provide information that is semantically organized has led to significant potential solutions for a variety of activities, including question-answering, recommendation, and information retrieval. This paper presents their use in the field of requirement engineering phases of software development. We have discussed the process of creating knowledge graphs, their functioning, and some major approaches discussed in the literature so far, especially in the field of requirement engineering. We also tried to present the positive and negative sides of the approaches and a comparative analysis of these approaches is also included.

Keywords Knowledge graph · Requirement engineering · Software development · Semantic graph · Ontology · Natural language processing

---

P. Malik (✉) · V. Mittal
Graphic Era (Deemed to be) University, Dehradun, India
e-mail: preetishivach2009@gmail.com

L. Nautiyal
University of Bristol, Bristol, UK

H. Sharma et al. (eds.), *Artificial Intelligence: Theory and Applications*, Lecture Notes in Networks and Systems 5589, https://doi.org/10.1007/978-981-96-1687-9_1

1 Introduction

The correctness and completeness of the requirement specification (RS) document serve as the foundation and source for the design and execution of software systems, and they are essential in assessing the quality of the software [1, 2]. Manual methods are frequently employed while producing the RSs for software projects, nevertheless. Between business people, requirements researchers, and software developers, there are frequent discrepancies and ambiguities in the realizing requirements during the requirements elicitation process, like gaps in the research or confusing business processes [3]. This substantially raises the anticipated research and development expenditure and causes fatal errors in the design of software systems [4].

The manually created software requirement specifications (SRSs) documents, however, are primarily intended for computer specialists. They are tough for business people to understand and have poor readability. Because of this, business people are unable to adequately verify if the RS accurately captures their needs. Currently, there are three primary ways to create RSs documents: manually creating UML documents, automatically creating RSs based on natural language processing (NLP), and extraction rule-based RSs [5, 6]. Though the process of preparing RS manually has a strong ability for personalized customization to produce RSs, it is typically challenging to address issues with complex manual preparation, being prone to error, and struggle while communicating with business people, which are simply impacted by the writer's subjective ability [7]. The extraction rules used by the RS generation technique based on extraction rules may successfully extract the information about the RS from the text, however, they will change differently for various functional needs. The creation of extraction rules theoretically can't account for every software scenario.

An innovative concept for the automatic extraction of RSs has emerged with the advancement of NLP technology. On the one hand, it may efficiently increase the effectiveness of creating SRSs; on the other hand, it can efficiently reduce the writer's human mistakes. User roles, functional requirements, data tables, and business processes make up the bulk of the SRS document.

1.1 Background

Knowledge Graphs (KGs) have grown to be the foundation of various information systems that need access to structured knowledge in recent years. It is possible to trace the origins of the Semantic Web back to Berners- Lee's [8] study from 2001. In his work, Berners-Lee advocated for the development and promotion of technical standards including the Web Ontology Language (WOL), Resource Description Framework (RDF), and Uniform Resource Identifier (URI). By using the RDF standard early on, certain studies helped to advance the graph-based representation of knowledge. These networks' nodes stand in for entities, and the edges connecting them denote the relationships between those things. The sets of relations can be arranged in an ontology or schema that specifies their correlativity and limitations on use.

In 2009, the idea of "Linked Data" [9] was introduced. To consider various datasets as a single, enormous, global knowledge network, it is suggested that they should be connected through the Semantic Web. In the Connected Open Data cloud, around 1,000 datasets have been linked to one another up until 2014; the majority of these linkages connect identical entities [10]. To exploit semantic knowledge in online searches, Google suggested a new technique in 2012 called KG. To improve the functionality of the search engine and improve user experience, Google's KG is used to recognize and decipher items in the text, enhance search results with semantically structured summaries, and give links to related entities in exploratory search. Many other businesses then began to create their own KGs. For instance, Satori, a related KG, has been connected with Microsoft's Bing search engine. In today's world, the term "KG" also refers to semantic web knowledge bases like DBpedia [11], YAGO [12], Wikidata, or Freebase [13]. In both the academic and industrial worlds, KGs have raised significant controversy. Many individuals believe that the information they supply holds enormous potential for the development of more intelligent machines since it is semantically organized and computer interpretable. As per, Gartner's 2022 Hype Cycle visualization places KGs are at the peak of inflated expectations region (refer Fig. 1).

1.2 Building Knowledge Graph

The building methods of KGs were the subject of many reviews [14, 15]. Knowledge acquisition and representation have become an important goal of AI-based systems which led to the development of knowledge-based systems. There must be a systematic process for the development and maintenance of these systems. KGs are the most promising solution to this problem. However, there are various methods for creating KGs in literature. This section describes the most general method for creating KGs, which is adopted from [17]. According to the authors, unlike traditional knowledge engineering methodologies, building KGs focus on large Assertional Boxes (ABoxes) and not complex Terminological Boxes (TBoxes). The author has divided the whole process is the following four major tasks (see Fig. 2):

- The first step is the ***Knowledge Creation step***. This step concentrates on the efficient creation of large Assertional Boxes and lightweight ontology engineering based on Schema.org for Terminological Boxes.
- The second step is the ***Knowledge Hosting step*** and which considers a source of the produced knowledge from diverse sources and contextualization of the same.
- The next step is the ***Knowledge Curation step***. The objective of this step is to evaluate, correct, and improve the KG. The evaluation task assesses the quality in different dimensions. Hereafter, the quality of KG is better mainly in two

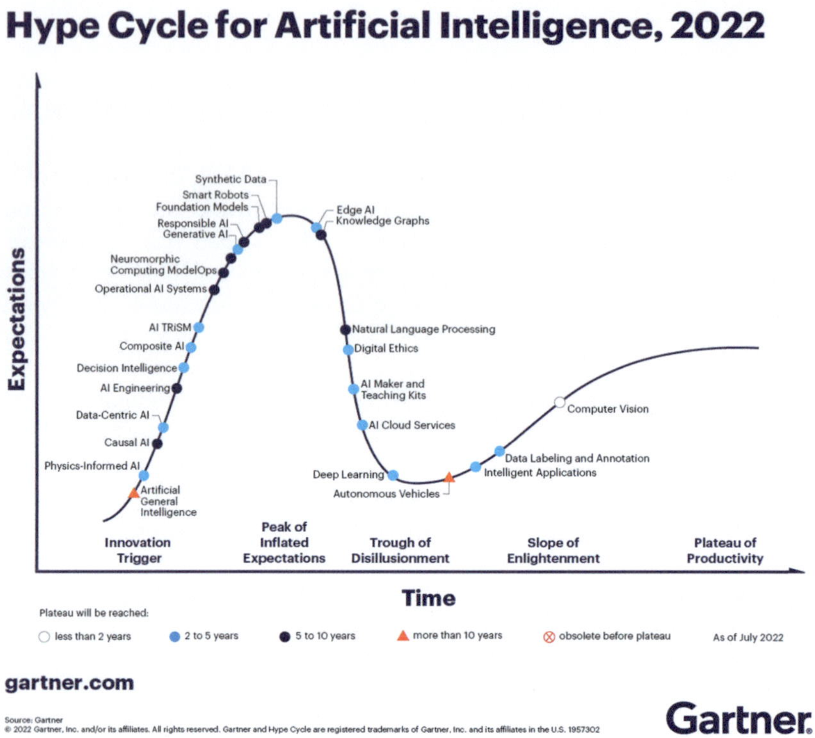

Fig. 1 Gartner hype cycle fetched from [16]

Fig. 2 Methodology adopted from [17]

dimensions, i.e., correctness and completeness by cleaning and enriching process, respectively.

- The last step is ***Knowledge Deployment step***. This step deploys the KG.

1.3 Knowledge Graph Functioning

We will discuss a broad architectural overview of the functioning of NLP-based KG which was proposed by Ergeta Muca [18] (see Fig. 3).

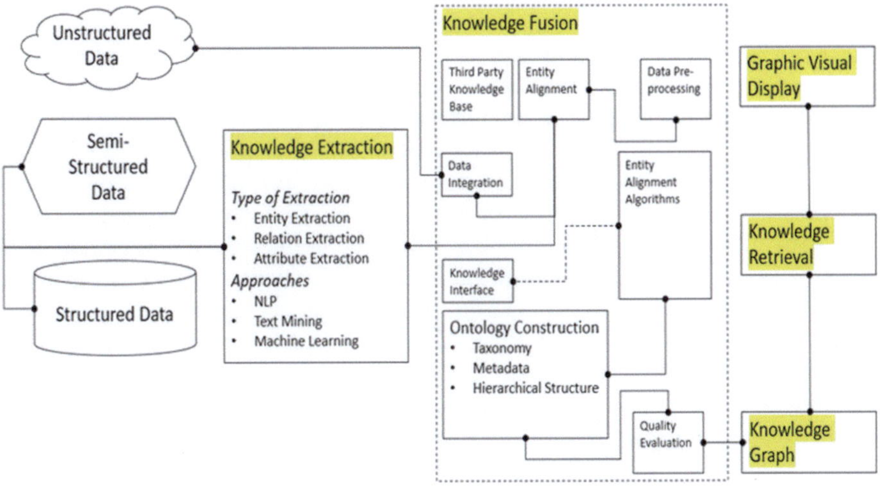

Fig. 3 KG Functioning according to Ergeta Muca [18]

Data Sources

There can be various data sources that produce different types of data like unstructured data, structured data, and semi-structured data. Unstructured data is data in the form of free text, photos, and documents, semi-structured data is data in the form of HTML files, JSON files, and XML files, and example of structured data is relational databases.

Knowledge Extraction

The process of extracting knowledge starts once the data has been absorbed. This step retrieves information like entities, relationships, and characteristics from the semi-structured and unstructured input data. For this purpose, NLP, text mining, and machine learning approaches are used where machine learning approaches can be both supervised and unsupervised learning). The entity extraction task recognizes the words that designate entities belonging to particular categories. An entity can be a person, thing, place, or object which can belong to a variety of classes, each of which has its subclasses. Information is obtained about the entities and their properties once all the entities have been removed. These may consist of both relationships between entities and their attributes. The relationship may be described using an entity attribute. The verb in the given sentence is typically, but not always, the entity attribute. Following the identification of the objects and topics, part of speech tagging is used to extract the entities.

Knowledge Fusion

The goal of this phase is to combine all the information bases from various sources to obtain a complete picture. Realizing entity alignment and ontology development are two of its unique objectives. Finding out whether "different entities refer to the same things in the actual world" is what entity alignment (or entity resolution) is all about. Data standardization is a crucial step in entity alignment since it creates a consistent language for the data. In this stage, any discrepancies or disintegrations in the data are fixed. A nice illustration of the value of data pre-processing and standardization is provided by LinkedIn's KG. Many of their entities have "meaningless names, erroneous or incomplete characteristics, outdated information, or no member linked to them" since they are user-generated organic entities. Several stages may be performed, such as producing entity candidates, de-duplicating entities using techniques like word2vec, grouping entities based on their contexts, and utilizing machine translation models to translate all entities into the identical language.

The records that pertain to the same object are merged using pairwise and collective alignment once the data has been fully integrated and consistent. Using various text similarity functions, such as cosine similarity, pairwise similarity comparisons are carried out. Deep learning approaches may also be integrated. Utilizing structural similarity algorithms like pattern recognition, collective alignment is accomplished. All of this effort results in the development of an ontology. Finally, the addition of a taxonomy, hierarchical structures, and metadata finishes the step and also improves the KG's quality. The produced ontology is compared to industry standards like schema.org to confirm overall KG quality, and if necessary, the process is iterated and enhanced. Given that the knowledge fusion stage is where the majority of the modeling takes place, it is crucial to emphasize the value of this step's iterative nature.

Storage, Retrieval, and Representation of KG

Once it has been generated, a KG is kept in a NoSQL database, either in a graph database or an RDF (resource description framework). KGs are represented by the triple (Subject, Predicate, Object) in RDF, and nodes, edges, and characteristics of graphs are stored in graph databases. The standard query language for obtaining data from extensive knowledge networks is SPARQL. One of the most extensively studied issues in this field, KG visualization is frequently carried out using browser-based programs.

2 Use of KG in Requirement Engineering Process

This section is dedicated to the previous work done in the field of using KG for the requirement engineering process.

2.1 SRS Construction Model Using NLP and KG: BiLSTM-CRF-KG

The proposed work presents an automatic generation model of SRS named Bi-Directional Long Short-Term Memory-Conditional Random Field-KG (BiLSTM-CRF-KG) model [19]. This model combines KGs with the RS and uses the KG's visual expression feature, which can significantly reduce the possibility of misunderstanding during the development of the RS and make it easier for business people and software engineers to communicate. The model considers the Chinese language mainly. Following are the primary elements that make up the BiLSTM-CRF-KG SRS graph:

1. *Pre-processing:* It mainly realizes the sentence segmentation (Chinese sentences), word segmentation, POS tagging, named entity, and relation tagging of the original requirements description corpus.
2. *Entity relationship extraction:* This process extracts the relationship between functional entities and functions, then generate a preliminary entity relationship set.
3. *Entity disambiguation and hidden relation learning:* To optimize the entity relationship set generated in the previous step, the hidden relationships between entities are identified by using context information.
4. *Generation of functional structure graph:* Use the hierarchical structure of the multi-tree to transform the optimized entity relationship set into the demanded relationship tree. Then, using the breadth-first search strategy, the requirement relationship tree is transformed into a (parent function, parent–child relationship, child function) function structure triplet, which is saved in the neo4j database to realize the automatic embedding of the functional structure graph.
5. Study the method of converting the traditional U/C matrix into the S/U/C matrix to reflect the real flow relationship between the creator/sender/user of the business data table.
6. Transform the business data table description into the S/U/C matrix.
7. Use the relationship between the S/U/C matrix mapping function and the data table to generate the (function-create/send/use relationship-data table) triplet and save it in the neo4j database to realize the automatic embedding of the function-data graph.
8. Integrate the function structure graph and function-data graph to generate the SRS graph.

Pros

- The BiLSTM-CRF-KG model is very stable.
- The precision rate and recall rate are stable at about 96%.

Cons

- The KG created only considers embedding of function, data, and their relationship but not user roles and other important entities.

- UML diagrams are created manually which is a time-consuming task.

2.2 System Requirement eXtraction with Chunking and Rule Mining (SRXCRM)

The study introduced SRXCRM, an NLP-based model to fetch, associate, and rank NL requirements of safety–critical systems [20]. Design specification of a Propulsion and Control railway subsystem (PCS) is used for the study. This model comprises three main phases viz; pre-processing, main chunking, and rule and knowledge extraction.

Pre-processing: Tokenization, stop word removal, and POS tagging are done in this phase.

Main Chunking: The goal of this phase is to identify PCS components and separate them from design specifications and customer requirements. To do this, the phase takes already tagged sentences from the previous phase, runs the Noun Phrase Chunking process for each sentence, and after this chunking process filtering of the results is done to get the most pertinent chunks.

Rule and Knowledge Extraction: The primary chunks, their derivative chunks, and the processed design specifications and customer requirements are inputs to the rule and knowledge extraction phase, which outputs the association rules found between the major chunks and the graph showing their relationships. In this stage, the weight of each chunk that appears in the dataset is used in the Weight Association Rule Mining (WARM) approach, which is an evolution of the ARM methodology.

The Weight Attribution and Frequent Itemset Extraction tasks, which compute the weight of involvement for each chunk and extract item sets when they appear in the dataset, respectively, support the WARM process. The rules are then retrieved by WARM.

Pros

- The proposed model was validated by domain expert analysis where almost 90% of the processed noun phrase chunks can be considered relevant to system engineers.

Cons

- A better processing technique is required for customer requirements because of their broader domain.

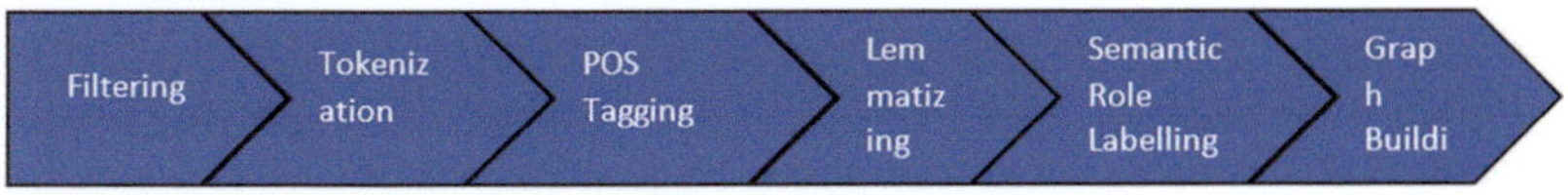

Fig. 4 NLP Pipeline to convert NL requirement to KGs

2.3 Knowledge Representation of Requirement Document Using NLP

In this study, the authors provide a pipeline for NLP that converts a set of natural language requirements into a KG [21]. This KG's goal is to organize and summarize all of the ideas and connections found in the requirements for all of the subsystem specs. RDF (Resource Description Framework)-inspired triples are used in the graph to describe this knowledge. The NLP pipeline (see Fig. 4) employs Semantic Role Labeling (SRL) to extract single triples before performing a variety of pre-processing procedures on the requirements to standardize the data to generate the graph. After that, the triples are combined into a single KG.

The initial steps of the pipeline have general meaning and the last step, i.e., graph building is detailed here. An exporter creates a graph from the annotated natural language as the final stage of the pipeline. The exporter looks for semantic role-labeling verbs and their arguments in the text. The subject is the first SRL argument, an object is the second, and the verb serves as the predicate in an RDF triple if at least a first and second argument is present.

Pros

- The author validated the proposed approach for one academic and one industrial requirement document. The validation demonstrates good results.

Cons

- The proposed pipeline is just a model which is unable to accurately extract all data or construct a reliable ontology or RDF graph.
- A causal relationship is not considered.
- Redundant nodes and edges in KG as the semantic similarity is not identified.

2.4 Trace Link Recovery Using Semantic Relation Graph

To enable additional engineering work, Trace Link Recovery (TLR) aims to find and connect linked current needs [22]. The ability to explain trace linkages is constrained by algebraic information retrieval techniques like the distance between tf-idf scores, which also operate on smaller datasets without training. In this paper, the authors describe a TLR strategy that leverages Spreading Activation to respond to trace inquiries across a semantic relation graph that explicitly represents the content of requirements.

The proposed method is entirely automated and includes an NLP pipeline to convert requirements expressed in unconstrained plain language into a graph. We believe that an explicit knowledge representation is an intriguing artifact to investigate in TLR techniques to produce explanations and refine findings, even though the predictive power.

Pros

- The proposed approach is scalable for various sizes of datasets in terms of a graph and the algorithm and is immediately applicable.
- It is a fully automated approach.
- The graph immediately adapts to the changes in requirements.

Cons

- The approach does not outperform the syntactical vector space model.

2.5 *Improving TLR Using Semantic Relation Graph*

The authors have improved the above approach by making considerable changes to the graph's construction and subsequent structure [23]. Unlike before, the process starts with single words and works our way up. All nouns are added to the graph as a vertex. Nouns are put together into phrases at the next level. To create a graph, any noun and person phrases that do not include a verb phrase are also added as vertices. They are then connected to the nouns they contain through edges. The same vertices are utilized when nouns or phrases are repeated. The lemma of words is utilized to facilitate this reuse. Additionally, sentences that include determiners or prepositions at the beginning are deleted depending on the POS tag. Finally, verb vertices are added to the graph. This time, instead of including additional argument vertices, the noun and sentence are used directly as argument vertices. There are two ways to assign an argument to its verb. The graph also has representations for adjectives and adverbs. Adverbs are deliberately inserted as components to the label of verb vertices, unlike adjectives, which are implicitly incorporated. Authors removed the identifier vertices to omit the separation between the short phrase and identifier vertices. Instead, all of a requirement's vertices are given the requirement identifiers as metadata. When a term or phrase appears more than once in a requirement, the identifier is also repeated more than once.

Pros

- The graph and its semantic relations are independent of the natural language.
- Semantic relations are considered.

Cons

- The approach only supports the English language.

2.6 *Identifying Inter-Related Requirements*

This strategy intends to provide fault-prone areas for both pre- and post-inspection to requirements engineers [24]. Inter-related requirements are used in this technique to assist remove unnecessary and redundant defects from related requirements prior to inspection, and after inspection, it helps engineers analyze the effects of a modification in one requirement on other related requirements which is also called change impact analysis.

Pros

- Studies change impact analysis for NL requirements.
- Early detection of fault results in cost effective software development.

Cons

- Manual evaluation of KG.

2.7 *Security Requirement Elicitation Using KG*

For security-critical software systems, it is crucial to specify security requirements (SR) during the requirement analysis process. To investigate SRs in depth, however, is challenging and complicated. In this study, the authors provide a KG-based SR acquisition technique [25]. First, a significant amount of security concept descriptions is retrieved as KG interactions among security ideas. Then, using triples as inputs, TransE is trained as a KG representation learning model. The KG can reason suitable Security Assurance Requirements (SAR) based on the same Evaluation Assurance Level (EAL) or reason-matching SFRs based on identical Security Threats/OSPs.

Pros

- The technique can obtain more accurate SRs by comparing them with the two SR recommendation models, the FT-SR model [26] and the co-occurrence model [27].

Cons

- Restricted to security requirements.
- Proposed only for small datasets.

3 Comparative Analysis

This section is dedicated the comparison of these approaches based on some common features. Detailed definition of these features is out of scope of this paper. Interested candidate can found the detailed description in [28]. We have done our best to fetch the information. Below, we will briefly define them and comparison results are shown in Table 1:

(1) ***Knowledge Graph Model:*** Model represents graph structure. It can be directed edge-labeled graph, heterogeneous/ homogeneous, property graph, or graph datasets.
(2) ***Deductive Knowledge:*** A person can deduce knowledge from KG. For deductive knowledge the approaches used are; ontologies, semantics and entailment and reasoning (rule or description logic).
(3) ***Inductive Knowledge:*** Inductive knowledge includes generalizing patterns for imprecise predictions. It can be gained through graph analytics, framework used, analytics with querying, KG embedding, symbolic learning etc.
(4) ***Quality Assessment:*** A KG can be assessed for quality parameters like; accuracy, coverage, coherency, and succinctness.
(5) ***Refinement:*** Knowledge graph completion and correction comes under this head like; link prediction, fact validation, and inconsistency repair.

4 Future Directions

Based on the above discussion, we present possibilities for future work in other dimensions as well.

(1) ***Other Requirement Engineering Phases:*** Requirement engineering process comprising of four phases; elicitation, specification, verification and validation, and requirement management. Researcher have used KG mainly for elicitation and analyzing requirements. The field of requirement management and validation can also be explored for betterment.
(2) ***Deductive Knowledge Approach:*** Semantic information is widely used. Ontology and description logic can also be good option.
(3) ***Inductive Knowledge:*** Node relevance can be studied to prioritize requirements. Connectivity of a node for requirement dependency and other metrics can also studied.
(4) ***Quality Assessment:*** There are many quality characteristics of SRS that can validated using KG like; accuracy, ambiguity, completeness, consistency, and validity, etc. One can work in this direction also.
(5) ***Refinement:*** A very little work has been done in fact validation and inconsistency repair. Hence this can also be good option.

Table 1 Comparative analysis of approaches

	RE phase	KG model	Deductive knowledge	Inductive knowledge	Quality assessment	Refinement
[19]	SRS construction and ambiguity removal	Graph dataset (function data and function structure graph)	Reasoning: rule-based	*Graph analytics:* framework (TensorFlow) *KG embedding:* tensor decomposition model	Consistency	*Completion:* hidden relationship learning
[20]	Req, analysis(req. similarity and variance)	Directed edge-labeled graph	Reasoning: rule-based	*Graph analytics:* requirement ranking *Analytics on data graph:* weighing	*Syntactic accuracy:* Human experts	–
[21]	Req. analysis (semantic embedding)	Graph dataset (main and sub graph)	Semantics	–	–	–
[22]	Req. analysis	Directed graph	Semantic	*Graph analytics:* techniques: path finding, ranking	Accuracy and coverage	Link prediction
[23]	Req. analysis	Property graph	Increased semantics	*Graph analytics:* techniques: path finding, ranking	Accuracy and coverage	Link prediction
[24]	Req. analysis (similarity detection)	Simple undirected graph	Semantics	*Graph analytics:* centrality, community detection	Accuracy, timeliness	Change impact analysis
[25]	Req. elicitation	Directed graph	No clear reasoning	*Graph analytics:* node similarity *Embedding:* TransE model *Analysis on data graph:* weighing	–	–

5 Conclusion

To the best of our knowledge, this paper is the first to review about applications of KGs requirement engineering field. Overall, we conclude that knowledge graphs have a great ability to provide semantically structured information and important advancements in applying such ability to specific domains have been made in recent years. Requirement specification language (English, Chinese, etc.) plays an important role while dealing with KGs as we have to perform tokenization and POS tagging, etc. techniques before generating KG. This paper presents some of the approaches which make use of KG for requirement engineering process with their pros and cons. A comparative analysis has also been done on some common features of the approaches. We found that size of SRS is also a matter of concern. Other issues have been discussed in the paper. This article gives researchers a direction to perform their research.

References

1. Adikara F, Anggarani A (2018) Qualitative requirements analysis process in organization goal-oriented requirements engineering (OGORE) for e-commerce development. Lect Notes Electr Eng 449:318–324
2. Boyarchuk A, Pavlova O, Bodnar M, Lopatto I (2020) Approach to the analysis of software requirements specification on its structure correctness. CEUR Workshop Proc Ser 2623:85–95
3. Pamuji R, Saputra YA, Ramdani F (2018) The relation of fit tables and textual requirements to understand software requirements. In: Proceedings of the 3rd international conference on sustainable information engineering and technology, Malang, Indonesia; pp 142–146
4. Bilal M, Gani A, Liaqat M, Bashir N, Malik N (2020) Risk assessment across life cycle phases for small and medium software projects. J Eng Sci Technol 15:572–588
5. Shimada H, Nakagawa H, Tsuchiya T (2018) Constructing a goal model from requirements descriptions based on extraction rules. Commun Comput Inf Sci 809:175–188
6. Murugesh S, Jaya A (2021) An integrated approach towards automated software requirements elicitation from unstructured documents. J Ambient Intell Humaniz Comput 12:3763–3773. [CrossRef]
7. Stanik C, Maalej W (2019) Requirements intelligence with OpenReq analytics. In: Proceedings of the 2019 IEEE 27th international requirements engineering conference, Jeju, Korea; pp 428–483
8. Berners-Lee T, Hendler J, Lassila O et al (2001) The semantic web. Sci Am 284(5):28–37
9. Bizer C, Heath T, Berners-Lee T (2011) Linked data: the story so far. In: Semantic services, interoperability and web applications: emerging concepts. IGI Global, pp 205–227
10. Schmachtenberg M, Bizer C, Paulheim H (2014) Adoption of the linked data best practices in different topical domains. In: International semantic web conference. Springer
11. Bizer C, Lehmann J, Kobilarov G, Auer S, Becker C, Cyganiak R, Hellmann S (2009) Dbpedia-a crystallization point for the web of data. Web Semant Sci Serv Agents World Wide Web 7(3):154–165
12. Suchanek FM, Kasneci G, Weikum G (2008) Yago: a large ontology from wikipedia and wordnet. Web Semant Sci Serv Agents World Wide Web 6(3):203–217
13. Bollacker K, Cook R, Tufts P (2007) Freebase: a shared database of structured general human knowledge. In: AAAI, vol 7, pp 1962–1963

14. Zenglin X, Yongpan S, Lirong H, Yafang W (2016) Review on knowledge graph techniques. J Univ Electron SciTechnol China 45(4):589–606
15. Zhao Z, Han SK, So IM (2018) Architecture of knowledge graph construction techniques. Int J Pure Appl Math 118(19):1869–1883
16. URL: https://www.gartner.com/en/articles/what-s-new-in-artificial-intelligence-from-the-2022-gartner-hype-cycle on 21-11-2022
17. Simsek U, Kärle E, Angele K et al (2023) A Knowledge graph perspective on knowledge engineering. SN Comput Sci 4:16
18. URL: https://medium.com/analytics-vidhya/introduction-to-knowledge-graphs-and-their-applications-fb5b12da2a8b on 21-11-2022
19. Wang Z, Pan J-S, Chen Q, Yang S (2022) BiLSTM-CRF-KG: a construction method of software requirements specification graph. Appl Sci 12:6016
20. Leitão V, Medeiros I (2021) SRXCRM: discovering association rules between system requirements and product specifications. In: REFSQ workshops
21. Schlutter A, Vogelsang A (2018) Knowledge representation of requirements documents using natural language processing. In: REFSQ 2018. Joint proceedings of REFSQ-2018 workshops doctoral symposium live studies track and poster track co-located with the 23rd international conference on requirements engineering: foundation for software quality (REFSQ 2018) Utrecht The Netherlands March 19 2018 ser. CEUR workshop proceedings, 2018, [online] Available: http://ceur-ws.org/Vol-2075/NLP4RE_paper9.pdf
22. Schlutter A, Vogelsang A (2020) Trace link recovery using semantic relation graphs and spreading activation. In: 2020 IEEE 28th international requirements engineering conference (RE). IEEE, pp 20–31
23. Schlutter A, Vogelsang A (2021) Improving trace link recovery using semantic relation graphs and spreading activation. In: International working conference on requirements engineering: foundation for software quality. Springer, Cham, pp 37–53
24. Singh M (2019) Using natural language processing and graph mining to explore inter-related requirements in software artefacts. ACM SIGSOFT Softw Eng Notes 44(1):1–7
25. Zhang Y, Li X (2020) The scheme of security requirement acquisition based on knowledge graph. In: 2020 International symposium on theoretical aspects of software engineering (TASE). IEEE, pp 225–231
26. Wang J, Li X, Feng Z et al (2017) Automated software security requirements recommendation based on FT-SR model. In: The international conference on software engineering and knowledge engineering, pp 382–385
27. Xu Y, Ge W, Li X, Feng Z, Xie X, Bai Y (2019) A co-occurrence recommendation model of software security requirement. In: International symposium on theoretical aspects of software engineering (TASE), Guilin, China, pp 41–48
28. Hogan A, Blomqvist E, Cochez M, d'Amato C, de Melo G, Gutierrez C, Gayo JEL, Kirrane S, Neumaier S, Polleres A, Navigli R, Ngomo A-CN, Rashid SM, Rula A, Schmelzeisen L, Sequeda J, Staab S, Zimmermann A (2020) Knowledge graphs. arXiv:2003.02320

Optimized Pathfinding for Autonomous Robots: Enhancing Operational Range and Safety in Varied Terrains

Surya Prakash, Akarshan P. Sami, and Bibhya Sharma

Abstract This research introduces a novel pathfinding framework designed for autonomous robots navigating diverse terrains. The framework leverages advanced Monte Carlo simulations to evaluate multiple potential paths dynamically under varying environmental conditions. Users can choose specific goals from Pareto optimal solutions, such as energy efficiency or safety. Our findings underscore the profound influence of path composition on the performance metrics of autonomous navigation systems. For instance, routes predominantly comprising unpaved sections culminating in a paved segment offer a balance between energy consumption and navigational risk. In contrast, fully paved routes escalate energy consumption but significantly mitigate navigational risk. Rigorous experimental validation demonstrates that our framework amplifies autonomous robots' operational range and safety across different terrains. This research propels the field of robotic navigation by providing a versatile decision-support tool, empowering users to make informed decisions that align with their operational strategies. Ultimately, our framework holds the potential to bolster the reliability and performance of autonomous robots, paving the way for safer, more efficient, and sustainable robotic operations.

Keywords Robotic navigation · Monte Carlo simulations · Multi-objective optimization · Pareto optimality · Autonomous systems

S. Prakash · B. Sharma
The University of the South Pacific, Suva, Fiji
e-mail: surya.prakash@usp.ac.fj

B. Sharma
e-mail: bibhya.sharma@usp.ac.fj

A. P. Sami (✉)
Monash University, Melbourne, Australia
e-mail: akarshan.sami@monash.edu

H. Sharma et al. (eds.), *Artificial Intelligence: Theory and Applications*, Lecture Notes in Networks and Systems 5589, https://doi.org/10.1007/978-981-96-1687-9_2

1 Introduction and Background

Effective pathfinding strategies are essential for navigating diverse terrains in autonomous robotics. Operating securely and efficiently in dynamically changing environments optimizes operational capabilities and ensures mission success. Autonomous robots encounter various environments, each presenting unique challenges and obstacles to path navigation. Path-planning algorithms typically aim to find the optimal path or a close approximation, minimizing one or more objective optimization functions. For instance, minimizing travel time is crucial in time-sensitive missions like search-and-rescue operations, where prompt response can be a matter of life or death. Additionally, minimizing the energy consumption of the robot is vital, especially in scenarios like planetary exploration where energy resources are limited. The path generated by the planner must adhere to imposed constraints, which may stem from the robot's limitations in adapting to certain terrains. Factors such as the robot's locomotion capabilities and terrain characteristics influence the range of maneuvers possible, thereby constraining the paths the planner can generate. Traditional pathfinding approaches, such as Dijkstra's algorithm or A^* search, may struggle to cope with the complexities of real-world terrain, especially when facing uneven surfaces, obstacles, or dynamic changes in environmental conditions. Researchers have developed specialized pathfinding techniques for autonomous robotics to address these challenges. One such method involves using Monte Carlo simulations to evaluate potential paths in various terrains. These algorithms identify optimal paths by simulating thousands of trajectories and considering terrain roughness, slope, and obstacle density. For example, a path that combines unpaved and paved sections may balance energy consumption and navigational risk, enabling the robot to traverse challenging terrain while minimizing the likelihood of accidents or breakdowns. Another approach to pathfinding in autonomous robotics utilizes machine learning algorithms to learn from past navigation experiences and adapt to changing environmental conditions. These algorithms refine their pathfinding strategies by analyzing sensor data and feedback from previous missions, enhancing the robot's ability to navigate autonomously in diverse terrain types.

This study aims to enhance autonomous robots' operational efficiency and safety by minimizing energy consumption and reducing navigational risks.

1.1 Problem Statement

Current robotics face challenges in balancing energy efficiency and navigational safety. Efficient energy management is crucial for the operational duration of battery-powered robots, while navigational safety is vital for avoiding obstacles and preventing damage. Existing pathfinding solutions often prioritize one aspect and struggle to adapt to changing environmental conditions. There is a need for a pathfinding approach that can optimally balance energy efficiency and safety in real time, especially in complex scenarios like an autonomous drone delivering medical supplies through risky areas.

1.2 Objectives

This study aims to improve autonomous robots' operational efficiency and safety by minimizing energy consumption and reducing navigational risks. Achieving these objectives will enhance robot performance, extend operational life, and ensure safer operations in unpredictable environments.

1.3 Contribution

This paper makes two primary contributions. First, it utilizes Monte Carlo simulations to explore a wide range of potential navigational paths under varied conditions, generating a comprehensive dataset for analysis. Second, it applies Pareto optimality principles to identify paths that optimally balance energy consumption and navigational safety. This approach enables the dynamic selection of optimal paths based on real-time assessments, significantly advancing the field of robotic navigation. The integration of these methods offers a sophisticated tool that adeptly handles the complexities of diverse and dynamic environments.

2 Literature Review

Path planning in robotics constitutes a vast and evolving research domain, tracing a long history, primarily focused on determining the most efficient route from a starting point to a destination while circumventing obstacles and other constraints [6]. This field has witnessed a resurgence of interest and significant expansion, especially with the proliferation of autonomous vehicles and drones [18]. A comprehensive review of path-planning algorithms is extensively covered in [4, 15]. Graph-based search algorithms emerge prominently among the myriad strategies employed in path planning due to their broad applicability. Techniques such as Dijkstra's algorithm and its heuristic-enhanced counterpart, the A^* algorithm, transform the environment into a graph to ascertain the shortest path between nodes. Dijkstra's algorithm is significant for its foundational role in operations research and computer science. It is often utilized to determine the shortest paths within a graph, assuming accessibility from the start node to all other nodes [1, 13]. Despite their usefulness, these algorithms are computationally intensive when applied to extensive environments [16] and are typically more suited to static rather than dynamic settings [17]. These algorithms primarily emphasize minimizing distance or cost, neglecting other crucial factors such as energy consumption, navigational risk, or the dynamic nature of the environment. In response to certain limitations of traditional graph-based algorithms, researchers have introduced sampling-based algorithms like rapidly exploring random trees (RRTs), probabilistic roadmaps (PRMs) [5], and adaptive path finding algorithm [7]. These techniques are praised for effectiveness in high-dimensional spaces and have demonstrated success in various complex environments, including dynamic ones. Despite these advancements, a significant gap persists in the conventional pathfinding algorithm's ability to holistically consider multiple objectives simultaneously. Most algorithms are not inherently designed to balance energy efficiency, safety, and operational constraints, which are crucial in many practical robotic applications. This limitation underscores the necessity for a more integrated approach to evaluate and optimize paths based on comprehensive performance criteria rather than merely focusing on the shortest route or time. In the literature, many path-planning approaches have continued to increase over the years.

Selecting the most appropriate approach based on specific requirements, such as locomotion, can be challenging. To address these multifaceted requirements, this research proposes a novel framework that transcends traditional pathfinding by integrating multi-objective optimization techniques, such as Pareto efficiency. Doing so aims to develop a more robust and versatile solution capable of navigating the complex trade-offs encountered in dynamic and unpredictable environments.

3 Methodology

This section outlines the comprehensive methodology adopted in this study, including the system model used for simulations, the Monte Carlo simulation parameters, the definition and calculation of objective functions, and the extraction of Pareto optimal solutions.

3.1 Network Model

The multi-objective path find problem involves choosing a combination of various available paths from a selected origin to a particular destination and considering the effects of implementing these combinations, such as navigational risks, energy consumption, time to destination, and path smoothness, to name a few.

The graph G is defined as a tuple (V, E, M), where V represents the set of vertices, E represents the set of edges, and M represents any additional information associated with the graph. Let V be the set of vertices that represent waypoints or nodes in the multimodal robotic path network. E is the set of edges that connect these vertices. M represents the various path types, such as paved, unpaved, or other specific terrain characteristics that are available within the network. Every edge e in the set E, which connects two vertices $\{u, v\}$ in the set V, is represented as (u, v) and has a weight $\delta(u, v; m)$ associated with it. Here, m belongs to the set M and denotes the type of path connected with the edge. The weight $\delta(u, v; m)$ represents the cost, which might encompass characteristics such as energy usage, trip duration, or risk, linked to moving from vertex u to vertex v over a path of type m.

Define $\phi(u^i, v^j)$ as the cost of transitioning from vertex u to vertex v when switching from path type i to path type j, where i, j are elements of the set M. This cost represents the possible need for extra resources or the hazards involved in modifying the robot's navigation approach to adapt to changes in path parameters.

It should be noticed that the kind of path for an edge represented by (u^i, v^j) is j. If i is equal to j, then the function ϕ applied to u raised to the power of i and v raised to the power of j equals zero. This implies no cost is associated with transitioning between vertices u and v if the path type remains the same. In addition, if u is the initial vertex of the route, the function $\phi(u^i, v^j)$ will be equal to 0. This indicates no cost is paid at the starting point, regardless of the path type.

Definition 1 u is adjacent to (or is a neighbor of) v if $(u, v) \in E$.

Definition 2 The set of all neighbors of u is the neighborhood of u and is denoted $N(u)$.

Definition 3 A path is defined as an ordered set of vertices $(v_1, \ldots, v_t), t > 1$ where t is an integer, such that $(v_t, v_{t+1}) \in E$ and the vertices are not repeated.

The multi-objective optimization problem can be formulated as follows:

$$F(x) = [F_1(x), F_2(x), \ldots, F_p(x)] \tag{1}$$

where

- $[F_1, F_2, \ldots, F_p]$ are the applicable objective functions which may include energy consumption, navigational risk, travel time, and cost, each function reflecting different operational priorities of the pathfinding system.
- x is a path as defined in Definition 3.
- $x = (N_O, N_2^m, N_3^m, \ldots, N_D^m)$ represents the sequence of vertices to be traversed to reach the destination N_D from the origin N_O, with m indicating the type of path used to arrive at each vertex. Each path type affects the evaluation of the objective functions differently.

3.2 Monte Carlo Simulations

In this step, we employ Monte Carlo simulation (MCS) to generate a diverse set of possible routes from the origin, N_O, to the destination, N_D. Monte Carlo simulations

(MCSs) are widely used across different fields due to their ability to handle complex systems and quantify uncertainty by generating numerous possible outcomes based on probability distributions. The applications of MCS are varied, including its use in benefit-cost analysis [12] and crash cost estimation [8]. Additionally, MCS is employed in project management for choosing risk response strategies [11], in planning for emergency relief transportation [9], and in creating synthetic schedules for detailed microsimulations of travel behavior [3]. Notably in transport research, MCS is celebrated for their role in modeling uncertainty and creating a range of scenarios, leading to significant insights into extensive routing options and the reliability of network travel times [10, 14]. In this instance, MCS is utilized to explore the stochastic nature of route selection and mode choice within the defined network.

The process begins at the origin vertex N_O during each simulation iteration ($k = 1$ to t). Here, the next vertex is randomly selected from the set of neighboring vertices $(N_2, \ldots, N_{D-1})$, which are identified through the network's adjacency matrix. This matrix specifies which vertices are directly accessible from any given vertex, thus facilitating a realistic simulation of possible movements. After selecting the vertex, a path type is also chosen randomly from the list of available options. This list corresponds to the types of paths, such as paved, unpaved, or rough terrain, that connect the vertex to the next. The process of selecting a vertex and path type continues iteratively until the destination N_D is reached.

If at any point a vertex has no neighboring vertices (indicating a dead-end within the network model), the simulation for that particular iteration terminates and restarts from N_O for the next iteration. This approach ensures that only feasible paths are considered in the analysis.

The outcome of this MCS process is a series of potential routes, each characterized by a sequence of vertices and associated paths of travel. These routes are then evaluated against the applicable objective functions, Eq. (1), to be screened in Sect. 3.3.

3.3 Pareto Optimal Solution Finding

In this methodology phase, we utilize the Pareto optimality principle [2] to identify the most efficient routes from the sets generated by Monte Carlo simulations. The Pareto optimality approach is well-suited for multi-objective problems where improving one objective without worsening another is desired. This approach allows decision-makers to:

- Assess the impact of different decision-making criteria directly.
- Choose from routes that represent the best compromise according to multiple criteria.
- Adapt decisions dynamically based on varying priorities or operational conditions.

To implement the Pareto optimality principle, we construct a Pareto frontier—a set of all Pareto optimal routes. This frontier represents the best possible trade-offs among the competing objectives.

Each route from the Monte Carlo simulation is evaluated against all others based on the defined objective functions. A route x is considered Pareto optimal if there does not exist another route y such that:

$$F_i(y) \leq F_i(x) \quad \text{for all } i \quad \text{and} \quad F_j(y) < F_j(x) \quad \text{for at least one } j, \tag{2}$$

where each F_i represents an objective function being minimized. This condition ensures that a Pareto optimal route x cannot be improved in any single objective without degrading it in at least one other.

3.4 Decision Support System

The Pareto optimal paths are then presented to the users, enabling them to make decisions based on their priorities in relation to the values of the applicable objective functions.

4 Applied Model

This section describes the experimental design used to validate the proposed methodology, detailing the simulation parameters, metrics for evaluation, and specific test scenarios used to assess the performance of the Pareto optimal pathfinding strategy under varying conditions. In this example, there are 20 vertices and 66 edges as shown in Fig. 1. There are at most two possible choices of path available which corresponds to paved path and unpaved between any vertices. Table 1 provides the edge weights which are associated costs between vertices for various paths if available. A dash (–) denotes that there is no connection between the associated vertices for the respective mode.

For our experiment, we have used only two objective functions, energy consumption, E and navigational risk, R, which are defined as follows:

Energy Consumption (F_1) Energy consumption is a critical metric for assessing the operational efficiency of robotic paths. It is influenced by several factors including distance traveled, the terrain type, speed, and acceleration. The general formula for calculating energy consumption (E) is given by:

$$E = \sum_{i=1}^{n} d_i \times c_i \times g(v_i, a_i, s_i) \tag{3}$$

Table 1 Vertex weights with respect to available paths

Edge	Paved	Unpaved	Edge	Paved	Unpaved
(1, 2)	2	1	(9, 14)	6	3
(1, 3)	10	5	(10, 11)	4	2
(1, 4)	3	1.5	(10, 14)	5	2.5
(2, 7)	3	1.5	(11, 12)	3	1.5
(2, 8)	9	4.5	(11, 15)	4	2
(3, 6)	7	3.5	(12, 16)	2	1
(3, 7)	1	0.5	(13, 18)	6	3
(4, 5)	2	1	(13, 19)	8	4
(5, 6)	9	4.5	(14, 15)	6	3
(5, 12)	2	1	(14, 17)	2	1
(6, 11)	3	1.5	(14, 18)	4	2
(7, 10)	7	3.5	(14, 20)	8	4
(7, 11)	3	1.5	(15, 16)	7	3.5
(8, 9)	5	2.5	(15, 17)	5	2.5
(8, 10)	1	0.5	(17, 20)	8	4
(9, 13)	10	5	(18, 20)	1	0.5
			(19, 20)	4	2

where

- d_i is the distance of the i-th segment of the path.
- c_i is the energy cost per unit distance for the i-th segment, which varies depending on the terrain type. Different terrains, such as asphalt, gravel, or grass, have different resistances and therefore different energy costs.

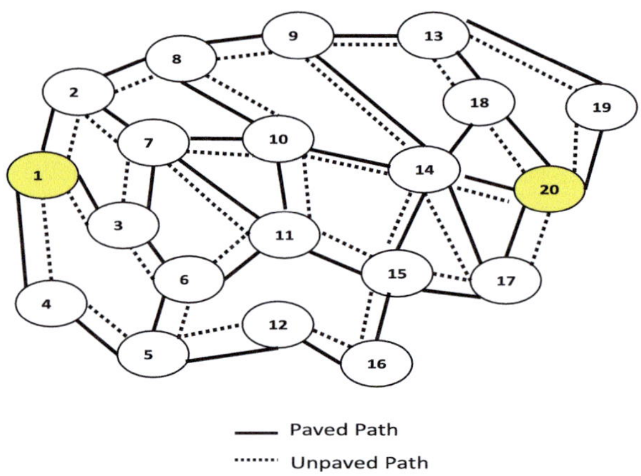

Fig. 1 Multi-path network

- $g(v_i, a_i, s_i)$ is a function that accounts for the speed (v_i), acceleration (a_i), and slope (s_i) on the i-th segment. This function reflects how changes in speed and acceleration, as well as the angle of ascent or descent, impact energy usage.

Navigational Risk (F_2) Navigational risk quantifies the potential hazards a robot may encounter along its path and the impact of these hazards on the mission's success. The calculation of navigational risk (R) involves evaluating both the probability of encountering specific hazards and the severity of their consequences. The general formula for navigational risk is given by:

$$R = \sum_{i=1}^{n} P_i \times C_i \tag{4}$$

where

- P_i represents the probability of encountering the i-th hazard on a given segment of the path.
- C_i is the consequence of the i-th hazard, quantified in terms of its impact on the robot's mission, which may include delays, damage, or total mission failure.

Types of Hazards Considered

- **Environmental Hazards**: Includes factors such as terrain ruggedness, weather conditions (e.g., rain, snow, fog), and natural obstacles (e.g., water bodies, rocky areas).
- **Operational Hazards**: Relates to the robot's interaction with dynamic entities such as pedestrians, other vehicles, or animals.
- **Technical Hazards**: Encompasses potential mechanical or software failures that could impede navigation.

Consequence Evaluation: The consequences (C_i) are typically rated based on their severity:

- **Minor**: Minimal impact, possibly causing slight delays or minor deviations from the planned route.
- **Moderate**: More significant impact, potentially causing damage that requires intervention but is not catastrophic.
- **Severe**: Major impact, likely leading to mission failure or extensive damage to the robot.

4.1 Simulation Parameters

Energy Consumption: For our purposes, we assume that the $g(v_i, a_i, s_i) = 1$ in Eq. (3). Paths are assigned energy consumption rates based on their type, with paved paths having a lower energy cost per unit distance ($c(m) = 2$) compared to unpaved paths ($c(m) = 3$). Paved paths generally offer less resistance to movement than unpaved paths. The smoother surface of paved paths reduces the friction and effort required by the robot to move, leading to lower energy consumption per unit distance. In contrast, unpaved paths may be composed of materials like gravel, dirt, or grass, which increase rolling resistance and, consequently, the energy needed to traverse these surfaces.

Navigation risk: For the experiment, we have used ($P = 0.1, C = 2.0$) for paved and ($P = 0.3, C = 1.5$) unpaved paths, in Eq. (4). The higher frequency of hazards on unpaved paths justifies a higher probability of encountering risks, whereas the lower severity aligns with the "Minor" (1.5) consequence. Conversely, the "Moderate" (2) consequence on paved paths suggests that while hazards are less frequent, their impact is more significant, demanding more robust mitigation strategies. The operational context significantly influences these assessments. For instance, in urban environments, paved areas may pose risks due to human activities and traffic, necessitating a moderate consequence rating due to the potential for more damaging interactions. In contrast, robots navigating in natural or less structured environments might frequently encounter minor terrain challenges that do not severely affect their operations.

Transition Costs Transition costs between path types are also considered, reflecting the additional energy required and potential hazards when switching from one path type to another. Specific transition costs in energy and risk are added when the path type changes:

- **Paved to unpaved**: $E_{\text{transition}} = 5$.
  This higher cost reflects the additional energy required to adapt the robot's systems for navigating less stable and rougher terrain.
- **Paved to unpaved**: $R_{\text{transition}}$: (Probability : 0.05; Consequence : 1.8) = 0.09.
  The increased consequence reflects the unexpected challenges and potential hazards (e.g., slipping, loss of control) when first encountering rougher surfaces.
- **Unpaved to Paved**: $E_{\text{transition}} = 3$.
  Transitioning back to a paved path involves less energy than the opposite transition but still requires adjustments, such as recalibrating for higher speeds and smoother navigation.
- **Unpaved to Paved**: $R_{\text{transition}}$: (Probability: 0.03; Consequence: 2.2) = 0.066.
  Despite a lower probability, the consequence is higher due to the increased potential for high-speed incidents when transitioning back to smooth, paved surfaces.

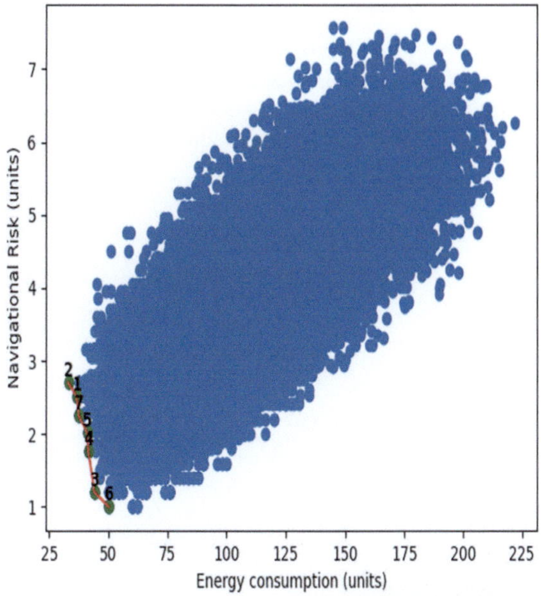

Fig. 2 Pareto optimal solutions

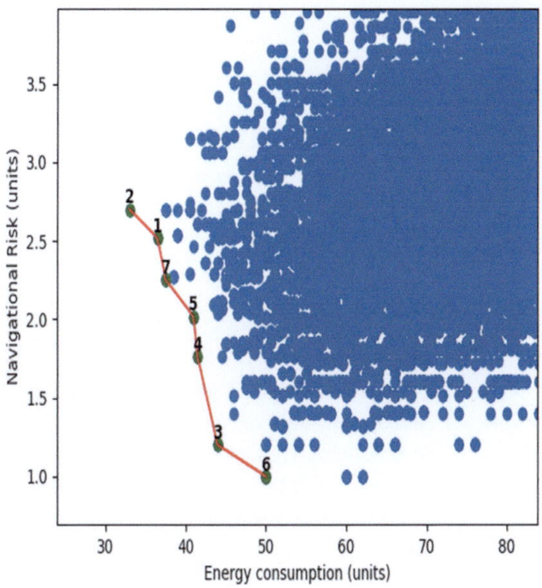

Fig. 3 Pareto solutions (magnified)

4.2 Results and Discussion

Once the Pareto front is identified, unique solutions are extracted. The plot of all solutions with Pareto front extracted is shown in Figs. 2 and 3. Table 2 depicts the details of the obtained Pareto optimal solutions obtained.

Table 2 Summary of Pareto optimal solutions with respective metrics

Solution #	Details: path, energy consumption, and navigational risk
1	Path: (1–2: unpaved)-(2–8: unpaved)-(8–10: unpaved)-(10–14: unpaved)-(14–18: unpaved)-(18–20: paved)
	Energy consumption: 36.5
	Navigational risk: 2.516
2	Path: (1–2: unpaved)-(2–8: unpaved)-(8–10: unpaved)-(10–14: unpaved)-(14–18: unpaved)-(18–20: unpaved)
	Energy consumption: 33.0
	Navigational risk: 2.7
3	Path: (1–2: paved), (2–8: paved)-(8–10: paved)-(10–14: paved)-(14–18: paved)-(18–20: paved)
	Energy consumption: 44.0
	Navigational risk: 1.2
4	Path: (1–2: unpaved)-(2–8: unpaved)-(8–10: paved)-(10–14: paved)-(14–18: paved)-(18–20: paved)
	Energy consumption: 41.5
	Navigational risk: 1.766
5	Path: (1–2: unpaved)-(2–8: unpaved)-(8–10: unpaved)-(10–14: paved)-(14–18: paved)-(18–20: paved)
	Energy consumption: 41.0
	Navigational risk: 2.016
6	Path: (1–2: paved)-(2–8: paved)-(8–10: paved)-(10–14: paved)-(14–20: paved)
	Energy consumption: 50.0
	Navigational risk: 1.0
7	Path: (1–2: unpaved)-(2–8: unpaved)-(8–10: unpaved)-(10–14: unpaved)-(14–20: unpaved)
	Energy consumption: 37.5
	Navigational risk: 2.25

The results presented in Table 2 demonstrate a diverse array of Pareto optimal solutions, each balancing energy consumption and navigational risk in different ways. Notably, the solutions vary significantly in their composition of path types, leading to important insights into how path selection impacts the performance metrics of autonomous navigation systems.

4.2.1 Impact of Path Type on Performance Metrics

Solutions #3 and #6, which consist entirely of paved paths, exhibit the highest energy consumption but the lowest navigational risks. This trend supports the hypothesis that paved paths, while energetically costly due to smoother and potentially faster travel, offer safer navigation conditions. Conversely, solutions like #1, #2, and #7, primarily utilizing unpaved paths, show reduced energy usage but increased risk, highlighting the trade-offs when prioritizing energy efficiency over safety.

The intermediate solutions (#4 and #5), which include transitions between paved and unpaved segments, represent strategic compromises that moderate both energy consumption and risk. These solutions are particularly noteworthy as they reflect real-world scenarios where a path might not uniformly consist of one type of terrain.

4.2.2 Strategic Implications for Robotic Pathfinding

The variation in energy and risk profiles across the solutions suggests that strategic planning of path type transitions can be critical in optimizing robotic operations. For instance, deploying robots in environments where safety is a higher priority might justify the higher energy costs associated with paved paths. In contrast, operations in energy-limited scenarios might necessitate accepting higher risks associated with unpaved paths.

One of the most significant advantages highlighted by this study is the empowerment of users to make informed decisions based on their specific operational priorities. The diverse array of Pareto optimal solutions derived from the Monte Carlo simulations and Pareto analysis offers users a menu of options, each balancing energy consumption and navigational risk differently. Each Pareto optimal solution represents a distinct navigation strategy, allowing users to customize their approach based on varying criteria such as safety, energy efficiency, or a balance of both. This flexibility is particularly valuable in environments where conditions can change rapidly, or different tasks may require different operational emphases.

4.2.3 Enabling Strategic Planning

The model's ability to provide multiple optimized routes also facilitates strategic planning over longer terms or across broader operational areas. Organizations can plan missions or deployments by selecting paths that align best with their strategic goals, operational constraints, and risk management policies.

4.3 Future Directions and Improvements

Several promising areas for further research still exist in exploring these as future directions; researchers can advance the field of autonomous robotics, developing navigation systems that are adaptive and efficient and capable of undertaking the complex challenges of real-world environments. This future works include:

- **Dynamic Path Adjustments**: One path worth exploring involves integrating real-time data into the pathfinding algorithm to enable dynamic path adjustments

based on evolving environmental conditions or operational needs. For instance, an autonomous drone could adapt its route to newly identified obstacles or emerging hazards in a search-and-rescue scenario, ensuring more efficient and adaptable navigation strategies.
- **Algorithm Efficiency**: Algorithm Efficiency: Another crucial area for future investigation is improving the algorithm's computational efficiency. Streamlining the process of generating and evaluating potential paths, particularly in complex environments, can significantly enhance autonomous navigation systems' overall performance and scalability. This could involve optimizing data structures and refining search heuristics to expedite the path-planning process.
- **Expanded Objective Functions**: Future studies could broaden the optimization capabilities of the pathfinding algorithm by incorporating additional objective functions. In addition to considering energy consumption and navigational risk, objectives such as time to destination or cost of operation could be integrated into the optimization framework. For instance, a path-planning algorithm in agricultural robotics could prioritize energy and time efficiency when navigating fields for crop monitoring or spraying operations, leading to more effective and cost-efficient farming practices.

5 Conclusion

In conclusion, this study has provided valuable insights into multi-objective pathfinding in autonomous robotics, focusing on strategic path selection considering energy consumption and navigational risk profiles. The findings confirm the effectiveness of the proposed model and highlight its significant potential in enhancing autonomous navigational strategies across diverse and dynamic environments. By integrating multi-objective optimization techniques, the study has demonstrated the feasibility of balancing competing objectives such as energy efficiency and navigational safety. Experimental validation has shown the model's ability to generate paths offering favorable trade-offs between these objectives, thereby improving autonomous robots' operational capabilities and reliability. An essential contribution of this research lies in its importance in adapting the pathfinding framework to diverse and incorporating changing environments. In real-world scenarios with unpredictable conditions, the ability to dynamically evaluate and adjust path options based on evolving factors like unevenness and obstacle density becomes crucial. The proposed model provides a versatile decision-support tool that permits users to make informed decisions aligned with operational priorities. Moreover, the study's implications extend to the broader field of autonomous robotics. Optimizing path selection to minimize energy consumption and navigational risk enables robots to operate more efficiently and safely across various applications, from search-and-rescue missions to planetary exploration. Additionally, insights gained can inform the development of advanced path-planning algorithms, paving the way for further advancements in autonomous navigation technology.

References

1. Andiwijayakusuma D, Mardhi A, Savitri I, Asmoro T (2019) A comparative study of the algorithms for path finding to determine the adversary path in physical protection system of nuclear facilities. J Phys Conf Ser 1198(9) (2019)
2. Arora JS (2017) Multi-objective optimum design concepts and methods. In: Arora JS (ed) Introduction to optimum design, fourth edn. Academic Press, Boston, pp 771–794 (Chapter 18). https://doi.org/10.1016/B978-0-12-800806-5.00018-4. https://www.sciencedirect.com/science/article/pii/B9780128008065000184
3. Balmer M, Meister K, Rieser M, Nagel K, Axhausen K (2008) Agent-based simulation of travel demand: structure and computational performance of MATSim-T. https://doi.org/10.3929/ethz-a-005626451
4. Karur K, Sharma N, Dharmatti C, Siegel JE (2021) A survey of path planning algorithms for mobile robots. Vehicles 3(3):448–468
5. Kavraki L, Svestka P, Latombe J, Overmars M (1996) Probabilistic roadmaps for path planning in high-dimensional configuration spaces. IEEE Trans Robot Automat 12:566–580
6. LaValle SM (2006) Planning algorithms. Cambridge University Press
7. Ng MK, Chong YW, Ko KM, Park YH, Leau YB (2020) Adaptive path finding algorithm in dynamic environment for warehouse robot. Neural Comput Appl **32**(17):13155–13171
8. Prakash S (2018) Alternative approach to estimating crash costs for cost-benefit analysis using Monte Carlo simulation. In: ATRF 2018—Australasian transport research forum 2018, proceedings
9. Prakash S (2019) Emergency relief goods transportation strategies—a Monte Carlo simulation approach. Australasian Transport Research Forum (ATRF)
10. Prakash S, Jokhan A (2016) An optimal cane delivery scheduling using the Monte Carlo method. In: ATRF 2016—Australasian transport research forum 2016, proceedings
11. Prakash S, Jokhan A (2017) Monte Carlo for selecting risk response strategies. In: ATRF 2017—Australasian transport research forum 2017, proceedings
12. Prakash S, Mitchell D (2015) Probabilistic benefit cost ratio—a case study. In: ATRF 2015—Australasian transport research forum 2015, proceedings
13. Prakash S, Mitchell D (2018) Estimating freight movements using Dijkstra's algorithm. In: Australasian transport research forum (ATRF). Darwin, Australia
14. Prakash S, Sharma B (2024) An optimized hybrid approach for path planning: a combination of lyapunov functions and high-level planning algorithms. In: Das S, Saha S, Coello Coello CA, Bansal JC (eds) Advances in data-driven computing and intelligent systems. Springer Nature Singapore, Singapore, pp 425–436. https://doi.org/10.1007/978-981-99-9524-0_32
15. Reda M, Onsy A, Haikal AY, Ghanbari A (2024) Path planning algorithms in the autonomous driving system: a comprehensive review. Robot Autonom Syst 174:104630. https://doi.org/10.1016/j.robot.2024.104630. https://www.sciencedirect.com/science/article/pii/S0921889024000137
16. Russell S, Norvig P (2009) Artificial intelligence: a modern approach. Prentice Hall
17. Stentz AT (1994) Optimal and efficient path planning for partially-known environments. In: Proceedings of (ICRA) international conference on robotics and automation, vol 4, pp 3310–3317
18. Thrun S, Montemerlo M et al (2006) Stanley, the robot that won the DARPA grand challenge. J Field Robot 9:661–692

Emotion Recognition Using GRU-Based Neural Networks on EEG Data

Ankit Tomar, Paras Jain, Vishan Kumar Gupta, Kamal Kant Verma, Ashish Kumar, and Vidisha

Abstract This paper will provide an update on the state of emotion recognition using EEG signals from 2016 to 2019, since the last thorough review was carried out between 2009 and 2016. This cutting-edge review concentrates on the components of study size, EEG hardware, machine learning classifiers, presentation strategy, and emotion stimuli type. Based on this state-of-the-art review, we propose various avenues for future research, such as taking a different tack when presenting the stimuli using virtual reality (VR). Considering this, a new section that reviews only VR studies in this field of study is provided, serving as inspiration for the suggested method of presenting stimuli through the use of virtual reality. Our proposed methodologies also highlight the integration of advanced neural network architectures, such as Gated Recurrent Unit network (GRU) that is a type of recurrent neural network (RNN), which has shown promising results in emotion recognition tasks. These approaches offer improved accuracy and robustness, setting the stage for more nuanced and reliable emotion recognition systems. We have gathered EEG dataset for emotion recognition from Kaggle.

Keywords EEG · Emotion · Convolutional neural network · GRU · Feature extraction · Data preprocessing

A. Tomar
Department of CSE, Graphic Era Deemed to be University, Dehradun, UK, India

P. Jain
SCEAI, VIT Bhopal University, Bhopal, MP, India

V. K. Gupta (✉) · Vidisha
Amity School of Engineering and Technology, Amity University Punjab, Mohali, India
e-mail: vishangupta@gmail.com

K. K. Verma
Department of CSE, Lovely Professional University, Phagwara, Punjab, India

A. Kumar
Department of CSE, Jaypee Institute of Information Technology, Noida, UP, India

H. Sharma et al. (eds.), *Artificial Intelligence: Theory and Applications*, Lecture Notes in Networks and Systems 5589, https://doi.org/10.1007/978-981-96-1687-9_3

1 Introduction

Emotions are essential to human nature and have a significant impact on thought processes. Emotion is frequently linked to perception, human interaction, reasoning, and to some extent, human intelligence itself. Reliable and implementable methods for identifying human emotional states are needed, as the research community's interest in creating some meaningful 'emotional' interactions between people and computers grows. The research community is very interested in the recent advancements in electroencephalography (EEG) for emotion recognition because consumer-grade wearable EEG solutions can offer an inexpensive, portable, and straightforward method of identifying emotions [1].

Even though human emotions are fundamental to daily life, science still knows very little about them. The development of society depends on advances in affective sciences, which is why the incorporation of machines that can detect emotions is necessary. This integration has the potential to increase efficiency and lower expenses in several areas, such as healthcare, education, and military training. Emotional assessments have historically relied on subjective accounts and self-assessment tools such as the Self-Assessment Manikin (SAM). However, people's reluctance or low self-awareness could make these techniques unreliable [2]. In response, physiological signals—especially brainwave signals—provide a quicker and more accurate assessment of people's emotional states [3]. Examining these signals offers insightful information for evaluating mental health issues.

The design and price of EEG headsets, which are impacted by the kinds of electrodes, their quantity, and where they are placed on the scalp, determine how well emotions can be recognized [4]. Issues like muscle movement artefacts, electrode sensitivity, and outside interferences must all be carefully taken into account when gathering data. For machine learning algorithms to accurately recognize emotions, filtering algorithms and signal transformations are necessary [5]. This paper updates a thorough review that covered a number of topics, such as electrode positions, brainwave bands, emotions, EEG, and a comparison between clinical and inexpensive EEG headsets. It also explores how emotions are processed by the brain, uses of virtual reality, and a review of earlier research on the categorization of emotions. It ends with a discussion and an overview of the results.

1.1 Problem Statement

With varied degrees of success, numerous electroencephalography studies on ER have been carried out. The various sample types used in the study are the cause of these discrepant results. Things like emotional selection, the testing environment, predesigned data, and feature selection frequently show the difference. It is challenging to compare and select the method that is unquestionably the best because of all these factors. As such, there is always space for improving classifiers that are appropriate for particular uses.

1.2 Tools Used

i. **Python:** The project is developed using Python, harnessing its expressive syntax and extensive library support.
ii. **NumPy:** NumPy, under the alias NumPy, is employed for its array and matrix manipulation capabilities, facilitating efficient mathematical operations crucial for the project's computations [6].
iii. **Scikit-learn:** Scikit-learn is a well-liked Python machine learning library with straightforward and effective tools for data mining and analysis. It offers different calculations for arrangement and relapse undertakings, which can help build feeling acknowledgment models.
iv. **OpenCV:** OpenCV assumes a basic part in different phases of facial feeling acknowledgment, from face discovery and milestone confinement to picture preprocessing and information expansion. It fills in as a major device related to AI strategies for building strong and exact feeling acknowledgment frameworks [7].

2 Literature Survey

Emotion recognition (ER) has emerged as a pivotal area in the realm of Human–Robot Interaction (HRI), particularly with the advent of Brain–Computer Interfaces (BCIs). Historically, emotion recognition methods primarily relied on observable cues like facial expressions and vocal patterns, achieving an accuracy range of 70–80% in controlled environments [8]. However, these approaches face limitations in capturing nuanced emotions that are not overtly expressed. In response, researchers have shifted their focus toward more objective and physiological indicators of emotions, with a specific emphasis on electroencephalography (EEG). EEG, capable of recording brain electrical activity, offers a promising avenue for real-time and non-invasive emotion recognition [9].

As applications of brain–computer interfaces continue to advance, so does research in ER. Allowing robots to exhibit social behaviors, social means being able to read the thoughts of others and act, accordingly, will lead to the benefits of HRI. In the past, attempts have been made to recognize different characters with varying degrees of accuracy using different techniques such as faces and voices [10]. Faces can be recognized with 70–80% accuracy in a controlled environment.

Emotions are not always shared, and sometimes people do not share them. In psychology, a clear distinction is made between physical arousal and behavior (emotion). The face and voice represent the second aspect of thought: expression,

Table 1 Literature survey table

Authors	Years	Methodology	Data set	Key findings
Calvo and D'Mello [5]	2010	Interdisciplinary review of affect detection	Various	Highlighted the importance of integrating multiple models and methods for affect detection
Koelstra et al. [8]	2012	Multimodal emotion analysis using EEG and other signals	DEAP	Demonstrated the feasibility of using physiological signals for emotion recognition
Casson [12]	2019	Wearable EEG technology review	Various EEG datasets	Discussed advancements in EEG technology and its applications beyond traditional settings
Gill and Singh [9]	2021	Deep learning approaches for emotion recognition	Various EEG datasets	Provided a comprehensive survey on the use of deep learning techniques for emotion recognition
Abadi et al. [10]	2015	Multimodal database for decoding affective responses	DECAF	Developed a multimodal database to facilitate research on decoding affective physiological responses

which can be easily modified and expressed [11]. This means there is no purpose. Therefore, heart rate, galvanic skin response, electroencephalogram, pupil dilation, etc. Physiological factors are taken into consideration. However, among these, many cognitive studies have been conducted using electroencephalography. Table 1 is showing the literature survey as per their author details, years in which this research conducted, methodology, dataset used, and various key findings regarding their work.

3 Methodology

The research design involves the selection of an appropriate pre-existing dataset that aligns with the study's objectives and the adoption of a quantitative approach for subsequent analysis using deep learning techniques. Participants are not directly involved in the data collection process since the EEG dataset is pre-existing. Figure 1 shows the methodology of the proposed work performed and discussion about these concerned steps are elaborated in corresponding sections.

3.1 Dataset Description

The dataset used in this study comprises EEG brainwave recordings associated with human emotional states. Each sample in the dataset consists of features extracted from the Fast Fourier Transform (FFT) of EEG signals, capturing patterns indicative of different emotional states [11]. The dataset includes a total of N samples, with

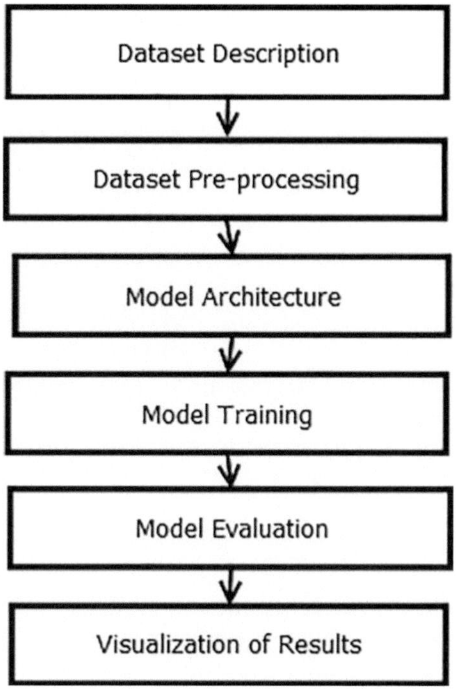

Fig. 1 Methodology of the proposed work

each sample containing M FFT features and a corresponding emotional label. The emotional labels are categorized into three classes: 'NEGATIVE,' 'NEUTRAL,' and 'POSITIVE.' A preliminary analysis of the dataset reveals the distribution of samples across different emotional categories as follows: The groups of samples are 'NEGATIVE,' 'NEUTRAL,' and 'POSITIVE' referring to N1 samples, N2 samples, and N3 samples, respectively.

3.2 Dataset Preprocessing

In this case feeding the data in the neural network model, the data undergoes several preprocessing steps. Firstly, the categorical emotional labels are converted from nominal format to numerical format using label mapping technique. After that, the data is divided into the training and testing data frames using the stratified train-test split that assigns 70% of data for training purpose and 30% for testing. This helps in making sure that both the training and the testing datasets have the right proportions for each emotional category [13]. Furthermore, due to variations in ranges of input features, normalization or standardization is done on features before feeding into the network in order to enhance model converge.

3.3 Model Architecture

The architecture used for emotion classification tasks is the neural network structure that contains multiple layers; the layers are aimed at capturing temporal dependencies from the given EEG signals. The model starts with an input layer designed to take the FFT features as input after passing through the preprocessing stage. This is then succeeded by a Gated Recurrent Unit (GRU) layer which consists of 256 units to help the model capture temporal patterns or sequence of EEG data [14]. The GRU layer is selected due to the stability of gradients, which is an essential factor when training deep networks, including recurrent models. After that, there is a flatten layer that changes the output of the GRU layer to single dimension and finally, the output layer with SoftMax activation function that generates class probability for emotion output.

3.4 Model Training

With the model architecture defined, the next step involves compiling and training the neural network on the preprocessed dataset. The model is compiled using the Adam optimizer and sparse categorical cross-entropy loss function, which are well-suited for multi-class classification tasks. During training, the model is trained on the training dataset for a specified number of epochs with a batch size of 32. To prevent overfitting, early stopping is implemented as a callback mechanism, monitoring the validation loss and restoring the best weights when the validation loss fails to decrease for a certain number of epochs.

3.5 Model Evaluation

Once training is completed, the trained model is evaluated on the held-out testing dataset to assess its performance. The evaluation metrics include test accuracy, which measures the proportion of correctly predicted emotional labels, and the generation of a confusion matrix and classification report [15]. These metrics provide insights into the model's ability to accurately classify emotional states and identify any potential biases or misclassifications across different emotional categories.

3.6 Visualization of Result

To aid in the interpretation of results, visualizations are utilized to illustrate key aspects of the model's performance. Figure 2 depicts a sample plot of FFT features

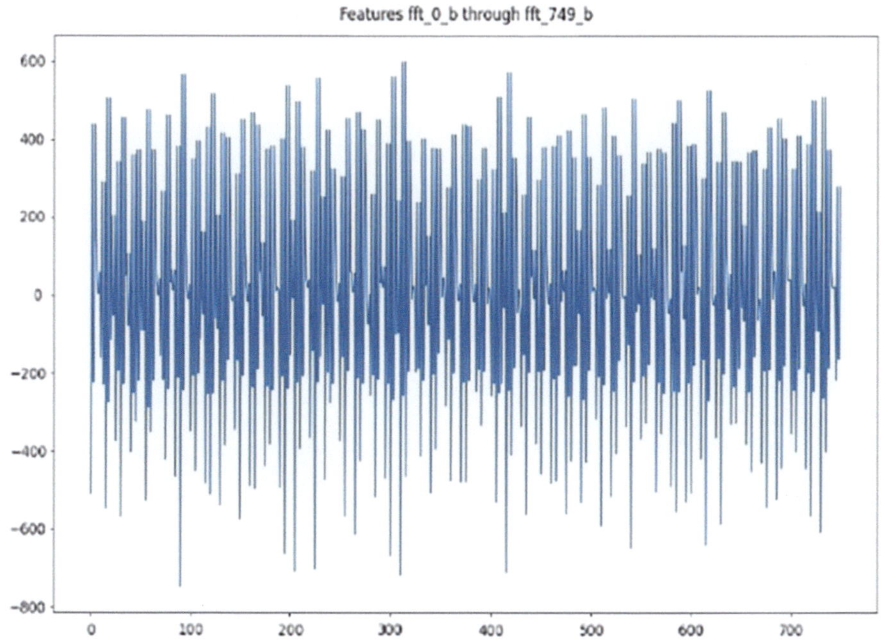

Fig. 2 Extracted FFT features

extracted from EEG signals, providing a visual representation of the input data. Additionally, Fig. 3 presents a heatmap of the confusion matrix generated during model evaluation, highlighting the distribution of predicted versus actual emotional labels [7]. These visualizations enhance the understanding of the model's behavior and its effectiveness in capturing patterns indicative of different emotional states. Table 2 is showing the stepwise methodology of working.

This methodology section provides a detailed overview of the processes involved in the study, including dataset description, data preprocessing, model architecture, training, evaluation, and visualization of results [16]. Every process is explained step by step to allow the reader to follow the sequence of the investigation and its outcomes.

4 Result Analysis

4.1 Model Performance Evaluation

i. **Classification Metrics:** The class wise report gives detailed information of how the model has done as per different emotions classes; Fig. 4 provides classification report of various classes in term of various performance metric.

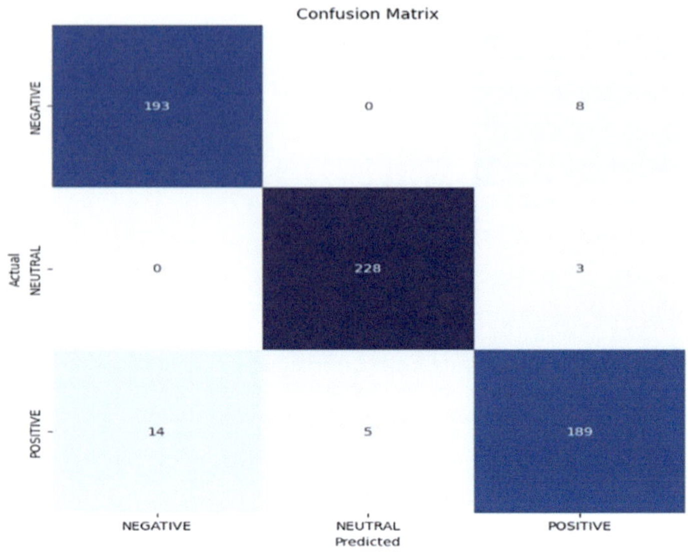

Fig. 3 Confusion matrix

Table 2 Stepwise methodology of working

Step	Description	Details
Data collection	Gathering EEG data for emotion recognition	Dataset: Emotions.csv from EEG brainwave dataset
Data preprocessing	Preparing data for model input	Label mapping: {'NEGATIVE': 0, 'NEUTRAL': 1, 'POSITIVE': 2} Train-test split: 70–30
Feature extraction	Selecting relevant EEG features	Features: fft_0_b to fft_749_b
Model architecture	Designing the GRU-based neural network	Input: Shape (number of features), GRU Layer: 256 units Output: Dense layer with SoftMax activation
Training parameters	Setting parameters for model training	Optimizer: Adam—Loss: Sparse categorical cross-entropy—Metrics: Accuracy—Batch size: 32—Epochs: 50—Early Stopping: Patience 5
Model training	Training the neural network on the preprocessed data	Validation split: 20%—Early stopping based on validation loss
Evaluation metrics	Assessing model performance on the test set	Confusion matrix: Classification report (Precision, Recall, and $F1$-score)
Visualization	Visualizing model performance	Plotting confusion matrix using Seaborn heatmap

Fig. 4 Classification report

Classification Report:

	precision	recall	f1-score	support
NEGATIVE	0.93	0.96	0.95	201
NEUTRAL	0.98	0.99	0.98	231
POSITIVE	0.94	0.91	0.93	208
accuracy			0.95	640
macro avg	0.95	0.95	0.95	640
weighted avg	0.95	0.95	0.95	640

ii. **Test Accuracy:** This indeed made the proposed neural network model have a good test accuracy of 95.312% on the EEG brainwave dataset for classifying emotions. A high accuracy means the model can objectively classify the kind of emotional state given EEG signals, proving its capacity to extract relevant patterns from the information.

iii. **Precision:** Precision measures the ability of the model in identifying positive samples by focusing on the identified positive samples. The higher precision means that whenever this model predicts an emotional class, the prediction is most often accurate. In our case the model achieves high accuracy in all the categories of emotions and the values read as follows 0.93–0.98. This means that the mathematics model is effective at making correct predictions while giving very few false positives.

iv. **Recall:** Recall measures how well the model can accurately classify a certain type of class out of all the classes in the testing data. Low specificity is also associated with high recall since low specificity means the model can capture all possible instances of a class. In our case, the model possesses high ranks of recall for all the emotional classes, these values are as follows: 91–0.98. It can be inferred from this that the model is capable of accurately classifying most instances of each emotion class.

v. ***F*1-Score:** The $F1$-score gives the balanced performance measure of model by being a harmonic means of precision and recall. Our model has high $F1$-scores across all emotional classes ranging from 0.93 to 0.98 meaning that there is an equilibrium in terms of precision and recall; this signifies strength in performance over various emotional categories.

The high-test accuracy and consistent performance metrics indicate that the model can easily fit into any new data, successfully identify EEG signals, and consequently group emotional states of the individual. On a wider perspective, the model's robustness is clearly depicted through its high precision plus recall and $F1$-scores. These values show how capable the model is at capturing complex patterns in the EEG data to make reliable predictions—thus regardless of the emotional classes presented for performance evaluation. The model always demonstrates strong precision, recall, and F1-scores which suggest that it does not deviate from consistency in performance.

5 Conclusion

We introduced an artificial neural network model for the classification of emotions based on EEG data. We used a dataset that contains FFT features extracted from EEG signals, and our model resulted in a test accuracy of 95.31%. This high level of accuracy demonstrates the capability of our model to effectively classify emotional states, which sets it apart from the rest. Additionally, the performance of our model was better than those reported in literature for similar tasks, which indicates that we have much more to offer toward the development of this field. By building upon previous work that has been done in affective computing and mental health monitoring with practical implications in different areas, we continue striving toward innovation despite having limitations within our study findings. In light of these promising results though, there are still some aspects that we can improve including looking at more preprocessing techniques as well as advanced neural network architectures; this will help us further increase the classification accuracy and robustness in future work that should be explored to strengthen our contributions. We found neural network models to be highly effective in emotion classification based on EEG, which emphasizes the considerable promise for improvements in emotion recognition technology. By continuing to refine and innovate in this area, we can unlock new opportunities for understanding human emotions and improving well-being in diverse contexts.

References

1. Liu Y, Sourina O, Nguyen MK, Jiang R (2010) Real-time EEG-based human emotion recognition and visualization. In: 2010 International conference on cyberworlds, Singapore. IEEE, Singapore, pp. 262–272
2. Picard RW (2000) Affective computing. MIT Press, Cambridge
3. Miotto R, Wang F, Wang S, Jiang X, Dudley JT (2018) Deep learning for healthcare: review, opportunities and challenges. Brief Bioinform 19(6):1236–1246
4. Bashivan P, Rish I, Yeasin M, Codella N (2015) Learning representations from EEG with deep recurrent-convolutional neural networks. arXiv preprint arXiv:1511.06448
5. Calvo RA, D'Mello S (2010) Affect detection: an interdisciplinary review of models, methods, and their applications. IEEE Trans Affect Comput 1(1):18–37
6. Gupta VK, Rana PS (2019) Activity assessment of small drug molecules in estrogen receptor using multilevel prediction model. IET Syst Biol 13(3):147–158
7. Singh MK, Singh P, Gupta VK, Mishra K, Gupta A (2023) Performance analysis of CNN models with data augmentation in rice diseases. In: 2023 3rd Asian conference on innovation in technology (ASIANCON), Ravet IN. IEEE, India, pp. 1–5
8. Koelstra S, Muhl C, Soleymani M, Lee JS, Yazdani A, Ebrahimi T, Patras I (2012) DEAP: a database for emotion analysis; using physiological signals. IEEE Trans Affect Comput 3(1):18–31
9. Gill R, Singh J (2021) A deep learning approach for real time facial emotion recognition. In: 2021 10th International conference on system modeling and advancement in research Trends (SMART). IEEE, pp 497–501

10. Abadi MK, Subramanian R, Kia SM, Avesani P, Patras I, Sebe N (2015) DECAF: MEG-based multimodal database for decoding affective physiological responses. IEEE Trans Affect Comput 6(3):209–222
11. Goel L, Gupta S, Gupta A, Rajan SN, Gupta VK, Singh A, Gupta P (2024) Advancing ASD detection: novel approach integrating attention graph neural networks and crossover boosted meerkat optimization. Int J Mach Learn Cybern 15(8):1–19
12. Casson AJ (2019) Wearable EEG and beyond. Biomed Eng Lett 9(1):53–71
13. Sharma D, Tiwari D, Singh V, Pandey P, Gupta VK (2022) Cloud-based occlusion aware intrusion detection system. In: International conference on cybersecurity in emerging digital era, Singapore: Springer Nature Singapore. Springer, India, pp 69–78
14. Agarwal S, Gupta A, Gupta VK, Shukla A, Sardana A, Pandey P (2022) Analysis of pulmonary fibrosis progression using machine learning approaches. In: International conference on cybersecurity in emerging digital era, Singapore: Springer Nature Singapore. Springer, India, pp 213–224
15. Gupta N, Agarwal S, Joshi K, Gupta VK, Shukla SK, Singh G (2023) Intelligent music recommendation system based on face emotion recognition. In: 2023 International conference on computing, communication, and intelligent systems (ICCCIS), Greater Noida. IEEE, India, pp 110–115
16. Gupta A, Tomar A, Gupta VK, Mishra SK, Shukla A, Tiwari H (2023) Phishing attacks Isolation via DNS prevention of 64-Bit time synchronized public key. In: 2023 IEEE international conference on blockchain and distributed systems security (ICBDS), New Raipur, India. IEEE, India, pp. 1–6

Language Models in Financial Fraud Detection: A Comparative Study

N. Neha and Vishalakshi Prabhu

Abstract Detecting financial frauds is crucial for maintaining the integrity and security of financial systems. This paper aims to compare the effectiveness of traditional machine learning models and advanced deep learning models, including large language models, in detecting financial fraud. Traditional models such as Random Forest and KNN were evaluated alongside deep learning models like Generative Adversarial Network (GAN), Artificial Neural Networks (ANNs), and large language models including GPT-2 and FinRoberta. Using financial filings from the U.S. Securities and Exchange Commission (SEC) and transaction statements, we assessed each model's performance based on accuracy, precision, recall, and F1-score. The study highlights the strengths and limitations of each approach in terms of accuracy and reliability in detecting fraudulent activities. Results indicate that while traditional models offer solid baseline performance, deep learning models, and large language models significantly enhance detection accuracy, with FinBERT and GPT-2 showing promise due to their advanced natural language processing capabilities. The insights from this research provide valuable guidance for researchers, practitioners, and policymakers on leveraging advanced machine learning techniques to combat financial fraud more effectively. This study underscores the potential of integrating cutting-edge deep learning and language models into financial fraud detection systems to improve their efficiency and robustness. Moreover, the practical implications of these findings suggest that adopting these advanced models could lead to more proactive and precise detection of fraudulent activities, thereby enhancing financial security.

Keywords Artificial Neural Networks (ANNs) · Deep learning models · FinBERT · Financial fraud detection · GPT-2 · Generative Adversarial Network (GAN) · Large language models · Machine learning models

N. Neha (✉) · V. Prabhu
Department of CSE, RV College of Engineering, Bangalore, India
e-mail: nehan.cs20@rvce.edu.in

V. Prabhu
e-mail: vishalaprabhu@rvce.edu.in

H. Sharma et al. (eds.), *Artificial Intelligence: Theory and Applications*, Lecture Notes in Networks and Systems 5589, https://doi.org/10.1007/978-981-96-1687-9_4

1 Introduction

Financial fraud encompasses a variety of illegal activities aimed at deceitfully acquiring financial assets from individuals or organizations. This type of fraud encompasses various forms such as corruption, asset misappropriation, and fraudulent financial statements. The impact of financial fraud is substantial, resulting in significant financial losses and undermining the trust and integrity of financial systems. The impact of financial fraud extends beyond direct financial losses, encompassing reputational damage, legal implications, and even organizational bankruptcy. As per the Association of Certified Fraud Examiners (ACFE), global businesses have been estimated to lose 5% of their annual revenues to fraud, which translates to trillions of dollars in losses. According to estimates, the global financial implications of fraudulent activities over the past two decades amount to approximately $5.127 trillion, with losses increasing by 56% in the last decade alone. Figure 1 demonstrates that the impact of financial fraud extends beyond merely financial losses. It shows the impact of financial frauds on various aspects of an organization including the reputational impact, operational impact, and revenue impact.

Detecting financial fraud is of paramount importance as it helps in safeguarding financial assets, maintaining consumer trust, and ensuring compliance with regulatory standards. Effective fraud detection systems not only protect businesses and individuals from financial losses but also deter potential fraudsters by increasing the risk of detection. Traditional methods of detecting fraud in financial systems often rely on rule-based systems and manual reviews, which can be time-consuming and less effective in identifying sophisticated fraudulent activities. In recent years, breakthroughs in machine learning and artificial intelligence have transformed the landscape of financial fraud detection. Machine learning models, such as Random Forest and K-Nearest Neighbors, have been widely adopted due to their ability to analyze enormous datasets and detect patterns indicative of fraudulent behavior. However, these models have limitations in handling complex and high dimensional data. Deep learning models, such as Generative Adversarial Network (GAN) and Artificial Neural Networks (ANNs), offer improved accuracy and robustness in

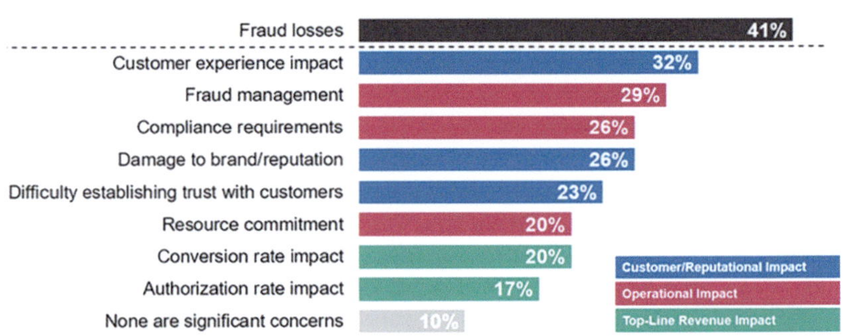

Fig. 1 Impact of financial fraud

detecting fraud by leveraging their ability to model intricate data structures. Additionally, the emergence of large language models like GPT-2 and FinRoBERTa has opened new avenues for detecting fraud through natural language processing (NLP) techniques, enabling the analysis of unstructured text data for signs of fraudulent activities. This paper seeks to present a comparative evaluation of the effectiveness of traditional machine learning models, deep learning techniques, and large language models in identifying financial fraud. The study involves a comparative evaluation of Random Forest, KNN, GAN, ANN, GPT-2, and FinRoBERTa in terms of their accuracy, precision, and recall in identifying fraudulent transactions. The findings of this research will offer valuable insights for researchers, practitioners, and policymakers in leveraging advanced techniques to enhance the effectiveness of financial fraud detection systems. The paper is structured as follows: Sect. 2 reviews related literature, Sect. 3 describes the methodology, Sect. 4 presents the experimental results and analysis, and Sect. 5 discusses the conclusions and future work.

2 Literature Survey

Dileep et al. [1] focuses on detecting fraudulent transactions using machine learning (ML) and deep learning (DL) techniques, including Local Outlier Factor (LOF), Isolation Forest (IF), and Autoencoders, achieving high accuracy with LOF at 99.0%.

Achary and Shelke [2] propose a machine learning-based approach for detecting financial fraud in banking transactions, using intelligent algorithms trained on public datasets to identify factors correlated with fraud and enhance early detection strategies.

Pathak and Singhal [3] address digital fraud in the financial sector, emphasizing the importance of machine learning algorithms like decision trees and XGBoost to detect fraud patterns and secure financial transactions.

Wang and Yang [4] explore ensemble learning for credit card fraud detection by combining Synthetic Minority Oversampling Technique (SMOTE) with random under sampling and Principal Component Analysis (PCA) to enhance data balance and improve recall by 2%.

Ranjan et al. [5] present a feasibility study of fraud detection methods for bank payments, focusing on individual transaction testing using machine learning models to identify fraudulent activities in real time and improve fraud detection efficiency.

Ashtiani et al. [6] conduct a systematic literature review on machine learning and data mining methods for detecting fraudulent financial statements, highlighting research gaps, and suggesting future directions in unsupervised and bio-inspired methods.

Kalbande et al. [7] propose an automated fraud detection system for online payment and credit card transactions, utilizing machine learning models to detect fraudulent activities in real time, offering a robust and cost-effective solution for financial sectors.

Priya and Saradha [8] review the importance of detecting and preventing digital fraud using machine learning and AI, discussing methods to capture customer information to identify fraud patterns and secure digital transactions.

Sánchez et al. [9] introduce FraudFind, a framework based on the fraud triangle theory to detect fraud in banking organizations by analyzing human behavior using semantic techniques, contributing to cybersecurity.

Turaba et al. [10] focus on credit card fraud detection using machine learning and deep learning techniques. They compare CNN, CNN with Gated Recurrent Units (GRUs), and Adaptive Boosting (AdaBoost) algorithms.

Mubalaike et al. [11] investigate the application of deep learning (DL) models for detecting fraudulent transactions with high accuracy. Using a dataset extracted from genuine financial logs of a mobile money service company in Africa, they compare machine learning techniques and deep learning techniques.

Kalbande et al. [7] address the need for an automated fraud detection system in financial services due to the increasing complexity and alarming rise in online frauds. They propose a cost-effective, efficient solution using a machine learning model to detect fraudulent and genuine transactions in real time.

3 Methodology

This study utilizes a thorough methodology to assess the performance of conventional machine learning techniques, deep learning frameworks, and large language models in identifying financial fraud. The methodology is structured into five key phases: Data Collection and Dataset Preparation, Data Visualization and Exploratory Data Analysis (EDA), Data Preprocessing, Training, Model Selection and Fine-Tuning, and Testing. Each phase is meticulously designed to ensure a robust and thorough analysis, starting from the initial collection of relevant financial data, through various stages of data processing and analysis, to the final evaluation of model performance. By systematically progressing through these phases, the study aims to identify the most effective techniques for enhancing fraud detection accuracy and efficiency. Figure 2 shows the steps followed in the proposed methodology.

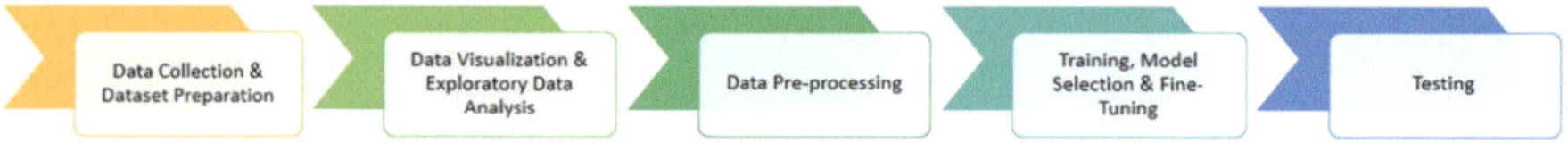

Fig. 2 Methodology for financial fraud detection

3.1 Data Collection and Dataset Preparation

The dataset utilized in this research comprises financial filings sourced from the U.S. Securities and Exchange Commission (SEC). These filings represent a diverse range of companies and contain critical information such as filenames, Central Index Keys (CIKs), filing years, textual sections, company names, and fraud labels. A meticulous selection process ensured the inclusion of filings from both fraudulent and non-fraudulent companies, with equal entries from each category. The SEC archives and enforcement data provided a reliable source for compiling this dataset. To streamline the dataset for analysis, specific sections like Management Discussion and Analysis (MD&A) and financial statements were prioritized. These sections are known to contain crucial information relevant to fraud detection. Leveraging a Python script and the Hugging Face's 10K-Filings dataset library, the data extraction process was facilitated, ensuring efficient retrieval of pertinent filings. The resulting dataset was meticulously structured, with designated text and label columns, aligning with the requirements of machine learning and large language models (LLMs) employed in subsequent analysis and model training.

3.2 Exploratory Data Analysis and Data Visualization

This section presents an Exploratory Data Analysis (EDA) of the dataset prepared from the SEC website. The dataset consists of filings from different companies, evenly split between fraudulent and non-fraudulent entities. An examination of the word counts in these filings reveals stark differences between the two categories. Fraudulent filings exhibit a notably higher word count as shown in frequency distribution Fig. 3.

Additionally, a word cloud representation as shown in Fig. 4 highlights the most frequently occurring words in both fraudulent and non-fraudulent filings.

A scatter plot analysis of the Gunning Fog Index scores shown in Fig. 5 offers insight into text readability The Gunning Fog Index is a readability metric developed to gauge the difficulty of a text by considering sentence complexity and word length. Fraudulent filings exhibit higher mean scores, indicating the use of more complex language, potentially to obfuscate information. In contrast, non-fraudulent filings demonstrate lower scores, indicative of greater transparency and accessibility in communication. The analysis of lexical diversity also shown in Fig. 5 reveals that non-fraudulent filings tend to use a more varied and rich vocabulary, whereas fraudulent filings often employ a more repetitive and less diverse vocabulary.

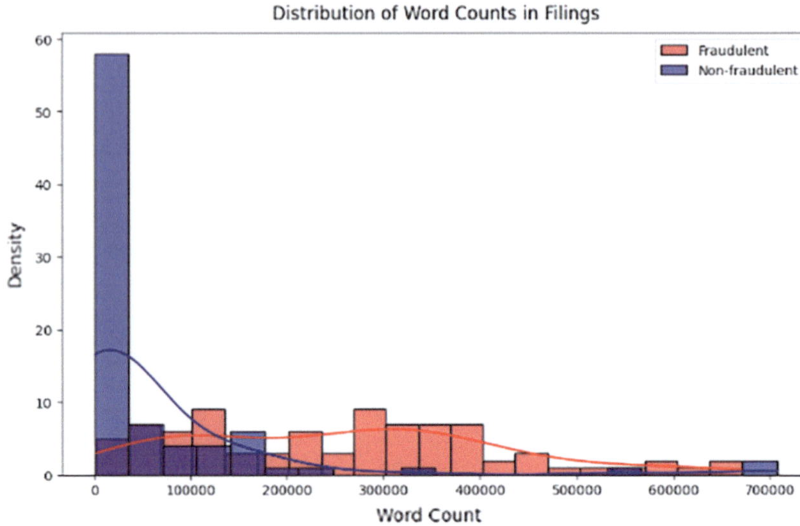

Fig. 3 Word count distribution

Fig. 4 Word cloud for frequent words in both the categories

3.3 Data Preprocessing

The data preprocessing phase of this research involved a series of steps aimed at preparing the dataset for training both traditional machine learning (ML) and deep learning (DL) models, including sophisticated language models like FinRoBERT and GPT-2. The process commenced with text cleaning, which entailed the removal of irrelevant characters, symbols, and punctuations, as well as standardizing the text to a consistent case format. Subsequently, tokenization was applied to segment the text into tokens, a prerequisite for converting textual data into numerical vectors. Rows containing null values were excluded from the dataset to prevent errors during model training, as imputation was not feasible for text data.

After conducting text cleaning and tokenization, the data was converted into a machine-readable format through word embedding techniques. Sequence padding was applied to ensure uniform input data by adjusting text sequences to a fixed

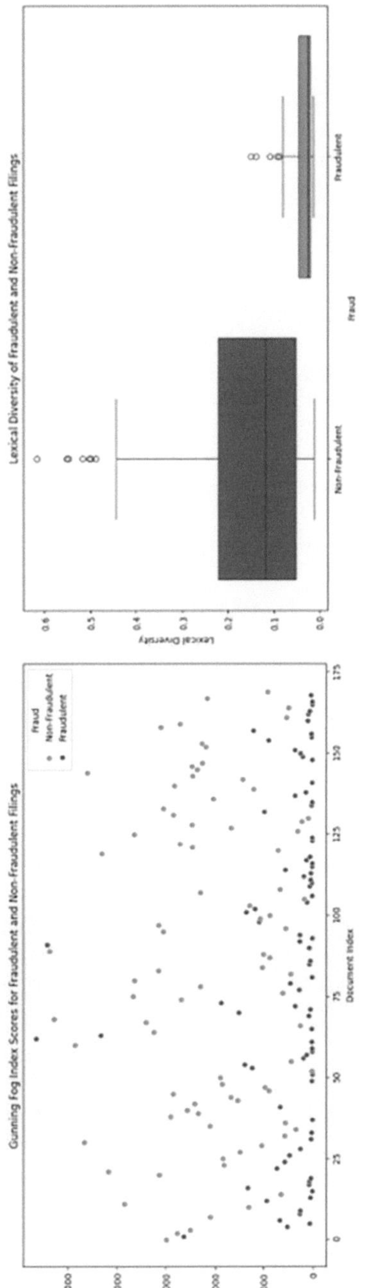

Fig. 5 Plots for readability and lexical diversity scores

length, either by truncating longer sequences or padding shorter ones. To maintain class balance and reduce potential bias in the model, an equal number of companies from both fraudulent and non-fraudulent categories were selected. For the analysis, only the most relevant features, such as the Management's Discussion and Analysis (MD&A) and financial statements, were retained to focus on the key elements. Special preprocessing steps were undertaken for language models, which included the incorporation of special tokens and adherence to specific formatting requirements mandated by models like FinRoBERTa and GPT-2. These steps were critical to ensuring that the data was appropriately structured for the subsequent modeling tasks.

3.4 Model Selection, Training, and Fine-Tuning

The training phase of this research involved implementing various machine learning (ML), deep learning (DL), and large language (LLM) models on the prepared dataset to detect financial fraud. Model selection was a critical aspect, focusing on comparing the effectiveness of traditional ML models compared to deep learning (DL) and large language models (LLMs) in fraud detection. Traditional ML models such as K-Nearest Neighbors and Random Forest (RF), as well as advanced DL models like Artificial Neural Network (ANN) and Generative Adversarial Network (GAN), along with LLM models like GPT-2 with attention and FinRoBERTa with attention, were chosen for training. These models were selected based on their proven performance in similar tasks and their potential applicability to the financial fraud detection domain.

The dataset was divided into three distinct sets: training, validation, and test sets, with a split ratio of 60-20-20, respectively. This partitioning ensured that the models had sufficient data for learning while also providing separate sets for tuning hyperparameters and evaluating performance. Random splitting was employed to minimize bias in the training and test data, with a fixed random state for reproducibility. Label encoding was applied to convert the labels into a binary format, with 0 representing non-fraudulent cases and 1 representing fraudulent cases. For transformer-based models like GPT-2 and FinRoBERTa, the input token length was set to a maximum of 512, with padding and truncation parameters used to handle sequences exceeding this length limitation.

Training and validation involved iterating through each model, feeding it with the training data, and adjusting model parameters to optimize performance on the validation set. Each model was trained using appropriate loss functions and optimization algorithms tailored to its architecture. The training process aimed to minimize the loss function and maximize model accuracy, ensuring robust performance in detecting financial fraud.

3.5 Testing

Once the models were trained and optimized on the validation set, they were evaluated on the test set to assess their generalization performance. The test set, which was kept separate from the training and validation sets, consists of unseen data instances that were not used during model training. Each model was applied to the test set, and its predictions were compared against the ground truth labels to calculate performance metrics. These metrics include accuracy, precision, recall, and $F1$-score, which provide insights into the model's ability to correctly classify instances of financial fraud. By analyzing these metrics, we can determine the effectiveness of each model in detecting fraudulent activities in financial data.

4 Results and Analysis

The performance of various machine learning (ML), deep learning (DL) models, and large language models (LLMs) in detecting financial fraud was evaluated using key metrics such as accuracy, precision, recall, and F1-score. Each model was trained and tested on a dataset prepared from SEC filings. The results provide insights into the strengths and limitations of each model in fraud detection, informing potential applications in real-world scenarios.

4.1 K-Nearest Neighbors

KNN demonstrated moderate performance, achieving an accuracy of 79.41% on both the training and test sets. While its recall rate was relatively high at 89.47% in the training set, indicating its ability to identify true positives, there was a slight decrease to 81.25% in the test set. Precision remained consistent, suggesting room for improvement in identifying false positives. Figure 6 shows the metrics obtained for KNN.

4.2 Random Forest

RF exhibited strong performance, with an accuracy of 91.18% on the training set and 94.12% on the test set. It demonstrated perfect recall on the training set, indicating its capability to identify all true positives. The precision remained high at 85.71%, showcasing its effectiveness in correctly identifying fraudulent cases while minimizing false positives. Figure 7 shows the metrics obtained for RF.

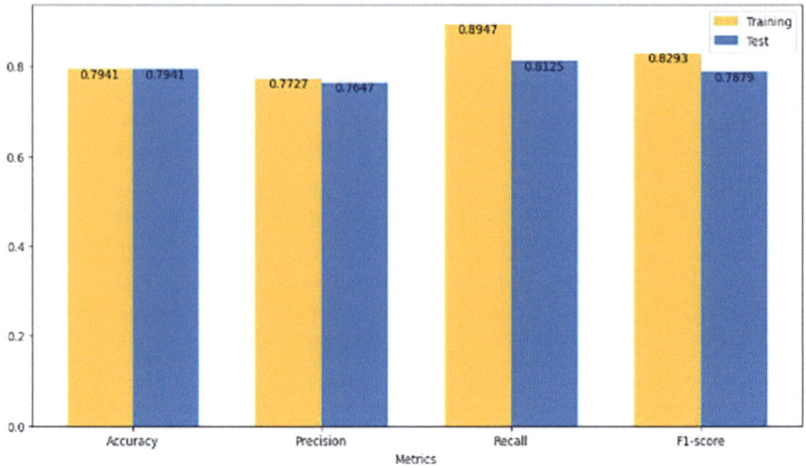

Fig. 6 KNN metrics

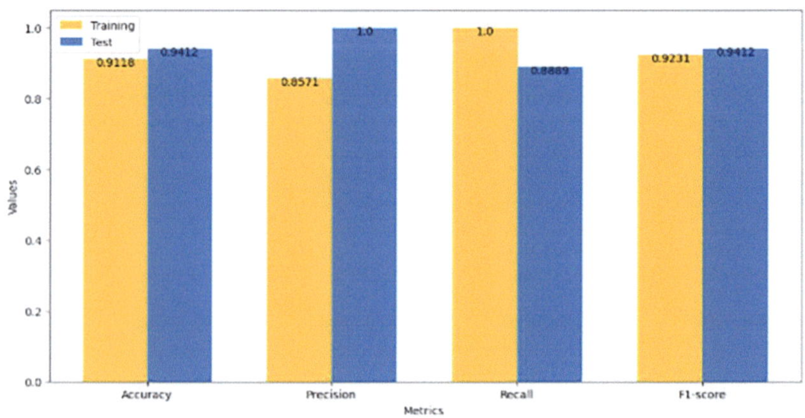

Fig. 7 RF metrics

4.3 Artificial Neural Network

ANN models showed robust performance, with validation accuracy of about 84% and test accuracies mirroring these results. These models demonstrated strong capabilities in fraud identification, with high recall rates indicating sensitivity to true positives. Figure 9 depicts the metrics obtained for ANN.

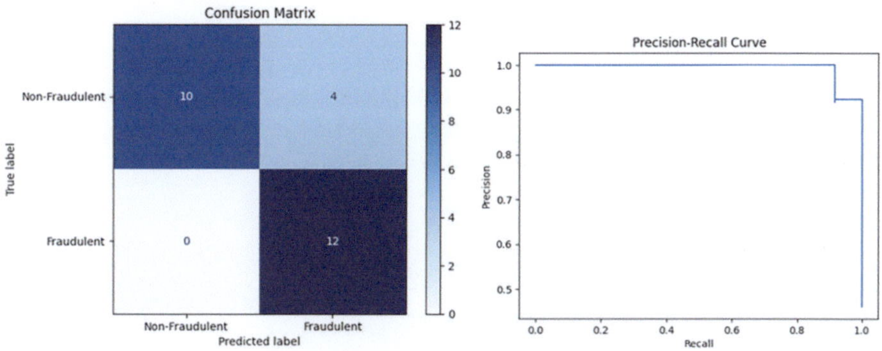

Fig. 9 ANN metrics

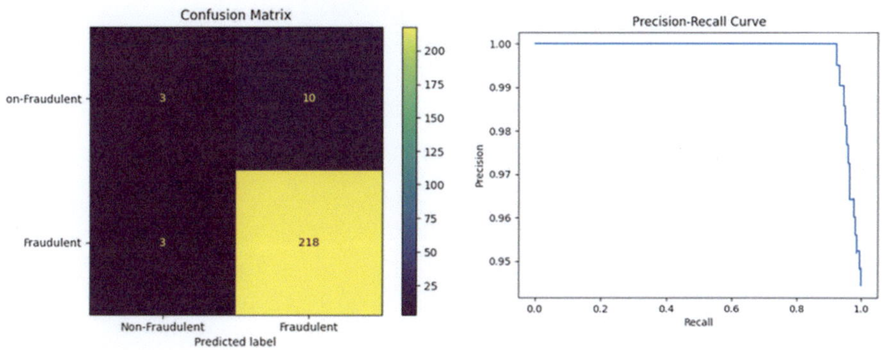

Fig. 10 GAN metrics

4.4 Generative Adversarial Network

GAN exhibited a validation accuracy at 95.06%. While it demonstrated perfect recall, indicating its ability to identify all true positives, its low precision raised concerns about false positives. This suggests limitations in its generalizability and effectiveness in capturing the dataset. Figure 10 shows the metrics obtained for GAN.

4.5 GPT-2 with Attention

GPT-2 showed promising results in the training set but performed poorly on the test set, indicating challenges in generalizability. While it demonstrated strong recall in the training set, this decreased significantly in the test set, suggesting difficulties in identifying actual fraud cases in new data. Figure 11 shows the metrics obtained for GPT-2.

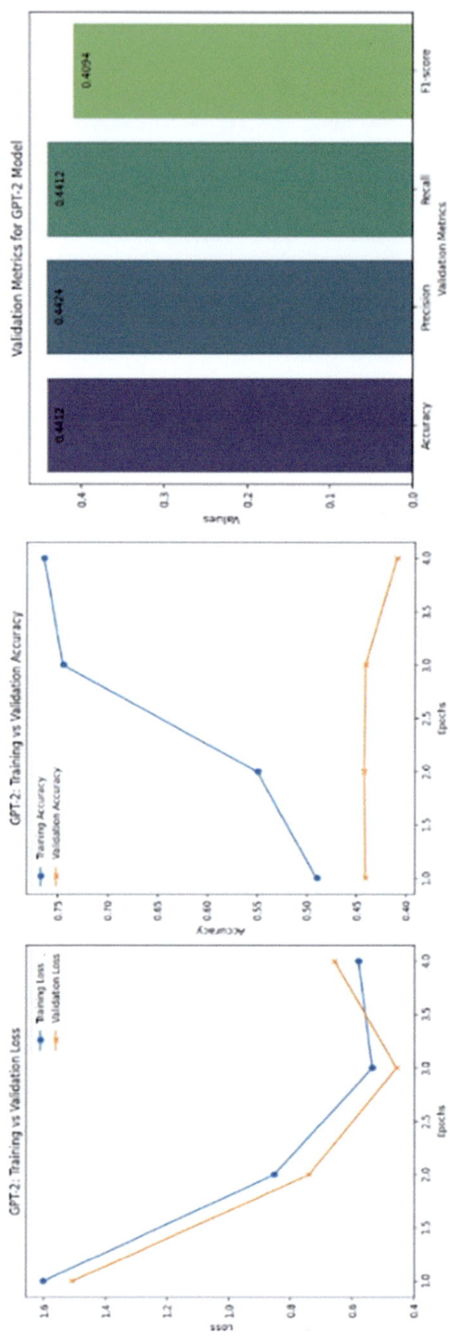

Fig. 11 GPT-2 metrics

4.6 *FinRoBERTa with Attention*

FinRoBERTa exhibited consistent and balanced performance across training and test sets, with accuracy and precision indicating good generalizability. Its ability to maintain high recall rates in both sets suggests effectiveness in identifying true positives, making it suitable for real-world applications in fraud detection. Figure 12 depicts the metrics obtained for FinBERT.

Table 1 represents a detailed overview of the performance metrics for each model.

5 Conclusion and Future Scope

In conclusion, this research conducted a comprehensive evaluation of machine learning (ML), deep learning (DL), and large language models (LLMs) in the context of financial fraud detection using textual data extracted from SEC filings. The study revealed that while traditional ML models such as Logistic Regression (LR), Support Vector Machine (SVM), and Random Forest (RF) demonstrated robust performance, LLMs showcased remarkable potential in detecting fraudulent activities. Specifically, FinBERT and GPT-2 with attention exhibited competitive performance, highlighting the effectiveness of pre-trained language models in capturing nuanced patterns within financial texts. Although challenges such as model interpretability and generalizability were observed, the findings underscore the transformative impact of LLMs in augmenting fraud detection capabilities.

Moving forward, there is significant potential for further advancements in leveraging LLMs for financial fraud detection. Continued research efforts should focus on enhancing the interpretability and explain ability of LLMs to facilitate trust and adoption in real-world applications. Additionally, exploring ensemble techniques that integrate the strengths of both traditional ML models and LLMs could lead to synergistic improvements in fraud detection accuracy. Furthermore, investigating the adaptation of LLMs to domain-specific financial contexts and exploring novel architectures tailored to fraud detection tasks hold promise for enhancing model performance. Real-world validation studies and deployment in operational environments will be crucial for assessing the practical utility and scalability of LLM-based fraud detection systems.

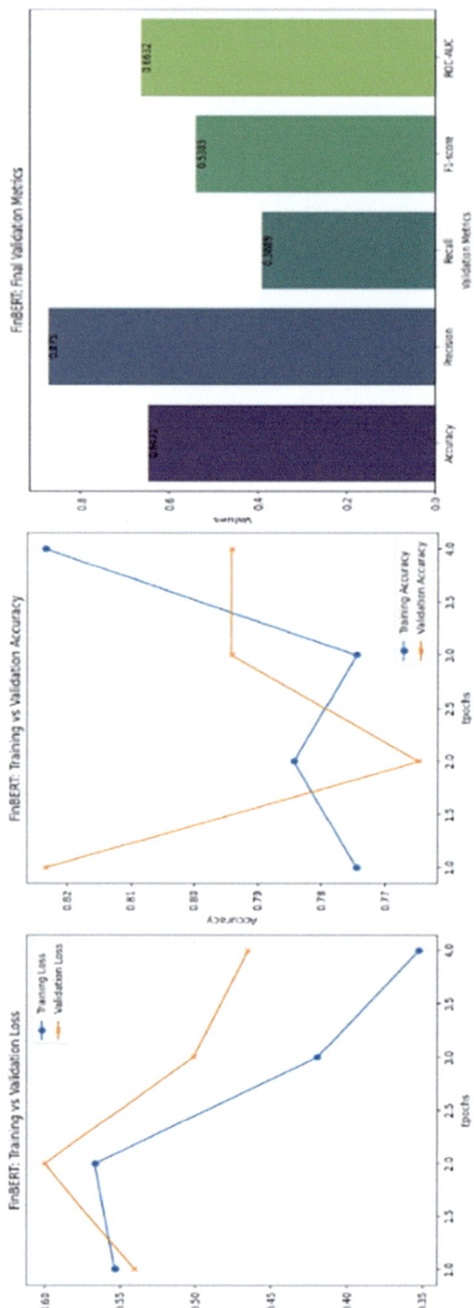

Fig. 12 FinBERT metrics

Table 1 Summary of model metrics

Model	Accuracy (%)	Precision (%)	Recall (%)	$F1$-Score (%)
KNN	79	77	89	83
Random Forest	91	100	88	94
ANN	84	85	100	92
GAN	91	88	100	95
GPT-2	43	44	44	41
FinRoberta	79	78	83	81

References

1. Dileep A, Karthik A, Krishna GS, Ganesh D, Hariharan S (2023) Financial fraud detection using deep learning techniques. In: 2023 International conference on distributed computing and electrical circuits and electronics (ICDCECE), Ballar, India, 2023, pp 1–6. https://doi.org/10.1109/ICDCECE57866.2023.10150467
2. Achary R, Shelke CJ (2023) Fraud detection in banking transactions using machine learning. In: 2023 International conference on intelligent and innovative technologies in computing, electrical and electronics (IITCEE), Bengaluru, India, 2023, pp 221–226. https://doi.org/10.1109/IITCEE57236.2023.10091067
3. Pathak N, Singhal S (2023) Fraud detection in financial domain using machine learning. In: 2023 International conference on artificial intelligence and smart communication (AISC), Greater Noida, India, pp 1449–1452. https://doi.org/10.1109/AISC56616.2023.10085181
4. Wang J, Yang C (2022) Financial fraud detection based on ensemble machine learning. In: 2022 IEEE international conference on dependable, autonomic and secure computing, international conference on pervasive intelligence and computing, international conference on cloud and big data computing, international conference on cyber science and technology congress (DASC/PiCom/CBDCom/CyberSciTech), Falerna, Italy, pp 16. https://doi.org/10.1109/DASC/PiCom/CBDCom/Cy55231.2022.9928001
5. Ranjan P, Santhosh K, Kumar A, Kumar S (2022) Fraud detection on bank payments using machine learning. In: 2022 International conference for advancement in technology (ICONAT), Goa, India, pp 1–4. https://doi.org/10.1109/ICONAT53423.2022.9726104
6. Ashtiani MN, Raahemi B (2022) Intelligent fraud detection in financial statements using machine learning and data mining: a systematic literature review. IEEE Access 10:72504–72525. https://doi.org/10.1109/ACCESS.2021.3096799
7. Kalbande D, Prabhu P, Gharat A, Rajabally T (2021) A fraud detection system using machine learning. In: 2021 12th International conference on computing communication and networking technologies (ICCCNT), Kharagpur, India, pp 1–7. https://doi.org/10.1109/ICCCNT51525.2021.9580102
8. Priya GJ, Saradha S (2021) Fraud detection and prevention using machine learning algorithms: a review. In: 2021 7th International conference on electrical energy systems (ICEES), Chennai, India, pp 564568. https://doi.org/10.1109/ICEES51510.2021.9383631
9. Sánchez M, Torres J, Zambrano P, Flores P (2018) FraudFind: financial fraud detection by analysing human behaviour. In: 2018 IEEE 8th annual computing and communication workshop and conference (CCWC), Las Vegas, NV, USA, pp 281–286.https://doi.org/10.1109/CCWC.2018.8301739
10. Turaba MY, Hasan M, Khan NI, Rahman HA (2022) Fraud detection during financial transactions using machine learning and deep learning techniques. In: 2022 International conference on communications, computing, cybersecurity, and informatics (CCCI), Dalian, China, pp 1–8. https://doi.org/10.1109/CCCI55352.2022.9926503

11. Mubalaike AM, Adali E (2018) Deep learning approach for intelligent financial fraud detection system. In: 2018 3rd International conference on computer science and engineering (UBMK), Sarajevo, Bosnia and Herzegovina, pp 598–603. https://doi.org/10.1109/UBMK.2018.8566574

AGCRNL: Automatic Glaucoma Classification Using Residual Network and LSTM

Sudeshna Pattanaik, Payal Mittal, Rosy Pradhan, and Santosh Kumar Majhi

Abstract This work presents the diagnosis of glaucoma in the early stage using deep learning models. Glaucoma is an irreversible disease that can cause blindness. Manual diagnosis depends on human skills, and it isn't easy to diagnose early. Glaucoma classification using deep learning is a challenging task. The concept of feature extraction using deep learning models has been utilized in the current study. An ensemble residual network, squeeze and excitation block, and Binary Long Short-Term Memory (BiLSTM) have been implemented for sequential feature extraction for faster and more accurate disease classification. A residual network is used for global feature extraction. The accuracy rate of the validation set was 92.3% for our model. The findings suggest that a cost-effective screening tool for early and cost-effective identification of glaucoma could be developed utilizing deep learning algorithms.

Keywords BiLSTM · Glaucoma classification · ResNet50 · SE block

1 Introduction

Glaucoma is the second progressive worldwide leading cause of eye diseases. The early stage detection of glaucoma is a challenging task. Currently, around 79.6 million individuals are affected by glaucoma [1] and it has been estimated that by 2040 the

S. Pattanaik · P. Mittal
Department of Computer Science and Engineering, Veer Surendra Sai University of Technology, Burla, Odisha, India

R. Pradhan (✉)
Department of Electrical Engineering, Veer Surendra Sai University of Technology, Burla, Odisha, India
e-mail: rosypradhan_ee@vssut.ac.in

S. K. Majhi
Department of Computer Science and Information Technology, Guru Ghasidas Vishwavidyalaya (Central University), Bilaspur, Chhattisgarh, India
e-mail: smajhi_cse@ieee.org

H. Sharma et al. (eds.), *Artificial Intelligence: Theory and Applications*, Lecture Notes in Networks and Systems 5589, https://doi.org/10.1007/978-981-96-1687-9_5

number of affected individuals will be around 111.8 million [2]. The eye is one of the most complex, sensitive and delicate organs responsible for passing visual information to the brain cells in the human body.

The root cause of glaucoma is the eye's elevated pressure, which can be associated with Intra-Ocular Pressure (IOP) [3]. Improper drainage of the liquid aqueous humor leads to increased pressure in the eye, subsequently affecting the retina, optic nerve, and human vision. The IOP value should generally be less than 21 mmHg to fight against glaucoma. Manual diagnosis is based on structural changes of the retinal nerve fiber layer and optic nerve head [4]. Glaucoma can broadly be classified into two categories: (1) Open-angle glaucoma and (2) Angle-closure glaucoma. The effect of open-angle glaucoma disease is only realized once it starts to impair human vision [2]. This type of glaucoma disease is painless and is caused when the drainage canal is not able to drain out the excess aqueous humor. People with high blood pressure and diabetes are more susceptible to this type of glaucoma. In the angle-closure glaucoma disease, the iris tends to block the drainage system by being extremely close to it. It requires immediate attention and can be severely painful.

Manual analysis of glaucoma can be time-consuming and cumbersome. The accuracy of parameter measurement for detection varies according to the doctors. In recent years, various automated techniques and diversity indices have been proposed to automatically detect the disease using retinal fundus images [5]. The Residual Network 50 (ResNet50) model has proven effective for extracting features and classifying the fundus image [6]. The accuracy of the model was found to be 93%. The residual network has been used in numerous research for feature extraction, including detection of colorectal cancer and insect pest recognition [7, 8]. Taking motivation from the research of Ovreiu et al. [9], residual with squeeze and excitation block has been combined to obtain channel-wise re-calibrated features. It reduces and discards features based on channel re-calibration and helps to train the model efficiently.

In this article, we propose a novel architecture for fully automated binary classification of glaucoma where the Fundus database has been utilized for training and testing purposes. Different databases (e.g., ORIGA, ACRIMA, and Fundus) can also be used for glaucoma classification. In the proposed framework, a residual network has been combined with a squeeze and excitation block followed by a BiLSTM block [10]. In addition, different available methods for glaucoma classification have been compared with the proposed model. Figure 1 represents diagram of normal eye and glaucoma affected eye.

The main objective of the work focuses on glaucoma disease, the second leading cause of blindness globally [12]. Identifying and categorizing glaucoma is challenging which motivated us to initiate this project. Early detection and treatment can stop irreversible vision loss and enhance the quality of life for those who are affected. Advances in deep learning and artificial intelligence approaches for glaucoma categorization are a major influence on the lives of millions of people at risk of glaucoma [5]. These techniques also provide intriguing potential for early identification and effective monitoring of the condition while preserving vision, enhancing patient lives, and advancement in ophthalmology.

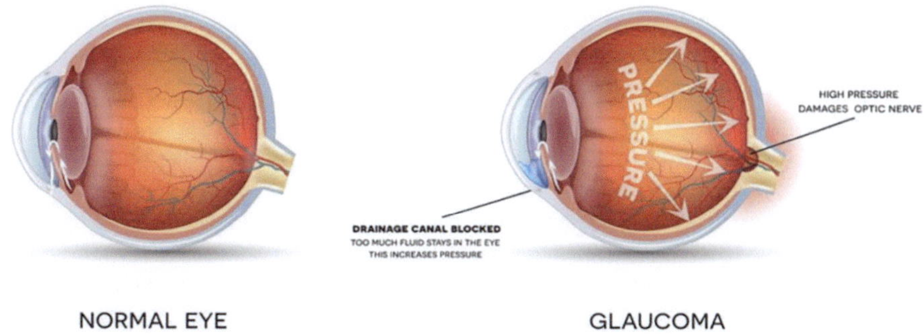

Fig. 1 Diagram of normal eye and eye with glaucoma [11]

2 Objective

The main objective of this study can be summarized as follows:

- The proposed model uses an ensemble method that deals with non-homogeneous and non-linear data elements while focusing on producing more accurate results than models with a single feature extraction level.
- Global feature extraction using the residual network combines squeeze and excitation block to perform dynamic channel-wise feature re-calibration. Sequential feature extraction using BiLSTM helps to extract every temporal feature of the image with accurate details.

The rest of the paper has been organized in the following manner: Sect. 2 includes the literature survey. Section 3 presents the materials and methods while Sect. 4 discusses results and various sample datasets. Finally, Sect. 5 concludes the work.

3 Literature Review

Previous researchers have defined many embedded models for feature extraction and classification. In 2007, Boland and Quigley [1] classified risk factors into six categories most likely to lead to the development of open-angle glaucoma in their study. In 2009, a significant addition to the field of ophthalmology was made through the study paper by Nayak et al. [2] which describes an automated process for feature extraction and SVM for classification. According to the authors, digital fundus images can extract the optic nerve head (ONH) and retinal nerve fiber layer (RNFL) features associated with glaucoma. Ramani et al. [3] described a novel method for classifying normal, glaucoma, and diabetic retinopathy cases using retinal image analysis and data mining methods. The authors strongly emphasize using computational methods to automatically forecast these eye diseases, which can be done by utilizing cutting-edge image analysis and data mining techniques. Ghosh et al. [4] proposed the Grid Color Moment method technique for obtaining color information from retinal pictures used in glaucoma classification. The image was divided into a grid of smaller sections, and color moment characteristics were calculated for each grid cell. The characteristics defining the statistical distribution of color values were subsequently fed into a neural network classifier. This approach demonstrated encouraging results in reliably distinguishing normal and glaucomatous retinal pictures, potentially aiding in the early detection and diagnosis of glaucoma. Chakravarty and Sivaswamy [5] introduced a glaucoma classification method that integrated segmentation- and image-based features for binary classification of glaucomatous and normal retinal pictures. The researchers used both segmentation-based and image-based approaches to extract characteristics from retinal pictures. In 2016, Dey and Bandyopadhyay [6] proposed a support vector machine (SVM) classification method for automated glaucoma detection. Principal component analysis (PCA) was used for feature extraction and SVM for classification. PCA was utilized to decrease the extracted features, and the smaller feature set was then used as input for SVM classification. The proposed method detected glaucoma with an impressive 96% accuracy rate. Hu et al. [8] introduced the SE block in their proposal for squeeze and excitation (SE) networks which models channel dependencies and automatically calibrates channel-wise feature responses. The SE block is a small and effective method that collects channel-wise information by adaptively weighting feature maps based on their channel-wise relevance. Ferreira and Vinicius [13] proposed a model using a convolution neural network and SVM classifier using textual descriptors. The CNN extracts the features, and the SVM stage is used for classification. A hybrid feature space for glaucoma classification that integrates texture data with transfer learning utilizing deep learning algorithms was proposed by Claro et al. [7]. The work used transfer learning, which uses deep learning models that have already been trained to extract essential features from big datasets and texture-based features that were recovered from retinal pictures. The classification of glaucoma using the hybrid feature space created by fusing texture information and transfer learning produced good results. The research paper entitled "Early Detection of Glaucoma Using Residual Networks" by Ovreiu et al. [9] *used residual networks to extract features and classify images into normal or glaucoma,* which had an accuracy of around 93%. Bisneto et al. [12] *proposed a model that* uses SVM and generative adversarial networks. The accuracy of the model was found to be 87.5%. Janani and Rajamohana [11] reviewed numerous methods for segmenting the optic disc and optic cup to detect glaucoma at an early stage. Oza et al. [14] proposed a model for classification using a convolution neural network. The accuracy of the model was found to be 95.44%. In 2022, ODGNet: a deep learning model for automatic optic disc localization and glaucoma classification using fundus images was proposed by Latif et al. [10]. The research introduces the ODGNet deep learning model, which uses fundus pictures to automate the optic disc localization and the categorization of glaucoma. The suggested model uses deep learning to identify the optic disc in fundus pictures and categories of glaucoma cases [10]. Moreover, different intelligent and deep learning techniques have been used to study glaucoma classification [15–23]. Recently, several attempts have been made for automatic classification of glaucoma [24–26].

4 Materials and Methods

4.1 Existing Model

In the existing model [9], the residual network used for classification achieved a classification accuracy of 93.28% for the validation data, where a total of 4 convolution blocks, each consisting of 3 convolution layers, were present. The Fundus database is used for training and validation purposes in this model. Figure 2 represents the existing residual network model.

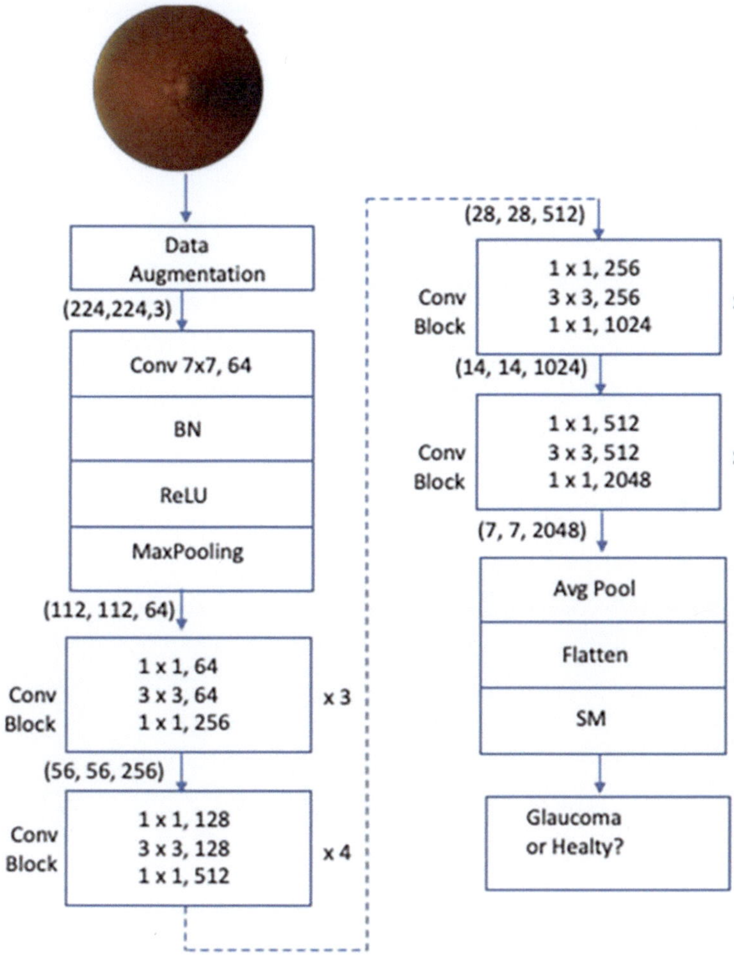

Fig. 2 Existing residual network model [9]

4.2 Proposed Model

Four crucial components make up the glaucoma detection strategy in use:

Step 1: Data Augmentation and Pre-processing.
Step 2: Global feature extraction: Residual network in combination with squeeze and excitation block (SE).
Step 3: Sequential feature extraction: BiLSTM.
Step 4: Classification.

The Image data generator is used to synthesize data using data augmentation and image quality is improved by re-sizing. The target image size is $224 \times 224 \times 3$. The images are passed to the residual and SE blocks for global feature extraction. It extracts 2048 features for each image. The 2D image is flattened to serve as input to the long short-term memory for sequential feature extraction. Global average pooling is applied to combine the effect of 2048 features in a single feature for classification. A fully connected layer is used for the classification of images. Figure 3 shows the block diagram of the proposed model.

Step 1: Data Augmentation

Data Augmentation generates synthetic data by rotating, scaling and shuffling the original data images. This technique is used to fulfill the requirements, the varied nature of the training data, and the volume of data. In addition to these, the class imbalance issue in classification jobs can also be solved using augmented data. In deep learning, models can be trained accurately and efficiently by training with augmented data.

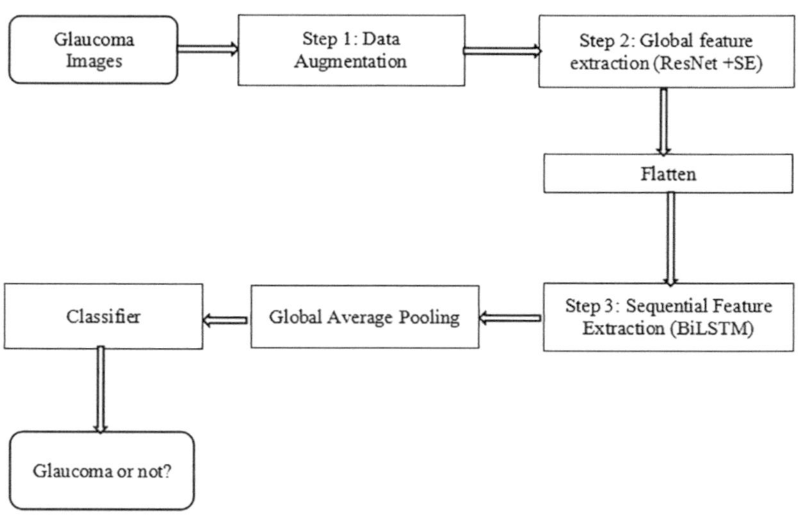

Fig. 3 Proposed model

Since deep learning models require a large amount of data for training and validation, ImageDataGenerator has been employed to generate synthetic data for training and validating the proposed model where the target image size is maintained at 224 × 224 × 3. In the proposed methodology from the Fundus dataset, 554 images were utilized for training, and 130 images were used for validation.

Step 2: Global Feature Extraction

Global feature is a term used for the entire image and associated attributes. Feature extraction is a part of dimensionality reduction to reduce the variables and computing resource usage. A residual network is used for feature extraction in deep learning. Residual block provides a skip connection to minimize the vanishing gradient problem, which can be considered a striking advantage. ResNet50 architecture has an extra convolution layer to reduce the dependencies, number of parameters, and matrix multiplication using bottleneck design for building blocks. ResNet 50 architecture has also been used for malicious software classification.

Squeeze and excitation blocks have two stages of action. SE block in the first 'Squeeze' step applies global average pooling to aggregate the feature map across the spatial dimension—an image with H × W × C aggregates to 1 × 1 × C. In the excitation stage, the channels are reduced by a factor $R = 16$ (reduction ratio) according to per-channel weights. The channels are re-calibrated, and activation functions are mapped to the output image to maintain consistency in the number of channels [8]. SE block helps discard the features that are not important for image classification and calculates the weight of each channel toward the classification during the training process.

SE block combined with the residual block provides the same accuracy with much lower computational cost. The input for our model is 224 × 224 × 3 and the output of the global feature extraction block is in the form of 55 × 55 × 2048. 2048 features were extracted in step 2 for each image. Figures 4 and 5 represent the block diagram of the residual network in combination with the SE block.

Step 3: Sequential Feature Extraction

An additional BiLSTM model has been ensembled that takes the global features as input and generates the sequential features which are more accurate from both directions [10]. BiLSTM learns the features while extracting sequential features from the global ones. Each BiLSTM block or cell is responsible for the extraction of a single global feature which also focuses on local patterns and features of each global feature extracted. A total of 2048 BiLSTM cells are being used to cover 2048 global features.

LSTM has been utilized to store long sequences of data for a particular amount of time or until the forget gate is activated, thereby reducing the long-term dependencies as it can remember the specific sequence and can be used for classification with reduced time. During training, it also learns the features and allows us to classify data with the least possible time. Figure 6 represents block diagram of sequential feature extraction.

Fig. 4 Residual network and SE block

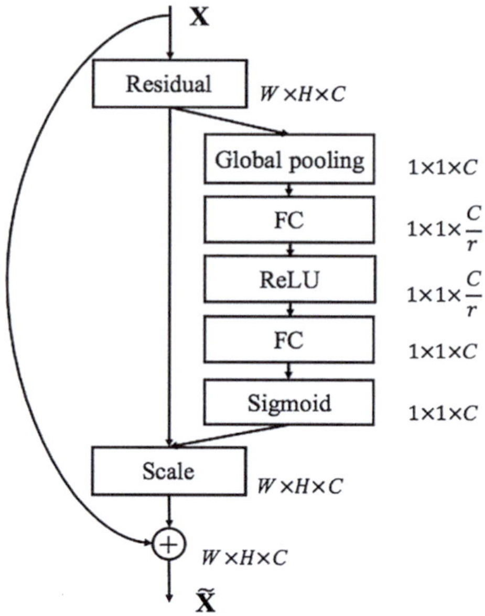

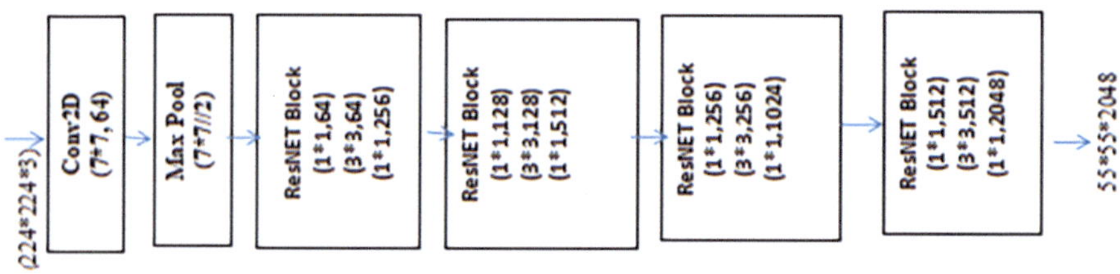

Fig. 5 Model of global feature extraction

Step 4: Classification

A dense or fully connected layer is placed after the global average pooling layer. It takes the aggregate feature map of 2048 features generated by the global average pooling. It produces an output for binary classification where standard images are denoted as one and a glaucoma affected image is denoted as 0.

5 Results and Discussion

In the proposed model, global features were extracted using a residual block which was then followed by sequential feature extraction. A single dense layer does classification. The results shown in Figs. 8 and 9 result from 50 epochs. Figure 7 represent the diagram of original glaucoma affected eye.

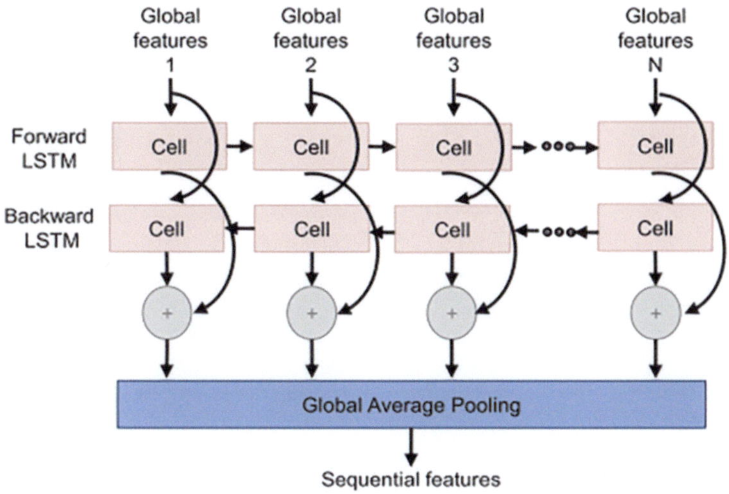

Fig. 6 Sequential feature extraction [10]

Fig. 7 Original image

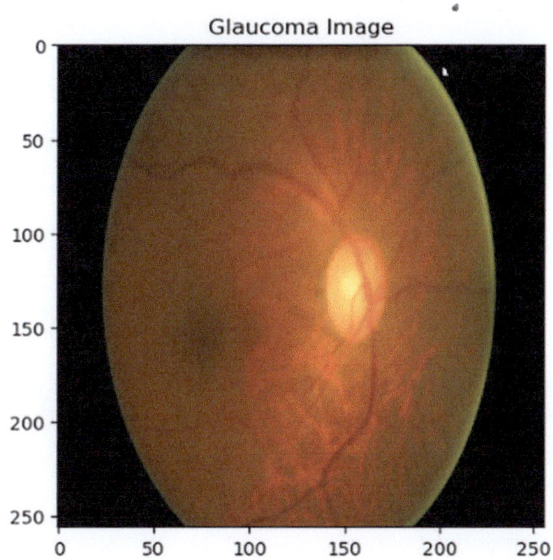

Global Features

Residual block, in combination with SE block, takes the input image of size 224 × 224 × 3 and extracts 2048 features for each in the output format 55 × 55 × 2048. After flattening the global features were fed to the BiLSTM block for sequential feature extraction. Figure 8 represents 8 features from the 2048 extracted features.

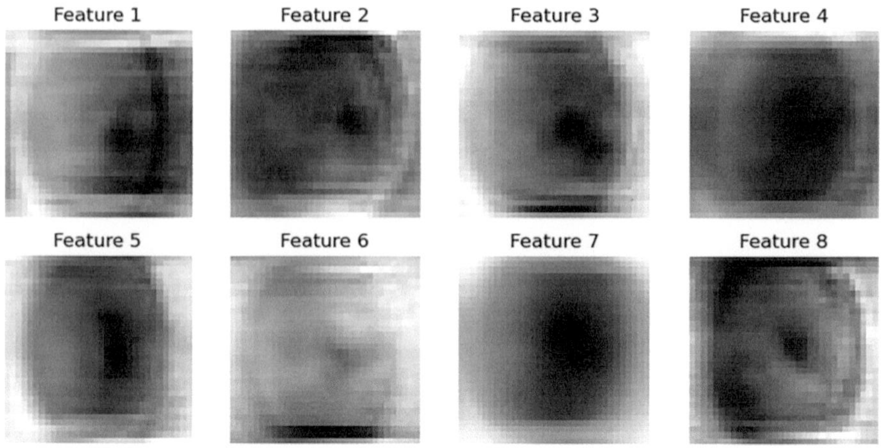

Fig. 8 Features extracted residual block

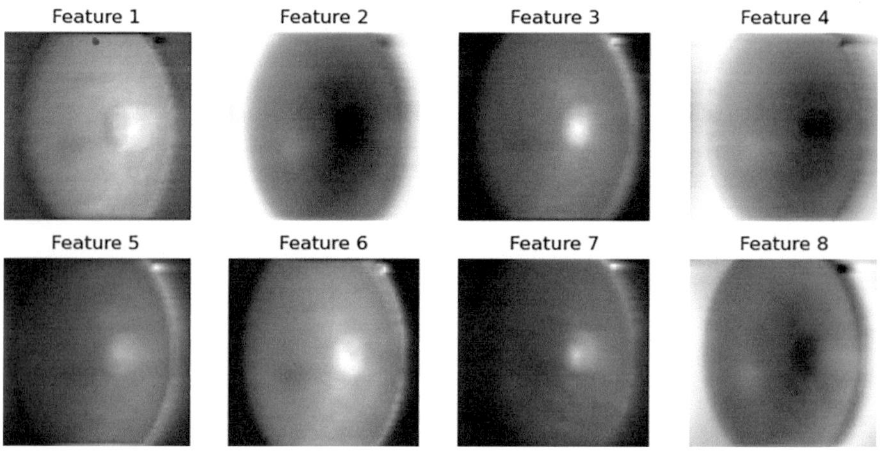

Fig. 9 Features after BiLSTM

Sequential Features

BiLSTM block takes the global features generated by the residual block and focuses locally on each feature to generate a sequential feature or pattern. Since BiLSTM can work on linear data, flatten is applied to change the dimension of the features generated by the residual block. Figure 9 shows the features after passing each global feature from the BiLSTM block.

Quantitative Result and Qualitative Result

Figure 10 shows the training and validation accuracy curves and the confusion matrices for the performance of the proposed model on the glaucoma datasets. These figures provide insight into how the model's accuracy changes as the training epoch progresses and how the model performs on different classes of images. The accuracy curves show that the performance of the proposed model improves significantly with increasing training epochs. At the same time, the confusion matrices provide a visual representation of the prediction made by the model as compared to the ground truth labels.

Table 1 summarizes the performance of the proposed model for glaucoma classification using three different datasets, i.e., ACRIMA, FUNDUS, and ORIGA. The proposed model is based on an advanced ensemble-based CNN architecture called BiLSTM along with ResNet50 and SE block, which has shown promising results in accurately detecting glaucoma from retinal fundus images. The study results demonstrate that the proposed method achieves higher accuracy than the existing methods for all three datasets, with an average accuracy of 94%. The precision of the proposed model is also high and ranges from 84 to 96%, indicating that the model has a low false positive rate. The recall rate, which measures the proportion of positive cases that the model correctly identifies, ranges from 92 to 98%, indicating that the model also has a low false negative rate.

Table 2 shows the comparative analysis of the performance of the proposed model to other architectures used in different literature. The architectures include CNN, ResNet50, transfer learning, and the proposed ResNet50 with SE block and BiLSTM

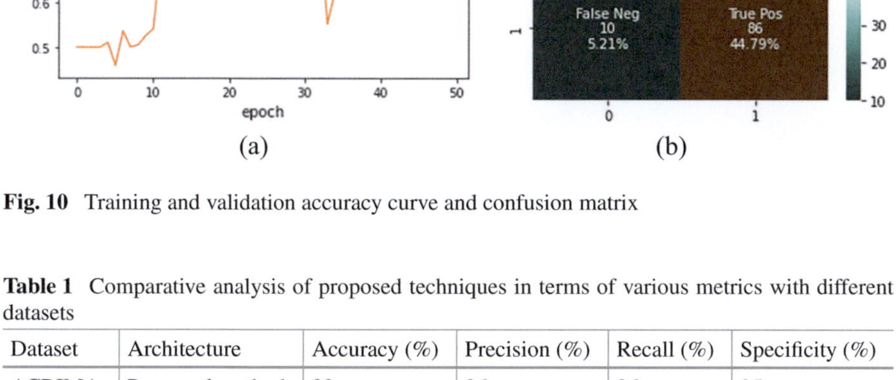

Fig. 10 Training and validation accuracy curve and confusion matrix

Table 1 Comparative analysis of proposed techniques in terms of various metrics with different datasets

Dataset	Architecture	Accuracy (%)	Precision (%)	Recall (%)	Specificity (%)
ACRIMA	Proposed method	99	96	96	95
FUNDUS	Proposed method	98	95	92	94
ORIGA	Proposed method	86	84	98	97

Table 2 Proposed model is evaluated comparatively with various existing methods

Architecture	Precision (%)
CNN	94
ResNet 50	93.28
Transfer learning	94
CNN with texture descriptor	92
Resnet50 + SEblock + BiLSTM	**98**

model. The results demonstrate that the proposed model achieved the highest precision at 98%, thereby outperforming all other architectures in the comparison. These findings suggest that the proposed model has the potential to be a valuable tool for accurately detecting glaucoma and preventing vision loss.

6 Conclusion

Glaucoma is a severe eye condition that can cause vision loss and even blindness if not detected early. To address this issue, a recent study proposed using ensemble-based CNN architectures, specifically BiLSTM with ResNet50 and SE block, as glaucoma classifiers. The study used only publicly available image datasets and data augmentation techniques to train and evaluate the performance of the proposed model. The results showed that the proposed architecture achieved the best performance, with an average accuracy of 0.98, an average specificity of 0.94, and an average recall of 0.92. The study also found that fine-tuning the proposed architecture led to these impressive results, achieved after analyzing 2594 images and implementing a sampling mechanism. These findings demonstrate the potential of using advanced CNN architectures to improve glaucoma diagnosis and ultimately prevent vision loss and blindness.

References

1. Boland MV, Quigley HA (2007) Risk factors and open-angle glaucoma: classification and application. J Glaucoma 16(4):406–418
2. Nayak J et al (2009) Automated diagnosis of glaucoma using digital fundus images. J Med Syst 33:337–346
3. Ramani, RG, Balasubramanian L, Jacob SG (2012) Automatic prediction of diabetic retinopathy and glaucoma through retinal image analysis and data mining techniques. In: 2012 International conference on machine vision and image processing (MVIP). IEEE
4. Ghosh A et al (2015) Grid color moment features in glaucoma classification. Int J Adv Comput Sci Appl 6(9):99–107
5. Chakravarty A, Sivaswamy J (2016) Glaucoma classification with a fusion of segmentation and image-based features. In: 2016 IEEE 13th international symposium on biomedical imaging (ISBI). IEEE

6. Dey A, Bandyopadhyay SK (2016) Automated glaucoma detection using support vector machine classification method. Br J Med Med Res 11(12):1
7. Claro M et al (2019) An hybrid feature space from texture information and transfer learning for glaucoma classification. J Vis Commun Image Representation 64:102597
8. Hu J, Shen L, Sun G (2018) Squeeze-and-excitation networks. In: Proceedings of the IEEE conference on computer vision and pattern recognition
9. Ovreiu S et al (2020) Early detection of glaucoma using residual networks. In: 2020 13th international conference on communications (COMM). IEEE
10. Latif J et al (2022) ODGNet: a deep learning model for automated optic disc localization and glaucoma classification using fundus images. SN Appl Sci 4(4):98
11. Janani R, Rajamohana SP (2021) Early detection of glaucoma using optic disc and optic cup segmentation: a survey. Mater Today: Proc 45:2763–2769
12. Bisneto TRV, de Carvalho Filho AO, Magalhães DMV (2020) Generative adversarial network and texture features applied to automatic glaucoma detection. Appl Soft Comput 90:106165
13. Ferreira S, Vinicius M et al (2018) Convolutional neural network and texture descriptor-based automatic detection and diagnosis of glaucoma. Expert Syst Appl 110:250–263
14. Oza MG, Rani G, Dhaka VG (2021) Glaucoma detection using convolutional neural networks. In: Handbook of research on disease prediction through data analytics and machine learning. IGI Global, pp 1–7
15. Singh PB, Singh P, Dev H (2023) Optimized convolutional neural network for glaucoma detection with improved optic-cup segmentation. Adv Eng Softw 175:103328
16. Chan WC, Morin JD, McCulloch C (1976) Optic disc observations in glaucoma. Can J Ophthalmol 11(2):134–139
17. Tanito M et al (2018) Different glaucoma types and glaucoma surgeries among different age groups. Graefe's Arch Clin Exp Ophthalmol 256:2013–2014
18. Sarwinda D et al (2021) Deep learning in image classification using residual network (ResNet) variants for detection of colorectal cancer. Procedia Comput Sci 179:423–431
19. Liu W et al (2020) DFF-ResNet: an insect pest recognition model based on residual networks. Big Data Min Analytics 3(4):300–310
20. Araújo JDL et al (2019) Glaucoma diagnosis in fundus eye images using diversity indexes. Multimedia Tools Appl 78:12987–13004
21. Ghosal P et al (2019) Brain tumor classification using ResNet-101 based squeeze and excitation deep neural network. In: 2019 Second international conference on advanced computational and communication paradigms (ICACCP). IEEE
22. Graves A, Graves A (2012) Long short-term memory. In: Supervised sequence labelling with recurrent neural networks, pp 37–45
23. Kumar PJ et al (2019) Intelligent detection of glaucoma using ballistic optical imaging. Adv Eng Inform 40:107–129
24. Pattanaik S, Chakraborty C, Behera S, Majhi SK, Pani SK (2024) A MIoT framework of consumer technology for medical diseases prediction. IEEE Trans Consum Electron 70(1):3754–3761
25. Pattanaik S, Behera S, Majhi SK, Dwibedy PK (2022) Stacked BiLSTM with ResNet50 for medical image classification. In: 2022 IEEE region 10 symposium (TENSYMP), Mumbai, India, pp 1–6. https://doi.org/10.1109/TENSYMP54529.2022.9864353
26. Pattanaik S, Behera S, Dwibedy PK, Majhi SK, Pradhan R (2022) AdaRes: ResNet50 with AdaSwarm for glaucoma classification & detection. In: 2022 3rd International conference for emerging technology (INCET), Belgaum, India, 2022, pp 1–5. https://doi.org/10.1109/INCET54531.2022.9825121

Intrusion Detection System in IOT Using Multiple Machine Learning Models

Santosh Kumar Majhi, Khushi Agrawal, Swarupa Panda, and Rosy Pradhan

Abstract The Internet of Things (IoT) is the connectivity of all physical things, and its objective is to link all physical devices. The number of Internet and IoT users is growing every day, as are security concerns, which predict security theft via IoT devices. To identify this harmful activity, several machine learning models are deployed in the IoT context to detect intrusive behavior. This research paper uses the NSL-KDD dataset to identify five distinct classes, four of which are attacks (DoS, U2R, R2L, Probe), and one is normal (the system is operating normally). NatureInspiredSearch performs machine learning models such as Decision Tree, Random Forest, AdaBoost, XgBoost, CatBoost, and hyper parameter. It has been found that hyper-tuned LightGBM outperforms other algorithms, with an accuracy of 99.86. It has also been found that performance indicators like as accuracy, recall, and $F1$ score decline without data sampling but improve with data sampling.

Keywords Intrusion detection · Ensemble machine learning techniques

1 Introduction

Over the years, the web has become an important aspect of people lives, with practically everyone utilizing it for daily tasks such as sending and receiving mails, apps, websites. As a result, a large quantity of data is exposed over the internet. Several

S. K. Majhi
Department of Computer Science and Information Technology, Guru Ghasidas Vishwavidyalaya (Central University), Bilaspur 495009, Chhattisgarh, India
e-mail: smajhi_cse@ieee.org

R. Pradhan (✉)
Department of Electrical Engineering, Veer Surendra Sai University of Technology, Burla 768018, Odisha, India
e-mail: rosypradhan_ee@vssut.ac.in

K. Agrawal · S. Panda
Department of Computer Science and Engineering, Veer Surendra Sai University of Technology, Burla 768018, Odisha, India

H. Sharma et al. (eds.), *Artificial Intelligence: Theory and Applications*, Lecture Notes in Networks and Systems 5589, https://doi.org/10.1007/978-981-96-1687-9_6

technologies have arisen over the years. One example is IoT (Internet of Things). IoT refers to the connectivity of networks of many devices for communication and control purposes. The goal is to link devices that lack communication or reception capabilities. These devices can communicate by connecting an additional module that generates and sends signals, as well as receives and acts on them. These additional modules include sensors and actuators. Automation is used in a variety of applications, including traffic automation, safe automation, and factory control, production control in industries, smart buildings, smart offices, and smart inventories. However, it also presents issues like scalability, privacy, and security. IoT devices are subject to a wide range of risks, including network attacks, software vulnerabilities, hardware failures, and possible data breaches. Additional security risks include Denial of Service attacks, jamming, data leakage, U2R, R2L, Probe, and so on. A Denial of Service (DoS) attack is a technique used by hostile persons to disrupt the operation of a computer or network, making it inaccessible to authorized users. This is accomplished by flooding the target with traffic or transmitting malicious data that causes it to crash. A User to Root (U2R) attack occurs when an attacker first gets access to a normal user account and then exploits system vulnerabilities to escalate their privileges and achieve root access. This may be accomplished using techniques such as Perl, Load Module, and Eject assaults. A Probe attack occurs when a hacker collects information about a network or system before launching an attack. Private Data Leaks are when sensitive personal information is unintentionally disclosed. This may include date of birth, social security numbers, email addresses, usernames, passwords, home addresses, phone numbers, and medical history [1, 2]. Spoofing is a method of tricking the intended recipient by presenting a message as if it comes from a known and trustworthy source, but it comes from an unknown source. This technique can be used to deceive people via a variety of channels, including emails, phone calls, and websites. It can also be accomplished using technological techniques such as IP address spoofing, ARP spoofing, or DNS server spoofing. Because IoT device makers have not incorporated robust security measures, these devices have become a significant concern, leaving enormous volumes of vulnerable devices accessible to the internet. The Internet of Things (IoT) is a fast growing field that connects physical things to the internet, allowing them to interact and share data with one another.

However, the networked nature of IoT devices poses security concerns, and intrusion detection has emerged as a key study area in IoT security. Wireless networks are the most common method of connecting IoT devices, and various precautions can be taken to protect them from intruders, such as not storing critical data and passwords on IoT devices or cloud servers, verifying software before installation, frequently changing passwords on IoT devices for better security, and so on. Furthermore, distributors and manufacturers should improve and expand IoT management methods by alerting users about out dated software, implementing secure passwords, limiting remote device access, and developing a simplified access control system for APIs. However, because IoT systems are sophisticated and linked, they are presenting new attack surfaces. Ensuring network security, encryption, access control, application safety, and other security measures is getting increasingly complex for big systems with many interconnected devices [3–10]. Furthermore, securing IoT systems requires taking into account a large volume of data created by IoT devices, making them useful to a wide range of sectors that may be difficult to handle using standard storage and processing approaches.

Machine learning (ML) is an effective method for analyzing data and studying normal and abnormal circumstances in IoT settings by investigating how IoT devices and components interact with one another. Training on current models can aid in the prediction of prospective attacks as well as the detection of changes in prior assaults. This paper addresses employing EL-based categorization, which is capable of processing a huge number of input attributes. Bagging, boosting, voting classifiers, AdaBoost, Gradient Boosted Machines, Cat Gradient Boosted Machines, and Light Gradient Boosted Machines are some of the regularly utilized EL methods. Bagging is a technique for improving the stability and precision of machine learning models and may be used to address classification and regression issues. Boosting is another EL strategy that reduces variation and bias by pairing poor learners with strong ones. The study uses a variety of machine learning models for intrusion detection and performs hyper parameter tweaking on them. Table 1 lists dome of the recent IDS using machine learning models.

2 Methodology

2.1 Dataset Description

The NSL-KDD dataset is widely used to analyze internet traffic and is regarded as a benchmark in the area. This dataset may be traced back to the KDD Cup, a competition in which traffic logs were collected to construct a network intrusion detection system. This competition resulted in the creation the KDD'99 dataset, from which the University of New Brunswick eventually built the NSL-KDD dataset. These datasets comprise records of internet traffic as monitored by a basic intrusion detection network, and each record has 43 characteristics, the last two of which indicate whether the traffic is normal or an attack, as well as its severity [8–15]. The data collection includes four types of attacks: Denial of Service (DoS), Probe, User to Root (U2R), and Remote to Local. A brief explanation of each attack is provided below:

- DoS: The most common assault in the NSL-KDD dataset is a Denial of Service (DoS) attack. During this assault, the target system receives excessive traffic that overwhelms the Intrusion Detection System (IDS), forcing it to shut down for protection. As a result, the network becomes unavailable to genuine users. During a sale day, an online merchant may see a spike in orders, causing the network to break down and preventing legitimate customers from making purchases.

Table 1 IDS employs machine and deep learning methods

Year	Reference	Methods	Security threats	Dataset type
2022	[1]	CNN	DDoS, DoS, Reconnaissance	NID, BoT and IoT
2022	[2]	DT, RF, NB, SVM	DoS, Probe, R2L	NSL-KDD
2022	[3]	CNN + DT	DoS, Probe, R2L	NSL-KDD
2022	[4]	Hybrid Bayesian Optimization hyper-tuned CatBoost approach	Normal, Wrong Setup, Spying, Scan, malicious operation, malicious control, DoS, dataProbing	DOS2DOS
2022	[5]	LGBM, RF, GNB, SVM, KNN, AB, Stacking of LGBM + RF	DoS	UNSW NB15, NSL-KDD, BoT
2021	[6]	LRE, RF, DT, CNN, LSTM, TCNN	DDoS, DoS, theft	BoT-IoT
2020	[7]	NB, SVM, DT, Adaboost	Malicious	Sensor 480
2020	[8]	DNN	DoS, Probe, R2L, U2R	KSS-Cup'99, NSL-KDD, UNSW NB15
2020	[9]	SVM, RF, MLP, CNN	DoS	BoT-IoT
2020	[10]	LightGBM	Image Forensic	Experimental
2020	[11]	GBM Technique	Anomaly and sensor attacks	IoT sensors
2019	[12]	XGB, DT, RF, GBR	Network and spam attacks	Public dataset of the smart home IoT device
2019	[13]	RF, AB, GBM, XGB, CART, MLP	DoS attacks	CIDDS-001, UNSW-NB15, NSL-KDD
2019	[14]	DT and DNN	DoS attacks	NSL-KDD
2019	[15]	SMOTE	Anomaly attacks	Publicly available
2019	[16]	DT, EL, LR, DN	Public health	Publicly accessible
2019	[17]	Decision Tree based Genetic Algorithm and AdaBoost	People's health	Diabetic Patients
2019	[18]	LR, SVM, DT, RF, ANN	DoS, Probing, Malicious control, Malicious operation, Scan, Spying, Wrong setup, Normal	DS2OS
2018	[19]	Ensemble of DBNs and SVMs	Network traffic	Benchmark

- The probe or surveillance attack aims to extract sensitive information from a network, similar to how a burglar takes important data like personal or banking information. The attacker wants to obtain as much information as possible.
- A U2R attack starts with a regular user account and seeks to get super-user or root access to a system or network. The attacker gains privileged access by exploiting system weaknesses.
- An R2L attack involves attempting to reach a distant workstation without local access to the system or network. The attacker attempts to bypass security measures and obtain unauthorized access to the network (Fig. 1).

The descriptions show that the DoS attack behaves differently from the other three. Figure 2 shows the control flow diagram for implementing a machine learning

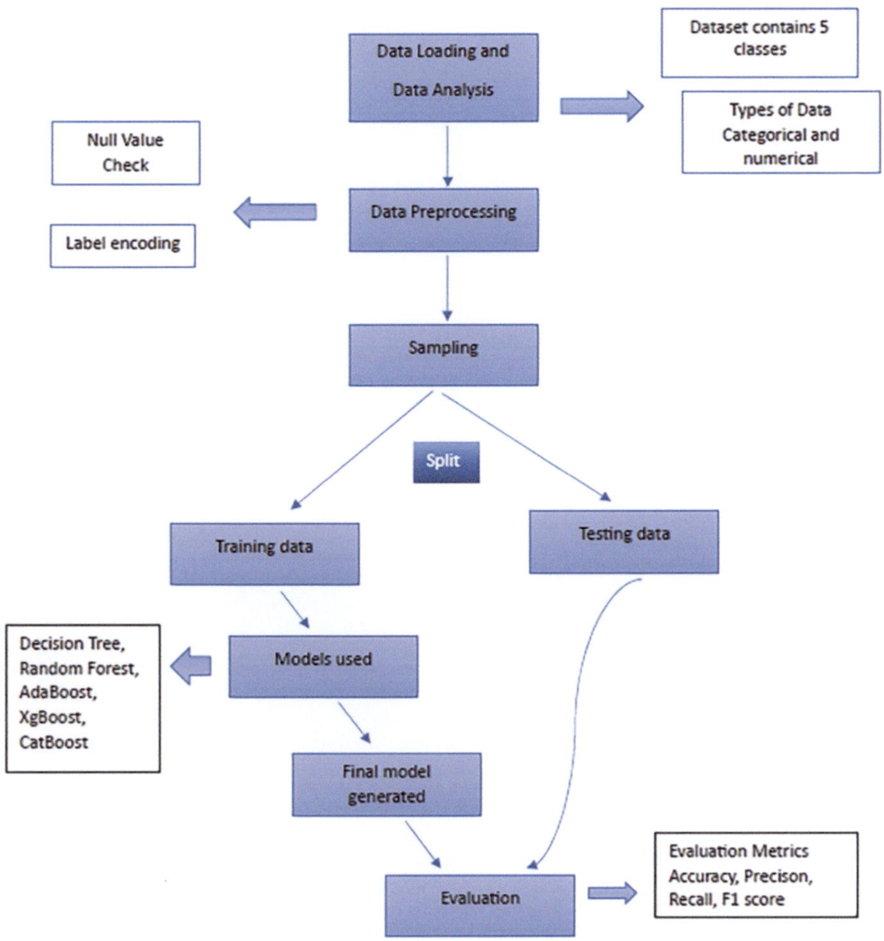

Fig. 1 Control flow diagram for implementing a machine learning model

model. While DoS seeks to disrupt the system and block traffic flow, the other three types of assaults try to enter the system undetected. Table 2 displays the numerous subcategories of each assault found in the dataset.

The types of characteristics in a traffic record offer information on the interaction between the Intrusion Detection System (IDS) with the traffic input. The four categories are: intrinsic, content, host-based, and time-based. These categories comprise several sorts of features, each with its own distinct properties that aid in distinguishing the nature of the traffic input. The data collection includes four distinct categories of characteristics, which are classified as follows:

- Four of them are categorical, denoted by characteristics 2, 3, 4 and 42.
- There are six binary characteristics, denoted by 7, 12, 14, 20, 21, and 22.
- The 23 distinct characteristics are represented by features 8, 9, 15, and 23–41 and 43.
- The last ten features are continuous: 1, 5, 6, 10, 11, 13, 16, 17, 18, and 19.

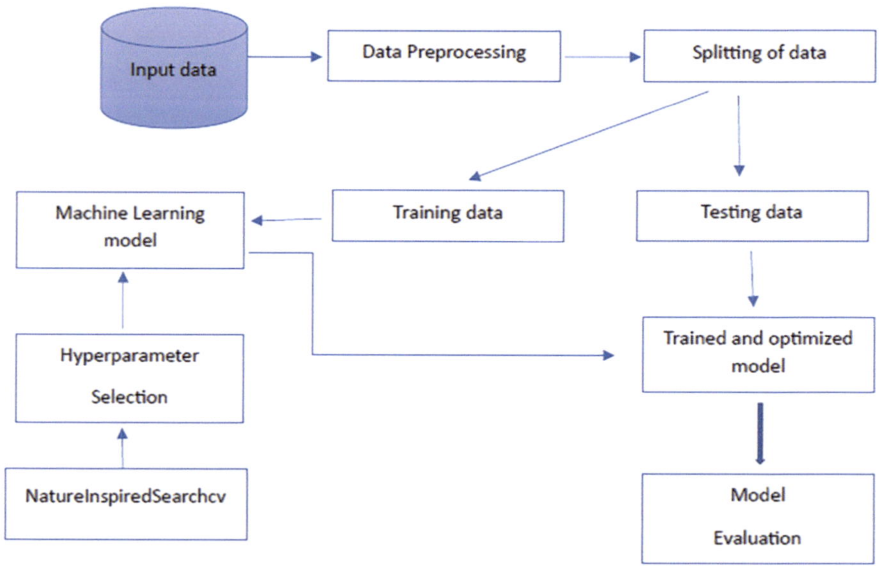

Fig. 2 Proposed model workflow

Table 2 Attacks and subclasses

Attacks	Subclasses	Count
DoS	Neptune, teardrop, smurf, pod, land, back	6
Probe	Ipsweep, portsweep, namp, satan	4
R2L	Warezclient, guess_passwd, ftp_write, multihop, imap, warexmaster, phf, spy	8
U2L	Rootkit, buffer_overflow, loadmodule, perl	4

There are three alternative values for the Protocol type feature, 60 for the Service feature, and 11 for the Flag feature.

2.2 Pre-processing

Data pre-processing is an important stage in developing a machine learning model since it includes cleaning and preparing raw data to make it ready for analysis. Real-world data might be noisy, missing numbers, or in an incompatible format, making it challenging to deal with. Thus, data pre-processing is required to prepare the data before training the machine learning model. Data preparation improves the model's accuracy and ensures that it receives high-quality data for training [17]. A real-world dataset may have noise, missing data, or be in an inappropriate format for further processing. As a result, it becomes an essential step before training the data; in this way, we may hope to obtain more accuracy while also providing quality data for training our model. It includes few steps:

- Null values check
- Handling missing data
- Encoding categorical data
- Handling imbalanced data
- Splitting of data into train and test set
- Data standardization

Handling missing data: Missing data poses major challenges for machine learning models. Therefore, it is critical to identify and handle any missing values in the data. There are primarily two techniques to deal with missing data, which are: the most frequent way to manage null data is to remove the rows or columns that contain them. However, this strategy is not particularly successful because eliminating data might result in a loss of information, which can lead to incorrect results; alternative way is to calculate the mean, which entails finding the average value of a column or row with missing data and replacing the null values. It is an excellent approach for dealing with missing data in numerical parameters such as age, salary, year, and so on.

Encoding categorical data: Machine learning models only work with numerical values; the existence of categorical variables in a dataset might complicate model construction. As a result, encoding becomes necessary for converting these classified variables into numerical values. Label encoding is one method for converting textual data to numeric data [17]. It essentially puts all the unique strings in alphabetical order, assigns a number to each unique data starting at 0, and replaces all of those string data with their equivalent label.

Handling imbalanced data: Imbalanced data is a dataset with considerably more or fewer observations for one class label than for another. This results in an unequal distribution of observations across the target classes. Suppose we have 1000 images to identify as apple or orange. There are 800 apple and 200 orange photos. If we classify all 1000 data points as apples and calculate accuracy, we get around 80%, which is a good score, but our model is unable to correctly predict even a single orange data point, resulting in inaccurate accuracy, and the data that is more in number is called the majority, and the data that is less in number is called the minority. These types of data must be managed carefully; otherwise, we may receive biased or erroneous results. Depending on the dataset, there are several approaches to tackle it: one by selecting the appropriate evaluation metric, and another by resampling.

Splitting the dataset into training and testing dataset: During the data preparation step of machine learning, we divided our dataset into two parts: training and test sets. This is an important phase in the process since it enhances the performance of our machine learning model [9]. If we train our model on one dataset and then test it on a completely different one, our model may struggle to recognize the relationships and patterns in the data. If our model is trained to be highly accurate on a certain dataset, it may not perform well on a new or unknown dataset. As a result, we hope to develop a machine learning model that performs well on both the training and test datasets. In this context, the training dataset refers to the data used to train the model, whereas the test dataset is a distinct dataset used to assess the model's performance on fresh and unknown data.

Feature scaling: In machine learning, feature scaling is the act of transforming a dataset's predictor variables to fit a certain range or scale. This is done to standardize the variables and guarantee that no one variable takes precedence over the others. The goal is to place all variables on an equal footing, allowing for more accurate comparison and analysis.

3 Proposed Work

This study uses machine learning techniques such as Decision Tree, Random Forest, AdaBoost, XgBoost, and CatBoost, as well as hyper tuning Random Forest, CatBoost, and LightGBM with NatureInspiredSearchCV, to identify invasive behavior in an IoT environment. Random Forest uses the following parameters: n_estimators, max_depth, min_samples_split, and max_features. CatBoost considers the following parameters: iterations, learning rate, depth, border count, and L2 leaf regularization. LightGBM considers the following parameters: boosting type, importance type, Lambda, and alpha values. The dataset utilized in this study is NSL-KDD, which has 23 classes representing harmful and regular behavior in an IoT framework. Furthermore, the 23 classes were classified into five categories based on the style of assault. It mainly comprises of four types of attacks: Denial of Service, U2R, R2L, Probe, and a regular class. The performance measures studied were accuracy, recall, precision, and $F1$ score. Figure 2 shows the workflow of the proposed model.

3.1 Classification Model Used

Decision Tree: Decision Trees (DTs) are supervised learning techniques used for classification and regression applications. The primary goal of DTs is to create a model that can predict the value of a target variable using data attributes and simple decision rules. A Decision Tree is a tree-like structure with each node representing a test on a specific property and each branch representing the result of the test [2]. The leaf nodes, also known as terminal nodes, contain the anticipated class label. To forecast the class of a given dataset, the method begins at the root node of the tree and progresses via the decision rules until it reaches the leaf node. In the Decision Tree algorithm, the value of an attribute in the root node is compared to the corresponding attribute value in the real dataset, and the algorithm selects the branch that corresponds to the matching outcome [3]. The algorithm then continues to the next node, repeating the process until it reaches the leaf node. To choose which branch to take at each node, the algorithm compares the attribute value to that of other sub nodes. One of the issues in developing a Decision Tree is deciding the characteristic to utilize for the root node and subsequent sub nodes. To overcome this issue, a method known as Attribute Selection Measure (ASM) is applied. This approach aids in finding the best appropriate characteristic for each node in the tree. There are two main ASM techniques: information and the Gini index.

Random Forest: Random Forest is a popular supervised machine learning method. This algorithm can solve both classification and regression issues. The primary idea behind Random Forest is to use the notion of ensemble learning, which entails mixing numerous classifiers to tackle complicated issues and improve the model's overall performance. Random Forest is a classification system that consists of many Decision Trees trained on different subsets of the supplied dataset. To increase forecast accuracy, it averages the findings from all trees rather than depending on a single tree. The ultimate production is forecasted using the majority vote of forecasts. The more trees in the forest, the better the accuracy and the smaller the likelihood of over fitting [18, 19].

AdaBoost: AdaBoost, or Adaptive Boosting, is an Ensemble Method used in Machine Learning. AdaBoost's principal estimator is a one-level Decision Tree. This refers to Decision Trees with only one split, known as Decision Stumps. It is a machine learning strategy that uses equal weights to create a model from all data points. The program then recognizes the improperly categorized data points and awards them larger weights. The following model concentrates on the data points with greater weights. The method continues to train models until it reaches a low error rate [17].

XGBoost: XGBoost is a sophisticated machine learning package that combines ensemble learning and gradient boosting to generate accurate models. It is intended to train models efficiently and manage huge datasets. The approach combines predictions from several weak models to get a more robust forecast. XGBoost is extensively utilized and has achieved cutting-edge performance in several machine learning applications, including classification and regression. One of XGBoost's key benefits is its ability to properly manage missing values, making it suited for real-world datasets that do not require extensive preparation. XGBoost offers the unique feature of parallel processing, which allows it to handle huge datasets effectively. It is an implementation of gradient boosted Decision Trees, in which Decision Trees are generated consecutively. XGBoost assigns weights to independent variables, which are subsequently utilized to train Decision Trees and produce predictions. When a Decision Tree predicts an inaccurate result in XGBoost, the weights of the variables involved are raised, and the same variables are passed on to the next Decision Tree. This technique is continued until all Decision Trees have been created. The different Decision Trees are then integrated to create a more robust and accurate model. XGBoost can tackle machine learning issues like regression, classification, ranking, and user-defined predictions [12, 13].

CatBooost: CatBoost is an open-source package that employs gradient boosting on Decision Trees to handle a variety of business challenges. It was created by Yandex researchers and engineers and can handle a variety of data kinds. CatBoost is utilized in a variety of applications, including search, recommendation systems, personal assistants, self-driving vehicles, and weather prediction. The term "CatBoost" is a mix of the phrases "Category" and "Boosting." It is a machine learning method that can successfully handle data from numerous categories, such as voice, text, images, and historical data [4].

Hyper parameter tuning: Hyper parameter tweaking is an important step in improving the performance of a machine learning model. Hyper parameters, which are user-defined values that the model does not learn during training, are crucial in deciding how well the model works with fresh data. Hyper parameters include the learning rate, number of hidden layers in a neural network, regularization coefficient, and number of trees in a Random Forest. To optimize performance, experiment with multiple hyper parameter settings to discover the ideal combination for each dataset and task [4]. There are numerous approaches to hyper parameter adjustment. Here are some typical techniques: Grid Search, Random Search, Bayesian Optimization, Gradient-Based Optimization, and Ensemble Optimization.

4 Results

In this work, the models such as DT, RF, AdaBoost, XgBoost, CatBoost, Random Forest (RF) are used along with NatureInspiredSearchCV. In Table 3, the corresponding parameters of each model is shown. Figure 3 represents the decision matrix for CatBoost. Figure 4 shows the confusion matrix for CatBoost along with NatureInspiredSearchCV. Figure 5 represents the confusion matrix of Decision Tree. Figure 6 shows the confusion matrix of Random Forest. In Table 4, the comparison of several models with accuracy, precision, recall, $F1$ score, and data sampling are shown (Figs. 7 and 8).

Table 3 Parameters optimized for all examined models

Models	Parameters optimized
DT, RF, AdaBoost, XgBoost, CatBoost	With default parameters (base model)
RF + NatureInspiredSearchCV	N_estimators=max_depth=, min_samples_split=, max_features=auto
CatBoost + NatureInspiredSearchCV	Iterations = 29, depth = 8, learning_rate = 0.7, border_count = 268, 12_leaf_reg = 4
LightGBM + NatureInspiredSearchCV	bt = goss, it = gain, reg_alpha = 0.01, reg_lambda = 0.03f

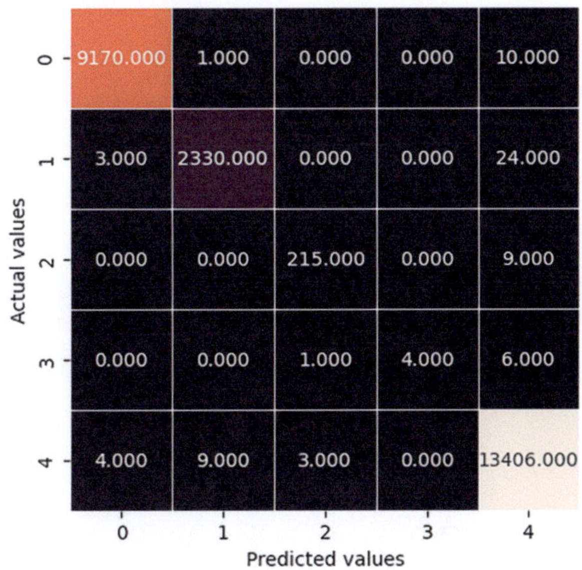

Fig. 3 Confusion matrix of Catboost

5 Conclusion and Future Scope

IoT networks are extremely vulnerable to attacks, and an intrusion detection system must be developed and tested to safeguard these devices and networks from attack. In this work, IDS was created utilizing different machine learning models such as Decision Tree, Random Forest, AdaBoost, XgBoost, CatBoost, and LightGBM, with hyper parameter tweaking conducted using nature-inspired search. Data sampling improves performance parameters such precision, recall, and $F1$ score. Additionally, LightGBM hyper tuned with nature-inspired search achieves an accuracy rate of 99.86. This study can be expanded by evaluating more datasets and applying the same strategies to those datasets.

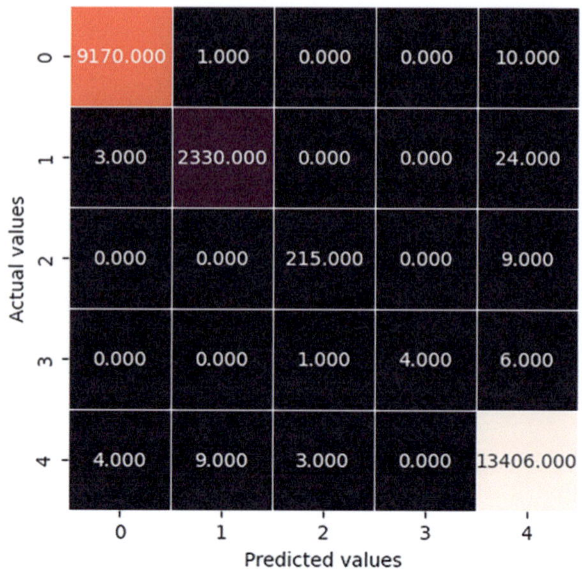

Fig. 4 Confusion matrix of CatBoost + NatureInspiredsearchcv

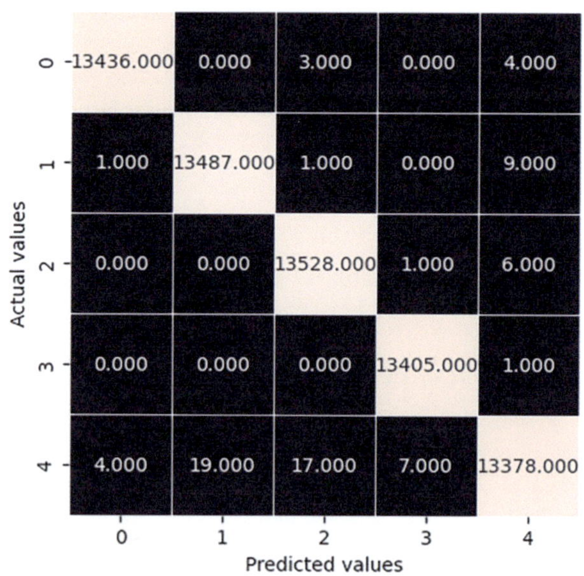

Fig. 5 Confusion matrix of decision tree

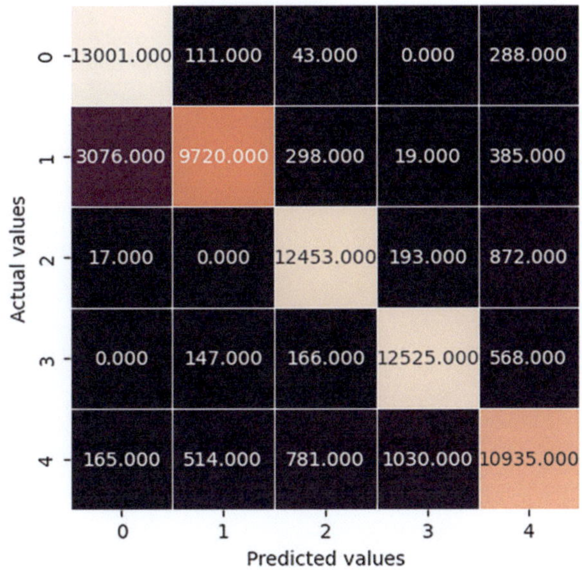

Fig. 6 Confusion matrix of random forest

Table 4 Comparison of several models based on accuracy, precision, recall, and $F1$ score without data sampling

Models	Accuracy	Recall	Precision	$F1$ score
Decision tree	99.71	91.43	88.85	89.97
Random forest	95.68	56.53	58.38	57.35
AdaBoost	89.41	61.04	52.18	53.51
XgBoost	99.71	83.56	91.74	86.39
CatBoost	99.31	74.84	79.37	76.85
Random forest + NatureInspiredSearchCV	99.84	84.47	99.80	87.97
CatBoost + NatureInspiredSearchCV	99.72	86.19	99.46	89.85
LightGBM + NatureInspiredSearchCV	99.86	90.56	91.56	91.03

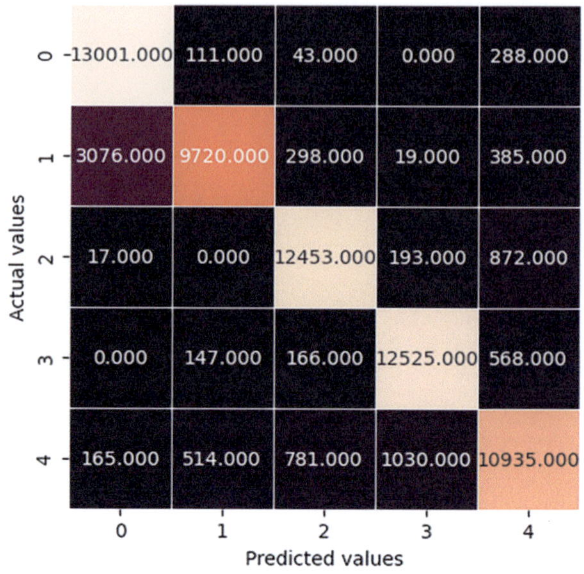

Fig. 7 Confusion Matrix of AdaBoost

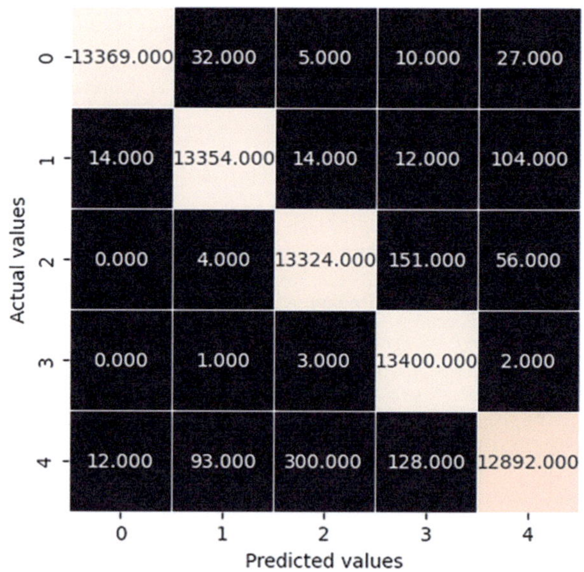

Fig. 8 Confusion matrix of XgBoost

References

1. Saba T et al (2022) Anomaly-based intrusion detection system for IoT networks through deep learning model. Comput Electr Eng 99:107810
2. Baich M et al (2022) Machine learning for IoT based networks intrusion detection: a comparative study. Procedia Comput Sci 215:742–751

3. Simon J et al (2022) Hybrid intrusion detection system for wireless IoT networks using deep learning algorithm. Comput Electr Eng 102:108190
4. Nayak J et al (2022) Hybrid Bayesian optimization hypertuned catboost approach for malicious access and anomaly detection in IoT nomalyframework. Sustain Comput Inform Syst 36:100805
5. Mishra AK, Paliwal S (2022) Mitigating cyber threats through integration of feature selection and stacking ensemble learning: the LGBM and random forest intrusion detection perspective. Cluster Comput 1–12
6. Aljumah A (2021) IoT-based intrusion detection system using convolution neural networks. PeerJ Comput Sci 7:e721
7. Kiran KS et al (2020) Building a intrusion detection system for IoT environment using machine learning techniques. Procedia Comput Sci 171:2372–2379
8. Choudhary S, Kesswani N (2020) Analysis of KDD-Cup'99, NSL-KDD and UNSW-NB15 datasets using deep learning in IoT. Procedia Comput Sci 167:1561–1573
9. Susilo B, Sari RF (2020) Intrusion detection in IoT networks using deep learning algorithm. Information 11(5):279
10. Xu G et al (2020) An ensemble learning-based prediction model for image forensics from IoT camera in smart cities. IEEE Access 8:222117–222125
11. Singh K, Singh N (2020) An ensemble hyper-tuned model for IoT sensors attacks and anomaly detection. J Inf Optim Sci 41(7):1715–1739
12. Zainab A, Refaat SS, Bouhali O (2020) Ensemble-based spam detection in smart home IoT devices time series data using machine learning techniques. Information 11(7):344
13. Verma A, Ranga V (2019) ELNIDS: ensemble learning based network intrusion detection system for RPL based Internet of Things. In: 2019 4th International conference on Internet of Things: smart innovation and usages (IoT-SIU). IEEE
14. Illy P et al (2019) Securing fog-to-things environment using intrusion detection system based on ensemble learning. In: 2019 IEEE wireless communications and networking conference (WCNC). IEEE
15. Tang S et al (2019) Smart home IoT anomaly detection based on ensemble model learning from heterogeneous data. In: 2019 IEEE international conference on big data (big data). IEEE
16. Yacchirema D et al (2019) Fall detection system for elderly people using IoT and ensemble machine learning algorithm. Pers Ubiquit Comput 23:801–817
17. Abdollahi J, Moghaddam BN, Parvar ME (2019) Improving diabetes diagnosis in smart health using genetic-based ensemble learning algorithm. Approach to IoT infrastructure. Future Gen Distrib Syst J 1:23–30
18. Hasan M et al (2019) Attack and anomaly detection in IoT sensors in IoT sites using machine learning approaches. Internet of Things 7:100059
19. Huda S et al (2018) Securing the operations in SCADA-IoT platform based industrial control system using ensemble of deep belief networks. Appl Soft Comput 71:66–77

Advancements in Cyber Security Using Deep Learning Techniques Attack Detection for Trojan Horses

S. Amutha, E. Sagin Sandoz Fernando, G. Jayanth, K. Ajay Kumar Reddy, K. Angel Jean Vincy, and K. Nivethika

Abstract Cybersecurity risks, particularly attacks using Trojan horses, represent a significant risk to modern computer systems. We suggest and evaluate improved deep learning methods for Trojan horse attack detection using the Kaggle-hosted Trojan Detection dataset. Deep learning may help cybersecurity identify Trojan horses. These neural network methods find bad trends in massive data. Deep learning models re more accurate and efficient than traditional detection approaches because they adapt to changing threats. Recent research demonstrates that new Trojan horse detection methods are quicker and more accurate. It prevents data breaches and unauthorized access, making it a vital cybersecurity improvement. This work makes use of these techniques. To achieve a high degree of accuracy in identifying malicious software, our technique is based on the classification of convolutional neural networks (CNNs). We preprocess and enrich the dataset to enhance the model's generalizability. As a result of extensive testing and optimization, Trojan horse attacks and other cyberattacks risk sensitive data and systems. Trojan horses and malicious software masquerading as legitimate, may be catastrophic. CNN-based classification solves this. From Kaggle's Trojan Detection dataset. Data preparation and enrichment increase feature representation. Models classify threats. This proposed Model achieved an impressive accuracy rate of 98.6% in identifying Trojan horse attacks in cybersecurity applications.

Keywords Cybersecurity · Trojan horses · Convolutional neural networks · Kaggle

S. Amutha (✉) · E. Sagin Sandoz Fernando · G. Jayanth · K. Ajay Kumar Reddy · K. Angel Jean Vincy · K. Nivethika
Department of Computer Science and Engineering, Vel Tech Rangarajan Dr. Sagunthala R&D Institute of Science and Technology, Chennai, Tamil Nadu, India
e-mail: amuthamukilraj@gmail.com

K. Nivethika
e-mail: knivethika@veltech.edu.in

H. Sharma et al. (eds.), *Artificial Intelligence: Theory and Applications*, Lecture Notes in Networks and Systems 5589, https://doi.org/10.1007/978-981-96-1687-9_7

1 Introduction

Hacker assaults are now the most important problem in the field of technology. Hackers are responsible for the majority of cyberattacks. In this context, the phrase relates to a destructive process using an exploitable vulnerability to get sensitive information, modify the code, or even bring the whole system to its knees. Additionally, cyberattacks might take the form of harmful software. Internet overlay networks like Trojan Attacks need certain software, setup, or authorization [1]. Its protocol for communication is often modified. Some examples of darknets include document-storing peer-to-peer social networks. Anonymous proxy services like Tor create a web of hidden links [2].

Despite its widespread use in the media, the phrase "Trojan Attack" lacks official recognition. Silk Road utilized it, which connected it to Tor's Onion services. Freenet, I2P, and Tor all have their uses, both legitimate and illicit. Security, anonymity, and resistance to censorship are the guiding principles in their work to protect digital rights. Witnesses, activists, journalists, and media organizations may all communicate anonymously on darknets using tools like SecureDrop [3]. Take into consideration a machine learning system that has identified broad principles and can explain them by itself. The analysis of training data has the potential to assist machine learning algorithms in becoming more accurate predictors illustrated Fig. 1.

To better secure IoT communication networks, the research takes on the challenge of identifying and classifying assaults, drawing attention to the need for a top-down machine learning architecture. None of the limitations or specific problems with this study are mentioned in the provided reference [4].

This research focuses on attack-aware ensemble learning traffic routing for IoT networks as a means to improve the performance of these systems by considering potential attacks at every stage of the routing process. There is no mention of the limitations and flaws of the proposed method in the reference [5].

The research proposes an anomaly-based approach that utilizes the Variational Autoencoder (VAE) and network flow characteristics to improve the accuracy of identifying anomalous network activity. None of the limitations or drawbacks of this work are mentioned in the provided reference [6].

1.1 Background

After going over all the studies on using neural network techniques for detecting intrusions and viruses, it's clear that people have accomplished a lot in this area. The bulk of research, however, has focused on deep learning techniques or has been developed to identify certain malware (e.g., malware targeting Android, malware targeting Windows, or network abnormalities). There aren't many surveys accessible on the subject, like comprehensive reviews that cover the full spectrum of neural network techniques for intrusion and virus detection.

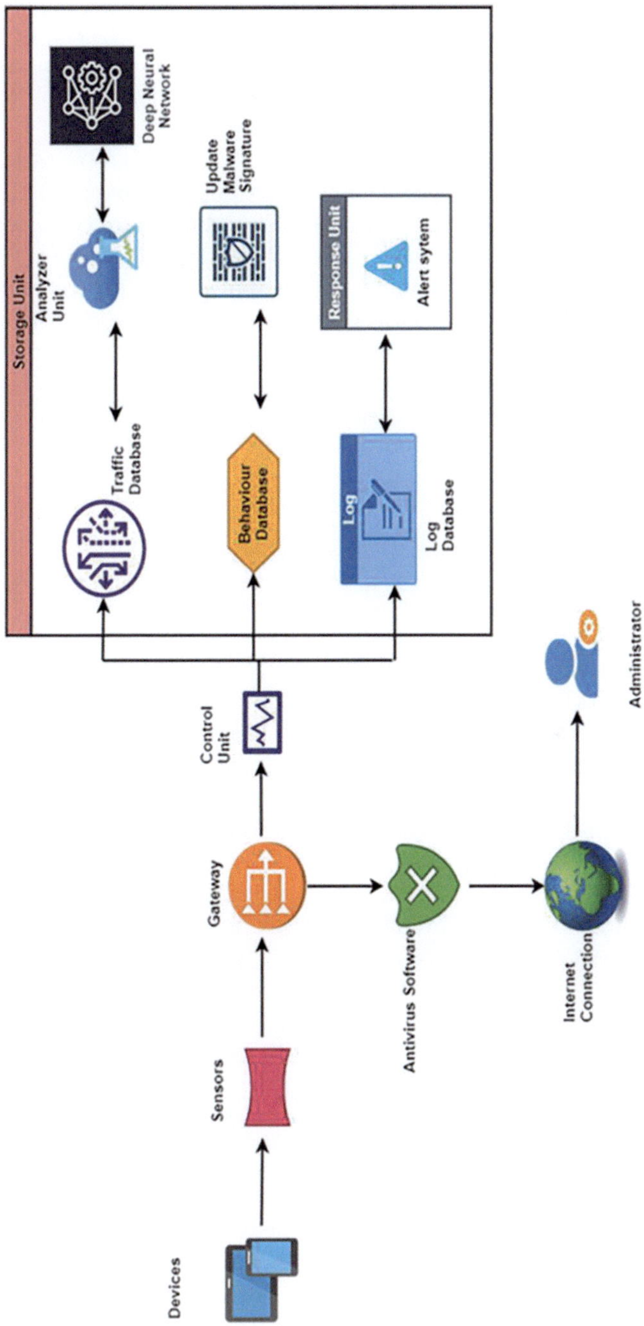

Fig. 1 Malware detection architecture

Malware (both Windows and Android), intrusion, network anomalies, and other security issues may each need a separate investigation. Ransomware is only one of many types of threats that should be the subject of a Systematic Literature Review (SLR). Nonetheless, we think it's important to provide new scholars in this area with a broad viewpoint and a thorough analysis. To do this, the conversation might be divided into portions according to the various platforms.

By distributing harmful malware and stealing personal information, cybercriminals endanger organizations, people, and governments throughout the world. Numerous cybercriminals break into systems, steal information, and conduct illegal financial activities daily through the use of harmful software. This is why protecting individuals' private data is quickly rising to the top of the scientific community's list of priorities. To detect harmful software and block its access to personally identifiable information, this research employs data analysis and artificial intelligence methods of classification. To do this, we analyze features based on signatures and features based on anomalies to provide a reliable and efficient method for malware identification and categorization. The suggested approach outperforms the alternatives, according to the experimental data.

1.2 Contribution

- The primary objective of this research is to identify malicious software in Internet of Things (IoT) devices using convolutional neural networks (CNNs).
- For accurate real-time malware detection and categorization, consider utilizing the proposed convolutional neural network (CNN) approach.
- Two subsets of the dataset were created by the author. The dataset was trained using the first subset, which accounts for 70% of the data.
- To test how well the CNN model performed, we used the second subset, which included 30% of the data.
- Without resorting to intricate feature engineering, with our deep neural network model, we achieved a 98.2% success rate.

2 Related Works

This study aims to address the challenge of understanding and categorizing Tor-related behaviors by analyzing Tor (The Onion Router) usage on mobile and PC platforms using multilayer identification and classification techniques. The citation fails to address the limitations and flaws of the proposed multilevel analysis [7].

Automating the real-time detection of harmful intent in darknet traffic is achieved via the study's usage of weight-agnostic neural network architecture and huge data analysis. There is no mention of the study's limitations or specific flaws in the provided reference [8].

Protecting cloud infrastructures using Software-Defined Networking (SDN) against distributed denial of service (DDoS) attacks is the topic of this research. The main focus is on enhancing security against DDoS attacks by using entropy-based protection mechanisms and the POX controller. All of the limitations and disadvantages of this research are not mentioned in the referenced article [9].

To understand and analyze the actions of IoT malware, this study makes use of association rule learning. To better detect and categorize malware associated with the Internet of Things, the research analyzes data collected from darknet sensors [10].

The limitations and flaws of the approach utilized to probe IoT malware activities are not mentioned in the reference.

This article details a preliminary study that examined Software-Defined Networking (SDN) DDoS attacks using Wireshark to examine network metrics. Improved detection and mitigation of distributed denial of service (DDoS) attacks is a primary focus of SDN technology. The provided reference [11] does not detail any particular drawbacks or restrictions of this pilot study.

This endeavor takes on the problem of identifying global cyber dangers in real time based on darknet activities. To enhance cyber threat awareness and rapid detection, the approach uses Graphical Lasso to assess anomalous synchronization. The reference [12] does not explicitly state the limitations or drawbacks of the proposed approach for real-time cyber threat detection.

Many people in the world succumb to coronary disease. There is a wealth of clinical data available for predictive analysis thanks to biomedical equipment and hospital apps. AI and ML can detect and decide upon heart problems. Recent advances have been made in medical machine learning. To use clinical test data for illness prediction, models are trained using case samples. Heart disease data has to be comprehended for better prediction. By removing superfluous features, feature selection helps to save both time and money while training predictive models. Prediction of coronary disease using artificial intelligence-based highlight determination is proposed. Using a variety of feature selection techniques, this research suggests a disease prediction model. Validating the model for prediction using data on cardiovascular disease. We assess the strategy's accuracy, precision, review, and f1-score. Heart conditions may be easier to diagnose with the use of these technologies [13].

Since cloud computing is accessible to everybody, millions of people store and access their data there. In such a situation, distributed denial of service assaults constantly jeopardize the availability of resources and services. An autonomous multi-agent system that uses particle swarm optimization can identify and counteract distributed denial of service attacks. Smart judgments are made by agents who communicate properly. In distributed denial of service attacks, many agents alert the coordinating agent. By analyzing entropy and covariance, the coordinating agent can detect DDoS attacks. At this level, monitoring agents keep an eye on the networking and resources in the cloud in real time. Unusualities notify those who can help. The results show that this approach improves the safety and performance of cloud platforms in trials [14].

Since the Internet of Things serves as the basis for these devices, the size of the wireless sensor network (WSN) has grown. The use of clustering enhances the activity of wireless sensor networks. Before choosing Cluster Heads, Cluster Nodes must be created. The information about the CH environment is provided by clustered wireless sensor networks. Recently, sink-selection methods have been working on the assumption that data can never be exhausted. Problems are likely to arise if many CHs utilize the same instance. There is a risk of parcels going missing due to the limited drop-off site. Rules governing sinks must be airy and light. Using SIS, the number of sink nodes may be reduced. The preferred option for regulators who take a more laid-back approach is CH. Having R-recommendations that are standardized would be a great help. Regulators use reductions as a means of regularization. These techniques reduce not only packet loss and delay but also residual energy and FHL production. Considerations such as power consumption, throughput, latency, dependability, packet loss, durability, and end-to-end delay are crucial. Researchers are now looking into clustering methods for deploying sensor networks to overcome design limitations and the fundamental principle of streamlining [15].

3 Methodology

3.1 Research Problem

It is possible to find malware threat components by use of both static and dynamic evaluation. For static analysis and the virus decompilation method to work, it is necessary to parse the corrupted files for detrimental strings. When malware is running in controlled conditions, like a computer running a program, dynamic analysis may detect it and stop it in its tracks. It would be best if you used both approaches while studying malware since they both have their advantages and disadvantages. Less hazardous materials would have been utilized in its design if virus detection had been improved. When a more reliable selection of characteristics might do the job for malware detection, we're concerned that far too many factors are being used. Determining which harmful traits to employ begins with identifying viable algorithms or methodologies. It is critical to detect previously unseen malware while drastically reducing the required criteria.

The research asserts that it may provide improved deep learning approaches. The CNN model's feature explanations are quite helpful, and they may help us learn more about Trojan horse assault traits. In addition to showing better effectiveness, this method makes the detection process more interpretable, which might lead to improvements in cybersecurity defenses against these types of assaults. The review does a good job of setting the stage for the present research and drawing attention to the holes in the current methods that the study intends to fill. Malware detection using convolutional neural networks (CNNs) is a well-described and relevant approach. Methods for feature extraction and classification, as well as the authors' framework and dataset, are detailed in the paper. Because of CNNs' usefulness in automatically learning characteristics from raw data, their usage is warranted, especially in virus detection. Several important metrics and visualizations are provided in the results section. These include a confusion matrix, graphs showing loss and accuracy, and a history of the model. Impressive as it may be, the 98.6% accuracy rate shows that the CNN model is great at detecting Trojan horses.

3.2 Framework

Deep learning can determine the next course of action without requiring any labeled input. Convolutional neural networks (CNNs) alleviate features are engineering, a technique that is essential for ML methods that rely on shallow data sets, but may not capture sufficient important features for job classification or address dependency issues, inconsistencies, or fundamental details in the data. CNNs can acquire knowledge from each other's representations and acquire novel characteristics at higher layers, even when they are trained on the same dataset. CNNs excel at gathering temporally sequential data and identifying localized spatial correlations. Convolutional neural networks (CNNs) are ideal for modeling API call sequences due to their skill to identify features at every step of the process and go through an entire series using filter-sequence convolutional techniques. CNN can infer connections from a set of apparently unconnected data points over a lengthy period, illustrated Fig. 2.

An overview of our advanced machine learning algorithm for detecting malware. This technique involves training a classifier, detecting sophisticated malware, and selecting features from promising datasets. The investigation's approach is further out Fig. 2.

3.3 Dataset

In the present age of modern technology, the menace of cyberattacks has emerged as a major worry for people, corporations, and governments alike. A particularly pernicious kind of cyber assault is the Trojan horse, a malevolent malware that masquerades as a benign file or application. It is essential to identify and thwart these assaults to protect the security of our computerized systems. Trojan horses may infiltrate computer systems or networks when users inadvertently welcome them by visiting unfamiliar and harmful websites or downloading software. The concept and functionality of a Trojan horse may be comprehensively grasped by constantly watching the network and collecting the network packets. The current dataset comprises both dangerous (specifically, Trojan category) and benign networked packets that may be used for Trojan horse detection.

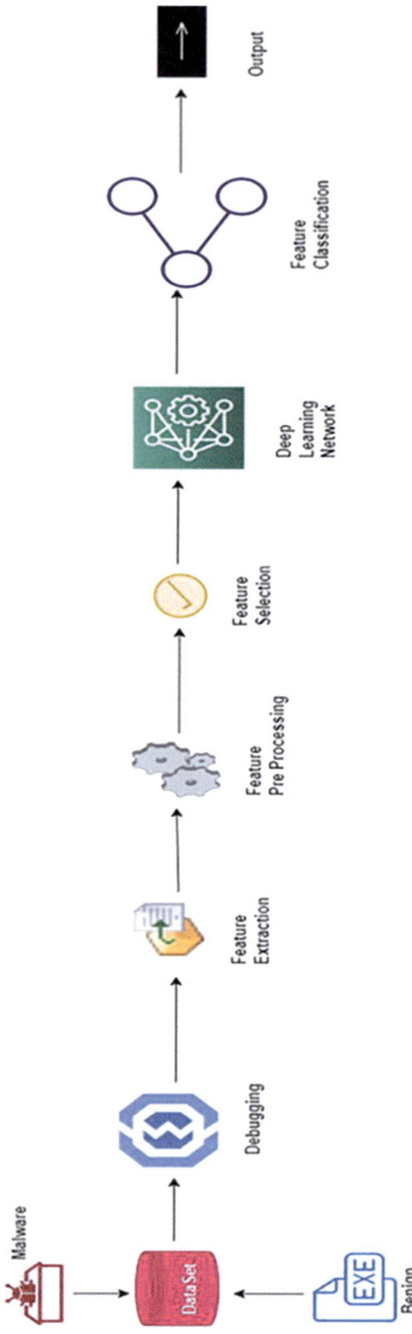

Fig. 2 CNN-based malware detection architecture

3.4 Feature Extraction

An example of a deep neural network is one with an incorporated convolutional neural network (CNN) layer, to extract attributes from the dataset. This setup is shown in Fig. 6. Hundreds of thousands of attributes are typical for datasets in the twenty-first century. With an ever-increasing amount of features put into ML models, overfitting has been a hot subject as of late. We narrow down the original long list of qualities to just the most important details to solve the problem. Figure 6 shows the present dataset; this study aims to improve it by removing unnecessary variables and focusing on the most important static and dynamic properties.

$$F_{i,j} = \sigma\left(\sum_{m=0}^{M-1}\sum_{n=0}^{N-1} W_{m,n} \cdot I_{i+m,j+n} + b\right) \tag{1}$$

3.5 Features Selection

After the process of feature extraction, which entails identifying more characteristics, the next step is feature selection. Feature selection is a critical procedure that involves choosing certain characteristics from a group of recently identified ones. This process is critical for minimizing overfitting, streamlining the model, and increasing accuracy. Experts discovered used a wide range of feature classification algorithms in their endeavor to detect malicious software. The feature rank approach is extensively used in this work because of its efficacy in picking relevant features for the development of malware identification models.

$$f_i = \frac{1}{N}\sum_{j=1}^{N} F_{i,j} \tag{2}$$

f_i represents the ith feature in the pooled feature vector, $F_{i,j}$ denotes the ith feature map at position j in the CNN layer output, N is the total number of elements in each feature map.

3.6 Feature Classification

The development of deep learning methods in cybersecurity, particularly in the detection of Trojan horse assaults using convolutional neural networks (CNNs), relies heavily on feature classification. When it comes to recognizing intricate patterns that might indicate malicious conduct, convolutional neural networks (CNNs) shine because of how effectively they acquire complicated and layered representations of information. To train the CNN model effectively, feature categorization involves taking input data, such as computer network activity or file content, and carefully selecting and extracting relevant features. When it comes to cybersecurity, feature classification is all about finding those unique things that Trojan horse attacks have but regular network traffic doesn't. Network packet headers, payload data, or patterns of behavior associated with known attack signatures are examples of characteristics that could be present. With proper feature extraction and selection, CNN models may potentially detect malicious network traffic from benign data with high accuracy. Convolutional neural network (CNN) detection systems can adapt to new threats and identify subtle variations in attack patterns with the help of better feature classification. Possible ways to improve feature categorization include data augmentation and transfer learning. To enhance classification accuracy, transfer learning makes use of models of convolutional neural networks that have already been trained. In contrast, data augmentation entails producing more training data to boost the model's efficiency. By doing so, we improve the model's ability to generalize its knowledge to other attack scenarios and lower the probability of false positives or missed detections. In addition, advancements in feature classification make CNN-based detection systems more understandable and clearer, which helps cybersecurity analysts understand the logic behind detection decisions and adjust their detection strategies accordingly. By developing and implementing better methods for classifying information into CNN designs, researchers may make Trojan horse intrusion detection systems stronger and more efficient. In a world where threats are always changing, this will make cyber security defenses stronger.

Table 1 illustrates the usual procedure of developing and using a convolutional neural network (CNN) to picture classification problems is shown by this pseudocode. To construct and train the model, it makes use of the TensorFlow library. A convolution neural network (CNN) consists of 4 layers: convolutional in nature, maximum pooling, flattening, and completely linked. Before training on the training data, the model is equipped with suitable optimization and loss functions. It is possible to make predictions on previously viewed pictures after training the model and then evaluating it on test data.

4 Results and Discussions

All methods of classification must include training and testing phases. The system has to be trained with both harmful and benign input for it to work properly. Machine learning techniques may train a classifier to make reliable predictions without human intervention. With each additional piece of labeled data, the CNN model's performance improves.

The Convolutional Neural Network (CNN) model employs dropout in the final fully connected layer. It seems that dropout is less of a regularization technique and more of a method to augment the model with additional layers. At this stage, we

Table 1 CNN Pseudocode

model = tf.keras.models.Sequential([tf.keras.layers.Conv2D(32, (3, 3), activation='relu', input_shape=(img_height, img_width, img_channels)),
tf.keras.layers.MaxPooling2D((2, 2)),
tf.keras.layers.Conv2D(64, (3, 3), activation='relu'),
tf.keras.layers.MaxPooling2D((2, 2)),
tf.keras.layers.Conv2D(64, (3, 3), activation='relu'),
tf.keras.layers.Flatten(),
tf.keras.layers.Dense(64, activation='relu'),
tf.keras.layers.Dense(num_classes, activation='softmax')])
model.compile(optimizer='adam', loss='sparse_categorical_crossentropy',
metrics=['accuracy'])
model.fit(train_images, train_labels, epochs=10, validation_data=(val_images, val_labels))
test_loss, test_acc = model.evaluate(test_images, test_labels)
print('Test accuracy:', test_acc)
predictions = model.predict(test_images)
sample_image = np.expand_dims(test_images[0], axis=0)
prediction = model.predict(sample_image)
predicted_class = np.argmax(prediction)
print('Predicted class:', predicted_class)

provide the actual results obtained from the test of how well our malware categorization and detection method worked. Trials make use of the created malware and cleanware datasets. In this research, we employ a controlled artificial intelligence strategy—more especially, the CNN classifier to analyze malware and group it into many classes.

Figure 3 shows a data collection used for training and testing the Trojan horse attack detection model. There are both good and bad examples of Trojan horse assaults in this collection of network traffic or file content samples. The labels on each sample show whether it shows typical network activity or an effort at malicious penetration.

Figure 4 shows the model history, which includes important parameters including factors that affect training accuracy, validation accuracy, validation loss, and training accuracy during the model training process. By showing how the model is learning to prevent overfitting and underfitting, this visualization sheds light on the model's convergence to optimum performance.

Figure 5 shows the loss result, which shows the decrease, in how much the loss component has changed throughout the training period. If the sum is going down, it means the model is doing a good job of learning to differentiate between Trojan horse assaults and reducing prediction mistakes. Model performance and detection accuracy are both improved with lower loss levels.

	Unnamed: 0	Flow ID	Source IP	Source Port	Destination IP	Destination Port	Protocol	Timest
0	73217	10.42.0.42-121.14.255.84-49975-80-6	10.42.0.42	49975	121.14.255.84	80	6	17/07/ 01:18:
1	72089	172.217.6.226-10.42.0.42-443-49169-17	10.42.0.42	49169	172.217.6.226	443	17	17/07/ 10:25:
2	96676	10.42.0.1-10.42.0.42-53-37749-17	10.42.0.42	37749	10.42.0.1	53	17	30/06/ 07:16:
3	42891	10.42.0.1-10.42.0.42-53-41352-17	10.42.0.42	41352	10.42.0.1	53	17	13/07/ 03:48:
4	169326	10.42.0.151-107.22.241.77-44353-443-6	10.42.0.151	44353	107.22.241.77	443	6	05/07/ 10:47:

Fig. 3 Sample dataset

```
Model: "sequential"
_________________________________________________________________
Layer (type)                 Output Shape              Param #
=================================================================
dense (Dense)                (None, 128)               11008
_________________________________________________________________
dense_1 (Dense)              (None, 64)                8256
_________________________________________________________________
dense_2 (Dense)              (None, 64)                4160
_________________________________________________________________
dense_3 (Dense)              (None, 64)                4160
_________________________________________________________________
dense_4 (Dense)              (None, 1)                 65
=================================================================
Total params: 27,649
Trainable params: 27,649
Non-trainable params: 0
_________________________________________________________________
```

Fig. 4 Model history

The model's accuracy result is shown in Fig. 6, which shows the proportion of dataset samples that were properly identified. The program is quite good at detecting Trojan horse assaults with few misclassifications; its stated accuracy is 98.6%. The outstanding stated accuracy of 98.6% shows that the suggested strategy is very successful. A confusion matrix, graphs showing accuracy and loss, and a history of the model are some of the visualizations that help to improve the presentation of

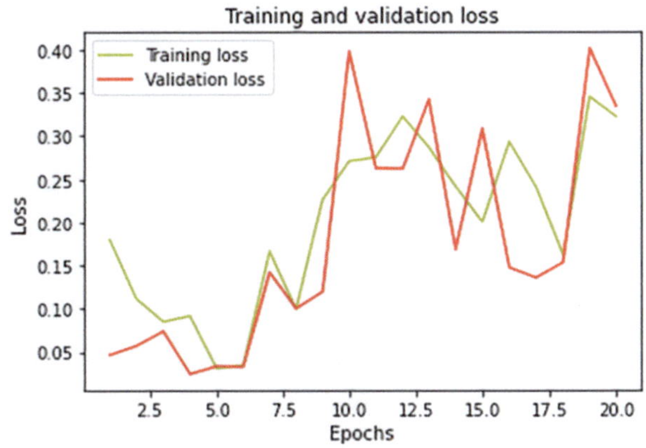

Fig. 5 Loss result

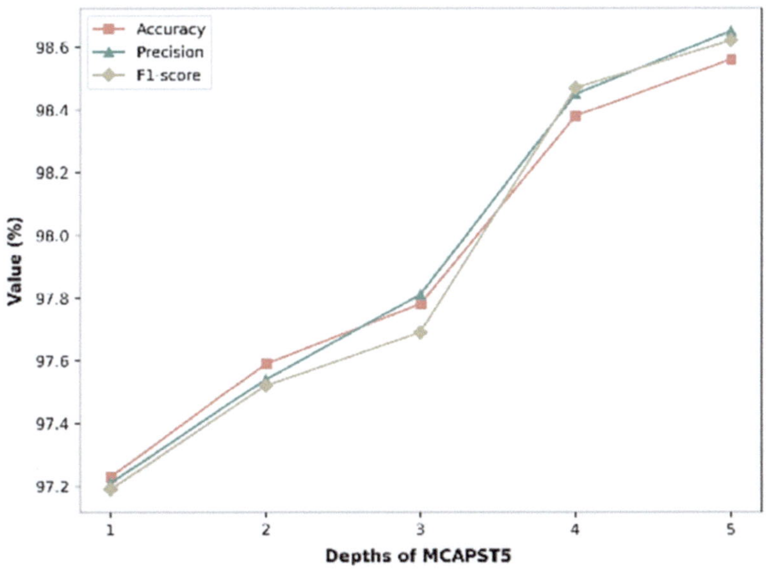

Fig. 6 Accuracy result

the data. Nevertheless, these findings may benefit from a more thorough analysis, especially concerning their practical implications. The method's dependability and practical significance in many real-world cybersecurity contexts should be better understood with this further examination.

The confusion matrix, a thorough visual representation of the distribution of the precision, accuracy, precision deficit, and precision deficit of the model's prediction, is shown in Fig. 7. By comparing the model's performance across classes, this matrix reveals where it might want improvement and where it could be vulnerable in

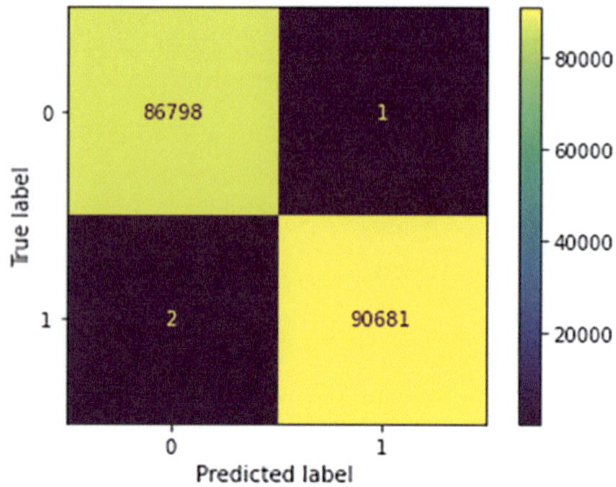

Fig. 7 Confusion matrix

terms of detection accuracy. Cybersecurity analysts may improve the model's training and parameters to maximize detection effectiveness and decrease false alarms by reviewing the confusion matrix.

5 Conclusion

The DL models are often implemented using Python and Python-based frameworks such as sci-kit-learn, TensorFlow, and Keras. However, there is little to no evidence on just how well those DL-based safety measures will work in actual use cases, and it is still unclear whether they could manage the massive volumes of data generated by businesses and on the internet. This study developed a new deep neural network by using CNN is 98.6% accuracy, to address key shortcomings in the recognition of malware, such as the limitations of existing learning methods and the inefficiency of human feature-building. For advanced malware detection that does not involve feature engineering, the approach is suggested. When it comes to detection accuracy, the CNN is on top. Training accuracy for malware detection using the suggested CNN model is increased by a value close to 1, and the accuracy of testing is almost identical to that of training. The combined strengths of a convolutional neural network (CNN) make this feasible by representing sequences and learning from long-term relationships, while simultaneously extracting spatially local correlations.

Limitations and Future Work

There are some limits to deep learning methods, even if they are very accurate in detecting Trojan horse assaults. Not having access to big, varied datasets, which are necessary for efficient model training, is a major obstacle. To add insult to injury, deep learning models may be very resource-intensive thanks to their computing demands. Efficient model optimization and new approaches to create or enhance training data should be the primary goals of future research. Integrating explainability to improve cybersecurity apps' trust and transparency and investigating how these models may adapt to identify new, changing threats are also important areas for future study. For validation and implementation in the actual world, it will also be essential to work with professionals in the field.

References

1. Nasiri E, Berahmand K, Li Y (2022) Robust graph regularization nonnegative matrix factorization for link prediction in attributed networks. Multimedia Tools Appl
2. Gayard L (2018) Darknet: geopolitics and uses. Wiley, Hoboken, NJ, p 158. ISBN 9781786302021
3. Pradhan S (2020) Anonymous. The darkest web: the dark side of the Internet, India, p 9. ISBN 9798561755668
4. Press Foundation, Freedom of the "SecureDrop". GitHub. Freedom of the Press Foundation. Retrieved 28 Jan 2019
5. Abu Al-Haija Q (2022) Top-down machine learning-based architecture for cyberattacks identification and classification in IoT communication networks. Front Big Data 4:782902
6. Abu Al-Haija Q, Al-Badawi A (2021) Attack-aware iot network traffic routing leveraging ensemble learning. Sensors 22:241
7. Zavrak S, Iskefiyeli M (2020) Anomaly-based intrusion detection from network flow features using variational autoencoder. IEEE Access 8:108346–108358
8. Wang L, Mei H, Sheng VS (2021) Multilevel identification and classification analysis of Tor on mobile and PC platforms. IEEE Trans Ind Inform 17:1079–1088
9. Demertzis K, Tsiknas K, Takezis D, Skianis C, Iliadis L (2021) Darknet traffic big-data analysis and network management for real-time automating of the malicious intent detection process by a weight agnostic neural networks framework. Electronics 10:781
10. Mishra A, Gupta N, Gupta B (2021) Defense mechanisms against DDoS attack based on entropy in SDN-cloud using POX controller. Telecommun Syst 77(1):47–62. https://doi.org/10.1007/s11235-020-00747-w
11. Ozawa S, Ban T, Hashimoto N, Nakazato J, Shimamura J (2020) A study of IoT malware activities using association rule learning for darknet sensor data. Int J Inf Secur 19(1):83–92. https://doi.org/10.1007/s10207-019-00439-w
12. Varghese JE, Muniyal B (2020) A pilot study in software-defined networking using wireshark for analyzing network parameters to detect DDoS attacks. In: Information and communication technology for competitive strategies (ICTCS 2020). Springer, Singapore, pp 475–487
13. Han C, Shimamura J, Takahashi T, Inoue D, Takeuchi JI, Nakao K (2020) Real-time detection of global cyberthreat based on darknet by estimating anomalous synchronization using graphical lasso. IEICE Trans Inf Syst E103-D:2113–2124
14. Ramathilagam A, Kaviya B, Thiyagarajan M, Amutha S (2024) Artificial intelligence methods for the prognosis of cardiovascular disease.https://doi.org/10.1109/icnwc60771.2024.10537248
15. Kesavamoorthy R, Soundar KR (2018) Swarm intelligence-based autonomous DDoS attack detection and defense using a multi-agent system. Clust Comput 22(S4):9469–9476. https://doi.org/10.1007/s10586-018-2365-y

Unlocking the Skies: Addressing the Role of Government Support and Policies in Fueling the Adoption of Drones in Last-Mile Deliveries

Kalya Lakshmi Sainath and C. Lakshmi Devasena

Abstract Artificial Intelligence and other technological developments improve the performance of business organizations worldwide. Technological development can also be utilized in logistics and the supply chain industry. Last-mile deliveries are crucial in logistics and supply chain management. The adoption of drones is an upcoming technological trend in last-mile delivery (LMD). Developed nations already started using drone technology and yielding presidential advantages out of it. Since drones are environmentally friendly and reduce carbon footprint, they reduce air pollution. It also consumes less fuel than diesel vehicles used for goods delivery. In India, technology has yet to start at a high pitch and reach levels in developed countries. The adoption of drones may be affected by various factors and conditions of the organization and its environment. Last-mile deliveries using drone technology affect the environment as the products are delivered using drones (a sort of unmanned air vehicle), and governmental rules and regulations might be forced to adopt and use it at any organization. This research examines the effect of government support and policies for adopting drone technology in last-mile deliveries by collecting data from the working professionals of the organizations involved in logistics and supply chain management in the Indian context using structured equation modeling. It was found that government support and policies positively affect the adoption of drone technology in last-mile deliveries, thus improving the operational performance of the business organization.

Keywords Adoption of drone technology · Drone in last-mile delivery · Government support and policies · Role of government support in drone adoption

K. L. Sainath (✉) · C. Lakshmi Devasena
ICFAI Business School (IBS) Hyderabad, IFHE University, Hyderabad, India
e-mail: kalyalakshmi.sainath21@ibsindia.org

C. Lakshmi Devasena
e-mail: ldevasena@ibsindia.org

H. Sharma et al. (eds.), *Artificial Intelligence: Theory and Applications*, Lecture Notes in Networks and Systems 5589, https://doi.org/10.1007/978-981-96-1687-9_8

1 Introduction

Customer need for increased logistics efficiency is on the rise. However, competitive price battles among e-commerce platforms pressure logistics firms and their revenue margins [1]. As a result, several restricted budgets are being used for vehicle purchases and repairs and running expenditures to enhance last-mile logistics efficiency [2]. The e-commerce business has substantial labor turnover and needs a driver shortage. In 2020, there was a shortage of 150,000 drivers, which shows the necessity of innovative and alternative methods to address this issue. Last-mile deliveries (LMD) expenses are accountable for 50% of the logistics service providers' variable costs [3–5]. Drones provide prospects for change in the last-mile deliveries in urban and rural logistics because of their flexibility, high speed, and low cost [6]. However, drone usage remains low, and the current expectations indicate that the number of delivery drones will continue to rise [7]. Gartner's report predicts that a million drones will be used in last-mile delivery by 2026 instead of 20,000 drones as predicted before [8]. End users prefer drone deliveries because of the product's increased value and speed [9]. Drones can provide a more targeted response, fly independently, and make decisions without the assistance of the pilot or the operator on the ground by sending communications to the warehouse [10]. In recent years, businesses like E-Commerce and Medicine delivery companies have been using drones for packet delivery. By conducting these tests in a real-world environment, companies like Amazon, UPS, DHL, and other logistics players are evaluating the options for safely and effectively delivering mechanisms to reach the packages to customers [11–13]. Government support is needed to implement drone technology in any state or country. Despite progressive regulations, obtaining necessary permissions for drone operations can take time and effort. Compliance with safety and privacy standards also poses significant challenges [14]. The widespread spread of drone technology has yet to be evident despite the potential benefits of drone usage in last-mile deliveries, which include lowering costs and increasing responsiveness. This paper addresses the role of government support and policy that influence the adoption of drones to improve operation performance.

2 Theory Foundation and Hypotheses

2.1 Drones in Last-Mile Deliveries

Using drones, particularly last-mile logistics, is gaining momentum globally to deliver packages faster to customers. In India, a country characterized by diverse geographical landscapes and significant urban–rural disparity, the potential to use and implement drone technology in last-mile deliveries is considerable. Recent trials from government and private bodies indicate a growing interest in the field. For example, Dunzo, a leading hyperlocal delivery firm, has conducted successful trials in Telangana to deliver medical supplies and groceries demonstarting the feasibility of drone deliveries [15]. A few more firms, like Swiggy and Zomato, have expressed interest in leveraging drone technology to enhance delivery efficiency and process [16]. Particular advantages are identified in the literature to address the importance of drone technologies in last-mile deliveries. According to [17], speed and efficiency are the primary factors that can reduce delivery times, especially in congested urban areas and remote rural areas, and this will help to complete the deliveries up to 70% faster than the traditional methods. The second advantage is cost-effectiveness; the initial investment in drone technology can be high, but operational costs are relatively low compared to conventional delivery vehicles. The cost advantage is particularly beneficial in areas with poor road infrastructure, as identified by [18]. Third, drones have a lower carbon footprint than conventional delivery vehicles, aligning with global sustainability goals. A study by [19] found that the electric propulsion systems in drone technology produce fewer emissions, contributing to greener logistics operations.

2.2 Government and Policy

The Indian Government has recognized the potential of drone technology. In 2018, the Directorate General of Civil Aviation (DGCA) released the Civil Aviation Requirements (CAR) for the operation of drones, providing a regulatory framework to facilitate drone operations [20]. The CAR, effective from December 1, 2018, categorized drones based on their weight and established guidelines for their operation, including the requirement for Unique Identification Numbers (UINs), operator permits, and the no-permission-no-takeoff (NPNT) protocol. In 2021, the Indian Government introduced more liberalized drone rules to boost the adoption of drone technology. The drone rules, 2021 reduced the number of forms required to operate drones from 25 to 5 and reduced the types of fees charged from 72 to 4. The government also abolished the requirement for a drone pilot license and replaced it with a remote pilot certificate issued by a DGCA-approved drone school. This regulatory support has encouraged several startups and established logistics companies to explore drone-based delivery solutions [20]. The state governments are also interested in using drones in different areas, such as the "Medicine from the Sky" initiative by the Government of Telangana, addressed in the World Economic Forum [21]. This project successfully used drones to deliver vaccines and medical supplies to remote areas, showcasing the potential of drones in improving healthcare logistics. Another project by "Andhra Pradesh's Agriculture Drone Project" highlighted that deploying drones to deliver agricultural inputs like seeds and fertilizers to farmers in remote areas helps support rural economies and agricultural productivity [22]. Third, The Indian Council of Medical Research (ICMR) and the Ministry of Civil Aviation permitted drone delivery of COVID-19 vaccines in Manipur and Nagaland, showcasing the potential for large-scale vaccine distribution using drones [23]. Collaborative efforts between government bodies, private sector stakeholders, and technological innovators are essential to harness the full benefits of drone technology. As advancements continue and regulatory frameworks evolve, drones will likely become a cornerstone of India's logistics and delivery ecosystem. This leads to the following hypothesis in this research work.

H1: Government Support and Policies Have Positive Effect on Adoption of Drone Technology in LMD.

2.3 *Intensity to Adopt Drones*

Technology adoption in last-mile deliveries is critical for improving efficiency, reducing costs, and enhancing customer satisfaction. The intensity to adopt technology refers to the degree of commitment and speed with which individuals or organizations embrace new technological innovations [24]. The surge in e-commerce has dramatically changed the landscape of last-mile deliveries in India. Consumers demand faster and more reliable delivery services, and firms are turning to advanced technologies to meet customer expectations [25]. Highlights that Artificial Intelligence applications in last-mile delivery can improve operational efficiency by up to 20–25%. Moreover, Machine Learning applications and their algorithms can analyze the data to forecast demand, enabling better inventory management and reducing stock-outs. Recent studies conducted as industry case studies and real-world examples showed how technology adoption can change the performance of operations. A study conducted by [26] in one of the leading e-commerce platforms in India addressed how Artificial Intelligence helped route optimization, real-time tracking, and automated sorting systems in last-mile deliveries. The second study conducted by [27] in one of the leading logistics service providers in Malaysia identified the use of technology helped the firm to reduce the delivery cost, accurate deliveries, inventory optimization in proper stacking, real-time tracking, and route planning of shipments. A third study by [28] in tech-enabled delivery integrators in food and grocery deliveries revalorized their last-mile deliveries by using advanced mechanisms and technologies and technology helped to understand the personalized customer experiences, real-time order tracking, and integration of digital payments to initiate contactless delivery and to enhance customer safety. Integration of drone technology aids in resource conservation, power efficiency, pollution reduction, and emergency readiness. There are numerous advantages to using drone technology, but the real question that needs to be addressed is how to incorporate drone technology into existing logistics systems [29]. These benefits not only reduce the carbon footprint or environmental air pollution but lead to improvement in the organization's operational performance, as explained in session 2.4, thus the following hypothesis:

H2: Adoption of Drone Technology Positively Improves the Operational Performance.

2.4 Operational Performance

Integrating drone technology in the last-mile delivery has significantly improved operational performance. Drone deliveries can lower operational costs by reducing the need for human labor and fuel consumption associated with traditional delivery vehicles. This is crucial in a price-sensitive market like India, where cost efficiency can provide a competitive advantage. Drone technology can cut delivery costs by 20–40% due to savings in labor, fuel, and vehicle maintenance [30]. Drones are generally more environmentally friendly than traditional delivery methods. They contribute to lower carbon emissions, aligning with global sustainability goals and appealing to environmentally conscious consumers. The CO_2 emissions from drones and conventional diesel vans were calculated and found that drones had significantly reduced CO_2 emissions per unit distance [31].

3 Methodology

3.1 Sample

The sample for the research work has been drawn through purposive sampling technique with working professionals from the logistics and supply chain industries having working experience from 1 to 30 years, including both genders and very few of the organizations (less than 6%) already started using drone technology for their last-mile deliveries. The data was gathered through an online survey by sending Google forms circulated through social media platforms.

G* Software [32] is used to determine the sample size. At a 10% significance level with a 0.95 power value, the minimum required sample is 132, and finally, 140 responses were collected, which is adequate to perform the analysis.

3.2 Measurement Instrument

The measurement instrument is constructed for government support and policies (3 items), intention to adopt drone technology (3 items), and operational performance (3 items), and a total of 9 items measured on a 5-point Likert Scale of Strongly Disagree (0) to Strongly Agree (5).

3.3 *Data Analysis*

PLS_SEM is employed in this research to investigate the associations between government support and policies and the intention to adopt drone technology and between the adoption of drone technology and operational performance. The WebPower analysis tool [33] for Multivariate Normality scrutinized the data. Mardia's multivariate skewness and kurtosis are significant, signifying the data is non-normal. Therefore, covariance-based structural equation modeling (CB-SEM) would not be appropriate for further analysis. Another reason for using PLS-SEM is that it has more predictive power through PLSPredict than CB-SEM [34].

4 Results and Discussion

4.1 *Common Method Bias*

Before further analysis, Common Method Bias (CMB) is checked for the data. CMB may exist if a similar type of measurement measures the constructs. One random variable is constructed, and the constructs are connected with that to check whether CMB exists; the variance inflation factor (VIF) values between them are calculated. Using the collinearity approach, variance inflation factor values of the constructs are verified compared to the endogenous random variable, and any construct's VIF value is greater than 3.3, indicating that CMB may occur [35]. As the constructs formed in this research work do not exceed 3.3, it is ensured that no CMB exists [36].

4.2 *Measurement Model Assessment*

The measurement model assesses the formulation of constructs and checks they are reliable to analyze. For that, the constructs' internal construct reliability, indicator reliability, convergent validity, and discriminant validity are examined. The construct reliability is calculated from composite reliability and Cronbach's alpha. It is calculated as Rho A (pA) in PLS-SEM software, as mentioned by [37], and the values between 0.7 0 and 0.95 are generally suggested for instituting the reliability of the construct [38]. Indicator reliability is achieved if the item's outer loading values are 0.707 or above [34, 39]. Construct's convergent validity is measured by average variance extracted (AVE) and accepted as more than 0.5 [40], as the constructs exhibit more than 50% of the total variance.

Discriminant validity of a construct exhibits "how the construct is distinct from other constructs taken for analysis." The Hetrotrait-Monotrait (HTMT) ratio is advocated to be superior to the Fornell-Larcker criterion [41] for validating the discriminant validity of the constructs [42]. Therefore, the data were analyzed using the

HTMT criterion, and its value of less than 0.85 establishes the discriminant validity. For understanding purposes, the Fornell-Larcker criterion is also given in Table 3. The validity and reliability of the constructs are shown in Tables 1 and 2 exhibits the discriminant validity with the HTMT criterion. The values-driven analysis ensures that the reliability and validity of all constructs are correctly established.

Table 1 Validity and reliability (measurement model)

Construct name	Factor loadings	Cronbach's alpha	pA	Composite reliability	AVE
Govt. support and policies		0.712	0.712	0.839	0.634
GOV1	0.797				
GOV2	0.817				
GOV3	0.775				
Drone adoption		0.807	0.807	0.886	0.722
INTADPT1	0.883				
INTADPT2	0.856				
INTADPT3	0.809				
Operational performance		0.738	0.739	0.851	0.656
OPEXC1	0.788				
OPEXC2	0.814				
OPEXC3	0.827				

Table 2 Discriminant validity—Heterotrait–Monotrait ratio (HTMT) matrix

	Government support and policies	Adoption of drone	Operational performance
Government support and policies			
Adoption of drone	0.667		
Operational performance	0.666	0.714	

Table 3 Discriminant validity—Fornell-Larcker criterion

	Government support and policies	Adoption of drone	Operational performance
Government support and policies	0.796		
Adoption of drone	0.506	0.850	
Operational performance	0.481	0.552	0.810

4.3 Measurement Model Assessment

First, in the structural (inner) model, collinearity analysis of inner constructs was calculated, and they were found to be less than 5; hence, any discrepancies due to multicollinearity issues were ruled out [34, 43]. Assessment of model fit in PLS-SEM is acclaimed through the "standardized root mean square residual" (SRMR) measurement [44]. However, values as high or equal to 0.08 are considered adequate for model fit [45].

The model established is verified using Smart PLS 4.03 software using a steady PLS algorithm. Ten thousand subsamples were run using bootstrapping to investigate the coefficients' significance. Table 4 depicts the established model outcomes for direct and indirect effects. Figure 1 demonstrates the model analysis outcomes (Constructs with Cronbach's Alpha, Inner model with path coefficients, and Outer model items with outer loadings). The path coefficients of the constructs were significant, and the indirect effect of government support and policies on organizational performance is also supported. Therefore, hypotheses H1, and H2, stand supported.

This research work (From Table 4) reveals that government support and policies positively support the adoption of drone technology in the logistics and supply chain organizations in India. Also, the adoption of drones in last-mile services improves organizational performance. At the same time, government support and policies

Table 4 Structural model results

Path	Coefficients	T-Statistics	*p*-value	Inference
GOV INTADPT	0.506	7.309	0.000	H1 is supported
GOV → OPEXC	0.279	4.846	0.000	Indirect effect of government support and policies for operational performance
INTADPT → OPEXC	0.552	8.345	0.000	H2 is supported

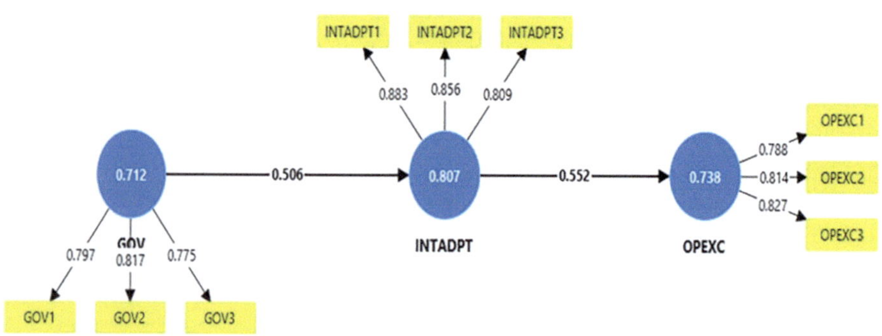

Fig. 1 Model estimation results (constructs: cronbach's alpha, inner model: path coefficients, outer model: outer loadings)

indirectly positively affect the organization's operational performance by adopting drone technology for last-mile services.

5 Implications and Conclusion

5.1 Managerial Implications

Logistics and supply chain organizations can adopt drone technology in last-mile services. The procedures for adopting drone technology are less compared to the earlier times when drones were used for last-mile services, and the government wants technological development to happen in the industry and provide tremendous support. A one-time investment in drone technology will reduce the upcoming cost of traditional delivery expenses. Managers can use this research outcome to suggest drone adoption in their organization, as this has been the first research work in the Indian context.

5.2 Conclusion

With rapid advancements in hardware technologies such as drones and better integration with refined software applications, major market players are increasingly considering the adoption of drones for their last-mile delivery operations. The primary goal behind this strategic shift is to enable faster deliveries while simultaneously reducing operational costs. The research work has analyzed the role of government support and policies in the Indian context to explore its relationship with the adoption of drone technology in last-mile services and further how it affects organizational performance. It is revealed that government support and policies have a positive effect on the adoption of drone technology, and the adoption of drone technology in last-mile services improves operational performance positively. India's logistics and supply chain organizations can consider adopting drone technology for their last-mile services. As explained in session 2.2, drone deliveries can reduce the need for human labor and fuel consumption associated with traditional delivery vehicles. This will drastically reduce the operational cost of an organization and make delivery faster.

References

1. Wei J, Chang M (2023) Are price matching and logistics service enhancement always effective strategies for improving profitability? Eur J Oper Res 307(1):103–115. https://doi.org/10.1016/j.ejor.2022.08.025

2. Zhen L, Wu Y, Wang S, Yi W (2021) Crowdsourcing mode evaluation for parcel delivery service platforms. Int J Prod Econ 235:108067. https://doi.org/10.1016/j.ijpe.2021.108067
3. Humphreys RM, Dumitrescu A, Biju NO, Lam YY (2020) COVID-19 and the maritime and logistics sector in Africa. https://documents1.worldbank.org/curated/en/830701594047530798/pdf/COVID-19-and-the-Maritime-and-Logistics-Sector-in-Africa.pdf
4. Ji-Hyland C, Allen D (2020) What do professional drivers think about their profession? An examination of factors contributing to the driver shortage. Int J Log Res Appl 25(3):231–246. https://doi.org/10.1080/13675567.2020.1821623
5. Ptock J (2018) Lkw-Fahrermangel: Bis 2020 werden mindestens 150.000 Fahrer fehlen. Pobrane z. https://www.logistik-watchblog.de/neuheiten/1543-lkw-fahrermangel-2020-mindestens-150000-fahrer-fehlen.html
6. Agatz N, Bouman P, Schmidt M (2018) Optimization approaches for the traveling salesman problem with drone. Transp Sci 52(4):965–981. https://doi.org/10.1287/trsc.2017.0791
7. Bain & Company (2018) Customer experience tools—delivery drones. https://www.bain.com/insights/customer-experience-tools-delivery-drones/. Accessed 4 June 2024
8. Goasduff L (2020) Why flying drones could disrupt mobility and transportation beyond COVID-19. https://www.gartner.com/smarterwithgartner/why-flying-drones-coulddisrupt-mobility-and-transportation-beyond-covid-19. Accessed 6 May 2024
9. Kim SH (2020) Choice model based analysis of consumer preference for drone delivery service. J Air Transp Manag 84:101785. https://doi.org/10.1016/j.jairtraman.2020.101785
10. Mehndiratta M, Kayacan E (2019) A constrained instantaneous learning approach for aerial package delivery robots: onboard implementation and experimental results. Auton Robot 43(8):2209–2228. https://doi.org/10.1007/s10514-019-09875-y
11. Heutger M, Kückelhaus M (2014) Unmanned aerial vehicle in logistics a DHL perspective on implications and use cases for the logistics industry. In: DHL customer solutions & innovation; DHL customer solutions & innovation: Bonn, Germany. https://www.dhl.com/discover/content/dam/dhl/downloads/interim/full/dhl-trend-report-uav.pdf
12. Amazon Prime Air. Available online: https://www.amazon.com/Amazon-Prime-Air/b?ie=UTF8&node=8037720011. Accessed on 18 Apr 2024
13. Wankmüller C, Truden C, Korzen C, Hungerländer P, Kolesnik E, Reiner G (2020) Optimal allocation of defibrillator drones in mountainous regions. OR Spektrum 42:785–814. https://doi.org/10.1007/s00291-020-00575-z
14. Mehrotra K (2024) Drone regulations: the Indian scenario. Available at SSRN 4817307. https://papers.ssrn.com/sol3/papers.cfm?abstract_id=4817307
15. Kashyaap S (2021) Dunzo teams up with Telangana govt to pilot drone delivery of medicines. YourStory. [online] 1 June. Available at: https://yourstory.com/2021/06/dunzo-startup-pilot-drone-delivery-medicines-telegana-government
16. Mathew AO, Jha AN, Lingappa AK, Sinha P (2021) Attitude towards drone food delivery services—role of innovativeness, perceived risk, and green image. J Open Innovation Technol Market Complexity 7(2):144. https://doi.org/10.3390/joitmc7020144
17. Park J, Kim S, Suh K (2018) A comparative analysis of the environmental benefits of drone-based delivery services in urban and rural areas. Sustainability 10(3):888. https://doi.org/10.3390/su10030888
18. Kumbhani C, Kant R (2024) Ranking the benefits of drone-based last-mile delivery due to adoption of its enablers. J Adv Manage Res ahead-of-print No(ahead-of-print). https://doi.org/10.1108/JAMR-03-2024-0103
19. Stolaroff JK, Samaras C, O'Neill ER, Lubers A, Mitchell AS, Ceperley D (2018) Energy use and life cycle greenhouse gas emissions of drones for commercial package delivery. Nat Commun 9(1):409. https://doi.org/10.1038/s41467-017-02411-5
20. Directorate General of Civil Aviation (DGCA) (2018) Civil aviation requirements (CAR) for remotely piloted aircraft systems. https://doi.org/10.1016/B978-0-12-420139-2.00007-1
21. World Economic Forum and NITI Aayog (2021) Medicine from the sky: transforming healthcare delivery through drones. https://www.weforum.org/publications/medicine-from-the-sky-india-taking-primary-healthcare-to-all/ Accessed on 4 June 2024

22. Division of Agriculture Engineering, ICAR (2022) Applications of drones in agriculture: status and scope. https://farmech.dac.gov.in/New_Folder/Drone_Booklet_Low.pdf
23. Indian Council of Medical Research (ICMR) (2021) Drone delivery of covid-19 vaccines. https://www.icmr.gov.in/idrone.html Accessed on 25 May 2024
24. Venkatesh V, Morris MG, Davis GB, Davis FD (2003) User acceptance of information technology: toward a unified view. MIS Q 27(3):425–478. https://doi.org/10.2307/30036540
25. Garg V, Niranjan S, Prybutok V, Pohlen T, Gligor D (2023) Drones in last-mile delivery: a systematic review on efficiency, accessibility, and sustainability. Transp Res Part D: Transport Environ 123:103831. https://doi.org/10.1016/j.trd.2023.103831
26. Lari HA, Vaishnava K, Manu KS (2022) Artificial intelligence in E-commerce: Applications, implications and challenges. Asian J Manage 13(3):235–244. https://doi.org/10.52711/2321-5763.2022.00041
27. Zainuddin N, Deraman N, Raman D, GANESAN L (2022) Efficiency of last mile delivery of logistics service providers (LSPS) in Malaysia: post-covid. Quantum J Soc Sci Humanit 3(6):88–104. https://doi.org/10.55197/qjssh.v3i6.197
28. Bhotvawala MA, Balihallimath H, Bidichandani N, Khond MP (2016) Growth of food tech: a comparative study of aggregator food delivery services in India. In: Proceedings of the 2016 international conference on industrial engineering and operations management, Detroit, Michigan, USA, pp 140–149. https://ieomsociety.org/ieomdetroit/pdfs/49.pdf
29. Ali SS, Khan S, Fatma N, Ozel C, Hussain A (2024) Utilisation of drones in achieving various applications in smart warehouse management. Benchmarking: An Int J 31(3):920–954. https://doi.org/10.1108/BIJ-01-2023-0039
30. Gabani PR, Gala UB, Narwane VS, Raut RD, Govindarajan UH, Narkhede BE (2021) A viability study using conceptual models for last mile drone logistics operations in populated urban cities of India. IET Collaborative Intell Manuf 3(3):262–272. https://doi.org/10.1049/cim2.12006
31. Figliozzi MA (2017) Lifecycle modeling and assessment of unmanned aerial vehicles (Drones) CO_2e emissions. Transp Res Part D: Transp Environ 57:251–261. https://doi.org/10.1016/j.trd.2017.09.011
32. Faul F, Erdfelder E, Buchner A, Lang AG (2009) Statistical power analyses using G* Power 3.1: tests for correlation and regression analyses. Behav Res Methods 41(4):1149–1160. https://doi.org/10.3758/BRM.41.4.1149
33. Zhang Z, Yuan K-H (eds) (2018) Practical statistical power analysis using WebPower & R. ISDSA Press. https://doi.org/10.35566/power
34. Hair JF, Risher JJ, Sarstedt M, Ringle CM (2019) When to use and how to report the results of PLS-SEM. Eur Bus Rev 31(1):2–24. https://doi.org/10.1108/EBR-11-2018-0203
35. Kock N, Lynn G (2012). Lateral collinearity and misleading results in variance-based SEM: an illustration and recommendations. J Assoc Inf Syst 13(7):546–580. https://ssrn.com/abstract=2152644
36. Kock N (2015) Common method bias in PLS-SEM: a full collinearity assessment approach. Int J e-Collaboration 11(4):1–10. https://doi.org/10.4018/ijec.2015100101
37. Dijkstra TK, Henseler J (2015) Consistent partial least squares path modeling. MIS Q 39(2):297–316. https://www.jstor.org/stable/26628355
38. Hair JF, Hult GTM, Ringle CM, Sarstedt M (2022) A primer on partial least squares structural equation modeling (PLS-SEM) (3rd ed). Sage Publications
39. Chin WW (2010) How to write up and report PLS analyses. In: Esposito Vinzi V, Chin W, Henseler J, Wang H (eds) Handbook of partial least squares. Springer Handbooks of computational statistics. Springer, Berlin, Heidelberg. https://doi.org/10.1007/978-3-540-32827-8_29
40. Hair JF, Black WC, Babin BJ, Anderson RE (2010) Multivariate data analysis (7th ed). Prentice Hall
41. Fornell C, Larcker DF (1981) Evaluating structural equation models with unobservable variables and measurement error. J Mark Res 18(1):39–50. https://doi.org/10.1177/002224378101800104

42. Henseler J, Ringle CM, Sarstedt M (2015) A new criterion for assessing discriminant validity in variance-based structural equation modelling. J Acad Mark Sci 43(1):115–135. https://doi.org/10.1007/s11747-014-0403-8
43. James G, Witten D, Hastie T, Tibshirani R (2013) An introduction to statistical learning: with applications in R. Springer. https://doi.org/10.1007/978-1-0716-1418-1
44. Henseler J, Dijkstra TK, Sarstedt M, Ringle CM, Diamantopoulos A, Straub DW, Ketchen DJ Jr, Hair JF, Hult GTM, Calantone RJ (2014) Common beliefs and reality about partial least squares: comments on Ronkko and Evermann (2013). Organizational Res Methods 17(2):182–209. https://doi.org/10.1177/1094428114526928
45. Hu LT, Bentler PM (1999) Cutoff Criteria for fit indexes in covariance structure analysis: conventional criteria versus new alternatives. Struct Equation Modeling 6(1):1–55. https://doi.org/10.1080/10705519909540118

A Simulative Study on the Pivotal Role of the Active Disturbance Rejection Controller (ADRC) in Feedback Control System Design

S. J. Lolu and M. A. Bazaz

Abstract The design of the feedback control system is profoundly influenced by the paradigm shift brought about by active disturbance rejection control. It effectively mitigates the impact of modeling errors and process parameter variations, thereby establishing itself as a valuable asset for control engineering professionals. Its tuning method is straightforward. This paper intends to provide foundational information on observer-based linear ADRC for the intended audience. In this paper for evaluation of ADRC, a simulative study is carried out using first- and second-order processes for the practitioners. This paper presents investigation into the impact of variation in process parameters, demonstrating that ADRC adapts severe parameter variations without relying on the exact mathematical model of the physical process. It illustrates the capabilities of the ADRC in the presence of varying process parameters or structural uncertainties. The simulation results demonstrate the robustness of linear ADRC in the presence of varying uncertainties. The performance comparison of the linear active disturbance rejection controller (LADRC), PI controller, and PID controller is presented in this paper showing that former is more superior. Also, linear ADRC achieves better performance in effectively rejecting external disturbances than PI and PID Controllers.

Keywords Linear active disturbance rejection controller (LADRC) · Extended State Observer (ESO) · Proportional–integral (PI) controller · Proportional–integral–derivative (PID) controller

---

S. J. Lolu (✉)
Department of Electrical Engineering, Islamic University of Science & Technology, Awantipora, J&K, India
e-mail: sami.jan@islamicuniversity.edu.in

M. A. Bazaz
Department of Electrical Engineering, National Institute of Technology, Srinagar, J&K, India
e-mail: abid@nitsri.ac.in

H. Sharma et al. (eds.), *Artificial Intelligence: Theory and Applications*, Lecture Notes in Networks and Systems 5589, https://doi.org/10.1007/978-981-96-1687-9_9

1 Introduction

It is difficult to think of a modern engineering process that does not involve feedback control mechanisms. Feedback control affects nearly all engineering disciplines. This technology was the driving force behind the industrial revolution. An early instance of an automatic control system is Watt's steam engine flyball governor. In feedback control, it is assumed that there is a causal relationship between the input and output of a physical process, and the controller manipulates the input variable in order to achieve the desired output change. Practitioners have demonstrated a staunch preference for the deductive control paradigm (based on theories) over the inductive control paradigm (based on experimentation) over the course of these years. This belief needs to be uplifted by the development of model-free control design methods.

Over these years, practitioners have preferred simple control theories over complex ones. About 90% of industrial control systems employ the proportional–integral–derivative (PID) control method (first proposed by Minorsky in 1922) [1, 2]. However, this hold may be broken by the new competent replacement, the Active Disturbance Rejection Controller (ADRC), due to its performance and practicality, leading to a new era.

ADRC is a productive emerging control technology [3–7]. ADRC has surpassed PID through the following enhancements:

- A trajectory profile generator helps to improve error computation.
- Using a noise-free tracking differentiator reduces the complications caused by derivative control.
- Linear control laws are replaced by nonlinear control laws.
- Estimation of the total disturbances and rejection.

Han introduced the ADRC concept to the 1990s [8, 9]. ADRC was first introduced in English in 2001 [10]. This concept of ADRC was expanded upon in [11]. Han spent close to twenty years of his life resolving the PID tuning difficulty problem [9, 12].

The ADRC principle effectively addresses the shortcomings of PID controller. Furthermore, ADRC adapts severe parameter variations without relying on the exact mathematical model of the physical process treating all modeling errors as the external disturbance [5, 13]. To show the capabilities of the ADRC in the presence of varying process parameters or structural uncertainties, a series of experiments are performed in this paper. Considering the ADRC's perspective, this paper's objective is as follows:

- **Robustness with respect to model uncertainties**: The stability and control performance of the system remain unaffected by variations of approximately one thousand percent in the K and T parameters of the first-order plant and by approximately five hundred percent in the K, D, and T parameters of the second-order plant, respectively.
- **Comparative Analysis**: The performance of ADRC is superior to that of other controllers.

- **Disturbance Rejection**: ADRC can effectively reject external disturbances, achieving better performance than traditional controllers like PI and PID controllers.
- **Overshoot and Settling Time**: It minimizes the overshoot and settling time for first and second-order systems while maintaining system stability.
- **Minimal system information needed**: ADRC requires minimal information about the processes being controlled, making it more practical in situations where the system parameters are difficult to obtain.

To achieve the above-mentioned objectives, the following design philosophy is followed:

Choose plant model.
⇓
Design ESO for plant model.
⇓
Design Controller for plant model.
⇓
Choose the tuning parameters.

In this paper, we discuss the weaknesses of a simple, traditional PID control scheme before introducing the ADRC control scheme followed by some simulation examples in Sect. 5. Finally, the concluding remarks are given in Sect. 6.

2 Traditional Proportional–Integral–Derivative (PID) Control

Traditional PID is a simple, easily understandable, primeval feedback compensator structure executed on the basic principle of error-feedback control, as shown in Fig. 1. Here, the output of a PID controller, which is equal to the control input to the plant, is calculated in the time domain from the feedback error as follows [12]:

$$u(t) = K_p \cdot e(t) + K_i \cdot \int e(t) + K_D \cdot \frac{de(t)}{dt}. \tag{1}$$

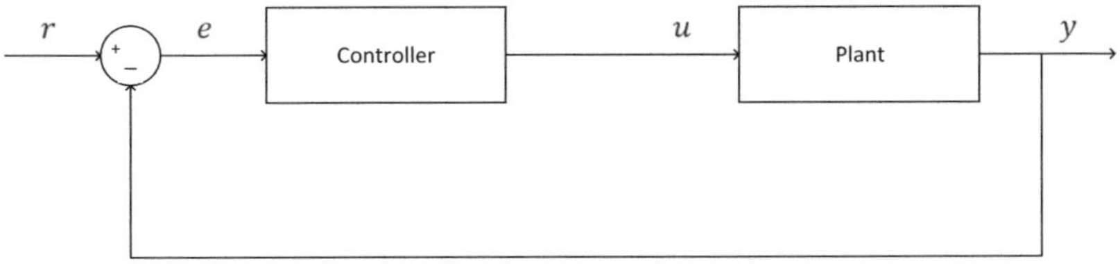

Fig. 1 PID control topology

In [9], the author addresses the PID limitations that prompted practitioners to adapt ADRC control, which are as follows:

- Step function set point, which is not appropriate for most dynamic systems.
- The D part of PID causing noise sensitivity.
- Control law based on a Linear weighted sum.
- I part of PID causing saturation and stability problems.

3 Active Disturbance Rejection Control (ADRC)

As mentioned by Z. Gao in [14–18], the trait that was missing in traditional feedback control system is "disturbance rejection" and for that a "disturbance rejector" or simply "rejector" is used. Consequently, the integrated controller-rejector combination exemplifies the fundamental architecture of disturbance rejection control (DRC), as illustrated in Fig. 2. Here, the controller's task of meeting the design specifications becomes a lot easier when it deals with the enforced plant, which tends to be much simpler and less uncertain than the actual process. And this is the concept missing from the controller-only paradigm of modern control. DRC provides a viable alternative to a single controller that would otherwise be unnecessarily complicated or even unmanageable when dealing with uncertainties and meeting performance specifications simultaneously.

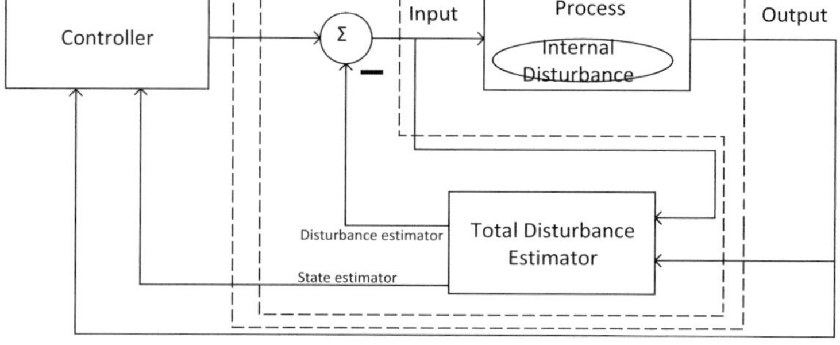

Fig. 2 Disturbance rejection control platform

DRC exemplifies the category of control systems whose primary objective is disturbance rejection. The rejector, as shown in Fig. 2, performs two key tasks:

- The collection, processing, and dissemination of disturbance and state information.
- Canceling, or eradicating, the disturbance.

The rejector performs the main task of rejecting disturbances, and controller forces the output to follow the command.

The Extended State Observer (ESO) was proposed specifically to estimate and cancel "total disturbance" in the context of ADRC. A survey of different types of state and disturbance observers can be found in [10, 19, 20].

3.1 Linear Active Disturbance Rejection Control

In this subsection, a Linear Active Disturbance Rejection Control (LADRC) for single-input, single-output systems is introduced [9, 21–24]. The LADRC loop mainly consists of three blocks, as shown in Fig. 3.

- Tracking differentiator: It is used to avoid setpoint jumps, and for that, it is necessary to construct a transient profile r_1 for the reference $\tilde{r}$ and the corresponding derivatives $\dot{r}_1$, $\ddot{r}_1$,..., $\ddot{r}_n$.
- Extended State Observer (ESO): It estimates the system states x_1, x_2, ..., x_n and the additional state x_{n+1}, representing non-modeled dynamics and perturbations.
- Controller: It provides a state feedback control law u_0 for the disturbance-free extended plant.

The control structure for a first-order process is shown in Fig. 4, as explained in [25–30]. Comparably, Fig. 5 shows the control structure for a second-order process, which is described in [26, 31–34].

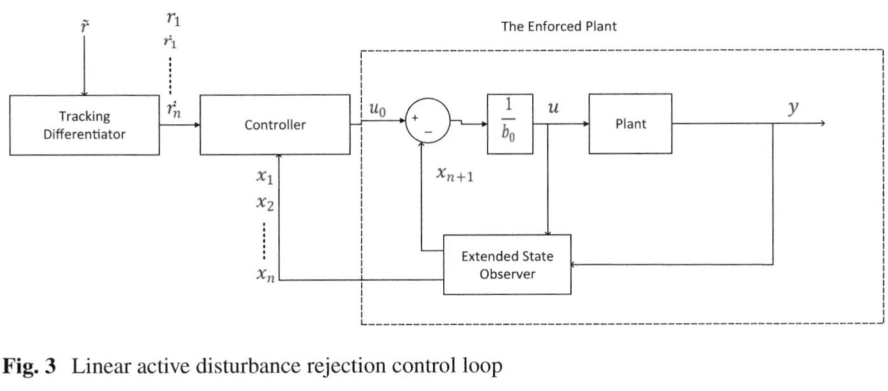

Fig. 3 Linear active disturbance rejection control loop

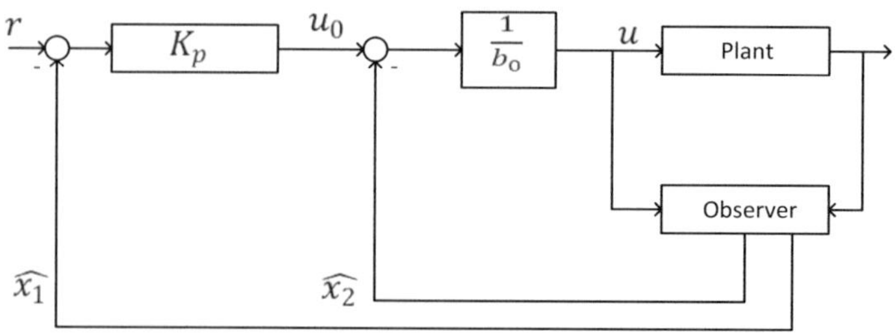

Fig. 4 Control model with ADRC for first-order process

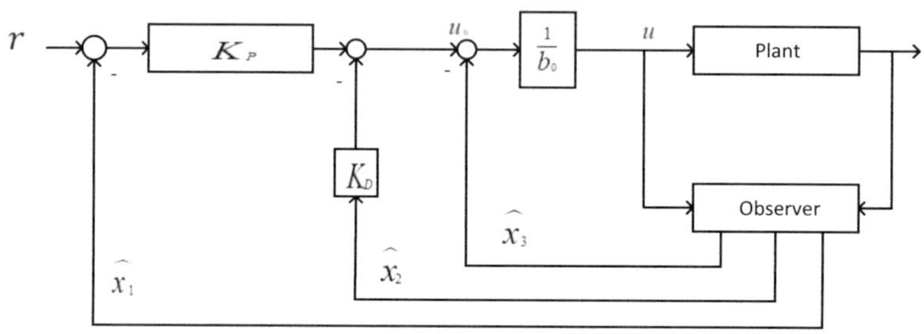

Fig. 5 Control model with ADRC for second-order process

3.2 *Time-Domain Formulation*

ADRC requires only an approximation of the plant dynamics to design controllers that provide robust disturbance rejection with no overshoot. In this work, ADRC uses a first-order or second-order model approximation of the known system dynamics along with the unknown dynamics and disturbances modeled as an extended state of the plant.

An Extended State Observer (ESO) is used to estimate $f(t)$ and implement disturbance rejection control by reducing the effect of estimated disturbances on the known part of model approximation [26].

4 Methodology

The modeling of first-order ADRC and second-order ADRC is carried out in this section.

The simulation model is made within the MATLAB/Simulink environment using MATLAB/Simulink library. MATLAB/Simulink's 'Edit Mask' command provides all settings for an edited module when packaging a subsystem. It can mask any subsystem.

4.1 Create Simulation Model

4.1.1 ADRC Control System Using a First-Order Process

A continuous-time ADRC operating on a first-order system will be examined in the first series of experiments. The controller was designed using the following process structure and nominal parameters:

$$P(s) = \frac{y(s)}{u(s)} = \frac{K}{Ts + 1}. \tag{2}$$

According to the Extended State Observer (ESO) algorithm, the simulation model of ESO is established and depicted in Fig. 6. Figure 7 represents the packaging subsystem. If a subsystem module is double-clicked, a parameter setting window depicted in Fig. 8 will appear allowing users to directly modify parameter values. To tune LADRC for a first-order plant in MATLAB/Simulink, the subsequent steps must be taken:

- Determine the desired ω_c from settling time T_s (assuming it is known in practical problem).
- Determine K_P from ω_c.

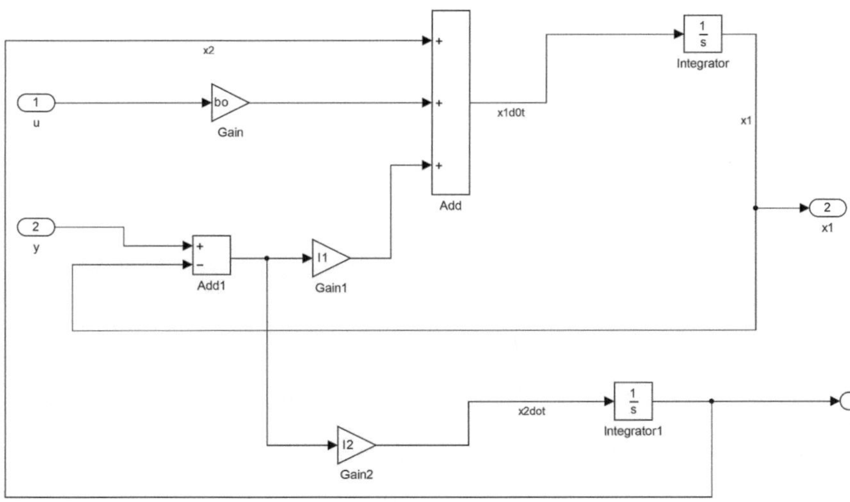

Fig. 6 Simulation model of an ESO subsystem

Fig. 7 Encapsulation module

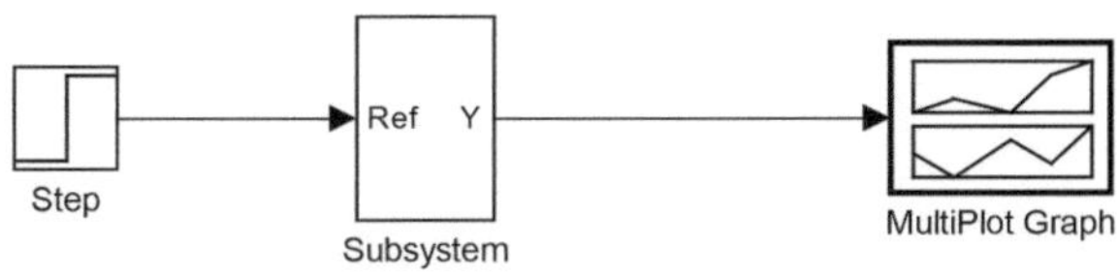

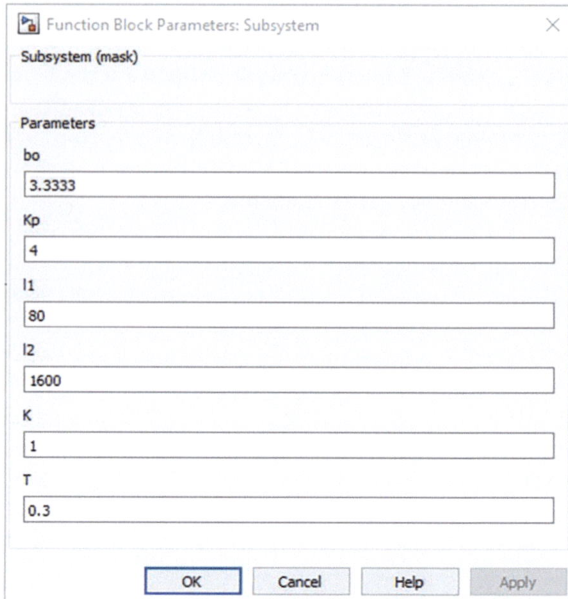

Fig. 8 Parameter setting window

- Assuming ω_0, determine l_1 and l_2.
- Determine the robustness of the controller by modifying the plant coefficients to nearly $\pm 1000\%$ and recording the result.

4.1.2 ADRC Control System Using a Second-Order Process

A continuous-time ADRC operating on a second-order system will be examined through a series of experiments. The controller was designed using the following process structure and nominal parameters:

$$P(s) = \frac{y(s)}{u(s)} = \frac{K}{T^2 s^2 + 2DTs + 1}. \tag{3}$$

According to the Extended State Observer (ESO) algorithm, the simulation model of ESO is established and depicted in Fig. 9. As depicted in Fig. 7 the packaging subsystem is the same as in first-order ADRC. If a subsystem module is double-clicked, the parameter setting window depicted in Fig. 10 will appear, allowing users to directly modify parameter values.

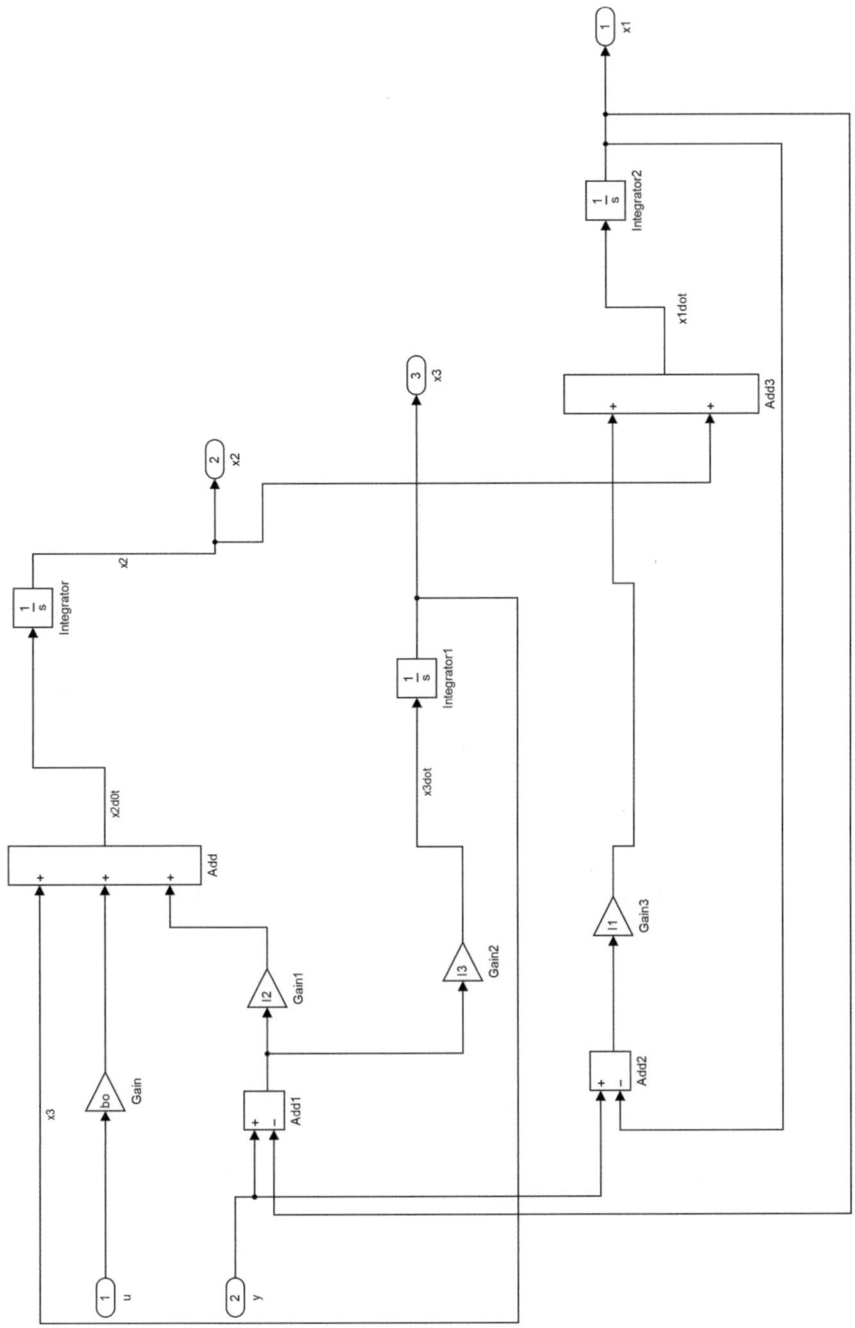

Fig. 9 Simulation model of an ESO subsystem

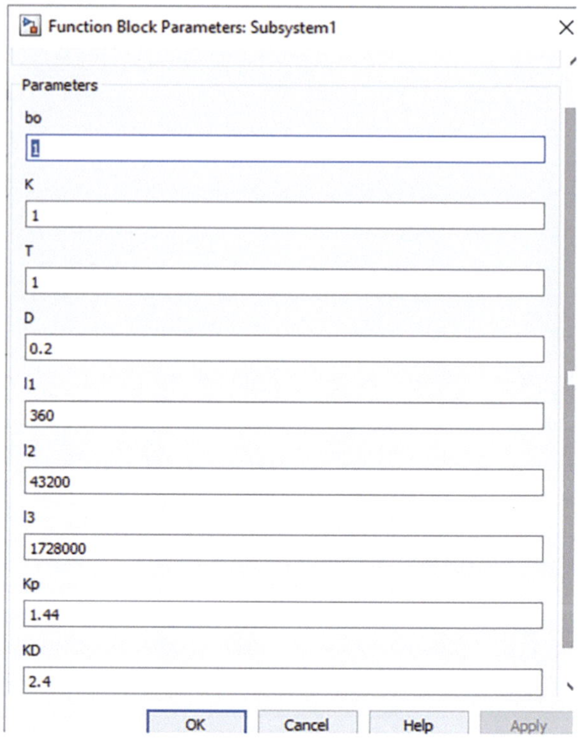

Fig. 10 Parameter setting window for second-order process

To tune LADRC for a second-order plant, the subsequent steps must be taken:

- Determine the desired ω_c from settling time T_s (assuming it is known in practical problem).
- Determine K_P and K_D from ω_c.
- Assuming ω_0, determine l_1, l_2, and l_3.
- Determine the robustness of the controller by modifying the plant coefficients to nearly $\pm 500\%$ and recording the result.

Since the intended settling time is specified, only one parameter, b_0, must be tuned.

5 Simulations

5.1 ADRC with a First-Order Process

We will now examine the experiments conducted on the controlled process:

$$P(s) = \frac{y(s)}{u(s)} = \frac{K}{Ts + 1} = \frac{1}{0.3\,s + 1}. \tag{4}$$

Here, $K = 1$ and $T = 0.3$. The ADRC is designed according to the tuning procedure described in Sect. 4. For now, we will assume that we have perfect knowledge of our process and set

$$b_0 = \frac{K}{T} = \frac{1}{0.3} = 3.3333$$

and leave this value unchanged for remaining experiments in this section. $T_{sett} = 1$ is chosen as the preferred closed-loop settling time. The other parameters chosen are

$$\omega_c = 4,$$

$$K_P = 4,$$

$$l_1 = 80,$$

$$l_2 = 1600,$$

the observer pole $\omega_0 = 10\omega_c$.

5.1.1 Vulnerability to Varying Process Parameters

On the model of ADRC depicted in Fig. 11, a series of simulations were executed with parameters listed in Tables 1 and 2. Figures 12 and 14 with Figs. 13 and 15 display the closed-loop response and the controller outputs, respectively. The values chosen for K and T ranged between 10% and 1000% of their nominal values.

The obtained results demonstrate that the closed-loop step response remains almost same regardless of parameter settings in Fig. 12. Only for 500% and 1000% parameter variations are larger overshoots discernible, as displayed in Fig. 14.

5.1.2 Influence of Observer Pole Locations

Here, simulations with varying T parameter are repeated for faster ($\omega_0 = 100\omega_c$) to slower observer poles ($\omega_0 = 5\omega_c$), as mentioned in Tables 3, 4, and 5.

Figures 16, 17, 18, 19, 20, and 21 display results in order from fastest to slowest observer setting. According to the results, the fastest observer poles show better results as compared to the slowest poles. For slower poles, closed-loop dynamics differ from desired dynamics.

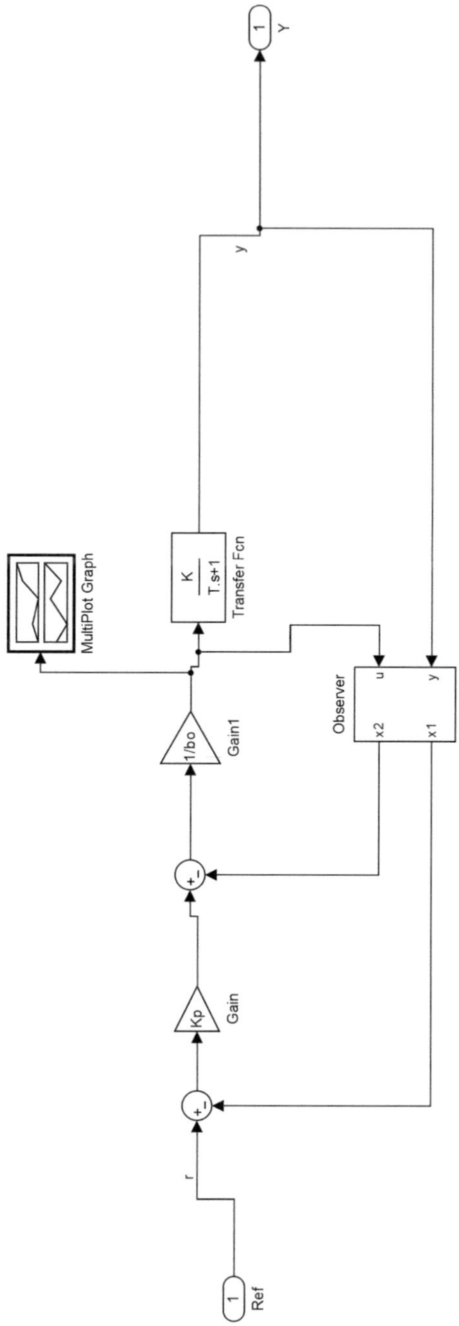

Fig. 11 Model of ADRC for first-order system

Table 1 First-order process with varying K parameter over a range from 10% to 1000%

b_0	K_P	l_1	l_2	K	T	Variation (%)
3.3333	4	80	1600	1	0.3	0
3.3333	4	80	1600	0.1	0.3	10
3.3333	4	80	1600	0.2	0.3	20
3.3333	4	80	1600	0.5	0.3	50
3.3333	4	80	1600	2	0.3	200
3.3333	4	80	1600	5	0.3	500
3.3333	4	80	1600	10	0.3	1000

Table 2 First-order process with varying T parameter over a range from 10% to 1000%

b_0	K_P	l_1	l_2	K	T	Variation (%)
3.3333	4	80	1600	1	0.3	0
3.3333	4	80	1600	1	0.03	10
3.3333	4	80	1600	1	0.06	20
3.3333	4	80	1600	1	0.15	50
3.3333	4	80	1600	1	0.6	200
3.3333	4	80	1600	1	1.5	500
3.3333	4	80	1600	1	3	1000

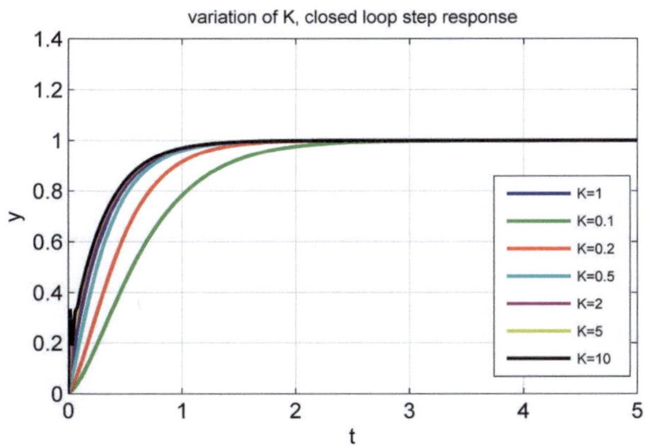

Fig. 12 Closed-loop step response of first-order process for varying K parameter

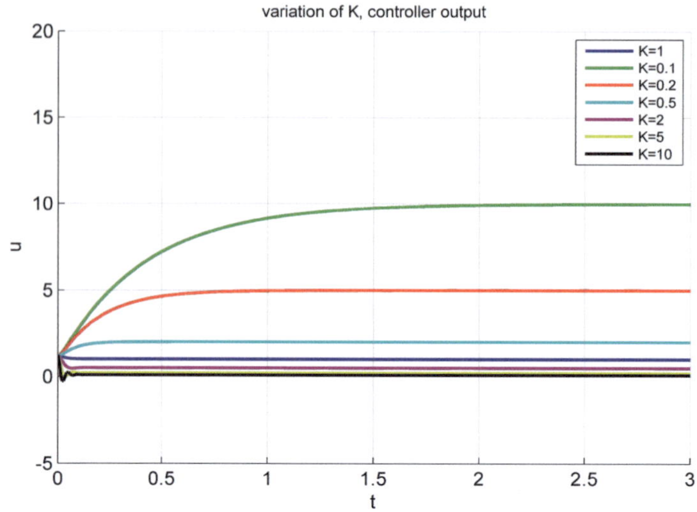

Fig. 13 Controller output u of first-order process for varying K parameter

Fig. 14 Closed-loop step response of first-order process for varying T parameter

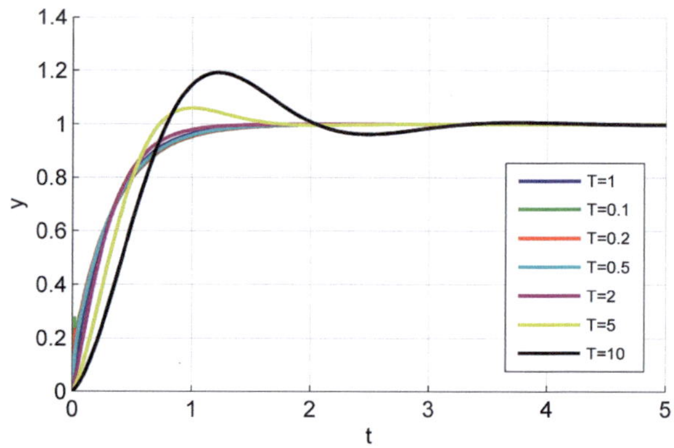

5.1.3 Effect of Dead Time

As many real-world processes exhibit dead time, a series of simulations were performed on the model depicted in Fig. 22.

In Table 6, various experiments with varying amounts of dead time (≤ 0.15) in the process were conducted, equating to nearly 50% of the process's time constant T.

Increasing dead time in process indicates a rise in oscillations, which can be mitigated by delaying the controller signal before it reaches the ESO. Since the process is already delayed by system dynamics (dead time), the delaying controller signal will synchronize the signals that go to the observer and provide accurate estimates. In Figs. 23 and 24, this approach was implemented using unknown (T_{dead}^{ESO}), demonstrating that the oscillations are more prominent compared to Figs. 25 and 26,

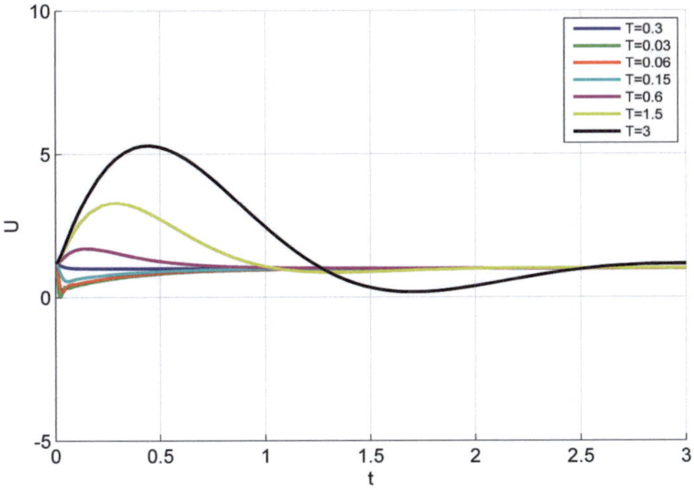

Fig. 15 Controller output u of first-order process for varying T parameter

Table 3 First-order process with varying T parameter over a range from 10% to 1000% for ESO pole location $\omega_0 = 100\omega_c$

b_0	K_P	l_1	l_2	K	T	Variation (%)
3.3333	4	800	160000	1	0.3	0
3.3333	4	800	160000	1	0.03	10
3.3333	4	800	160000	1	0.06	20
3.3333	4	800	160000	1	0.15	50
3.3333	4	800	160000	1	0.6	200
3.3333	4	800	160000	1	1.5	500
3.3333	4	800	160000	1	3	1000

where $(T_{dead}^{ESO}) = 0.06$ results in oscillations that are less prominent even if (T_{dead}^{ESO}) does not match the actual time. Above the $(T_{dead}^{ESO}) = 0.06$, the oscillations are more prominent.

5.1.4 Effect of Structural Uncertainties

To demonstrate the behavior of plant if higher-order dynamics become apparent in the plant, let's consider the second pole in the process. Figure 27 demonstrates that the addition of a second pole results in the addition of another time constant T_2.

$T_2 \leq 0.2$ or up to 70% of the dominant time constant T as shown in Table 7. The results in Figs. 28 and 29 show that oscillations became prominent as higher-order pole approaches the dominant pole.

Table 4 First-order process with varying T parameter over a range from 10% to 1000% for ESO pole location $\omega_0 = 10\omega_c$

b_0	K_P	l_1	l_2	K	T	Variation (%)
3.3333	4	80	1600	1	0.3	0
3.3333	4	80	1600	1	0.03	10
3.3333	4	80	1600	1	0.06	20
3.3333	4	80	1600	1	0.15	50
3.3333	4	80	1600	1	0.6	200
3.3333	4	80	1600	1	1.5	500
3.3333	4	80	1600	1	3	1000

Table 5 First-order process with varying T parameter over a range from 10% to 1000% for ESO pole location $\omega_0 = 5\omega_c$

b_0	K_P	l_1	l_2	K	T	Variation (%)
3.3333	4	40	400	1	0.3	0
3.3333	4	40	400	1	0.03	10
3.3333	4	40	400	1	0.06	20
3.3333	4	40	400	1	0.15	50
3.3333	4	40	400	1	0.6	200
3.3333	4	40	400	1	1.5	500
3.3333	4	40	400	1	3	1000

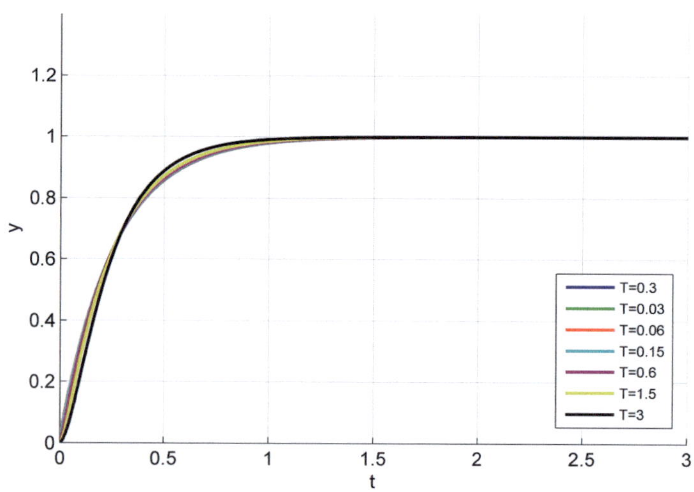

Fig. 16 Closed-loop step response of a first-order process with varying T parameter when pole locations for ESO ($\omega_0 = 100\omega_c$)

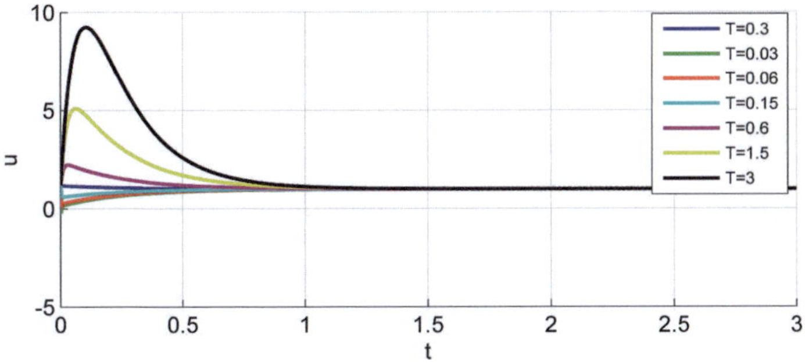

Fig. 17 Controller output u of a first-order process with varying T parameter when pole locations for ESO ($\omega_0 = 100\omega_c$)

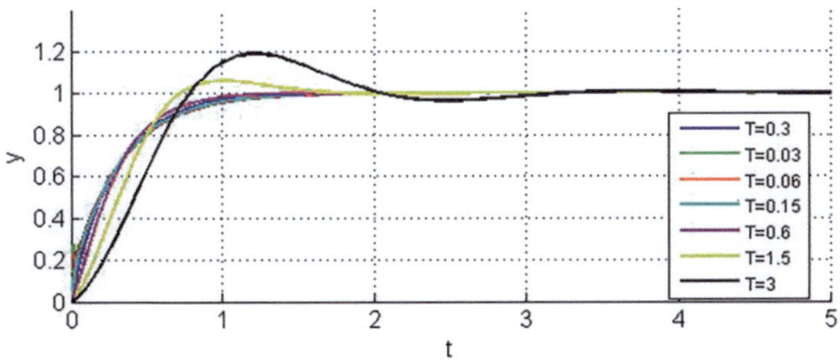

Fig. 18 Closed-loop step response of a first-order process with varying T parameter when pole locations for ESO ($\omega_0 = 10\omega_c$)

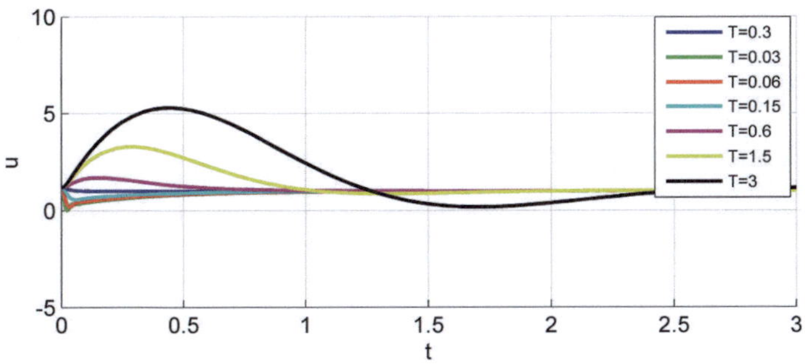

Fig. 19 Controller output u of a first-order process with varying T parameter when pole locations for ESO ($\omega_0 = 10\omega_c$)

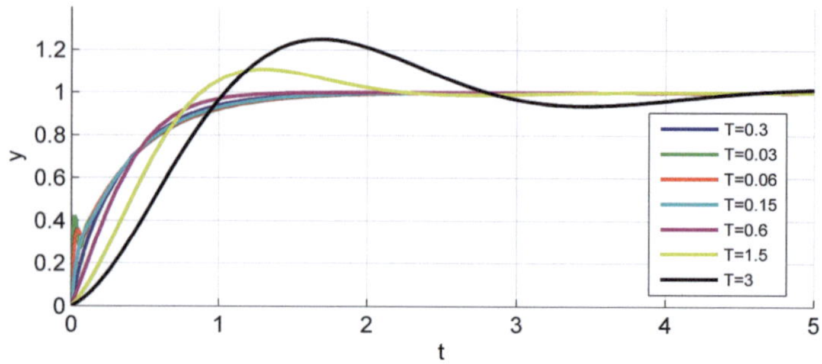

Fig. 20 Closed-loop step response of a first-order process with varying T parameter when pole locations for ESO ($\omega_0 = 5\omega_c$)

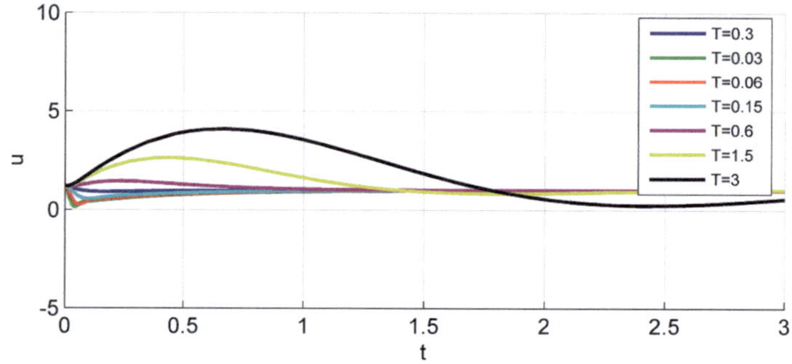

Fig. 21 Controller output u of a first-order process with varying T parameter when pole locations for ESO ($\omega_0 = 5\omega_c$)

5.1.5 Comparison with PI Control

The PI controller is designed for same first-order process as the ADRC, maintaining the same settling time.

$$K_p = 1.1693$$

and

$$K_I = 8.183$$

are the PI controller parameters obtained when adjusting the controller for same-order dynamics while considering the settling time. Figure 30 depicts the model of the PI controller that controls the first-order process.

The comparison of simulation results in Figs. 31, 32, 33, and 34 clearly shows that ADRC controller has high ability to maintain similar closed-loop dynamics despite significant parameter variations.

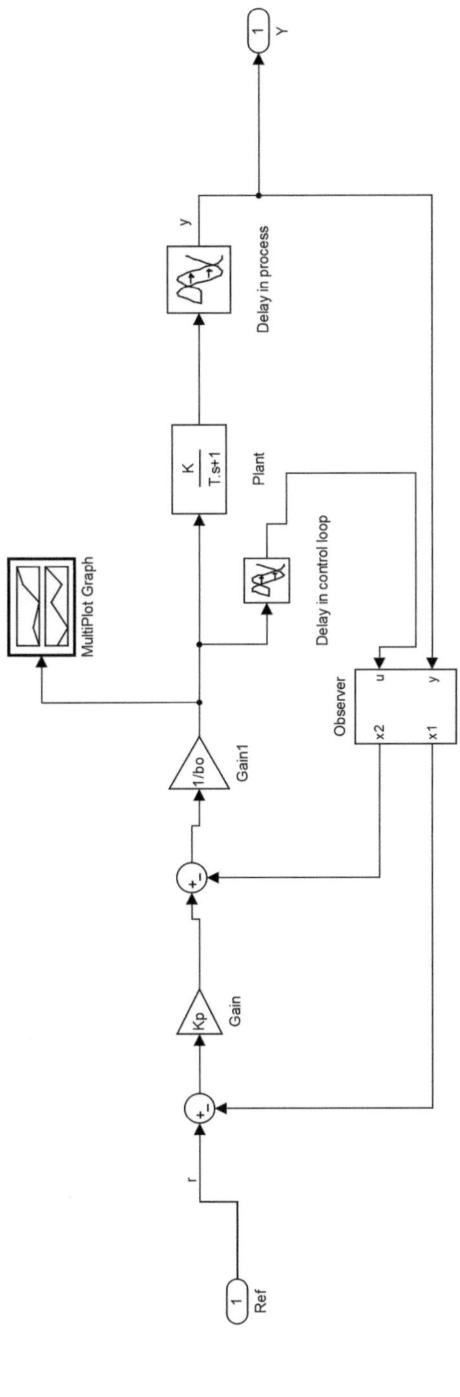

Fig. 22 Control model using ADRC on first-order process with dead time

Table 6 First-order process for ESO pole location $\omega_0 = 5\omega_c$ with varying process dead time

b_0	K_P	l_1	l_2	K	T	Variation (%)	Dead time
3.3333	4	40	400	1	0.3	0	0
3.3333	4	40	400	1	0.3	10	0.03
3.3333	4	40	400	1	0.3	20	0.06
3.3333	4	40	400	1	0.3	50	0.15

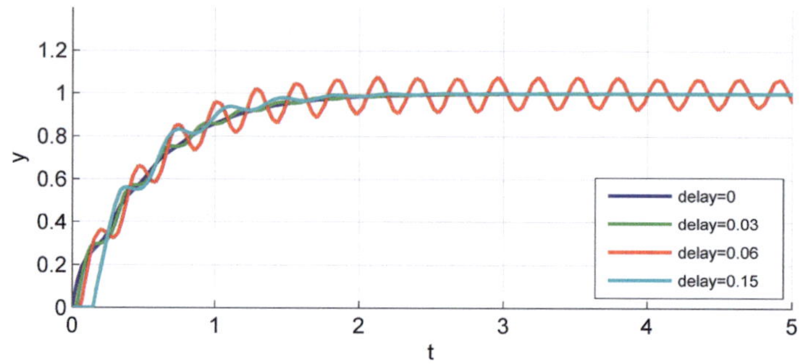

Fig. 23 Closed-loop step response of a first-order process with varying T parameter when pole locations for ESO ($\omega_0 = 5\omega_c$) and delay in ESO is unknown

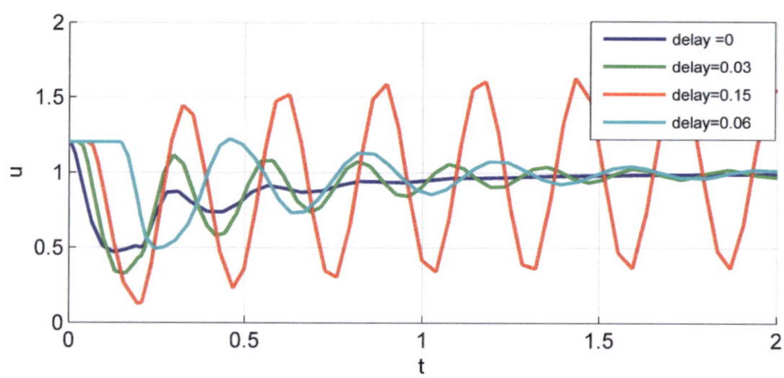

Fig. 24 Controller output u of a first-order process with varying T parameter when pole locations for ESO ($\omega_0 = 5\omega_c$) and delay in ESO is unknown

5.1.6 Disturbance Rejection of ADRC and PI

To test the disturbance rejection capabilities, we inoculate an internal disturbance into the process for both ADRC and PI controllers. The input disturbance operates from $t = 2$ to $t = 4$. Both controllers have the same reaction to setpoint changes. However, the ADRC controller compensates for effect of disturbances much faster than the PI controller, as depicted in Fig. 35 and 36.

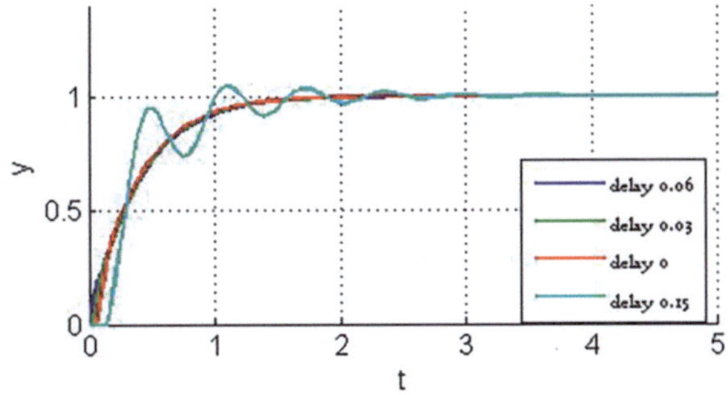

Fig. 25 Closed-loop step response of a first-order process with varying T parameter when pole locations for ESO ($\omega_0 = 5\omega_c$) and delay in ESO = 0.06

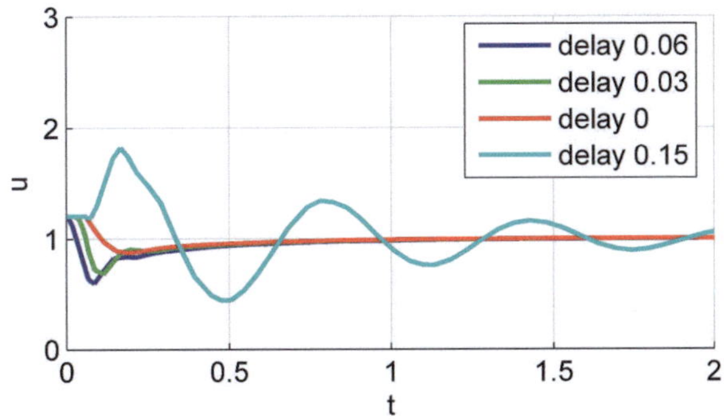

Fig. 26 Controller output u of a first-order process with varying T parameter when pole locations for ESO ($\omega_0 = 5\omega_c$) and delay in ESO = 0.06

5.2 *ADRC with a Second-Order Process*

We will now examine the experiments conducted on the controlled process:

$$P(s) = \frac{y(s)}{u(s)} = \frac{K}{T^2 s^2 + 2DTs + 1} = \frac{0.5}{0.5\, s^2 + s + 1} \tag{5}$$

with $K = 0.5$, $D = 0.71$, and $T = 0.71$. The ADRC is designed according to the tuning procedure outlined in Sect. 4. Assuming perfect knowledge of our process at this time, we have set

$$b_0 = \frac{K}{T^2} = \frac{0.5}{0.71^2} = 1$$

and will maintain this value for rest of the experiments in this section. $T_{sett} = 4$ is chosen as the favored closed-loop settling time. The other parameters chosen are

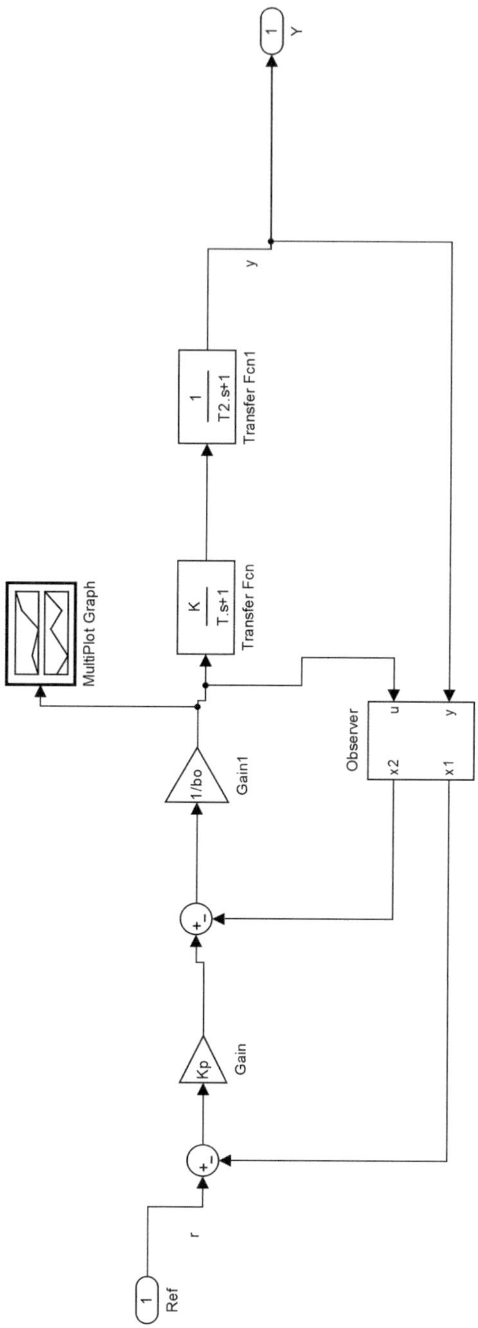

Fig. 27 Model of ADRC for first-order system with structural uncertainties, i.e., with dominating pole T and unknown second pole T_2

Table 7 First-order process for ESO pole location $\omega_0 = 5\omega_c$ with varying second pole T_2

b_0	K_P	l_1	l_2	K	T	T_2
3.3333	4	40	400	1	0.3	0
3.3333	4	40	400	1	0.3	0.001
3.3333	4	40	400	1	0.3	0.002
3.3333	4	40	400	1	0.3	0.01
3.3333	4	40	400	1	0.3	0.02
3.3333	4	40	400	1	0.3	0.03
3.3333	4	40	400	1	0.3	0.2

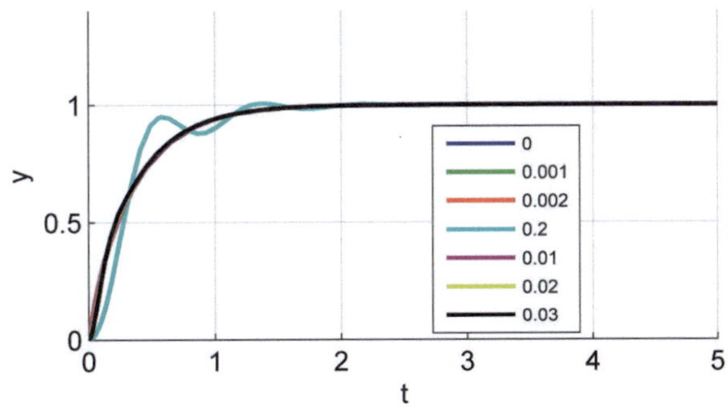

Fig. 28 Closed-loop step response of first-order process with varying T parameter, $\omega_0 = 5\omega_c$ and second pole $T_2 \leq 0.2$

$$\omega_c = 1.5,$$

$$K_P = 2.25,$$

$$K_D = 3,$$

$$l_1 = 45,$$

$$l_2 = 675,$$

and

$$l_3 = 3375.$$

The observer pole chosen is $\omega_0 = 10\omega_c$.

5.2.1 Vulnerability to Varying Process Parameters

On the ADRC model shown in Fig. 37, a series of simulations were performed using the parameters listed in Tables 8, 9, and 10.

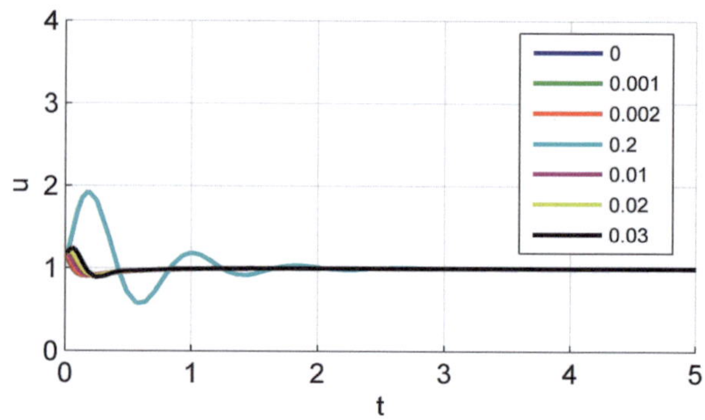

Fig. 29 Controller output u of first-order process with varying T parameter, $\omega_0 = 5\omega_c$ and second pole $T_2 \leq 0.2$

In Figs. 38, 40, and 42, the closed-loop responses are depicted and controller outputs are depicted in Figs. 39, 41, and 43.

As shown in Table 8, variations for K parameter were chosen between 10% and 500% of their nominal values. Figures 38 and 39 illustrate that the closed-loop behavior is almost unaffected by such changes. Only for small values of K, the step response slows down. Similarly, for varying D parameters as in Table 9, the results of Figs. 40 and 41 show that the closed-loop performance over a wide range is not affected. Only for larger values of D, some overshoot is visible.

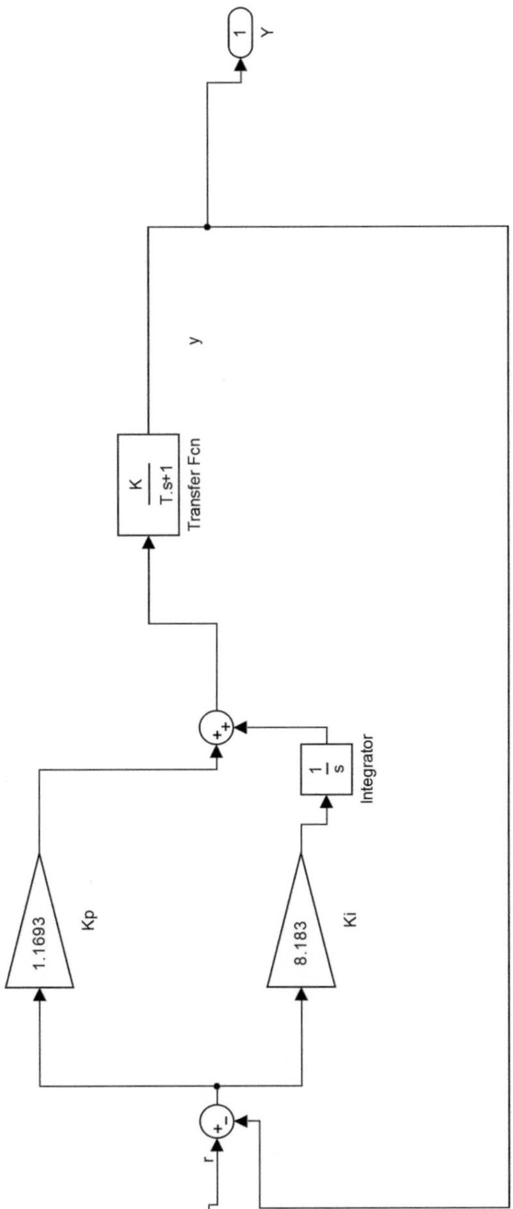

Fig. 30 Simulation model of first-order process using PI controller

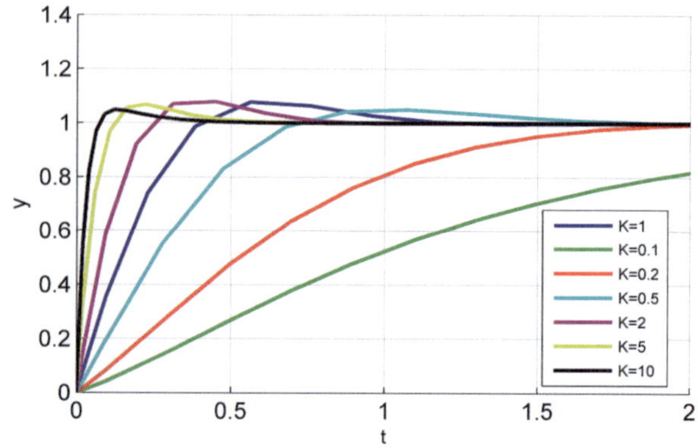

Fig. 31 Closed-loop step response of PI controller for first-order process with varying K process parameter

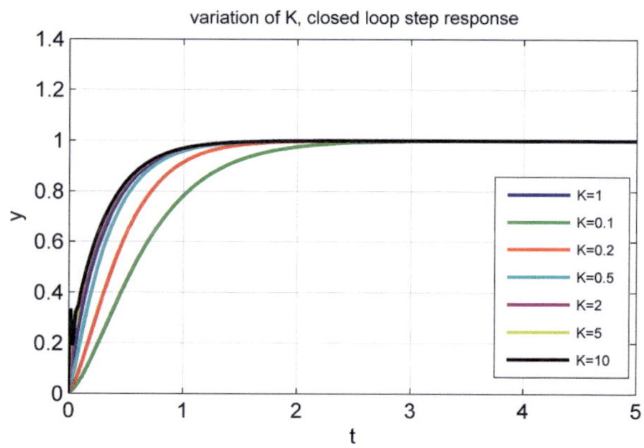

Fig. 32 Closed-loop step response of ADRC controller for first-order process with varying K process parameter

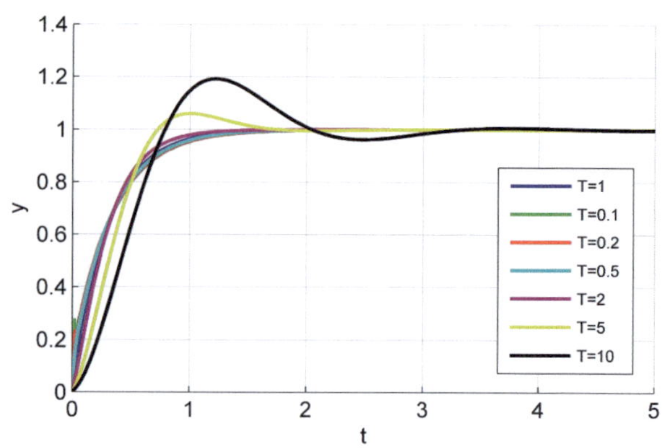

Fig. 33 Closed-loop step response of ADRC controller for first-order process with varying T process parameter

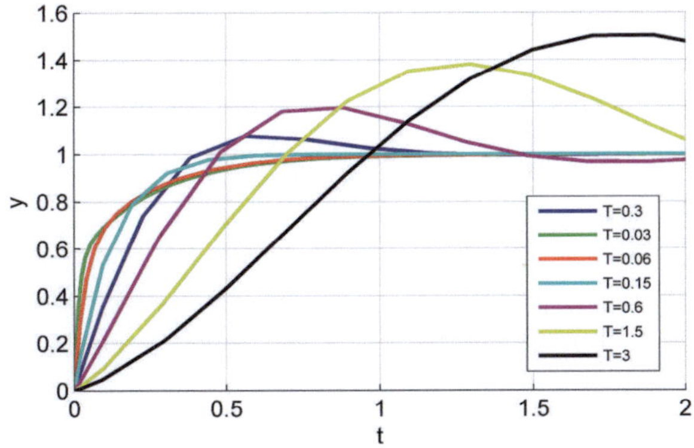

Fig. 34 Closed-loop step response of PI controller for first-order process with varying T process parameter

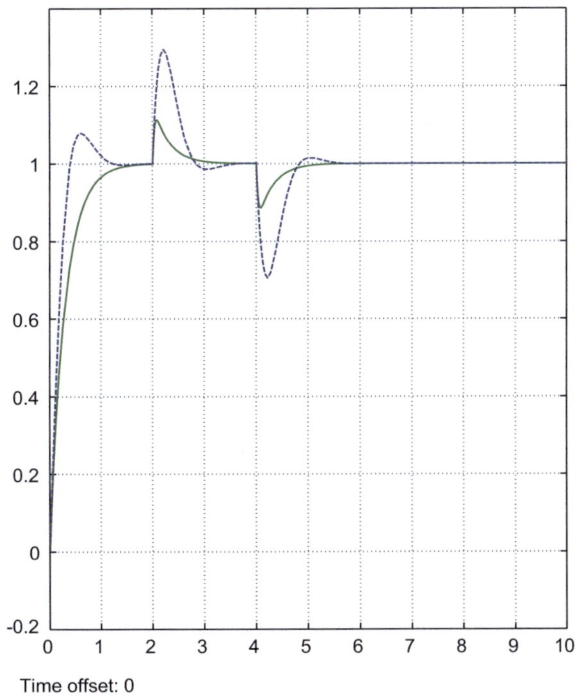

Fig. 35 Closed-loop step response on comparison of disturbance rejection behavior—ADRC versus PID controller, input disturbance was effective from $t = 2$ uptil $t = 4$

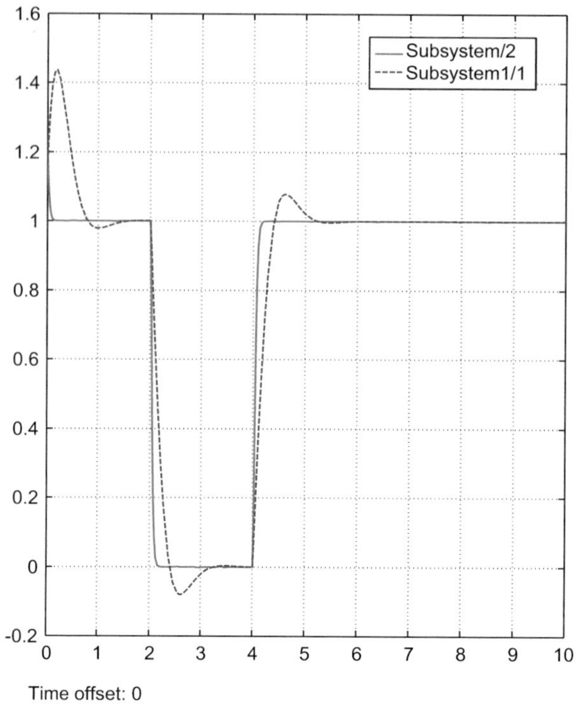

Fig. 36 Controller output u on comparison of disturbance rejection behavior—ADRC versus PID controller, input disturbance was effective from $t = 2$ uptil $t = 4$

Changes in values of T as in Table 10 show increased oscillations and overshoot for larger values as shown in Figs. 42 and 43 .

5.2.2 Influence of Observer Pole Locations

Previously, we observed that changes in T affect the closed-loop behavior than changes in K. Therefore, here we exemplify the influence of observer pole locations on varying T parameter only. These changes are shown in Tables 11, 12, and 13. Both faster and slower observers will be evaluated by setting $\omega_c = 100\omega_0$ and $\omega_c = 5\omega_0$.

Figs. 44, 45, 46, 47, 48, and 49 illustrate the closed-loop dynamics. The slow observer settings exhibit a large overshoot than the faster observer settings.

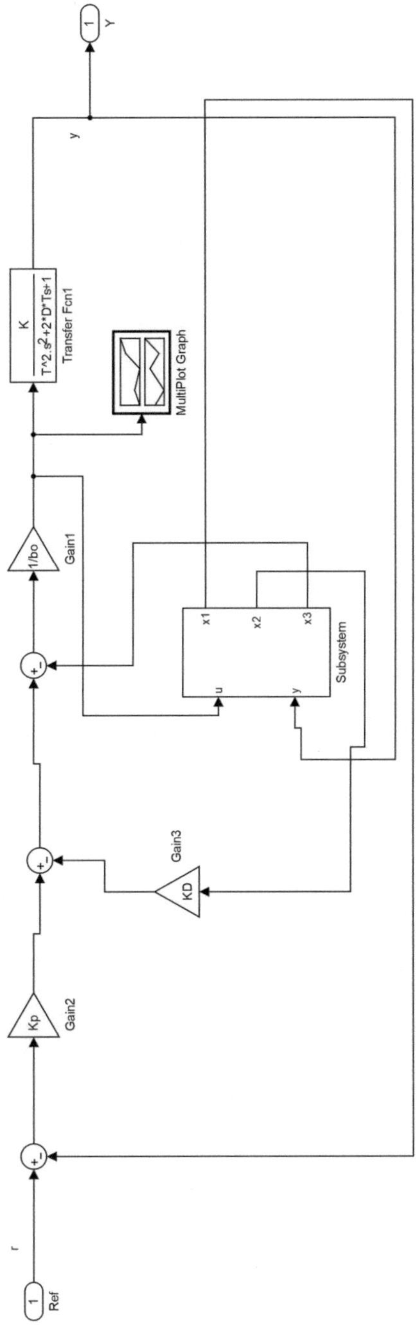

Fig. 37 Model of ADRC for second-order system

Table 8 Second-order process with varying K parameter over a range from 10% to 500%

b_0	K_P	K_D	l_1	l_2	l_3	K	T	D	Variation (%)
1	2.25	3	45	675	3375	0.5	0.71	0.71	0
1	2.25	3	45	675	3375	0.05	0.71	0.71	10
1	2.25	3	45	675	3375	0.1	0.71	0.71	20
1	2.25	3	45	675	3375	0.25	0.71	0.71	50
1	2.25	3	45	675	3375	1	0.71	0.71	200
1	2.25	3	45	675	3375	2.5	0.71	0.71	500

Table 9 Second-order process with varying D parameter over a range from 10% to 1000%

b_0	K_P	K_D	l_1	l_2	l_3	K	T	D	Variation (%)
1	2.25	3	45	675	3375	0.5	0.71	0.71	0
1	2.25	3	45	675	3375	0.5	0.71	0.07	10
1	2.25	3	45	675	3375	0.5	0.71	0.14	20
1	2.25	3	45	675	3375	0.5	0.71	0.36	50
1	2.25	3	45	675	3375	0.5	0.71	1.42	200
1	2.25	3	45	675	3375	0.5	0.71	3.55	500
1	2.25	3	45	675	3375	0.5	0.71	7.1	1000

5.2.3 Effect of Dead Time

To understand the effect of dead time, consider the second-order ADRC with an unknown dead time, as shown in Table 14. Figures 50 and 51 demonstrate that until 40% of variation, or $T_{dead} \leq 0.28$, output is hardly affected by dead time. However, with large values of dead time, oscillations increase dramatically.

These oscillations are reduced by delaying the input to observer, as depicted in Figs. 52 and 53, where $(T_{dead}^{ESO}) = 0.071$, and oscillations are significantly reduced even if (T_{dead}^{ESO}) does not equal actual dead time.

5.2.4 Effect of Structural Uncertainties

In addition to unknown dead times, the controller may be facing high-order dynamics in the process. To examine such effect, third pole is introduced in the process within the range of $0.001 \leq T_3 \leq 1$ as shown in Table 15. The results as in Figs. 54 and 55 depicted that second-order ADRC has good robustness against an unknown higher-order dynamics.

Table 10 Second-order process with varying T parameter over a range from 50% to 350%

b_0	K_P	K_D	l_1	l_2	l_3	K	T	D	Variation (%)
1	2.25	3	45	675	3375	0.5	0.71	0.71	0
1	2.25	3	45	675	3375	0.5	0.36	0.71	50
1	2.25	3	45	675	3375	0.5	1.42	0.71	200
1	2.25	3	45	675	3375	0.5	1.78	0.71	250
1	2.25	3	45	675	3375	0.5	2.13	0.71	300
1	2.25	3	45	675	3375	0.5	2.49	0.71	350

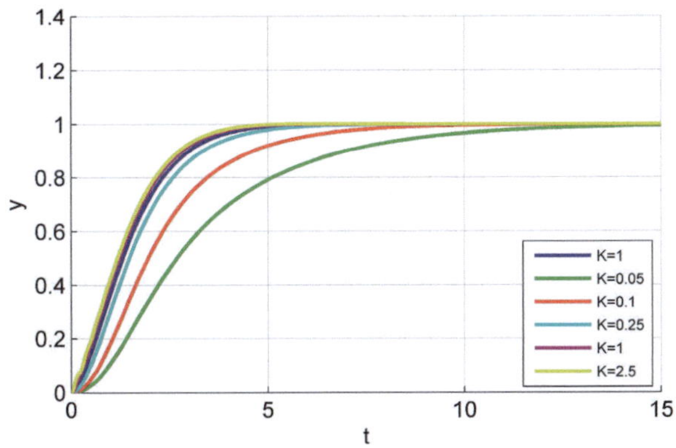

Fig. 38 Closed-loop step response of second-order process for varying K parameter

5.2.5 Comparison with PI and PID Controller

The PI and PID Controllers are designed for second-order process, maintaining the settling time same as for ADRC. Tuning PI and PID Controllers for the same-order dynamics while considering the settling time yielded $K_p = 3.0859$ and $K_I = 2.1014$ for the PI controller and $K_p = 3.618$, $K_I = 3.6663$, and $K_D = 0.89256$ for the PID controller. Figures 56 and 57 depict the model of the PI and PID controllers controlling the second-order process.

The comparison of simulation results in Figs. 58, 59, 60, 61, 62, and 63 clearly demonstrates that the ADRC controller is highly capable of maintaining similar closed-loop dynamics even in the presence of significant parameter variations (Figs. 64, 65, and 66).

5.2.6 Rejecting ADRC, PI, and PID Disturbances

The same closed response must be tuned for ADRC, PI, and PID controllers in order to assess their disturbance rejection capabilities. To each controller, an input

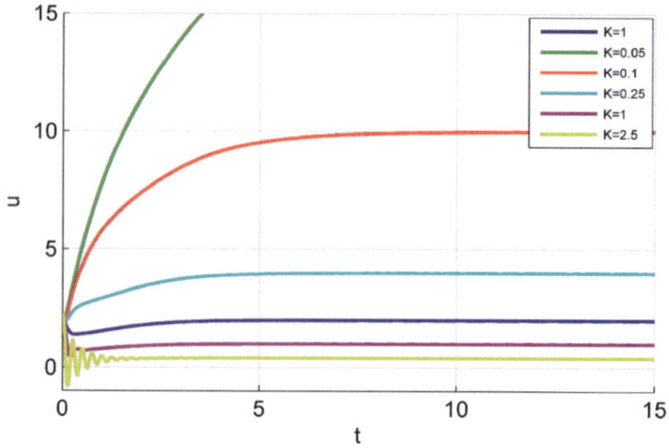

Fig. 39 Controller output u of second-order process for varying K parameter

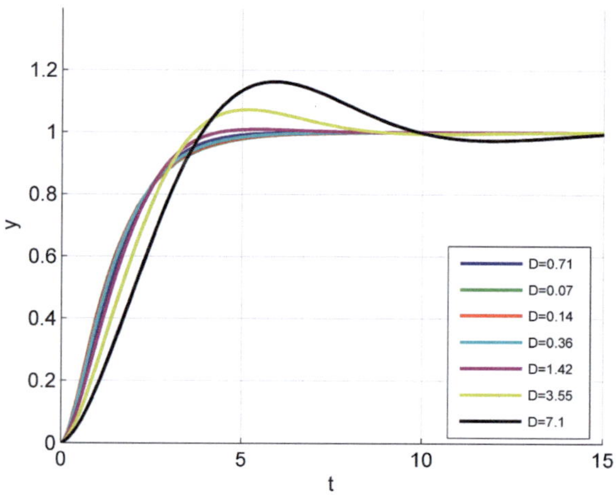

Fig. 40 Closed-loop step response of second-order process for varying D parameter

disturbance, $d = 0.5$, is applied for an interval of ten seconds as depicted in Figs. 67 and 68 .

ADRC compensates for the disturbance significantly faster than the other two controllers.

Table 11 Second-order process with varying T parameter over a range from 50% to 350%, $\omega_c = 100\omega_0$

b_0	K_P	K_D	l_1	l_2	l_3	K	T	D	Variation (%)
1	2.25	3	450	67500	3375000	0.5	0.71	0.71	0
1	2.25	3	450	67500	3375000	0.5	0.36	0.71	50
1	2.25	3	450	67500	3375000	0.5	1.42	0.71	200
1	2.25	3	450	67500	3375000	0.5	1.78	0.71	250
1	2.25	3	450	67500	3375000	0.5	2.13	0.71	300
1	2.25	3	450	67500	3375000	0.5	2.49	0.71	350

Table 12 Second-order process with varying T parameter over a range from 50% to 350%, $\omega_c = 10\omega_0$

b_0	K_P	K_D	l_1	l_2	l_3	K	T	D	Variation (%)
1	2.25	3	45	675	3375	0.5	0.71	0.71	0
1	2.25	3	45	675	3375	0.5	0.36	0.71	50
1	2.25	3	45	675	3375	0.5	1.42	0.71	200
1	2.25	3	45	675	3375	0.5	1.78	0.71	250
1	2.25	3	45	675	3375	0.5	2.13	0.71	300
1	2.25	3	45	675	3375	0.5	2.49	0.71	350

Table 13 Second-order process with varying T parameter over a range from 50% to 350%, $\omega_c = 5\omega_0$

b_0	K_P	K_D	l_1	l_2	l_3	K	T	D	Variation (%)
1	2.25	3	22.5	168.75	421.88	0.5	0.71	0.71	0
1	2.25	3	22.5	168.75	421.88	0.5	0.36	0.71	50
1	2.25	3	22.5	168.75	421.88	0.5	1.42	0.71	200
1	2.25	3	22.5	168.75	421.88	0.5	1.78	0.71	250
1	2.25	3	22.5	168.75	421.88	0.5	2.13	0.71	300
1	2.25	3	22.5	168.75	421.88	0.5	2.49	0.71	350

Table 14 Second-order process with varying dead time over a range from 10% to 40%, observer input $(T_{dead})^{ESO} = 0$, $\omega_c = 5\omega_0$

b_0	K_P	K_D	l_1	l_2	l_3	K	T	D	Dead time	Var wrt T (%)
1	2.25	3	22.5	168.75	421.88	0.5	0.71	0.71	0.071	10
1	2.25	3	22.5	168.75	421.88	0.5	0.71	0.71	0.14	20
1	2.25	3	22.5	168.75	421.88	0.5	0.71	0.71	0.21	30
1	2.25	3	22.5	168.75	421.88	0.5	0.71	0.71	0.28	40

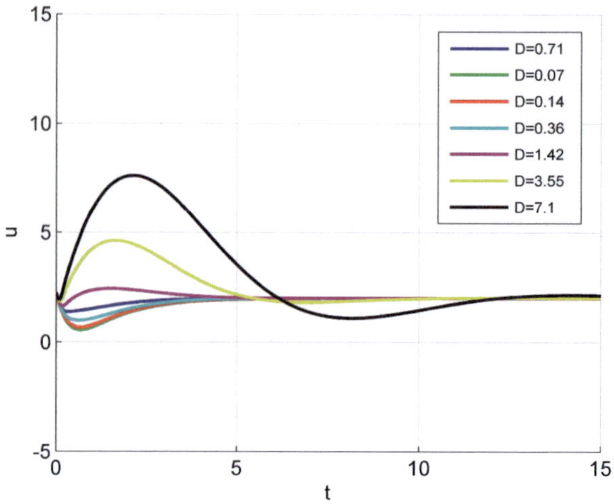

Fig. 41 Controller output u of second-order process for varying D parameter

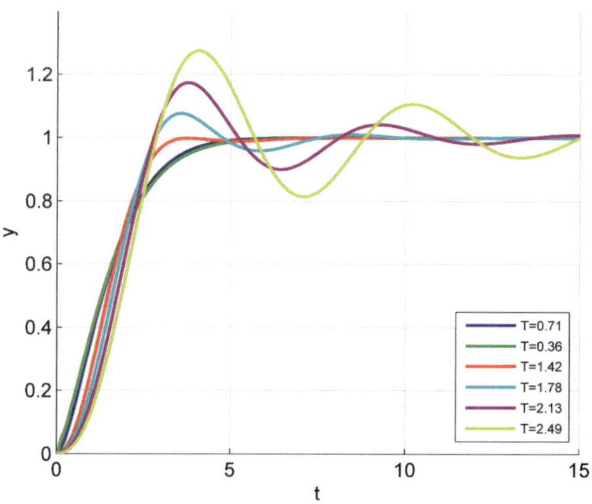

Fig. 42 Closed-loop step response of second-order process for varying T parameter

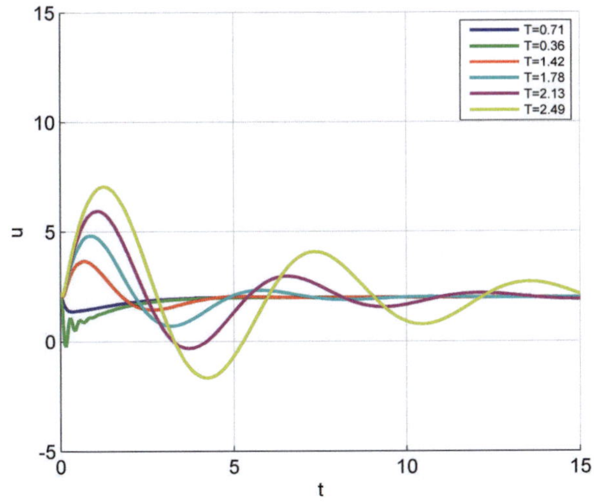

Fig. 43 Controller output u of second-order process for varying T parameter

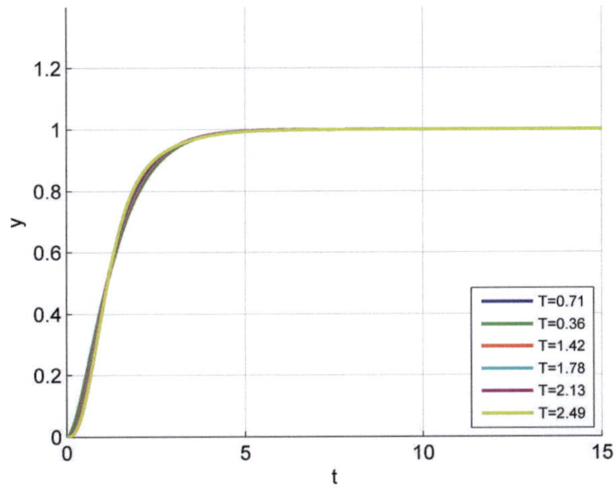

Fig. 44 Closed-loop step response of second-order process for varying T parameter, $\omega_c = 100\omega_0$

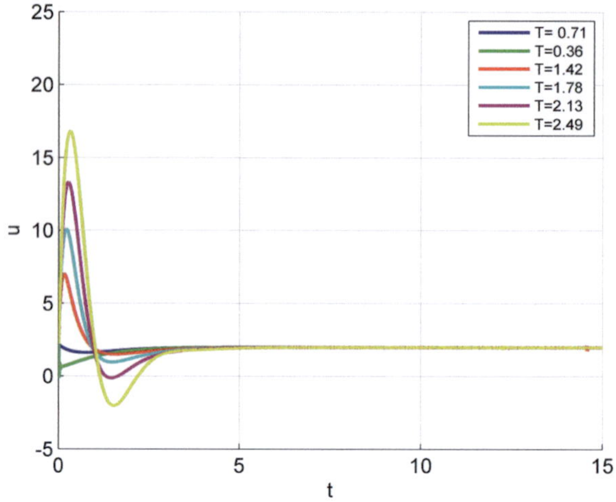

Fig. 45 Controller output u of second-order process for varying T parameter, $\omega_c = 100\omega_0$

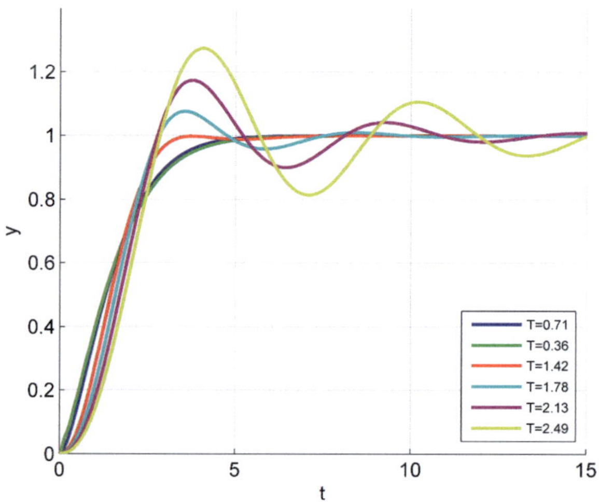

Fig. 46 Closed-loop step response of second-order process for varying T parameter, $\omega_c = 10\omega_0$

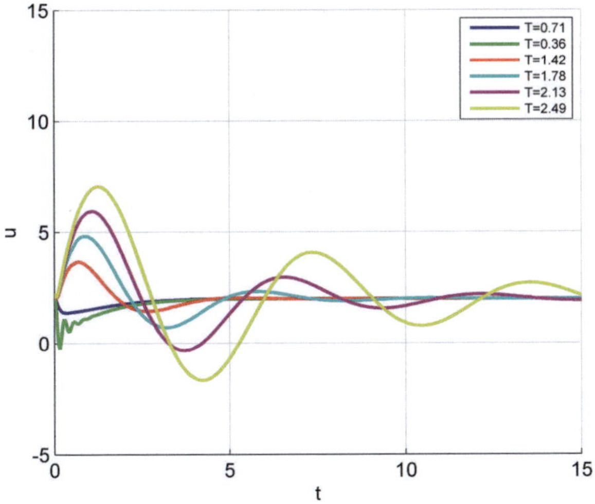

Fig. 47 Controller output u of second-order process for varying T parameter, $\omega_c = 10\omega_0$

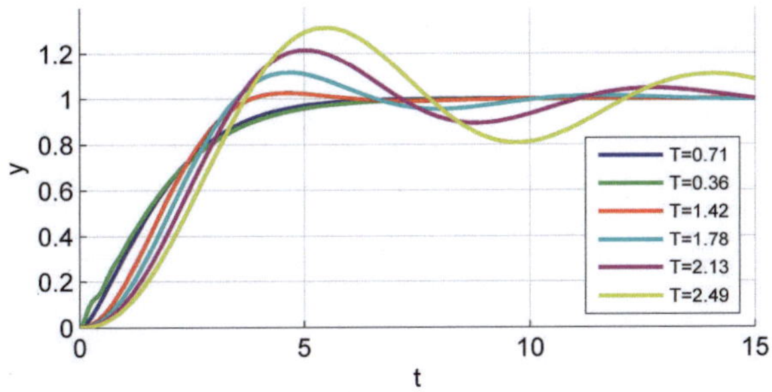

Fig. 48 Closed-loop step response of second-order process for varying T parameter, $\omega_c = 5\omega_0$

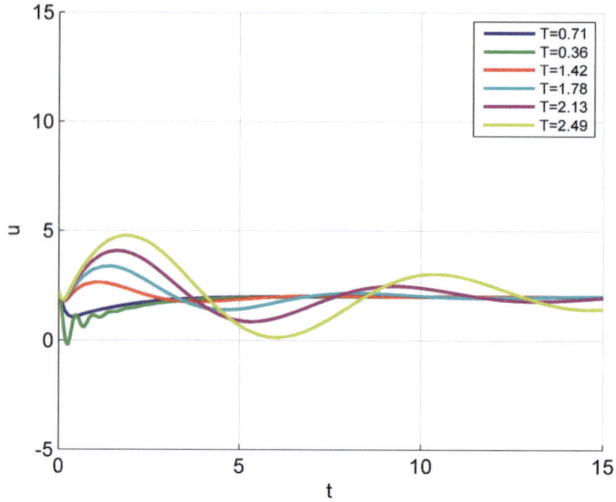

Fig. 49 Controller output u of second-order process for varying T parameter, $\omega_c = 5\omega_0$

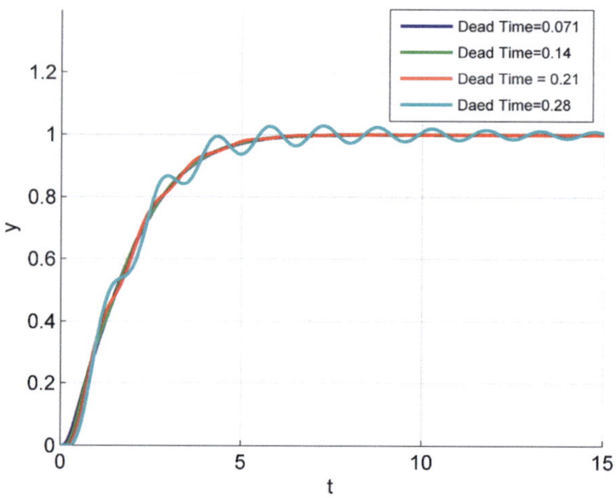

Fig. 50 Closed-loop step response of second-order process with varying dead time for observer input ($T_{dead}^{ESO} = 0$, $\omega_c = 5\omega_0$)

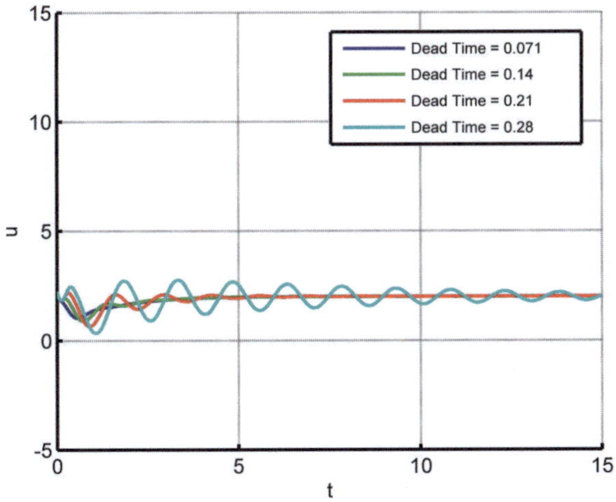

Fig. 51 Controller output u of second-order process with varying dead time for observer input ($T_{dead}^{ESO} = 0, \omega_c = 5\omega_0$)

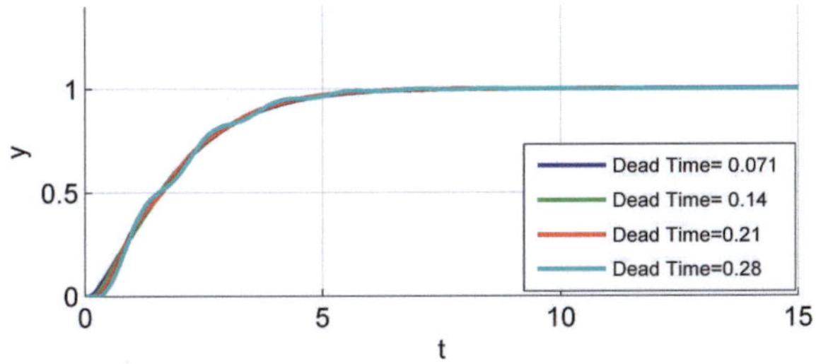

Fig. 52 Closed-loop step response of second-order process with varying dead time for observer input ($T_{dead}^{ESO} = 0.071, \omega_c = 5\omega_0$)

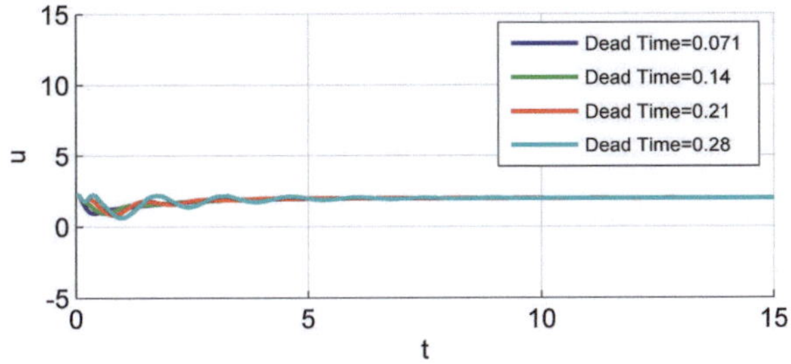

Fig. 53 Controller output u of second-order process with varying dead time for observer input ($T_{dead}^{ESO} = 0.071, \omega_c = 5\omega_0$)

Table 15 Second-order process for $\omega_c = 5\omega_0$ with extra pole T_3

b_0	K_P	K_D	l_1	l_2	l_3	K	T	D	T_3
1	2.25	3	22.5	168.75	421.88	0.5	0.71	0.71	0.001
1	2.25	3	22.5	168.75	421.88	0.5	0.71	0.71	0.01
1	2.25	3	22.5	168.75	421.88	0.5	0.71	0.71	0.1
1	2.25	3	22.5	168.75	421.88	0.5	0.71	0.71	1

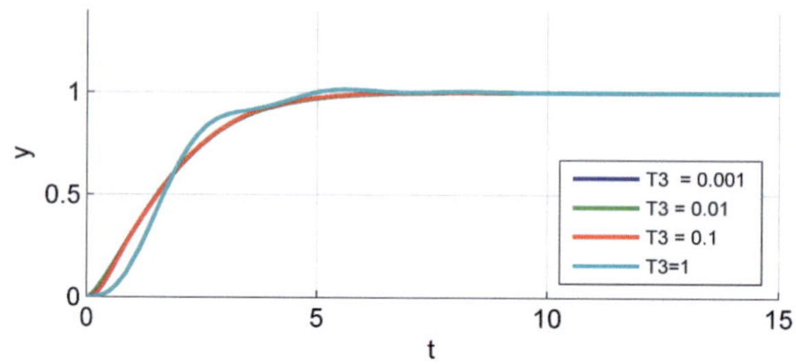

Fig. 54 Closed-loop step response of second-order process with extra unknown pole T_3, $\omega_c = 5\omega_0$

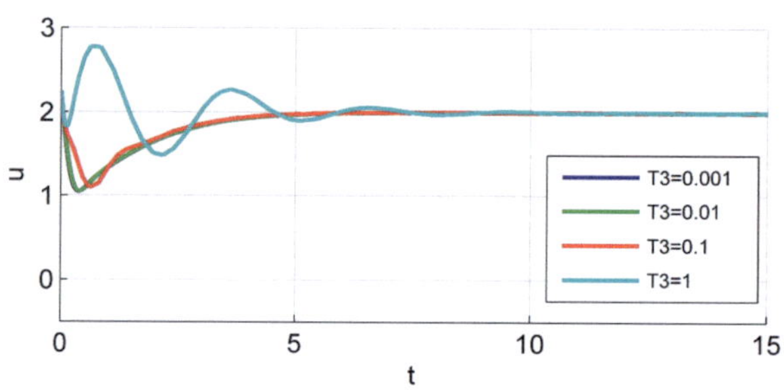

Fig. 55 Controller output u of second-order process with extra unknown pole T_3, $\omega_c = 5\omega_0$

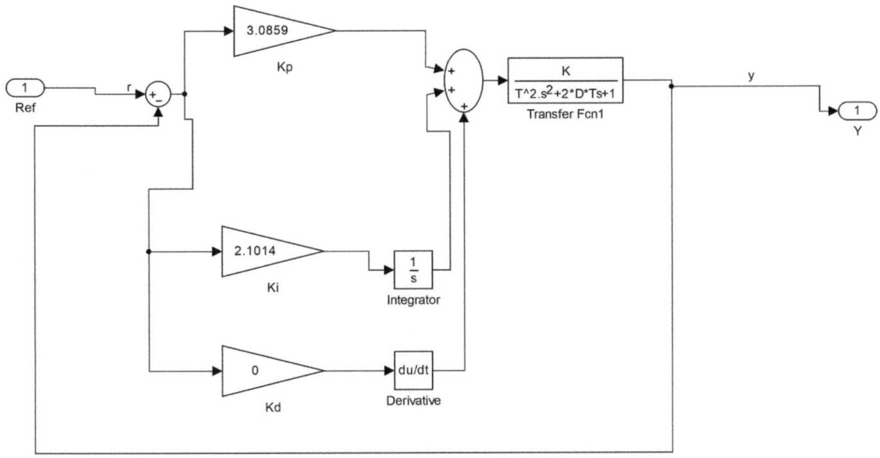

Fig. 56 Smulation model of second-order process using PI controller

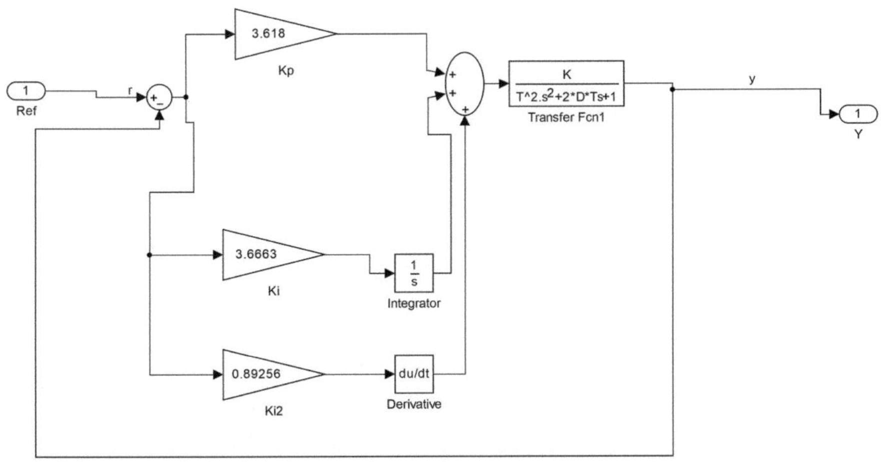

Fig. 57 Smulation model of second-order process using PID controller

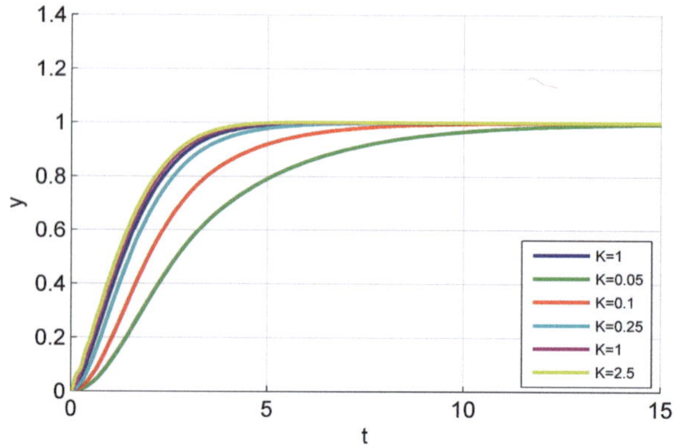

Fig. 58 Closed-loop step response of second-order ADRC controller with varying K process parameter

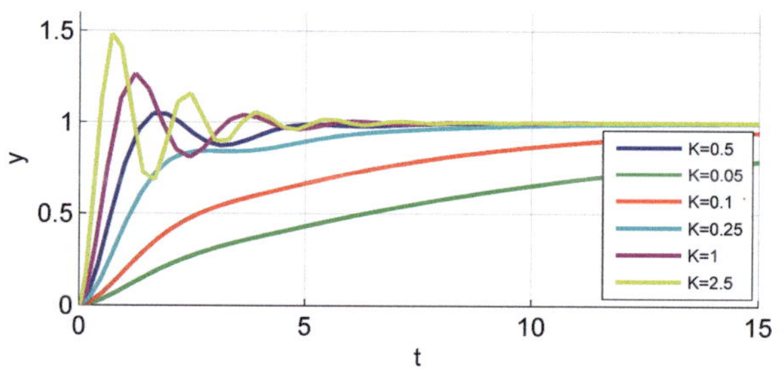

Fig. 59 Closed-loop step response of second-order PI controller with varying K process parameter

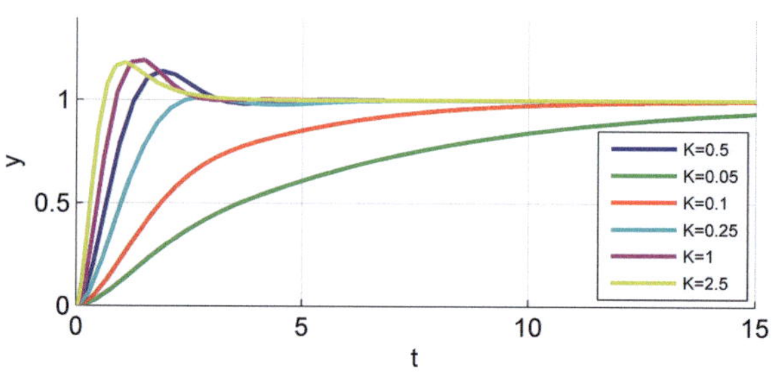

Fig. 60 Closed-loop step response of second-order PID controller with varying K process parameter

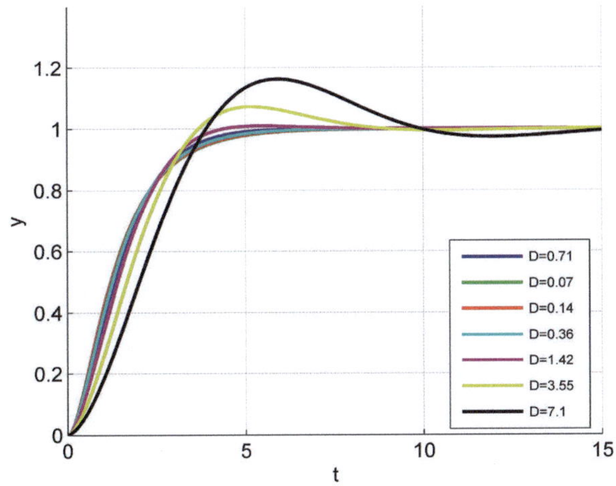

Fig. 61 Closed-loop step response of second-order ADRC controller with varying *D* process parameter

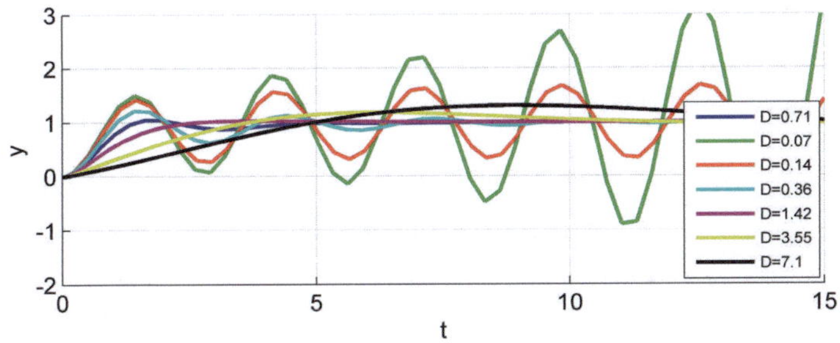

Fig. 62 Closed-loop step response of second-order PI controller with varying *D* process parameter

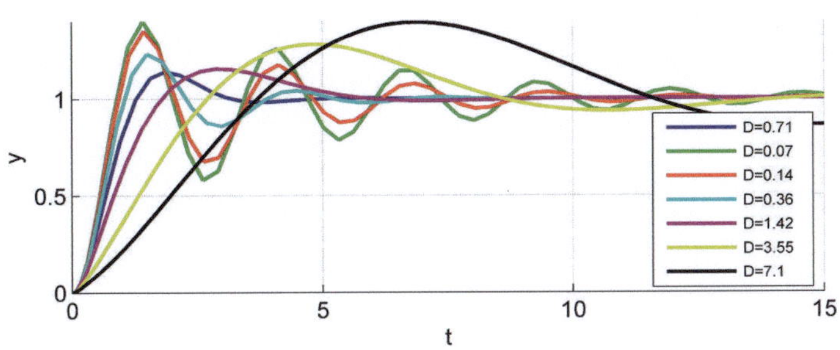

Fig. 63 Closed-loop step response of second-order PID controller with varying *D* process parameter

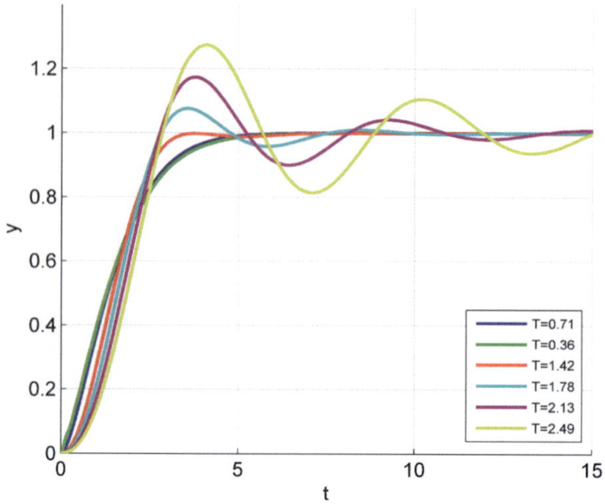

Fig. 64 Closed-loop step response of second-order ADRC controller with varying T process parameter

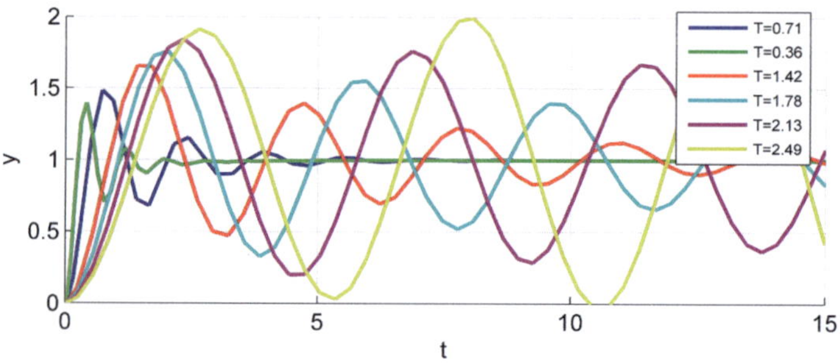

Fig. 65 Closed-loop step response of second-order PI controller with varying T process parameter

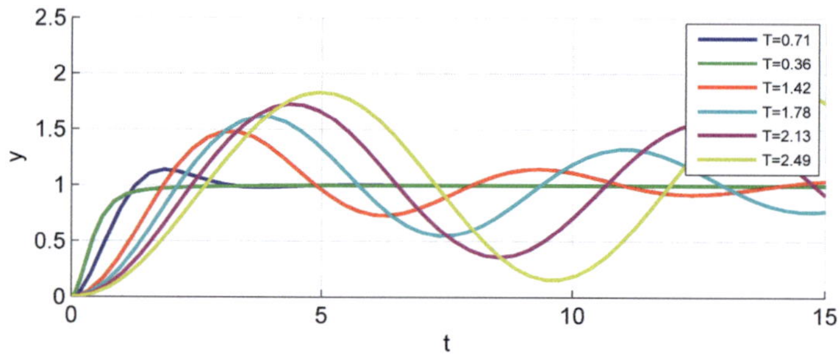

Fig. 66 Closed-loop step response of second-order PID controller with varying T process parameter

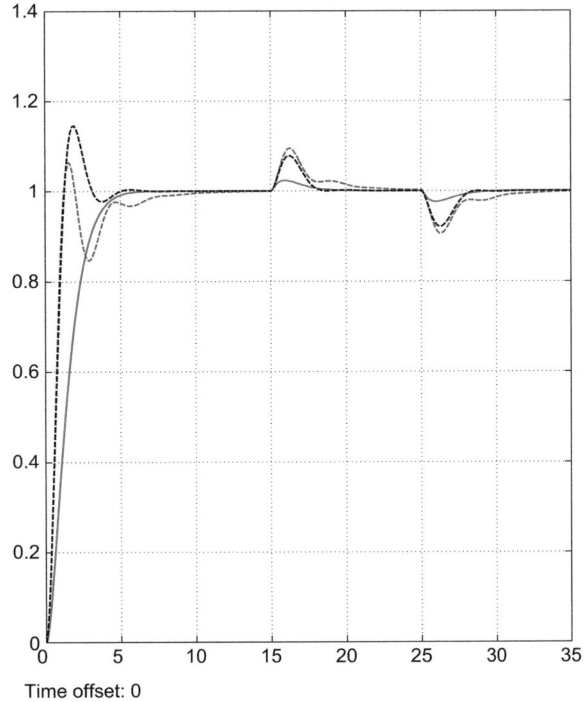

Fig. 67 Closed-loop step response on comparison of disturbance rejection behavior, input disturbance $d = 0.5$ effective from $t = 15$ to $t = 25$

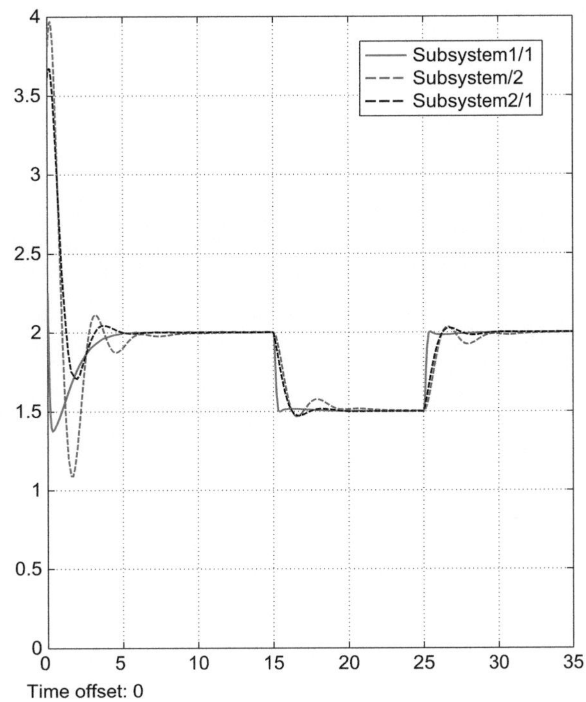

Fig. 68 Controller output on comparison of disturbance rejection behavior, input disturbance $d = 0.5$ effective from $t = 15$ to $t = 25$

6 Conclusion

The primary focus of this paper is to demonstrate that ADRC is an effective control method in mitigating the impact of modeling errors and process parameter variations. This paper focused on the linear scenario and the processes used in this paper are first-order and second-order plants which are investigated in Simulink/MATLAB environment. The results show that ADRC can adapt high process "parameter" variation without relying on the exact mathematical model of process unlike PI or PID controllers. Furthermore, ADRC requires minimal information about the processes being controlled, making it more adaptive in situations where the system parameters are difficult to obtain. The dynamic responses suggest that the controller performance deteriorates for slower observer settings. Also, the disturbance rejection comparison of ADRC with PI and PID controllers has been examined showing that former is more superior than others.

References

1. O'Dwyer A (2006) Pi and pid controller tuning rules: an overview and personal perspective. In: 2006 IET Irish signals and systems conference, pp 161–166
2. Wang Q (2005) Handbook of PI and PID controller tuning rules, Aidan O'dwyer, Imperial College Press, London, 375pp. ISBN 1-86094-342-x, 2003 Autom 41(2):355–356. https://doi.org/10.1016/j.automatica.2004.09.012
3. Elameen MHE, Li J, Zhang B, Tan W (2019) Active disturbance rejection control design and analysis for first-order plus time delay systems. In: 2019 Chinese Control Conference (CCC), pp 71–75. https://doi.org/10.23919/ChiCC.2019.8865719
4. Gao Z (2015) Active disturbance rejection control: From an enduring idea to an emerging technology. In: 2015 10th international workshop on Robot Motion and Control (RoMoCo), pp 269–282. https://doi.org/10.1109/RoMoCo.2015.7219747
5. Herbst G (2019) A simulative study on active disturbance rejection control (ADRC) as a control tool for practitioners. CoRR http://arxiv.org/abs/1908.04596
6. Li D, Yu W, Zhou J, Jin Q (2016) Active disturbance rejection control with adaptive delayed input and its performance monitoring for time-varying delay system. In: 2016 35th Chinese Control Conference (CCC), pp 5964–5969. https://doi.org/10.1109/ChiCC.2016.7554293
7. Zhao S, Gao Z (2014) Modified active disturbance rejection control for time-delay systems. ISA Trans 53(4):882–888
8. Han J (2013) An anthology on active disturbance rejection control, in memory of prof. Jingqing Han on the 5th anniversary of his passing. In: Key laboratory of systems and control. Chinese Academy of Sciences
9. Han J (2009) From pid to active disturbance rejection control. IEEE Trans Ind Electronics 56(3):900–906. https://doi.org/10.1109/TIE.2008.2011621
10. Gao Z, Huang Y, Han J (2001) An alternative paradigm for control system design. In: Proceedings of the 40th IEEE conference on decision and control (Cat. No.01CH37228), vol 5, pp 4578–4585. https://doi.org/10.1109/CDC.2001.980926
11. Gao Z (2006) Active disturbance rejection control: a paradigm shift in feedback control system design. In: 2006 American control conference, p 7. https://doi.org/10.1109/ACC.2006.1656579
12. Inc TM (2021) Control tutorials for matlab and simulink—introduction: Pid controller design. https://ctms.engin.umich.edu/CTMS/index.php?example=Introduction&section=ControlPID
13. Hamid PS, Shafi S, Nahvi SA, Bazaz MA, Koul MH (2023) Experimental investigation of the effectiveness and robustness of active disturbance rejection control using a dc motor-driven piston. In: 2023 7th international conference on Computer Applications in Electrical

Engineering-Recent Advances (CERA), pp 1–6. https://doi.org/10.1109/CERA59325.2023.10455106
14. Gao Z (2003) Scaling and bandwidth-parameterization based controller tuning. In: Proceedings of the 2003 American control conference, vol 6, pp 4989–4996
15. Gao Z (2003) Scaling and bandwidth-parameterization based controller tuning. In: Proceedings of the 2003 American control conference, vol 6, pp 4989–4996. https://doi.org/10.1109/ACC.2003.1242516
16. Gao Z (2013) On the centrality of disturbance rejection in automatic control. ISA Trans 53. https://doi.org/10.1016/j.isatra.2013.09.012
17. Zhao S, Gao Z (2013) Modified active disturbance rejection control for time-delay systems. ISA Trans 53. https://doi.org/10.1016/j.isatra.2013.09.013
18. Zheng Q, Gao Z (2013) Predictive active disturbance rejection control for processes with delay. ISA Trans 53. https://doi.org/10.1016/j.isatra.2013.09.021
19. Gao Z, Hu S, Jiang F (2001) A novel motion control design approach based on active disturbance rejection. In: Proceedings of the 40th IEEE conference on decision and control (Cat. No.01CH37228), vol 5, pp 4877–4882. https://doi.org/10.1109/CDC.2001.980980
20. Gao Z, Huang Y, Han J (2001) An alternative paradigm for control system design. In: Proceedings of the 40th IEEE conference on decision and control (Cat. No.01CH37228), vol 5, pp 4578–4585. https://doi.org/10.1109/CDC.2001.980926
21. Martínez Carvajal B, Sanchis J, Garcia-Nieto Rodriguez S, Martínez M (2021) Tuning rules for active disturbance rejection controllers via multiobjective optimization—a guide for parameters computation based on robustness. Mathematics 9:517. https://doi.org/10.3390/math9050517
22. de Mello F (1991) Boiler models for system dynamic performance studies. IEEE Trans Power Syst 6(1):66–74. https://doi.org/10.1109/59.131048
23. Seborg DE, Edgar TF, Mellichamp DA, Doyle III FJ (2016) Process dynamics and control. Wiley
24. Tan W, Fang F, Tian L, Fu C, Liu J (2008) Linear control of a boiler-turbine unit: analysis and design. ISA Trans 47(2):189–197
25. Fu C, Tan W (2017) Control of unstable processes with time delays via ADRC. ISA Trans 71:530–541
26. Inc TM (2023) Active disturbance rejection control. https://in.mathworks.com/help/slcontrol/ug/active-disturbance-rejection-control.html
27. Li X, Ai W, Tian S (2018) ADRC with feedfoward control for time-delay systems. In: 2018 IEEE 7th Data Driven Control and Learning Systems Conference (DDCLS), pp 736–742. https://doi.org/10.1109/DDCLS.2018.8516030
28. Tan W, Fu C (2015) Analysis of active disturbance rejection control for processes with time delay. In: 2015 American Control Conference (ACC), pp 3962–3967. https://doi.org/10.1109/ACC.2015.7171948
29. Wu X, Chen Z, Sun M, Sun Q (2017) Active disturbance rejection generalized predictive control and its application on large time-delay systems. In: 2017 6th Data Driven Control and Learning Systems (DDCLS), pp 1–8. https://doi.org/10.1109/DDCLS.2017.8067716
30. Zheng Q, Gao Z (2016) On active disturbance rejection for systems with input time-delays and unknown dynamics. In: 2016 American Control Conference (ACC), pp 95–100. https://doi.org/10.1109/ACC.2016.7524898
31. Radke A, Gao Z (2006) A survey of state and disturbance observers for practitioners, p 6. https://doi.org/10.1109/ACC.2006.1657545
32. Taguchi H, Araki M (2000) Two-degree-of-freedom pid controllers—their functions and optimal tuning. In: IFAC digital control: past, present and future of PID control, vol 33. https://doi.org/10.1016/S1474-6670(17)38226-5
33. Wang L, Li Q, Tong C, Yin Y, Gao Z, Zheng Q, Zhang W (2015) On control design and tuning for first order plus time delay plants with significant uncertainties. In: 2015 American Control Conference (ACC), pp 5276–5281. https://doi.org/10.1109/ACC.2015.7172163
34. Wu JA, Tian C, Yan P (2018) A predictor-based ADRC for input delay systems subject to unknown disturbances. In: 2018 Chinese Automation Congress (CAC). IEEE, pp 231–236

Early Investigations into Music-Induced Neural and Cardiovascular Responses

Prasad Purnaye and Sanket Salvi

Abstract This study aims to explore the neural correlation to heartbeat in response to music, unraveling the mechanisms by which musical stimuli influence the human nervous and cardiovascular systems. Through empirical research involving 30 participants aged 18–22, we observed that slow music significantly accelerates heart rate recovery post-exercise, with an average resting heart rate of 75.6 bpm and an average elevated heart rate of 103.8 bpm. Utilizing a K-Nearest Neighbors (KNN) algorithm, our machine learning analysis demonstrated a notable predictive accuracy with an R^2 score of 0.77, indicating a substantial correlation between predicted and actual recovery times. Additionally, EEG data revealed distinct neural patterns corresponding to different musical stimuli, suggesting that slow music enhances parasympathetic activation and promotes a faster return to homeostasis. Through empirical research, the study strives to contribute valuable insights that can inform the development of innovative applications and interventions, harnessing the power of music to promote happier and healthier lives.

Keywords EEG · Arduino · Music · Neural response · Cardiovascular function · Data analytics

1 Introduction

Regular physical activity is crucial for overall health, and understanding the recovery phase post-workout is essential for optimizing fitness levels. This project investigates the impact of music genre on heart rate recovery following a workout, comparing scenarios with and without music. By investigating whether different music genres

P. Purnaye (✉) · S. Salvi
Dr. Vishwanath Karad MIT World Peace University, Kothrud, Pune, India
e-mail: prasad.purnaye@mitwpu.edu.in

S. Salvi
e-mail: sanketsalvi.salvi@gmail.com

H. Sharma et al. (eds.), *Artificial Intelligence: Theory and Applications*, Lecture Notes in Networks and Systems 5589, https://doi.org/10.1007/978-981-96-1687-9_10

expedite heart rate recovery to baseline levels, we aim to contribute to the growing understanding of music's potential therapeutic applications in exercise science and sports medicine [1, 2]. Through empirical investigation, our goal is to uncover insights that may inform the development of personalized exercise regimens and recovery strategies, ultimately enhancing physical fitness and performance. Research literature highlights that music can indeed play a role in post-exercise recovery, with findings highlighting the potential of music as a tool to enhance the post-workout experience and optimize recovery [1, 2].

The paper is organized as follows: In Sect. 2, we review the related work in the field of neural correlation to heartbeat in response to music. Section 3 outlines the proposed work and research methodology adopted for this study. We discuss the requirement analysis and implementation details in Sect. 4, providing insights into the experimental setup and results in Sect. 5. Finally, we conclude our findings and discuss future directions in Sect. 6.

2 Related Work

The ability of music to evoke emotions has sparked interest in its neural correlates. Various methodologies, including EEG-based studies and multimodal approaches, have been used to explore this phenomenon. Sun et al. [3] utilized EEG signals and deep learning to recognize music-induced emotions, while Zhang et al. [4] combined EEG with other modalities for better emotion recognition. These studies highlight neural mechanisms, but they face challenges such as limited ecological validity and individual differences in emotional processing.

The influence of music on post-exercise recovery is another area of interest. Music positively impacts mood, motivation, and perceived exertion. Guzzo et al. [5] reviewed music's effects on exercise performance, noting its benefits. However, diverse study designs and individual differences make it difficult to draw definitive conclusions. The synchronization of heart rate induced by music relies on the consistent physiological responses to the music itself, rather than being influenced by the listener's mood or level of motivation [6].

Music-based meditation has shown promise in managing stress and anxiety. Hassed et al. [7] conducted a meta-analysis on mindfulness-based stress reduction with music, finding strong evidence for its efficacy. Despite the effectiveness, the mechanisms behind music-based meditation are not well understood, and the diversity of interventions requires further research. Bhattacharya et al. [8] and Wang et al. [9] have demonstrated music's impact on brain function and stress reduction. In contrast, Lo et al. [10] highlighted music's potential for cognitive enhancement in older adults. Another study by researchers [11] found that listening to different emotional music pieces affected motor sequence learning and movement kinematics. Facilitating changes in emotional and physiological states, music could potentially accelerate relaxation and recovery processes following physical exertion.

3 Proposed Work

In this section, the proposed system and research methodology are discussed, which employs a multi-faceted approach to investigate the intricate relationship between music, physiological responses, and relaxation. The subsections that follow delve into specific methodologies utilized to explore EEG signals, the application of machine learning techniques, and a proposed approach to merge Spotify data with real-world EEG experiments.

The study employed a comprehensive approach by integrating both EEG signals and pulse rate data collected through Arduino sensors to monitor physiological changes in real-time during and after exercise. Participants' baseline heart rates and neural responses were first recorded before engaging in a controlled cardiovascular activity, such as climbing stairs, to elevate their heart rates. Post-exercise, participants were exposed to slow music, and the time taken for their heart rates to return to baseline was measured and compared to times without music to evaluate the music's impact on recovery. The integration of Spotify data with EEG experiments allowed for a nuanced analysis of how specific musical tracks influenced neural responses and relaxation, bridging subjective experiences with measurable physiological effects.

3.1 EEG Signals

The OpenBCI EEG signals, illustrated in Fig. 1, provide a comprehensive view of brain activity by decomposing the signals into various frequency bands. These range from very low frequencies, such as Delta waves, to high frequencies like Gamma waves. The figure showcases the OpenBCI system's capability to capture and display EEG signals, facilitating detailed analysis of brain wave patterns.

3.2 Proposed Approach

Figure 2 shows the proposed system under which experiments are conducted to assess the effects of music on neural and cardiovascular responses. First, we will record the baseline heart rate and neural responses of the subjects. Subsequently, the subjects will engage in a cardiovascular activity by climbing a flight of stairs within a fixed duration. Post-exercise, the elevated heart rates, and neural responses will be documented. Then, a slow music track will be played, and the time for the heart rate to return to baseline will be measured. The same procedure will be repeated without the music, recording the time taken to return to baseline heart rate. A comparative analysis of these times, with and without music, will be performed to determine whether music facilitates a quicker return to baseline heart rate, thereby aiding in faster recovery.

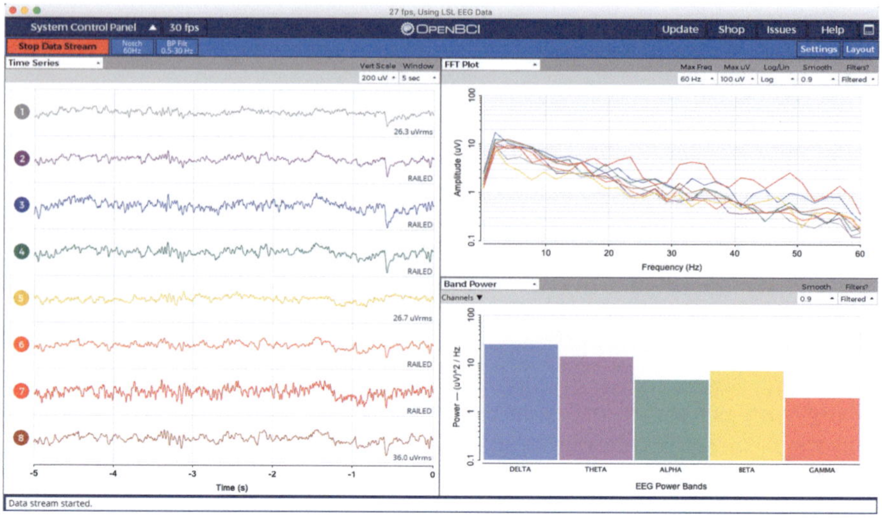

Fig. 1 Open BCI EEG signals

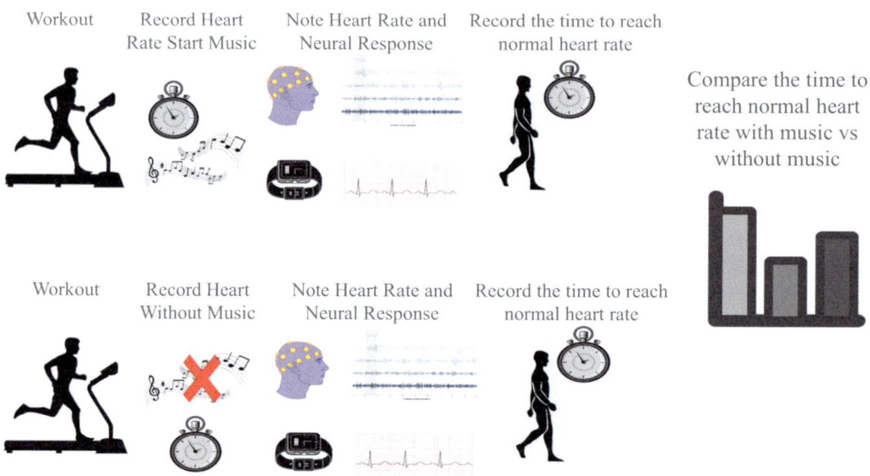

Fig. 2 Proposed system to test music-induced neural and cardiovascular responses

Our approach merges Spotify data with EEG experiments to validate music's effectiveness in inducing relaxation and quantify its extent. We aim to bridge the gap between subjective relaxation experiences and measurable physiological responses by combining music streaming data with neuroscientific measurements. In this study, we will leverage machine learning, specifically the K-Nearest Neighbors (KNN) algorithm, to explore the complex relationship between physiological responses and music features in post-exercise recovery. Integrating physical exertion, heart rate monitoring, EEG measurements, and music during cooldown phases, our research dataset will include individual attributes and features derived from physiological responses and song characteristics. Through a user-friendly Flask web server, participants will input specific songs to gauge their predicted effectiveness for recovery.

4 Requirement Analysis and Implementation

Table 1 presents the hardware and software requirements essential for the implementation of the proposed system and conducting the experiment.

The data acquisition process involved three primary methods to gather comprehensive information about the subjects: EEG data collection, real-time monitoring of pulse rates using an Arduino sensor, and obtaining additional subjective insights through a Google Form. Figure 3 illustrates the steps involved in the proposed workflow. Starting with the bottommost layer, the sensor data is captured from the human subject, which is then logged as raw data. This raw data undergoes preprocessing where missing values and timestamps are normalized. Subsequently, the data from the two independent sensors is merged based on the timestamps, and additional subjective information from Google Forms is appended to generate a multimodal dataset.

Table 1 Hardware and software requirements for the experiment

Component	Quantity	Justification
Arduino UNO	1	Arduino UNO is used as the microcontroller to interface with the pulse sensor and communicate with the computer. It provides analog-to-digital conversion and digital input/output capabilities
Pulse sensor	1	The pulse sensor measures heart rate by detecting changes in blood volume in the fingertip. It is interfaced with the Arduino UNO to monitor heart rate during the experiment
EEG headset with OpenBCI	1	The EEG headset from OpenBCI is used to record neural responses during the experiment. It provides a non-invasive method to monitor brain activity, allowing for the assessment of the effects of music on neural responses
Laptop	1	The laptop is equipped with an Intel Core i5 processor, 8 GB RAM, and a 256 GB SSD. It serves as the interface for running the experiment software, processing data, and analyzing results
Google Form	1	Google Form is used to collect subjective feedback from participants regarding their relaxation experiences with different music tracks. It provides an easy-to-use and accessible platform for data collection

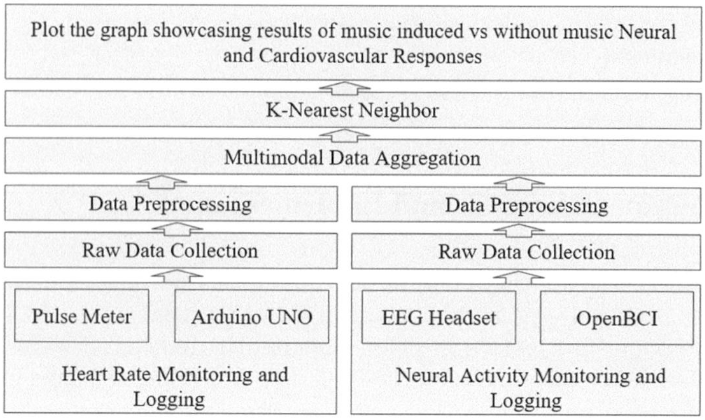

Fig. 3 Functional block diagram

By applying K-Nearest Neighbors (KNN) on this dataset, a graph showcasing the difference between the two sets of experiments is plotted. This comprehensive approach ensures the integration of physiological and subjective data, providing a holistic view of the subjects' responses. The multimodal dataset enables the exploration of correlations between neural and cardiovascular responses, enhancing the robustness of the analysis. Moreover, the preprocessing steps ensure data integrity and consistency, facilitating accurate and reliable model predictions.

4.1 EEG Data Collection

EEG data is collected using EEG headsets, with electrodes FP1 and FP2 strategically positioned on the forehead to capture electrical signals from the brain. This data provides insights into the neural activities of the subjects during various states and activities.

4.2 Pulse Rate Monitoring with Arduino

Real-time heartbeat data was acquired using a pulse rate monitoring sensor connected to an Arduino device as shown in Fig. 4. Simultaneously recorded EEG data, along with timestamps, provided a synchronized dataset for analysis. The time taken for subjects to reach a relaxed state, both with and without music after a workout, was calculated allowing for the assessment of the impact of music on relaxation.

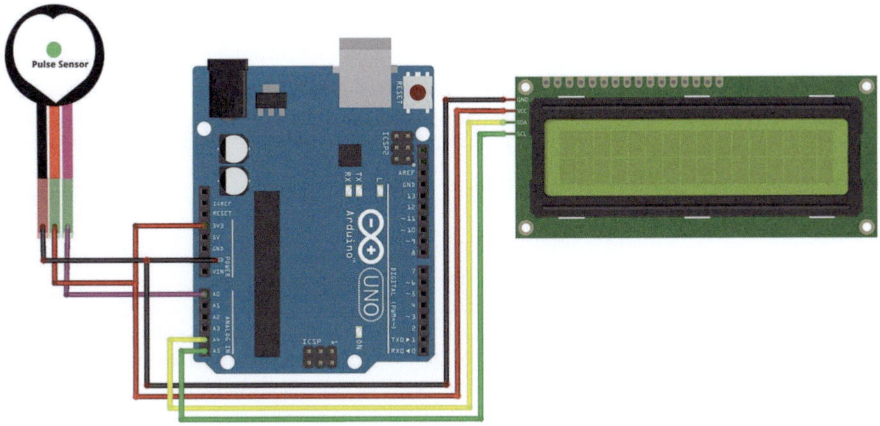

Fig. 4 Pulse sensor Arduino UNO circuit diagram

4.3 Subjective Information Through Google Form

Subjects were requested to fill out a Google Form to gather additional information, including user details such as name, age, weight, and gender. Preferences related to soothing music, a crucial factor in the study, were also collected through this form.

The data collection process is structured into three distinct phases, each designed to capture different physiological states and responses from the subjects as shown in Table 2.

5 Experimental Setup and Results

The sample of 30 humans with a balanced set of 15 females and 15 males was used to test the hypothesis. The age group range considered was from 18 to 22 years old. The average heart rate was 75.6 bpm for the subjects at rest. The subjects were asked to climb 50 stairs within 1 min. The average elevated heart rate post-exercise was 103.8 bpm. It was observed that the average time to reach baseline heart rate with music was 247 s while without music it was 298 s.

Regression analysis, a supervised machine learning method, forecasts a continuous outcome variable based on input variables. The aim of regression analysis is to establish a connection between the input and output variables to make predictions about new and unknown data. Various evaluation metrics are employed to assess the accuracy of the model. Mean Squared Error (MSE) as shown in Eq. 1 computes the average of the squares of the errors. The MSE value should be lower for better performance of the model.

Table 2 Summary of data collection phases

Phase	Objective	Procedure
Phase 1: Baseline benchmark reading	Obtain a benchmark reading to establish the average physiological state of the subject	Subjects remain relaxed without any specific activities. EEG and heartbeat data are recorded continuously for 2 min. This phase serves as a baseline for comparison in subsequent phases
Phase 2: Intense workout and relaxation	Assess the impact of intense physical activity on brain waves and heartbeat, followed by relaxation	Subjects perform a vigorous activity, running up and down stairs, to elevate heart rate. Continuous monitoring of EEG and heart rate data during and after the intense workout. Data collection continues until the subject's readings return to the baseline observed in Phase 1
Phase 3: Music-enhanced intense workout and relaxation	Evaluate the influence of music on relaxation during and after an intense workout	Subjects engage in a similar intense workout session as in Phase 2. Music, chosen by the subject and provided through the Google Form, is introduced during this phase. Continuous monitoring of EEG and heartbeat data while the subject engages in the activity. Data collection extends until the subject reaches a relaxed state post-exercise, comparing the impact of music on relaxation

$$\text{MSE} = \frac{1}{N}\sum_{i=0}^{N}(y_i - \hat{y}_i)^2. \tag{1}$$

Root Mean Squared Error (RMSE) measures the square root of the Mean Squared Error, providing insight into the gap between the predicted and actual values. Similar to MSE, a lower RMSE value indicates a better model as shown in Eq. 2.

$$\text{RMSE} = \sqrt{\frac{\sum_{i=1}^{N}(y_{\text{pred}} - y_{\text{ref}})^2}{N}}. \tag{2}$$

Mean Absolute Error (MAE), as shown in Eq. 3, computes the average absolute difference between the predicted and actual values, emphasizing more on the average prediction error, irrespective of the magnitude of the error.

Table 3 Performance of machine learning models on regression task

Model	MSE	RMSE	MAE	MAPE	R^2
Random forest	68.380	8.269	4.991	0.103	0.711
KNN_euclidean	67.395	8.209	2.822	0.062	0.716
KNN_manhattan	52.740	7.262	2.586	0.055	0.777
KNN_mahalanobis	55.572	7.455	2.917	0.060	0.765
SVM	193.943	13.926	11.026	0.213	0.181
AdaBoost	62.681	7.917	2.827	0.061	0.735

$$\text{MAE} = \frac{1}{n}\sum_{i=1}^{n} |y_i - \hat{y}_i|. \tag{3}$$

The coefficient of determination (R^2 score) measures the proportion of the variance in the dependent variable using Eq. 4. Its values range from 0 to 1, where 1 implies that the model's predictions are accurate.

$$R^2 = 1 - \frac{\sum_{i=1}^{n}(y_i - \bar{y})^2}{\sum_{i=1}^{n}(y_i - \hat{y}_i)^2}. \tag{4}$$

The R^2 score, a metric indicating the goodness of fit, revealed a strong correlation between the predicted and actual values. The obtained R^2 score, 0.77, signifies the model's capability to capture and explain variations in the effectiveness of recovery based on the selected features as shown in Table 3.

In the presented study, the selected Manhattan machine learning model, based on the K-Nearest Neighbors (KNN) algorithm, demonstrated a commendable performance in predicting the effectiveness of recovery strategies. The model was trained on a dataset encompassing individuals' weight, gender, and experimental data collected during a series of drills as shown in Table 4.

Figure 5 shows one of the human subjects undergoing the proposed experiment. One key aspect of our investigation involved the consideration of physiological responses, specifically BPM recovery time difference and EEG recovery time difference. The model successfully predicted EEG recovery time difference, providing valuable insights into the impact of music-induced relaxation during the cooldown phase as shown in Fig. 6.

Furthermore, the analysis of the effectiveness of recovery strategies, expressed as a percentage, allowed us to gauge the practical implications of the model's predictions. Users interacting with the analysis dashboard could input a specific song and receive a quantifiable measure of its predicted effectiveness for recovery. The incorporation of song features such as danceability, valence, speechiness, and loudness enriched the dataset, enabling a holistic exploration of the interplay between music characteristics and individual physiological responses. These results underscore the

Table 4 Features and significance in music track analysis

S. No.	Features	Significance
1	Time stamp	Important for temporal analysis
2	EXG Channel_0	Measures electrical activity of the brain
3	EXG Channel_1	Measures electrical activity of the brain
4	BPM	Reflects heart rate and emotional state
5	Age	May influence music preferences
6	Gender	May influence music preferences
7	Weight	May influence music preferences
8	Delta_tEeg	Measures sleepiness and relaxation
9	Acousticness	Reflects the musicality of the track
10	Danceability	Reflects the suitability for dancing
11	Energy	Reflects the intensity and liveliness of the track
12	Speechiness	Reflects the presence of spoken words
13	Valence	Reflects the emotional tone of the track

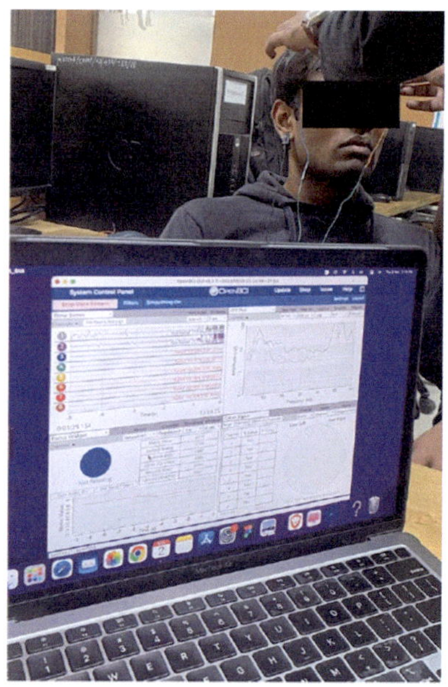

Fig. 5 Human subject undergoing proposed experiment

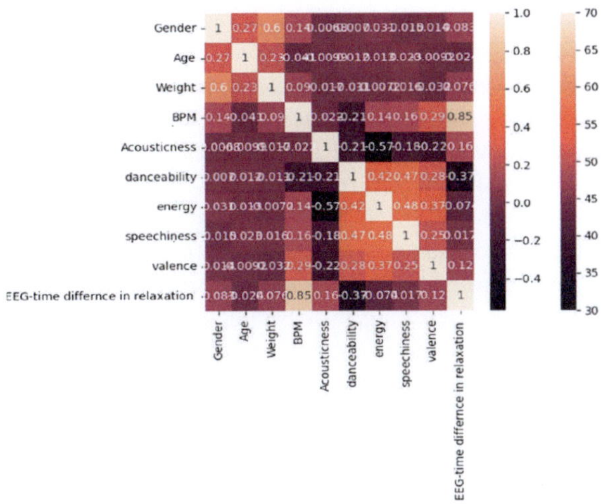

Fig. 6 Final confusion matrix

potential of our approach in tailoring recovery strategies to individual preferences and physiological profiles. The integration of machine learning, physiological data, and music features opens avenues for personalized interventions in post-exercise recovery. Future iterations of this research could explore additional factors influencing recovery effectiveness and refine the model to enhance predictive accuracy.

6 Conclusion

Our study on the potential of music-induced relaxation as a post-physical exertion recovery strategy yielded insightful findings and identified avenues for future research. Through meticulous experimentation involving 30 participants aged 18–22, we observed promising outcomes, including an average resting heart rate of 75.6 bpm and an average elevated heart rate of 103.8 bpm post-exercise. It was observed that heart rate returns to normal faster with slow music compared to no music. Leveraging a machine learning model based on the K-Nearest Neighbors (KNN) algorithm, we achieved notable predictive accuracy, with an R^2 score of 0.77 indicating a substantial correlation between predicted and actual values. This analysis highlighted the significance of physiological responses, particularly heart rate and EEG recovery time differences, in understanding the impact of music-induced relaxation during the cooldown phase.

Furthermore, the EEG data analysis supports the hypothesis that music-induced relaxation positively influences neural and cardiovascular systems, highlighting the importance of integrating music into recovery protocols to optimize overall well-being. Specifically, slow music likely enhances parasympathetic activation, facilitating a reduction in heart rate variability and promoting a faster return to homeostasis.

These findings underscore the potential of slow music in optimizing cardiovascular recovery processes post-exercise.

However, our study also encountered limitations that warrant consideration. The sample size, though balanced in gender and age distribution, may not fully capture the diverse range of factors influencing individual responses to music therapy. Additionally, the scope of our investigation was confined to a specific age group and recovery scenario, limiting the generalizability of our findings. Future research endeavors could address these limitations by expanding the dataset to include a more diverse population and exploring a broader range of recovery contexts.

There are exciting possibilities for further exploration in this field. Enhancing our understanding of the long-term effects of music therapy on recovery outcomes and overall well-being could provide valuable insights into its sustained benefits. Moreover, integrating advanced machine learning techniques and investigating the synergistic effects of music therapy with complementary modalities hold promise for optimizing personalized intervention strategies. By addressing these avenues for future research, we can continue to unlock music's full potential as a therapeutic tool for promoting relaxation and enhancing recovery from physical exertion, ultimately contributing to improved health and well-being.

References

1. Reddy NM, Rao C, Savitha D (2014) Effect of different musical tempo on post-exercise recovery in young adults. Int J Res Med Sci 2(2):409–414
2. Desai RM, Thaker RB, Patel JR, Parmar J (2017) Effect of music on post-exercise recovery rate in young healthy individuals. Int J Res Med Sci 3(4):896–898
3. Zhao X, Sun Y, Li J, Yan S (2019) Music-induced emotions recognition based on eeg signals using deep learning. In: 2019 IEEE International Conference on Acoustics, Speech and Signal Processing (ICASSP), pp 1893–1897
4. Zhang Y, Zhang Y, Reilly RB (2019) Music emotion recognition using multi-modal deep learning framework. In: 2019 IEEE International Conference on Acoustics, Speech and Signal Processing (ICASSP), pp 2079–2083
5. Fulton JM, Guzzo MS, Ely JE, Ugulava TD (2015) The effects of music on exercise performance. Sports Med 45(4):587–602
6. Ryota N (2024) Reliability for music-induced heart rate synchronization. Sci Rep 14(1):12200
7. Kabat-Zinn JD, Hassed CK, Santorelli MB, Segal RB (2010) Mindfulness-based stress reduction: a comprehensive meta-analysis. J Clin Psychiatry 71(3):120–131
8. Kumar S, Kumar A, Bhattacharya S (2022) Neural correlates of music-induced emotions: a functional magnetic resonance imaging study. Frontiers Human Neurosci 16
9. Li Y, Yang W, Wang Y (2023) Effects of music-based meditation on stress and anxiety: a randomized controlled trial. Complementary Therapies Med 60:103063
10. Chen CC, Lin CH, Lo AH (2023) Relationship between music listening and cognitive function in older adults: a longitudinal study. The Gerontologist 63(4):e141–e148
11. Gaia B, Giovanna L, Marco B, Ambra B, Alessandro B, Martina P, Carola C, Susanna M, Elisa P, Laura A (2023) How music moves us: music-induced emotion influences motor learning. Neuroscience 526:246–255

Exploring the Dynamics: Analyzing Trends in Wind and Solar Energy Production

Milind Shah, Dweepna Garg, Kinjal Gandhi, Jigar Sarda, Adit Jindal, Kush Jindal, Ankita Bhatt, and Jay Mochi

Abstract This research comprehensively analyzes energy production data from 2020 to 2023, focusing on generating energy from wind and solar sources within a particular nation. The use and influence of renewable energy sources within the nation's energy landscape may be better understood by analyzing trends and patterns that have manifested throughout this four-year period. To shed light on the oscillations and development trajectories of wind and solar energy generation, the approach involves collecting extensive data and statistical analysis of that data. According to the findings, there are noticeable changes in the energy generation from wind and solar, highlighting the changing roles these two sources play in the country's energy mix. The implications of these patterns are analyzed regarding the planning of future energy, sustainability, and policy.

M. Shah
Department of Computer Engineering, Sardar Vallabhbhai Patel Institute of Technology (S.V.I.T), Vasad, Gujarat, India

D. Garg (✉) · K. Gandhi
Department of Computer Engineering, Devang Patel Institute of Advance Technology and Research (DEPSTAR), Faculty of Technology and Engineering (FTE), Charotar University of Science and Technology (CHARUSAT), Anand, Gujarat, India
e-mail: dweeps1989@gmail.com

J. Sarda
Electrical Engineering, CSPIT, CHARUSAT, Anand, Gujarat, India
e-mail: jigarsarda.ee@charusat.ac.in

A. Jindal
Georgia Institute of Technology, Atlanta, USA

K. Jindal
Strawberry Field High School, Chandigarh, Punjab, India

A. Bhatt
Accenture Pvt Ltd, Bhubanneswar, India

J. Mochi
ControlCase LLC, Mumbai, India

H. Sharma et al. (eds.), *Artificial Intelligence: Theory and Applications*, Lecture Notes in Networks and Systems 5589, https://doi.org/10.1007/978-981-96-1687-9_11

Keywords Wind energy · Solar energy · Renewable energy · Energy generation · Trends · Sustainability

1 Introduction

By the mid-century, solar and wind energy utilization is anticipated to increase globally faster than any other energy source. Wind and solar energy are renewable, ubiquitous, environmentally friendly, and peaceful energy sources. Consequently, solar and wind energy are progressively being integrated into a wide range of applications and are currently dominating the alternative energy sector regarding renewable energy sources.

The social progress of every country is closely linked to its energy consumption and sustainability. Consequently, numerous nations have witnessed a substantial surge in energy requirements over the past few decades; this development has been fueled by industrialization, urbanization, and population expansion. The worldwide electricity generation demand was recorded at 8.35 TW in 2020, and forecasts indicate it will increase to 13.7 TW by 2050. Global electric energy demand is projected to increase by nearly 50 percent over the next three decades, according to the International Energy Agency (IEA). This increase will be predominantly driven by population and economic development, especially in developing Asian countries. Environmental contamination and the possibility of energy shortages are consequences of this escalating energy demand. Notably, the combustion of fossil fuels to generate electricity contributes significantly to air pollution. Moreover, the dependence on electricity generated from fossil fuels is untenable in light of the limited availability of these resources. Therefore, environmentally sustainable renewable energy sources, including solar and wind power, are urgently required to tackle these challenges effectively.

The utilization of renewable energy (RE) sources has experienced substantial growth in recent times, propelled by the escalating worldwide energy requirement and increased consciousness regarding climate change. Over the last two decades, solar energy has witnessed significant expansion on a global scale, with particular emphasis on countries located within the solar belt, including India and China. Photovoltaic (PV) and concentrated solar power (CSP) systems have become the predominant solar technologies. Solar photovoltaic (PV) has become increasingly prominent due to its ability to capture diffuse horizontal irradiance and direct normal irradiance and its comparatively low energy production costs. In addition, wind energy has become the second most prevalent form of renewable energy, contributing 7% to the electricity supply in the United States and accounting for 4% of global electricity generation. Due to their abundance and cleanliness, solar and wind energies are essential to the transition to RE sources. As the expenses of solar and wind energy technologies decrease, they have emerged as progressively more feasible alternatives for fulfilling energy needs.

Panel, including grime, air pollution, and shadow. Several photovoltaic technologies in the Middle East's Taklimakan and Gobi arid regions are adversely affected by dust, solar radiation, and relative humidity. Understanding these environmental challenges in a specific geographic area is critical for implementing renewable energy systems and estimating their efficiency and output. Furthermore, understanding these obstacles is crucial for designing and modeling wind and photovoltaic systems, as well as atmospheric energy balance, climatology, and air quality research. Given the prevalence of these challenges in arid environments, it is probable that both passive and active solutions can be implemented, contingent upon the unique conditions of such regions. Concerning solar energy development in the area, social community engagement and sustainable development are topics of discussion, as is the need to enhance the human capital necessary to address the social and technical challenges associated with alternative solar energy. Furthermore, potential challenges to wind energy necessitate a meticulous evaluation of wind attributes, including the mean wind speed and frequency distribution.

Increasing energy demand in the Gulf Cooperation Countries (GCC)—Kuwait, Qatar, Bahrain, Saudi Arabia, the United Arab Emirates, and Oman—increases the likelihood of electricity shortages, load peaks, and the depletion of hydrocarbon resources. Implementing alternative and ecologically sustainable energy systems in Kuwait will result in improved power supply efficiency, an updated electrical infrastructure, increased employment prospects, greener surroundings, and a more favorable international reputation.

The paper consists of six sections. Section 1 is the Introduction; Sect. 2 discusses about Research Objectives; Sect. 3 focuses on Survey of Literature; Sect. 4 deals with Methodology; Sect. 5 includes Results and Analysis; Sect. 6 is Conclusion which concludes the paper.

2 Research Objective

The objective of this paper is to offer a thorough analysis of the historical development of energy generation from various sources. We will analyze not only the yearly and monthly variations, but also scrutinize the data to comprehend how output varies on different weekdays and during the hours of the day.

3 Survey of Literature

Wind energy generates around two times as much electrical energy as solar radiation on average in Germany. Furthermore, there is a discernible annual cycle whereby the proportion of generated power is balanced during the summer and 70/30 wind/solar in the winter [1].

This research demonstrates how multivariate and data visualization methods can be utilized to depict the intricate network of relationships between the development of wind and solar energy and a variety of socioeconomic, political, and environmental variables, as well as to determine the positioning of economies based on the dynamic interplay between these variables. The proposed method is founded on an approach to reducing dimensionality that is capable of simultaneously processing ordinal and numeric variables and can account for nonlinearities in their relationships. By employing this methodology, the present study additionally seeks to offer researchers an alternative means of discerning crucial attributes that influence the placement of economies [2].

The research employs a holistic methodology, which includes spatial analysis to determine the suitability of land, techno-socio-economic feasibility analyses, and machine learning-based forecasts of the variability and trends of renewable energy (RE) resources. Using Geographic Information Systems (GIS), spatial analysis is utilized to determine appropriate locations for constructing solar photovoltaic (PV) and wind farms, utilizing multi-criteria evaluation techniques. Using the System Advisor Model (SAM), a techno-socio-economic feasible assessment for solar PV and wind energy sources under different setups and scenarios is performed. Additionally, a machine learning algorithm known as Support Vector Machine, or SVM, is created and implemented to predict significant meteorological variables, including wind speed, Global Horizontal Irradiance (GHI), and wind speed, at various locations throughout South Africa [3].

Integrating large-scale solar and wind energy sources into the power grid fundamentally depends on studying their combined application. Consequently, by analyzing various methodologies and locations, this work aims to critically review the most recent approaches to comprehend and evaluate the complementary nature of connected to the grid solar and wind power systems. Integrating wind and solar energy sources typically yields a more consistent power provision, a critical factor in ensuring the functionality and security of electrical infrastructure globally [4].

The principal objective of this investigation is to enhance the current renewable energy provision to fortify the dependability of power grid components. The V2G concept expands the intelligent charging system, enabling electric vehicles to discharge battery energy into the power infrastructure. In doing so, these vehicles function as distributed generators or energy storage systems. This analysis demonstrates how battery storage, as well as parallel V2G storage, contribute to the power grid. Moreover, decentralized V2G battery storage will be incorporated into future renewable energy systems, according to the review [5].

This study examines various algorithms for deep learning (DL) implemented in solar and wind energy resources. It assesses the effectiveness of these algorithms using an original taxonomy. Additionally, it provides an exhaustive review of the current body of literature that supports an evaluation and performance assessment of DL techniques. Furthermore, it addresses significant obstacles and prospects for comprehensive research. Variations in precision, accuracy, and robustness, in addition to the capacity for generalization, appear to be the most prevalent obstacles to implementing DL techniques, according to the findings. When confronted with large datasets, DL techniques exhibit a markedly superior performance compared to alternative CI techniques. Nonetheless, emphasis should be placed on developing and implementing hybrid DL techniques in conjunction with additional optimization methods to improve the structure of the techniques. Hybrid networks consistently exhibit superior performance compared to single networks because hybrid techniques leverage the strengths of two or more prediction methods. Mixed methods are advised for use in DL techniques [6].

As a result of the aforementioned environmental challenges, the research found that photovoltaic, or PV, cells are not the most suitable source of energy for Kuwait; thus, alternative forms of renewable energy are regarded as more feasible. The results of one year of operation at wind fields and solar units in Kuwait demonstrate that wind energy generates more than the industry average. This technology's sustained high capacity factors led to a yearly power production that was 2.3 times greater than that of photovoltaics (PV). It could supply energy to 450 households, whereas PV could only power 199 homes [7].

Since 2021, research in this domain has been advancing rapidly in China. Four topics will be prominent in this discipline between 2013 and 2022: The utilization of power-to-gas (P2G) technology in integrated energy systems and renewable energy consumption; the integration of heat and power through fuel cell combined heat and power systems; the preparation, performance, and application of hydrogen storage materials (mainly hydrogen storage alloys); and the catalyst employed in the electrolysis of water to produce hydrogen and the hydrogen evolution reaction [8].

This study determines that a 30 kW solar project would generate 150 energy per day, given that a 5 kW solar plant can produce five daily units. Although some units are lost due to systemic losses, the system as a whole can generate 27 kW of power and supply it to the grid. Voltage loss on both the AC and DC sides does not exceed the 2% limit allowed on either side, as the voltage drop on the DC side is 0.5% under the design parameters, while the voltage drop on the AC side is 0.7%. MPPT accepts a maximum input current of 26.85 A, close to the rated input current of 33A and, therefore, appropriate for higher efficiencies. The rated voltage of the string is approximately 580 V, which falls between 600 V and the midpoint of the input voltage range. All bus voltages at the Breshnakot and distribution substations fall within the permissible range of 95 to 100%. The simulation above illustrates the percentage loads on each bus; no bus is overburdened, under voltage, or overwhelmed [9].

The objective was to assess the advantages of solar and wind power in the Texas Panhandle region. To achieve this, the AOC 15/50 50 kW wind turbine system at AEI WTC and the PV system at PDRC served as the research data collection and analysis foundations. By examining the gathered data in conjunction with the PV WATTS calculator, it was possible to ascertain that the annual energy outputs of both systems amounted to 228,531 kWh for the solar system and 81,581 kWh in total. The return periods for wind and solar energy sources were approximated to be 13 years and 19 years, respectively [10]

In this paper, the operational efficacy of both RESs is technically evaluated concerning rotor angle stability. MATLAB/Simulink was implemented as the modeling environment. The findings indicate that wind turbines exhibit superior performance compared to solar PV systems, but the latter can supply the grid with more consistent power. This is in contrast to wind turbines, which necessitate implementing battery energy storage systems (BESS) and supplementary frequency support to generate stable power [11].

The electric heater can significantly mitigate wind curtailment and enhance the system's overall stability. Additionally, the findings indicate that the hybrid system featuring a more extensive thermal storage system's capacity and smaller solar multiple exhibits superior efficacy in mitigating wind curtailment. Moreover, since the solar considerable remains constant, the average power change rate diminishes as the capacity of the thermal storage system increases. As the solar multiple increases, the rate at which the average power change rate decreases also increases [12].

This study examines the necessity for adequate resource modeling to incorporate snow by analyzing the present and anticipated PV capacity of utilities. At the state level, mid-case and high-penetration usefulness PV scenarios were analyzed and compared to the annual average snowfall. With utility PV compound annual growth rates of 5% or greater,

Colorado, Michigan, New York, Wisconsin, and Minnesota stand out as high-snowfall states in both scenarios.

Despite the findings indicating an inverse correlation between snowfall and predicted photovoltaic (PV) expansion by state, certain states characterized by substantial snowfall may still witness substantial growth in utility PV capacity throughout the subsequent 28 years. A service-level analysis was conducted in conjunction with the state-level analysis, utilizing NSIDC snow data and utility IRP forecasts. According to the findings of this stage, utility PV projections in the northern Midwest could reach 6500 MW by 2040. Even utilities that have lesser projections for photovoltaics exhibit higher cumulative yearly growth rates [13].

This research presents evidence of reduced mean wind velocities as reported by nearby meteorological stations, as well as superior energy production from a 160W solar panel in comparison with a 500W wind turbine. The prognostic study reveals that there is not much agreement among the models regarding the anticipated change; however, the results underscore the necessity of conducting more comprehensive analyses of future scenarios in order to accurately model both regional and local circulation. A comprehensive social, economic, and political analysis reveals the distinct ramifications of climate change on various stakeholders, including those associated with micro-turbine systems. The findings of this research culminate in suggestions for enhancing energy and wind speed prognostic studies, as well as adjustments to the stakeholder interactions [14].

4 Methodology

4.1 Dataset Description

During the period beginning in the year 2020, this dataset is comprised of wind and solar energy production records, which are recorded on an hourly basis for the French grid. The major objective of this initiative is to provide the Commission de Régulation de l'Énergie (CRE) with the ability to compute the reference price that is utilized in the computation of supplementary compensation for the wind and solar manufacturing industries [15].

4.2 Data Cleaning

The dataset that will be used for our research will be prepared in this phase.

4.2.1 Data Attribute

When acquiring the dataset, make sure that the "Date" attribute is identified as a date data type by using the "parse_dates" argument for the "Date" attribute.

At one point in time, the dataset contained information such as the date and the hour. We removed the "Date and Hour" feature in order to simplify the analysis and minimize down on unnecessary repetition within it. We have instead divided the "Date" attribute into three components: the "Year," the "Month," and the "Day" components. The analysis of annual and monthly patterns will be further simplified as a result of this separate.

4.2.2 Handling Missing Values

Specifically, we have found two entries that are missing from this dataset, which is a relatively small amount of missing values. In light of the fact that these missing data have minimal impact on our research, we shall proceed with the removal of these numbers.

5 Results and Analysis

In this research, the comparative analysis of wind and solar energy production has been implemented and it has been found that only 19% energy has produced by solar whereas wind has produced 81% and this analysis was done on Renewable Daily

Electricity Production data which was collected in France country in the year 2020. This research undertakes an examination of energy production data encompassing the years 2020 to 2023, intending to clarify trends and patterns in wind and solar energy generation within the nation. Providing a detailed examination of the development of energy production from various sources throughout the span of time is the objective of the research. Not only did we analyze the yearly and monthly fluctuations, but we analyzed the data in order to gain an understanding of how production varies on different days of the week and during the hours of the day.

We want to discover significant findings that can add to our knowledge of the dynamics of renewable energy sources in The Netherlands by comparing energy production from wind and solar power side by side. These findings can be used to influence sustainable energy initiatives.

Figure 1 shows pie chart showing the distribution of produced energy by wind and solar.

Figure 2 illustrates the analysis of energy production from 2020 to 2023. It displays two parameters: production and date. The blue color represents solar analysis, while wind analysis is represented in red.

Figure 3 illustrates the analysis of annual energy production from the year 2020–2023. The graph displays two parameters: production and date. The blue color represents solar analysis, while wind analysis is represented in red.

Figure 4 illustrates the analysis of average energy production by month from 2020 to 2023. The graph displays two parameters: average production and month. Solar analysis is represented in blue, while wind analysis is depicted in red.

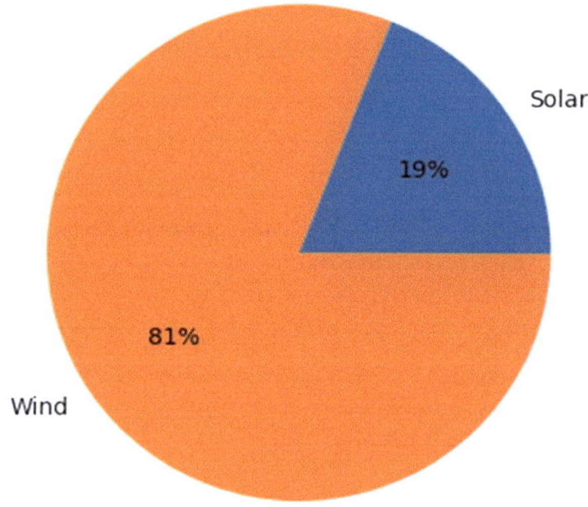

Fig. 1 Analysis showing energy produced by wind and solar

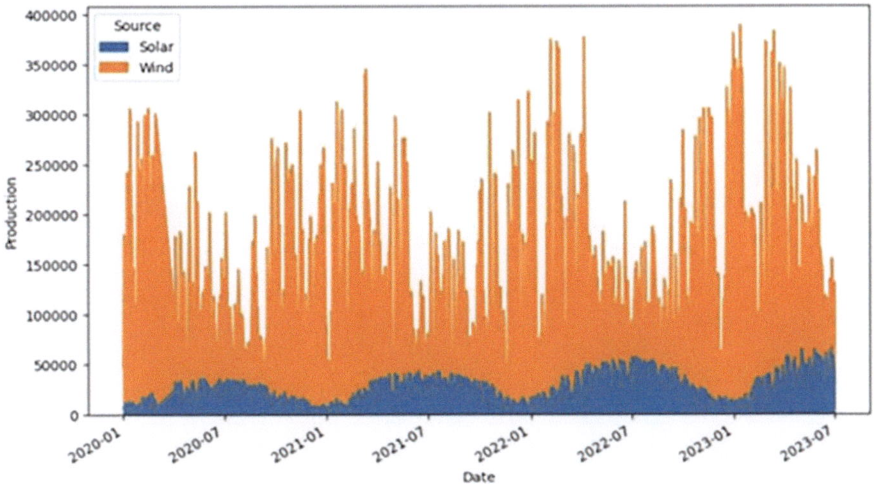

Fig. 2 Energy production source 2020–23

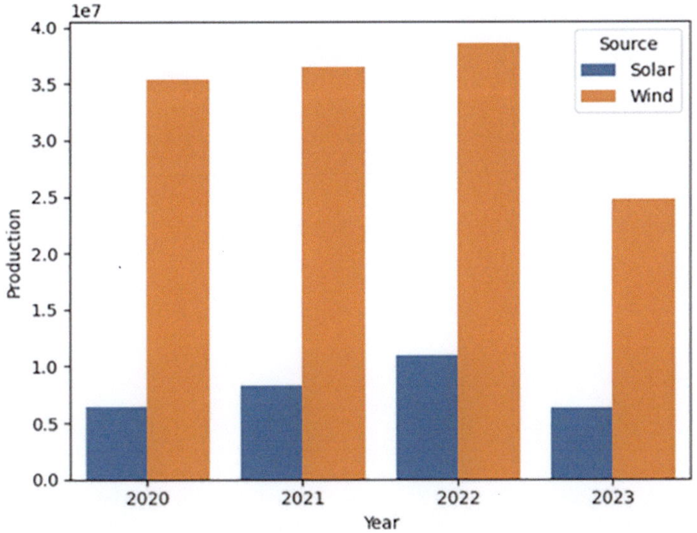

Fig. 3 Annual energy production 2020–23

Figure 5 illustrates the analysis of average daily energy production from 2020 to 2023. The graph displays two parameters: average production and month. Solar analysis is represented in blue, while wind analysis is depicted in red.

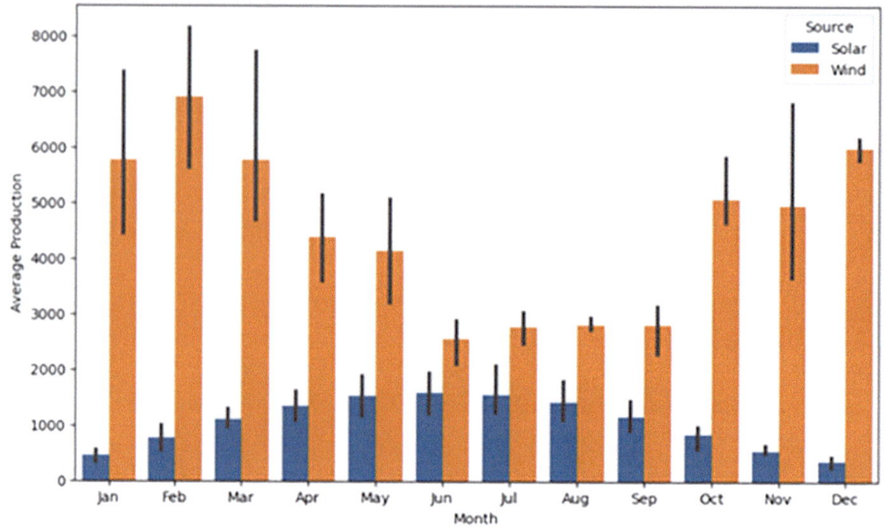

Fig. 4 Average energy production by month 2020–23

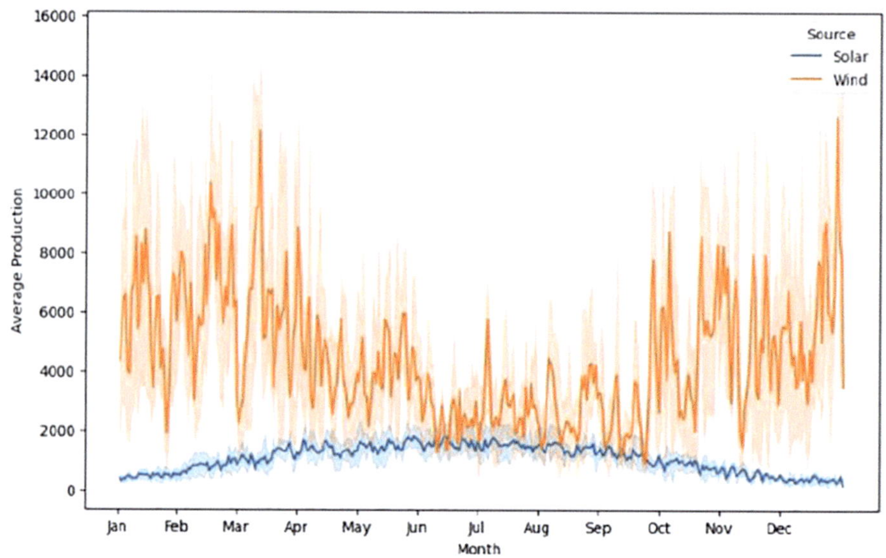

Fig. 5 Average daily energy production 2020–23

6 Conclusion and Future Work

In conclusion, the findings of our analysis of the data related to energy production from the years 2020 to 2023 have provided us with substantial insights into the patterns and trends of wind and solar energy collection. Over the past four years, we have witnessed a significant rise in the production of energy from both wind and solar sources, which seems to indicate a good trend toward the utilization of renewable energy sources. In addition, research in the future could analyze the possibility of incorporating cutting-edge technology, such as energy storage systems and smart grids, to improve the effectiveness and dependability of renewable energy systems. Overall, this research contributes to the existing body of knowledge on the dynamics of renewable energy and offers significant insights that can be used to influence decision-making and drive progress toward a more sustainable energy future in The Netherlands and worldwide.

References

1. Drücke J et al (2021) Climatological analysis of solar and wind energy in Germany using the Grosswetterlagen classification. Renew Energy 164:1254–1266
2. Carty C, Claveria O (2024) A synergistic analysis of solar and wind energy deployment in Europe. Environ Dev 49(December 2023):100967
3. Imam AA, Abusorrah A, Marzband M (2024) Results in engineering potentials and opportunities of solar PV and wind energy sources in Saudi Arabia: land suitability, techno-socio-economic feasibility, and future variability internal rate of return. Results Eng 21(December 2023):101785
4. Weschenfelder F et al (2020) A review on the complementarity between grid- connected solar and wind power systems. J Clean Prod 257:120617
5. Lehtola T, Zahedi A (2019) Solar energy and wind power supply supported by storage technology: a review. Sustain Energy Technol Assessments 35(February):25–31
6. Shamshirband S, Rabczuk T, Chau KW (2019) A survey of deep learning techniques: application in wind and solar energy resources. IEEE Access 7:164650–164666
7. Al-Dousari A et al (2019) Solar and wind energy: challenges and solutions in desert regions. Energy 176:184–194
8. Xiao Z, Zhang Z, Yu Y, Ran M (2023) Analysis of the development trend of China's electricity-hydrogen energy technology research based on bibliometrics. In: 2023 6th International conference on energy, electrical power and engineering CEEPE 2023, pp 1511–1516
9. Sherzay S, Perveen R (2021) Analysis of tie grid 30 kW solar project. In: 2021 IEEE 6th International conference computing on communication and automation ICCCA 2021, pp 725–729
10. Chang B, Starcher K (2019) Evaluation of wind and solar energy investments in Texas. Renew Energy 132:1348–1359
11. Jhumka AA, Ah AK, Khoodaruth A, Ramasawmy C (2022) Comparative performance analysis of solar energy and wind energy systems using rotor angle stability. In: ENERGYCON 2022–2022 IEEE 7th international energy conference on proceedings, no Energycon, pp 1–5
12. Ding Z, Hou H, Yu G, Hu E, Duan L, Zhao J (2019) Performance analysis of a wind-solar hybrid power generation system. Energy Convers Manag 181(September 2018):223–234
13. Wickett S, Dyreson A (2023) Trends in solar PV growth in snowy climates and impact on resource adequacy. In: 2023 IEEE 50th Photovoltaic Specialists Conference, pp 1–10

14. Benzo C, Singh P, Hunt I, Ortega A (2020) Whole systems assessment of current and future wind speed and energy trends in rural Peru. In: 2020 IEEE global humanitarian technology conference (GHTC)
15. Upton H (2023) Wind and solar daily power production, Kaggle. [Online]. Available: https://www.kaggle.com/datasets/henriupton/wind-solar-electricity-production. Accessed: 20 July 2024

Cyberbullying Detection in Low Resource Code-Mixed Languages Using ML and NLP Techniques

Nidhi Saralaya, P. P. Jashma Suresh, and Sridevi Saralaya

Abstract Cyberbullying has become a pervasive issue on social media platforms, necessitating effective detection methods across diverse linguistic contexts. This study proposes a methodology for cyberbullying detection in Kannada, Kannada written in English, and English texts. Leveraging machine learning and natural language processing techniques, we develop and evaluate three types of models: traditional (Random Forest, Support Vector Machine), neural network (Single-Layer Perceptron, Multi-layer Perceptron), and deep neural network (Long Short-Term Memory, BERT, IndicBERT). Results show that deep learning models, particularly those leveraging pre-trained language models, exhibit superior performance in detecting offensive language, contributing to the ongoing efforts to combat cyberbullying in contemporary discourse. The model BERT outperforms the Random Forest model by 5.40 in accuracy and 13.79 in $F1$-score. Similarly, the Semi-Supervised model surpasses the Logistic Regression model by 10.62 in accuracy and 19.37 in $F1$-score, highlighting significant improvements. Additionally, an analysis of the underlying mechanisms highlights the inherent strengths of deep learning architectures, such as their ability to capture intricate linguistic patterns and contextual nuances, as well as model long-range dependencies within textual data, thereby enhancing their discriminatory capabilities in identifying cyberbullying behavior across multilingual and cross-script contexts.

Keywords Natural language processing · Cyberbullying detection · Machine learning · Neural network · Deep neural network models · Kannada language detection · Offensive language detection

N. Saralaya (✉) · P. P. Jashma Suresh
Department of Computer Science and Engineering, Manipal Institute of Technology, Manipal Academy of Higher Education, Manipal, Karnataka, India
e-mail: nidhisaralaya@gmail.com

P. P. Jashma Suresh
e-mail: jashma.suresh@manipal.edu

S. Saralaya
Department of Computer Science and Engineering, St. Joseph Engineering College, Mangaluru, Karnataka, India
e-mail: sridevis@sjec.ac.in

H. Sharma et al. (eds.), *Artificial Intelligence: Theory and Applications*, Lecture Notes in Networks and Systems 5589, https://doi.org/10.1007/978-981-96-1687-9_12

1 Introduction

As social media usage has evolved, so has the unfortunate capability for cyberbullying through these platforms. The anonymity and virtual nature provided by platforms like Twitter, Snapchat, and Instagram have led to the proliferation of cyberbullying [1]. In the past ten years, cyberbullying has increased a lot in different languages, including those spoken in Asian countries like India, where many people use mobile phones. Given the broad target audience susceptible to cyberbullying, it is imperative to detect such behavior across diverse linguistic contexts. The literature review uncovered substantial research efforts in cyberbullying detection, particularly in English and local languages spoken in different parts of the world. However, there is scarcity of studies focusing on local languages written in English. One such Indian language is Kannada spoken in the state of Karnataka. During literature review it was found that there is absence of studies which emphasize on pure Kannada and Kannada text written in English. Consequently, this study proposes a methodology for cyberbullying detection in Kannada, Kannada written in English, and English. In this study, we create a model that can identify both cyberbullying and non-cyberbullying text. Given the escalating prevalence of cyberbullying alongside the expanding use of social media, the developed model employs machine learning and natural language processing techniques to identify instances of cyberbullying in text conversations. It is anticipated that the implementation of this model will prove invaluable in combating cyberbullying in contemporary discourse.

Additionally, code-mixed language, where two or more languages are used in the same discourse, presents a unique challenge for natural language processing tasks. This phenomenon is particularly prevalent in multilingual societies like India, where speakers frequently switch between languages. Code-mixing occurs due to various sociolinguistic factors, including cultural identity, social context, and language proficiency. In such cases, conventional language processing models may struggle to accurately interpret and classify text, as they are usually trained on single-language datasets. However, recent advancements in natural language processing, including models like BERT and IndicBERT, have shown potential in handling code-mixed data by capturing intricate linguistic patterns and contextual nuances present in multilingual communication. This study seeks to enhance cyberbullying detection in diverse linguistic environments by addressing the complexities of code-mixed language, ultimately contributing to more inclusive and accurate models for combating online harassment.

The remainder of the paper is structured as follows. In Sect. 2 we discuss related work, methodology is discussed in Sect. 3, results in Sect. 4 and we conclude in Sect. 5.

2 Related Work

In recent years, the escalation of cyberbullying has spurred the exploration of methodologies integrating natural language processing (NLP) and machine learning (ML) to detect and mitigate abusive content online. Notably, the study by Mehendale, Ninad, [2] utilize real-time social media data to train ML algorithms, including Linear SVC, Logistics Regression, and Multinomial Naive Bayes, for cyberbullying detection across multilingual contexts, particularly in languages like Hindi and Marathi. These efforts underscore the importance of diverse linguistic datasets and multilingual approaches in enhancing cyberbullying detection system accuracy and efficiency. Pawar and Raje present a Multilingual Cyberbullying Detection System focusing on Hindi and Marathi [3], essential languages given the surge in cyberbullying incidents, notably in regions like India with a large mobile user base. Achieving up to 97% accuracy and 96% $F1$-score, the system addresses the scarcity of research in Indian languages, emphasizing tailored approaches. Leveraging ML techniques for classification, the system employs data scraping, manual labeling, and synthesis to rectify imbalances in bullying instances, with Logistics Regression emerging as the top-performing algorithm. The study highlights the system's potential in combating cyberbullying across languages, proposing future extensions such as larger dataset validation and sentiment analysis integration.

Moreover, advancements in deep learning models, such as BERT and MuRIL, have further propelled research in hate speech and offensive language identification, particularly in code-mixed texts. Pawar and Raje [4] demonstrates the effectiveness of models like MuRIL in detecting offensive language in Tamil and Malayalam code-mixed texts, emphasizing the importance of tailored approaches for diverse linguistic contexts. Additionally, initiatives like KanCMD, as highlighted in [5], contribute to the development of datasets for sentiment analysis and offensive language detection in under-resourced languages like Kannada, fostering research in real-world code-mixed social media text. Despite these advancements, challenges persist, including dataset scarcity and the need for nuanced feature extraction techniques, necessitating continued research efforts to effectively combat cyberbullying and online harassment.

Significant progress has been made in natural language processing (NLP) for Indian languages, aided by libraries like iNLTK, Indie NLP, and Stanford NLP, which offer valuable tools and resources. However, challenges such as the lack of corpora and the need for better communication between linguists and programmers persist, hindering the development of robust NLP tools [6]. Efforts toward machine translation from Sanskrit to Hindi highlight the importance of leveraging innovative approaches like zero-shot translation to overcome translation challenges and foster inclusivity in digital communication platforms.

The study by Shannaq, Fatima [7] focuses on developing an optimized offensive language detection model for classifying Arabic tweets into offensive and non-offensive categories. The methodology comprises two stages of optimization: fine-tuning word embedding models and applying GA-based algorithms (a hybrid of GA and SVM or XGBoost). The dataset preparation involved collecting tweets from various domains using the Twitter streaming API and annotating them manually. Text normalization was performed to remove noise, followed by feature extraction using various non-contextualized pre-trained word embedding models. The first optimization stage involved fine-tuning the models with the ArCybC training dataset, while the second stage optimized hyperparameters using GA.

Experimental results showed performance improvements across all phases, with SVM consistently outperforming other models and XGBoost showing enhancements in specific scenarios. The study underscores the effectiveness of feature engineering and model optimization in offensive language detection and suggests future exploration of transfer-based models like AraBERT for further improvements.

Yang's study [8] explores natural language processing (NLP) with convolutional neural networks (CNNs) and semi-supervised algorithms. The research defines NLP, examines CNN and semi-supervised methods, and evaluates word vector feature recognition in deep learning optimization. Results show improved prediction accuracy with merged word vectors and semi-supervised learning. The study reviews relevant research, emphasizing NLP's evolving landscape and challenges like Chinese question analysis. Additionally, Yang discusses CNN's structural components, highlighting its depth and weight sharing strategy to address parameter explosion. The training methodology, predominantly based on Backpropagation (BP), is outlined, along with the learning process of semi-supervised algorithms. This comprehensive approach offers insights into NLP optimization using deep learning techniques.

Despite these advancements, challenges persist, including dataset scarcity and the need for nuanced feature extraction techniques, necessitating continued research efforts to effectively combat cyberbullying and online harassment.

3 Methodology

The methodology presents a structured framework guiding the research process, encompassing key stages from data acquisition to model evaluation, ensuring reliability, and reproducibility. The stages include data collection, preprocessing, feature selection, model training, integration, and implementation, adhering to best practices in data science and machine learning. This framework is illustrated in Fig. 1, which visually represents the flow and connection between each stage. In this study, three types of models have been implemented: traditional models (e.g., Random Forest (RF), Support Vector Machine (SVM)), neural network models (e.g., Single-Layer Perceptron (SLP), Multi-layer Perceptron (MLP)), and deep neural network models (e.g., Long Short-Term Memory (LSTM), BERT, IndicBERT). Each model type offers distinct advantages and addresses specific research objectives, providing a comprehensive exploration of the data [9] and enabling a thorough evaluation of performance across different methodologies. The models were selected after analyzing the studies obtained from prior research works [10–37].

Fig. 1 Process flow

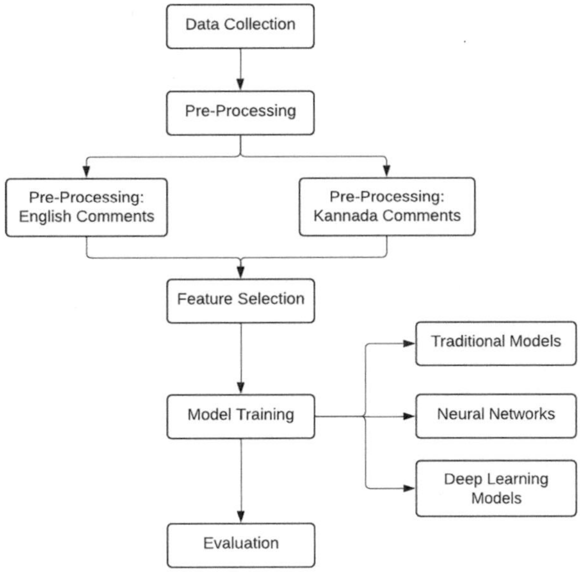

3.1 Data Collection and Preprocessing

The dataset was obtained by extracting comments from YouTube using the YouTube Data API. Python code utilizing the API key was written to retrieve comments and replies associated with a specific video ID. The comments were then stored along with their corresponding replies in a structured format. Each comment was manually labeled based on the presence of offensive language, with a label of 1 indicating the presence of offensive words and 0 indicating their absence. The comments were in English and also comments written in Kannada, requiring special handling during preprocessing.

The process of gathering data entailed loading the dataset from CSV files, which included labels and comments. Preprocessing stages for traditional classifiers included managing missing values and using text cleaning methods such eliminating stop words and punctuation. Tokenization was used to separate the text into discrete tokens, or words. Lemmatization and stemming were used to break words down into their most basic forms. Identical preprocessing procedures were used for neural networks, along with extra tokenization and padding to guarantee consistent input size. To ensure comprehensive preparation of the textual data, deep learning models such as BERT used specific tokenizers to handle complicated text features and language nuances.

3.2 Feature Selection and Encoding

Text data was quantitatively represented by traditional classifiers using feature engineering techniques like word embeddings and TF-IDF vectorization. TF-IDF vectors were developed to draw attention to the importance of words in the dataset. Additional preparation measures, including transliteration or translation to English, were needed for Kannada comments in order to resolve linguistic disparities. Word embeddings were used in neural networks to map letters into a continuous vector space. Word2Vec and GloVe, two pre-trained embeddings, were used to capture the semantic links between words. Contextual embeddings were utilized by deep learning models to comprehend the meaning of each word in a phrase. The input formats for tokenized sequences were customized to fit the model's framework.

3.3 Model Training

Traditional classifiers used preprocessed data to train a variety of algorithms, including Random Forests (RF), Decision Trees (DTs), and Naive Bayes (NB). To maximize performance, hyperparameters were adjusted through the use of strategies like grid search and cross-validation. TensorFlow or PyTorch systems were used to build model architectures for neural networks, such as feedforward neural networks and recurrent neural networks (RNNs). To improve the efficiency of the model, hyperparameters including the number of layers, units per layer, and activation functions were adjusted. Transfer methods for learning were used to fine-tune models that were previously trained on the labeled dataset, which were then used by deep learning models like IndicBERT. Model parameters were updated during model training using optimization techniques such as Adam or RMSprop, which were trained using training data.

3.4 Integration and Model Evaluation

After model training, integration and implementation involved evaluating the models on the test dataset to assess their performance. Traditional classifiers were evaluated using metrics such as accuracy, precision, recall, and $F1$-score. Neural networks were assessed based on similar metrics, with additional attention to training and validation losses. Deep learning models like BERT were evaluated on various performance metrics including accuracy, precision, recall, and $F1$-score, along with regression metrics like RMSE and R-squared. Semi-supervised learning techniques, such as pseudo-labeling, were explored to leverage unlabeled data for improved model performance.

4 Results

The evaluation of offensive word detection models in Kannada texts encompasses three categories: traditional models, neural networks, and deep learning models. Key performance metrics, such as accuracy, precision, recall, and $F1$-score—all of which are extensively accepted and applied in the literature on machine learning—are used to evaluate the performance of each category. These parameters provide an in-depth analysis of the models' capacities by offering a thorough grasp of how well the models identify and classify data.

4.1 Accuracy

The integrated approach demonstrates commendable accuracy in detecting and categorizing offensive words within Kannada texts. As the main measure of the detection system's effectiveness, accuracy is calculated as the ratio of successfully identified offensive instances to all instances.

4.2 Precision and Recall

Metrics such as precision and recall provide more insights into the effectiveness of the combined strategy. By calculating the ratio of accurately recognized offensive cases to the total anticipated offensive cases, precision measures the model's ability to reduce false positives. Recall, on the other hand, evaluates the model's ability to identify genuine offending incidents while avoiding the detection of offensive content. Robust detection systems with few false alarms and missed detections are indicated by high precision and recall values.

4.3 F1-Score

A comprehensive assessment of the model's total performance is given by the $F1$-score, which is the harmonic mean of precision and recall. A detection method that effectively balances precision and recall can minimize false alarms and reduce missed detections, as indicated by a high $F1$-score.

4.4 Comparison and Analysis

The comparative analysis reveals distinct performance characteristics among the three categories of models. Table 1 presents and compares all of the data, giving a thorough rundown of the model performance indicators. Traditional models exhibit a modest level of accuracy and F1-score, with the highest scores going to Support Vector Machine (SVM) and Logistic Regression (LR). However, these models exhibit limited performance in capturing complex linguistic nuances (see Fig. 2).

Neural networks showcase improved performance metrics, with Multi-layer Perceptron and Artificial Neural Network exhibiting higher accuracy and $F1$-score compared to traditional models (see Fig. 3). Nevertheless, these models still struggle with handling contextual variations in offensive language.

Deep learning models, particularly Bert and IndicBert, demonstrate superior performance across all metrics, outperforming both traditional models and neural networks (see Fig. 4). Pre-trained language models greatly improve the model's capacity to identify complex patterns in offensive language. Furthermore, the best

Table 1 Performance metrics of offensive word detection models

Model	Accuracy (in %)	Precision (in %)	Recall (in %)	$F1$-score (in %)
Multinomial Naive Bayes	71.30	77.05	35.61	48.70
Support vector machine	71.88	79.66	35.61	49.21
Random forest	71.30	73.24	39.39	51.23
K-nearest neighbors	64.93	92.31	9.09	16.55
Decision Tree	66.96	58.04	49.24	53.28
Logistic regression	71.88	81.82	34.09	48.13
Single-layer perceptron	61.45	20.48	33.33	25.37
Multi-layer perceptron	69.75	49.48	41.33	40.37
Artificial neural network	70.56	49.48	41.33	41.37
Feedforward neural network	68.11	49.48	40.03	38.62
Recurrent neural network	62.02	31.40	50.37	38.58
Gated recurrent unit	62.89	54.84	35.73	28.58
LSTM	61.45	20.48	33.33	25.37
BERT	66.70	64.35	65.70	64.94
IndicBERT	66.67	57.03	55.30	56.15
Semi supervised	82.50	83.03	57.35	67.50

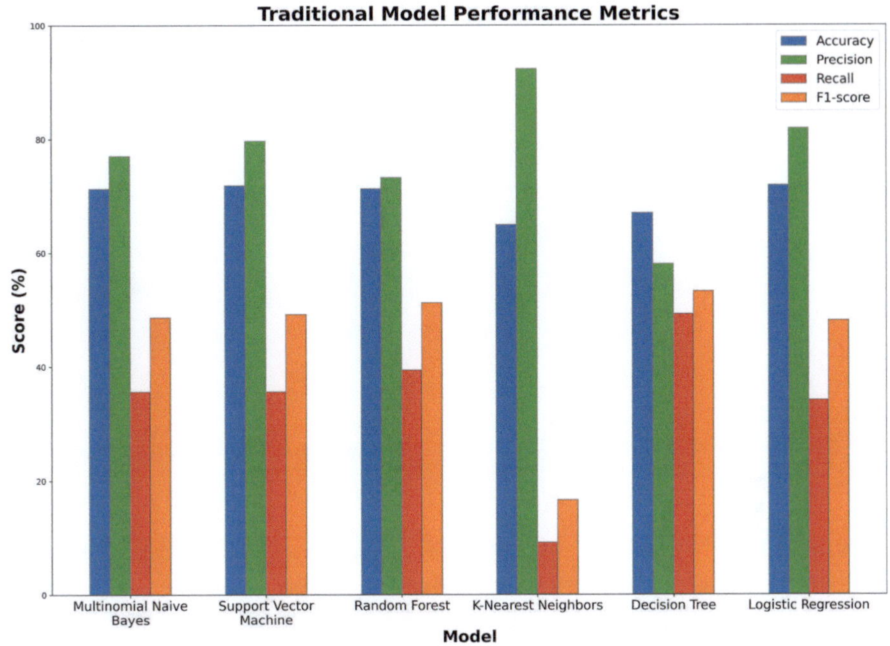

Fig. 2 Accuracy, precision, recall, and $F1$-score of traditional models

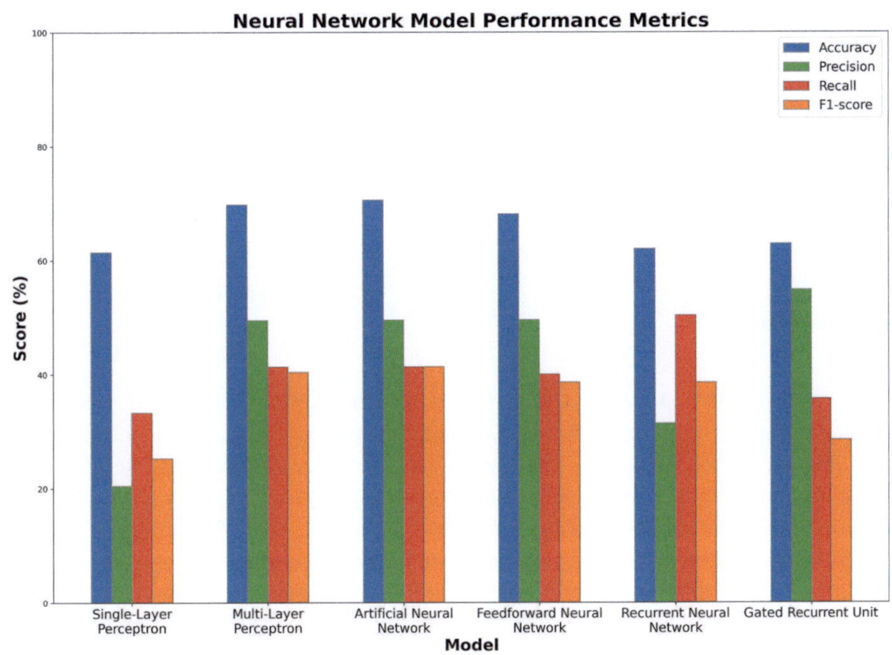

Fig. 3 Accuracy, precision, recall, and $F1$-score of neural network models

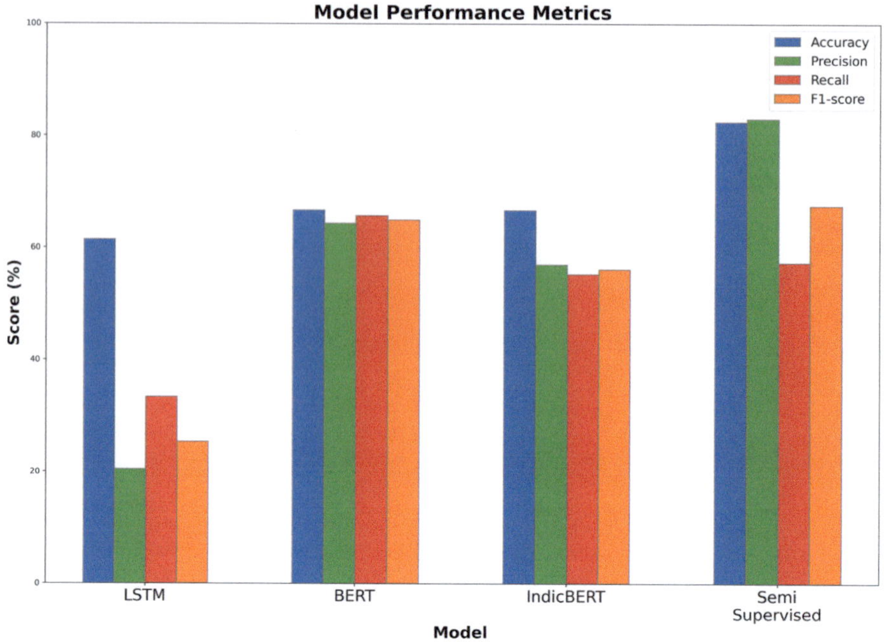

Fig. 4 Accuracy, precision, recall, and $F1$-score of deep learning models

accuracy and F1-score are obtained by semi-supervised learning, which uses Bert in combination with unsupervised methods. This shows how successful the model can be to employ both supervised and unsupervised learning strategies.

Through comprehensive analysis and comparison, it is evident that deep learning models, particularly those leveraging pre-trained language models, offer the most promising approach for offensive word detection in Kannada texts.

5 Conclusion

This study presents a systematic investigation into offensive word detection in Kannada texts, employing a structured methodology encompassing data acquisition, preprocessing, model training, and evaluation. Traditional classifiers, neural networks, and deep learning models were rigorously explored, each offering distinct advantages and performance characteristics.

The evaluation of performance metrics, including accuracy, precision, recall, and $F1$-score, reveals that while traditional models demonstrate moderate effectiveness, neural networks show improvement but struggle with contextual variations in offensive language. However, deep learning models, particularly those leveraging pre-trained language models such as BERT and IndicBERT, exhibit superior performance across all metrics.

Specifically, traditional models like Multinomial Naive Bayes and Support Vector Machine achieve an accuracy of around 71% and $F1$-score of approximately 49%. Because of their ease of use and capacity for managing high-dimensional data, these models are especially useful for preliminary baseline comparisons. However, they struggle with capturing the complexity of linguistic patterns in offensive language.

Neural networks like Multi-layer Perceptron and Artificial Neural Network demonstrate slightly higher accuracy around 70–72% and $F1$-score around 40–41%. These models benefit from their ability to learn non-linear relationships and patterns in the data. However, despite their improved performance over traditional models, they still exhibit limitations in capturing contextual nuances and handling large variations in offensive language.

Deep learning models, especially those utilizing pre-trained language models like BERT and IndicBERT, show significant improvement, with accuracies hovering around 66–67% and $F1$-scores ranging from 56 to 65%. The strength of these models lies in their capacity to understand and model complex linguistic structures and contextual information due to their extensive pre-training on diverse textual data. This pre-training allows them to capture intricate patterns and contextual variations more effectively than traditional and simpler neural network models.

Moreover, the incorporation of semi-supervised learning techniques, coupled with pre-trained language models, yields further enhancements in accuracy and $F1$-score, showcasing the effectiveness of hybrid supervised-unsupervised approaches. Specifically, the semi-supervised approach attains a remarkable 82.50% accuracy and a 67.50% $F1$-score. This method makes use of both labeled and unlabeled data, improving the model's capacity to generalize from a small number of labeled examples and gaining from the unlabeled data's wider contextual understanding.

In summary, our study emphasizes the significance of deep learning models as the most effective method for detecting offensive words in Kannada literature, particularly those that employ pre-trained language models. The results not only enhance offensive language detection systems but also offer insightful information for future machine learning and natural language processing research.

This project's future scope includes multiple directions for development and application. Initially, adding more frequently spoken languages to the mix beyond Kannada could improve the offending word identification system's applicability and inclusivity. Moreover, considering the prevalence of code-mixed languages in informal communication, adapting the model to detect offensive language in such contexts could further improve its utility. Additionally, there is potential for integrating the system into social media applications to enable real-time detection and reporting of bullying instances, thereby fostering safer online environments.

References

1. Hindustan Times: 85% of Indian children have experienced cyberbullying (2022). https://www.hindustantimes.com/technology/85-of-indian-children-have-experienced-cyberbullying-report-101659881741150.html
2. Reynolds K, Kontostathis A, Edwards L (2011) Using machine learning to detect cyberbullying. In: 2011 10th International conference on machine learning and applications and workshops, Honolulu, HI, USA, pp 241–244. https://doi.org/10.1109/ICMLA.2011.152
3. Mehendale N, Shah K, Phadtare C, Rajpara K (2022) Cyber bullying detection for Hindi-English language using machine learning. SSRN 4116143. https://doi.org/10.2139/ssrn.4116143
4. Pawar R, Raje RR (2019) Multilingual cyberbullying detection system. In: 2019 IEEE international conference on electro information technology (EIT), Brookings, SD, USA, pp 040–044. https://doi.org/10.1109/EIT.2019.8833846
5. Bhawal S, Roy P, Kumar A (2021) Hate speech and offensive language identification on multilingual code mixed text using BERT. Working notes of FIRE 2021-forum for information retrieval evaluation (Online). CEUR
6. Hande A, Priyadharshini R, Chakravarthi BR (2020) KanCMD: Kannada CodeMixed dataset for sentiment analysis and offensive language detection. In: Proceedings of the third workshop on computational modeling of people's opinions, personality, and emotions in social media, pp 54–63
7. Kumar R, Sahula V (2021) Intelligent Approaches for natural language processing for Indic languages. In: 2021 IEEE international symposium on smart electronic systems (iSES). IEEE, pp 331–334
8. Shannaq F, Hammo B, Faris H, Castillo-Valdivieso PA (2022) Offensive language detection in Arabic social networks using evolutionary-based classifiers learned from fine-tuned embeddings. IEEE Access 10:75018–75039
9. Yang S (2022) Natural language processing based on convolutional neural network and semi-supervised algorithm in deep learning. In: 2022 International conference on artificial intelligence in everything (AIE). IEEE, pp 174–178
10. Kulkarni H, Kulkarni P, Khandate V, Lohkare C, Dhore M (2022) Online hate classifier for social media platform. In: 2022 6th international conference on computing, communication, control and automation (ICCUBEA). IEEE, pp 1–6
11. Hani J et al (2019) Social media cyberbullying detection using machine learning. Int J Adv Comput Sci Appl 10(5)
12. Mahmud T, Ptaszynski M, Eronen J, Masui F (2023) Cyberbullying detection for low-resource languages and dialects: review of the state of the art. Inf Process Manage 60(5):103454
13. Chia ZL, Ptaszynski M, Masui F, Leliwa G, Wroczynski M (2021) Machine Learning and feature engineering-based study into sarcasm and irony classification with application to cyberbullying detection. Inf Process Manage 58(4):102600
14. Achuthan K, Nair VK, Kowalski R, Ramanathan S, Raman R (2023) Cyberbullying research—alignment to sustainable development and impact of COVID-19: Bibliometrics and science mapping analysis. Comput Hum Behav 140:107566
15. Singh S, Thapar V, Bagga S (2020) Exploring the hidden patterns of cyberbullying on social media. Procedia Comput Sci 167:1636–1647
16. Akhter A, Acharjee UK, Talukder MA, Islam MM, Uddin MA (2023) A robust hybrid machine learning model for Bengali cyberbullying detection in social media. Nat Lang Process J 4:100027
17. Nandhini BS, Sheeba JI (2015) Online social network bullying detection using intelligence techniques. Procedia Comput Sci 45:485–492
18. Sadiq S, Mehmood A, Ullah S, Ahmad M, Choi GS, On BW (2021) Aggression detection through deep neural model on Twitter. Futur Gener Comput Syst 114:120–129
19. Jahan MS, Oussalah M (2023) A systematic review of hate speech automatic detection using natural language processing. Neurocomputing 126232

20. Mahajan E, Mahajan H, Kumar S (2024) EnsMulHateCyb: multilingual hate speech and cyberbully detection in online social media. Expert Syst Appl 236:121228
21. Subramanian M, Sathiskumar VE, Deepalakshmi G, Cho J, Manikandan G (2023) A survey on hate speech detection and sentiment analysis using machine learning and deep learning models. Alex Eng J 80:110–121
22. Tiwari RS (2024) Hate speech detection using LSTM and explanation by LIME (local interpretable model-agnostic explanations). In: Computational intelligence methods for sentiment analysis in natural language processing applications. Morgan Kaufmann, pp 93–110
23. Sharma D, Gupta V, Singh VK (2024) Abusive comment detection in Tamil using deep learning. In: computational intelligence methods for sentiment analysis in natural language processing applications. Morgan Kaufmann, pp 207–226
24. Kumar A (2022) A study: hate speech and offensive language detection in textual data by using RNN, CNN, LSTM and Bert model. In: 2022 6th international conference on intelligent computing and control systems (ICICCS). IEEE, pp 1–6
25. Li Z, Shimada K (2022) Combining pre-trained language models and features for offensive language detection. In: 2022 13th International congress on advanced applied informatics winter (IIAI-AAI-Winter). IEEE, pp 5–10
26. Kanessa LG, Tulu SG (2021) Automatic hate and offensive speech detection frame-work from social media: the case of afaan oromoo language. In: 2021 International conference on information and communication technology for development for Africa (ICT4DA). IEEE, pp. 42–47
27. Preetham J, Anitha J (2023) Offensive language detection in social media using ensemble techniques. In: 2023 International conference on circuit power and computing technologies (ICCPCT). IEEE, pp 805–808
28. Hajibabaee P, Malekzadeh M, Ahmadi M, Heidari M, Esmaeilzadeh A, Abdolazimi R, James H (2022) Offensive language detection on social media based on text classification. In: 2022 IEEE 12th Annual computing and communication work-shop and conference (CCWC). IEEE, pp 0092–0098
29. Prasad D, Kadambari KV, Mukati R, Singariya S (2023) Real-time multi-lingual hate and offensive speech detection in social networks using meta-learning. In: TENCON 2023–2023 IEEE region 10 conference (TENCON). IEEE, pp 31–35
30. Santhiya S, Uma S, Abinaya N, Jayadharshini P, Priyanka S, Dharshini MN (2024) A comparative exploration in text classification for hate speech and offensive language detection using BERT-based and GloVe embeddings. In: 2024 2nd International conference on disruptive technologies (ICDT). IEEE, pp 1506–1509
31. Yadav Y, Bajaj P, Gupta RK, Sinha R (2021) A comparative study of deep learning methods for hate speech and offensive language detection in textual data. In: 2021 IEEE 18th India council international conference (INDICON). IEEE, pp 1–6
32. Panchala GH, Sasank VVS, Adidela DRH, Yellamma P, Ashesh K, Prasad C (2022) Hate speech and offensive language detection using ML and NLP. In: 2022 4th International conference on smart systems and inventive technology (IC-SSIT). IEEE, pp 1262–1268
33. Mehta M, Gada D, Sharma R, Chavan K, Kanani P (2022) Offense detection using BERT and CNN. In: 2022 IEEE 3rd global conference for advancement in technology (GCAT). IEEE, pp 1–5
34. Oriola O, Kotzé E (2020) Evaluating machine learning techniques for detecting offensive and hate speech in South African tweets. IEEE Access 8:21496–21509
35. Sreelakshmi K, Premjith B, Chakravarthi BR, Soman KP (2024) Detection of hate speech and offensive language CodeMix text in dravidian languages using cost- sensitive learning approach. IEEE Access
36. Bhardwaj M, Sundriyal M, Bedi M, Akhtar MS, Chakraborty T (2023) HostileNet: multilabel hostile post detection in Hindi. IEEE Trans Comput Soc Syst
37. Acar AN, Omurca Sİ (2023) Offensive language detection in social media for Turkish language. In: 2023 8th International conference on computer science and engineering (UBMK). IEEE, pp 267–270

Machine Learning Based Hybrid Approach to Detect Intrusion in Wireless Communication

Bobby Sharma, Google Kakati, and Vikramjit Kakati

Abstract Recognizing the crucial role of communication, data delivery, and access across various sectors, including government, business, and individual domains, it becomes imperative to pinpoint faults and vulnerabilities in cyber communication. To safeguard personal, governmental, and business data from the growing threats of advanced attacks on wireless communications, the implementation of Intrusion Detection System is essential. Information security serves as a robust defense for both host machines and networks. Analytical methods, particularly machine learning, are employed for the identification and prevention of diverse wireless network threats. In this paper, a novel approach is presented based on machine learning model, is integrated with different cross-validation techniques to find few performance matrix parameters in identifying and eliminating intrusions from a wireless network. This hybrid methodology employs various attributes, such as the Gini Index and Entropy of the decision tree (DT) model, different DT methodologies, including those using Gini Index, train-split method, and information entropy, along with their respective subdivisions like K-Fold validation and Stratified K-Fold validation, that are implemented and evaluated on a recent dataset containing DDoS network attack activities. The proposed algorithm mentioned in the paper establishes its performance by evaluating network performance parameters including packet delivery ratio and network throughput.

Keywords Decision tree · Gini index · Information entropy · Cross-validation · Stratified K-fold validation

---

B. Sharma (✉)
Department of CSE, School of Technology, Assam Don Bosco University, Guwahati, India
e-mail: bobby.sharma@dbuniversity.ac.in

G. Kakati
Department of CSE, Vellore Institute of Technology, Vellore, Tamil Nadu, India
e-mail: google.kakati2023@vitstudent.ac.in

V. Kakati
Department of MNE, School of Technology, Assam Don Bosco University, Guwahati, India
e-mail: vikramjit.kakati@dbuniversity.ac.in

H. Sharma et al. (eds.), *Artificial Intelligence: Theory and Applications*, Lecture Notes in Networks and Systems 5589, https://doi.org/10.1007/978-981-96-1687-9_13

1 Introduction

The realm of cyber security marks a significant advancement in shielding against a myriad of sophisticated attacks that have proliferated in the age of the Internet. It encompasses a series of protocols specifically crafted to safeguard personal, governmental, and business data from potential misuse [1]. The significant demand for wireless communication security focuses extensively on notable advancements in protecting programs and network data. The swift evolution of social networks, cloud services, and web technologies [2] has prompted the implementation of numerous security measures. In [3], author mentioned that to cope with security in the network, certain challenges are to be faced by machine learning technique.

Security of wireless communication over Internet, i.e., cyber security techniques are composed of computer and network security system [4] which has enhanced the integrity to protect the data from the susceptible cyber-attacks and threats [5].

The Intrusion Detection System (IDS) plays a vital role in network security as well as cyber security, as it possesses the capability to identify new misuse attacks [6] within the network. It safeguards the system administrator from malicious activities by effectively detecting and protecting against them. IDS operates on two levels: one at the individual machine level as a host, referred to as a Host-Based Intrusion Detection System (HIDS), and another level that monitors malicious attacks capable of compromising confidentiality, integrity, or causing violations [7] within the network, known as Network-Based Intrusion Detection System (NIDS) [8].

The innovative application of Data Science has progressed in analytical techniques and methods to counter the increasing complexity of cyber security attacks, particularly when applied to real-time data. Decision tree algorithm is considered as one of the algorithm for classification in case attack detection [9]. This application incorporates Artificial Intelligence (AI) and its derivatives, including machine learning (ML), deep learning (DL), and others [10]. Machine learning (ML) serves as the intermediary link between Artificial Intelligence (AI) and computational statistics, providing computers with the capability to enhance their learning from provided datasets [2]. The application of ML presents challenges to the methodological and theoretical approaches for addressing and resolving unknown cyber security situations, which are becoming progressively crucial in a potential manner [7]. The new automated ML techniques resolve the problem of solving by the traditional human analyst [11] or conventional signature-based system [5] as it has the capability to learn from the past experience for better future performance [12]. ML models can be broadly classified into supervised, unsupervised, and semi-supervised learning. In supervised learning perspective, any labeled data can be used for training and map the new output whereas in the unsupervised, the learning is done from unlabeled dataset where there is exploration, clustering of data. But the semi-supervised is the act of two learning technique for high achievement. The methods undergoes through three main stages: training, validation, and testing [4]. The collection of raw security data sets can be effective in analyzing, with the use of tools for preprocessing classification, regression, clustering, association rules, and visualization [13] the feature which helps in building learning-based security model that provides productive services [14].

2 Literature Review

In this section, there is a brief discussion of the related works based on machine learning techniques:

Alqahtani et al. [15] uses of different machine learning algorithms for intrusion detection from KDD'99 cup datasets. The datasets have classified four main cyber-attacks. The effectiveness is measured by evaluating the performance metrics, precision, recall, f1-score, and accuracy. The experiment shows a comparison among the machine learning algorithms. With numerous kinds of attacks the machine learning algorithms for Intrusion Detection System (IDS) may vary in the accuracy or sometimes in other prediction area, which may show the performing ability for some is less compare to other classifiers.

Sabar et al. [16] discussed the idea of using the new proposed framework Hyper-Heuristic with the traditional Support Vector Machine (SVM) algorithm. The traditional Kernel SVM model has bi-objection formulation showcasing the accuracy and complexity of the model as conflicting objectives. The new framework consist two levels namely, high-level strategy, and low level heuristics (LLHs).The Hyper-Heuristic Support Vector Machine (HH-SVM) model aims to establish the benefits with different effects of using multiple LLHs on the search performance. The HH-SVM even compared with other ML algorithms to check the accuracy performance.

Moustafa et al. [17] proposed the use of new Network Intrusion Detection System (NIDS) dataset, UNSW-NB 15 [18] that challenges the existing threats in the cyber security field. The existing benchmarks datasets has limited amount of attacks and information packets that was produced a decade ago which can be an obstacle in the platform of research. To overcome such limitation of previous datasets, UNSW-NB15 is developed IXIA Perfect Storm tool, by the cyber security research lab of the Australian Centre for Cyber Security (ACCS) at UNSW in Canberra. NIDS dataset quality points out two major features that are a comprehensive demonstration of modern threats and the new range of traffics. KDDCups99 dataset lacks the first feature of NIDS dataset. Though NSL-KDD is the improve version but it lacks in pointing out the zero-footprint attacks. UNSW-NB15 dataset has 9 major families and 49 features, setting up as a new benchmark in the dataset community.

Bagui et al. [19] generates the comparison of two different classifiers of machine learning using the hybrid feature selection process. The UNSW-NB15 dataset undergo through the feature selection process that are k-means clustering and a Correlation-Based Subset evaluation (CFS). After the process of feature techniques, the data undergoes through the two classifiers. The probabilistic Naïve Bayes (NB) classifier gives a higher classification accuracy rate, i.e., 99% and lowers FAR, i.e., 1% for most attack families using the selection of feature process. However, when it comes to J48 decision tree, the classification rate was not impacted by the hybrid process, though it perform almost same with or without the selection factors.

Sarraf et al. [20] detects and classify the kinds of intrusions. To detect the intrusions, binary classification was performed in the network traffic. After the classification, machine learning algorithms are performed to determine the accuracy and other performance metrics. Among them, XGBoost performed quite well then the rest with 83.92% accuracy.

Singh et al. [21] compares between SVM and iSVM. SVM algorithm is dependent on Kernel and each Kernel function has capability to map the non-linearly separable data points into different dimensions. By modifying the traditional Gaussian Kernel, the paper proposed the idea of modified Gaussian Kernel making the SVM into Improved SVM (iSVM). The traditional SVM shows less execution in the detection of cyber-attack compare to iSVM in KDDCUP2009 dataset. The comparison of iSVM is not limited; it is even compared with Bayesian Network (BN) and Classification and Regression Tree (CART) but the performance doesn't show satisfying result for R2L and U2r, respectively. But still the iSVM accuracy has remarkably improved for the attacks class like normal and DOS than the traditional SVM algorithm.

Sasan et al. [22] analyze to build a hybrid model with the feature selection method to observe the behavior of the attacks of network packets in the standard version of NSL-KDD data sets. The hybrid model is made up of J48 and CART algorithm. As the proposed model is used for achieving high accuracy in the Intrusion Detection System with 29 features is about 88.23%. The author further talks about working on more parameters to improve the intrusion detection.

Kumari et al. [18] proposed the usage of logistic regression and Naive Bayes and logistic regression for differentiating network scenarios using data packet throughput and the CAIDA 2007 dataset. Logistic regression demonstrated more accuracy (99–100%) over Naive Bayes (98–99%). The paper efficiently identified Distributed Denial of Service (DDoS) attacks, conducted a performance comparison and emphasized the efficacy of machine learning. The conclusion reported a high accuracy of 100% for the machine learning model and 99.75% for the mathematical model, highlighting efficient DDoS identification. It suggests live attack testing, applications beyond DDoS such as memory management and firewalls, and acknowledges a single dataset limitation, and proposes examining distributed datasets for enhanced robustness and applicability in future research.

In their work, Gebremeskel et al. [23] recommended a solution leveraging software-defined networks (SDNs) capabilities for early-stage Distributed Denial of Service (DDoS) attack detection at the multi-controller level. They used the CICDDoS2019 dataset, applying PCA, decision tree, random forest regressor, and the chi-squared (x^2) test for feature selection. The Keras framework, using regularization, recurrent, activation, and dense layer shapes, facilitated deep learning model development. The research highlighted the importance of using distributed controller designs for future SDN systems, introducing an entropy-based and deep learning model for enhanced attack classification. The chi-square (x^2) test served as the feature selection algorithm, and categorical classification addressed the binary classification limitations of the baseline model. Successfully implementing a multi-controller-based SDN solution, the study demonstrated improved efficiency and accuracy with the Long Short-Term Memory (LSTM) model achieving a 99.42% accuracy. The conclusion highlighted the benefits of a logically centralized and physically distributed architecture, coupled with feature selection, for enhanced reliability, efficiency, and availability in SDN environments.

3 Proposed Methodology

The proposed methodology (Fig. 1) is an approach to identify and mitigate malware-related activities within a network using artificial intelligence-based agents and machine learning technique. The process basically involve with the following modules:

- **Data Collection**: Artificial intelligence-based agents are deployed within the network to collect and manage data concerning malware activities. Any node in the network that suspects malicious activity from other nodes reports it to these agents.
- **Threshold Analysis**: The agents analyze the reported data and calculate two key parameters
  - Packet Delivery Ratio (PDR)
  - Network throughput

  If these parameters exceed a predetermined threshold value, it suggests a potential issue with the reported node.

- **Suspected Malicious Nodes**: Nodes that trigger the threshold are moved to a "suspected malicious nodes" archive for further investigation.
- **Packet Capture**: Packets are captured from the suspected malicious nodes for further analysis using ML.

Machine Learning Model: This step involves gathering the raw data from network traffic, which then undergoes data preprocessing and normalization. This is essential because the feature values of network packets often vary in range, requiring conversion into numeric values to ensure their utility and comprehensibility for the machine learning model. Once the data is appropriately refined, it becomes ready for processing in the "decision tree-based machine learning model (DTML)." This model is responsible for various tasks, including data preprocessing, feature extraction, and both training, testing and validation of the data.

Here, DTML had been extended by implementing Gini Index in the decision tree algorithm. The different DT methodologies including DT using Gini Index:

- K-Fold validation,
- Using Stratified K-Fold validation.

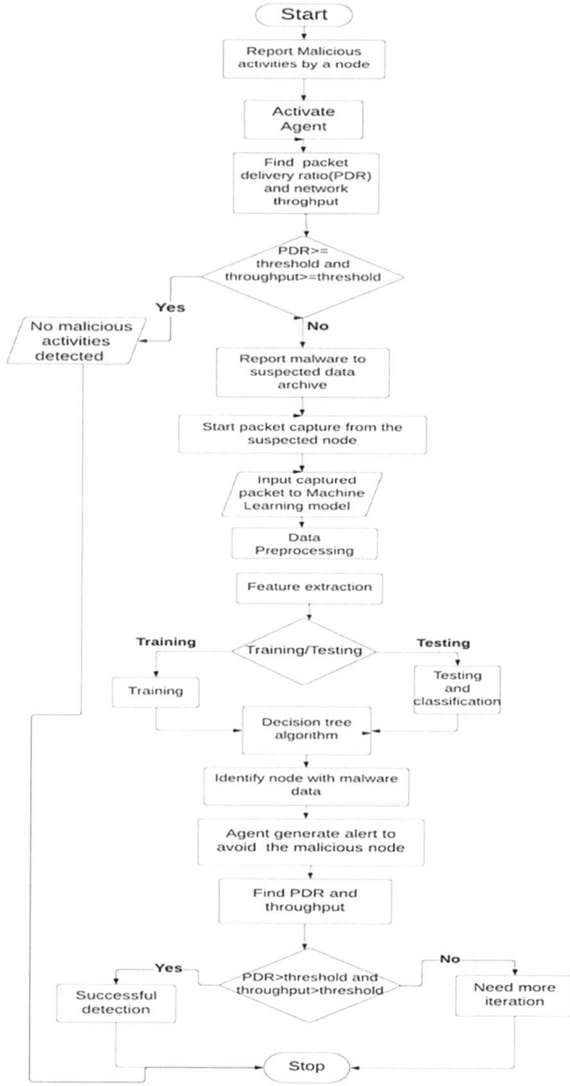

Fig. 1 Flowchart for proposed methodology

Where Gini Index is the measurement of the probability that a variable that will not be classified correctly if it is randomly chosen.

a. **Malicious Activity Detection:** DTML generates an output that determines whether the captured data represents malicious activities or not. If it identifies malicious activities, an alert will be generated within the network to warn other nodes in the network to avoid communication with the suspicious nodes.
b. **Monitoring and Improvement**: If the alert is generated, the network is closely monitored. If, over time, the PDR and network throughput improve and surpass

the threshold values, the detection process can be stopped. The network continues to operate normally unless new suspicious activities are reported.

The network packet data are gathered and subjected to data preprocessing and normalization to address variations in the ranges of feature values. This involves converting the data into numeric values that are both useful and interpretable for the machine learning model. Once the data has undergone refinement, it is prepared for processing according to the flowchart shown in Fig. 1.

4 Implementation

Wireless network had been designed using network simulator NS 3 with the various network parameters as mentioned in Table 1.

In Table 1, it is mentioned that number of nodes in the simulation process for wireless communication is 50 with "*random walk 2D*" mobility model. Similarly speed of each node is determined by an exponential random variable in which parameter specifies mean speed on 1 unit to maximum speed up to 2 units. Simulation time is considered as 2 s. The network communication is based on 802.11n wireless standard with an operational speed of 5 GHz. The wireless channel considered in the simulation is "*YansWifiChannelHelper*". "*YansWifiPhyHelper*" determines the process of transmission and receiving of signal from and to physical layer. "*ArfWifiManager*" is remote station manager, used to specify rate control algorithm to adjust transmission rate. For this simulation the type of attack simulated is considered as "Distributed *Dinial of Service (DDoS)*" attach. Transmission range is considered as default transmission range of 802.11n standard and PHY layer. For TCP connection, data rate is

Table 1 Network parameters

Number of nodes	50
Mobility model	Random Walk 2D
Mobility parameters	Speed: ExponentialRandomVariable[Mean = 1|Bound = 2]
Time	2 s
Wireless standard	802.11n (5 GHz)
Wireless channel	YansWifiChannelHelper::Default()
PHY layer	YansWifiPhyHelper::Default()
Remote station manager	ArfWifiManager
Type attack impose	DDoS attack
Transmission range	Default
TCP connection parameters	Data Rate: 5 Mbps Packet Size: 1024 bytes
TCP connection duration	2 s
Simulation duration	2 s

considered as 5Mbps and packet size is 1024 bytes. TCP connection duration and simulation duration are considered as 2 s.

Packets are captured for two scenarios including, *without attack* and *with attack* (*Specially DDoS attack*).

Captured packets are then utilized for data preprocessing and feature extraction for further analysis using DTML. Model as shown in Fig. 1.

DTML will identifies the node involved in malicious activities and generate an alarm to avoid communication through suspected malicious node. Then PDR and network throughput will be measured again, if it finds the PDR and network throughput is above threshold value, the process of finding malicious activities will be stopped unless any new malicious activities is reported.

Following table shows the different network parameters while testing the same using NS 3.

5 Results and Analysis

The performance of the proposed hybrid approach was evaluated based on key network performance parameters. Out of several performance evaluation parameters, packet delivery ratio as well as network throughputs are considered as two key parameters to evaluate performance accuracy of the proposed system. The results demonstrated that the hybrid approach performed well in identifying and mitigating network attacks which can be visualized based on Figs. 2 and 3. This implicate that hybrid methods can significantly improve the robustness and reliability of wireless network security systems.

Packet delivery ratio and network throughput on periodic basis had been calculated for both the scenarios, i.e., with and without intrusion and plotted in Fig. 2.

In Fig. 3, there is a comparison of network throughput with or without implanting proposed system for induced attack in the network.

From the above results, it is established that proposed system is an efficient system to mitigate DDoS attack in a wireless network.

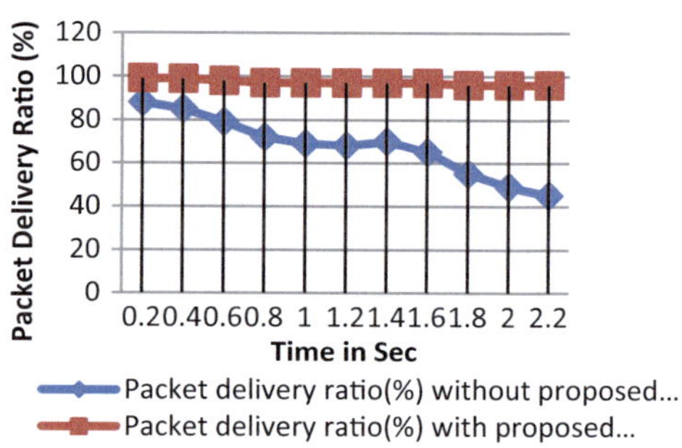

Fig. 2 Comparison of packet delivery ratio with or without proposed system for DDoS attack

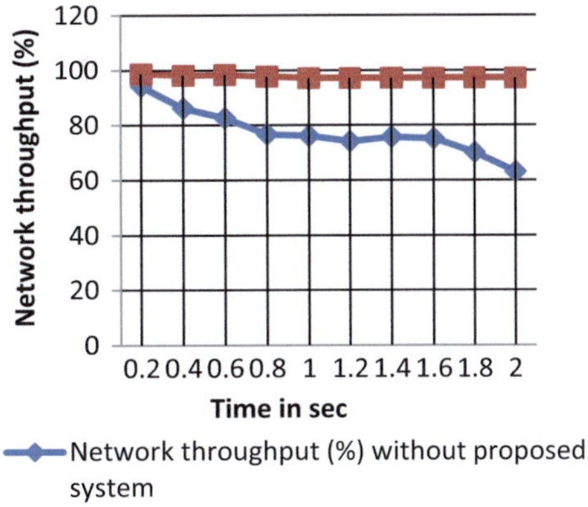

Fig. 3 Comparison n of network throughput with or without proposed system for DDoS attack

6 Conclusions

Wireless network security has grown more and more important in today's information-driven environment. Network vulnerability has increased as a result of the recent spike in attacks. Computers must quickly evolve in order to properly detect or prevent these attacks. Thus, in order to classify threats quickly, learning models need to adapt their methods. The proposed hybrid approach to wireless network attack detection shows good performance in identifying and eliminating the same from the network. The results in terms of network performance parameters had been evaluated, and thus establishes the performance of the proposed system.

By looking the importance of threat detection in various network scenarios, the proposed hybrid approach to attack detection offers a promising solution, demonstrating good performance in identifying, and eliminating network threats.

Although the current results are promising, yet there is scope for further improvement. Implementation of additional machine learning algorithms may be an addition to the existing work for better results. A comparative analysis of various algorithms could provide deeper insights into their effectiveness in different scenarios.

References

1. Torres JM, Comesaña CI, Garcia-Nieto PJ (2019) Machine learning techniques applied to cybersecurity. Int J Mach Learn Cybern 10(10):2823–2836
2. Xin Y, Kong L, Liu Z, Chen Y, Li Y, Zhu H, Gao M, Hou H, Wang C (2018) Machine learning and deep learning methods for cybersecurity. IEEE Access 15(6):35365–35381
3. Amit I, Matherly J, Hewlett W, Xu Z, Meshi Y, Weinberger Y (2018) Machine learning in cyber-security-problems, challenges and data sets. arXiv preprint arXiv:1812.07858
4. Buczak AL, Guven E (2015) A survey of data mining and machine learning methods for cyber security intrusion detection. IEEE Commun Surv Tutorials 18(2):1153–1176
5. Shaukat K, Luo S, Varadharajan V, Hameed IA, Xu M (2020) A survey on machine learning techniques for cyber security in the last decade. IEEE Access 2(8):222310–222354

6. Abdulraheem MH, Ibraheem NB (2019) A detailed analysis of new intrusion detection dataset. J Theor Appl Inf Technol 97(17):4519–4537
7. Ford V, Siraj A (2014) Applications of machine learning in cyber security. In: Proceedings of the 27th international conference on computer applications in industry and engineering 2014 Oct 13, vol 118. IEEE Xplore, Kota Kinabalu
8. Sonule AR, Kalla M, Jain A, Chouhan DS (2020) UNSWNB15 dataset and machine learning based intrusion detection systems. Int J Eng Adv Technol 9:2638–2648
9. Sharma H, Kumar S (2016) A survey on decision tree algorithms of classification in data mining. Int J Sci Res (IJSR) 5(4):2094–2097
10. Khan A. Data science in action: key to cyber security
11. Sinclair C, Pierce L, Matzner S (1999) An application of machine learning to network intrusion detection. In: Proceedings 15th annual computer security applications conference (ACSAC'99). IEEE, pp 371–377
12. Das K, Behera RN (2017) A survey on machine learning: concept, algorithms and applications. Int J Innovative Res Comput Commun Eng 5(2):1301–1309
13. Revathi S, Malathi A (2013) A detailed analysis on NSL-KDD dataset using various machine learning techniques for intrusion detection. Int J Eng Res Technol (IJERT) 2(12):1848–1853
14. Sarker IH, Kayes AS, Badsha S, Alqahtani H, Watters P, Ng A (2020) Cybersecurity data science: an overview from machine learning perspective. J Big Data 7(1):1–29
15. Alqahtani H, Sarker IH, Kalim A, Hossain SM, Ikhlaq S, Hossain S (2020) Cyber intrusion detection using machine learning classification techniques. In: International conference on computing science, communication and security. Springer, Singapore, pp 121–131
16. Sabar NR, Yi X, Song A (2018) A bi-objective hyper-heuristic support vector machines for big data cyber-security. IEEE Access 6(6):10421–10431
17. Moustafa N, Slay J (2015) UNSW-NB15: a comprehensive data set for network intrusion detection systems (UNSW-NB15 network data set). In: 2015 military communications and information systems conference (MilCIS). IEEE, pp 1–6
18. Kumari K, Mrunalini M (2022) Detecting denial of service attacks using machine learning algorithms. J Big Data 9(1)
19. Bagui S, Kalaimannan E, Bagui S, Nandi D, Pinto A (2019) Using machine learning techniques to identify rare cyber-attacks on the UNSW-NB15 dataset. Secur Priv 2(6):e91
20. Sarraf J, Chakraborty S, Pattnaik PK (2020) Detection of network intrusion and classification of cyberattack using machine learning algorithms: a multistage classifier approach. In: International conference on smart computing and cyber security: strategic foresight, security challenges and innovation. Springer, Singapore, pp 285–295
21. Singh S, Agrawal S, Rizvi MA, Thakur RS (2011) Improved support vector machine for cyber attack detection. In: Proceedings of the world congress on engineering and computer science, vol 1
22. Sasan HP, Sharma M (2016) Intrusion detection using feature selection and machine learning algorithm with misuse detection. Int J Comput Sci Inf Technol 8(1):17–25
23. Gebremeskel TG, Gemeda KA, Krishna TG, Ramulu PJ (2023) DDoS attack detection and classification using hybrid model for multicontroller SDN. Wirel Commun Mob Comput 2023:1–18

Smart Asthma Inhaler: Childhood Asthma Management Through Real-Time Monitoring and Personalized Care

Advik Narendran, Anantha Hothri Inuguri, Hemanth Saga, and S. Thangam

Abstract There are numerous respiratory diseases which are known to affect children, one of those is Asthma. Asthma is a long-term illness that mainly impacts children and tends to be brought on by environmental concerns, and volunteers involve repeated attacks with signs that are not obvious. This work presents a new smart inhaler that can help children, especially those who are conversant with smart gadgets, in monitoring their asthma conditions. There are several features in the proposed system such as, real-time inhaler usage, medication reminder, location tracking, heart rate measurement, and environmental quality measurement all based on the Arduino technology. This two-circuit device, one for physiological monitoring and the other for inhaler management, aims to improve self-care and health management, thus reducing minor medical visits and asthma attacks. The smart inhaler represents a significant advancement in pediatric asthma care, promoting better self-management and improving the quality of life for children with asthma.

Keywords Smart inhaler · Pediatric asthma · Real-time monitoring · Medication alerts · Environmental quality · Arduino technology · Health management

1 Introduction

Asthma, a common chronic respiratory illness affecting millions worldwide, imposes considerable burdens on both sufferers and healthcare systems, significantly impacting children. Children with asthma often suffer from ineffective management of their condition: their asthma attacks occur repeatedly, and their lives are frequently interrupted. Despite improvements in clinical care, one of the biggest challenges for

A. Narendran · A. H. Inuguri (✉) · H. Saga · S. Thangam
Department of Computer Science and Engineering, Amrita School of Computing, Amrita Vishwa Vidyapeetham, Bengaluru, India
e-mail: ananthahothri@gmail.com

S. Thangam
e-mail: s_thangam@blr.amrita.edu

H. Sharma et al. (eds.), *Artificial Intelligence: Theory and Applications*, Lecture Notes in Networks and Systems 5589, https://doi.org/10.1007/978-981-96-1687-9_14

caregivers continues to be aftercare management of childhood asthma as poor compliance and inadequately designed monitoring methods lead to a poor prognosis. In response to these issues, this work seeks to depict an innovative solution: an advanced inhaler system, whose core purpose is to cater to the needs of children with asthma, scrutinize medications, and provide individualized reminders.

One of the primary limitations of existing solutions is their inability to provide continuous, real-time monitoring and personalized care. Traditional inhalers and asthma management systems lack the integration of smart technology that can offer medication reminders, real-time health monitoring, and environmental assessments. Consequently, children often forget to take their medication or fail to recognize early warning signs of an asthma attack, leading to frequent hospital visits and a diminished quality of life.

The proposed smart inhaler system is designed with two circuits to support better asthma control in children by monitoring the use of inhalers. The first of them is built into a smartwatch and monitors vital health indicators 24/7, including heart rate, body temperature, humidity, and air quality. These sensors offer an instant response to the situation, which enables avoiding asthmatic triggers. The second circuit is designed to be incorporated in smart inhaler, it keeps track of inhaler utilization, reminds the patient when it is time to use the inhaler and has a buzzer to help locate the inhaler. The information is relayed to a central unit for further analysis in the detection of patterns, generation of health reports, and recommendation suited to the individual. This dual-circuit system of IoT and data analytics is therefore a holistic solution that would assist in improving self-care, decreasing medical visits, and improving the quality of life of children with asthma.

The system operates on the principles of personalized medicine in conjunction with the use of real-time medication reminders and health data to ensure that asthma is treated successfully, and its adherence and control are met. In 'Asthma medication adherence among children' [1], it has been observed that reminder and electronic monitoring techniques have led to increased adherence rates among asthma patients. The system operates on the model of the smart inhaler in which the Arduino Nano microcontroller with the necessary sensors to measure the patients' heart rate, temperature, humidity, and air quality, all fitted onto an inhaler. This tries to find weight from results in books such as 'Low-cost wearable asthma monitoring system with a smart inhaler' [2], in which there was an observation that continuous monitoring of asthma ascertains proper disease control. Inspired by disruptive innovation in the Internet of Things, precision medicine, data analytics, and power, this system is designed to support the asthmatic in managing health, reducing infection rates, and hence improving the quality of life. The IoT system described is 'Smart Asthma', which is similar to the concept of 'Smart Inhalers: The implementation of IoT for better Asthma management' [3] contributing to the holistic aims of personalized medicine and patient-centered care. In addition, the system has a provision of features to locate the asthma inhaler. The asthma inhaler can be made to beep on the smartwatch by using a buzzer. The design thinking is having a user-friendly interface for easy availability to the patients Users can plan for the medication and receive alert corrections and get detailed information of the health status. This is similar to the concept in 'Modernizing Asthma Management: Personalized Asthma Action Plans Using Smartphone Applications' [4].

As a contribution to ongoing asthma care efforts, this project aims at fostering collaboration among healthcare professionals, researchers, and technology developers toward a smarter and more advanced future.

2 Literature Survey

Asthma continues to be a major health problem worldwide, even in children, with noncompliance to therapeutic regimens and hypersensitivity to environmental stimuli being the major concern. Also, features for real-time monitoring of existing interventions are lacking regarding children. Current advances in the field of IoT technologies are paving new ways for improving asthma management. It is to collect the information needed for designing a Smart Asthma Inhaler system specifically for children that this literature survey is being carried out.

Antonia et al. [5] proposed an AR approach to demonstrate the proper way of using the inhaler to children and adolescents. This approach is unconventional yet proves efficiency in the interaction with young patients with asthma which, in turn, will help in the overall management of the condition. The strength of this study therefore resides in the fact that the SMART inhaler technology used encouraged patients to practice correct inhaler usage. But it is more centered on the educational aspect than repeated observation of the usage of inhalers or even immediate feedback.

Esther et al. [6] introduced a device that provides real-time feedback to children with uncontrolled asthma, aiming to improve inhalation techniques and therapy adherence. This study highlights the importance of immediate feedback in managing asthma. However, it does not address environmental triggers or integrate comprehensive health monitoring, which limits its scope. Lavanya et al. [7] developed a low-cost wearable system for continuous monitoring of patient's vital signs and environmental parameters. This proactive approach shifts asthma care toward prevention by continuously monitoring potential triggers. The limitation of this study is its focus on vital signs and environmental parameters without integrating inhaler use monitoring or medication reminders.

It is to be noted that Naudiyal et al. [8] highlighted the cost implications of IoT in healthcare. While IoT can enhance convenience and simplify healthcare processes, it can also increase costs. This study underscores the need for cost-effective solutions in developing asthma management systems. A successful implementation of an IoT-alert system for asthma attacks in Rwanda was highlighted by [9]. This study demonstrated the potential of IoT for immediate response in asthma management. However, it does not address long-term management or continuous monitoring.

Yuqing et al. [10] conducted a meta-analysis on the efficacy of inhaled corticosteroids in pediatric asthma. This comprehensive analysis provides valuable insights into the effectiveness and side effects of corticosteroids but does not incorporate technological advancements for improved asthma management. Murugesan et al. [11]

designed an IoT framework for monitoring asthma patients, integrating vital signs, inhaler tracking, and environmental sensing. This holistic approach is promising but may face challenges in implementation and user adoption.

Tippannavar et al. [12] integrated a gas sensor with a smart inhaler to continuously scan the atmosphere, improving asthma management by reducing triggers. This system focuses on environmental monitoring but does not provide comprehensive patient health monitoring. Lijina and Nippun [13] explored the use of Bluetooth RSSI for collision avoidance in robots, which could be adapted for tracking inhaler use and environmental triggers in asthma management. However, this application is theoretical and requires further adaptation for practical use in healthcare.

Megalingam et al. [14] developed intelligent crutches using IoT technologies for real-time feedback and monitoring, which can serve as a guideline for creating smart asthma inhalers. However, direct applicability to asthma management requires further exploration. Rajeshkumar and Mathi [15] designed an IoT-based solution for reducing COVID-19 risk, integrating health metrics and environmental monitoring. The methodologies can be adapted for asthma management, though specific adjustments are needed for asthma-specific parameters.

Reddy and Thangam [16] incorporated machine learning for predicting myocardial infarction relapses. The predictive models and data analysis methodologies can be adapted for predicting asthma attacks, enhancing proactive care. Thangam et al. [17] demonstrated that cloud-based monitoring systems can be cost-effective for remote monitoring and precision farming. Similar approaches can be applied to asthma management, providing real-time data to caregivers and healthcare providers.

The most critical problem that Yadav et al. [18] discuss is the security vulnerabilities on current interconnected systems like IoT, IIoT, SCADA, and Android applications. In actual sense we explain how these systems are increasingly being depended on for varied services that can be accessed through android phones as well as other gadgets used in Internet of Things. This leaves room for defects from bugs or even factory set options that renders them attackable. Furthermore, essentials regarding security are covered including privacy maintenance, prevention of unauthorized access, and maintainability.

Mwangi et al. [19] explore the growing importance of blockchain in automating processes across finance, supply chain, education and healthcare. Manual methods for accessing and verifying large datasets are time-consuming. To address this the paper proposes a new Private Blockchain-Cloud System (PBCS) for healthcare services. This will secure patient records through e-block access platform so only authorized individuals with valid keys can access the data. Security is further enhanced by blockchain block.

The reviewed literature collectively highlights the potential and challenges of using IoT and related technologies for asthma management. The proposed smart asthma inhaler system aims to integrate several advanced features to improve asthma management in children. Building on the work of Esther et al. [6], the system will track inhaler usage in real-time, providing immediate feedback on inhalation techniques and adherence. Inspired by Antonia et al. [5], it will include reminders and alerts to ensure timely medication use, enhancing compliance. Incorporating environmental monitoring aspects from Lavanya et al. [7] and Tippannavar et al. [12], the system will continuously monitor environmental parameters and provide alerts for potential triggers. Adopting the holistic IoT framework from Murugesan et al. [11], it will integrate vital signs monitoring, inhaler tracking, and environmental sensing into a single platform. Utilizing predictive models discussed by Reddy and Thangam [16], the system will analyze data to predict potential asthma attacks, allowing for proactive interventions. Addressing cost concerns highlighted by Naudiyal et al. [8], the system will aim to use cost-effective technologies and open-source frameworks to ensure affordability and scalability. Following cloud-based approaches suggested by Thangam et al. [17], the system will store and analyze data on the cloud, providing real-time access to caregivers and healthcare providers. By integrating these features, the proposed Smart Asthma Inhaler system will provide a comprehensive solution for managing asthma in children, improving their quality of life and health outcomes through effective symptom tracking and management. This approach builds upon the strengths of existing studies while addressing their limitations, offering a more holistic and child-centric solution for asthma care.

3 Methodology

The methodology states the stages of development and the parts of the data collection, monitoring, and analysis techniques to be used in the implementation of the real-world treatment of asthma in children. The proposed smart inhaler system entails the use of a dual circuit to address the issues of asthma in children. In this way, there is a real-time record and control of both physiological data and inhaler usage. The two circuits include one for physiological monitoring of the patient and another circuit that deals with the inhaler. Here is a detailed explanation of each circuit:

The Smart Inhaler Project is a combination of real-time monitoring, medication reminders, and assessment of environmental quality in the management of asthma in children using IoT technologies. As shown in Fig. 1 the current approach integrates sensors from different micro-controllers to collect, process, and analyze data on health conditions further and transmit timely alerts and feedback to users and attendants.The design was focused on developing a system for continuous monitoring of the user's heart rate, environmental conditions, usage of an inhaler, as well as giving timely reminders of medication or emergency alerts.

3.1 Physiological Monitoring Circuit

This circuit is incorporated into a child's wrist-worn gadget. It has the ability to check and display the main health indicators of the child as and when necessary. The primary components of this circuit include:

Fig. 1 Working of smart asthma inhaler

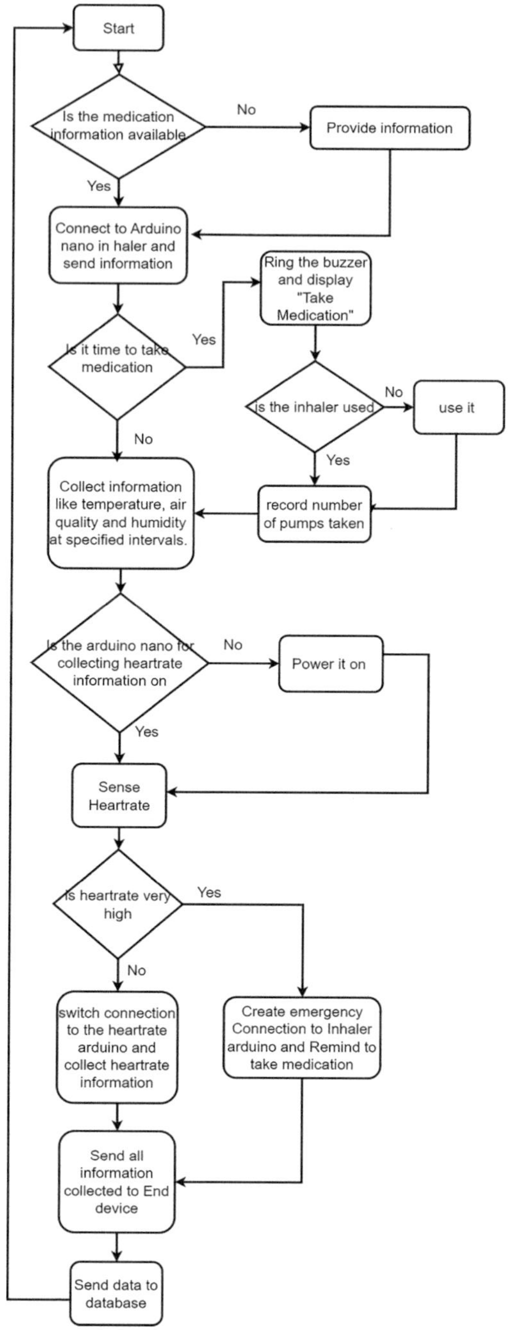

- **Heart Rate Sensor (Pulse Sensor)**: Measures the childs' heart rate continuously to detect any irregularities that may indicate an impending asthma attack or other health issues.
- **Temperature Sensor (DHT11)**: Monitors the body temperature to identify fever or other conditions that might affect asthma.
- **Humidity and Air Quality Sensors (MQ135)**: These sensors assess the environmental conditions around the child, including humidity levels and air quality. Poor air quality and high humidity can trigger asthma symptoms, so real-time monitoring helps in taking preventive actions.

These sensors located in the smartwatch gather data and send it to a central unit which may be a smartphone or a cloud server for further processing and storage. It is applied for alerting and reminding, so children and caregivers can better manage asthma with the help of real-time data.

3.2 Inhaler Management Circuit

This circuit is built into the actual smart inhaler itself. Its main purpose is to oversee the inhaler and make sure that the medication is given properly and at the right time. The main components of this circuit include:

- **Usage Tracking Sensor**: Detects when and how often the inhaler is used. This helps in tracking medication adherence and identifying usage patterns that could indicate misuse or overuse.
- **Medication Reminder System**: Sends alerts and reminders to the child and caregiver when its' time to take the medication. This system ensures that the child follows the prescribed medication schedule without missing doses.
- **Location Monitoring**: Helps locate the inhaler using a buzzer that can be activated from the smartwatch or smartphone. This feature is particularly useful for children who might misplace their inhaler frequently.
- **Connectivity Module**: Facilitates communication between the inhaler and other devices (like the smartwatch and smartphone), ensuring seamless data transfer and integration.

3.3 Integration and Data Analysis

Together, these two circuits offer a complete solution for asthma control. The data obtained from the circuit of monitoring of the child physiological parameters is compared with the data obtained from the circuit of tracking the usage of inhalers to get the information about the state of health of the child and the efficiency of the employed method of the treatment of asthma. Advanced data analytics are applied to:

- Detect patterns and predict potential asthma attacks.
- Provide personalized recommendations for medication and environmental adjustments.
- Generate comprehensive health reports for caregivers and healthcare providers.

3.3.1 Integration with Arduino Nano and Data Transmission

1. **Arduino Nano**: Acts as the central processing unit, collecting data from sensors and managing communication protocols.
2. **Data Transmission**: Sensor data is transmitted to a central unit (e.g., smartphone or cloud server) via Bluetooth module for real-time monitoring and storage.

3.3.2 Software Algorithms and Data Analysis

1. **Data Collection**: Arduino Nano gathers sensor readings at regular intervals, ensuring continuous monitoring.
2. **Monitoring Processes**: Algorithms analyze sensor data to detect anomalies (e.g., abnormal heart rate or poor air quality) that may trigger alerts for caregivers.

This approach is strong, efficient, and easily applicable using IoT technology and data analysis in the management of childhood asthma while enhancing the overall quality of life of children.

Much of the methodology entails a sequential and extensive process of managing asthma in children by monitoring the conditions and symptoms frequently, administering medications on time, and determining the quality of the environment in which the child spends most of their time. Thus, we choose trusted parts like Arduino Nano, Pulse Sensor, DHT11 Sensor, MQ135 Sensor, OLED Display, Bluetooth modules, Buzzer, and so on.

4 Implementation

This section explains the system components necessary for the successful implementation of the project. Each component plays a key role in enabling real-time monitoring, data collection, and interaction with the user, contributing to the overall effectiveness of the smart asthma inhaler. The required components are briefly described in Table 1.

Table 1 Hardware components

S. No.	Item	Model	Description	Qty
1	Arduino	Nano, ATmega328P, 16 MHz clock, 5 V input	Microcontroller board for data processing and management. It provides GPIO pins, analog inputs, and various communication interfaces	2
2	Wi-Fi module	ESP8266 12E, IEEE 802.11 b/g/n WiFi network	Enables Wi-Fi connectivity for IoT projects, and divices to communicate with each other and data transmission	1
3	Temperature and humidity sensor	DHT11, 3.5–5.5 V power, measures 0–50 °C, 20–90% RH	Measures ambient temperature and humidity	1
4	Air quality sensor	MQ135, 5 V power, detects 200–10,000 ppm VOCs	Detects air quality and pollutants, useful for detecting gases that can affect health and trigger asthma attacks	1
5	Heart rate sensor	3.3 V power, analog output (ADC dependent)	Measures heart rate by detecting changes in blood volume in the fingertip, essential for monitoring health conditions and detecting anomalies like irregular heartbeats	1
6	GPS module	Neo 6M, 3.3 V typical, serial communication	Provides accurate positioning and time information	1
7	Breadboard	Connection platform, 0.1″ spacing, 1 A current	Solderless platform for prototyping circuits	2
8	Blue OLED display	0.96 in., SPI/I2C-4Ppin	OLED display for visual output	2
9	Buzzer	3–12 V power, 85 db sound output	Audible alarm for alerts and notifications	1
10	Buttons	3.3 V/5 V, digital output	Input devices for user interaction	2

4.1 System Architecture

This system can be split into two major modules as explained in the methodology

- **Bracelet Module**: This module is worn on the wrist to track the user's heart rate. It displays the data on an OLED screen and then transfers the information to the inhaler module using Bluetooth.
- **Inhaler Module**: Collects environmental data, displays the information to the patient, provides reminders for medication intake, retrieves the heart rate data from the bracelet, and triggers alerts if necessary.

The system architecture in Fig. 2 explains the logical steps and decision points of operation of the system that will ensure asthma in children is monitored and managed efficiently. The first comes as the setting of the asthma management process, along with initial checks regarding the asthma management process. From that, it becomes medication information, and in case this information is absent, it will ask the user to enter the required information. After entering, the system, on its own, will connect to the Arduino Nano on the inhaler and send the drug details. The system checks, at intervals, the time of medication and, after checking, rings a buzzer with a message 'Take Medication' from the display. It also checks the number of times the inhalers have been used, and if not, it sends the user a reminder to do so. Environmental data, which includes temperature, air quality, and humidity, are, on the other hand, collected at the time interval users will set using DHT11 and MQ135 sensors. Another critical process the system, here, performs is monitoring the heart rate. It checks the power of the heart rate monitoring Arduino, then senses and continuously monitors the heart rate. In cases where the rate is extremely high, indicating danger, the system builds a connection to the Arduino inhaler and sends the user a reminder to take medication. Otherwise, in cases where the rate is ordinary, monitoring will persist. The final collected data—medication use data, data on conditions of the environment, and heart rate data—are then transmitted to the end device—mobile/laptop—using Bluetooth and are stored in a local database for further analysis and monitoring.

The implementation outlines the system architecture, that is divided into Bracelet, facilitates continuous monitoring of heart rate and environmental conditions while providing timely alerts and reminders to the user. This integrated approach as shown in Fig. 3 ensures a holistic solution for asthma management, aiming to improve the quality of life for children suffering from asthma.

5 Results and Discussion

The Smart Asthma Inhaler Project focuses primarily on improved asthma management along with real-time monitoring, medication alerts, and the assessment of environmental quality. It improves asthma treatment for children by organizing individual care, monitoring the condition, and intervening before an acute episode occurs. The

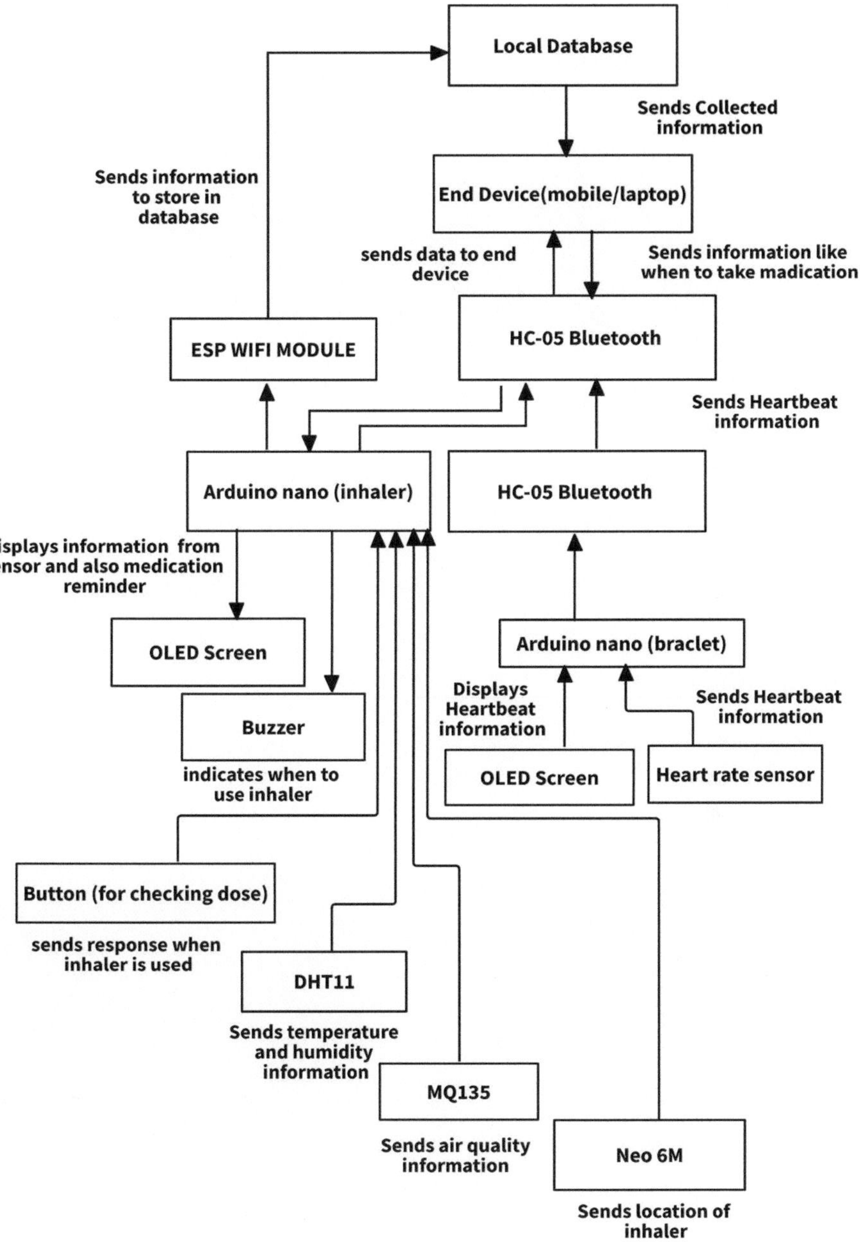

Fig. 2 System architecture diagram

Fig. 3 Anduino nano inhaler implementation

system focuses on addressing the patient's medical needs using IoT and data analysis that involve prescription medications and the control of symptoms. The system was tested under different circumstances, assessed in performance, and tested its efficiency in monitoring health parameters to trigger the desired alert. The inhaler was tested in different scenarios to confirm its efficiency in real-time monitoring and alerting in ensuring effective asthma management. It can identify changes in health and surrounding environments and trigger appropriate alerts for caregivers to act upon.

The response of the system is evaluated with the deteriorating air quality, indicating a potential trigger of asthma as shown in Fig. 4. The MQ135 sensor present in the

Fig. 4 Displayed results of the gas sensed by MQ135

Fig. 5 Heart rate sensed by the pulse sensor

inhaler module detects harmful gases; hence, the measurement of air quality is taken. In this case, the testing of the module presents experiences where volatile organic compounds and other harmful gases are increased. The MQ135 sensor within the inhaler module detects these changes before relaying the data back to the inhaler. The system alerts the user whenever the gas values exceeds the predefined safety threshold of 120 ppm, and the buzzer in the inhaler sounds up an alarm to the child to use the inhaler. The system has been able to monitor the air quality and raise the alert as gas values increase to mitigate the rise of asthma.

The system detects and alerts abnormal cases of deviation in heart rates from the normal heart rate as shown in Fig. 5. The pulse sensor is integrated into the bracelet module and is continuously detecting the heartbeat of the user. When heart rates deviate from the normal range of 60–120 BPM, the system raises emergency alerts via Bluetooth transmission to the inhaler module, prompting inhaler use. The buzzer of the inhaler will buzz, and the child will be instructed to take his inhaler in use. It can be inferred from these results that the system effectively detects abnormal rates of the heart and raises emergency alerts to make corrective interventions to prevent complications of health.

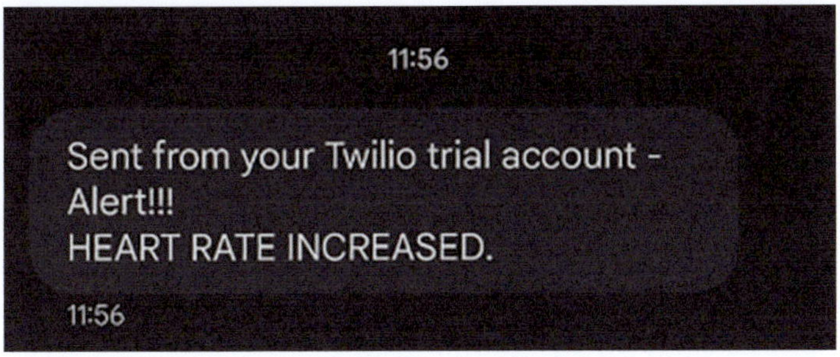

Fig. 6 Alert message

The system's Wi-Fi module transmits gathered data to caregivers' devices, enabling real-time health monitoring. Abnormal conditions trigger alerts, as shown in Fig. 6. This feature ensures caregivers can promptly respond to changes in the child's health, enhancing safety and security. The system was evaluated under various conditions to assess its ability to communicate critical health alerts. The Wi-Fi module connects to caregivers' devices, sending alerts when gas sensor values or heart rates exceed safety limits. Caregivers receive detailed alert messages, enabling timely response and monitoring.

Existing asthma management systems are made up of scheduled check-ups and data input on patients' condition, they infrequently include real-time alerting. These limitations are solved in this study that has continuous monitoring, automatic data transmission, and alerts in real time. This integration greatly enhances asthma by allowing the intervention to occur before a severe incident occurs.

5.1 Impact on Asthma Management

The smart asthma inhaler system improves asthma self-care, monitoring, and intervention of children with asthma using technology. They aimed to use IoT solutions and big data to adjust the treatment strategy based on the needs of patients and promote correct use of medications and control of symptoms. Monitoring existing vital parameters of the environment, including heart rate, temperature, humidity, and air quality, makes it possible to identify the probability of asthma attacks and prevent them, which, in its turn, dramatically decreases the possibility of severe forms of the disease. The system also enables the device to send messages to the caregivers instantly and this includes alert messages and reminders to ensure that parents are aware of their childs' health status.

This proactive approach not only improves the idea that children and caregivers can play an active role in the disease, but also improves the quality of life of the child. Constantly monitoring and informing the system also prompts the proper behavior and minimizes asthma attacks occurrences, as well as ensuring parents the feeling of protection for their children. Moreover, future development, including the integration of machine learning for the purpose of alerting asthma attack and having an application for real-time data analysis, lends credibility and reliability to the system. That being said, the Smart Asthma Inhaler creates a new benchmark for pediatric asthma care and promising the world a smarter and more advanced asthma management solutions.

5.2 Impact of Cost on Implementation

The smart asthma inhaler system transforms the manner in which asthma is managed particularly in children since it provides customized treatment, consistent tracking, timely intervention, and constant care to both the patient and the parent. Its contribution towards asthma management is quite compelling, since it helps in disease controlling, increasing the quality of life, and encouraging users to adopt preventive strategies in the occurrence of an asthma attack.

The cost issue is highly relevant when it comes to the implementation of the system and, more specifically, the choice of materials needed for the accomplishment of the project. The acquisition of IoT components and sensors, connectivity methods, and storage impact the cost of the system. For operational dependability, tools have to be made of long-lasting high-quality materials, which could be very costly at first. It is now important to maintain the quality of these materials while at the same time making the system affordable to the larger population, especially those who live in the low-income areas.

Hence, strategies involve analyzing the cost effects on the materials used for the project and appropriating the findings to determine the cost of the project and the feasibility of the project in delivering sustainable management of asthma. This consists of providing for the increased and efficient, effective and dependable hardware and at the same time being conscious of the costs of acquisition of such equipment. Other possibilities for cost reduction in the future could be obtained by purchasing in larger quantities, manufacturing technology improvement as well as the creation of cheaper yet equally effective sensors. In addition, it is possible to investigate funding options for the service, such as cooperation with healthcare providers, insurance companies, and government health programs to make the price more affordable and reach as many people as possible. In the long run, improvements can be made in technology bringing down costs of production even more and therefore increase the coverage.

6 Conclusion and Future Scope

The enhanced system of the smart asthma inhaler is a demonstration of its decent behavior in the improvement of control measures for childrens' asthma through the integration of different technologies. It provides a high level of detail in monitoring childrens' health and the environment, encouraging proactive behavior, decreasing the risk of serious asthma attacks, and significantly improving the quality of life of the patient.

The potential for future enhancements, such as incorporating machine learning algorithms for predicting asthma attacks and introducing a mobile application for real-time data monitoring, further strengthens the system's capabilities. Overall, the smart asthma inhaler sets a new standard in pediatric asthma care, paving the way for smarter and more advanced asthma management solutions.

References

1. Aparna P (2023) Adherence to inhaler technique among children with asthma. Int J Res Paediatr Nurs. Available from: https://doi.org/10.33545/26641291.2023.v5.i1b.121
2. Kavyashree, Lavanya K, Nikitha K, Sowmya, Deepthi SR (2021) Low-cost wearable asthma monitoring system with a smart inhaler. In: 2021 Asian conference on innovation in technology (ASIANCON), Pune, India, pp 1–7. https://doi.org/10.1109/ASIANCON51346.2021.9544970
3. Margam R (2024) Smart inhalers: harnessing IoT for precise asthma management. Int Educ Res J 10
4. Isaac N, Sampath N, Gay V (2018) Modernising asthma management: personalised asthma action plans using a smartphone application. In: 12th international symposium on medical information and communication technology (ISMICT). Sydney, NSW, Australia, pp 1–5. https://doi.org/10.1109/ISMICT.2018.8573707
5. O'Connor A, Tai A, Kopsaftis Z, Carson-Chahhoud K (2021) Qualitative study protocol: augmented reality technology to deliver asthma inhaler technique training for children and adolescents with asthma. Int J Qual Methods 20:16094069211042228
6. Sportel ET, Wolcherink MJO, van der Palen J, Lenferink A, Thio BJ, Movig KLL, Brusse-Keizer MGJ (2020) Does immediate smart feedback on therapy adherence and inhalation technique improve asthma control in children with uncontrolled asthma? A study protocol of the IMAGINE I study. Trials. Available from. https://doi.org/10.1186/S13063-020-04694-4
7. Shahida M, Choudhury S, Venugopal V, Parhi C, Suganya J (2021) A smart monitoring system for asthma patients using IoT. In: 2021 fifth international conference on I-SMAC (IoT in social, mobile, analytics and cloud) (I-SMAC), Palladam, India, pp 130–135. https://doi.org/10.1109/I-SMAC52330.2021.9641059
8. Naudiyal A, Aswal S, Anandaram H, Joshi K, Sidhu KS, Singh A (2023) Effect of cost on Internet of things-based integrated remote electronic health surveillance, self-regulating home, and security alert system: a review analysis. In: 2023 6th international conference on contemporary computing and informatics (IC3I), Gautam Buddha Nagar, India, pp 998–1003. https://doi.org/10.1109/IC3I59117.2023.10398027
9. Mwangi A, Ndashimye E, Karikumutima B, Ray SK (2020) An IoT-alert system for chronic asthma patients. In: 11th IEEE annual information technology, electronics and mobile communication conference (IEMCON). Vancouver, BC, Canada, pp 0012–0019. https://doi.org/10.1109/IEMCON51383.2020.9284816
10. Qin Y, Wang J, Qin J, Yang N, Li S, Xu L, Han Y (2022) Correlation meta-analysis of the efficacy of inhaled corticosteroids in children with asthma based on smart medical health. J Healthc Eng. Available from. https://doi.org/10.1155/2022/6220774
11. Murugesan D, Kaviyarasan M, Matheswari S, Preethika M, Tamilarasan B, IOT based assist device for pulmonary diseased patients monitoring framework
12. Tippannavar SS, Yashwanth SD, Madhu Sudan MS, Puneeth KM, Chandrashekar Murthy BN, Prasad V (2023) BreatheFree—a smart inhaler for a smart, safe & healthier world. In: IEEE renewable energy and sustainable E-mobility conference (RESEM). Bhopal, India, pp 1–6. https://doi.org/10.1109/RESEM57584.2023.10236222
13. Lijina P, Nippun KA (2016) Bluetooth RSSI based collision avoidance in multirobot environment. In: 2016 international conference on advances in computing, communications and informatics (ICACCI), Jaipur, India, 2016, pp 2168–2174. https://doi.org/10.1109/ICACCI.2016.7732373
14. Megalingam RK, Greeshma MG, Pillai SS (2019) Design and implementation of intelligent crutches for medical applications. In: 2019 international conference on communication and signal processing (ICCSP), Chennai, India, pp 0926–0929. https://doi.org/10.1109/ICCSP.2019.8697940
15. Rajeshkumar A, Mathi S (2022) Smart solution for reducing COVID-19 risk using Internet of things. Indonesian J Electr Eng Comput Sci 25(1):474–480

16. Lokeswar Reddy K, Thangam S (2022) Predicting relapse of the myocardial infarction in hospitalized patients. In: 3rd international conference for emerging technology (INCET). Belgaum, India, pp 1–7. https://doi.org/10.1109/INCET54531.2022.9825439
17. Tejaswi TSM, Supritha V (2023) Low cost precision farming and remote monitoring using cloud. In: 2023 14th international conference on computing communication and networking technologies (ICCCNT), Delhi, India, 2023, pp 1–6. https://doi.org/10.1109/ICCCNT56998.2023.10306633
18. Yadav CS, Gupta S (2022) A review on malware analysis for IoT and android system. SN Comput Sci 4(2):118
19. Gupta S (2022) Private blockchain-cloud system (PBCS) for healthcare services. Int J Softw Innov (IJSI) 10(2):1–8

Automatic Speech Recognition with Customized Bi-GRU Hybrid Model for Real-Time Speech Analysis

Anushka Tonk, Arsh Akhtar, and P. Pankaja Lakshmi

Abstract Automatic Speech Recognition (ASR) is an emerging technology in today's digital world, allowing for seamless human–computer interaction via voice commands, transcription services, and other features. This work presents an end-to-end model capable of transcribing the real-time captured speech into text. A Customized Bi-directional Gated Recurrent Unit (Bi-GRU) with convolutional neural network (CNN) along with the technique of layer normalization. The model incorporates convolutional layers to robustly extract features from raw audio signals, combined with layer normalization and Bi-GRUs to capture long-range temporal dependencies and contextual information from both past and future frames. Significant improvements were observed over five training epochs. The training and validation loss have been reduced from 311.9084 to 70.7906 and 303.5839 to 46.6886, respectively. The dataset provides comprehensive metadata in transcript file, assisting an extensive number of research initiatives in speech synthesis and synthetic speech detection. The audio clip encoded as a single-channel 16-bit PCM.wav file at a sample rate of 22,050 Hz. The experiment was done to design more accurate and reliable speech recognition systems and highlight the effect of layer normalization in reducing the CTC loss.

Keywords ASR · Deep learning · Bi-GRU · Mel spectrogram

A. Tonk (✉) · A. Akhtar · P. Pankaja Lakshmi
Vellore Institute of Technology, Chennai, Tamil Nadu, India
e-mail: anushka.tonk2021@vitstudent.ac.in

A. Akhtar
e-mail: arsh.akhtar2021@vitstudent.ac.in

P. Pankaja Lakshmi
e-mail: pankajalakshmi.p@vit.ac.in

H. Sharma et al. (eds.), *Artificial Intelligence: Theory and Applications*, Lecture Notes in Networks and Systems 5589, https://doi.org/10.1007/978-981-96-1687-9_15

1 Introduction

ASR technology has advanced rapidly in recent years, enabling seamless human–computer interaction through voice commands, transcription services, and other applications. Traditional ASR systems usually contain a complex pipeline of tasks like feature extraction, acoustic modeling, and language modeling, each requiring substantial expertise and manual tuning. The recent advances in deep learning have enabled end-to-end models that can jointly learn all required components from raw audio and transcribed text data. The paper by Hannun et al. [1] presents an end-to-end speech recognition model inspired by the DeepSpeech2 framework. The model incorporates convolutional layers to robustly extract features from raw audio signals, combined with layer normalization as discussed in Kiros et al. [2] and about Bi-GRUs as discussed in paper by Saxena et al. [3] to capture long-range temporal dependencies and contextual information from both past and future frames. This architecture enhances the model's ability to accurately recognize and transcribe speech in English. Figure 1 depicts the overall flow diagram of the model. The end-to-end model discussed in the paper is trained on the LJ Speech Dataset, a public domain collection of 13,100 audio recordings with corresponding transcripts from a single speaker reading non-fiction texts. Each 16-bit mono-channel audio clip is sampled at 22,050 Hz, with metadata provided to assist speech synthesis and recognition research Arora et al. [4], since understanding the number of audio channels is crucial for properly processing the input data. After training for 5 epochs on this large audio dataset, the model achieved impressive training and validation losses, demonstrating its capability to generalize and transcribe unseen speech data accurately, which can be referred to the study by Hannun et al. [1] and Reza et al. [5]. The field of ASR has been shaped by foundational works such as Levis et al. [6], which provided a comprehensive overview of ASR technology and its key components, and Brown et al. [7], which tackled the acoustic modeling. Their research highlighted the challenges of transcribing diverse speech patterns and introduced ASR models to handle speaker variations and background noise. Feature extraction using Mel-Frequency Cepstral Coefficients (MFCC) and a hybrid architecture that combines transformers and CNN was given by Kadam et al. [8]. Through the development of complex end-to-end neural architectures and basic Hidden Markov Models (HMMs) as described by Rabiner et al. [9], research has greatly increased recognition efficiency and accuracy.

The challenges in ASR includes:

1. Dataset with different accents, background noise, and speaking styles.
2. Due to their reliance on manually created features and statistical models, which struggled to generalize across different datasets and environments.
3. Deep learning and neural networks have completely changed the field and made it possible to create models that are more resilient and flexible.

The primary objective of this research paper is:

1. To develop an end-to-end ASR model that integrates convolutional layers, layer normalization, and Bi-GRUs for improved feature extraction and temporal

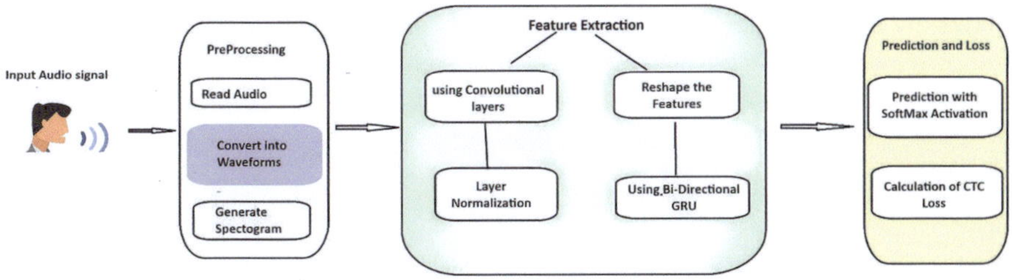

Fig. 1 Flow diagram of the model

context capture, in order to compare the proposed model's performance with existing state-of-the-art ASR systems.

2. To highlight the effect of layer normalization in reducing the CTC loss in the final stages.

Section 2 provides a comprehensive review of the literature, emphasizing significant developments and current research directions in the field of automatic speech recognition. Section 3 presents the methodology along with a brief discussion of Mel spectrum patterns, Gated Recurrent Units (GRUs) and the layer normalization procedure along with a few additional details. After that, we describe the experimental design, dataset gathering, and preprocessing methods in Sect. 4. Section 5 concludes by discussing the result obtained and limitations of our study and suggesting directions for further research in the area, with final conclusion depicted in Sect. 6.

The contributions in this research work include:

1. An end-to-end model capable of transcribing the real-time captured speech into text is proposed.
2. The model has been implemented to extract crucial features for accurate transcription and learn input-output alignments without pre-aligned data.

2 Literature Survey

Previous studies, as discussed in Table 1, have explored various approaches for ASR, with diverse model architectures like Residual Convolutional Neural Networks (ResCNN) and Bi-GRU, in Reza et al. [5], proposed a customized ResCNN and Bi-GRU-based ASR model that incorporates layer normalization instead of batch normalization, improving model training and performance. Leveraging these advancements, their model achieved a Character Error Rate (CER) of 4.7% and 3.61% on the LibriSpeech corpus and LJ Speech dataset, respectively, Raghu et al. [10]. This model, trained without the need for external language models, demonstrates the importance of dataset size and novel architectural choices in enhancing ASR performance, Reza et al. [5]. The above study and research work performed by Raghu et al. [10] tackled the computational intensity and inefficiency of traditional models by incorporating ResCNN and Bi-GRU architectures with layer normalization, achieving significant performance improvements without relying on external language models.

Table 1 Key literature in the field of ASR, presenting significant data from a range of influential papers

S. No.	Author details	Title	Model used	Inference
1	Reza et al. [5]	A customized residual neural network and bi-directional gated recurrent unit-based automatic speech recognition model	Customized Residual Neural Network, Bi-GRU	Improved ASR accuracy and robustness, showing significant performance gains over traditional models
2	Lee et al. [12]	Intermediate Loss Regularization for CTC-Based Speech Recognition	CTC, Intermediate Loss Regularization	Enhanced ASR performance by mitigating overfitting and improving generalization, especially in noisy environments
3	Saxena et al. [3]	Introduction to Gated Recurrent Unit	GRU	Provided foundational understanding and practical insights into the implementation and benefits of GRU in sequence modeling
4	Wang et al. [13]	Residual Convolutional CTC Networks for Automatic Speech Recognition	Residual Convolutional Networks, CTC	Demonstrated that residual convolutional networks with CTC achieve state-of-the-art performance in ASR tasks
5	Raghu et al. [10]	A Perspective Study on Speech Emotion Recognition: Databases, Features, and Classification Models	Various databases, feature extraction techniques, classification models	Reviewed existing SER systems, highlighted challenges, and proposed future research directions
6	Malik et al. [11]	Automatic speech recognition: a survey	Overview of ASR technologies, challenges, applications	Provided a comprehensive overview of ASR evolution, core technologies, challenges, applications, and future research directions
7	Rabiner et al. [9]	An introduction to hidden Markov models	Hidden Markov Models (HMM)	Introduced HMMs as a powerful statistical tool for modeling temporal sequences, significantly impacting ASR development

The paper by Malik et al. [11] provides an in-depth survey of the field of ASR, as it follows the evolution of ASR from its early stages to modern advancements, highlighting significant milestones and technological breakthroughs. The survey discusses core technologies such as feature extraction, acoustic models, language models, and decoding strategies, as well as traditional methods such as Hidden Markov Models (HMMs) and modern deep learning techniques. HMMs, outlined in Rabiner et al. [9], are statistical models used to represent systems that transition between a set of hidden states in a probabilistic manner.

The ReduceLROnPlateau callback is used to reduce the learning rate when a metric has stopped improving. This helps the model to converge faster and potentially reach a better final state. In the context of mental health assessment, this callback could be beneficial in fine-tuning the model's performance and ensuring that it is learning effectively from the emotion analysis and scales data, discussed in the paper of Lee et al. [12]. This paper focuses on enhancing Connectionist Temporal Classification (CTC) models for ASR by proposing an intermediate CTC loss function incorporated into an intermediate layer of the CTC encoder network. By regularizing the training process, this strategy aims to optimize model performance at a minimal computational cost. Additionally, the paper by Saxena et al. [3], explores combining the intermediate CTC loss with stochastic depth training, which further enhances the model's robustness and performance. The approach in the study by Wang et al. [13] demonstrated notable improvements in word error rates (WER) and CER when tested on datasets like the Wall Street Journal (WSJ) and AISHELL-1. It was able to achieve competitive performance without the need of beam search algorithms or external language models. [13] The latest advancements in ASR, as highlighted in the paper by Kheddar et al. [14], include the development of end-to-end models and hybrid architectures. Models like those combining convolutional neural networks (CNNs) with transformers, have significantly improved the accuracy and efficiency of ASR systems by integration the attention mechanisms and the Transformer architecture. Despite these advancements, challenges remain, such as ensuring model scalability, managing computational costs, and maintaining accuracy across diverse datasets.

3 Materials and Methods

3.1 Dataset

The dataset used in this work is the LJ Speech dataset which is a public domain dataset comprising 13,100 short audio clips of a single speaker reading passages from seven non-fiction books, making up to a total of 2.6 Gigabytes. Each audio clip varies in length from 1 to 10 s, totaling approximately to 24 h of recorded audio, each accompanied by an audio transcription. The texts were published between 1884 and 1964, whereas the audio was recorded in 2016–2017 under the LibriVox project, both available in the public domain. The dataset provides audio files in single-channel 16-bit PCM WAV format with a sample rate of 22,050 Hz. Metadata for each clip, including the transcription and normalized transcription, is provided in the transcripts file.

3.2 Preprocessing

The dataset was processed by converting it into spectrograms along with respective label from the transcript file and divided into train set and validation set each having of size 11,790 and 1310, respectively, with a train test split ratio of 9:1.

To prepare the dataset for training and validation, a vocabulary is constructed to represent the characters present in the transcriptions. The vocabulary includes lowercase alphabets (a-z) along with common punctuation marks—[“.”, “,”, “?”, “&”, “!”, “;”, “-”, “’”, “ ”] and a space character. Each character in the transcription is then mapped to its corresponding index in the vocabulary. Out-of-vocabulary tokens are handled by replacing them with an empty string. For the audio data, each audio file is read and decoded into a waveform. The waveform is converted into a spectrogram using the STFT, whose absolute value gives the required spectrogram. The spectrograms are then padded to ensure uniform length across examples, with a batch size of 32. All dataset processing steps were implemented using the TensorFlow framework. Finally, the processed batches are prefetched to optimize training performance.

3.3 Proposed Methodology

The end-to-end ASR approach as depicted in Fig. 2 begins with data preprocessing, in which input audio files are read and converted into waveforms using TensorFlow. The Short-Time Fourier Transform (STFT) technique is then used to transform these waveforms into spectrograms, which provide a time–frequency representation of the audio signals. Using spectrogram transformation and CTC loss, the model was able to extract crucial features for accurate transcription and learn input–output alignments without pre-aligned data. The spectrograms, as shown in Fig. 3, are subsequently normalized to ensure consistency and enhance model performance. The normalized spectrograms are passed through a series of convolutional layers in the feature extraction phase. Each layer applies filters to extract hierarchical features from the spectrograms, capturing key patterns and attributes of the audio signal. Each convolutional layer is followed by layer normalization to enhance convergence and stabilize the learning process, the proposed architecture of the model can be referred in Fig. 5.

These reshaped feature maps are then fed into a stack of Bi-GRU layers. The GRU layers are designed to capture temporal dependencies in the feature maps, allowing the model to understand the sequential nature of speech, as discussed in the study by Maas et al. [15]. A dense layer with a softmax activation function

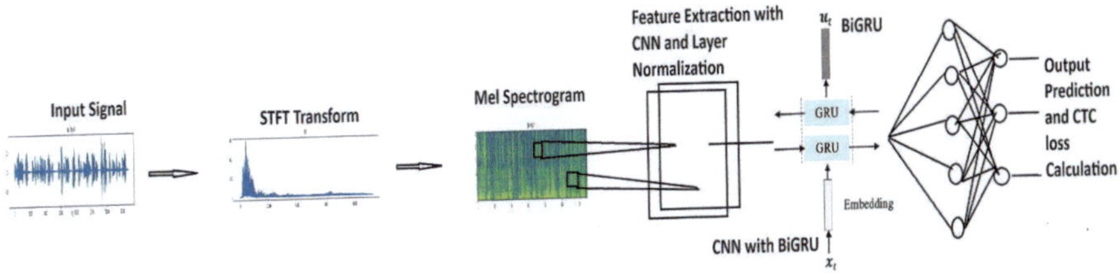

Fig. 2 Proposed end-to-end ASR model

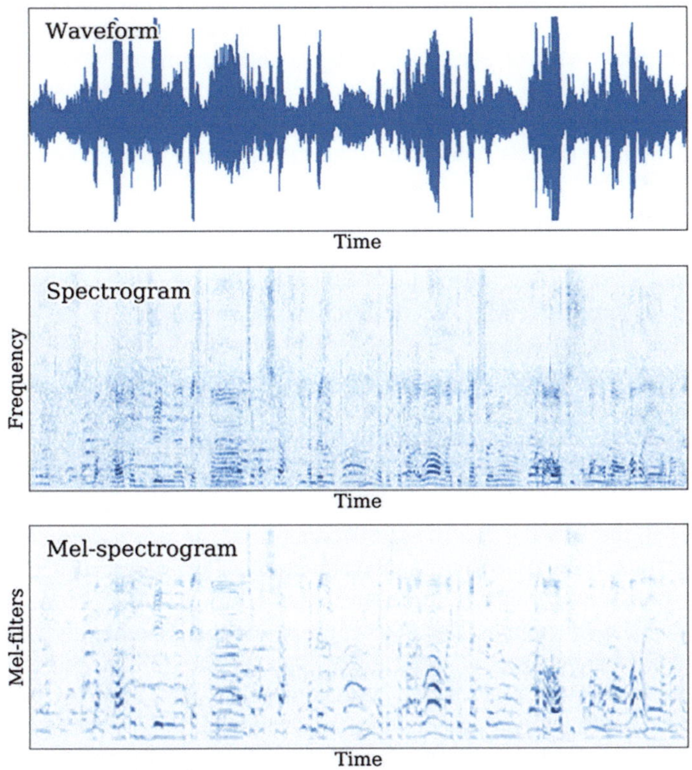

Fig. 3 Mel spectrogram for an input signal

forms the final output layer. It generates probabilities for every character in the vocabulary, along with a unique token for unknown characters. The model is compiled using the CTC loss function, which is particularly effective for sequence-to-sequence tasks where the input and output lengths can vary. The training process utilizes callbacks for early stopping and learning rate reduction to prevent overfitting and ensure optimal performance. The dataset, which includes sections from seven non-fiction books published between 1884 and 1964, contains nearly 24 h of speech data, whose in-depth description can be found in Sect. 4.

3.4 Mel Spectrogram

Designed to convert English audio speech into text, DeepSpeech2 is a popular end-to-end speech recognition model that was first released in December 2015 [2]. Its architecture consists of converting input speech into Mel Spectrograms as shown in Fig. 3 by De Benito-Gorron et al. [16], using a combination of CNNs and Recurrent Neural Networks (RNNs) with Long Short-Term Memory (LSTM) units, and using the CTC loss function for alignment and transcription. The end-to-end model further integrates language models, which improve the authenticity of the generated text. The model described in the paper by Hannun et al. [1], was trained on datasets such as AN4, LibriSpeech, and TEDLIUM, demonstrating its ability to handle different voice input and generate accurate transcriptions. The ASR discussed in the above paper works by converting the audio signals into Mel Spectrograms which can be passed into the model. A Mel spectrogram as shown in Fig. 3 is an audio signal representation that focuses on frequencies that are more perceptually relevant to human hearing.

For creating a Mel spectrogram, the audio signal is first divided into short, overlapping frames. For each frame, the STFT is applied to obtain the power spectrum, which represents the distribution of signal energy across different frequencies as discussed in paper by Bhangale et al. [17]. In the context of this research article Yin et al. [18], the STFT is likely used for analyzing the spectral content of audio signals over time. It involves dividing the signal into short segments and computing the Fourier transform for each segment, providing a time-varying representation of the signal's frequency content. The Fourier transform, on the other hand, is a mathematical tool used to decompose a signal into its constituent frequencies. It provides a static representation of the frequency content of a signal, making it suitable for analyzing stationary signals or obtaining an overall frequency profile of a signal as discussed in the study by Yin et al. [18]. The Mel spectrogram is produced by mapping the power spectrum onto the Mel scale, yielding a two-dimensional representation with one axis representing time and the other representing frequency (in mel scale), in Bhangale et al. [17].

In audio signal processing, the frame length refers to the duration of each segment of audio data that is analyzed at a time. It is typically measured in milliseconds (ms) or number of audio samples. A longer frame length captures more temporal information but may result in less frequency resolution. Shorter frame lengths offer higher frequency resolution but may lose some temporal context. The frame step, also known as hop size or frame shift, determines the overlap between consecutive frames in audio signal processing. It specifies the amount of time or number of samples between the starting points of adjacent frames. A smaller frame step increases temporal resolution by providing more overlap between frames, allowing for smoother transitions between segments. However, it also increases computational complexity. Conversely, a larger frame step reduces computational load but may lead to information loss, especially in transient or rapidly changing audio signals. Adjusting the frame step is crucial for balancing temporal resolution and computational efficiency in audio feature extraction. The frame step value for the end-to-end model discussed in this paper is 128, whereas the frame length value has been kept as 225 based on the observation, since it is a crucial parameter in the feature extraction method in STFT.

3.5 Gate Recurrent Unit (GRU)

After getting the spectrograms for each audio wave file along with the labels, the model architecture contains a stack of convolutional layers and bi-directional layers embedded with GRU layers. Based on the referenced paper Reza et al. [5], GRUs are a type of RNN that are used to capture temporal dependencies in sequential data. GRUs are similar to LSTM units shown in Fig. 4a, but have a simpler architecture, combining the forget gate and input gate into a single update gate. However, in the context of this paper [5] on speech recognition, GRUs shown in Fig. 4b were chosen for their balance of efficiency and performance, making them well-suited for the temporal nature of speech signals present in the dataset. GRU, like LSTM, uses gates to control the flow of information. They are relatively new in comparison with LSTM. This is the reason why they outperform LSTM while having a simpler architecture [13].

Unlike LSTM, a separate cell state (C_t) is absent in GRU, instead it only has a hidden state (H_t). This also makes GRUs faster to train due to its simpler architecture. The inputs for a GRU are an X_t which is taken from timestamp t, and an H_{t-1} being the hidden state input from the previous timestamp $t-1$ as per Saxena et al. [3]. Later it outputs a new hidden state H_t which again passes to the next timestamp. There are primarily two gates in a GRU as opposed to three gates in an LSTM cell. The first gate is the reset gate and the other one is the update gate. The reset gate is responsible for the short-term memory of the network, i.e., the hidden state (H_t).

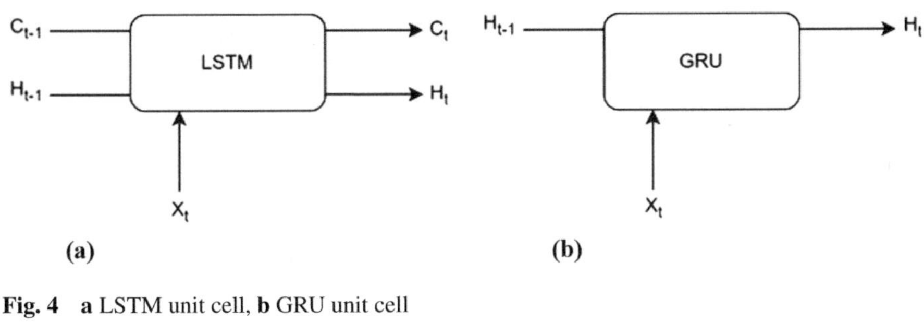

Fig. 4 **a** LSTM unit cell, **b** GRU unit cell

Fig. 5 End-to-end ASR model architecture

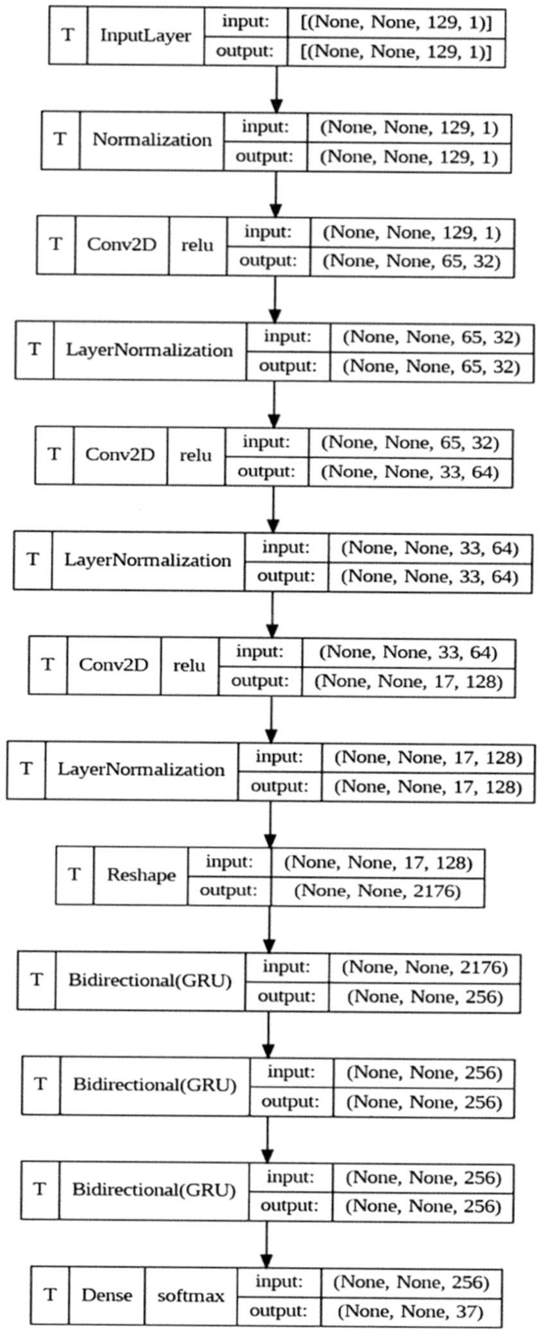

Here is the equation of the reset gate:

$$r_t = \sigma(X_t * U_r + H_{t-1} * W_r) \tag{1}$$

The update gate is responsible for the long-term memory and its equation is given below:

$$r_t = \sigma(X_t * U_u + H_{t-1} * W_u) \tag{2}$$

After evaluation of sigmoid, r_t ranges from 0 to 1. Here, $\mathrm{U_r}$, U_u, U_g, W_u, $W_{g,}$ and W_r are weight matrices. GRUs operates through a two-step process to compute the hidden state H_t. The candidate hidden state equation:

$$\hat{H}_t = \tanh\big(X_t * U_g + (r_t \odot H_{t-1}) * W_g\big) \tag{3}$$

Initially, a candidate hidden state is generated by incorporating the input and the previous timestamp's hidden state, scaled by the reset gate output r_t, and then applies the tanh function to produce the candidate's hidden state. The influence of the previous hidden state on the candidate state is controlled by the reset gate, where $r_t = 1$ signifies full consideration and $r_t = 0$ indicates complete disregard. Subsequently, the update gate regulates the incorporation of both historical information from H_{t-1} and new information from the candidate state and generates the current hidden state H_t. The hidden state equation:

$$H_t = U_t * H_{t-1} + (1 - U_{t-1}) * \widehat{H}_t) \tag{4}$$

This mechanism allows GRU to adaptively update its hidden state based on the input and the previous state, facilitating effective sequence modeling as described in the study by Wang et al. [13].

3.6 *Layer Normalization*

The architecture of the end-to-end model described in the paper have a stack of convolutional layers along with layer normalization technique. Layer normalization as discussed in the paper by Shanthi et al. [19], unlike batch normalizing, performs the same computation at both the training and testing stages. It is also straightforward to apply to recurrent neural networks by collecting the normalization statistics individually for each time step. Layer normalization is a highly effective method for stabilizing the hidden state dynamics in recurrent networks. Empirically, it can be observed from the study by Shanthi et al. [19], that layer normalization can substantially reduce the training time compared to previously published approaches. In a standard RNN, the summed inputs in the recurrent layer are computed from the current input X_t and previous vector of hidden states H_{t-1} which are computed as $a_t = W_{hh}H_{t-1} + W_{xh}X_t$. The layer normalized recurrent layer re-centers and re-scales its activations using the extra normalization terms as shown below:

$$h_t = f\left[\frac{g}{\sigma^t}(a_t - \mu_t) + b\right] \tag{1}$$

$$\mu_t = \frac{1}{H}\sum_{i=1}^{H} a_t^i \tag{2}$$

$$\sigma_t = \sqrt{\frac{1}{H}\sum_{i=1}^{H}(a_t^i - \mu_t)^2} \tag{3}$$

3.7 Callbacks Used

In deep learning, callbacks are specialized functions that can be inserted at various stages of the training process to monitor and optimize the model's performance. They enable dynamic interventions during training, such as adjusting the learning rate, saving the model's best weights, or stopping the training early if needed, for optimizing the training process, preventing overfitting, and ensuring that the model generalizes well to unseen data. For this research, the following callbacks were utilized:

1. The ReduceLROnPlateau callback, which reduces the learning rate when the monitored metric, such as validation loss, has stopped improving. This is particularly useful for fine-tuning the learning process. The callback was configured with the attributes:

   a. *monitor = 'val_loss'*
   b. *factor = 0.2*
   c. *patience = 5*
   d. *min_lr = 0.001*

   This setup ensures that if the validation loss does not improve for 5 consecutive epochs, the learning rate is reduced by a factor of 0.2, without dropping below 0.001.
2. The EarlyStopping callback halts the training process when the validation loss ceases to improve, preventing overfitting. Configured with:

   e. *monitor = 'val_loss'*
   f. *patience = 10*
   g. *restore_best_weights = True*

> It stops the training if there is no improvement in the validation loss for 10 consecutive epochs, restoring the weights from the epoch with the best validation performance. These callbacks enhance the training process, making it more efficient and effective by ensuring the model does not overfit and continues to improve or stabilize, leading to better performance on unseen data.

It stops the training if there is no improvement in the validation loss for 10 consecutive epochs, restoring the weights from the epoch with the best validation performance. These callbacks enhance the training process, making it more efficient and effective by ensuring the model does not overfit and continues to improve or stabilize, leading to better performance on the unseen data.

3.8 Model Architecture

The end-to-end model outlined in the research paper starts with spectrogram inputs and applies normalization before going through three convolutional layers with strides and increasing filter sizes, each of which is then followed by layer normalization described in Fig. 5. To capture temporal dependencies bi-directionally, the resulting feature maps are reshaped and fed into three Bi-GRU layers. The final output layer uses a softmax activation to predict the probability distribution over the vocabulary size adding one for the '*[LINK]*' token. The architecture is illustrated below:

The final layer which is giving the output of the probabilities over the character set is a dense layer with a softmax activation. It is a mathematical function that converts a vector of arbitrary real values into a probability distribution, with each value representing the likelihood of a corresponding class or category. Its output values range between 0 and 1, and the sum of all output values is equal to 1. The model whose architecture is depicted in Fig. 5, is trained using the CTC loss function and Adam optimizer. The Adam optimizer is a gradient-based optimization algorithm for updating the parameters of the neural networks by efficiently adjusting the learning rate for each parameter based on exponentially decaying averages of past gradients and their squares. In model training, the batch size refers to the number of samples processed by the model in each iteration during training. Instead of updating the model's parameters after each individual sample (which can be computationally inefficient), training is typically performed in batches.

The end-to-end model is trained with a batch size of 32, which determines how many samples are propagated through the network before the model's parameters are updated based on the computed gradients. The learning rate is a hyperparameter that controls the size of the step taken during the optimization process of training a machine learning model. It determines how much the model parameters are adjusted with respect to the gradient of the loss function. The learning rate is $1 \times e^{-3}$ in the final epoch. It achieves accurate transcription by effectively capturing the complexities of speech signals and learning from unaligned sequences.

Table 2 Hyperparameters used in this model

Hyperparameters used in end-to-end model	
Learning rate	$1 \times e^{-3}$
Batch size	32
Frame length	225
Frame step	128
Epochs	5
Training time	48,900 s (13 h approx)

4 Experimental Setup

The training was done on a system with Linux Mint 20.3 Cinnamon operating system, powered by an Intel® Core™ i7-11700 Processor, with 64 GB of RAM and a 2 TB ROM. The system provided ample resources for training purposes. Graphics processing was facilitated by a 12 GB NVIDIA GeForce RTX 3060, ensuring smooth visualization during model training and evaluation. This setup offered a reliable computing environment for conducting research and experimentation effectively. The loss values have dramatically dropped over the course of the five training epochs, indicating an improvement in the model's performance.

In the first epoch, the training loss started at 311.9084, while the validation loss was 303.5839. By the second epoch, the training loss had dropped to 263.7853 and the validation loss to 195.7637. In the third epoch, this tendency continued as the training loss decreased to 139.0291 and the validation loss to 87.8569. There was higher progress in the fourth epoch, which had a training loss of 89.1267 and a validation loss of 59.0740. In the end, the training loss was discovered to be 70.7906 in the fifth epoch, while the validation loss decreased to 46.6886, see Fig. 7 for reference. Learning was the last set of hyperparameters utilized to train the model, for more details about the hyperparameters refer to Table 2.

5 Results and Discussion

5.1 Spectrogram Obtained

An essential step in the ASR model is transforming the audio input shown in Fig. 6a into a spectrogram, depicted in Fig. 6c, which transforms the time-domain data into a time–frequency representation using Short-Time Fourier Transform as shown in Fig. 6b. We can observe the unique patterns and structures that correlate to various phonetic aspects of speech by examining the spectrogram Fig. 6c. This visualization highlights the model's ability to extract meaningful features from the raw audio signal, laying the foundation for subsequent layers to process and decode the speech

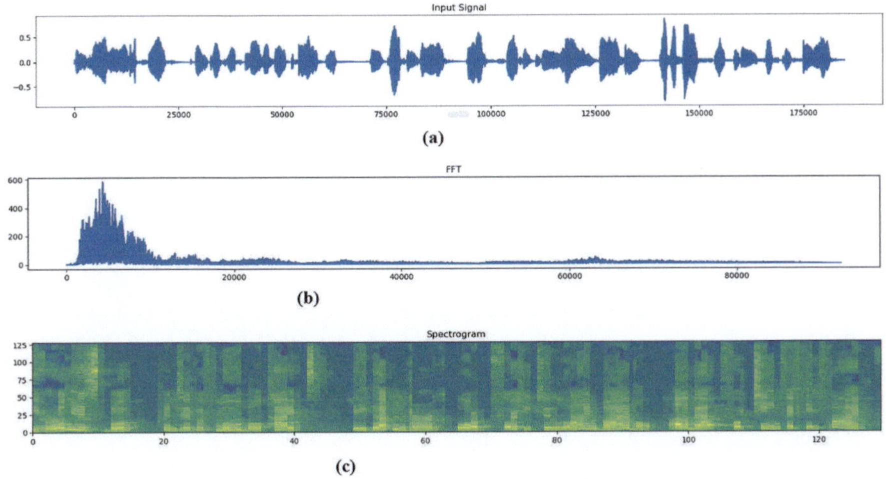

Fig. 6 **a** Input signal, **b** short time Fourier transform on input signal, **c** Mel spectrogram for an input signal

into text. The spectrogram's detailed representation ensures that the model captures the variations in speech, leading to more accurate and robust transcription results.

5.2 Connectionist Temporal Classification (CTC) Loss

CTC is a loss function used in ASR and other sequence-to-sequence tasks where the alignment between the input and output sequences is unknown. Traditional approaches to sequence labeling rely on having pre-aligned input and output sequences, but CTC allows the model to learn this alignment on its own. This makes it particularly useful for tasks like speech recognition, where the exact timing of each word or phoneme in the input signal is not explicitly labeled as per the knowledge gained in the information obtained from neuralearn source [20].

CTC works by introducing a special "blank" token and defining a many-to-one mapping from the input sequences to the output sequences. In this work, while defining the vocabulary, we have named that empty/blank token as "[LINK]." During

Fig. 7 Training CTC loss and validation CTC loss over 5 epochs

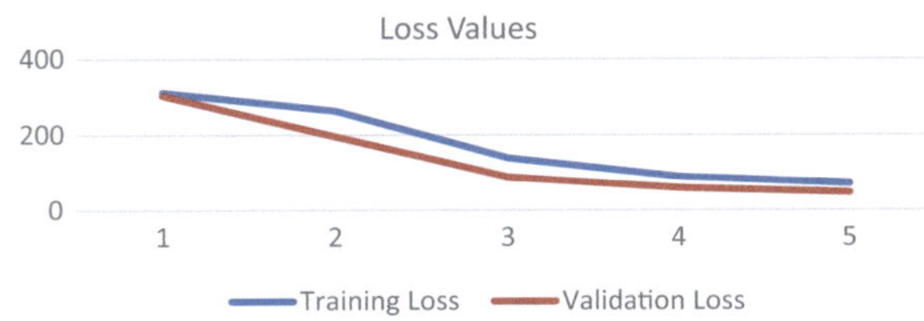

training, it considers all possible alignments of the input sequence that could produce the output sequence, summing the probabilities of these alignments to calculate the loss. This approach enables the model to handle variable-length input sequences and outputs, making it robust and flexible for real-world ASR applications. The final results from this research work show a significant decrease in CTC loss over five training epochs, with the training loss dropping from 311.9084 to 70.7906 and the validation loss decreasing from 303.5839 to 46.6886. This indicates substantial improvements in the model's accuracy and robustness, validating the effectiveness of using CTC loss in ASR applications.

The end-to-end model for Automatic speech recognition was first tested without implementing layer normalization technique with 3 GRU layers having 128 filters each, which gave the value of CTC loss on the 5th epoch as 301 and loss on validation dataset as 275, trained with a batch size of 64. The model was further implemented, with three GRU layers having 32, 64 and 128 number of filters, respectively, without layer normalization, which gave the final training CTC loss of around 289.6263 and validation loss of 291.23 with batch size of 32. Finally, the model was implemented using the layer normalization technique after every Convolutional layer and the number of filters as 128 in each of the three GRU layers, converging the CTC loss.

Over the course of the five epochs of training, the loss values have significantly decreased, showing that the model's performance has improved. The training loss began at 311.9084 in the first epoch, and the validation loss was 303.5839. The validation loss decreased to 195.7637 and the training loss to 263.7853 by the second epoch. The training loss dropped to 139.0291 and the validation loss to 87.8569 in the third epoch, following this trend. The fourth epoch, which had a validation loss of 59.0740 and a training loss of 89.1267, demonstrated more improvement. Ultimately, the validation loss dropped to 46.6886 while the training loss was found to be 70.7906 in the fifth epoch as depicted in Fig. 7. In the paper by Reza et al. [5], their model achieved a CER of 4.7% and 3.61% on the LibriSpeech corpus and LJ Speech dataset, respectively. The final hyper parameters used for training the end-to-end model described in this paper were learning rate as $1 \times e^{-3}$ with frame length of 255 and frame size of 128. These outcomes validate the effectiveness of the model by showing its rising accuracy and robustness in speech recognition. The total time taken for training the model for 5 epochs was around 48,900 s approximating to 13 h, with average time per epoch as 9764.6 s. The comparison of CTC losses over various models and techniques can be seen in Table 3.

5.3 Limitations and Future Scope

The concept of transfer learning can be used to train the model on different loss functions to make it a more reliable model, instead of CTC, which will help improve the model performance in certain areas. CER can be used to evaluate the model's performance as it outputs the percentage of wrongly predicted characters from a generated

Table 3 Comparison of CTC losses for different models and techniques

S. No.	Model	Batch size	Layer normalization	Training loss	Validation loss
1	3 GRU Layers with Filters = (128, 128, 128)	64	Absent	301	275
2	3 GRU Layers with Filters = (32, 64, 128)	32	Absent	289.6263	291.2300
3	3 GRU Layers with Filters = (128, 128, 128)	32	Present	70.7906	46.6886

transcript to the actual transcript. A low value implies an accurate prediction with the ideal value being 0. It is calculated based on the following formula:

$$\text{CharErrorRate} = \frac{S + D + I}{N} = \frac{S + D + I}{S + D + C} \tag{4}$$

S number of substitutions
D number of deletions
I number of insertions
C number of correct characters
N number of characters in the reference ($N = S + D + C$)

The current model has a high CER which can be improved using the Resnet 50 layers or other transfer learning models that can be implemented along with the current model layers for better results. Furthermore, an increase in the number of epochs will result in a more optimized model and lower the loss values, which are generally caused by the noise in the input audio. One effective method involves using an autoencoder for signal noise removal. The future scope can include various method to improve the overall robustness of the model. Implementing an autoencoder can effectively remove noise from input audio signals. By training an autoencoder on noisy speech data, a clean speech features can be extracted and then the above-described ASR model. Integrated attention mechanisms, such as those found in transformer models, can enhance the ASR system's ability to focus on relevant parts of the input sequence. Also, integrating attention mechanisms into your Bi-GRU layers can help the model better handle complex linguistic patterns and improve overall performance. This approach leverages neural networks to learn a compressed representation of the input data, which can then be used to reconstruct the signal with reduced noise.

6 Conclusion

Using spectrogram transformation and CTC loss, we developed and assessed an ASR model in this work. The model transforms the raw audio signal into a spectrogram which is allowed to extract crucial features for accurate transcription. The CTC loss enabled the model to learn input–output alignments without pre-aligned data. The significant improvement was observed across five training epochs in which the training loss was decreased from 311.9084 to 70.7906, while validation loss decreased from 303.5839 to 46.6886. Further lowering CTC loss with the use of enhanced Bi-GRU layers and layer normalization improved the accuracy and resilience of the model. Further improvements can combine ResNet50 layers and use transfer learning with different loss functions, such as CER, to better optimize the model. The results mentioned above confirm the efficacy of our model and emphasize the necessity of continuous enhancements to attain more accurate and reliable speech recognition systems.

Abbreviations

ASR	Automatic speech recognition
Bi-GRU	Bi-directional gated recurrent unit
CER	Character error rate
CNN	Convolutional neural network
CTC	Connectionist temporal classification
GRU	Gated recurrent unit
HMM	Hidden Markov model
LSTM	Long short-term memory
ResCNN	Residual convolutional neural networks
STFT	Short-time Fourier transform

References

1. Hannun A, Case C, Casper J, Catanzaro B, Diamos G, Elsen E, Prenger R et al (2014) Deep Speech: scaling up end-to-end speech recognition. arXiv preprint arXiv:1412.5567
2. Ba JL, Kiros JR, Hinton GE (2016) Layer normalization. arXiv preprint arXiv:1607.06450
3. Saxena S (2023) Introduction to gated recurrent unit (GRU). Analytics Vidhya, https://www.analyticsvidhya.com/blog/2021/03/introduction-to-gated-recurrent-unit-gru/
4. Arora SJ, Singh RP (2012) Automatic speech recognition: a review. Int J Comput Appl 60(9)
5. Reza S, Campos Ferreira M, Machado JJ, Tavares JM (2023) A customized residual neural network and bi-directional gated recurrent unit-based automatic speech recognition model. Expert Syst Appl 215:119293
6. Levis J, Suvorov R (2012) Automatic speech recognition. The encyclopedia of applied linguistics

7. Brown PF (1987) The acoustic-modeling problem in automatic speech recognition. Carnegie Mellon University
8. Kadam S, Jani J, Kudtarkar A, Koshy R (2024) Speech emotion recognition using mel-frequency cepstral coefficients & convolutional neural networks. In: 2024 2nd International conference on intelligent data communication technologies and Internet of Things (IDCIoT). IEEE, pp 1595–1602
9. Rabiner L, Juang B (1986) An introduction to hidden Markov models. IEEE ASSP Mag 3(1):4–16
10. Raghu K, Sadanandam M (2021) A Perspective study on speech emotion recognition: databases, features and classification models. Traitement du Signal 38(6)
11. Malik M, Malik MK, Mehmood K, Makhdoom I (2021) Automatic speech recognition: a survey. Multimedia Tools Appl 80:9411–9457
12. Lee J, Watanabe S (2021) Intermediate loss regularization for ctc-based speech recognition. In: ICASSP 2021–2021 IEEE international conference on acoustics, speech and signal processing (ICASSP). IEEE, pp 6224–6228
13. Wang Y, Deng X, Pu S, Huang Z (2017) Residual convolutional CTC networks for automatic speech recognition. arXiv preprint arXiv:1702.07793
14. Kheddar H, Hemis M, Himeur Y (2024) Automatic speech recognition using advanced deep learning approaches: a survey. Information Fusion 2024:102422
15. Maas AL, Le QV, O'Neil TM, Vinyals O, Nguyen P, Ng AY (2012) Recurrent neural networks for noise reduction in robust asr. In: Interspeech, vol 2012, pp 22–25
16. de Benito-Gorron D, Lozano-Diez A, Toledano DT, Gonzalez-Rodriguez J (2019) Exploring convolutional, recurrent, and hybrid deep neural networks for speech and music detection in a large audio dataset. EURASIP J Audio Speech Music Proc 2019(1):9
17. Bhangale K, Mohanaprasad K (2020) Speech emotion recognition using mel frequency log spectrogram and deep convolutional neural network. In: International conference on futuristic communication and network technologies. Springer Nature, Singapore, pp 241–250
18. Yin H, Nadeu C, Hohmann V (2010) Pitch-and formant-based order adaptation of the fractional Fourier transform and its application to speech recognition. EURASIP J Audio Speech Music Proc 2009:1–14
19. Shanthi N, Stonier AA, Sherine A, Devaraju T, Abinash S, Ajay R, Prasath VA, Ganji V (2022) An integrated approach for mental health assessment using emotion analysis and scales. Healthc Technol Lett
20. Master deep learning with tensorflow 2 with computer vision, natural language processing, sound recognition & deployment, https://neuralearn.ai/. Last accessed 12/2023

Estimating the Dynamic Properties of Steel Fiber-Reinforced Ultra-High-Performance Concrete

Chi Trung Nguyen and Ngoc Thanh Tran

Abstract The establishment of steel fiber-reinforced ultra-high-performance concrete (SFUHPC) signifies a significant breakthrough in concrete technology, showcasing superior performance compared to conventional fiber-reinforced cementitious composites. SFUHPCs not only boast higher tensile strength but also exhibit greater tensile ductility under both static and dynamic loads, attributed to the unique strain-hardening behavior accumulated in multiple microcracks. Consequently, this material holds promise for structural applications under quasi-static and dynamic loading conditions. While several empirical models have been developed to forecast the dynamic increase factor (DIF) of SFUHPC tensile strength, their limitations arise from being based on a fixed set of individual test data, potentially lacking the necessary breadth to ensure accuracy. This study introduces a novel approach by employing a machine learning model to estimate the DIF of SFUHPC tensile strength. Drawing from a collection of 140 experimental test results, an artificial neural network (ANN) model has been developed, accounting six input parameters characterized by intricate relationships. The prediction results underscore the excellent performances of the developed ANN model in forecasting the DIF of SFUHPC tensile strength across both training and testing sets. Through a sensitivity analysis, it was revealed that the strain-rate emerged as the greatest contribution parameter, significantly impacting the DIF of SFUHPC tensile strength, while fiber length exhibited the least importance among the variables.

Keywords SHFRCC · Dynamic increase factor · Tensile strength · Machine learning model · Prediction

C. T. Nguyen · N. T. Tran (✉)
Institute of Civil Engineering, Ho Chi Minh City University of Transport, Ho Chi Minh City, Vietnam
e-mail: ngocthanh.tran@ut.edu.vn

H. Sharma et al. (eds.), *Artificial Intelligence: Theory and Applications*, Lecture Notes in Networks and Systems 5589, https://doi.org/10.1007/978-981-96-1687-9_16

1 Introduction

The establishment of steel fiber-reinforced ultra-high-performance concrete (SFUHPC) represents a significant advancement in concrete technology, as SFUHPCs demonstrate superior performance compared to conventional fiber-reinforced concrete (FRC). SFUHPCs not only exhibit much higher tensile strength but also greater tensile ductility under direct tension than FRCC due to the unique strain-hardening behavior accumulated in multiple microcracks [1]. Notably, this material maintains these excellent properties even under high-rate loading [2]. Among the various properties of SFUHPC, tensile strength becomes a crucial parameter at both static and high strain rates for recognizing the superiority of this material over other concretes [3, 4].

Given its importance, numerous studies have aimed to investigating the tensile strength of SFUHPC at both static and dynamic loads. Consequently, most studies have investigated the dynamic increase factor (DIF), which is the ratio of tensile strength at dynamic rate to that at static rate. The DIF of SFUHPC tensile strength has been reported to depend strongly on various factors [5–10]. Tran et al. [5] demonstrated that the DIF is influenced by factors such as fiber type, fiber volume content, fiber length, and specimen size. Similarly, fiber properties like type, diameter, and length were found to affect the DIF of SFUHPC tensile strength [6]. Park et al. [7] concluded that matrix strength significantly influences the DIF of SFUHPC tensile strength. Conversely, Park et al. [8] observed an increase in the DIF as strain-rates rose from quasi-static to high rates more than 150 s^{-1}. Wille et al. [9] and Kim et al. [10] agreed that the DIF of SFUHPC tensile strength increases slowly under low strain rates.

As mentioned earlier, the DIF of SFUHPC tensile strength is influenced by numerous factors, making accurate estimation challenging. To predict the DIF precisely, a model must consider the effects of multiple parameters simultaneously. Despite several attempts with empirical models proposed by various authors [11], these models have limitations as they are based on a fixed set of individual test data, which may not be extensive enough, raising questions about their accuracy. Therefore, it is crucial to propose a more accurate model for estimating the DIF of SFUHPC tensile strength. One promising method is the utilization of the machine learning approach, as it has been demonstrated to predict the mechanical resistance of concrete accurately utilizing a multitude of input variables with complex relationships [12, 13]. However, there is minimal research on utilizing a machine learning (ML) method to estimate the DIF of SFUHPC tensile strength.

To address this gap, this research recommends the utilization of a machine learning approach to predict the DIF of SFUHPC tensile strength, leveraging a substantial amount of collected test results. The main objectives are as follows: (1) proposing a ML-based approach to estimate the DIF of SFUHPC tensile strength and (2) determining the contribution of all input parameters to the DIF of SFUHPC tensile strength.

2 Database and Statistical Metrics

2.1 Experimental Database

A comprehensive compilation of 140 experimental test results on the dynamic properties of SFUHPC was assembled from 14 prior studies [5–10, 14–21]. Six input variables were taken into consideration: compressive strength of matrix (MS), fiber type (FT), volume fiber content (FV), diameter of fiber (FD), length of fiber (FL), and strain-rate (SR). The output variable focused on the DIF of SFUHPC tensile strength. Considerable variability was observed in these variables: MS (20–200 MPa), FT (twisted, hooked, and straight), FV (0.5–3%), FD (0.1–0.375 mm), FL (10–30 mm), and SR (0.0001–300 s^{-1}). Moreover, the DIF of SHUHPC tensile strength ranged from 0.5 to 15.

2.2 Metrics

To measure the effectiveness of the models, several metrics were employed: the correlation coefficient (R) is calculated by Eq. (1), while root mean squared error (RMSE) and mean bias error (MBE) have been formulated by Eqs. (2) and (3), respectively.

$$R = \frac{s\sum_{i=1}^{s}\hat{z}_i z_i - \left(\sum_{i=1}^{s}\hat{z}_i\right)\left(\sum_{i=1}^{s}z_i\right)}{\sqrt{s\sum_{i=1}^{s}\left(\sum_{i=1}^{s}\hat{z}_i^2\right) - \left(\sum_{i=1}^{s}\hat{z}_i\right)^2}\sqrt{s\sum_{i=1}^{s}\left(\sum_{i=1}^{s}z_i^2\right) - \left(\sum_{i=1}^{s}z_i\right)^2}} \tag{1}$$

$$\text{RMSE} = \sqrt{\frac{\sum_{i=1}^{s}(\hat{z}_i - z_i)^2}{s}} \tag{2}$$

$$\text{MBE} = \frac{\sum_{i=1}^{s}(\hat{z}_i - z_i)}{s} \tag{3}$$

where s is number of test dataset, $\hat{z}_i$ represents the prediction values, and z_i is the test values.

Table 1 Performance of current model

Metrics	Value
R	0.62
RMSE	0.67
MBE	− 0.03

3 Performance of Current Empirical Model

3.1 Current Empirical Model

Recently, Noh et al. [22] withdrew an empirical model for estimating the DIF of SHUHPC tensile strength as follows:

$$\text{DIF} = 0.0508\log(\dot{\varepsilon}) + 1.3049 \text{ for } \dot{\varepsilon} < 10\ \text{s}^{-1} \tag{4}$$

$$\text{DIF}_{pc} = 1.7139\log(\dot{\varepsilon}) - 0.4846 \text{ for } \dot{\varepsilon} \geq 10\ \text{s}^{-1} \tag{5}$$

where $\dot{\varepsilon}$ represents the strain rate varying from 0.000001 to 300 s^{-1}

3.2 Performance of the Current Model

The metrics for the prediction of DIF from the current method are summarized in Table 1. The correlation from the prediction of the current method demonstrated weak performance, with a R value of 0.62. Furthermore, the RMSE and MBE values were 0.67 and 0.03, respectively, demonstrating the low reliability of the current method in estimating the DIF of SHUHPC tensile strength. This is due to the fact that the current model considered only the influence of SR in the prediction, while other parameters such as the MS, FT, FV, FD, and FL were not included. Moreover, the current model was obtained by fitting limited test data.

4 Machine Leaning-Based Model

4.1 Back-Propagation Technique

In this research, a machine learning method, specifically an artificial neural network model based on back-propagation technique (ANN), was developed to forecast the DIF of SFUHPC tensile strength. The typical structure of this model is illustrated in Fig. 1. Generally, the ANN model consists of one input layer, one or more hidden

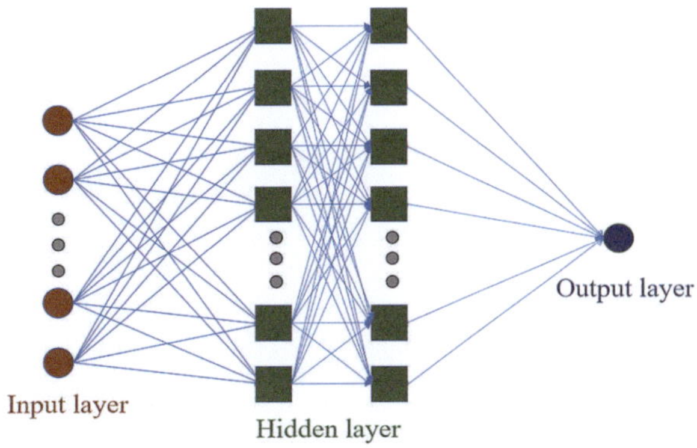

Fig. 1 Structure of an ANN model

layers, and one output layer. Each input layer and hidden layer can consist of multiple nodes, but the remained output layer contains a single neuron. The data starts from input layer and then flow to hidden layers. Subsequently, the hidden layers capture the complex nonlinear behavior of the data and process them. Finally, the results are generated by output layer.

The ANN model included 1 input layer, 1 hidden layer, and 1 output layer. There was 6 neurons in input layer and they are: MS, FT, FV, FD, FL, and SR. Through tuning and optimizing parameters, the hidden layer consisted of 9 neurons. Additionally, the DIF of SFUHPC tensile strength served as the single node in the output layer. To calculate the total of the weighted inputs and the bias term, converted into the output layer, the Sigmoid activation function was employed. Furthermore, gradient descent was selected as the optimization algorithm for training neural networks to optimize the function, which measures the deviation of the neural network's predictions from the actual values. The learning rate, utilized to scale the magnitude of variable updates during gradient descent, was set to 0.001. The data was classified into two segments: 80% and 20% for train and test sets, respectively.

4.2 Performance of the ANN Model

Figure 2 illustrates the correlation between forecasted and tested database in the train set of the ML model. Additionally, the metrics for the train set are summarized in Table 1. The correlation in the training dataset demonstrated strong performance, with a correlation coefficient value of 0.99. Furthermore, the RMSE and MBE values were 0.11 and 0.01, respectively, demonstrating the suitable reliability of the ANN model in estimating the DIF of SFUHPC tensile strength. Specifically, the RMSE value was approximately 6% of the average value of DIF for SFUHPC tensile strength, ranking it as excellent according to the guide for the assessment of concrete strength test results recommended by ACI Committee 214.

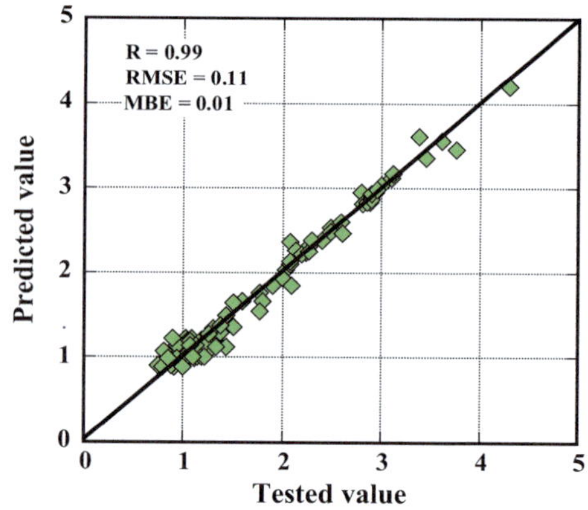

Fig. 2 Accuracy of the machine learning model in training set

The ability of the machine learning model in predicting the DIF of SFUHPC tensile strength in the testing set is depicted in Fig. 3, with corresponding metrics provided in Table 2. Similarly, the correlation between two variables including estimated and tested data presented excellent performance even in the testing set when the correlation coefficient value of prediction was 0.98. Besides, the RMSE value was 0.14 while the MBE value was 0.01. The RMSE value was approximately 9% of the average value of DIF for SFUHPC tensile strength, indicating a good ranking according to the guide for the assessment of concrete strength test results recommended by ACI Committee 214. Hence, the proposed ANN model showcased its ability to estimate the DIF of SFUHPC tensile strength with excellent performances in both sets. Compared to the current model, ANN model exhibited outstanding prediction ability in both coefficient and root mean squared error values.

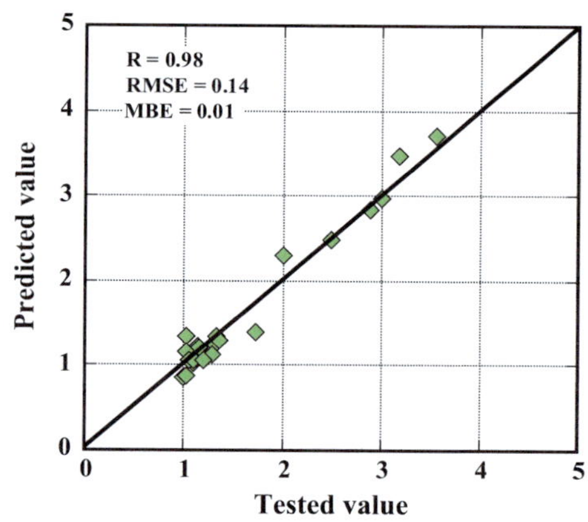

Fig. 3 Accuracy of the ANN model in testing set

Table 2 Performance of ANN model in training and testing sets

Metrics	Training set	Testing set
R	0.99	0.98
RMSE	0.11	0.14
MBE	0.01	0.01

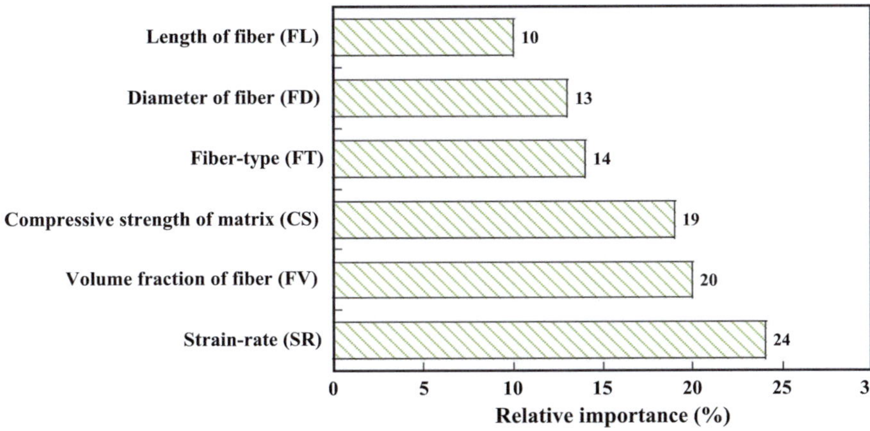

Fig. 4 Contribution of each input variable

4.3 *Sensitivity Analysis*

A sensitivity analysis based on the method proposed by God [23–25] was performed to find out the contribution of all input parameters to the DIF of SFUHPC tensile strength. Figure 4 illustrates the contribution of all input parameters based on the results of sensitivity analysis. The findings from the ANN model highlighted that the strain rate held the highest importance among the variables, significantly influencing the final DIF of SFUHPC tensile strength. Conversely, fiber length emerged as the least influential variable. Additionally, the order of performance was as follows: SR > FV > MS > FT > FD > FL.

5 Conclusion

This study proposes the use of an ANN model to estimate the DIF of SFUHPC tensile strength. Through the collection of 140 experimental test results, an ANN model was developed, taking into account six input parameters characterized by intricate relationships. Base on the prediction of ANN model and contribution of each input parameter, the following conclusions emerge:

- The current empirical model showed low performance in predicting the DIF of SFUHPC tensile strength. The *R* value of predictions was 0.62 and the RMSE value was 0.67.
- The ANN model effectively estimates the DIF of SFUHPC tensile strength by simultaneously considering multiple input parameters.
- The developed ANN model demonstrates excellent performances in estimating the DIF of SFUHPC tensile strength across both sets. The correlation coefficient values of predictions were greater than 0.98 and the RMSE values were less than 0.14.
- The strain-rate emerged as the greatest contribution parameter, significantly impacting the DIF of SFUHPC tensile strength, while fiber length displayed the least importance among the variables.

References

1. Tran NT, Nguyen DL, Kim DJ, Ngo TT (2021) Sensitivity of various fiber features on shear capacities of ultra-high-performance fiber-reinforced concrete. Mag Concr Res 74(4):190–206
2. Tran TK, Nguyen TK, Tran NT, Kim DJ (2022) Improving the tensile resistance at high strain rates of high-performance fiber-reinforced cementitious composite with twisted fibers by modification of twist ratio. Structures 39:237–248
3. Tran NT, Nguyen DL, Vu QA, Kim DJ, Ngo TT (2022) Dynamic shear response of ultra-high-performance fiber-reinforced concretes under impact loading. Structures 41:724–736
4. Tran TK, Tran NT, Kim DJ (2021) Enhancing impact resistance of hybrid ultra-high-performance fiber-reinforced concretes through strategic use of polyamide fibers. Constr Build Mater 271:121562
5. Tran TN, Tran TK, Kim DJ (2015) High rate response of ultra-high-performance fiber-reinforced concretes under direct tension. Cem Concr Res 69:72–87
6. Tran NT, Kim DJ (2017) Synergistic response of blending fibers in ultra-high-performance concrete under high rate tensile loads. Cem Concr Compos 78:132–145
7. Park JK, Kim SW, Kim DJ (2017) Matrix-strength-dependent strain-rate sensitivity of strain-hardening fiber-reinforced cementitious composites under tensile impact. Compos Struct 162:313–324
8. Park SH, Kim DJ, Kim SW (2016) Investigating the impact resistance of ultra-high-performance fiber-reinforced concrete using an improved strain energy impact test machine. Constr Build Mater 125:145–159
9. Wille K, Xu M, El-Tawil S, Naaman AE (2016) Dynamic impact factors of strain hardening UHP-FRC under direct tensile loading at low strain rates. Mater Struct 49:1351–1365
10. Kim DJ, El-Tawil S, Naaman AE (2009) Rate-dependent tensile behavior of high performance fiber reinforced cementitious composites. Mater Struct 42:399–414
11. Thomas RJ, Sorensen AD (2017) Review of strain rate effects for UHPC in tension. Constr Build Mater 153:846–856
12. Ngo TT, Le QH, Nguyen DL, Kim DJ, Tran NT (2023) Experiments and prediction of direct tensile resistance of strain-hardening steel-fiber-reinforced concrete. Mag Concr Res. Ahead of Print. https://doi.org/10.1680/jmacr.22.00060
13. Tran NT, Nguyen TK, Nguyen DL, Le QH (2022) Assessment of fracture energy of strain-hardening fiber-reinforced cementitious composite using experiment and machine learning technique. Struct Concr Early View. https://doi.org/10.1002/suco.202200332

14. Tran TK, Tran NT, Nguyen DL, Kim DJ, Park JK, Ngo TT (2021) Dynamic fracture toughness of ultra-high-performance fiber-reinforced concrete under impact tensile loading. Struct Concr 22:1845–1860
15. Tran TK, Kim DJ (2013) Investigating direct tensile behavior of high performance fiber reinforced cementitious composites at high strain rates. Cem Concr Res 50:62–73
16. Tran TK, Kim DJ (2014) High strain rate effects on direct tensile behavior of high performance fiber reinforced cementitious composites. Cement Concr Compos 45:186–200
17. Pyo S, Wille K, El-Tawil S, Naaman AE (2015) Strain rate dependent properties of ultra high performance fiber reinforced concrete (UHP-FRC) under tensions. Cement Concr Compos 56:15–24
18. Pyo S, El-Tawil S, Naaman AE (2016) Direct tensile behavior of ultra high performance fiber reinforced concrete (UHP-FRC) at high strain rates. Cem Concr Res 88:144–156
19. Fujikake K, Senga T, Ueda N, Ohno T, Katagiri M (2006) Effects of strain rate on tensile behavior of reactive powder concrete. J Adv Concr Technol 4:79–84. https://doi.org/10.3151/jact.4.79
20. Cadoni E, Meda A, Plizzari GA (2009) Tensile behaviour of FRC under high strain-rate. Mater Struct 42:1283–1294
21. Caverzan A, Cadoni E, Di Prisco M (2012) Tensile behaviour of high performance fibre reinforced cementitious composites at high strain rates. Int J Impact Eng 45:28–38
22. Noh HW, Truong VD, Cho JY, Kim DJ (2022) Dynamic increase factors for fiber-reinforced cement composites: a review. J Build Eng 56:104796
23. God ATC (1995) Back-propagation neural networks for modeling complex systems. Artif Intell Eng 9:143–151
24. Tran NT, Nguyen DH, Tran QT, Le HV, Nguyen DL (2024) Experimental and machine learning based study of compressive strength of geopolymer concrete. Mag Concr Res 76:723–737
25. Tran NT, Nguyen DH, Tran TK, Kim DJ, Nguyen DL (2024) Synergy performance of hybrid fiber-reinforced ultra-high-performance cementitious composites with low fiber contents. Compos A Appl Sci Manuf 187:108423

Enhancing Scholarship Opportunities: A Multi-label Classification Approach

Vaishnavi Padiya, Vidit Gala, Shubham Mehta, Pratham Shah, and Lynette DMello

Abstract Scholarships are critical in making education accessible to students from diverse socio-economic backgrounds. This research aimed to develop a multi-label classification-based scholarship recommendation system for engineering students in India. Utilizing a dataset of dummy students and scholarships, various multi-label classification models were evaluated, including Binary Relevance, Classifier Chains, Label Powerset, Multi-Output Classifier (MOC) with Random Forest, Multi-Output Classifier (MOC) with KNeighbors and Multi-Label k-Nearest Neighbors. Models were assessed based on $F1$ Scores, Jaccard Scores, Accuracy, and Hamming Loss. The Random Forest model emerged as the top performer with the highest accuracy of 88% and lowest Hamming Loss of 0.009, followed by KNeighbors and Multi-Label KNN as strong alternatives. The study highlights the effectiveness of multi-label classification in enhancing scholarship recommendation systems, providing a robust tool to support students in identifying suitable scholarship opportunities.

Keywords Multi-label classification · Scholarships machine learning · Deep learning

1 Introduction

Scholarships are financial aid awards designed to support students in funding their education, which they do not need to repay. These awards are crucial as they provide financial relief, reduce the burden of educational expenses, and encourage academic and extracurricular excellence. Scholarships are awarded based on various criteria, including academic achievement, athletic talent, artistic skills, financial need, and

V. Padiya (✉) · V. Gala · S. Mehta · P. Shah · L. DMello
Computer Engineering, D. J. Sanghvi College of Engineering, Vile Parle, Mumbai, Maharashtra, India
e-mail: vaishnavipadiya@gmail.com

L. DMello
e-mail: lynetteDMello@djsce.ac.in

H. Sharma et al. (eds.), *Artificial Intelligence: Theory and Applications*, Lecture Notes in Networks and Systems 5589, https://doi.org/10.1007/978-981-96-1687-9_17

other specific attributes or accomplishments. The necessity of scholarships lies in their ability to make education accessible to students from diverse socio-economic backgrounds, ensuring that financial constraints do not hinder talented individuals from pursuing higher education. They also offer emotional and psychological support, boosting students' motivation and morale.

Ramadhianti and Soegoto [1] studied the influence of scholarships on students' learning, motivation, and academic achievement, highlighting the challenges they faced. Findings indicated that scholarships played a crucial role in enhancing motivation and performance by providing both financial aid and emotional support. Scholarship recipients exhibited higher motivation, consistent attendance, and better academic achievements compared to non-recipients. The research recommended continuous enhancement of scholarship support through increased funding, improved selection processes, and better mentoring programs. It emphasized the need for greater transparency regarding scholarship eligibility and selection criteria to help students and their guardians better navigate available opportunities.

A field survey of students aged 14–22 was conducted which revealed low awareness of major scholarships, with significant portions of the population unaware of key opportunities. This lack of awareness resulted in missed opportunities and benefits. The study called for the development of a comprehensive platform to provide updated information on scholarships and career options, making it easier for students to access and benefit from these resources. Traditional scholarship search methods relied on manual filtering, resulting in missed relevant opportunities and limited access. The process was also tedious, overwhelming, and time-consuming, requiring students to match their profiles with scholarship requirements, remember due dates, and cross-check documents for applications.

To address these challenges, a scholarship recommendation system was developed to save time and effort, offering personalized suggestions and targeted selections that automatically cross-checked user details to recommend eligible scholarships. The research primarily focused on Scholarships for Engineering Students in India post-12th grade but it could be expanded further. To achieve this, a dataset of dummy scholarships and dummy users along with their details was prepared. A script was used to match users to scholarships based on their details manually. This dataset was then utilized to train multi-label classification models such as Binary Relevance, Classifier Chains, Label Powerset, Multi-Output Classifier (MOC) with Random Forest, Multi-Output Classifier (MOC) with *K*Neighbors and Multi-Label *k*-Nearest Neighbors.

2 Literature Review

A scholarship is a form of financial aid awarded to students to support their education. Scholarships are typically granted based on various criteria, which can include academic merit, athletic ability, artistic talent, financial need, or other specific attributes or achievements. Alvarol and Gabayan [2] introduced a special issue on digital scholarship and highlighted the significance of research in digital libraries and digital scholarship. The paper underscored the importance of communicating ideas and enhancing scholarly communication within the digital realm. Khatri et al. [3] proposed an online portal, DESTINY, utilizing Object-Oriented Analysis and Design (OOAD) methodology and technologies such as React and MongoDB to tackle the lack of awareness surrounding scholarships and courses in India, their research underscored the significance of centralizing resources for improved accessibility. A web application was designed using the Rapid Application Development (RAD) model that leveraged web scraping techniques which provided personalized scholarship recommendations to students by Sulaiman and Ahmad [4], their future scope suggested expanding the scholarship database, introducing additional search and filter criteria, and enhancing the personalization algorithm. Patro et al. [5] propose a Hybrid Action-Related K-Nearest Neighbor (HAR-KNN) recommender system to enrich the user behavior matrix and enhance the recommendation system. A smart scholarship system was created by Khelifi et al. [6] for Abu Dhabi University (ADU) students to apply for scholarships, verify their scholarship status, and expedite the scholarship office's operations. Nugroho and Hermawan [7] developed a scholarship recommendation system utilizing content-based filtering, achieving a 100% success rate in recommendation accuracy, the system, employed the term frequency-inverse document frequency weighting, which aids in decision-making for scholarship selection in higher education institutions. Using more advanced means Ginting et al. [8] introduce a scholarship advice system utilizing clustering with K-Means and TOPSIS algorithms to group scholarships and provide customized tips with an 88% accuracy rate. The study by Setiawan [9] introduced a recommendation system for scholarships utilizing the Fuzzy TOPSIS algorithm. By integrating a client-server application and advanced algorithmic techniques, the system efficiently processes scholarship applications, providing objective recommendations tailored to each student's qualifications and preferences. Zeeshan et al. [10] address the complexity of recommendation systems, focusing on multi-criteria recommendation (MCRS). They proposed a feature-based approach that combines content-based and collaborative filtering techniques to enhance recommendation accuracy, demonstrating the effectiveness of the algorithm across multiple datasets and providing insights into improving recommendation systems' performance and efficiency. The literature suggested that a digital and personalized approach is essential and that combining multiple algorithms can significantly improve recommendation accuracy also incorporating detailed user profiles and continuously updating them with new data can enhance the personalization aspect.

3 Methodology

The methodology for building the scholarship recommendation system follows a structured process of data preparation, feature selection, model training, and evaluation. Figure 1 visually represents the steps involved, starting from data preparation to model training and evaluation.

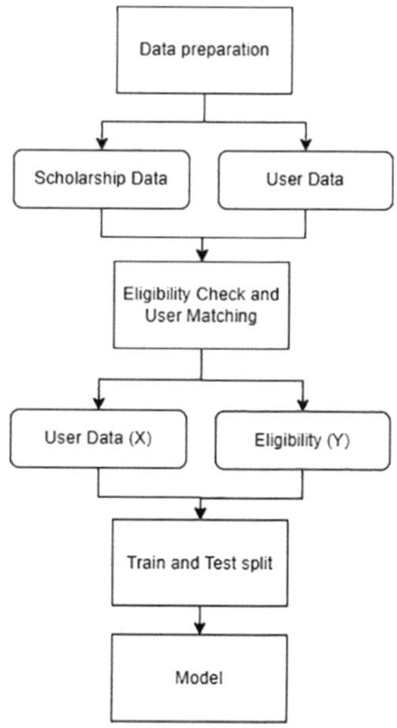

Fig. 1 Methodology flowchart. The flowchart illustrates the stages of building the machine learning model

Dataset Preparation: To develop the scholarship recommendation system, a synthetic dataset of 1000 dummy students and 15 scholarships was created, informed by observations of existing scholarship websites. The student dataset included attributes such as student ID, category, family income, academic performance (10th and 12th percentages), and entrance exam scores (State CET and JEEMains). The scholarship dataset comprised scholarship ID, name, due date, amount, category, income eligibility, required documents, and academic performance criteria. Each student was mapped to eligible scholarships, generating a new dataset where students were represented as rows and scholarship IDs as columns, with binary values indicating eligibility.

Normalization and Feature Selection: Irrelevant features, such as students' names, email addresses, contact information, and addresses, were removed from the dataset as they did not contribute to the scholarship recommendation process. The remaining student data was then normalized. Label Encoding was applied to encode students' documents and categories. Normalization was performed to ensure that all features were on a consistent scale, which facilitated accurate comparisons and improved the performance of the recommendation algorithm.

Train-Test Split of the Dataset: To prevent overfitting and ensure proper model training and accuracy, the dataset was divided into an 80:20 ratio. Specifically, 80% of the student data, totaling 800 students, was used for training the model, while the remaining 20% was reserved for testing and evaluating the model's accuracy.

Model Comparison: Given that each student could be eligible for multiple scholarships, multi-label classification models were employed. These models allowed for the assignment of multiple labels to a single data point, as the labels were not mutually exclusive, and a data point could be relevant to several categories simultaneously. The models utilized included Binary Relevance, Classifier Chains, Label Powerset, Multi-Output Classifier (MOC) with Random Forest, Multi-Output Classifier (MOC) with KNeighbors, and Multi-Label k-Nearest Neighbors. A comparison of the results was conducted using various evaluation metrics to determine which model demonstrated the best performance.

Evaluation Metrics: To evaluate the results, various metrics were employed which included accuracy, $F1$ Score (weighted, micro, and macro), Hamming loss, and Jaccard Score (weighted, micro, and macro). True Positive (TP): The number of instances correctly predicted. False Positive (FP): The number of instances incorrectly predicted as positive. True Negative (TN): The number of instances correctly predicted as negative. False Negative (FN): The number of instances incorrectly predicted as negative. Precision: It is the ratio of correctly predicted positive observations to the total predicted positives. High precision indicates a low false positive rate.

$$\text{Precision} = \frac{\text{TP}}{\text{TP} + \text{FP}} \tag{1}$$

Recall: It is the ratio of correctly predicted positive observations to all observations in the actual class. High recall indicates a low false negative rate.

$$\text{Recall} = \frac{\text{TP}}{\text{TP} + \text{FN}} \tag{2}$$

***F*1 Score**: The $F1$ Score balances precision and recall, providing a harmonic mean that is especially useful for handling imbalanced datasets.

$$F1_{\text{weighted}} = \sum \left(\frac{N_c}{N} \cdot \frac{2 \cdot \text{Precision}_c \cdot \text{Recall}_c}{\text{Precision}_c + \text{Recall}_c} \right) \tag{3}$$

$$F1_{\text{micro}} = \frac{2 \cdot \sum(\text{True Positives})}{2 \cdot \sum(\text{True Positives}) + \sum(\text{False Positives} + \text{False Negatives})} \tag{4}$$

$$F1_{\text{macro}} = \frac{1}{C} \sum_{c=1}^{C} \frac{2 \cdot \text{Precision}_c \cdot \text{Recall}_c}{\text{Precision}_c + \text{Recall}_c} \tag{5}$$

N is the total number of instances, N_c is the number of instances belonging to class c and C is the total number of classes. Precision_c is the precision for class c, defined as $\frac{\text{TP}_c}{\text{TP}_c + \text{FP}_c}$. Recall_c is the Recall for class c, defined as $\frac{\text{TP}_c}{\text{TP}_c + \text{FN}_c}$.

Hamming Loss: Hamming loss measures the fraction of labels that are incorrectly predicted. Lower values indicate better performance.

$$\text{Hamming Loss} = \frac{1}{N \cdot L} \sum_{i=1}^{N} \sum_{j=1}^{L} \mathbf{1}_{\{y_{ij} \neq \hat{y}_{ij}\}} \tag{6}$$

N is the total number of instances, L is the total number of labels, y_{ij} is the true value of the j-th label for the i-th instance, $\hat{y}_{ij}$ is the predicted value of the j-th label for the i-th instance and $\mathbf{1}_{\{y_{ij} \neq \hat{y}_{ij}\}}$ is the indicator function that is 1 if $y_{ij} \neq \hat{y}_{ij}$ and 0 otherwise.

Jaccard Score: The Jaccard Score evaluates the similarity between predicted and true labels. Higher values indicate better performance.

$$\text{Jaccard}_{\text{weighted}} = \sum \left(\frac{N_c}{N} \cdot \frac{|\text{True Positives}_c|}{|\text{True Positives}_c + \text{False Positives}_c + \text{False Negatives}_c|} \right) \tag{7}$$

$$\text{Jaccard}_{\text{micro}} = \frac{\sum \text{True Positives}}{\sum(\text{True Positives} + \text{False Positives} + \text{False Negatives})} \tag{8}$$

$$\text{Jaccard}_{\text{macro}} = \frac{1}{C} \sum_{c=1}^{C} \frac{|\text{True Positives}_c|}{|\text{True Positives}_c + \text{False Positives}_c + \text{False Negatives}_c|} \tag{9}$$

N is the total number of instances, and N_c is for instances belonging to class c. True Positives$_c$ are the number of correctly predicted positive instances for class c, False Positives$_c$ are the number of incorrectly predicted positive instances for class c, and False Negatives$_c$ are the number of positive instances that are incorrectly predicted as negative for class c.

4 Results

The preprocessed dataset was subsequently employed to train a variety of multi-label classification models. Each model was evaluated for its efficacy in predicting multiple labels simultaneously.

Binary Relevance Binary Relevance is a straightforward approach that treats each label as an independent binary classification problem. For each label, a separate binary classifier is trained. The predictions for all labels are then combined to form the final multi-label output. While computationally efficient, this method often neglects label dependencies. An accuracy of 62.5% along with hamming loss of 0.0343 was obtained, it also has moderate $F1$ and Jaccard Scores on testing. $F1$ and Jaccard Scores are mentioned in Table 2. It indicates a balance between precision and recall but not the best performer.

Classifier Chains Classifier Chains address the limitations of Binary Relevance by considering label dependencies. A chain of classifiers is constructed, where the prediction of a label is influenced by the predictions of preceding labels in the chain. The order of labels in the chain can significantly impact performance. An accuracy of 45% with a hamming loss of 0.056 was obtained which suggests it didn't perform

Table 1 Comparison of multi-label classification models

Model	Accuracy	Hamming loss
Binary Relevance	0.625	0.0343
Classifier Chains	0.45	0.056
Label Powerset	0.67	0.0237
MOC Random Forest	0.88	0.009
MOC *K*Neighbors	0.845	0.011
Multi-label *K*NN	0.845	0.01

well compared to other models on testing. The Jaccard Scores and $F1$ Scores are mentioned in Table 2. It has lower $F1$ and Jaccard Scores compared to other models.

Label Powerset Label Powerset transforms the multi-label problem into a multi-class problem by considering all possible label combinations as distinct classes. A single multi-class classifier is trained on this transformed dataset. However, this method suffers from computational challenges as the number of classes grows exponentially with the number of labels. An accuracy of 67% was obtained on testing along with a hamming loss of 0.024. The Jaccard Scores and $F1$ Scores are mentioned in Table 2. It showed good $F1$ and Jaccard Scores especially in Micro and Weighted Metrics therefore shows a better performance compared to Binary Relevance and Classifier Chains.

Multi-output Classifier (MOC) with Random Forest Similarly, a MOC was utilized with Random Forest as the base estimator. Random Forest's ability to handle complex interactions between features was leveraged for multi-label prediction. Each label was assigned a Random Forest classifier, and their predictions were combined to produce the final output. An accuracy of 86.5% was obtained on testing along with a hamming loss of 0.01. The Jaccard Scores and $F1$ Scores are mentioned in Table 2. It showed the highest Jaccard Scores among all the models and high $F1$ Scores especially Micro and Weighted, therefore its the best performer across all metrics.

MultiOutputClassifier with *K*Neighbors The MultiOutputClassifier with *K*NeighborsClassifier treated each label as an independent classification problem using the K-Nearest Neighbors algorithm as the base classifier. This method involved training a separate *K*NeighborsClassifier for each label, where the majority vote among the k-nearest neighbors in the feature space determined the prediction for each label. An accuracy of 84.5% was obtained on testing along with a hamming loss of 0.011. The Jaccard Scores and $F1$ Scores are mentioned in Table 2 are quite high and comparable to those of Random Forest, its a strong performer but not better than Random Forest.

Multi-label *k*-Nearest Neighbors The Multi-Label k-nearest Neighbors (ML-*k*NN) algorithm extended the k-nearest Neighbors approach to multi-label classification by identifying the k-nearest neighbors of an instance and using their label information

Table 2 Comparison of multi-label classification models ($F1$ and Jaccard Scores)

Model	$F1$ Score			Jaccard Score		
	Micro	Macro	Weighted	Micro	Macro	Weighted
Binary Relevance	0.5960	0.4916	0.6265	0.4245	0.3535	0.4659
Classifier Chains	0.4716	0.3895	0.5488	0.3086	0.2615	0.3945
Label Powerset	0.6432	0.5356	0.6680	0.4740	0.4407	0.5450
MOC Random Forest	0.7999	0.4697	0.7316	0.654	0.404	0.6571
MOC KNeighbors	0.7692	0.4461	0.7389	0.6250	0.3737	0.6587
Multi-label KNN	0.7692	0.4054	0.6983	0.6708	0.4170	0.6718

to make predictions. For each label, it calculated the frequency of occurrence among the neighbors and assigned the label based on a threshold. This method leveraged label relationships and distribution in the feature space for improved classification. An accuracy of 84.5% was obtained on testing along with a hamming loss of 0.009. The Jaccard Scores and $F1$ Scores are mentioned in Table 2. It performs similar to KNeighbors because it has High $F1$ and Jaccard Scores comparable to KNeighbors.

In Table 1 when comparing the performance of various multi-label classification models and evaluating them based on $F1$ Scores, Jaccard Scores, Accuracy, and Hamming Loss. Random Forest emerged as the best overall performer since it achieved the highest accuracy (0.88) and the lowest Hamming Loss (0.01). Additionally, it performed exceptionally well in terms of $F1$ and Jaccard Scores across all metrics, making it the most reliable model for this multi-label classification task. KNeighbors and Multi-Label KNN also demonstrated strong performance, both models had high accuracy (0.845) and low Hamming Loss (0.011 and 0.01, respectively), and their $F1$ and Jaccard Scores were close to those of the Random Forest, making them strong alternatives. Label Powerset showed good performance with a balanced approach in handling multiple labels and Binary Relevance achieved a good balance between accuracy and other performance metrics (Table 2).

To evaluate the performance of the recommendation model, custom user input was tested using the Binary Relevance model. Based on the input provided-category: OBC, maxinc: 400,000, 10th percentage: 85.5, 12th percentage: 82.0, State CET percentile: 95.0, JEE Mains percentile: 90.0, and documents: OBC Certificate-the model recommended three scholarships: Nehru Scholarship Scheme for Farmers 2024–2025 (NSSF), HFFC Scholarship Scheme for Scientists 2024–2025 (HFFC-SSS), and Ambedkar Scholarship Scheme for Farmers 2024 (ASSF) Table 3. This outcome demonstrated that the model performed accurately.

Table 3 Scholarship recommended (output)

Name	LastDate	Maxinc	For	Document	Amt	10th	12th	CET	JEE
NSSF	2024/05/24	215,004	OBC	OBC doc	21,410	83	98	77	66
HFFCSSS	2024/12/13	729,345	OBC	OBC doc	21,589	93	87	88	72
ASSF	2024/11/28	454,780	OBC	OBC doc	96,590	74	71	92	98

5 Conclusion

This study investigated the development of a scholarship recommendation system for engineering students in India using various multi-label classification models. Among the models tested, Random Forest demonstrated the highest accuracy (88%) and the lowest Hamming Loss (0.009), making it the most reliable model for predicting multiple scholarship eligibilities. Its performance across $F1$ and Jaccard Scores further affirmed its suitability for this multi-label classification task. KNeighbors and Multi-Label KNN also showed strong performance, with accuracies of 84.5% and comparable $F1$ and Jaccard Scores, establishing them as viable alternatives to Random Forest. Label Powerset, while achieving a lower accuracy of 67%, balanced precision and recall effectively, indicating its potential utility in certain scenarios. The model was also tested on custom user output to recommend scholarships which implied the accuracy and usage of the model. This research mainly focuses on Scholarship Recommendations for Indian students who want to pursue Engineering, but to enhance the system further, future research could focus on integrating additional data sources, and other domains, refining feature selection, and exploring hybrid models that combine the strengths of various algorithms. Additionally, expanding the dataset and including real-world scholarship data could provide more robust insights and improve the system's applicability in diverse contexts. Overall, the study demonstrates the potential of multi-label classification models in enhancing scholarship recommendations and provides a foundation for future advancements.

References

1. Ramadhianti RH, Soegoto DS (2024) The role of scholarships on learning motivation. In: Proceeding of international conference on business, economics, social sciences, and humanities, vol 7, pp 252–258
2. Alvaro PA, Gabayan V (2021) Scholarship information management system with academic performance predictor and decision support system. Linker J Comput Technol 2(2)
3. Vishal K, Kanish G, Yash A, Raghav G, Hrithik D (2023) Research on the awareness of various scholarships and courses available in India and development of an online courses and scholarships portal. IRJMETS Int Res J Mod Eng Technol Sci 5:01
4. Sulaiman NAMNH (2023) ScholarsAid: a personalized scholarship web application based on web scraping. Math Sci Inf J 4:11–22

5. Patro SGK, Mishra BK, Panda SK, Kumar R, Long HV, Taniar D, Priyadarshini I (2020) A hybrid action-related k-nearest neighbour (har-$k$nn) approach for recommendation systems. IEEE Access 8:90978–90991
6. Khelifi A, Ehtesham H, Al-Mansoori M, Hasan AT, Tamim SB (2023) Smart scholarship system. In: From Industry 4.0 to Industry 5.0: mapping the transitions, vol 470, pp 821–830
7. Nugroho DK, Hermawan H (2018) Development and quality analysis of decision support systems as software for scholarship recommendation in higher education. J Phys Conf Ser 1140:012045
8. Ginting DSB, Sihombing P, Nasution NA (2024) Analysis of combination k-means clustering and technique for order preference by similarity in recommendation of scholarship. AIP Conf Proc 2987
9. Setiawan H (2021) Implementation of the fuzzy TOPSIS algorithm in the scholarship recommendation system. JTECS Jurnal Sistem Telekomunikasi Elektronika Sistem Kontrol Power Sistem dan Komputer 1:157
10. Zeeshan Z, Ain QU, Bhatti UA, Memon WH, Ali S, Nawaz SA, Nizamani MM, Mehmood A, Bhatti MA, Shoukat MU (2021) Feature-based multi-criteria recommendation system using a weighted approach with ranking correlation. Intell Data Anal 25:1013–1029

A Multimodal Approach for English-to-Telugu Translation with FNet and Transformer Models

Joshita Malla, Ujwala Kanumuri, Swathi Mytreye Chaganti, and J. Swaminathan

Abstract This research paper develops a multimodal translation system that is versatile in nature and specifically designed for converting English-to-Telugu language, which addresses accessibility issues across various applications. Our method involves using ResNet50 to ensure that we have robust image feature extraction while using word tokenization to generate English and Telugu tokens. The translation is done through FNet encoder-decoder architectures with the multimodal encoder acting as the de facto combiner of textual and visual information and the decoder generating the Telugu translation to ensure the results are both accurate and contextually appropriate. Based on extensive experimental observations, our model achieved a BLEU score of 54.73, significantly outperforming unimodal approaches such as LSTM (28.97), GRU (35.61), and transformer (46.22) models. Our framework goes beyond assistive devices for the blind and has the possibility of being used as a navigation aid, communication tool, or educational resource thereby improving access and inclusion among individuals who speak different languages.

Keywords Multimodal translation · ResNet50 · Transformer-based encoder-decoder · Accessibility · Inclusivity

1 Introduction

Due to the increasing globalization in today's world, there is an increasing demand for easily accessible technology especially in areas with several languages. Countries like India, with a large number of languages and dialects, present unique challenges in making technology accessible to all as the translation of these languages is not an extensively researched area compared to more dominant languages. The syntactical and semantical differences between dominant languages such as English and

J. Malla (✉) · U. Kanumuri · S. M. Chaganti · J. Swaminathan
Department of Computer Science and Engineering, Amrita School of Computing, Amrita Vishwa Vidyapeetham, Amritapuri, India
e-mail: joshitamalla9@gmail.com

H. Sharma et al. (eds.), *Artificial Intelligence: Theory and Applications*, Lecture Notes in Networks and Systems 5589, https://doi.org/10.1007/978-981-96-1687-9_18

regional languages like Telugu result in a translation gap, which is one of the primary challenges preventing the proper usage of devices. Overcoming this gap is crucial for encouraging inclusion and equitable access to technology among different types of end users.

Existing translation models have limitations [1], such as a lack of high-quality training data and syntactic differences between English and Telugu. This process is further complicated by structural differences in word arrangement and sentence structure between these two languages. This demonstrates that effective translation algorithms capable of handling multilingual settings are required because they involve a wide range of assistive devices, educational tools, navigation systems, and communication devices.

Traditional approaches frequently miss out on visual cues because they rely on text-only-based input. Different machine translation techniques like statistical and neural machine translation [2] have been explored to deal with these challenges. Attention mechanisms have also gained significant attention because of their ability to improve translation quality [3, 4]. However, incorporation of multimodal elements such as images has started being viewed as a promising approach toward improving the quality of translations generated. While textual inputs just represent the linguistic content, images provide context, cultural references, and visual cues that can help understand the meaning and enhance translation accuracy. Nonetheless, the exploration of multimodal translation techniques in less-resourced languages like Telugu is still relatively limited. In fact, most research on this topic has been done using English, Spanish, and Chinese as target languages thus not taking into account language diversity and richness of languages like Telugu.

So in this research paper, we aim to address these problems by proposing a new translation strategy based on a multimodal approach as seen in Fig. 1. Based on the studies [5–7]. We first created a dataset consisting of different types of content and language complexities represented by texts in Telugu and English with their corresponding images. This dataset is used to train multimodal translation models which are then evaluated with respect to their performance at generating the Telugu translations. Multimodal machine translation models can effectively use visual input even in cases with limited textual context [8]. So, we developed this algorithm by combining textual features with visual elements in order to improve translation accuracy and meet the linguistic needs of a wide range of users. Although there have been tremendous improvements in machine translation systems, the diversity of language nuances and intricate visual contexts present new challenges that require better approaches. The effectiveness of the algorithm in generating precise translations and

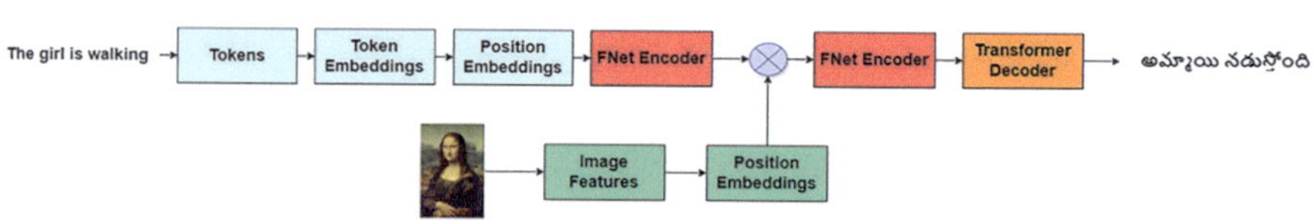

Fig. 1 Proposed model flow diagram

improving technology accessibility will be thoroughly evaluated through an analysis of its performance.

This research has major implications for the development of translation technology with special emphasis on reducing the gap between languages that differ by linguistic resources. The approach we have taken is more adaptable than mere translation. Including visual cues in the process of translation expands opportunities for various applications such as aiding the visually impaired as seen in the study [9] education, navigation, and communication tools thus improving user experiences across domains.

2 Related Works

The study [10] by Lekshmy O. H. and Swaminathan J. employs a multimodal machine translation framework guided by images and text to enhance translation correctness into Malayalam. The research employs the image-guided Multimodal Machine Translation methodology with LSTM models as a means to enhance translation accuracy and scene distributions in Malayalam. However, it should be noted that using sequential models such as Long Short-Term Memory (LSTM) has a drawback because of the vanishing gradient problem leading to reduced efficiency in translation.

The authors further reflect on major data resources, evaluation campaigns as well as state-of-the-art approaches for both end-to-end and pipeline research designs. Performance measurement issues are also highlighted within this paper, along with future research needs including larger datasets and more input–output multimodality. However, ethical concerns especially on privacy and confidentiality in education have received little focus in this study.

The study [11] by Paul Pu Liang et al. focuses on Multimodal machine learning. In this area, several other areas, such as video understanding and multisensor fusion in healthcare and robotics, have emerged, bringing up computational challenges due to data heterogeneity. It provides insights into advancements and open research directions through historical and recent perspectives by identifying key principles and technical challenges.

The study [12] by Suxia Lei et al. involves the development and application of a machine English translation evaluation system using a BP neural network algorithm to improve translation efficiency and quality. This model will help to get rid of some limitations inherent in conventional MT systems, including poor data utilization as well as huge model parameters through optimizing translation performance via applying neural network algorithms that may be hampered by such matters as inadequate information processing and immense model sizes.

The study [13] by Loitongbam Sanayai Meetei et al. proposes a method of multimodal English-to-Hindi machine translation that uses news images and captions. These are different from normal NLP images, which show significant events to give context. The text and image inputs are processed in parallel by separate encoders.

Two tests were done to assess the influence of the image contents on translation systems: BPE for text representation and VGG-19 for visual feature extraction.

The study [14] by Sahinur Rahman Laskar et al. focuses on participation in the WAT2022 workshop. This work examines neural machine translation (NMT) for low-resource language pairs and multimodal approaches that improve translation accuracy. The transliteration-based phrase pairs in English-to-Hindi multimodal translations were improved through an approach developed by Team CNLP-NITS-PP during the WAT2022 workshop which led to a significant performance increase. However, this research only uses transliteration, thus it ignores much of what translated text is all about.

In comparison with existing studies in multimodal machine translation, our research represents a significant advancement in English-to-Telugu translation that includes both images and texts. Many of the previous inquiries in this area had to overcome several challenges. The most usual stumbling blocks were small datasets, vagueness in language, and lexical and semantic differences. Such issues have always prevented accurate translations of both text and visual information.

To tackle these problems, we propose a multimodal approach for translation tasks from English-to-Telugu architecture. Conventional multimodal translation models have been based on traditional encoders like convolutional neural networks (CNNs) for image processing and recurrent neural networks (RNNs) for text. However, we implemented an FNet encoder-transformer decoder architecture for our multimodal machine translation model. Instead of complicated sequential models making use of sequential encoders and decoders, the FNet offers a simpler yet effective alternative for encoding images using its Fourier-based transformation method. The model then utilizes a transformer decoder that has proved useful in many natural language processing tasks for capturing long-range dependencies effectively through self-attention mechanisms to merge information from images and texts seamlessly. The outcome of the new combination not only advances translation accuracy but also assists in deepening the comprehension of how visuals and text interact during translations.

3 Dataset

There are no existing datasets for multimodal machine translation from English to Telugu. Hence, the dataset used in this task is a Telugu Visual Genome which we translated using the existing Hindi Visual Genome. In this dataset, the train data was created using 28,928 parallel English-to-Telugu sentences and 28,928 images. Each dataset item comprises an image, a rectangular region in the image, a source text in English describing the image region, and its translation in Telugu. After removing duplicate sentences having ID numbers 2391240, 2385507, 2328549 from parallel data and one image having ID number 2326837 (since the corresponding text is not present in parallel data), the parallel and image train data was reduced to 28,927. The selected rectangular portion of the image is represented using image coordinates

(X, Y Width, Height). The testing and validation sets consist of 1595 and 998 images respectively. Another text dataset containing English-to-Telugu text translation with 155798 sentences was also employed and added to the token dictionary to generate more tokens and include all possible variations of Telugu sentences and words.

4 Proposed Methodology

In order to improve translation accuracy and contextual relevance, we propose a multimodal translation approach in which we leverage both visual and textual information for translation.

As shown in Fig. 2, we start by tokenizing the text and integrating position embeddings, which contain the information on the position of each token in the sentence, to capture sequential patterns. Simultaneously, the pre-trained ResNet-50 model is used to extract rich visual features from the images. The textual and visual features are then fed into our multimodal encoder comprising two FNet layers, where they undergo a fusion. The fused representations are then passed through a transformer decoder translating these representations, which includes attention mechanisms that enable

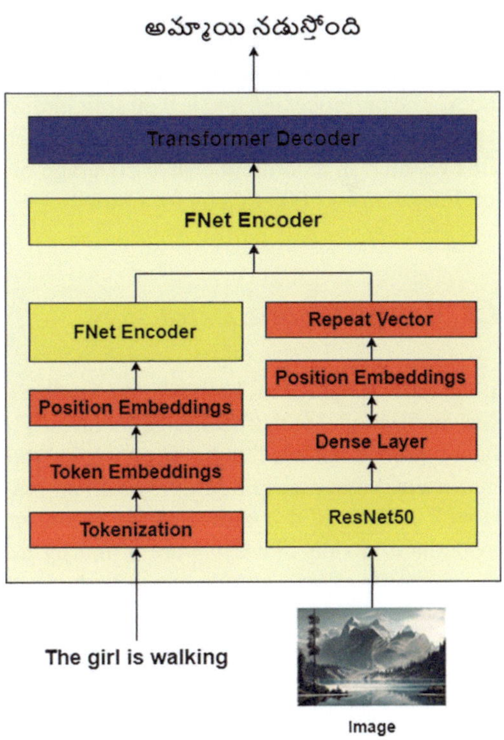

Fig. 2 Model architecture

the model to focus on relevant parts of the input sequence and feed-forward networks that provide additional nonlinearity and computational depth. This decoder can thus handle complex linguistic structures while producing contextually appropriate Telugu translations.

4.1 Text Processing

Tokenization is the process of breaking down text into smaller units called tokens. The types of tokenization are word tokenization which breaks down the text into individual words based on spaces, character tokenization which breaks text into individual letters and subword tokenization which creates units that are larger than characters but smaller than the entire word.

As character-level tokenization may result in loss of semantic information and context subword tokenization was employed initially. However, current algorithms couldn't accurately perform subword tokenization for Telugu due to its complex structure. So, as a practical solution word tokenization has been implemented.

Word Tokenization Word tokenization effectively handled the Telugu vocabulary within the translation framework. It made sure that the model captures the semantic meaning of words. The tokenizer class from the TensorFlow Keras library was used to implement word tokenization. This involves initializing the Tokenizer and fitting it to the corpus. This will create a dictionary where each word is mapped to a unique index. The preprocessed text is then split into words based on spaces and each word is represented by its corresponding index converting the text into tokenized sequences. This resulted in a comprehensive coverage of vocabulary of around 50,000 tokens for Telugu and 17,000 tokens for English.

4.2 Image Feature Extraction

In the current approach, based on the studies [15–17] we have used the pre-trained ResNet-50 model to extract image features. The pre-trained ResNet-50 is one of the most powerful models that has been trained thoroughly on the ImageNet dataset. We have trained this pre-trained CNN with an image dataset containing 28,928 images as well as their ROI coordinates for optimum results by removing noise. Usually, pre-trained models like CNN are primarily used for image classification, and they are composed of mainly two parts—feature extraction and classification. In the case of a ResNet-50, the architecture has a global average pooling layer which is followed by a single fully connected layer without activation, used for classification tasks. As displayed in Fig. 3, we have adopted this architecture for feature extraction by using layers up to the fully connected stage. This model accepts the input of shape (224, 224, 3) and returns the feature vectors of shape (, 2560). These feature vectors are

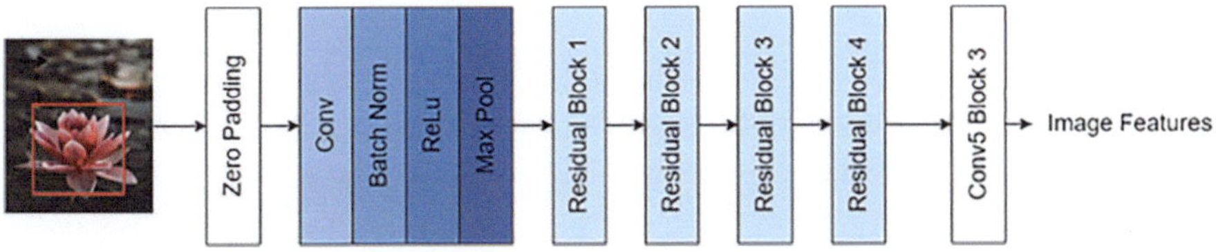

Fig. 3 Image feature extraction using ResNet50

further reshaped by passing through fully connected layers. Finally, a repeat vector layer is introduced to add one more dimension to the model, which is equal to the length of the maximum sentence in the English language which is 34; therefore our output vector will be of shape (, 34, 512).

4.3 FNet Encoder

Taking the tokens and image features as input, our encoder architecture comprises two FNet encoder layers, which effectively extract features from the sentences, providing higher-quality translations. In the FNet architecture seen in Fig. 4, the self-attention mechanism of the traditional transformer models is replaced with a leaner mechanism: a parameterized Fourier transformation-based linear mixer for input tokens.

This approach is more effective and scalable, while also ensuring that the loss of information during the transformation is minimal. Before passing the textual data into the FNet layer, the text data is first passed through an embedding and position encoding layer where all the input tokens are mapped to their corresponding dense vector representations along with their positional information. These textual embeddings are then passed as input to the first FNet encoder layer. An early multimodal fusion of the resulting embeddings and image features is then performed, and this is fed as input into the subsequent FNet layer. So the encoder extracts all the meaningful features and encodes them into a comprehensive representation for subsequent translation.

4.4 Transformer Decoder

Our intended pipeline contains a decoder for generating the translations from the embedded sequences obtained from the encoder output. The decoder part works with attention mechanisms and feed-forward networks utilizing a transformer-based architecture which can be visualized in Fig. 4. The decoder's ability to attend to all parts of the input sequence simultaneously enhances its capability to handle complex

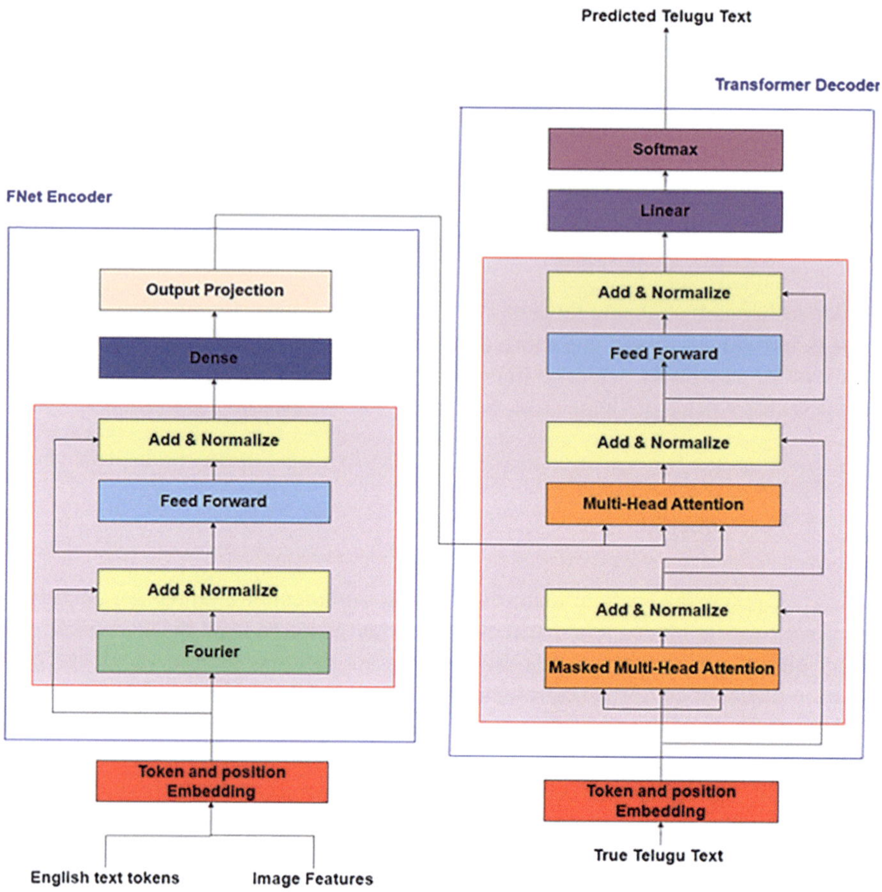

Fig. 4 Model architecture

linguistic structures and idiomatic expressions, resulting in more accurate translations, The decoder also uses attention mechanism to identify important parts of the source text, which includes long-range dependencies and context more efficiently. Due to this, the decoder model does not suffer from the vanishing gradient problem that sequential models face. Furthermore, the integration of feed-forward networks into this model enables it to learn intricate relationships between input and output sequences that in turn promote the production of coherent as well as fluent translations. In this way, our decoder goes beyond legacy sequential decoders thereby offering improved translation performance and guaranteeing much more accurate results given their context.

Table 1 Result of LSTM, GRU, transformer encoder-decoder, and FNet encoder-transformer decoder model using BLEU score

Metric	LSTM model	GRU model	TED	FETD
Bleu score	28.9677	35.6099	46.2206	54.7341

TED transformer encoder-decoder model; *FETD* FNet encoder-transformer decoder model

4.5 Sampling

In order to generate the translations from the resulting probability distribution sampling is used. At first, in order to achieve proper translation within our pipeline, a greedy sampling technique was used. Though it is simple and efficient, this method often produced more than one sentence that meant the same thing, thus reducing the quality of translations as a whole. In response to this, we tried beam sampling as our next step toward improvement. By considering several competing translations at once and identifying the best ones using a predefined beam width, repetitions were effectively eliminated from beam sampling and the translated texts became more fluent with high coherence level.

5 Experiments and Results

The efficiency of various encoder-decoder architectures was explored in our experiments on multimodal English-to-Telugu translation. The main objective was to improve translation quality from classic LSTM, GRU, and transformer encoders. We found that by using a FNet encoder and a transformer decoder we got the best results as shown in Table 1.

The model was trained using a training dataset which contains 28,928 rows of data and monitored using a validation dataset containing 998 rows of data, respectively.

5.1 Test Dataset

To evaluate our model, we used a comprehensive test dataset that contains 1595 rows of English sentences, their corresponding Telugu translations, and images. The dataset contained unseen data that allowed us the effectively train our model. We used the BLEU score metric to evaluate our model.

The results made by our model showed significantly better results in comparison with previous sequential models achieving a BLEU score of 54. This can be seen in Fig. 5. Which compares the translation displayed by our model on comparison to that of Google Translate. Conversely, other traditional types of sequential models

Image Id : 1593079		
	Source Text (English)	The batter swinging the bat
	Predicted Text (Telugu)	బ్యాట్‌ను బంతి కొట్టే ఆటగాడు బ్యాట్‌ను ఊపుతున్నారు
	Google Translation	ఒక కొట్టు బ్యాట్ ఊపుతున్నాడు

Fig. 5 Results of the translation

like LSTM and GRU registered much lower BLEU scores at 29 and 35, respectively. Even though transformer models are well-known for their effectiveness in natural language processing tasks, they were outperformed with a BLEU score of 46.

6 Conclusion and Future Works

This research is an attempt to aid the translation of a low-resource language like Telugu to improve the accuracy as well as preserve the context which may often be lost in unimodal approaches. So our multimodal English-to-Telugu translation approach extracts meaningful features from the text and images and integrates these features to generate more accurate and contextually rich translations as seen in Table 1 when compared to existing sequential models like LSTMs, GRU, or traditional unsequential transformer encoder-decoder models. It finds its vast potential applications in navigation aids, communication tools, or assistive technology. As part of future work, more analysis and research can be done to improve the performance of the machine translation model which can include but is not limited to multimodal fusion strategies such as early fusion, late fusion, or dynamic fusion mechanisms as well as experimenting with other architectures.

References

1. Kumari D (2013) Translation an attempt to deconstruct historiography in the Brande. Criterion 4:1–8
2. Premjith B, Kumar MA, Soman KP (2019) Neural machine translation system for English to Indian language translation using MTIL parallel corpus. J Intell Syst 28:387–398

3. Singh S, Kumar A, Soman KP (2018) Attention based English to Punjabi neural machine translation. J Intell Fuzzy Syst 34:1551–1559
4. Vaswani A, Shazeer N, Parmar N (2017) Attention is all you need. In: Proceedings of the 31st international conference on neural information processing systems (NIPS'17). Curran Associates Inc., Red Hook, NY, USA
5. Laskar SR, Dadure P, Manna R (2022) English to Bengali multimodal neural machine translation using transliteration-based phrase pairs augmentation. In: Proceedings of the 9th workshop on Asian translation
6. Parida S, Panda S, Biswal SP (2021) Multimodal neural machine translation system for English to Bengali. In: Proceedings of the first workshop on multimodal machine translation for low resource languages (MMTLRL 2021)
7. Kumar VV, Lalithamani N (2022) English to Tamil multi-modal image captioning translation. In: 2022 IEEE world conference on applied intelligence and computing (AIC). IEEE
8. Caglayan O, Madhyastha P, Specia L, Barrault L (2019) Probing the need for visual context in multimodal machine translation
9. Abhishek S, Sathish H, Kumar A, Anjali T (2022) Aiding the visually impaired using artificial intelligence and speech recognition technology. In: 2022 4th international conference on inventive research in computing applications (ICIRCA). IEEE
10. Lekshmy HO, Jayaraman S (2022) English-Malayalam vision aid with multi modal machine learning technologies. In: 2022 6th international conference on intelligent computing and control systems (ICICCS). IEEE
11. Liang PP, Zadeh A, Morency LP (2023) Foundations & trends in multimodal machine learning: principles, challenges, and open questions
12. Lei S, Li Y (2023) English machine translation system based on neural network algorithm. Proc Comput Sci 228::409–420
13. Meetei LS, Singh SM, Singh A (2023) Hindi to English multimodal machine translation on news dataset in low resource setting. Proc Comput Sci 218:2102–2109
14. Laskar SR, Singh R, Karim MF (2022) Investigation of English to Hindi multimodal neural machine translation using transliteration-based phrase pairs augmentation. In: Proceedings of the 9th workshop on Asian translation
15. Li Z, Gu T, Li B (2022) ConvNeXt-based fine-grained image classification and bilinear attention mechanism model. Appl Sci 12:9016
16. Kamal K, Hamid EZ (2023) A comparison between the VGG16, VGG19 and ResNet50 architecture frameworks for classification of normal and CLAHE processed medical images
17. Dhanya R, Paul IR, Akula SS (2019) A comparative study for breast cancer prediction using machine learning and feature selection. In: International conference on intelligent computing and control systems (ICCS)

Prediction of Inter-ocular Rice Leaf Disease Using Deep Learning Approach

Namala Shiva Prasad and Kiran L. N. Eranki

Abstract Agriculture is a major part of many nations' economies, and in Telangana, it plays a crucial role. Rice, also referred to as paddy is cultivated as a staple food cash crop, as rice is consumed three times a day by the people of Telangana. However, various factors such as diseases affecting paddy leaves, temperature fluctuations, and pest attacks can significantly impact paddy production, potentially reducing yields by 40–50%. The prevalence of paddy diseases varies by region and season, with the most common ones in Telangana being bacterial leaf blight, brown spot, and rice blast. Early detection of these diseases is essential to prevent widespread damage to farmland. We address this through our proposed system that uses preprocessing and data augmentation to expand the dataset. It employs a pretrained Inception_v3 algorithm and a deep learning-based CNN algorithm, which efficiently classifies paddy diseases. This approach not only enhances the accuracy of disease prediction but also reduces the time required for classification, helping farmers take timely action to protect their crops.

Keywords Deep learning · Paddy diseases prediction · Classification · Image recognition · CNN

1 Introduction

Agriculture is a cornerstone of Telangana's economy, and paddy cultivation holds particular importance due to rice being a dietary staple consumed three times daily by the local population. This crop underpins livelihoods both directly and indirectly. But paddy production causes loss every year due to some diseases in paddy leaf as well as insert pest attacks which will directly affect our country's economy. Farmers should

N. Shiva Prasad (✉) · K. L. N. Eranki
Department of Computer Science and AI, SR University, Warangal, India
e-mail: nspd1304@gmail.com

K. L. N. Eranki
e-mail: erankikiran@gmail.com

H. Sharma et al. (eds.), *Artificial Intelligence: Theory and Applications*, Lecture Notes in Networks and Systems 5589, https://doi.org/10.1007/978-981-96-1687-9_19

adopt appropriate precautionary measures during the early stages of paddy cultivation to prevent the spread of diseases. Since paddy diseases can be transmitted through air and water, they may quickly affect the entire field. Once the infection spreads, it becomes very difficult for farmers to control, potentially leading to significant crop loss. Great reduction in the cultivation of paddy. Scope of this project's goal is to find the diseases in the paddy leaf and to notify the farmer about the affected disease types so that farmer can take early precaution to protect the paddy diseases in initial stage. If the diseases are identified in early stage, there is no need to spray a high-dose fertilizer on the paddy crops [1]. The paddy crops are affected by bacteria, fungus, and virus diseases. Most of the critical diseases affect paddy which are blast, bacterial leaf blight, tungro, and paddy brown spot; it will cause great economic loss in the field of agriculture.

1. **Brown Spot Rice Leaf Disease**: A brown spot rice leaf disease is referred to as "Helminthosporiumoryzae" which is a fungal leaf disease that occurs mainly in Kharif season. The symptoms of these diseases are leaves with small brown spots; once they develop, the leaves look like round or oval brown spots [2]. The diseases are transferred from the infected paddy seeds. It causes fewer seedlings. This disease is found in Orissa, Tripura, Tamil Nadu, Andhra Pradesh, and Telangana. As shown in Fig. 1a, the brown spot leaf disease can be observed.
2. **Paddy Blast Rice Leaf Disease**: Paddy blast occurs in Kharif and Rabi season: November to February; it is caused by "Magnaporthe Oryzae"; it is fungus disease, and its symptoms are elliptical spots with light color center and reddish edges. The yield reduces up to 50%. This disease is found in Telangana, Andhra Pradesh, Kerala, Karnataka, and Tamil Nadu [1]. We can observe the paddy blast rice leaf disease in the snapshot view shown in Fig. 1b.
3. **Bacteria Rice Leaf Blight**: The primary cause of bacterial infections like bacterial leaf blight is "Xanthomonas Oryzae", and it is a monsoon season disease (Kharif); these types of diseases can be identified by the symptoms of water-soaked lesson begin at the leaf edges and extend to the leaf base. Once it's not identified in early stage, it will become pale brown to yellowish brown. Having a yellowish border separating the green portion of the paddy leaf from the dead portion. The diseases are caused by rainy, cloudy, and stormy climates as well as high nitrogen fertilizer [2]. This disease causes significant yield losses in Telangana, Kerala, coastal Andhra Pradesh, Odisha, and Tamil Nadu, as illustrated in Fig. 2a.
4. **Rice Tungro Disease**: Caused by *Rice tungro bacilliform virus* (RTBV), this disease leads to stunted growth, leaf yellowing, and reduced grain yield in rice plants. The virus is transmitted by green leafhoppers, which infect the crop [3]. RTBV disrupts nutrient absorption, compromising plant health and productivity. Effective management—such as planting resistant varieties and controlling leafhopper populations—is crucial to minimize damage. This disease is prevalent in Telangana, Andhra Pradesh, Kerala, Karnataka, and Tamil Nadu (Fig. 2b).

The remaining sections of paper are organized as follows: Section 2 covers the relevance of rice leaf disease images and their role in diagnosis and research review.

(a) Snapshot view of Rice Brown Spot.

(b) Snapshot of Paddy Blast.

Fig. 1 Snapshot view of rice leaf disease types—**a** Brown spot and **b** Blast

(a) Snapshot view of Bacterial Leaf Disease.

(b) Snapshot view of Leaf Tungro.

Fig. 2 Snapshot view of other rice leaf diseases: **a** Bacterial, **b** Leaf Tungro

Followed by Sect. 3 discusses the possible recommendations for premature leaf disease diagnosis. And finally we conclude the study with Sect. 4 which details the findings and concludes the study with pointers to future work.

2 Relevance of Rice Leaf Disease Images in Disease Diagnosis and Prognosis

Earlier studies revealed the importance of deep transfer learning to detect the paddy plant disease. Usage of DenseNet as a neural network while training the model, similarly, ImageNet and the Inception model-based methods were applied. However, some of these models failed to diagnose the leaf blast area precisely [1]. Studies also propose the segmentation methods to predict the disease, using feature extraction and classification but drawbacks here are still lacking to predict the diseases [4]. Kathiresan et al. suggest a procedure for identifying data using individual transfer learning paddy diseases. Using generative adversarial network here the advantage produces a good accuracy rate, and the demerits are lack of samples [1]. Utilizing a machine learning technique like *K*NN, deep models-based paddy leaf disease detection system obtained the set utilized for training had a success rate of 75.463%, whereas the test set had an average of 70.8333% [3]. Liang et al. suggest CNN algorithm to recognize the paddy blast diseases including LBPH and Haar-WT model; this method needs to optimize its parameter [5]. Kawakita et al. propose CNN algorithm for plant disease detection and diagnosis problem faced here is it doesn't accurately detect the plant diseases [6]. Jiang et al. propose a CNN algorithm incorporating GoogleNet Inception and Rainbow concatenation methods to train the model, achieving 78.80% detection accuracy for five apple leaf diseases. However, a key limitation of this approach is its inability to detect multiple objects simultaneously [7]. Lu et al. presence looks at founded on deep learning like convolutional neural network here the pictures can be given straightforwardly to the model to train the network [8]. Bhattarai et al. propose deep learning with machine learning method; here the disadvantage is resulting in dimensionality reduction [9]. Similarly, other researchers also used video to convert into still image, then detected the paddy diseases using faster R-CNN, and used transfer learning VGG16, ResNet-50, and YOLOv3 to train neural network [6].

3 Research Methodology and Proposed Disease Prediction Framework

The proposed framework has preparing and testing stage. The dataset was prepared with a significant number of images depicting paddy leaves affected by various diseases. There are three illnesses ordinarily impacted in paddy in particular bacterial leaf scourge, paddy impact, and earthy colored spot. The dataset pictures are gathered from the site Kaggle. These illness datasets are put away in isolated envelope. The dataset was split into 70% for training (including validation) and 30% for testing. This division resulted in 12,040 training images and 5,160 test images, with the entire dataset comprising 17,200 images in total (covering training, validation, and test sets).

1. **Preprocessing** First, step is pre-handling which is utilized to eliminate the undesirable commotion from the loaded infected paddy leaf image, here using image resizing ($50 * 50$) pixel to make the entire image into the same size so that it's easy for training the CNN network.

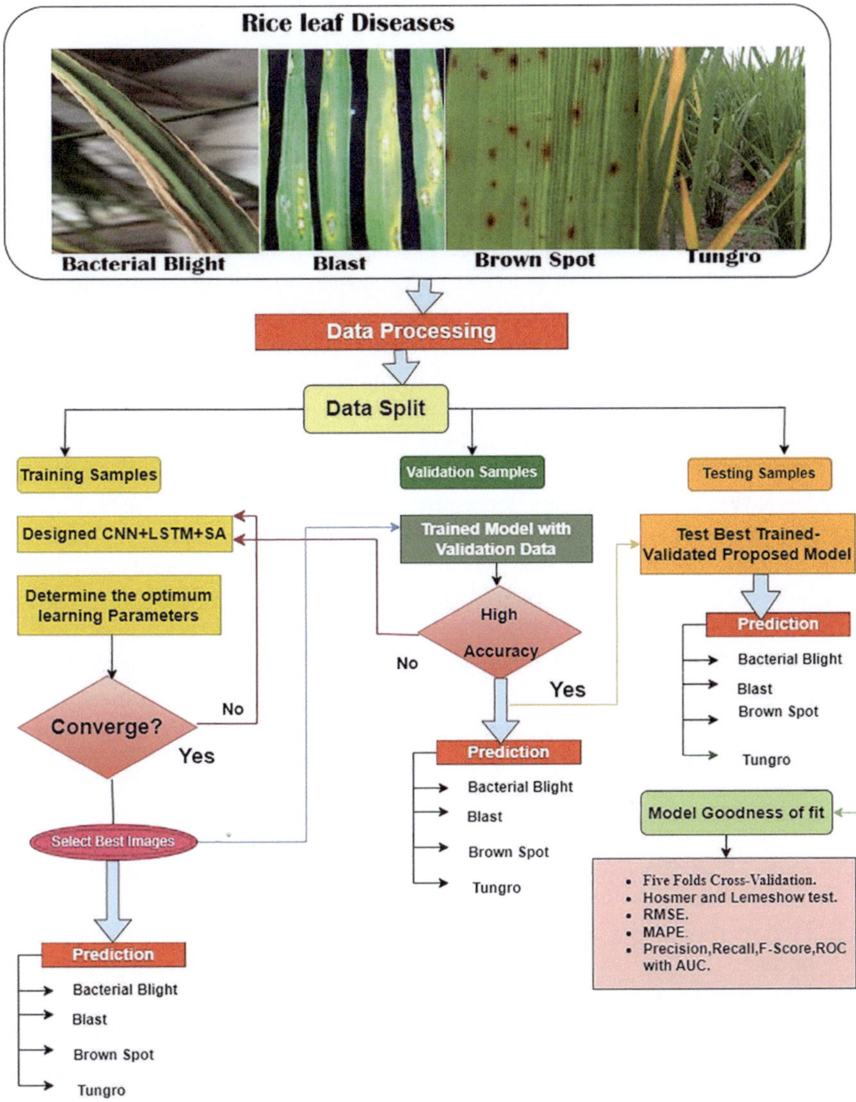

Fig. 3 Research approach to rice leaf disease diagnosis and its phases using CNN model

2. **Image augmentation** The image augmentation is the method to increase the original dataset. Here image rotation, image blurring, and image flipping of augmentation method are used to extend the dataset artificially. The extended original paddy diseases image dataset to 628 images.

3. **GANs augmentation** Still the dataset is a little small; This approach may lead to overfitting during training. To mitigate this issue, we employ Generative Adversarial Networks (GANs) to augment the dataset, facilitating more effective training of the CNN algorithm. Using GANs, the dataset was expanded to 17,200 images for neural network training. By using GANs method extend the dataset up to 12,000 images for training the neural network.
4. **Classification training module: transfer learning Inception_V3** Inception_V3 is a widely used technique for image classification. The primary advantage of using this method is the ability to reuse a pretrained model on a new task, which significantly reduces training time, improves performance, and facilitates loss function computation and epoch optimization. The convolutional and pooling layers act as filters to extract the feature from the image dataset. Fully connected layer or dense layer is used for classification purpose here using unsupervised classification algorithm like convolution neutral network (CNN). Here directly using the RGB image pixel value used for classification. CNN algorithm, using various layers like keras layers dense, dropping out, layer, become flattened, and dense layer of total params 22,853,411: here trainable params is 1,050,627 and non-trainable params is 21,802,784. This method trains the neural network by teaching various diseases in paddy leaf. Activation function SoftMax is used to predict the disease type.
5. **Testing module** Testing is a method of predicting the output of classifier as healthy leaf, as well as types of paddies affected conditions such as paddy blast, brown spot, and bacterial leaf blight, is common. These are tested in testing phase to check whether the given input is predicting the correct paddy leaf disease types or not. Phases of proposed framework:

   (a) Collect dataset from the rice cultivation center and preprocess the images.
   (b) Import libraries—keras, scklearn, and numpy to apply the necessary augmentation to datasets.
   (c) Prepare dataset for training and testing ensuring homogeneous distribution of data.
   (d) Visualize the training sample and validate the image details.
   (e) Build the CNN model, identify the layers and batch normalization to be performed.
   (f) Compute the gradient loss, error metrics, apply the optimizer to reduce model overfit issues.
   (g) Run the trained CNN model and validate the test sample to evaluate the model performance and accuracy.
   (h) Predict the accuracy of disease and compute the misclassification error to examine the model performance.

All the phases of proposed framework have been shown in Fig. 3.

Fig. 4 Preprocessed snapshots of four variants of rice leaf disease

Table 1 Evaluation metrics for CNN Inception V3 model

Metric	Precision	Recall	$F1$-score	Support
	1.00	0.67	0.80	3
	0.96	0.96	0.98	261
Accuracy			0.98	264
Macro avg.	1.00	0.83	0.90	264
Weighted avg.	1.00	1.00	1.00	264

4 Results and Discussion

The preprocessing result of paddy blast, brown spot, tungro, and bacterial leaf blight has been shown in Fig. 4.

Evaluation metrics of the proposed model performance show 98% accuracy in predicting the type of rice leaf disease, provided to the model. Comparative analysis of other selective models which are very popular in the rice leaf detection have also been examined, and the results of our observations have been shown in Table 1. Our proposed model seems to outperform the other neural network models considering the fine-tuning of hyper-parameters and selective normalization applied to control model overfit issues.

CNN Inception V3 model is applied to predict rice leaf disease diagnosis to classify the leaf disease categories and provide the necessary recommendations for the disease. Proposed model applies 30 epochs to train the model with 90% accuracy and test the leaf disease. Current implementation gives prediction accuracy of 95.7%

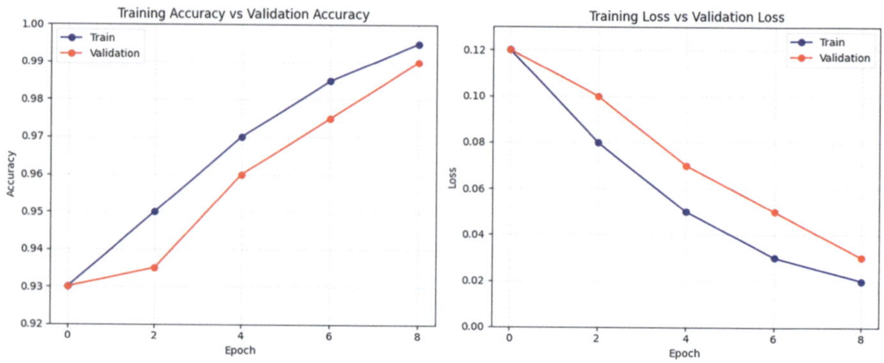

Fig. 5 Results of model performance on training and testing data with accuracy

Table 2 Comparative analysis of selective models performance on rice leaf disease

Model	Train-accuracy	Test-accuracy
RestNet-50	56.86	52.67
DenseNet-201	91.14	62.67
VGGNet-16	77.71	60.67
Proposed Inception_V3	95.7	75.00

on training with loss rate of 0.1081, val_loss 1.1025, and the val_acuuracy is 0.7500 as shown in Fig. 5. Batch normalization with ADAM optimizer has been applied to improve the performance of the model and control model overfit issues (Table 2).

5 Conclusion and Future Scope of Work

The proposed system for predicting paddy leaf disease applies data preprocessing, data augmentation, and a pretrained Inception_v3 algorithm, using deep learning-based CNN Model. The proposed model has demonstrated exceptional performance. The results indicate high precision and recall rates, particularly for the most prevalent disease class, which gave a perfect precision, recall, and $F1$-score of 1.00. The overall accuracy of the system stands at 1.00, underscoring its effectiveness in accurately classifying paddy leaf disease. The macro-average precision, recall, and $F1$-score are 1.00, 0.83, and 0.90, respectively, reflecting the system's balance performance across different disease categories. Additionally, the weighted average precision, recall, and $F1$-score show 1.00 further highlighting the robustness and reliability of the model in handling a diverse dataset. The high accuracy and balanced performance metrics ensure that farmers can rely on this tool for early detection and intervention, ultimately contributing to better crop health and higher yields. Future research should focus on optimizing the model further and expanding the dataset to maintain and enhance its predictive capabilities. This study significantly advances the use of machine learning in agriculture, offering a powerful solution for protecting paddy crops in Telangana.

References

1. Kathiresan G, Anirudh M, Nagharjun M, Karthik R (2021) Disease detection in rice leaves using transfer learning techniques. J Phys Conf Ser 1911:012004
2. Deepak Kumar K, Jeevithashree P, Dhanalakshmi M, Alfred Daniel J, Kalaikumaran T (2019) Gui based detection of paddy unhealthy leaves and to increase the growth using image processing techniques. Int J Res Eng Sci Manag 3:352–355
3. Stephen A, Punitha A, Chandrasekar A (2024) Optimal deep generative adversarial network and convolutional neural network for rice leaf disease prediction. Visual Comput 40(2):919–936
4. Sridevi S, Kiran Kumar K (2024) Optimised hybrid classification approach for rice leaf disease prediction with proposed texture features. J Control Decis 11(1):84–97
5. Lingwal S, Bhatia KK, Singh M (2024) A novel machine learning approach for rice yield estimation. J Exp Theoret Artifi Intell 36(3):337–356
6. Kawakita S, Yamasaki M, Teratani R, Yabe S, Kajiya-Kanegae H, Yoshida H, Fushimi E, Nakagawa H (2024) Dual ensemble approach to predict rice heading date by integrating multiple rice phenology models and machine learning-based genetic parameter regression models. Agric For Meteorol 344:109821
7. Sherline Jesie R, Godwin Premi MS, Jarin T (2024) Comparative analysis of paddy leaf diseases sensing with a hybrid convolutional neural network model. Measur Sens 31:100966
8. Lu Y, Tao X, Feng J, Du J, Li G, Liu Y (2024) Image recognition of rice leaf diseases using atrous convolutional neural network and improved transfer learning algorithm. Multimed Tools Appl 83(5):12799–12817
9. Bhattarai Y, Prasun K, Sharma G, Rai P (2024) Enhancing rice disease binary classification. An analysis of deep neural network models. Int J Res Publ 147(1):9

Scalable and Real-Time Sentimental Analysis Using NLP and Distributed Concepts

Kodati Bhanusri, Koti Leela Sai Praneeth Reddy, Julakanti Sai Yaswanth, Sreebha Bhaskaran, and Nalini Sampath

Abstract This paper describes a system that uses NGINX as the server to improve performance, combining advanced natural language processing (NLP) models with distributed computing. Fast and precise sentiment classification is provided by the system, which specializes in real-time sentiment analysis of massive amounts of text data. Scalability and dependability are guaranteed by distributed computing, and system responsiveness and dependability are boosted by NGINX's improved load balancing and traffic management. Different load balancing algorithms are used to distribute network traffic among multiple servers so that no single server is overloaded. These algorithms include Round Robin, Least Connections, IP Hashing, Geographic Load Balancing, and Consistent Hashing. The algorithms random forest (RF), support vector machine (SVM), and decision tree were tested; random forest yielded the best results in terms of accuracy. The flask-developed backend of the system manages requests effectively and interfaces with NLP models. To sum up, this work presents a real-time, scalable sentiment analysis system that leverages NGINX, distributed computing, and sophisticated NLP models to achieve better performance. The system produces accurate sentiment classification and is very good at quickly analyzing large amounts of text data. This work provides a thorough framework for creating scalable and effective real-time sentiment analysis systems, setting the stage for further developments in the field. For applications like social media monitoring and customer feedback analysis, integrating cutting-edge NLP and ML algorithms into a distributed framework provides helpful advice on how to implement dependable and robust sentiment analysis.

Keywords Distributed systems · Emotional categorization · Flask web applications · Load balancing algorithms · Real-time sentiment analysis

---

K. Bhanusri · K. L. S. P. Reddy · J. S. Yaswanth · S. Bhaskaran (✉) · N. Sampath
Department of Computer Science and Engineering, Amrita School of Computing, Amrita Vishwa Vidyapeetham, Bengaluru, India
e-mail: b_sreebha@blr.amrita.edu

H. Sharma et al. (eds.), *Artificial Intelligence: Theory and Applications*, Lecture Notes in Networks and Systems 5589, https://doi.org/10.1007/978-981-96-1687-9_20

1 Introduction

Sentiment analysis is an important field in NLP that identifies sentiments in textual data, whether neutral or positive-negative. It recognizes emotions such as happiness and sadness, which helps businesses and governments analyze consumer input and market trends. The combination of ML and NLP has transformed sentiment analysis, improving the ability to understand the complexity of language. Scalability and reliability are critical components for long-term success in the complex ecosystem of real-time sentiment analysis online apps. The critical function of load balancing algorithms is at the core of ensuring flawless operation. These algorithms coordinate incoming requests, carefully dispersing them among multiple servers or computational resources to enhance resource efficiency and reduce response times.

This paper presents a novel approach to real-time sentiment analysis through the smooth integration of state-of-the-art NLP and ML algorithms into a distributed system framework. Taking advantage of the Flask framework's flexibility for web development, the system leverages advanced sentiment analysis libraries to efficiently parse textual data. Furthermore, load balancing algorithms that are carefully set up with Nginx are thoroughly tested and contrasted with Apache Bench. Their complex influence on system scalability and performance is clarified by this thorough analysis. Through the integration of cutting-edge NLP, ML, and distributed systems techniques, this research aims to provide invaluable insights into building robust and efficient sentiment analysis systems for practical use.

Hence, this paper focuses on developing a scalable and real-time sentiment analysis system using distributed computing and advanced NLP models. The main contributions of the paper in this direction are:

- Provide a brief explanation about scalable and real-time sentiment analysis using NLP techniques.
- Provide a comparative analysis of various NLP models, including SVM, Naive Bayes, LSTM, and Transformer-based models, in handling sentiment analysis tasks, and evaluate their performance using metrics such as accuracy, $F1$-score, and latency.
- Propose a hybrid sentiment analysis model that integrates the strengths of transformer-based models and LSTM to enhance the scalability and accuracy of real-time sentiment analysis. The comparative analysis is enhanced by including this new hybrid model. This approach aims to demonstrate the potential improvements in sentiment analysis efficiency offered by the proposed hybrid model.

The paper is organized as follows: Sect. 2 contains a literature review of the various studies examined. Section 3 describes the proposed methodology as well as how the algorithm simulations were implemented. Section 4 displays the obtained results. Section 5 concludes the paper.

2 Related Works

Sharma [1] and Pooja Pathak highlight sentiment analysis's role in social media content assessment, employing SVM, Naive Bayes, and LSTM for nuanced analysis including language use and sarcasm. They focus on enhancing accuracy through robust models capable of handling social media's language variability. Meanwhile, Lyu [2], Qi Zhu, and Yunfan Cheng emphasize transformer load management in power grids, employing load prediction curves and incident matrices to optimize load distribution and prevent overloads, ensuring reliable grid operations. Their approach reveals techniques that ensure balance and optimality of the loads over the structure paths to support the consistent grid.

Nikolaevich et al. [3] also examined the static and dynamic load balancing in data centers as a way to facilitate information processing. These are dynamic that respond to the current conditions of the system for instance the least connections dynamic method outperforms the static round robin method which is based on rules. They suggest that their work provides substantial evidence showing that techniques in dynamic load distribution which ensures varying network loads, boost up server farm processing capabilities. At the same time, another work proposed by Jadeja [4] and Kiran Parmar, and Dr. Chintan Thacker created a sentiment analysis (SA) system using deep learning methods and neural networks to analyze the social sentiments. Thus, their SA system based on the attention mechanisms and POS tagging allows providing cost-effective and highly effective sentiment analysis for startups and small-scale indutries where acquiring precise data on consumers' feedback is crucial.

Prananda et al. [5] highlighted some findings and issues on load balancing in the transporter power transformers for avoiding the equipment stress and lesser efficiency in power networks, for which the Indonesian PLN has observed the Federal DSO's recommendation for imbalances control less than 10%. Scholars have also developed fuzzy logic algorithms in the aspect of arranging loads for the transformers, cutting the processing time and ensuring all transformers are evenly loaded in terms of phases. Meanwhile another work proposed by Savaghebi [6], Masoud Hajian, and Muhammad Sadegh Golsorkhi—investigated the electrical control systems of islanded AC MGs and pointed toward shortcomings of the conventional low-voltage variation methods, which result in imprecise load-sharing among MGs. This is achieved through combination of distributed secondary control schemes with event-based messages that is enhanced by MATLAB Simulink simulations to enhance the organization and obtain a steady voltage level.

Activeness in computer science education pointed out by Smith [7] and Srishti Srivastava say that educators should incorporate PDC as soon as possible because of its complexity. Other practical methods of retention such as learning games and flashcards that make learning more effective are also looked into. Wang [8], LiPing Yuan, Yi Zhong discuss on load balancing techniques that helps to manage large number of users and concurrency in multiple servers. They support dynamic load balancing over the traditional load balancing approach such as seeding arguing ability to handle concurrency, throughput, and response time with the use of Siege tool.

The study by Handur [9], and Prakash R. Marakumbi still address the issue of efficient high-load applications in cloud computing to avoid latency as the use of cloud computation increases. Comprehensive throughput analyses of different load balancing proposed techniques and their impact on response times are detailed, with some based on power-aware particle swarm optimization (PSO) algorithms, analyzed and assessed through simulators such as CloudSim. Various researchers like [10], Akila V., B. Sravya, Harshitha T. delineate that web hosting has shifted from a traditional method to cloud hosting because of things like; scalability, high up time, and automated backup, opposition to things like high startup costs, vulnerability to malware. Resource distribution is always a critical factor in cloud application, which is why load balancing is so important; this is backed up by the likes of Azure which can help to increase load balancing software to further improve the performance and security of applications.

Juyal [11] emphasized the importance of sentiment analysis in Amazon reviews in gaining insight into customer opinions. Although real feedback provides real insights, fake reviews make it difficult to determine sentiment accurately. In order to improve accuracy, recent research has made use of NLP, ML, and DL techniques, especially hybrid approaches like SVM and NB. However, sentiment analysis tools enable brands to effectively navigate the constantly changing digital landscape and understand consumer sentiment. The growing significance of machine learning-driven sentiment analysis in NLP is emphasized by Athar et al. [12]. It provides information on how opinions are formed in a variety of contexts, including social media. Reaching high accuracy and efficiency is one of the challenges. Sentiment analysis is intended to be improved by ensemble learning models, especially soft voting ensembles that incorporate classifiers such as LR, NB, XGB, RF, and MLP.

Rai [13] introduces a convolutional neural network (CNN) for fault classification in distribution networks with distributed generators (DGs), utilizing raw and sampled three-phase voltage and current data. The CNN achieved 99.52% average accuracy across fault and no-fault classes, surpassing traditional methods in accuracy and efficiency. Similarly, a study on a mixed transmission and distribution network with photovoltaic DGs demonstrated accuracies of 99.92% and 99.97%, respectively. Pratheeba [14] focuses on managing large volumes of high-velocity data in real-time applications, advocating distributed frameworks like Hadoop and Apache Spark. Comparing speed and storage tradeoffs, Apache Spark is preferred for handling big data in real-time scenarios, enhancing emergency response systems for road travel using its superior performance capabilities.

Sreelekshmi [15], maintaining fault tolerance and quality of service (QoS) in the cloud relies heavily on effective load balancing, which involves distributing incoming user requests across multiple virtual machines (VMs). To address the issue of a single point of failure in traditional load balancer systems, this paper proposes a model utilizing multiple load balancers that share or borrow services to correct load estimations, enhancing fault tolerance and improving cluster capacity management. Another work proposed by Bhaskaran [16] integrates smart meters with a peak load management system (PLMS) and theft detection for cost-effective Smart Grid interoperability in India. Using IoT, it monitors real-time electricity and manages blackouts remotely. By distinguishing between light and power loads, it cuts power only to high-power loads during peak times, reducing inverter use. Consumers can pay extra for high-power loads during peak times via an online portal or manual breaker system, ensuring flexible power usage.

To summarize sentiment analysis was carried out using NLP and ML techniques such as SVM, Naive Bayes, and LSTM. These are also used to better analyze Amazon reviews, detecting false reviews, and use ensemble models for greater accuracy. Researchers investigated transformer load balancing, discovering that dynamic approaches such as least connections and fuzzy logic algorithms effectively prevent overloading and maintaining balance. Dynamic load balancing solutions in Nginx, and other servers, such as particle swarm optimization, enhanced speed, reduced latency, and managed concurrency more effectively. Research gaps include enhanced handling of context and sarcasm in sentiment analysis, optimization and real-time adjustments in data center load balancing, scalability and efficiency in transformer load management, integration of fuzzy logic with machine learning for improved load balancing, and improvement in fault tolerance and quality of service in cloud environments.

3 Methodology

3.1 *User Interface*

To gain access to the system, you must enter your credentials on the login screen found in the user interface. You will see two options after logging in: Fine-Grain Sentiment Analysis and Standard Sentiment Analysis. A text box where you can enter a message or paragraph will open when you choose one of these options. To view the sentiment analysis findings, click the "Analyze" button (Fig. 1).

3.2 *Sentiment Analysis Using NLP*

NLP techniques enable computers to efficiently perceive, analyze, and respond to human language, making them useful for applications like social media monitoring and customer feedback analysis. Sentiment analysis using NLP techniques involves analyzing text to detect the emotional tone of the words and categorizing them as positive, negative, or neutral as well as into excited, fearful, hopeful, grateful, etc. Using this method helps researchers and corporations assess public opinion and make more informed decisions.

Standard Sentiment Analysis: Classifying text as positive, negative, or neutral is a standard sentiment analysis technique. Without considering more subtleties or intensity of feeling, it gives a general impression of the sentiment of a text. Uses for

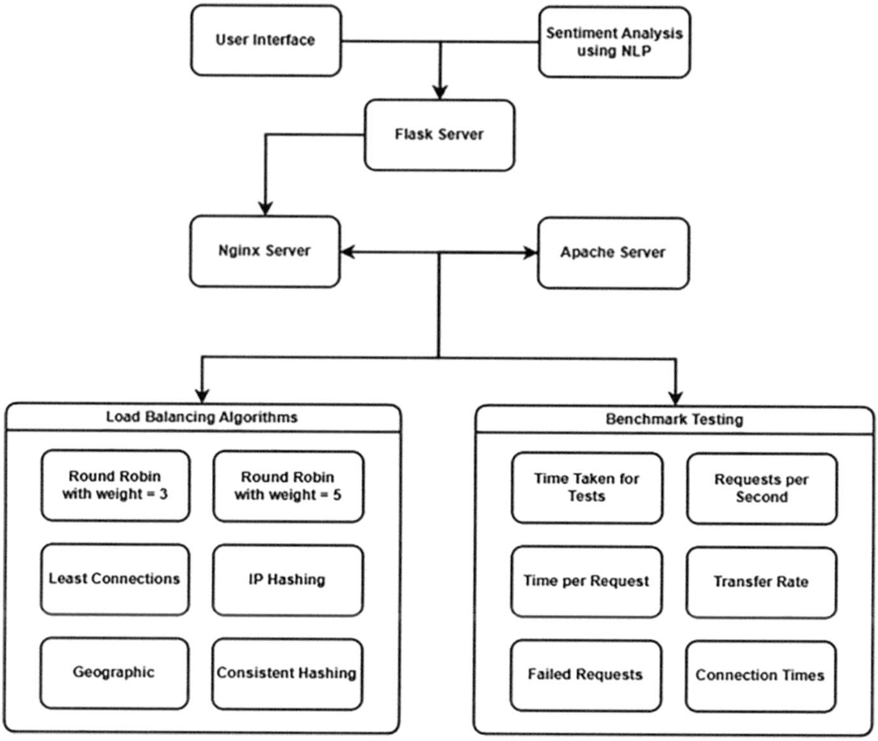

Fig. 1 Proposed architecture diagram

this technique include opinion mining, customer feedback analysis, and sentiment analysis in social media monitoring.

Fine-Grain Sentiment Analysis: Fine-grain sentiment analysis offers a comprehensive understanding of the sentiment present in a text. It is able to identify multiple sentiments within a single sentence, capturing the intensity and context of feelings. This approach is especially helpful for evaluating complex customer feedback, social media posts, and other situations where minute sentiment variations matter.

3.3 Proposed Methods

Support Vector Machine (SVM): SVM is used for sentiment classification because it performs well in high-dimensional spaces. The implementation entails training the SVM on a labeled dataset, with each text input assigned a sentiment category. The SVM then uses the learned patterns to predict the sentiment of new, previously unseen text.

Decision Tree (DT): A decision tree is a simple but effective algorithm that divides a dataset into subsets based on the most important attributes. It is simple to understand and visualize, providing valuable insights into the decision-making process behind sentiment classification.

Random Forest (RF): Random forest is an ensemble learning method that uses multiple decision trees during training to output the mode of the classes for classification. It improves prediction accuracy while controlling overfitting, making it ideal for sentiment analysis tasks.

NLP Techniques: Tokenization, stopword removal, stemming, and lemmatization are the preprocessing steps employed in this work. These steps ensure that the text data is properly cleaned and normalized before being fed into sentiment analysis models. Feature extraction methods like Term Frequency-Inverse Document Frequency (TF-IDF) are used to convert text into numerical vectors that the models can handle.

Flask Web Applets: A Flask server is used to build a lightweight web application that allows users to enter text and view sentiment analysis results. Flask was chosen because it is simple to use and versatile, allowing for rapid web application development and deployment.

Scalability and Real-Time Processing: To handle real-time processing and scalability, the Flask server works in tandem with Nginx, a fast and dependable web server and reverse proxy. Nginx efficiently manages and distributes incoming requests to the Flask server, ensuring peak performance and resource utilization.

3.4 Nginx Server

NGINX is a fast and reliable web server and reverse proxy server with efficient performance, essential functions, configurable configurations, and low use of resources. Originally, it has been designed to avoid the installation of virtual servers, speeding up the serving of static content and handling concurrent connections.

3.5 Load Balancing Algorithms

By dividing up network traffic among several servers to avoid overloading any one of them, load balancing algorithms increase performance and reliability. An efficient and ideal use of server resources is ensured by common algorithms like Round Robin, Least Connections, IP Hashing, Consistent Hashing, and Geographic Load Balancing. A simple and balanced distribution of incoming network traffic is ensured by the round robin technique, which assigns a sequential number of requests to each server in a cyclical order. By dynamically allocating requests based on the current workload, Least Connections routes traffic to the server with the fewest active connections. This is particularly helpful when server loads fluctuate greatly. In order to maintain session durability and equitable distribution, IP Hashing routes requests based on the client's IP address, guaranteeing that traffic from the same client is consistently directed to the same server. By distributing traffic among servers located in various regions and directing users to the closest or most suitable server based on their location, geographic load balancing enhances performance and reliability while reducing latency and improving user experience. Consistent Hashing divides requests among servers using a hash function, minimizing the effect of server pool changes on the system as a whole, minimizing redistribution when servers are added or removed, and guaranteeing dependable and effective resource allocation.

3.6 Apache Server

An Apache server is a widely used open-source web server software that provides robust and flexible features for delivering web content. Known for its stability and rich module ecosystem, Apache supports various functionalities such as SSL/TLS encryption, Benchmark Testing, URL redirection, and authentication. It is highly configurable and can serve dynamic and static web pages, making it suitable for a wide range of web applications and services.

3.7 Benchmark Testing

By comparing results to pre-established benchmarks, benchmark testing analyzes the speed, stability, and efficiency of hardware or software and provides insights for optimization and performance improvement. The duration of tests, which indicates the performance of the system, requests per second, which measures the throughput and capacity of the system, time per request, which indicates the responsiveness of the system and the user experience, transfer rate, which indicates the speed at which data is transmitted and the network's efficiency, and connection times, which examine different facets of connection establishment to offer insights into the effectiveness and performance of the network under varying loads, are important metrics.

4 Results and Analysis

Using the Flask framework, a real-time sentiment analysis application was created that uses sophisticated natural language processing (NLP) techniques to interpret and categorize text and provide instant feedback. This app provides specific emotions like happy, sad, and excited in addition to general sentiment categories like mixed, neutral, negative, and positive. In a distributed system, Nginx dispersed incoming requests

Table 1 Performance metrics without load balancing

Name of the performance metric	Value
Time taken for tests	36.212 s
Requests per second	243.99
Time per request (average)	3721.236 ms
Time per request (concurrent)	3.729 ms
Transfer rate	104.27 KB/s
Connection times (ms) (min/mean/median/max)	113/3326/3465/4006 ms

among several Flask instances. To simulate concurrent users, Apache Bench (ab) was used to evaluate performance metrics. Various load balancing techniques were evaluated based on metrics such as transfer rate, time taken for tests, and requests per second.

The Flask application's performance metrics without load balancing are shown in Table 1. Figure 2 presents the minimum, mean, median, and maximum connection times of the Flask application without load balancing, measured in milliseconds.

The performance measures for the different load balancing algorithms—Round Robin, Least Connections, IP Hashing, Weighted RR (3), Weighted RR (5), Geographic, and Consistent Hashing—are shown in Table 2. A visual comparison of the performance of various load balancing algorithms is presented in Fig. 3, which shows the connection times associated with each algorithm. Tables 2 and 3 show that load balancing significantly enhances performance, improving test times, requests per second, and time per request compared to the "No Load Balancing" scenario.

Fig. 2 Connection times without load balancing

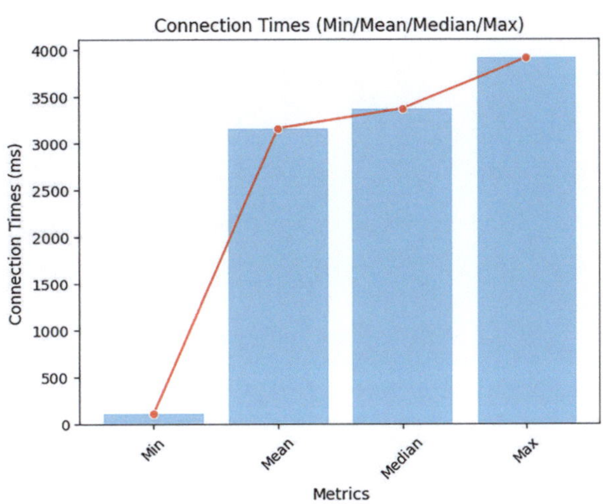

Table 2 Performance metrics for different load balancing algorithms

Metric	Round robin	Least connections	IP hashing	Weighted round robin	Weighted round robin	Geographic	Consistent hashing
				Weight = 3	Weight = 5		
Time taken for tests (s)	35.336	35.241	35.818	35.265	35.492	35.279	35.951
Requests per second	283	283.76	279.19	283.57	281.75	283.46	278.16
Time per request (average)	3533.591	3524.103	3581.828	3526.495	3549.224	3527.856	3595.058
Time per request (concurrent)	3.534	3.524	3.582	3.526	3.549	3.528	3.595
Transfer rate	108.89	109.18	107.42	109.11	108.41	109.07	107.03
Connection times (min/mean/median/max) (ms)	101/3165/3374/3923	112/3154/3369/3909	121/3215/3379/3924	100/3158/3366/3946	130/3176/3389/3932	132/3161/3374/3885	111/3228/3397/3928

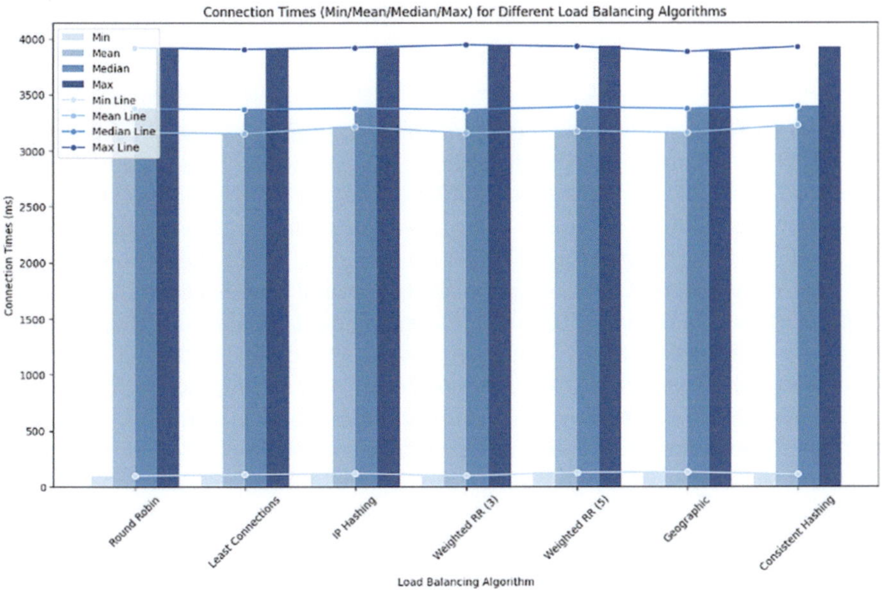

Fig. 3 Connection times with different load balancing algorithms

According to the analysis, all load balancing algorithms outperform the scenario in which load balancing is not used. By handling 283.76 requests per second and cutting the test time to 35.241 s, the Least Connections method consistently outperformed other algorithms. Additionally, this approach showed the best results in terms of transfer rate (109.18 KB/s), average and concurrent request times, and overall performance.

Even though the variations in performance metrics are negligible, they have the potential to become substantial in situations of high load. Even though the Consistent Hashing algorithm yielded the highest times, applications that need session persistence based on particular keys—like user IDs—might still favor it.

Table 3 Overall summary of comparative analysis

Metric	No load balancing	Best load balancer	Worst load balancer
Time taken for tests (s)	36.212	35.241	35.951
Requests per second	243.99	283.76	278.16
Time per request (avg)	3721.236	3524.103	3595.058
Time per request (concurrent)	3.729	3.524	3.595
Transfer rate	104.27	109.18	107.03
Connection times (min/mean/median/max) (ms)	113/3326/3465/4006	112/3154/3369/3909	111/3228/3397/3928

5 Conclusion

In conclusion, this work presents a scalable, real-time sentiment analysis system that improves performance by utilizing distributed computing, advanced NLP models, and NGINX. Accurate sentiment classification is produced by the system, which is quite good at analyzing massive amounts of text data quickly. In addition to improving responsiveness, NGINX's load balancing and traffic management provide scalability and reliability when coupled with distributed computing. In order to prevent overload and improve performance, the study evaluated several load balancing algorithms, such as Consistent Hashing, IP Hashing, Geographic Load Balancing, Round Robin, and Least Connections. Significant contributions include integrating NLP and ML techniques like decision tree (DT), random forest (RF), and support vector machine (SVM) for sentiment analysis (RF demonstrating superior accuracy), a scalable distributed computing framework, efficient load balancing with Least Connections achieving the best results, and a robust Flask-based backend. Limitations include variations in performance across different NLP models or datasets, as well as controlled evaluations of load balancing algorithms. Future research should include more NLP model evaluations, real-world testing, hybrid models, and advanced load balancing strategies. This study establishes a comprehensive framework for developing scalable and efficient real-time sentiment analysis systems, laying the groundwork for future advances in the area.

References

1. Sharma R, Pathak P (2021) A comparative study on various approaches of sentimental analysis. In: 2021 international conference on simulation, automation & smart manufacturing (SASM). IEEE, Aug 2021, pp 1–4
2. Lyu Z, Cheng Y, Zhu Q (2022) A regional load balancing method based on the average value of peak load rates for associated transformers. In: 2022 China international conference on electricity distribution (CICED). IEEE, Sept 2022, pp 421–425
3. Nikolaychuk NN, Galiaskarov DF, Stepanov SN, Slatina OA, Maslov AA (2024) Analysis of information flows distribution with a load balancer in data centers. In: 2024 systems of signals generating and processing in the field of on board communications. IEEE, Mar 2024, pp 1–5
4. Jadeja H, Thacker C, Parmar K (2022) Categorizing data for sentimental analysis by auto-focusing mechanism using natural language processing. In: 2022 second international conference on computer science, engineering and applications (ICCSEA). IEEE, Sept 2022, pp 1-6
5. Prananda R, Dini HS, Putri TWO (2021) Design of load balancing method on secondary distribution network using artificial intelligence based on fuzzy logic. In: 2021 3rd international conference on high voltage engineering and power systems (ICHVEPS). IEEE, Oct 2021, pp 372–377
6. Golsorkhi MS, Hajian M, Savaghebi M (2022) Distributed event-triggered control strategy based on adaptive VI droop characteristic for accurate load sharing in AC microgrids. In: 2022 IEEE 13th international symposium on power electronics for distributed generation systems (PEDG). IEEE, June 2022, pp 1–6

7. Smith ML, Srivastava S (2023) Introducing parallel and distributed computing concepts through the use of flashcards and a card game. In: 2023 IEEE international parallel and distributed processing symposium workshops (IPDPSW). IEEE, May 2023, pp 278–283
8. Qin E, Wang Y, Yuan L, Zhong Y (2020) Research on NGINX dynamic load balancing algorithm. In: 2020 12th international conference on measuring technology and mechatronics automation (ICMTMA). IEEE, Feb 2020, pp 620–624
9. Vidya SH, Prakash RM (2020) Response time analysis of dynamic load balancing algorithms in cloud computing. In: 2020 fourth world conference on smart trends in systems, security and sustainability (WorldS4). IEEE, July 2020, pp 371–375
10. Sheela T, Akila V, Sravya B, Harshitha T (2021) Secure online examination system on azure cloud with load balancer. In: 2021 4th international conference on computing and communications technologies (ICCCT). IEEE, Dec 2021, pp 285–292
11. Juyal P (2022) Sentimental analysis of amazon customers based on their review comments. In: 2022 international interdisciplinary humanitarian conference for sustainability (IIHC). IEEE, Nov 2022, pp 914–919
12. Athar A, Ali S, Sheeraz MM, Bhattachariee S, Kim HC (2021) Sentimental analysis of movie reviews using soft voting ensemble-based machine learning. In: 2021 eighth international conference on social network analysis, management and security (SNAMS). IEEE, Dec 2021, pp 01–05
13. Rai P, Londhe ND, Raj R (2021) Fault classification in power system distribution network integrated with distributed generators using CNN. Electr Power Syst Res 192:106914. ISSN 0378-7796
14. Pratheeba B, Prathilothamai M (2016) A study of distributed systems in real-time applications. Int J Control Theory Appl 9(10):4233–4240
15. Sreelekshmi S, Remesh Babu KR (2018) Fault-tolerant multiple synchronized parallel load balancing in cloud. In: HIS 2017 conference. Springer International Publishing AG, part of Springer Nature
16. Bhaskaran S, Ullas S (2016) Advanced metering infrastructure (AMI) with combination of peak load management system (PLMS) and theft protection. In: 2016 online international conference on green engineering and technologies (IC-GET). Coimbatore, India

Design of a Robust Load Frequency Controller Using PID, GA and IMC for Hydro Turbine Power Systems with Non-minimum Phase Dynamics

Supriya Y. Bhuran and Sharad P. Jadhav

Abstract This paper presents a robust load frequency controller using Internal Model Control (IMC) and Genetic Algorithm (GA) for a single-area hydro turbine power system with non-minimum phase dynamics. These dynamics often cause instability and oscillations. The study explores PID, IMC, and GA optimization techniques for Load Frequency Control (LFC) in Hydro Turbine Power Systems (HTPS) with Non-Minimum Phase (NMP) behavior. IMC outperforms conventional PID and modern optimization methods by leveraging system dynamics, maintaining stability and minimizing overshoot, settling time, and steady-state error. MATLAB simulations demonstrate significant improvements in disturbance rejection and system reliability. Comparative analysis highlights the limitations of traditional tuning methods for HTPS with NMP characteristics. IMC offers flexibility in adjusting internal model parameters, optimizing performance and shows potential for soft computing optimization to tailor system responses. While GA iteratively refines solutions, IMC excels in accommodating intricate dynamics and built-in tuning rules, showcasing superior precision and responsiveness, making it highly effective for real-world applications.

Keywords Genetic algorithm · Internal model control · Non-minimum phase system · Hydro power · Single area power system

1 Introduction

Our lives rely heavily on gadgets, driving a constant surge in electrical energy demand. Yet, the supply isn't keeping pace. This growing mismatch is nothing but load frequency control which calls for enhanced control mechanisms to ensure

S. Y. Bhuran (✉) · S. P. Jadhav
Ramrao Adik Institute of Technology, Navi Mumbai, India
e-mail: supriya.bhuran@rait.ac.in

S. P. Jadhav
e-mail: sharad.jadhav@rait.ac.in

H. Sharma et al. (eds.), *Artificial Intelligence: Theory and Applications*, Lecture Notes in Networks and Systems 5589, https://doi.org/10.1007/978-981-96-1687-9_21

the power grid can handle sudden spikes in demand without disruption. In power generation, particularly in non-minimum phase hydro turbine systems, this becomes critical. The non-minimum phase (NMP) systems are sluggish, unstable, sensitive to minor changes in disturbance which leads to susceptibility to disturbance/very less disturbance rejection, and inverse response for some inputs like step response. These systems inherently possess uncertainty and rapidly varying delay characteristics, which leads to design controllers effectively [1]. Recent trends point to the use of nature-inspired algorithms for optimization. Genetic algorithms, a subset of evolutionary algorithms, offer adaptive heuristic search techniques ideally suited to address these complex control challenges. A robust power supply generation system hinges on the implementation of high-quality controllers capable of maintaining stability even in fluctuating demands.

Genetic algorithms have emerged as a powerful tool for tuning PID controllers, as demonstrated by various researchers. For instance, P. Shiva Krishna utilized this optimization technique to fine-tune fractional PID controllers tailored for medical applications [2]. Similarly, in another study, PID controllers were designed to effectively reject disturbances by leveraging optimal tuning strategies [3]. Notably, Tianyu Hui ventured into the realm of nuclear engineering, employing a GA-PID approach to regulate core power [4]. Moreover, the realm of robotics has benefited from this synergy as demonstrated in research where a PID controller optimized through a multi-objective genetic algorithm effectively governed a robotic manipulator [5]. Authors of [6] employed a GA-PID approach to control the Electric Power-Assisted Steering (EPAS) system. The efficacy of genetic algorithms in PID tuning is underscored by their superiority over traditional methods, as indicated by the literature. However, a significant gap exists regarding the application of GA-PID in non-minimum phase systems. Addressing this gap, researchers have explored alternative control methodologies, such as Internal Model Control (IMC), which exhibits robustness against time delays and system complexities [7]. A noteworthy study by Pushkar Prakash Arya and Sohom Chakrabarty showcased the effectiveness of IMC controllers specifically tailored for non-minimum phase systems [8]. Their research demonstrated promising outcomes in set-point tracking and disturbance rejection. Additionally, efforts have been made to enhance the performance of IMC-based PID controllers, resulting in improved closed-loop performance [9].

The study in [10] explores a robust PID controller design method based on maximum peak criterion for single area power systems, integrating IMC-based PID control for load frequency control. The application of intelligent controllers for load frequency control in interconnected power systems investigated, focusing on hydro units with non-minimum phase characteristics using Bat algorithm [11]. A robust fractional IMC-based PID controller design for non-reheated thermal power systems, which include non-minimum phase elements could achieve a robust controller which is good at set-point tracking and disturbance rejection [12]. The [13] focuses on the design of robust PID controllers for load frequency control in non-minimum phase hydro power plants using PSO-enabled automated Quantitative Feedback Theory. The comprehensive review covers various PID controller design methods for load frequency control, highlighting challenges and advancements in the context of power systems with non-minimum phase behavior [14]. The researcher presents a novel quasi-oppositional harmony search algorithm for automatic generation control in hydropower systems, comparing its performance with QOHS-PID and IMC methods [15]. In his study the author applies an AHP-based optimized tuning method to Modified Active Disturbance Rejection Control for load frequency control, highlighting its robustness against non-minimum phase dynamics [16]. The study in this paper introduces a PID-based load frequency control system for a stochastic microgrid power system, addressing non-minimum phase systems [17]. This paper introduces an active disturbance rejection control strategy optimized via quantum genetic algorithms for load frequency control in power systems with non-minimum phase elements [18]. The researcher discusses a fuzzy-PID controller design based on logarithmic approximations for IMC-based load frequency control models [19]. The paper [20] presents PID controller tuning rules for non-minimum phase hydro turbines using IMC principles. The author presented a PID controller design via the IMC scheme utilizing a 2DoF-IMC technique for robust load frequency control in multi-area power systems [21]. The paper presents a fractional order PID (FOPID) based on IMC and CRONE techniques for non-minimum phase elements in power systems [22]. This study investigates the internal model control (IMC) for load frequency control of power systems with non-minimum phase characteristics of hydro turbines [23].

In summary, while genetic algorithms have demonstrated remarkable success in PID tuning across various applications, further exploration is warranted to assess their efficacy in non-minimum phase systems. Meanwhile, IMC-based approaches offer a promising alternative, exhibiting robustness and enhanced control performance in such intricate systems.

Rest of the paper is structured as in Sect. 2 non-minimum phase hydro turbine power system's mathematical representation is presented. In Sect. 3 the control strategies of GA-PID and IMC are discussed. Section 4 covers the simulation of the SAHTPS with GA-PID and IMC and performance analysis in terms of time domain, frequency domain and different error calculations and then results and discussion covered in 5th section; Conclusion is in Sect. 6.

2 Single Area Hydro Turbine Power System (SAHTPS)

A single area hydro turbine power system is a simplified model used to oversee the production of electrical power from a hydro turbine within a specific geographic region. This setup typically consists of a hydro turbine, generator, transmission lines and related control mechanisms. The hydro turbine harnesses the kinetic energy of water, converting it into mechanical energy to propel the generator and generate electricity. This power is then transmitted through the grid to satisfy the energy needs of consumers in the designated area. Effective control of this system is vital for ensuring stability and a dependable power supply.

The mathematical representation of the load frequency control (LFC) of a single area hydro turbine power system (SAHTPS) is derived from the block diagram as

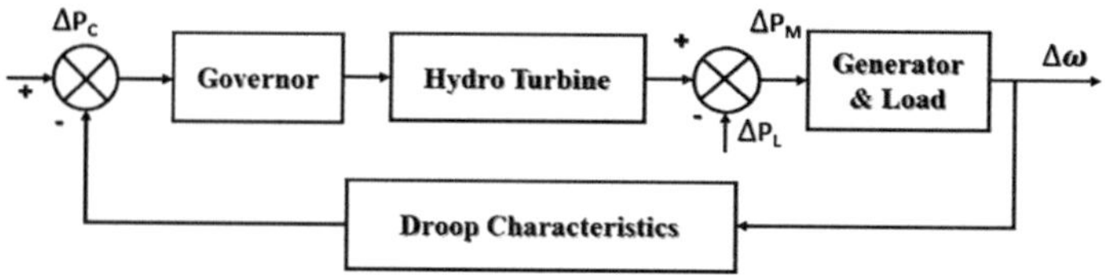

Fig. 1 Load frequency control of single area hydro turbine power system (SAHTPS)

shown in Fig. 1. In this, the mathematical model of a hydro turbine is non-minimum phase dynamics means it possesses at-least one zero in the right half of *s*-plane (RHS). The presence of RHS zero leads to instability, exponential growth, time delay, oscillatory behavior and decreased robustness. The entire system's functionality is encapsulated in the transfer function denoted as Eq. (1) [7].

$$\frac{\Delta\omega}{\Delta Pc} = \frac{(1 - Tws)(Kp)}{(1 + Tgs)(1 + 0.5Tws)(1 + Tps) + \frac{1}{R}}. \tag{1}$$

The mathematical model for the plant, incorporating droop characteristics with specific parameter values as outlined by $Kp = 1$, $Tw = 4$, $Tg = 0.2$, $Tp = 6$, and $R = 0.05$, is formulated [24]. Upon substituting these characteristic values into the equation, the resulting mathematical transfer function is presented in Eq. (2).

$$\frac{\Delta\omega}{\Delta Pc} = \frac{(1 - 4s)}{\left(2.4s^3 + 13.6s^2 + 8.2s + 21\right)}. \tag{2}$$

After examining Eq. (2), it can be seen that the system has a zero in the RHS of the *s*-plane, showcasing its non-minimum phase attributes.

3 Controller Approach

Various control strategies, such as PID controllers and sophisticated optimization techniques like genetic algorithms, can be deployed to regulate turbine speed, fine-tune power output and handle disturbances within the system. Optimizing the performance of a single area hydro turbine power system is crucial for achieving efficient and sustainable energy generation, particularly in regions where hydroelectric power plays an important role in the overall energy landscape.

3.1 Genetic Algorithm for PID Tuning (GA-PID)

Genetic Algorithms (GAs) are intelligent search algorithms inspired by nature's evolutionary processes. They belong to a broader category of techniques called evolutionary algorithms. GAs leverages the principles of natural selection and genetics to solve optimization and search problems effectively. In essence, GAs mimics the concept of natural selection, where species that can best adapt to their environment thrive and reproduce. Similarly, in GAs, individuals within a population compete for survival and reproduction, with only the fittest individuals being selected to pass their traits to the next generation. Each generation comprises a group of individuals, each representing a potential problem solution. These individuals are encoded as strings of characters, integers, floats, or bits, akin to chromosomes in biological organisms. Through a process of crossover, mutation and selection, GAs iteratively refine and evolve these populations toward optimal or near-optimal solutions [25]. The genetic algorithm optimizes the proportional, integral and derivative gain of the PID controller as shown in Fig. 2. The flowchart diagram Fig. 3 shows the process of the algorithm.

The steps involved in the PID tuning using GA optimization are:

1. Representation: Defining a suitable encoding scheme to represent potential solutions as individuals in the population. This encoding could be in the form of binary strings, integer arrays or any other appropriate representation depending on the problem domain.
2. Objective Function: Establishing an objective function which quantitatively estimates the fitness or quality of each individual solution. This function provides a measure of how well a solution performs with respect to the problem's goals.

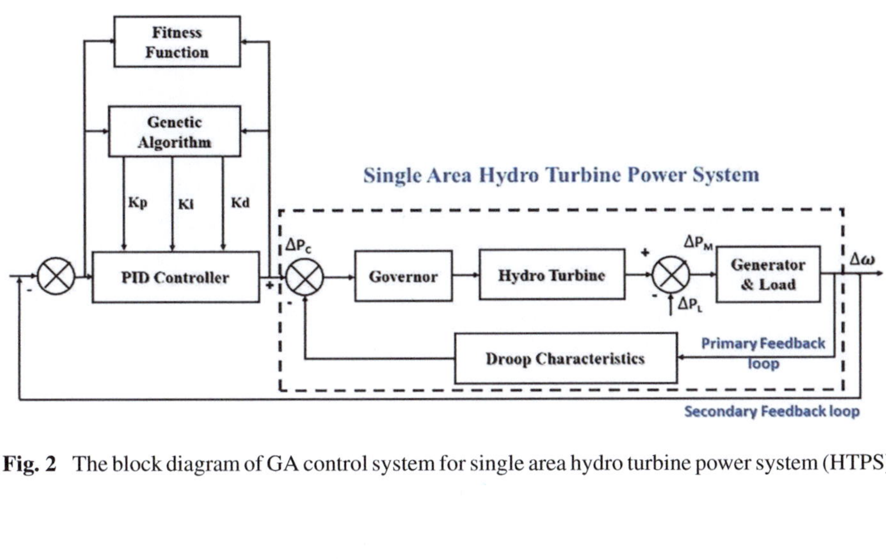

Fig. 2 The block diagram of GA control system for single area hydro turbine power system (HTPS)

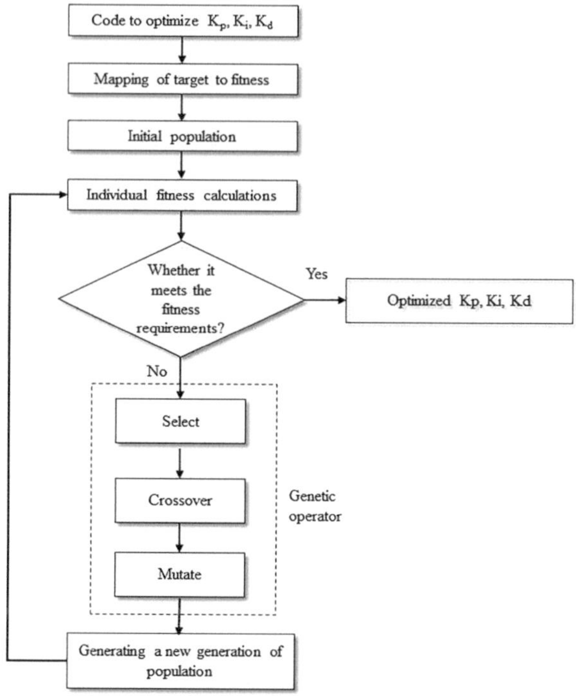

Fig. 3 Flowchart of GA optimization of PID parameters

3. Population Initialization: Initial population of individuals are generated randomly or using some heuristic method. This population represents the starting point for the evolutionary process.
4. Parent Selection Mechanism: The selection of parents from the current population who assist to create next generation offsprings. The selection process can be based on fitness proportionate selection, tournament selection or other strategies aimed at favoring individuals with higher fitness values.
5. Variation Operator—Crossover (Recombination): Perform crossover or recombination operations on selected pairs of parents to produce offspring. This involves exchanging genetic material between parents to create new solutions that potentially inherit favorable characteristics from both parents.
6. Variation Operator—Mutation: Introduction of arbitrary deviations or mutations to the offspring generated through crossover. Mutation contributes to preserving diversity within the population and prevents premature convergence toward suboptimal solutions.
7. Termination Condition: Define a termination criterion to determine when the algorithm should stop iterating and return the best solution found so far.

These steps collectively form the core framework of a Genetic Algorithm, guiding the iterative process of evolving a population toward increasingly better solutions to

Table 1 Initialization parameters of genetic algorithm for optimization

GA parameters	Value/method	GA parameters	Value/ method
Population size	25	Selection method	Uniform
Maximum number of generations	50	Mutation method	Uniform
Variable bounds (*Kp*, *Ki*, *Kd*, *dt*)	$[(-100:100), (-100:100), (-100:100), (0:1)]$	Crossover function	Constraint dependent

the optimization or search problem at hand. The parameters mentioned in Table 1 are used to design the PID controller by using a genetic algorithm in MATLAB.

3.2 Internal Model Control (IMC) Controller

The IMC systems have a unique setup comprising a controller and a simulated version of the process called the internal model. Within this system, the internal model loop calculates the disparity between the outputs of the actual process and those of the internal model. This difference indicates the impact of disturbances and discrepancies in the model. The IMC control structure is as shown in Fig. 4. Studies have demonstrated that internal model control devices exhibit robustness against disturbances and model inconsistencies, particularly when dealing with linear process models.

The IMC controller is designed stepwise as:

Identify Positive Plant Dynamics: Start by isolating the parts of the plant model that contribute to positive responses, such as delays and positive zeros.

Identify Negative Plant Dynamics: Then pinpoint the elements of the plant model that contribute to negative responses, including negative zeros and other components.

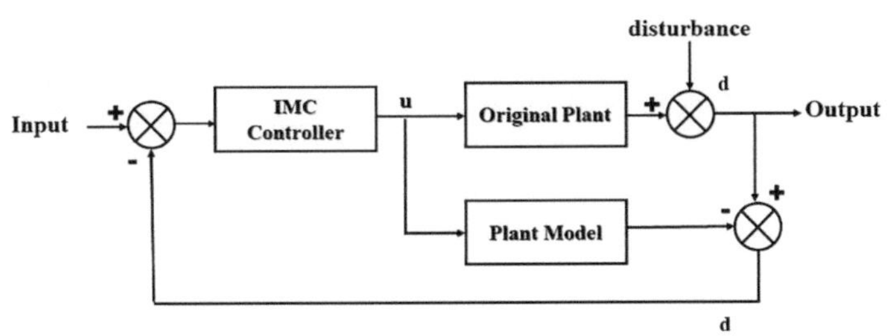

Fig. 4 IMC control structure

Select Suitable Low Pass Filter: Choose a low pass filter that ensures the IMC controller functions properly and responds quickly. This filter helps maintain stability and effectiveness in the control system.

This stepwise procedure enables the systematic design of an IMC controller, ensuring it effectively manages both positive and negative dynamics while incorporating a filter for optimal performance.

IMC controller is given as in (3),

$$\text{IMC\_ controller}\,(s) = \left(\frac{1}{\text{Plant Model}}\right)\text{LPF}\,(s), \tag{3}$$

where, Plant Model = the factor $\tilde{G}_{-}(s)$ which has no delays and all of its zeros are negative. LPF$(s) = \frac{1}{(1+T_f s)^{\eta}}$ is a low pass filter; selected to maintain properness of IMC controller [7]. To ensure the properness of the IMC controller, η is selected as 3. Additionally, a value of $T_f = 2.0$ is chosen to regulate the speed response.

The final transfer function of the IMC controller is obtained in (4).

$$\text{IMC\_ controller} = \frac{2.4s^3 + 13.6s^2 + 8.2s + 21}{8s^3 + 12s^2 + 6s + 1}. \tag{4}$$

This controller is used for closed-loop simulation.

4 Controller Simulation

Analytically, the IMC controller was designed, followed by its integration with the plant model in Simulink for simulation purposes, where graphical outputs were generated. Similarly, a Genetic Algorithm (GA) algorithm was implemented in an online-MATLAB environment to optimize the values of *Kp*, *Ki*, and *Kd*. These optimized values were then utilized in Simulink blocks for direct comparison. Additionally, the performance of a PID controller, tuned using built-in tuning applications within Simulink, was assessed. The combined blocks were simulated for 100 s with a step input provided to analyze the transient response of the system as depicted in Fig. 5.

Figure 6 illustrates the transient response of the Single Area Hydro Turbine Power System (SAHTPS) under various control strategies. When employing either a PID controller or a GA-optimized PID controller, the system exhibits oscillatory behavior before reaching stabilization. Furthermore, the time taken to achieve stability is notably longer compared to the IMC-based controller. Table 2 presents the time domain analysis results with the optimal values highlighted [26].

It is observed that the IMC controller outperforms other controllers demonstrating superior stability 3.5 times higher and 3 times faster response than PID as well as GA controller.

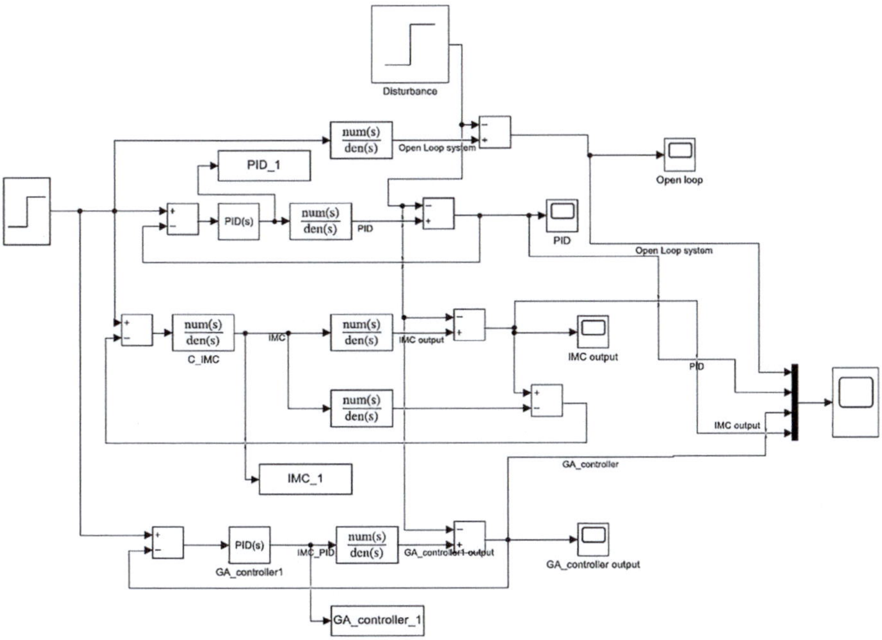

Fig. 5 Simulation of the closed-loop system with different control strategies

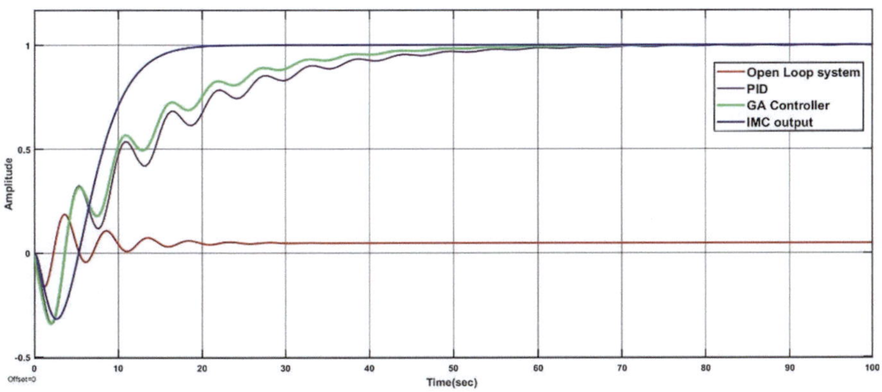

Fig. 6 Closed-loop response of the NMP HTPS when step input is given in NO-disturbance condition

Following a sudden 20% increase in load, initiated after 30 s, the transient responses of implemented controllers are evaluated. Figure 7 illustrates that post-disturbance, both the PID and GA-PID controllers exhibit oscillations and fail to attain the desired level promptly. In contrast, the IMC controller maintains stability, showcasing a brief dip but swiftly recovering. Once more, the IMC-based controller

Table 2 Time domain analysis with no load condition

Transient response	Open loop	PID controller	GA controller	IMC controller
Rise time (s)	0.1729	26.9175	26.8078	**7.3331**
Settling time (s)	23.9371	47.0059	46.4253	**16.7570**
Overshoot	291.7782	0	0	**0**
Undershoot	340.7513	33.2084	33.9472	**31.7474**

The bold values signifies the best values comapred to other values

demonstrates its superior performance. The corresponding time domain analysis tabulated in Table 3, highlights the advantageous values of the IMC controller.

Calculating performance metrics like ITSE (Integrated Time Squared Error), ITAE (Integrated Time Absolute Error), and ISE (Integrated Squared Error) provides valuable insights into the overall control system performance [27]. These metrics quantify the cumulative error over time, considering both the magnitude and duration of deviations from the desired setpoint. This metric places emphasis on larger errors that persist over time. It penalizes sustained deviations from the setpoint, making it useful

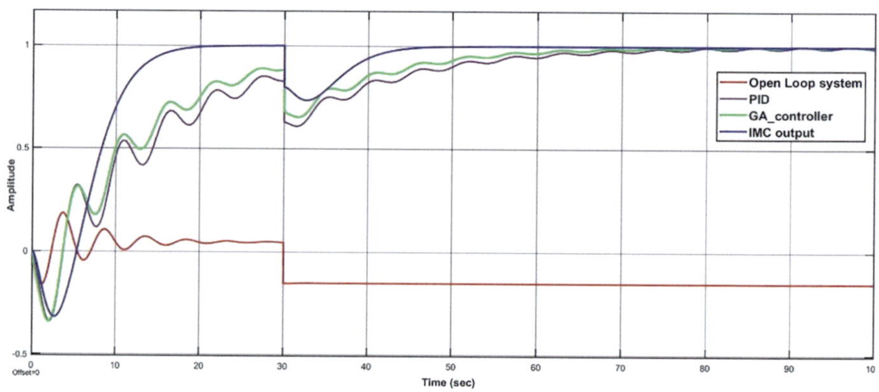

Fig. 7 Transient response of the NMP HTPS when step input is given with 20% load disturbance condition after 30 s

Table 3 Time domain analysis with 20% load change

Transient response	PID controller	GA controller	IMC controller
Rise time (s)	36.4	35.0	**7.380**
Settling time (s)	85.481	82.4	**14.726**
Overshoot	0	0	**0**
Undershoot	4.622	3.242	**6.695**

The bold values signifies the best values comapred to other values for systems where long-term accuracy is crucial [28]. ITAE gives weight to the absolute magnitude of errors over time. By considering both the error magnitude and duration, it provides a balanced assessment of control system performance. ITAE is often preferred when rapid response to disturbances is critical, as it penalizes both large errors and quick changes. ISE is a fundamental metric that quantifies the total squared error over time. It provides a straightforward measure of overall system performance, with larger errors contributing more significantly to the total. ISE is valuable for assessing the overall efficiency of a control.

IAE stands for Integrated Absolute Error. It is a performance metric used in control systems to quantify the cumulative absolute value of the error over time. Unlike the squared error metrics like ISE and ITAE, IAE considers only the absolute magnitude of the error without squaring it. The significance of IAE lies in its ability to provide a measure of the overall accuracy of a control system's response to setpoint changes or disturbances. By integrating the absolute error over time, IAE gives equal weight to both positive and negative deviations from the desired setpoint. This makes it particularly useful for applications where overshoot and undershoot need to be minimized equally.

Table 4 presents the error values, showcasing that the IMC controller exhibits minimal errors compared to traditional PID and GA-PID controllers. This observation underscores the superior performance of the IMC controller in achieving tighter control and more accurate tracking of the setpoint, affirming its effectiveness in practical applications.

Calculating the 2-norm and infinity norm in controller design serves several purposes and offers insights into the performance and robustness of the control system. The 2-norm and infinity norm are commonly used to analyze the robustness of a control system against uncertainties, disturbances and variations in system parameters. By evaluating the sensitivity of the closed-loop system to perturbations, researchers can assess how well the controller maintains stability and performance under different operating conditions. These norms provide quantitative measures of the control system's performance relative to its desired specifications. Researchers can use them to fine-tune controller parameters to achieve specific performance objectives, such as minimizing overshoot, settling time or steady-state error. Stability analysis is a crucial aspect of controller design. The 2-norm and infinity norm help determine the stability margins of the closed-loop system, indicating how close the system is to instability under various conditions [29]. This information guides controller design to ensure robust stability across a range of operating scenarios. By evaluating

Table 4 Performance indices

Error calculations	PID controller	GA controller	IMC controller
ISE	11.3005	10.7301	9.6250
IAE	16.3743	15.5767	10.0
ITSE	8.1131×10^3	7.0594×10^3	1.7825×10^3
ITAE	195.9649	179.3235	47.9992

Table 5 Control efforts

Norms	PID controller	GA controller	IMC controller
2-norm	0.4941	0.5757	**0.4677**
∞-norm	1.1383	1.6521	**1.0264**

The bold values signifies the best values comapred to other values

the norms of the closed-loop transfer function, researchers can assess the overall performance of the control system including its ability to reject disturbances, track setpoints accurately and maintain desired performance specifications.

Table 5 presents the control efforts required to achieve optimal values in which the IMC Controller showcasing its superior performance compared to other controllers. These values highlight the effectiveness of the IMC Controller in achieving desired control objectives such as minimizing errors, reducing settling time and enhancing system stability. The selection of these optimal values is indicative of the IMC Controller's ability to deliver robust and precise control in practical applications.

The random load variation of 40% [30] as shown in Fig. 8 applied and the robustness of the controller analyzed; which is depicted in Fig. 9. The dynamic response provides insight into how effectively the IMC controller handles random load changes. This suggested controller is effective and adaptable for non-minimum phase single area hydro turbine power systems.

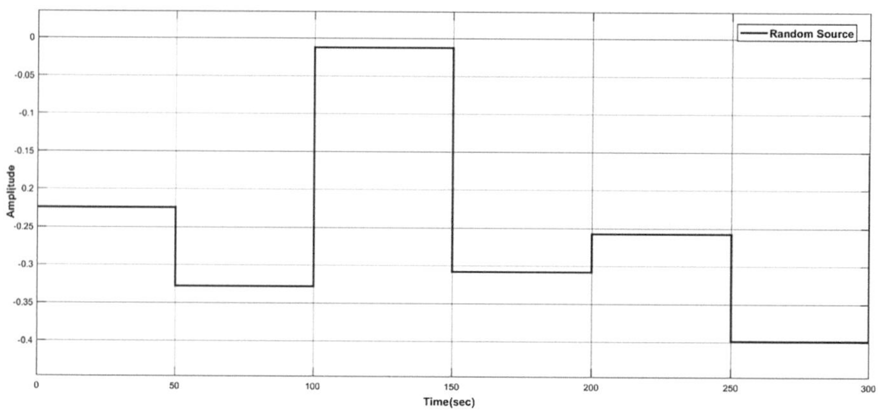

Fig. 8 Random 40% load disturbance

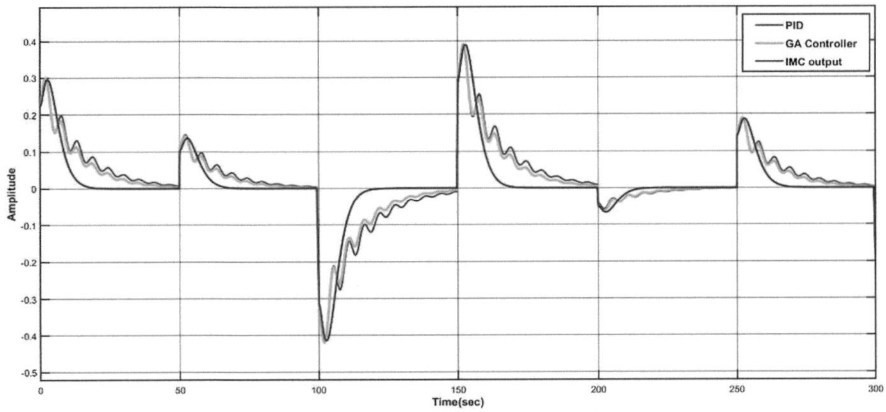

Fig. 9 Closed-loop response with 40% random load disturbance

5 Results and Discussion

The online-MATLAB simulation unequivocally demonstrates the effectiveness of PID, IMC and GA controllers in managing non-minimum phase systems. While the genetic algorithm operates by iteratively refining a population of individual solutions—selecting parent candidates from the current population to produce offspring for the subsequent generation—the IMC stands out as particularly well-suited for such systems. Its superiority lies in its capability to accommodate the intricate dynamics of systems with zeros in the right half-plane, coupled with built-in tuning rules that streamline the design process.

In comparison to the genetic algorithm's iterative approach, the IMC-based controller showcases superior precision and responsiveness in system response. By inherently addressing the complexities of non-minimum phase systems, the IMC offers a compelling advantage, underscoring its efficacy in real-world applications.

6 Conclusion

In this paper the analysis of a Hydro Turbine Power System (HTPS) exhibiting Non-Minimum Phase (NMP) behavior is presented for various control strategies. It effectively tackles the challenging task of Load Frequency Control (LFC) by devising controllers based on PID, Internal Model Control (IMC) and Genetic Algorithm (GA) optimization techniques. The performance of these controllers is thoroughly reviewed across both time domains and performance indices. The study reveals that IMC, leveraging an internal model of system dynamics, adeptly manages the NMP dynamics of HTPS, outperforming conventional PID controller and GA optimization method. Through comparative analysis using performance indices and control effort norms, the paper emphasizes the limitations of traditional tuning methods in meeting LFC requirements for HTPS with NMP characteristics. IMC design offers a crucial advantage by providing flexibility in adjusting internal model parameters, thereby optimizing controller performance. Furthermore, the study identifies a promising avenue for applying soft computing optimization approaches to tailor the system's response characteristics according to specific requirements. In future the study can be expanded with the other optimization methods as well as if permission granted by authorities, investigation the requirements and implications of real-time implementation.

References

1. Tong W, Wang S, Kiomars S (2024) What are the key stability challenges in high-bandwidth, non-minimum phase systems with time-varying, and non-smooth delays? Heliyon 10(6):e26949. ISSN 2405-8440. https://doi.org/10.1016/j.heliyon.2024.e26949
2. Siva Krishna P, Gopi Krishna Rao PV (2024) Fractional-order PID controller for blood pressure regulation using genetic algorithm. Biomed Sig Process Control 88(Part B):105564. ISSN 1746-8094. https://doi.org/10.1016/j.bspc.2023.105564
3. Krohling RA, Rey JP (2001) Design of optimal disturbance rejection PID controllers using genetic algorithms. IEEE Trans Evol Comput 5(1):78–82. https://doi.org/10.1109/4235.910467
4. Hui T, Zeng W, Yu T (2020) Core power control of the ADS based on genetic algorithm tuning PID controller. Nucl Eng Des 370:110835. ISSN 0029-5493. https://doi.org/10.1016/j.nucengdes.2020.110835
5. Ayala HVH, dos Santos Coelho L (2012) Tuning of PID controller based on a multiobjective genetic algorithm applied to a robotic manipulator. Exp Syst Appl 39(10):8968–8974. ISSN 0957-4174. https://doi.org/10.1016/j.eswa.2012.02.027
6. Hassan MK, Azubir NAM, Nizam HMI, Toha SF, Ibrahim BSKK (2012) Optimal design of electric power assisted steering system (EPAS) using GA-PID method. Proc Eng 41:614–621. ISSN 1877-7058. https://doi.org/10.1016/j.proeng.2012.07.220
7. Bhuran SY, Jadhav SP (2024) Design of PID, IMC and IMC based PID controller for hydro turbine power system of non-minimum phase dynamics. J Robot Control (JRC) 5(2):416–426. ISSN: 2715-5072. https://doi.org/10.18196/jrc.v5i2.21342
8. Arya PP, Chakrabarty S (2018) IMC based fractional order controller design for specific non-minimum phase systems. IFAC Pap Online 51(4):847–852. ISSN 2405-8963. https://doi.org/10.1016/j.ifacol.2018.06.123
9. Ghousiya Begum K, Seshagiri Rao A, Radhakrishnan TK (2017) Enhanced IMC based PID controller design for non-minimum phase (NMP) integrating processes with time delays. ISA Trans 68:223–234. ISSN 0019-0578. https://doi.org/10.1016/j.isatra.2017.03.005
10. Gupta A, Gupta Y, Gupta AD, Gautam RB, Anan SP (2019) A novel PID controller design technique for load frequency control. Int J Rec Sci Res 10(01(A)):30208–30213
11. Hussein AA (2018) Load frequency control for twoarea multi-source interconnected power system using intelligent controllers. Tikrit J Eng Sci 25(1):78–86. ISSN: 1813-162X (Print); 2312-7589 (Online)
12. Kasireddy I, Nasir AW, Singh AK (2018) Non-integer IMC based PID design for load frequency control of power system through reduced model order. Int J Electr Comput Eng (IJECE) 8(2):837–844. ISSN: 2088-8708
13. Satpati B, Bandyopadhyay I, Das G, Koley C (2008) Robust controller design for load frequency control of non-minimum phase hydro power plant using PSO enabled automated quantitative feedback theory. In: 2008 annual IEEE India conference, Kanpur, India, pp 349–354. https://doi.org/10.1109/INDCON.2008.4768748

14. Hote YV, Jain S (2018) PID controller design for load frequency control: past, present and future challenges. IFAC Pap Online 51(4):604–609. ISSN 2405-8963. https://doi.org/10.1016/j.ifacol.2018.06.162
15. Shiva CK, Mukherjee V (2016) Automatic generation control of hydropower systems using a novel quasi-oppositional harmony search algorithm. Electr Power Compon Syst 44(13):1478–1491. https://doi.org/10.1080/15325008.2016.1147103
16. Srikanth MV, Yadaiah N (2018) An AHP based optimized tuning of modified active disturbance rejection control: an application to power system load frequency control problem. ISA Trans 81:286–305. ISSN 0019-0578. https://doi.org/10.1016/j.isatra.2018.07.001
17. Dhasagounder M, Kaliannan J, Shah P, Sekhar R (2023) Optimization of load frequency control gain parameters for stochastic microgrid power system. J Robot Control (JRC) 4:726–742. https://doi.org/10.18196/jrc.v4i5.19772
18. Huang Z, Chen Z, Zheng Y, Sun M, Sun Q (2021) Optimal design of load frequency active disturbance rejection control via double-chains quantum genetic algorithm. Neural Comput Appl 33. https://doi.org/10.1007/s00521-020-05199-6
19. Arunsai V, Srividya Devi P, Vijaya Laxmi U (2021) Design of fuzzy-PID controller using logarithmic approximations for IMC-LFC model. In: E3S web of conferences. 3rd international conference on design and manufacturing aspects for sustainable energy (ICMED-ICMPC 2021), vol 309
20. Bošković MČ, Šekara TB, Rapaić MR (2020) Novel tuning rules for PIDC and PID load frequency controllers considering robustness and sensitivity to measurement noise. Int J Electr Power Energy Syst 114:105416. ISSN 0142-0615. https://doi.org/10.1016/j.ijepes.2019.105416
21. Kumar A, Anwar M, Huba M (2022) Load frequency controller design based on the direct synthesis approach using a 2DoF-IMC scheme for a multi-area power system. Symmetry 14:1994. https://doi.org/10.3390/sym14101994
22. Saxena S (2019) Load frequency control strategy via fractional-order controller and reduced-order modeling. Int J Electr Power Energy Syst 104:603–614. ISSN 0142-0615. https://doi.org/10.1016/j.ijepes.2018.07.005
23. Shukla RR, Panda AK, Garg MM et al (2023) Performance analysis of diverse-source interconnected power system with internal model controller in the presence of EVs. Arab J Sci Eng 48:14295–14312. https://doi.org/10.1007/s13369-022-07527-5
24. Bose S, Hote YV, Hanwate SD (2018) Analysis of effects due to right half plane zeros in PI controller based hydro turbine. IFAC Pap Online 51(4):633–638. ISSN 2405-8963
25. Pillai RP, Jadhav SP, Patil MD (2013) Tuning of PID controllers using advanced genetic algorithm. In: IJACSA special issue on selected papers from international conference and workshop on advance computing, pp 1–6
26. Poljak D, Sesnic S, Susnjara A, Paric D, Doric V, Antonijevic S (2018) Frequency domain and time domain analysis of the transient field radiated by GPR antenna. In: 2018 3rd international conference on smart and sustainable technologies (SpliTech), Split, Croatia, pp 1–6
27. Ansari AQ, Ibraheem, Katiyar S (2014) Application of ant colony algorithm for calculation and analysis of performance indices for adaptive control system. In: 2014 innovative applications of computational intelligence on power, energy and controls with their impact on humanity (CIPECH), Ghaziabad, India, pp 466–471. https://doi.org/10.1109/CIPECH.2014.7019078
28. Kristanto JA, Danang Wijaya F, Firmansyah E, Setiawan N (2019) Optimization of PLL loop filter control to reduce time delay using genetic algorithm on single phase grid tie inverter application. In: 2019 5th international conference on science and technology (ICST), Yogyakarta, Indonesia, pp 1–6. https://doi.org/10.1109/ICST47872.2019.9166164
29. Mustafa D, Glover K (1988) Controllers which satisfy a closed-loop H/sup infinity /-norm bound and maximize an entropy integral. In: Proceedings of the 27th IEEE conference on decision and control, vol. 2. Austin, TX, USA, pp 959–964. https://doi.org/10.1109/CDC.1988.194456
30. Sariki M, Shankar R (2022) Optimal CC-2DOF(PI)-PDF controller for LFC of restructured multi-area power system with IES-based modified HVDC tie-line and electric vehicles. Eng Sci Technol Int J 32:101058. ISSN 2215-0986. https://doi.org/10.1016/j.jestch.2021.09.004

Robust Control of Time Delay Systems Using Particle Swarm Optimized Fractional-Order PID Controller

Diptee S. Patil and Sharad P. Jadhav

Abstract Time delays in industrial, biological, and chemical processes can significantly hinder system performance, which conventional PID controllers may not adequately address. This paper presents Fractional-Order Proportional Integral Derivative (FOPID) controller using particle swarm optimization (PSO) to achieve robust control for time delay systems (TDSs). The proposed PSO-FOPID method optimizes controller parameters by minimizing the Integral of Squared Error (ISE). Simulations conducted using MATLAB/Simulink demonstrate that the PSO-FOPID controller outperforms Integer-Order PID (IOPID), Fractional M_s Constrained Integral Gain Optimization Method (FMIGO), and Fractional-order Proportional Integral (FOPI) controllers in terms of overshoot, rise time, settling time, and disturbance rejection. The PSO-FOPID controller achieves the lowest ISE value, exhibiting enhanced performance in time-domain parameters and robustness against gain variations.

Keywords Fractional calculus (FC) · Fractional-order controller · Integer-order controller · Time delay system · Particle swarm optimization algorithm

1 Introduction

Time delay occurs in all real-world engineering applications. It can occur on input, measurement, or output side and can cause instability. As a result, stability and closed-loop response of control systems with delays are important considerations [1]. Over the past decades, many researchers have invested in the analysis, control, and applications of time delay systems [2]. Control systems are designed to perform specific tasks

D. S. Patil (✉) · S. P. Jadhav
Department of Instrumentation Engineering, Ramrao Adik Institute of Technology, Navi Mumbai, Maharashtra, India
e-mail: diptee.patil@rait.ac.in

S. P. Jadhav
e-mail: sharad.jadhav@rait.ac.in

H. Sharma et al. (eds.), *Artificial Intelligence: Theory and Applications*, Lecture Notes in Networks and Systems 5589, https://doi.org/10.1007/978-981-96-1687-9_22

and actions [3, 4]. The control systems are studied with their performance specifications [5]. They generally relate to accuracy, stability, and speed of response. There are numerous different types of Integer-order (IO) and Fractional-order (FO) controllers used in various applications depending on the application-specific requirements. The various types of controllers documented in the literature are IO controllers, FO controllers, fuzzy controllers, adaptive controllers, internal model control (IMC) controllers, etc. Proportional Integral Derivative (PID) controllers [6, 7] are the prominent controllers applied in industry for structure simplicity, robustness, and easy tuning methods based on minimum plant model knowledge [8–10].

In the last decade, FC and applications to control theory [11] were researched focus. A FO controller is a type of controller that uses FC, which is a branch of mathematics that generalizes the traditional calculus to non-integer order derivatives and integrals [12]. The idea behind FO control [13] is to use a controller with a FO transfer function instead of the traditional IO transfer function, which can lead to better performance and stability in some control problems [14, 15]. Compared to PID controller [16], there are five parameters (K_p, K_d, K_i, λ, μ) where FO operators are λ and μ. The intelligent concept of control was first introduced somewhere two decades ago. Intelligent control evolved as productive research and development field in various industrial engineering applications in recent years. Soft computing paradigms enhance artificial intelligence (AI) and incorporate human expert knowledge in computing processes. Among various soft computing algorithms, this paper discusses swarm intelligence (SI) that focuses on the coordination of individuals within various natural and artificial systems through decentralized control and self-organization [17]. The different swarm intelligence techniques include colonies of ants, fish school, and birds flocks. All algorithms have merits and demerits and therefore are suited for specific set of problems. Optimization algorithm becomes effective when it converges to global minima in less number of iterations. PSO is a popular heuristic method and effectively deals with various optimization problems. Evolutionary computation (EC) methods are among the popular optimization techniques, known for their effectiveness in solving optimization problems [18, 19]. Some EC algorithms inspired by biological behaviors include genetic algorithm (GA), ant colony optimization (ACO), and particle swarm optimization (PSO). These algorithms widely adopted in engineering research areas [20, 21].

This paper aims to provide thorough and timely review of PSO algorithms and applications. Section 1 introduces time delay system, IO and FO controllers, PSO applications across various fields. Section 2 presents brief on time delay system, its modeling, and characteristics. Section 3 summarizes various existing controllers including the latest developments in PSO algorithms and its tuning strategy for FOPID controllers. Section 4 describes simulation of the proposed PSO algorithm for FOPID controllers applied to a first-order time delay (FOPTD) system. The results and analysis are given in Sect. 5 and conclusion is in Sect. 6.

2 Time Delay System

The time delay system (TDS) system exhibits a delayed response to changes in input, leading to a sluggish dynamic performance. Time delay is a typical aspect in all the branches of engineering. The various industrial processes involve subsequent phenomena within their internal dynamics. The theoretical research on delay systems is crucial, and is definitely a growing field for researchers [3]. These delays cause instabilities of original systems and make it difficult for system analysis and controller design. These disadvantages must be worked upon for necessary application in design process and highlighting good properties of time delays can bring about new advances in practical design [4].

2.1 *Modeling of Time Delay System*

The time delay systems can be approximated as FOPTD model and Second-Order Plus Time Delay (SOPTD) Model [11]. This paper focuses on FOPTD model. The FOPTD has the following mathematical model:

$$P(s) = \frac{K}{1 + Ts} \mathrm{e}^{-Ls}, \tag{1}$$

where gain of Process is (K), Time constant is (T), Delay is (L). A crucial property of the FOPTD models is relative dead time of system, which is defined as

$$\tau = \frac{L}{L + T} \tag{2}$$

The range of parameter τ is between 0 and 1. Systems in which $L \gg T$ are called delay dominant, $L \ll T$ are called lag dominant, $L \approx T$ are called balanced lag, delay system.

2.2 *Characteristics of Time Delay System*

Time delay systems, also known as dead-time systems, are characterized by a delay between input and observable effect on output. The main characteristics of time delay systems are as follows:

- The presence of a delay reduces system's responsiveness to control actions.
- The time delay can cause a phase lag which can destabilize the system.
- The delay introduces a phase lag, reducing phase margin and making the system more prone to oscillations and instability.

- Performance metrics rise, settling time, overshoot are adversely affected by the delay.

3 Controllers for Time Delay System

There are numerous different types of IO and FO controllers used in various applications as specific requirements [11, 14]. The majority of controllers are always difficult to design. Among them, in FOPID controller, there are five boundaries for tuning the controller. This controller's tuning is difficult due to the increased number of parameters. The block diagram of proposed tuning strategy is given in Fig. 1.

3.1 Structure of IO and FO Controllers

The structure of various controllers [6, 11] for IO and FO are given below:

$$C_{\mathrm{IOPID}}(s) = K_{\mathrm{p}} + \frac{K_{\mathrm{i}}}{s} + K_{\mathrm{d}}s. \tag{3}$$

The corresponding generalized FO controller is given by,

$$C_{\mathrm{FOPI}}(s) = K_{\mathrm{p}} + \frac{K_{\mathrm{i}}}{s^{\lambda}}, \tag{4}$$

$$C_{\mathrm{FOPID}}(s) = K_{\mathrm{p}} + \frac{K_{\mathrm{i}}}{s^{\lambda}} + K_{\mathrm{d}}s^{\mu}. \tag{5}$$

The paper presents design of the parameters of FOPID controller using PSO algorithm. For comparative analysis, IOPID, FOPI, and FMIGO controllers are also designed and implemented [22–24].

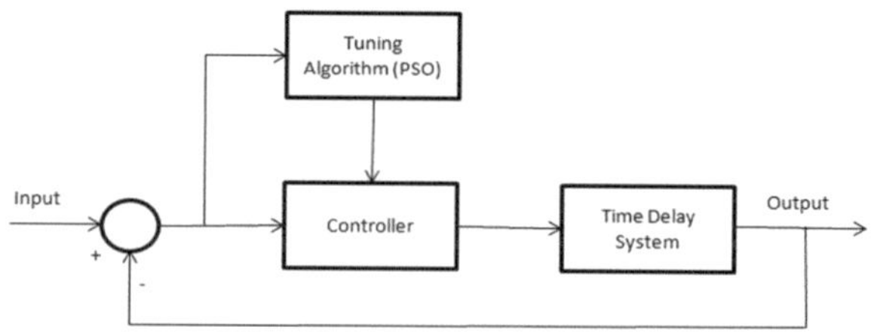

Fig. 1 Structure of PSO-based controller with time delay system [17]

3.2 Design Parameters of IO and FO Controllers

The design and implementation of FOPI controller uses Gain and Phase Margin Method [22]. In this method, firstly, complete stabilizing region of FOPI along with IOPID controller for FOPTD is found. Secondly, design steps [22] are carried out for satisfying gain crossover frequency (ω_{gc}), phase margin (PM), and flat phase constraint. Lastly, the complete feasible region of (ω_{gc}) and (PM) is collected. The equations given below are used for finding FOPI parameters.

$$K_p = \frac{-(B_1 S_1 + B_2 C_1)}{A K S_2 \omega^r} \tag{6}$$

$$K_i = \frac{B - B_1 S_1 C_1 + B_2 C_1^2}{A K S_1} + \frac{B_1 S_1 C_2 + B_2 C_1 C_2}{A K S_2} \tag{7}$$

For the FMIGO controller, the tuning is based on M_s constrained integral gain optimization. The design of FOPTD is found using fractional order α. The τ is used to determine best fractional order α gains and PI^{λ} gains [11]. The rejection of load disturbances is this method's primary design objective. The load disturbance at output will be minimal if integral gain K_i is maximized. The FMIGO design rules allow to find controller gains at any α. K_p is calculated using the following equation:

$$K_p = \frac{1}{k}\left(\frac{0.2978}{\tau + 0.00307}\right). \tag{8}$$

The integral gain K_i is calculated using following equation:

$$K_i = \frac{K_p(\tau^2 - 3.4022\tau + 2.405)}{0.8578T}. \tag{9}$$

3.3 Particle Swarm Optimization Algorithm

Particle swarm optimization (PSO), introduced by Kennedy and Eberhart in 1995, is inspired by the social behavior of bird flocks [17]. Evolutionary computing methods emulate the collective behaviors of swarms, which move and act intelligently to locate the most fertile feeding areas. A “swarm” refers to chaotic group of individuals that cluster together while each individual changes direction randomly. In PSO, a group of particles form a swarm to explore search space and identify optimal solution to optimization problem. Particle adjusts own trajectory in an n-dimensional space based on own experiences and experiences of other particles in the swarm. Each particle tracks its best position, known as *p*best, while the best position found by any particle in swarm is called *g*best. PSO algorithm updates each particle's velocity based on its *p*best and *g*best positions at each time step. The current position and velocity of each particle are adjusted; considering distances to both *p*best and *g*best. This approach models the collaborative behaviors observed in insects, birds, fish, and other animals [25, 26].

$$V_i^{k+1} = wV_i^{k+1} + c_1\text{rand}_1(\ldots)x(p\text{best}i - s_i^k) + c_2\text{rand}_2(\ldots)x(g\text{best}i - s_i^k) \tag{10}$$

where w, c_1, c_2 denote inertia weight coefficient, cognitive coefficient, social coefficient, resp., and rand_1, rand_2 are both random values between 0 and 1. The value Vrange $[-v_{\max}, +v_{\max}]$. The boundaries limit the particle velocity within the search space. Thus, preventing it from transitioning rapidly between points. This constraint enables the ith particle to effectively explore *p*best and *g*best. Therefore, each particle transitions from its current position to new position-updating its velocity according to Eq. (11).

$$P_i(n) = P_i(n-1) + V_i(n). \tag{11}$$

These algorithms emulate swarm behavior to find solutions to optimization problem. Specifically, this work combines studies of PSO and FOPID controllers, which have potential widespread industrial applications. The goal is to optimize FOPID controllers using PSO across various multidisciplinary fields [27].

PSO offers several advantages: it requires relatively few parameters to adjust, has a relatively fast convergence rate, and is straightforward to implement [21]. Due to technical merit, ease implementation, PSO is widely used technique for solving optimization problems. However, like many population-based evolutionary computing algorithms, PSO faces the challenge of potential solutions becoming easily trapped in local optima [19, 28]. Therefore, developing new PSO methods has become approach to addressing issue of premature convergence.

3.4 Tuning Method for Proposed PSO-Based FOPID Controllers

The PSO process Algorithm is presented in Table 1. The parameters selected for tuning FOPID controller using proposed PSO algorithm are shown in Table 2. The flowchart of proposed PSO FOPID is presented in Fig. 2.

Table 1 Proposed PSO algorithm

Proposed PSO algorithm
1. Initialize parameters of PSO algorithm population size P, w, c_1, c_2, and $V_{\max}$
2. Set swarm with-P particles
3. Initialize position x_i, 1, velocity v_i, 1, pi_i, 1 of each particle ($i = 1, 2, \ldots, P$); and initialize pg_1 of swarm
4. Calculate each particle's-fitness value
5. Update $\text{pi}_{i,k}$ of each particle and pg_k of swarm
6. Update $v_{i,k}$ and $x_{i,k}$ of each particle based on (10)
7. Confirm, maximum iterations met or fitness value reaches threshold, if not, go-to step 4

Table 2 Controller parameters

Controller	K_p	K_i	λ	K_d	μ
ZN-PID	7.4315	23.1613	–	–	–
IOPID	0.3852	0.7783	1	0.032	–
FOPI	3.3367	4.64	1.21	–	–
FMIGO	3.26	7.99	0.7	–	–
PSO-FOPID	4.717	4.115	1.01	0.0559	0.2

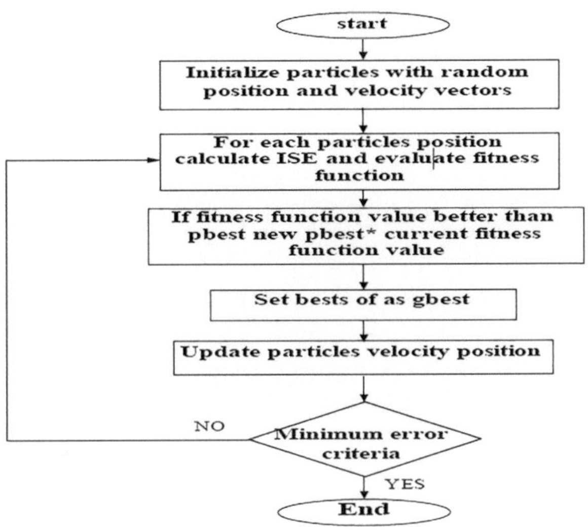

Fig. 2 Flowchart for PSO-based FOPID controller

4 Simulation of Closed-Loop System with Controllers

Firstly, the controller structures above obtain optimal parameters of the IO and FO controllers for FOPTD. The IO order PID controller was tuned using Ziegler–Nichols method [16] and conventional method [6]. The FOPI controller is tuned using stability region criteria as presented in [22]. The FMIGO controller [11] uses M_s constrained integral gain optimization, tuning rules for FOPTD dynamic systems. The performance is compared based on robustness of controllers. In this paper, a lag-dominant FOPDT model is chosen and the design methods mentioned above are tested for the process in (1). A simulated FOPDT lag-dominant processes for $K = 1$ is given in (12) as,

$$P(s) = \frac{1}{1+s} \mathrm{e}^{-0.1\,s} \tag{12}$$

The range of five variables K_p, K_i, K_d, λ, μ of controller parameters are 0–10 for K_p, K_i, K_d and 0–2 for λ, μ. The number of iterations is 50, population size is 50, $w_{\max}$ is 0.9, cognitive coefficient $c1$—1.2, and social coefficient $c2$—0.12. The primary function considered is based on the error criterion. In proposed work, the controller's performance index is evaluated using Integral of Squared Error (ISE) as $\mathrm{ISE} = \int |\mathrm{e}^2(t)|\mathrm{d}(t)$. The challenges and potential limitations of implementing PSO-based fractional-order controllers in real-time systems require specialized hardware to handle the computational load, which can increase cost and complexity. To address these limitations, it is necessary to carefully consider the specific application requirements, available computational resources, and potential trade-offs in controller performance. Adopting ISE index to optimize reduces significant deviations from the set point, leading to improved accuracy and performance of the FOPID controller. The implementation is done in MATLAB R2019b environment. The controller values obtained are shown in Table 2.

5 Results and Analysis

The simulation studies on FOPTD are presented. A step input is applied to closed-loop system with lag-dominant time delay system and five controllers, namely ZN-PID, IOPID, FOPI, FMIGO, PSO-FOPID (proposed) controller. The closed-loop performance of all the controllers is examined. To illustrate tracking, disturbance rejection, step response is shown in Fig. 3 and disturbance rejection is shown in Fig. 4. The observations in Table 3 show that PSO-FOPID has lesser overshoots across all controllers. The rise time 0.2116 s and settling time 1.289 s for proposed PSO-FOPID which is less compared to other controllers. A minimum value 0.1726 of ISE index is obtained for PSO-FOPID controller. For robustness analysis, a gain

Table 3 Time domain parameters and performance error index

Tuning method	Mp (%)	T_r (s)	T_s (s)	ISE
ZN-PID	57.937	0.9487	1.362	0.1994
IOPID	–	0.4572	5.726	0.8922
FOPI	9.341	0.3872	2.449	0.223
FMIGO	24.275	0.1887	1.546	0.197
PSO-FOPID	0.415	0.2116	1.289	0.1726

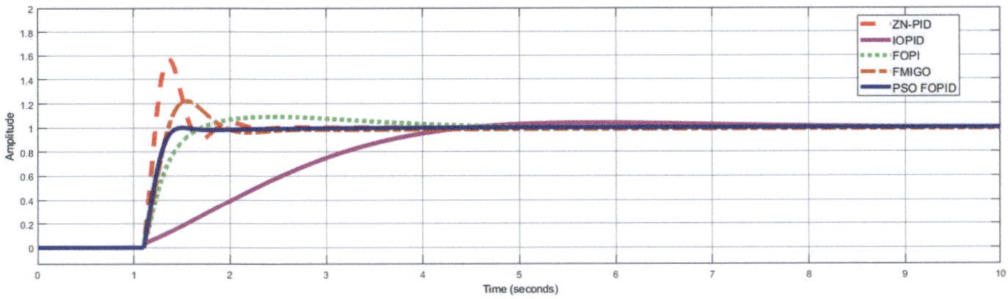

Fig. 3 Step responses of controllers with time delay system

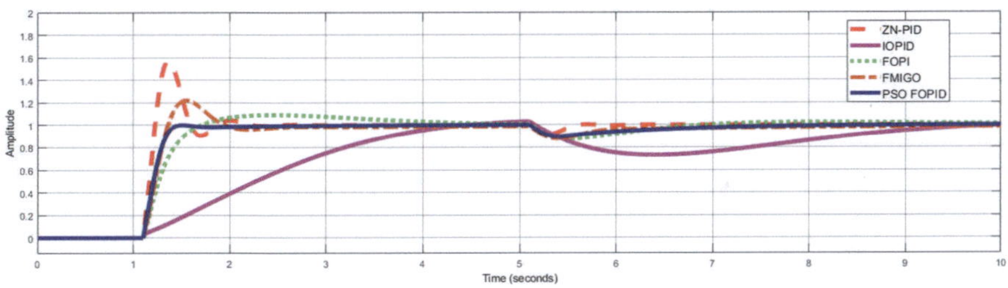

Fig. 4 Step responses with disturbance rejection of controllers with time delay system

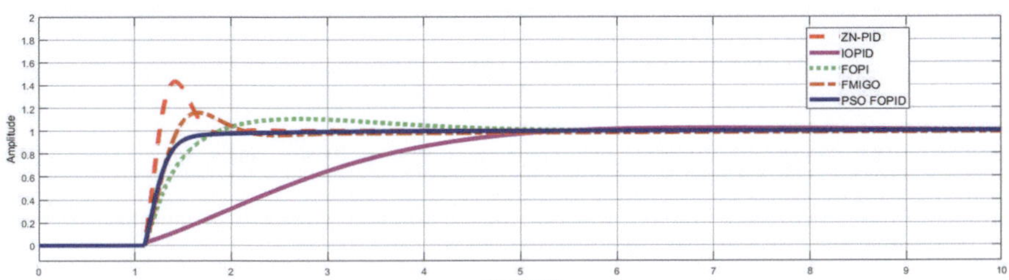

Fig. 5 Step responses of controllers with − 20% gain variation for time delay system

variation of ± 20% is presented in Figs. 5 and 6. Thus, it is observed proposed PSO-FOPID controller outperforms its counterpart controller in time-domain parameters, ISE, disturbance rejection, and gain variation.

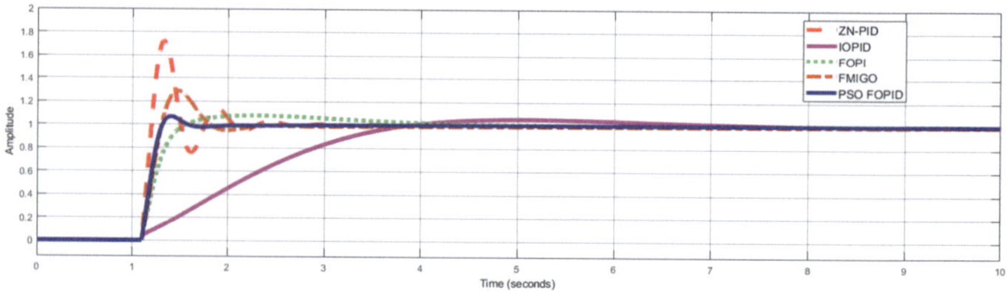

Fig. 6 Step responses of controllers with +20% gain variation for time delay system

6 Conclusion

This paper presents PSO-FOPID controller tuning method for time delay systems. A time-delay closed-loop process is used for control system design. Robust stability is analyzed in disturbance rejection and gain variations. Finally, effectiveness of proposed PSO-FOPID controller is illustrated by comparisons with various IO and FO controllers. The further work is to design method for more dynamic real-time applications such that proposed PSO-FOPID controller can be applied in a real control system.

References

1. Zavarei MM, Jamshidi M (1987) Time-delay systems: analysis, optimization and applications. Engineering
2. Richard JP (2003) Time-delay systems: an overview of some recent advances and open problems. Automatica 39(10):1667–1694
3. Zhong QC (2006) Robust control of time-delay systems, vol 1. Springer, London
4. Boubaker O, Balas VE, Benzaouia A, Chaabane M, Mahmoud MS, Zhu Q (2017) Time-delay systems: modeling, analysis, estimation, control, and synchronization. Math Probl Eng
5. Patil D, Jadhav S (2024) Design and analysis of IO and FO controllers to investigate the effects of process parameter perturbations on lag-dominant time delay systems. J Robot Control (JRC) 5(2)
6. Astrom KJ, Hagglund T (1995) PID controllers: theory, design and tuning. Engineering, Computer Science
7. Astrom KJ, Hagglund T (2006) Advanced PID control. ISA—The Instrumentation, Systems, and Automation Society
8. Astrom KJ, Panagopoulos H, Hagglund T (1998) Design of PI controllers based on non-convex optimization. Automatica 34(5):585–601
9. Panagopoulos H, Astrom KJ, Hagglund T (1999) Design of PID controllers based on constrained optimization. In: Proceedings of the 1999 American control conference (Cat. No. 99CH36251), vol 6, pp 3858–3862
10. Bharti J, Phadke G, Patil D (2020) Optimization of DC motor speed control using LQR technique. In: Kumar A, Paprzycki M, Gunjan V (eds) ICDSMLA 2019. Lecture notes in electrical engineering, vol 601. Springer, Singapore
11. Monje CA, Chen YQ, Vinagre BM, Feliu V (2010) Fractional order systems and controls, fundamentals and applications, vol 1

12. Chen YQ, Petras I, Xue D (2009) Fractional order control—a tutorial. In: American control conference, pp 1397–1411
13. Podlubny I (1999) Fractional-order systems and $PI^{\lambda}D^{\mu}$ controllers. IEEE Trans Autom Control 44(1):208–214
14. Shah P, Agashe S (2016) Review of fractional PID controller. Mechatronics 38:29–41
15. Patil DS, Patil MD, Vyawahare VA (2015) Design of fractional-order controller for fractional-order systems using Bode's ideal loop transfer function method. In: 2015 international conference on industrial instrumentation and control (ICIC), Pune, India, pp 490–495
16. Ziegler JG, Nichols NB (1993) Optimum settings for automatic controllers. J Dyn Syst Meas Control 115:220–223
17. Kennedy J, Eberhart R (1995) Particle swarm optimization. Paper presented at proceedings of IEEE international conference neural networks, Perth, WA, Australia
18. Bingul Z, Karahan O (2018) Comparison of PID and FOPID controllers tuned by PSO and ABC algorithms for unstable and integrating systems with time delay. Optim Control Appl Methods
19. Fang J et al (2023) A survey of algorithms, applications and trends for particle swarm optimization. Int J Netw Dyn Intell 2(1):24–50
20. Levy DC, Lu Y, Yan D, Zhou M, Chen S (2019) Particle swarm optimization of real-time PID controllers. In: Tian YC, Levy DC (eds) Handbook of real-time computing. Springer, Singapore
21. Khare A, Rangnekar S (2013) A review of particle swarm optimization and its applications in solar photovoltaic system. Appl Soft Comput 13:2997–3006
22. Luo Y, Chen Y (2019) Stabilizing and robust fractional order PI controller synthesis for first order plus time delay systems. Automatica 48(9):2159–2167
23. Ben Hmed A, Amairi M, Aoun M, Hamdi SE (2017) Comparative study of some fractional PI controllers for first order plus time delay systems. In: 18th international conference on sciences and techniques of automatic control and computer engineering, pp 278–283
24. Chen SY, Huang HX (2019) Design of fractional order proportional integral controller using stability and robustness criteria in time delay system. Meas Control 52(9–10)
25. Cao J-Y, Cao B-G (2006) Design of fractional order controllers based on particle swarm optimization. In: 1st IEEE conference on industrial electronics and applications, Singapore, pp 1–6
26. Daftari MA, Nekoui MA (2018) Analysing stability of time delayed synchronous generator and designing optimal stabilizer fractional order PID controller using particle swarm optimization technique. In: 2nd IEEE international conference on power electronics, intelligent control and energy systems
27. Xu L, Song BY, Cao MY (2021) An improved particle swarm optimization algorithm with adaptive weighted delay velocity. Syst Sci Control Eng 9:188–197
28. Rahimi A, Dev Kumar K, Alighanbari H (2013) Particle swarm optimization applied to spacecraft reentry trajectory. J Guid Control Dyn 36:307–310

Enhancing Crop Image Classification: Comparative Analysis of Augmentation Techniques for Small Datasets

Shruti Singh, Shantanu Kumar, Purnachandra Mandadapu, and Satish Mandavalli

Abstract In crop image classification, the sets of data often aren't very big, aren't balanced well, and don't exactly match what happens in real life. When models are trained using this data, they end up getting too good at the data they were trained on—and don't work so well on real data. The potency of three methods—transformative augmentation, data augmentation, and generative augmentation, in improving a small dataset of pests was estimated in this study. These methods were compared to a setup where the dataset wasn't changed at all. In the first method, random changes were made to things like brightness, rotation, and colors in the images to create more of them. In the second method, Generative Adversarial Network (GAN) is used to create fake images of pests to add to the small dataset. Three different models were then trained for each method to see how well they could identify the pests. This experiment was repeated a thousand times to ensure genuine results. It was found that at a certain level of confidence, the three methods made a big difference in how well the models could identify pests. The method using transformative augmentation worked the best. It helped make the images clearer by reducing things like background noise and differences in lighting and angles.

Keywords Generative Adversarial Network · Generative augmentation · Pests · Transformative augmentation

S. Singh
Washington State University, Seattle, WA, USA
e-mail: sshruti.connect@gmail.com

S. Kumar (✉)
Amazon, Seattle, WA, USA
e-mail: reach.shantanuk@gmail.com

P. Mandadapu
Deloitte, Dallas, TX, USA
e-mail: mpchandra39@gmail.com

S. Mandavalli
Microsoft, Dallas, TX, USA
e-mail: sat.manda@microsoft.com

H. Sharma et al. (eds.), *Artificial Intelligence: Theory and Applications*, Lecture Notes in Networks and Systems 5589, https://doi.org/10.1007/978-981-96-1687-9_23

1 Introduction

The potential of Deep Learning (DL) for pest detection, however, is hampered by limited datasets. Yet, current pest datasets fail on one or both criteria [1–5]. A dominant paradigm in addressing crop pests and their effects on livelihood is Integrated Pest Management (IPM). On the one hand, pest datasets are incomplete relative to standard-sized datasets. A promising avenue to automate pest detection is the use of DL techniques [6–11]. This application is particularly relevant in pest detection where pests may be classified based on their distinct physical traits. However, the number of samples per pest category of current datasets is remarkably low, averaging 160 samples per category. IPM seeks to combine all available pest control methods to manage pests in an economical and environmentally sustainable way. Automation should be used to augment local knowledge in pest management to support the precise and early diagnosis of pests. With climate change and sudden weather shifts favorable to pest populations—the ill effects of outbreaks on local economies are only expected to increase. An important application in this domain is image classification, whereby DL models are trained to extract relevant features of images and classify them given various classes. In the Philippines alone, invasive crop pests have cost the economy 600 million annually. Pests account for up to 4070 billion each year. In the Philippines, for example, some farmers are unable to make judicious use of pesticides due to uncertainty in identifying pests. Yet, its sustained adoption by local economies across the world remains tentative in part due to a lack of technical knowledge among farmers.

To build robust classification models, it is important to obtain datasets that are complete and representative. Therefore, a viable alternative to transformative augmentation is generative augmentation in which DL is used to synthesize new image data that may be used for expanding limited datasets. An important use case for Generative Adversarial Network (GAN) is augmenting datasets for which data is expected to be scarce as in medical images or images of rare plant disease. Hence, GANs represent an opportunity for artificially diversifying limited datasets in pest recognition to improve classification models. In this regard, a relevant approach to diversifying datasets is data augmentation. Data augmentation is the process of artificially increasing the size and quality of datasets. Several studies across different crops testify to the ability of basic data augmentation to improve model accuracy relative to models with no data augmentation at all—have shown that adding GAN-generated pest images to a limited dataset improves classification accuracy by as much as 3%. GANs are a family of DL frameworks that are used to generate synthetic data by learning the distribution of a given dataset. Yet, a study on crop disease detection has shown that models trained under homogeneous images report less than 50% accuracy when tested on real-life samples, which is a concern being addressed by recent curators of pest datasets. However, an important caveat in this finding is that IPM have used homogeneous images in GAN training, which we have previously shown contributes to poor generalizability to field condition images. This concern also highlights the need for datasets that not only have a high quantity of data but are also diverse enough to cope with the wide range of conditions found in the field. On the other hand, some datasets also fail to represent the vast array of field conditions. Such manipulations are indispensable to crop pest classification where visual diagnoses may be done from different angles, with varying backgrounds, and at different times of the day. Therefore, transformative augmentation, for example, adds new images simply through image manipulations such as flipping, rotating, scaling, adjusting contrast and brightness, and color transformations.

Showing how well transformative and generative augmentation work with field datasets is an important step in making automatic crop pest identification better. If these methods work well, it could mean we don't need to collect as much data manually. Instead, we could create different kinds of data artificially, making the training process for pest identification models faster and more accurate. It will help the farmers more effectively to figure out what's wrong with their crops as this is a big part of managing crop diseases, which is important for keeping food production steady over time. This knowledge can help us make existing methods better and, eventually, improve how good our data and models are for identifying crop pests. So, the goal of this study is to see how well transformative and generative data augmentation can improve models that identify tomato pests in real field conditions. Particularly, we wants to:

1. Use transformative and generative augmentation for different kinds of tomato pests.
2. Make models that identify tomato pests with no extra data, with transformative augmentation, and with generative augmentation.

The study is as follows; the materials and methods are shown in the next section. The experimental analysis is carried out in Sect. 3, and we wrap up the investigation with some conclusions and ideas for future work in Sect. 4.

2 Materials and Methods

A dataset of tomato pests curated by [12] contains 8 classes of tomato pests. Unlike datasets like plant village,[1] where the images were taken in labs with controlled conditions, [12] dataset has images with more complex backgrounds and different numbers and poses of pests, making it more realistic. Only the classes in the dataset with at least 50 image samples were considered, resulting in the use of 6 out of the 8 classes for the study as shown in Fig. 1. The dataset is somewhat limited and slightly unbalanced, which may lead to overfitting. However, after testing three different models for classifying images of pests (refer to Table 1) where each model uses a different method to change the images we solve the problem of overfitting.

[1] https://plantvillage.psu.edu/

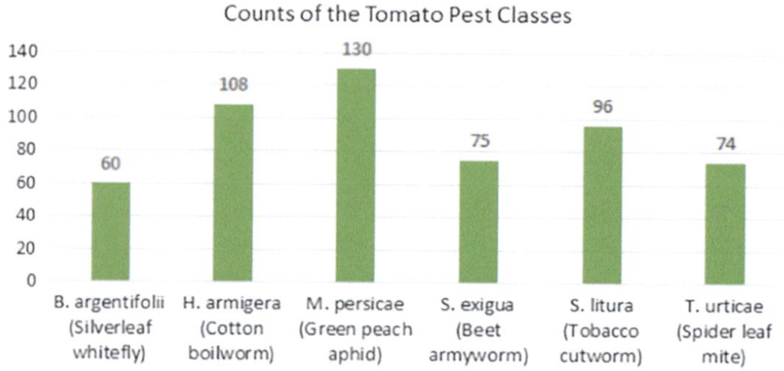

Fig. 1 Frequency distribution of images in the tomato pest dataset across different types

Table 1 Overview of experimental design

Component setup	Control setup (no augmentation)	Experimental setup 1 (transformative augmentation)	Experimental setup 2 (generative augmentation)
Training data	Original pest images	Original images + transformed images	Original images + transformed images + generated images
Model	MobileNet + image classification model	MobileNet + image classification model	MobileNet + image classification model

2.1 Augmentation

No Augmentation: **Experiment 1**. To see how well the model works without any changes to the images, the original dataset was used for training without any adjustments. The dataset was divided into three parts—training, validation, and testing as shown in Table 2. By only training the latter layers of the model—we preserve model learning of general objects such as leaves and insects, which is instrumental for learning the overall features of tomato leaves and pests. The image classification model employs a Convolutional Neural Network (CNN) [13–16], which includes a dense layer for classification purposes. Convolutional layers within the CNN are responsible for extracting high-level features from the images. Initially, these layers identify relatively simple features such as edges and shapes, while later layers identify more significant parts such as leaves and stems. Pooling layers take the convolved features of previous layers and reduce their dimensions, usually through max-pooling, to decrease the complexity of and heighten relevant features for classification. In fine-tuning, we excluded the last 5 layers of the model, appended a dense layer for classification, and froze the first 22 layers. To compensate for the lack of image data, we take advantage of transfer learning. However, the candidate class or category with the highest probability is inferred to be the subject of the image. Therefore, the MobileNet model was trained to convergence with early stopping,

Table 2 Train, validation, and test split of original dataset

Classes	Train	Validation	Test
B. argentifolii	43	5	12
H. armigera	79	7	22
M. persicae	96	8	26
S. exigua	54	6	15
S. litura	70	7	19
T. urticae	53	6	15

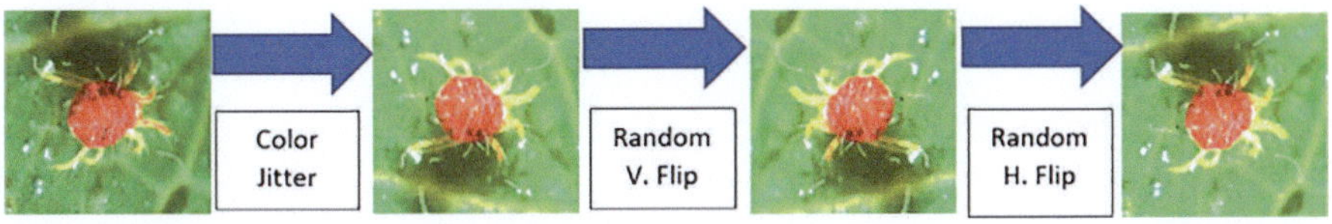

Fig. 2 Various transformations applied on a pest image

i.e., when the validation accuracy did not significantly improve after 3 epochs. For this experiment, therefore, we fine-tune a MobileNet model pre-trained on ImageNet available from the Keras API. The MobileNet architecture possesses a small number of parameters and file sizes, making it suitable for edge computing by farmers. A CNN requires a large amount of training data to be effective. This architecture entailed preprocessing the images to a size of 128 by 128 pixels and normalized pixel values from a range of -1 to 1. The activation function then transforms the weighted sum of the inputs through a non-linear function that gives corresponding probabilities to each of the candidate classes.

Transformative Augmentation: **Experiment 2**. Expanding the original training set by applying random transforms on each image. In contrast to the first experiment, an augmented version of the original dataset is used for model training. PyTorch—TorchVision[2] transforms are used to sequentially apply color jitter, random vertical flip, and random horizontal flip on each image. A round of data augmentation applied to an image as shown in Fig. 2—the set of configurations for the model is identical to that of Experiment 1. Table 3 shows the counts for the training, validation, and test split of each pest class. The augmented dataset is then used as training data for a separate MobileNet model. Note that the validation and test sets are not altered, which are identical to those of Experiment 1.

Generative Augmentation: **Experiment 3**. The goal of the generator is to synthesize fake images that are practically indistinguishable from real images by learning the distribution of the original dataset. Table 4 shows the counts of the real pest images fed to each GAN. Given the difficulty in training and evaluating GANs, it is not guaranteed that the synthetic images would be of comparative quality to that of the

[2] https://pytorch.org/vision/

Table 3 Train, validation, and test split of dataset with transformative augmentation

Classes	Train			Validation	Test
	Original	Transformed	Total		
B. argentifolii	43	957	1000	5	12
H. armigera	79	921	1000	7	22
M. persicae	96	904	1000	8	26
S. exigua	54	946	1000	6	15
S. litura	70	930	1000	7	19
T. urticae	53	947	1000	6	15

real images. On the other hand, for the real input, we first randomly transform the real training images and then feed them to the GANs. GANs require both fake and real images as input [17–22]. For this purpose, we use GANs to synthesize new images from the original pest dataset. The adversarial relationship is more formally represented by the mini-max loss function. For optimal training, we use a subset of GANs called the Wasserstein Generative Adversarial Network with Gradient Penalty (WGAN-GP) framework to generate synthetic pest images from the original dataset. After training, we add synthetic images generated by the GANs to the base dataset until images of each class in the classification dataset reach 1000. It is based upon the WGAN which boosted the model training stability of the original GAN through an improved loss function. On the other hand, the goal of the discriminator is to correctly classify images as either real or fake, outputting a binary value to indicate either real or fake. We use a batch size of 128 for training as per [23] implementation as shown in Table 5; therefore, all of the images were processed in one step per epoch. Initially, we explored transforming the images every several thousand epochs, but after significant training coming up with low-quality inputs, we halted the transformations. At times, therefore, researchers do not evaluate the model itself but simply judge the quality of the generated images. While the previous experiment relied on transformative augmentation, Experiment 3 utilizes generative augmentation, whereby DL is used to create fake tomato pest images.

However, we kept the training data and transformations to a minimum, hoping to reduce overall training time and complexity. WGAN training does not always

Table 4 Number of real training images used to train gan per class

Pest classes	Number of transformed images
B. argentifolii	43
H. armigera	79
M. persicae	96
S. exigua	54
S. litura	70
T. urticae	53

Table 5 Generator and discriminator model architecture

Discriminator		Generator	
Layer	Output shape	Layer	Output shape
Conv2d-1	[64, 64, 64]	ConvTranspose2d-1	[1024, 4, 4]
LeakyReLU-2	[64, 64, 64]	BatchNorm2d-2	[1024, 4, 4]
Dropout2d-3	[64, 64, 64]	ReLU-3	[1024, 4, 4]
Conv2d-4	[128, 32, 32]	ConvTranspose2d-4	[512, 8, 8]
BatchNorm2d-5	[128, 32, 32]	BatchNorm2d-5	[512, 8, 8]
LeakyReLU-6	[128, 32, 32]	ReLU-6	[512, 8, 8]

lead to local convergence for common use cases and GAN evaluation remains an open research problem. However, we did not have a stopping criterion for training inherent in the GAN model, unlike the early stopping mechanism of the classification model. We rely on a combination of objective and subjective measures such as the inception score and manual visual inspection of images to judge image quality. For this, we slightly altered Marko's GAN architecture by removing the sigmoid activation function since the discriminator/critic of WGAN-GP outputs a continuous rather than binary value to indicate the realness or fakeness of an image. Therefore, for each of the tomato pest classes, we train a separate GAN, producing 6 GAN models in total. For the initial fake input, we feed a noise vector of size 100 to each of the GANs. WGAN-GP adds a gradient penalty that regularizes the gradient norm for improved training. We just apply the transformations and do not expand the dataset, keeping the original counts. Table 3 shows the counts for the training, validation, and test splits (same as transformative augmentation). WGAN and WGAN-GP were originally designed for noise-to-image synthesis where the generator starts with random noise and gradually morphs it into an acceptable image across several epochs. As in Experiment 2, we do not alter or augment the original validation and test sets. Hence, the generator and discriminator are participating in a zero-sum adversarial game.

2.2 Evaluation Metrics

In particular, we use a well-known metric called the inception score (IS). The goal is for the IS of the synthetic images to approximate that of the real images. For reference, synthetic images generated from the CIFAR-10 dataset gained an IS of 8.09 compared to the IS of the real images of the same dataset of 11.24. A high IS indicates that salient objects were found in the images and that the images were significantly different. The IS for the generated dataset is arrived at by first predicting the class probabilities of the images using the Inception Model pre-trained on 1000 image classes. In addition to evaluating the classifiers, also assesses the quality of the generated images of the GAN to support the accuracy score obtained in the third experiment. Then, the conditional and marginal probabilities of the images are derived as a measure of their quality and diversity, respectively. The primary metric used to evaluate the skill of the three models is classification accuracy. However, given that model training and evaluation are stochastic procedures, adopt an experimental repeats approach for a more reliable estimate of accuracy. Then, obtain a grand mean of the accuracy scores per model, which is used to compare their general performance against each other. The IS measures two aspects of the image: (1) the "objectness" of the images, or the degree to which a salient object is perceivable in each image and (2) the variation among images. Applying this approach, repeat the training and evaluation of the models 1000 times, getting each of their accuracy scores at each repeat.

3 Experimental Analysis

3.1 *Image Classification*

The classification performance of the three models falls well within the 70% range, as shown in Table 6. Consistent with the hypothesis, the experimental models with augmentation performed better in terms of classification accuracy than the control model with no augmentation as shown in Table 1. In particular, the setup with transformative augmentation performed 2.1% better than the control setup while the generative augmentation performed 1% better than the control setup. Furthermore, the confidence intervals of the classification accuracies of the three models reveal that these differences were statistically significant ($\alpha = 0.05$) with no overlap among the intervals. Also, confirm the hypothesis that the augmented setups possess higher mean classification accuracy scores than the control setup ($\alpha = 0.033$). Among the three models, the model with transformative augmentation performed the best ($\alpha = 0.033$). In addition to having the highest classification accuracy, it also possessed a more compact distribution and fewer low-scoring outliers. On the other hand, both the control model and the GAN-assisted model had slightly wider spreads and more low-scoring outliers, as shown in Fig. 3. The experimental models performed better as augmentation increased the quantity and diversity of the training set. In particular, transformative augmentation reduced overfitting as shown in its compact range and minimal deviation of outliers. The transformative approach was less about creating new training images than it was about resisting the noise of existing images by reducing biases to background, color, and idiosyncratic best features.

On the other hand, the generative approach outperformed than the basic approach but did no better than the transformative one. This result may be attributed to the fact that while the GANs produced some images similar to those of the training set, the synthesized images were largely of low and variable quality. A close inspection of the model errors reveals commonalities and slight differences germane to the type of data augmentation applied, if any, as shown in Fig. 4. An important general

Table 6 Mean classification accuracy scores of the three models

Setup	Mean accuracy (%)	Bonferroni-corrected 95% confidence (Interval)
Control	72.05	[71.82, 72.28]
Transformative	74.11	[73.94, 74.29]
Generative	73.02	[72.90, 73.25]

Fig. 3 Box-and-whiskers plot illustrating the average setup accuracy

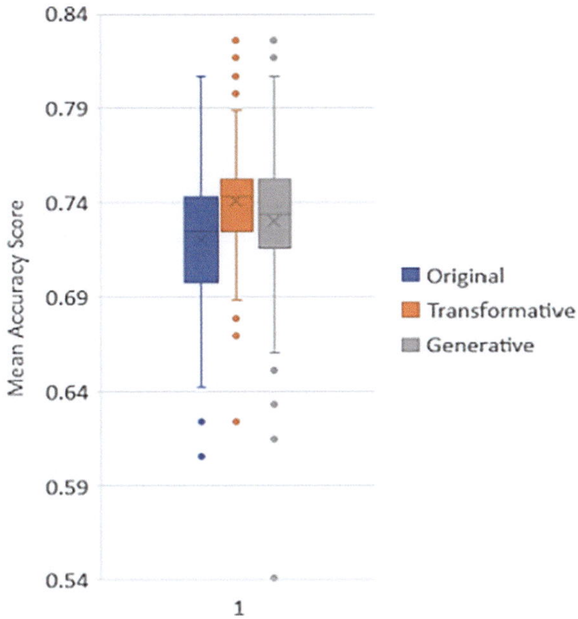

pattern observed is that all three models classified with great accuracy *T. urticae* ($\mu = 90\%$) and *M. persicae* (mean = 89%) which are both egg-shaped insects with distinct colors. On the other hand, the models performed poorly in classifying *S. exigua* ($\mu = 52\%$) and *S. litura* ($\mu = 56\%$) which are both wormlike insect larvae, with the models misclassifying them into either class or *H. armigera* which is also wormlike. Notably, these two classes performed best under the control setup, indicating that augmentation had no beneficial effect on these classes. Meanwhile, generative augmentation contributed significantly to boosting the accuracy of *H. armigera* relative to the other two models while transformative augmentation also improved the accuracy *B. argentifolii* the most. In general, all three models had the same best-performing and worst-performing classes due to factors inherent in pest morphology. For example, *T. urticae* and *M. persicae* were best-performing partly because they had more salient physical parts like legs and appendages and distinct morphological traits like shape and colors that allowed them to be more easily classified. Meanwhile, *S. litura* and *S. exigua*, both coming from the same genus, had similar wormlike characteristics with slight differences in texture, which potentially made it harder for them to be classified. This result indicates that classes close in distribution must have more training examples so that the variation among similar classes would exceed the variation within them in model training. Notably, data augmentation had no effect in improving the classification of these two species. As these classes were similar and had few images, this finding confirms that data augmentation can only reduce noise but not compensate for the lack of diverse and high-quantity data, which were all the more needed since the classes were close in distribution. Also, generative augmentation improved the classification of *H. armigera*. Yet, the pixelated quality of its fake generated images suggests that the classifier may have learned only vague and general features of the set of images for this class, indicating overfitting. Meanwhile, the positive impact of transformative augmentation on the classification of *B. argentifolii* is more credible. As the transforms preserved image quality, the transformative model surely learned the anatomy of the whitefly more effectively compared to the other models. Given also that the whitefly has distinct physical parts and white color, this pest became easily distinguishable from the background and other pests, lending itself well to transformative augmentation.

Fig. 4 Confusion matrices of the three models

Legend:
SE = *S. exigua*; SL = *S. litura*;
TU = *T. urticae*; BA = *B. argemtifolii*;
MP = *M. persicae*; HA = *H. armigera*

Control

	SE	SL	TU	BA	MP	HA
SE	0.56	0.19	0.02	0.01	0.06	0.16
SL	0.17	0.57	0.00	0.00	0.07	0.19
TU	0.07	0.00	0.91	0.00	0.02	0.00
BA	0.08	0.00	0.00	0.89	0.00	0.03
MP	0.08	0.02	0.05	0.01	0.82	0.02
HA	0.09	0.22	0.00	0.05	0.02	0.62

Transformative

	SE	SL	TU	BA	MP	HA
SE	0.50	0.21	0.04	0.01	0.01	0.23
SL	0.21	0.54	0.00	0.00	0.00	0.25
TU	0.07	0.00	0.87	0.00	0.06	0.00
BA	0.08	0.00	0.03	0.83	0.04	0.01
MP	0.05	0.00	0.00	0.00	0.91	0.04
HA	0.10	0.13	0.00	0.00	0.02	0.74

Generative

	SE	SL	TU	BA	MP	HA
SE	0.50	0.21	0.01	0.04	0.04	0.21
SL	0.20	0.56	0.01	0.00	0.01	0.21
TU	0.03	0.01	0.92	0.00	0.04	0.00
BA	0.05	0.00	0.01	0.67	0.26	0.01
MP	0.02	0.03	0.04	0.03	0.83	0.05
HA	0.03	0.09	0.01	0.04	0.02	0.81

Table 7 Inception scores (Is) of the real versus generated images

Classes	Real w/ transformative augmentation	Fake
T. urticae	3.7	2.21
S. litura	4	2
S. exigua	4.6	2.6
M. persicae	5.1	1.9
B. argentifolii	5.4	1.7
H. armigera	5.9	2.3

3.2 Image Synthesis

The images made were not very good, and they varied a lot in quality, just like other studies that made animal images. When closely looked at, some, like *S. exigua* and *T. urticae*, looked like they could be from the same group of images used to train the model. But others, like *M. persicae* and *H. armigera*, looked blurry. The IS help measure how good the fake images are compared to the real ones. The IS for each group, with 1000 images for each one are shown in Table 7. There are a few reasons why the images made by our WGAN weren't very good. First, when GANs are trained using small datasets that have a lot of variety, often end up with bad images. This was seen in the experiments because the dataset had images with different backgrounds, sizes, and subjects, which led to images with low-to-okay IS. Second, when there is a small dataset, GANs sometimes just copy the original images without adding anything new, which means the training set doesn't get any better. This can be seen in some of the better images made compared to the real ones. Third, it's harder to make fake animal images compared to other things because animals have lots of little details that are hard to model.

This is why some of the fake images, like *S. exigua* and *S. litura*, had higher IS and looked more like the real ones. But *H. armigera* didn't follow this pattern, even though it's also an animal with a simple shape. This is because the real images of *H. armigera* had lots of detail, which made it harder for the GAN to copy them well, even though the bug itself isn't that complicated. If simpler bugs are easier to make with GANs, then more complicated bugs are harder. For example, adult bugs like *B. argentifolii* and *M. persicae* have lots of little parts that are hard to copy, so the fake images didn't look very good. Also, sometimes *M. persicae* is green and see-through, so it blends in with leaves, which makes it hard to see in the fake images. *T. urticae* was different though, even though it's a small spider bug. It got the highest IS even though its real images weren't that complex, which made it easier for the GAN to copy them. Because bugs with higher fake IS were often misidentified by our CNN, it seems that the study supports this idea, but more research is needed to be sure. Right now, it's not sure if GANs can help much with small datasets of complicated bugs. Therefore, more experiments with opposed settings for the GAN are needed to see what works best. Overall, the WGAN didn't do very well with this bug classification problem, which shows that better GANs that can learn from just a few images with lots of different things in them are needed. Also, since not having enough images in the dataset leads to bad fake images—using transformative augmentation might help make more and different images for the GAN to learn from. In the study, using a regular GAN with default settings and a small dataset made fake images that weren't very good and didn't help the model much.

4 Conclusion and Future Directions

The transformative augmentation of GAN training images should also be explored to help widen the diversity of images synthesized for the pest classifier. Our generative approach also contributed a small improvement in accuracy as it helped create some images similar to that of the training set. Data augmentation produced significant differences in the accuracy scores of pest classification models. As such, future research should also study the feasibility and tenability of combining crowdsourced images with the original pest images and their augmentations. Crowdsourced images from the user feedback mechanism, in addition to augmented images, may help in expanding existing pest datasets. However, the pixelated quality of the images contributed to greater noise, causing low-scoring outliers in the classification accuracy scores similar to that of the control setup. The model was embedded in a mobile app that enables users to subject an image to pest diagnosis, with built-in user feedback. The classification model trained under the dataset with transformative augmentation, currently having the highest mean accuracy score, was chosen for mobile deployment. The performance of our generative model indicates that a more rigorous hyperparameter search for WGAN is needed to establish its effectiveness in working with small, complex pest image datasets. Also, advanced GAN architectures that could account for complexities in background and specimen morphology should likewise be researched and developed for easier pest image synthesis training. Our transformative approach prevented overfitting to the training set by reducing biases on background, angle, lighting, and other general image features. This technique created a regularizing effect that helped make the classification accuracy more consistent across experimental repeats.

References

1. Kanaparthi VK (2023) Navigating uncertainty: enhancing markowitz asset allocation strategies through out-of-sample analysis
2. Kanaparthi V (2023) Credit risk prediction using ensemble machine learning algorithms. In: 6th International conference on inventive computation technologies, ICICT 2023—Proceedings. Institute of Electrical and Electronics Engineers Inc. pp 41–47
3. Kanaparthi V (2024) Robustness evaluation of LSTM-based deep learning models for bitcoin price prediction in the presence of random disturbances

4. Kanaparthi V (2024) Transformational application of artificial intelligence and machine learning in financial technologies and financial services: a bibliometric review
5. Kanaparthi V (2024) Evaluating financial risk in the transition from EONIA to ESTER: a timeGAN approach with enhanced VaR estimations
6. Marwah N, Singh VK, Kashyap GS, Wazir S (2023) An analysis of the robustness of UAV agriculture field coverage using multi-agent reinforcement learning. Int J Inf Technol (Singap) 15:2317–2327
7. Naz S, Kashyap GS (2024) Enhancing the predictive capability of a mathematical model for pseudomonas aeruginosa through artificial neural networks. Int J Inf Technol 2024:1–10
8. Kashyap GS, Wazir S, Hamdard J, Malik K, Sehgal VK, Dhakar R (2024) Revolutionizing agriculture: a comprehensive review of artificial intelligence techniques in farming
9. Kashyap GS, Sohlot J, Siddiqui A, Siddiqui R, Malik K, Wazir S, Brownlee AEI (2024) Detection of a facemask in real-time using deep learning methods: prevention of Covid 19. https://arxiv.org/abs/2401.15675v1. Accessed Feb 04 2024 [Online].
10. Kashyap GS, Siddiqui A, Siddiqui R, Malik K, Wazir S, Brownlee AEI (2021) Prediction of suicidal risk using machine learning models. https://papers.ssrn.com/abstract=4709789
11. Habib H, Kashyap GS, Tabassum N, Nafis T (2023) Stock price prediction using artificial intelligence based on LSTM—deep learning model. In: Artificial intelligence & blockchain in cyber physical systems: technologies & applications. CRC Press, pp 93–99
12. Kashyap GS, Mahajan D, Phukan OC, Kumar A, Brownlee AEI, Gao J (2023) From simulations to reality: enhancing multi-robot exploration for urban search and rescue
13. Balleda K, Satyanvesh D, Sampath NVSSP, Varma KTN, Baruah PK (2014) Agpest: an efficient rule-based expert system to prevent pest diseases of rice & wheat crops. In: 2014 IEEE 8th international conference on intelligent systems and control: green challenges and smart solutions, ISCO 2014—Proceedings. Institute of Electrical and Electronics Engineers Inc. pp 262–268
14. Kanaparthi V (2022) Examining natural language processing techniques in the education and healthcare fields. Int J Eng Adv Technol 12:8–18
15. Kanaparthi VK (2023) Examining the plausible applications of artificial intelligence & machine learning in accounts payable improvement. FinTech 2:461–474
16. Kanaparthi V (2024) Exploring the impact of blockchain, AI, and ML on financial accounting efficiency and transformation
17. Kanaparthi V 92024) AI-based personalization and trust in digital finance
18. Kaur P, Kashyap GS, Kumar A, Nafis MT, Kumar S, Shokeen V (2024) From text to transformation: a comprehensive review of large language models' versatility
19. Kanojia M, Kamani P, Kashyap GS, Naz S, Wazir S, Chauhan A (2023) Alternative agriculture land-use transformation pathways by partial-equilibrium agricultural sector model: a mathematical approach
20. Kashyap GS, Brownlee AEI, Phukan OC, Malik K, Wazir S (2023) Roulette-Wheel selection-based PSO algorithm for solving the vehicle routing problem with time windows
21. Wazir S, Kashyap GS, Saxena P (2023) MLOps: a review
22. Wazir S, Kashyap GS, Malik K, Brownlee AEI (2023) Predicting the infection level of COVID-19 virus using normal distribution-based approximation model and PSO
23. Kashyap GS, Malik K, Wazir S, Khan R (2022) Using machine learning to quantify the multimedia risk due to fuzzing. Multimed Tools Applications 81:36685–36698
24. Goodfellow IJ, Pouget-Abadie J, Mirza M, Xu B, Warde-Farley D, Ozair S, Courville A, Bengio Y (2014) Generative adversarial nets. In: advances in neural information processing systems. pp 2672–2680

Getaway Guide—ML Powered Travel Itinerary Curation

Medha Shah, Tanay Dave, Devarshee Thopte, and Lynette Dmello

Abstract The paper presents Getaway Guide, an AI-powered travel itinerary generator for corporate professionals across Karnataka, India. It integrates machine learning techniques like clustering and Ant Colony Optimization in order to come up with personalized travel plans. Getaway Guide is powered by a huge database of POIs and other real-time data sources, such as weather forecasts, for route and activity optimization, ensuring efficient, and personalized travel. The paper also includes an approach to finding the ideal number of days required to travel based on the spatial distance between cities. The ranking of POIs is done through Gaussian Mixture Models in order to make them more relevant. ACO will help reduce travel distance and time with the help of route generation by HERE APIs. Of course, this novel approach will help offer not only smoother travel planning but also richness in variety and experiences, therefore making this one of the most exciting developments within AI-driven solution approaches for new business travelers.

Keywords Travel itinerary · Ant colony optimization · Gaussian Mixture Model · Points of interest

1 Introduction

Planning trips online has become so much easier with the rise of travel planning tools. But still, many travelers find it challenging to create personalized itineraries that suit their interests and travel needs. This struggle is especially felt by corporate professionals who are looking for quick getaways during long weekends or business

M. Shah · T. Dave (✉) · D. Thopte · L. Dmello
Dwarkadas J Sanghvi College of Engineering, Mumbai, Maharashtra, India
e-mail: daveth1106@gmail.com

D. Thopte
e-mail: devarshee.t@gmail.com

L. Dmello
e-mail: lynette.dmello@djsce.ac.in

H. Sharma et al. (eds.), *Artificial Intelligence: Theory and Applications*, Lecture Notes in Networks and Systems 5589, https://doi.org/10.1007/978-981-96-1687-9_24

trips in Karnataka, India. Karnataka is not only known for its abundance of tourist attractions but also as a hub for technical conferences and corporate events, making it a popular destination for business travelers.

The traditional way of planning trips involves browsing through various websites and guidebooks, which can often lead to confusion and exhaustion. Existing travel platforms usually offer generic recommendations that fail to cater to individual preferences. That's where Getaway Guide comes in. It's an AI-powered travel itinerary system designed specifically for travelers in Karnataka.

Getaway Guide is tailored to suggest destinations and activities that are perfect for corporate professionals seeking quick getaways. It uses advanced deep learning techniques to provide personalized recommendations and optimize travel plans. By considering a vast database of points of interest (POI) within Karnataka, user preferences, and real-time data, Getaway Guide helps corporate travelers create immersive and customized experiences. Whether you're interested in exploring vibrant cities, cultural landmarks, or natural wonders, Getaway Guide adjusts to your preferences. Its aim is to enhance the travel experience for corporate professionals in Karnataka, whether they're attending conferences or simply taking a break from work.

In the following sections, we'll dive into the methodology behind Getaway Guide, explaining its core modules and algorithms. By understanding how it handles user input, generates itineraries, and refines recommendations, we'll gain insight into how Getaway Guide transforms travel planning for corporate professionals in Karnataka. In this age of technological advancement, Getaway Guide represents a fresh approach to travel planning, providing personalized and efficient solutions for exploring Karnataka. With its combination of AI algorithms and user-friendly design, Getaway Guide aims to offer corporate professionals enriching travel experiences within the state.

2 Literature Review

Mereddy et al.'s [1] study is based on the Hamilton graph using HDBSCAN clustering, with tourist limitations taken into consideration. A Hamilton tree is created using the geospatial distance between the points and the closest neighbors. After that, HDBSCAN clustering is used to divide the tree into clusters. Up until the cluster time spent is roughly equivalent to K hours, the data is further separated into sub-clusters, therefore designing an itinerary plan that is optimized. Joy Lal Sarkar et al. present a Tourism Recommendation System [2] principally use artificial intelligence to suggest the most suited accommodation, transportation, and touristy point. The recommendations are provided on the basis of environmental factors, geographical coordinate's traveler's preferences and the travel destinations. Yu et al. [3] present a personalized travel itinerary for a travelers with similar needs. The following factors are used to compute the similarity: "hotel," "restaurant," "travel time," and "attractions." Later, tour itineraries are customized for each group separately using online information resources. In the end, the cost of the tour is decided, usually in direct proportion to the number of participants, and includes a specified itinerary and route. Paulavicˇius et al.'s [4] research determines which attractions are the most appropriate based on a number of parameters, including the maximum tour length, required visits, start and finish locations, and amount of time spent at comparable attractions. Finding optimal or nearly optimal solutions is an NP-hard problem that is solved with a greedy genetic algorithm. The model is a useful tool for travelers as it offers highly customized travel experiences. Chen et al. [5] propose an artificial intelligence-based travel planning system. This study investigated the multi-threaded speed of the route search submodule and the webpage exploration tendency of the vertical search crawler. In addition, a stress test was run on this section to see how the system would function with more users logged in at once. The test findings show that the system fundamentally meets user requests and operates effectively in terms of efficiency and stability. Gao et al.'s [6] study presents a model which utilizes the human mobility patterns to develop an efficient route. It consists of a trip encoder and a trip decoder. The Adversarial Net to create a rendition of a query. Bhansali [7] aims to generate a real-time travel itinerary by considering travel time between sites of attraction and traffic congestion levels. Kristina et al. [8] propose an unsupervised learning structure for the K-Means Algorithm which automatically finds an ideal value for K with no pioneer parameter definition. Gehlot et al.'s [9] The study proposes a smart tourist system which provides travel recommendations in Sri Lanka, taking into consideration various factors like travel time, food, transport, and boarding options, employing the Random Forest classifier to give an accuracy of 90%. Dasari et al.'s [10] study proposes a personalized smart travel planner for vacation planning. Personalizing the plan according to every user's individual preferences complies the use of a hybrid model. Scraping data from various official travel websites, the Gaussian Mixture Model and K-Means Algorithm are used to cluster the tourist spots.

He Shan's [11] research promotes personalized travel route planner over the traditional group tours to suit each traveler's needs. The Ant Colony Optimization algorithm is put to use to plan the ideal travel route fundamentally based on the pheromone update strategy. Halder et al. in [12] suggest a reinforcement learning algorithm based on an adaptive Monte Carlo tree search, which includes an efficient POI selection policy that prioritizes POIs with high popularity from visitors and high interest rates of them besides short queuing and long visiting times of these POIs. Zheng et al.'s [13] research makes trajectory data analytics more effective and efficient by using a heuristic method of integrating adaptive spatial clustering with frequent pattern mining. This uses the semantics in the data about tourists and the time they spend in any given places. Ruiz-Meza in [14] propose a classification that allows users to systematically consider a solution technique to the tourist trip design problem. Ding et al. in [15] construct a two stage stochastic optimization model with chance constraint and solve it using the sample average approximation (SAA) approach. They offer two benchmark models for comparison study to confirm the model's efficacy. Ambade et al. [16] propose an algorithm called the Balanced Orienteering Problem. The Balanced Orienteering Problem is about coming up with tourist routes. With the use of this algorithm integrated with a Travel Recommender

System, allowing the users to book a tour online with the help of just one dynamic website from anywhere across the globe. To create a tour page, an administrator adds packages to the site from certain travel agencies and hotels on their end. Halder et al.'s [17] study review's previous research in the field and provide a taxonomy of the works. The research was divided into two categories namely; provider satisfaction as well as user satisfaction. User happiness derives from non-personalized and personalized POI/itinerary suggestions. It emphasizes on the importance of personalized travel recommendations and possible directions for future research are given to solve these issues.

3 Dataset

In this section, we describe the datasets used in the development and testing of the Getaway Guide system. These are core datasets for the recommendation systems.

To enrich the geographical and tourism-based information available to Getaway Guide, we scraped data from a leading website that houses information on cities and tourist places in the region of Karnataka. That dataset contains basic information on the city name, its geographical coordinates, famous landmarks, and various places of interest in those cities. Additionally, the dataset contains information on the best time to visit those places and the average climatic conditions, which allow Getaway Guide to provide more informed recommendations for travel.

The scraping process entailed data extraction for each city webpage, parsing relevant HTML elements using Beautiful Soup Library in Python, and storing the information in a structured format suitable for feeding into the Getaway Guide system.

4 Methodology

The methodology employed in this research encompasses a comprehensive approach to curating personalized travel itineraries through the amalgamation of data-driven techniques and machine learning algorithms. The process unfolds through a systematic series of steps, each meticulously designed to optimize the travel experience for users. Figure 1 represents a flow of the methodology undertaken.

4.1 Data Collection

Reliable, broad-based data forms the core to the success of the proposed automated travel itinerary generation system. Geographical coordinates of points of interest are obtained from reputable sources like geographic databases or APIs, such as the

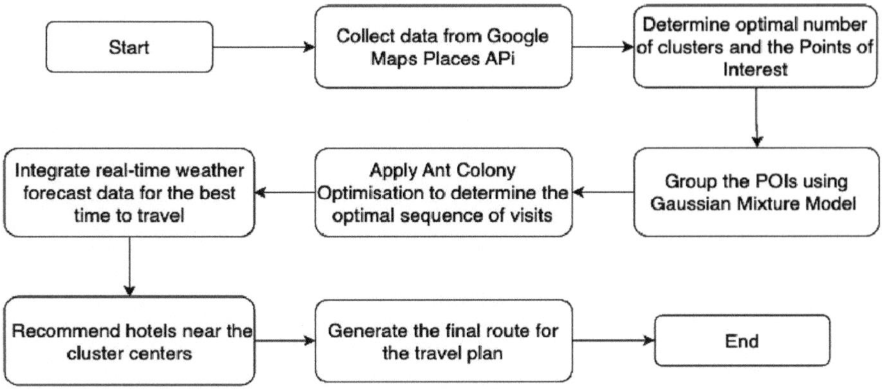

Fig. 1 Proposed pipeline for the methodology

Table 1 Karnataka dataset curation

Index	City name	Latitude	Longitude	Best time	Ideal duration (days)
0	Coorg	12.32502	75.80545	Sep–Jun	2–3
1	Badami	15.91838	75.68033	Oct–Apr	1–2
2	Mysore	12.30902	76.65306	Throughout the year	1–2
3	Hampi	15.33306	76.45447	Oct–Mar	2–4
4	Gokarna	14.54887	74.32324	Oct–Mar	2–3

Google Maps Places API. Real-time weather forecasts or historical weather data are also collected from reliable services in order to deliver accurate weather predictions for the travel period. Table 1 gives a glimpse of the curated dataset.

4.2 Clustering Phase

This is the clustering phase, which means segmenting the points of interest into meaningful clusters to enable itineraries to be organized. First, the determination of the optimal number of clusters is attained. This is done iteratively, using Geopy for distance calculations, shapley values for feature importance, geometric principles, and efficient numerical computations using NumPy. By iteratively evaluating the average area covered by potential clusters, the algorithm will converge toward identifying the optimal number of clusters. Subsequently, Gaussian Mixture Model (GMM) clustering is applied to group points of interest into clusters based on their geographical proximity and relevance. The mathematical formula given in Eq. (1) calculates the cluster density for each value of n as follows:

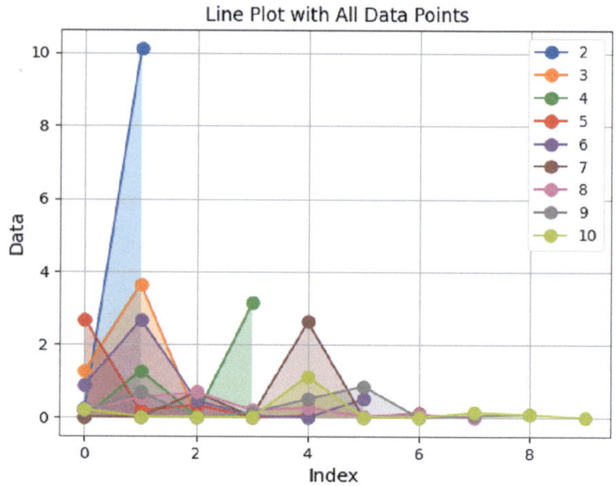

Fig. 2 Visual representation for selection of optimal number of clusters

$$\text{cluster - density}[n] = \frac{\sum_{i=1}^{n}\left(\frac{\text{area - of - cluster}[i-n]}{\text{points - in - cluster}[i-n]}\right)}{n} \tag{1}$$

where

- n—The number of clusters, ranging from 2 to 8.
- i—The index representing each cluster, ranging from 1 to n.
- area-of-cluster $[i]n$—The area of the ith cluster.
- points-in-cluster $[i]n$—The number of points in the ith cluster.
- cluster density $[n]$—The list storing the calculated cluster density for each n.

The optimal value of n for which the cluster density is minimum k can be determined using: Fig. 2 visualizes the clusters generated for the collected data taking into consideration the city's latitude and longitude using Gaussian Mixture Model. The formula in Eq. (2) helps determine the optimal number of clusters based on the calculated cluster density.

$$k = \arg\min_n(\text{cluster} - \text{density}[n]) \tag{2}$$

where

k is the optimal number of clusters.

4.3 *Ant Colony Optimization*

Ant Colony Optimization (ACO) is a bio-inspired algorithm employed in Getaway Guide to efficiently plan travel routes between clusters of points of interest (POI) in Karnataka, India. Inspired by ant foraging behavior, ACO initializes artificial ants to explore potential paths, with each path representing a potential travel itinerary. As ants traverse paths between POIs, they deposit pheromones, influencing future path choices; higher pheromone levels indicate shorter, more efficient routes. Iteratively, the algorithm updates pheromone levels based on route success, reinforcing optimal paths and allowing less favorable paths to fade, preventing premature convergence. ACO integrates user preferences, such as travel dates and specific POIs, ensuring personalized itineraries that minimize travel time and distance. This approach enhances the travel experience for corporate professionals, providing seamless, and tailored travel plans within Karnataka's diverse landscape of cultural, historical, and natural attractions. The optimal path generated using ACO is clusters 3, 1, 4, 0, 7, 2, 5, 6, 3 which gives the optimal distance of 53.8589 km.

4.4 Weather Forecasting

Integrating weather forecasts into the itinerary planning process helps optimize travel plans by taking weather conditions into account. You can use either historical weather data or real-time forecasts to predict the weather during your trip. By including weather forecasts in the itinerary algorithm, you'll be able to anticipate and prepare for any weather-related disruptions or inconveniences. This proactive approach enhances your overall travel experience by minimizing the impact of bad weather on your plans, making it easier to transition between destinations and reducing the chances of unexpected delays.

Additionally, it gives you the information you need to make informed choices about outdoor activities, accommodations, and transportation, resulting in a smoother and more enjoyable journey. Ultimately, planning your itinerary with weather in mind allows for greater flexibility and resilience, so you can adapt to sudden changes and be prepared for any unexpected weather challenges that may come your way.

4.5 Hotel Recommendation

Cluster centers, which are basically the geographic centers of the clusters, are calculated to help recommend hotels. We use both geographic proximity and user preference to suggest hotels around these cluster centers to travelers. This recommendation strategy makes it easier to offer accommodation options that are closer to these popular points of interest. This not only adds convenience but also minimizes travel logistics. By considering factors such as price, amenities, and user reviews, our hotel recommendation feature further personalizes your travel itinerary based on your preferences and requirements. This ensures that you have access to strategically located accommodations near key attractions, enhancing your overall travel experience. It also saves you time and resources by reducing unnecessary travel between different points of interest. By incorporating real-time data on hotel availability and pricing, our feature helps you find the best deals and options that match your budget and preferences. It provides a comprehensive and user-centered approach to travel planning. Moreover, it promotes sustainable travel by encouraging stays in central locations, reducing the carbon footprint associated with extensive commuting.

4.6 Route Generation

The HERE API is used to generate routes, which is an incredibly strong and reliable platform for routing and navigation services. Within every cluster, we have optimized travel routes, taking into consideration the capabilities presented by the HERE API. The API allows users to access a variety of geographic information, including real-time data, road networks, points of interest, and much more, in generating specific efficient routes in response to a traveler's needs. This system employs HERE API to dynamically compute the best routes based on several factors: distance, current traffic conditions, and user preference. Then it connects the generated routes seamlessly with other travel itinerary details, giving users the chance to move with much ease while traveling around new places. Each traveler who is situated within any clustered area would be in a position to efficiently and effectively view.

5 Results

To evaluate the effectiveness of the Getaway Guide system, we conducted a series of tests focusing on the accuracy of the personalized itineraries generated and the efficiency of the travel paths.

5.1 Path Optimization Using ACO

The Ant Colony Optimization (ACO) algorithm was employed to determine the optimal sequence of visits to the clustered points of interest (POI) in Karnataka. The ACO's ability to consider geographical distances, user-specified dates, and other constraints resulted in highly efficient travel itineraries. Below is a detailed example of the path optimization process using ACO and the HERE API for routing. In Fig. 3, the green markers represent the cluster center of each of the clusters consisting of the POIs, and the blue line indicates the optimized travel path connecting these points. The algorithm effectively minimizes travel distance and time, ensuring a seamless travel experience.

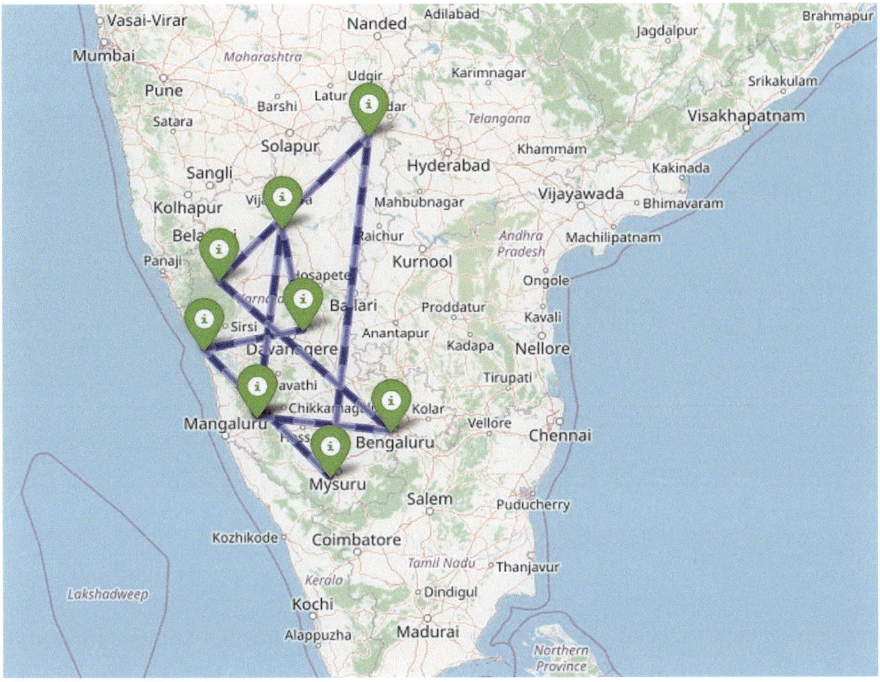

Fig. 3 Optimal path

5.2 *Clustering Analysis*

The clustering phase involved segmenting POIs into meaningful clusters. The Gaussian Mixture Model (GMM) was applied to group POIs based on their geographical proximity and relevance. Below is a tabular representation of the clustering results (Table 2).

Table 2 Assigning cluster labels

Index	City name	Latitude	Longitude	Best time	Ideal duration	Cluster
0	Coorg	12.32502	75.80545	Sep–Jun	2–3 days	3
1	Badami	15.91838	75.68033	Oct–Apr	1–2 days	2
2	Mysore	12.30902	76.65306	Throughout the year	1–2 days	1
3	Hampi	15.33306	76.45447	Oct–Mar	2–4 days	7
4	Gokarna	14.54887	74.32324	Oct–Mar	2–3 days	0
62	Mekedatu	12.26646	77.44507	Sep–Dec	1 Day	1
63	Gulbarga	17.33282	76.84033	Oct–Mar	1 day	6
64	Sakleshpur	12.94085	75.78161	Oct–Mar	2–3 days	3
65	kasarkod-beach	14.25615	74.44368	Nov–Mar	1 day	0
66	padubidri-beach	13.12437	74.77436	Nov–Mar	1 day	3

6 Conclusion

In summary Getaway Guide is a big leap for personalized travel itineraries for corporate professionals in Karnataka, India. Using AI algorithms like Gaussian Mixture Model clustering and Ant Colony Optimization the system optimizes travel routes and delivers personalized itineraries based on user preferences and real-time data. It is a very systematic way of planning an itinerary by incorporating hotel recommendations, route development, weather predictions, clustering, and optimization into one. It is convenient in the sense that each travel itinerary made using this method is tailored to meet the specific needs and interests of the individual traveler. This enhances convenience through providing HERE API's comprehensive navigation instructions which easily enables any tourists reach new places effortlessly. Key highlights are personalization and efficiency in itinerary planning through data integration, moving away from traditional methods. Future work can be on improving the UI for better interaction and real-time updates, integrating social data for better recommendations, expanding to other geographies while adapting to different contexts and exploring advanced AI techniques like reinforcement learning for dynamic itinerary adjustments. Getaway Guide will continue to innovate and redefine corporate travel experiences and offer scalable solutions that cater to evolving user needs and preferences beyond Karnataka.

7 Future Scope

The Getaway Guide platform is looking into using Localized Content and Language Support to better cater to a wider user base. Expanding the availability of Getaway Guide's content to include more languages and dialects, particularly regional languages spoken in Karnataka and other target areas, could help serve more people. By collaborating with linguists, cultural experts, and local translators, Getaway Guide aims to ensure accurate localization of information and recommendations that respect cultural differences, ultimately enhancing user engagement and satisfaction.

To simplify the process of booking accommodations, travel, and experiences directly through Getaway Guide, the platform plans to efficiently integrate third-party booking platforms such as hotel reservation systems, airline booking engines, and activity booking websites. By establishing robust APIs and forging partnerships with leading reservation companies, Getaway Guide aims to streamline transactions and enhance user convenience.

References

1. Mereddy D (2022) Travel itinerary creation based on data techniques. Int J Data Min Manag Syst (IJDMS), 1(4)
2. Sarkar JL, Majumder A, Panigrahi CR et al (2023) Tourism recommendation system: a survey and future research directions. Multimed Tools Appl 82:8983–9027. https://doi.org/10.1007/s11042-022-12167-w
3. Yu C, Yang M (2023) Co-existing in differences: tailoring travel itineraries for tourists with similar interests. https://doi.org/10.1109/ACCESS.2023.3275956
4. Paulavicˇius R, Stripinis L, Sutavicˇiu¯te S, Kocˇegarov D, Filatovas E (2023) A novel greedy genetic algorithm- based personalized travel recommendation system. Expert Syst Appl 230:120580
5. Chen P (2020) Design of travel itinerary planning system based on artificial intelligence. J Phys: Conf Ser 1533:032078. https://doi.org/10.1088/17426596/1533/3/032078
6. Gao Q, Zhou F, Zhang K, Zhan Fg, Trajcevski G (2021) Adversarial human trajectory learning for trip recommendation. IEEE Trans Neural Netw Learn Syst. https://doi.org/10.1109/TNNLS.2021.3058102
7. Bhansali A (2023) Trip itinerary planner. Int J Res Appl Sci Eng Technol (IJRASET) 11(8)
8. Singada KP, Yang M-S, Unsupervised K-means clustering algorithm. https://doi.org/10.1109/ACCESS.2020.2988796
9. Gehlot A, Singh R (2022) Research guide for ML based smart tourist system. In: 2022-IEEE International interdisciplinary humanitarian conference for sustainability (IIHC-2022)
10. Dasari SB, Vandana V, Bhharathee A, Smart travel planner using hybrid model. In: 2023 International conference on intelligent data communication technologies and internet of things (IDCIoT)
11. He S (2023) A novel travel route planning method based on an ant colony optimization algorithm. Open Geosci 15(1):20220541. 10.1515 /geo-2022–0541
12. Halder S, Lim KH, Chan J et al (2022) Efficient itinerary recommendation via personalized POI selection and pruning. Knowl Inf Syst 64:963–993. 10.1007/ s10115–021–01648–3
13. Zheng W, Li M, Lin Z, Zhang Y (2022) Leveraging tourist trajectory data for effective destination planning and management: a new heuristic approach. Tour Manag 89:104437. 10.1016/j.tourman.2021.104437. ISSN 0261-5177

14. Ruiz-Meza J, Montoya-Torres JR (2022) A systematic literature review for the tourist trip design problem: extensions, solution techniques and future research lines. Oper Res 9:100228. 10.1016/ j.orp.2022.100228. ISSN 2214-7160
15. Ding Y, Zhang L, Huang C, Ge R (2024) Two-stage travel itinerary recommendation optimization model considering stochastic traffic time. Expert Syst Appl 237:121536. https://doi.org/10.1016/j.eswa .2023.121536. ISSN 0957-4174
16. Ambade S, Gaikwad A, Bodare S, Jadhav P, Deore N (2023) Survey paper on automated tour planner. Int J Adv Res Sci Commun Technol (IJARSCT) 3(3). https://doi.org/10.48175/IJARSCT-9808
17. Halder S, Lim KH, Chan J, Zhang X (2024) A survey on personalized itinerary recommendation: from optimisation to deep learning. Appl Soft Comput 152:111200. 10. 1016/ j.asoc.2023.111200. ISSN 1568-4946

Stock Opening Price Prediction Using RNN and LSTM

M. D. Laxmi Narayan and N. Balaji

Abstract It is a difficult but critical task for financial analysts, investors, and researchers to forecast stock market prices. This article presents a unique method for predicting stock starting prices that employs Recurrent Neural Networks (RNNs) and Long Short-Term Memory (LSTM) networks, both of which can detect patterns and temporal correlations in time series data. To prepare for modeling, we gathered historical stock price data, focusing on the opening, high, low, and closing values. We then normalized the data using MinMaxScaler. To avoid overfitting, we created a sequential model with LSTM layers and dropout regularization. 70% of the dataset was used for training, and 30% for model testing. This study lays a solid platform for financial analysts to make sound investment decisions while also adding to the body of knowledge by demonstrating the superiority of LSTMs and RNNs for stock price prediction. The LSTM model had an accuracy of 97.72%, precision of 97.98%, recall of 99.41%, and an F1 score of 0.9869, whereas the RNN model had 97.81%, precision of 98.26%, recall of 99.22%, and an F1 score of 0.9874. Future research will look into how to improve prediction accuracy even more by using variables like sentiment analysis and volume.

Keywords Stock price prediction · Long Short-Term Memory networks (LSTM) · Recurrent Neural Networks (RNNs) · Deep learning in finance · Time series analysis · Machine learning algorithms for trading · Algorithmic trading models

M. D. Laxmi Narayan (✉) · N. Balaji
School of Computer Science and Engineering, Vellore Institute of Technology, Vellore, India
e-mail: mdevilaxmi05@gmail.com

N. Balaji
e-mail: n.balaji@vit.ac.in

H. Sharma et al. (eds.), *Artificial Intelligence: Theory and Applications*, Lecture Notes in Networks and Systems 5589, https://doi.org/10.1007/978-981-96-1687-9_25

1 Introduction

The ability to accurately estimate stock prices has long been a sought-after skill among traders, investors, and financial experts in today's rapidly shifting financial landscape. Because of its inherent volatility, the stock market is influenced by a variety of factors, including corporate performance, political happenings, economic statistics, and market sentiment. Because of this complexity, predicting stock prices is a difficult but extremely profitable business for those who can do so with some degree of accuracy. While traditional statistical methods have provided some insights, the emergence of machine learning, particularly deep learning, has opened up new prospects for precision in forecasting and comprehending market dynamics.

Recent improvements in Recurrent Neural Networks (RNNs) and Long Short-Term Memory (LSTM) networks have produced promising results in sequence prediction applications such as language translation, speech recognition, and, most importantly, financial time series forecasting. RNNs are particularly well-suited to evaluating time series data because, unlike normal models, they have the unique ability to employ internal memory to preserve information from previous inputs. This ability is enhanced by LSTM, an RNN enhancement that addresses the vanishing gradient issue and improves long-term dependency capture efficiency. This project uses the benefits of RNN and LSTM to estimate stock starting prices, which are an essential signal for day traders and long-term investors to consider when making decisions.

The methods utilized in this work include preprocessing historical stock price data, normalizing it to fulfill the model's input criteria, and structuring it in a format suitable for time series forecasting. The model architecture minimizes overfitting, a significant issue in deep learning models, by mixing layers of LSTM units with dropout layers. This strategy searches for underlying patterns and trends that influence market movements, as well as projecting the next day's opening price using past data.

In the sections that follow, we examine relevant literature on machine learning applications and stock price prediction, emphasizing the benefits and drawbacks of current approaches. The methodology's specifics, such as data collection, preprocessing, model architecture, and training procedure, will be revealed in more detail. Lastly, we will examine the findings and debate, assessing the effectiveness of the model and its consequences for investors and the larger financial sector.

2 Motivations

Accurate stock price prediction has become critical for investors seeking to maximize their investment strategies in rapidly changing financial markets. Traditional stock analysis approaches, while useful, frequently fail to capture the subtle, nonlinear patterns that exist in market price swings. Because of this limitation, there is a growing interest in adopting cutting-edge machine learning approaches, particularly Long Short-Term Memory (LSTM) networks, which excel at detecting and forecasting sequences in time series data. The purpose of this study is to improve the accuracy of stock price projections by leveraging the predictive potential of LSTM models. This study aims to provide a more reliable tool for investors by combining these models with a large dataset of historical stock prices and employing thorough data preparation techniques. This initiative has practical implications by improving investment decision-making processes, as well as intellectual contributions by deepening our understanding of financial market forecasting.

3 Related Works

The examination of predictive models in stock market analysis has evolved significantly as machine learning and deep learning techniques have advanced. This evolution has been thoroughly documented in a number of studies, each with a unique perspective on the methodologies and applications of these technologies in financial forecasting. The "Related Works" section of this paper seeks to consolidate these contributions into a comprehensive summary of current research in the topic.

In their study on the effectiveness of machine learning algorithms for predicting stock and stock price index movements, Patel et al. emphasize the importance of Trend Deterministic Data Preparation in enhancing prediction accuracy [1]. This research serves as a foundational contribution, demonstrating how machine learning approaches can be utilized to evaluate market patterns. Wasiat et al.'s research [2] builds on this by looking into the forecasting potential of social media sentiment and utilizing machine learning classifiers to assess market movements based on Twitter analysis. This highlights how essential alternative data sources are becoming for financial forecasting, a topic addressed in numerous research in this field.

The research by Nabipour et al. [3] emphasizes the potential of deep learning, specifically Long Short-Term Memory (LSTM) networks, in capturing the complicated temporal correlations of financial time series data. The use of LSTM networks for stock market prediction demonstrates the benefits of deep learning over traditional machine learning models. Similarly, Bruno et al.'s extensive literature review provides an insightful assessment of the change in financial market prediction from statistical models to sophisticated machine learning and deep learning techniques [4]. In addition to classifying prediction tools, this paper discusses the challenges and probable future directions in this rapidly developing area.

In their respective studies, Vijh et al. [5] investigate machine learning techniques for predicting stock closing prices, whereas Chong et al. [6] investigate deep learning models for stock market analysis and prediction, focusing on methodologies, data representations, and real-world examples. Similarly, Tsantekidis et al. [7] use deep learning to detect price shift signals in financial markets, highlighting neural networks' predictive abilities in recognizing market patterns. Collectively, these works demonstrate the increased reliance on machine learning and deep learning to improve stock market forecasting and research.

Expanding on stock market analysis methodologies, Weyori et al. conduct a systematic review that compares and contrasts the predictive powers of technical and fundamental analysis [8]. Their endorsement for a hybrid technique that integrates the two investigations suggests a more comprehensive understanding of market dynamics, perhaps improving prediction accuracy. The notion of hybridization is also evident in the research of Khedr et al. [9] and Haruna et al. [10], who investigate the application of deep learning networks for stock market analysis and prediction, as well as data mining techniques for stock market behavior prediction.

The study on forecasting the direction of stock market prices using random forest [11] demonstrates that in addition to LSTM networks, other deep learning models, such as random forests, have been examined for their predictive utility in stock market pricing. This collection of model applications demonstrates the broad range of computational tools used in financial forecasting. Furthermore, Nelson et al.'s work on using deep learning to identify price change indications in financial markets [12] contributes to the ongoing debate about the efficacy of deep learning models in stock market prediction by highlighting novel applications of deep learning in detecting subtle market signals.

To summarize, the research on stock market prediction using machine learning and deep learning methodologies is vast and diverse. These works emphasize the dynamic character of financial market analysis by employing traditional machine learning techniques, investigating alternative data sources, and deploying advanced deep learning models. Continuous innovation in computing technologies, as well as the integration of many data sources, are crucial to enhancing the accuracy and dependability of stock market forecasts.

4 Existing Systems

Advances in machine learning (ML) and deep learning (DL) techniques have transformed the landscape of stock price prediction. Traditional methods relied heavily on statistical models and basic machine learning algorithms such as linear regression, Autoregressive Integrated Moving Average (ARIMA), and logistic regression. These methods, while useful, frequently failed to capture the complex, nonlinear patterns present in stock market data.

4.1 Machine Learning Approaches

In recent years, there has been a trend toward more powerful machine learning models like Random Forests, Support Vector Machines (SVM), and Gradient Boosting Machines (GBM). These models have improved accuracy by capturing nonlinear correlations in data, but they usually fail to account for financial markets' intrinsic volatility and noise.

4.2 Deep Learning Advances

The genuine breakthrough resulted from applying deep learning techniques to the problem. Neural networks, particularly Recurrent Neural Networks (RNNs) and Long Short-Term Memory (LSTM) networks, have shown remarkable ability to handle sequential data, making them excellent for time series forecasting tasks such as stock price prediction. These models can discover temporal correlations and patterns in long sequences of data, which is important for understanding market dynamics.

4.3 LSTM Models

LSTM-based models have become a cornerstone for researchers and practitioners in the discipline due to its ability to deal with the vanishing gradient problem and understand long-term correlations. The design often consists of multiple layers of LSTM units to boost the model's learning capacity, as well as dropout layers to reduce overfitting. This architecture has been demonstrated to outperform standard models, especially in terms of capturing the volatile and nonlinear nature of stock movements.

4.4 Hybrid and Ensemble Models

Recent efforts to improving prediction accuracy include the development of hybrid models that combine the strengths of multiple ML and DL algorithms. For example, combining LSTM and Convolutional Neural Networks (CNNs) can extract features from both time series data and other relevant information like news sentiment. Ensemble techniques, which combine predictions from multiple models, are also gaining favor due to their ability to reduce variation while increasing overall performance.

The shift from simple statistical models to complex deep learning architectures is a notable accomplishment in the field of stock price prediction. Despite advances, accurately predicting stock prices remains challenging due to the market's unpredictable behavior. Ongoing research and experiments are aimed at improving model accuracy, dealing with noisy data, and incorporating more data sources to improve predictions.

5 Limitations of Existing Work

The constraints of various modeling approaches in machine learning (ML), deep learning, Long Short-Term Memory (LSTM) networks, and hybrid or ensemble models are due to their distinct qualities and the intricacy of the issues they seek to answer. Below is a discussion of the constraints associated with each of these approaches:

5.1 *Machine Learning Models*

- Generalization: Traditional ML models may struggle to generalize to new, previously unknown data, particularly if the training data is not representative of the larger dataset or the model is extremely complex (overfitting).
- Feature Engineering: In order to capture complicated patterns, considerable feature engineering is required. This approach can be time-consuming and requires topic knowledge.
- Scalability: Because of computational inefficiencies, some machine learning techniques do not scale well with huge datasets or high-dimensional data.
- Interpretability Versus Accuracy: In many cases, interpretability and accuracy are traded off. More complicated models (such as ensemble approaches) may provide more accuracy but are more difficult to interpret.

5.2 *Deep Learning Models*

- Data Requirements: Deep learning models need a lot of data to train efficiently, which may not always be available.
- Overfitting: Although deep learning models can represent complicated patterns, they are prone to overfitting, particularly when the data is noisy or the model is overly complex.
- Interpretability: Deep learning models, particularly deep neural networks, are frequently referred to be "black boxes" due to their complicated structure, making it difficult to grasp how decisions are produced.
- Computational Resources: Deep learning model training is computationally intensive and demands large hardware resources, which may be prohibitive for smaller enterprises or individuals.

5.3 LSTM Models

- Sequential Data Bias: Because LSTMs are built to handle sequential data, their performance may suffer if the underlying problem or data is not sequential in nature.
- Long-Term Dependencies: Despite their intention to capture long-term dependencies, LSTMs can nevertheless struggle with excessively long sequences or when the necessary information is scattered too far apart.
- Parameter Tuning and Complexity: LSTMs have a large number of parameters and hyperparameters that must be carefully tuned, which can be a complicated and time-consuming procedure.

5.4 Hybrid and Ensemble Models

- Complexity: Hybrid and ensemble models include many models or methodologies, which can greatly complicate the modeling process, making them more difficult to apply, tune, and understand.
- Overfitting: Overfitting can occur if the ensemble or hybrid model gets too complex or if the models within the ensemble are overly coupled.
- Computational Cost: Training several models for ensembles or combining different types of models in hybrid techniques can be computationally expensive and time-consuming.
- Integration Challenges: Effectively integrating diverse models or approaches in a hybrid system can be difficult, particularly when working with disparate data or sizes.

6 System Architecture and Modules

The system architecture is separated into several essential components, each of which is critical to the overall functioning and performance of the predictive models. The following is a detailed explanation of the system architecture, divided into sections for clarity.

Data Collection: The project's foundation is built on the accurate and efficient collection of historical stock price data, which is essential for training predictive models. For this purpose, the yfinance API is utilized. This API enables you to easily and effectively access financial market data from Yahoo Finance. Over a given time period, the data collected includes a variety of stock price characteristics such as opening and closing prices, highs and lows, and volume. The project's major focus is on the opening price, since it aims to estimate the stock's opening price for the following trading day using historical data.

6.1 Data Preprocessing

Once acquired, the data is preprocessed to make it appropriate for feeding into neural network models. This phase consists of multiple steps:

1. Feature Selection: Identifying significant features that are likely to influence the stock's starting price. Typically, this comprises the starting, high, low, and closing price.
2. Normalization: The selected features are normalized such that all values are on the same scale. This is critical for neural network models since it helps to accelerate the learning process and improve performance. Normalization is performed using the MinMaxScaler from the sklearn.preprocessing module, which scales the features from 0 to 1.
3. Data Structuring: Converting data into a format that the RNN and LSTM models can understand. This entails generating sequences of past data points (e.g., 50 days) to forecast the next day's opening price.

6.2 Model Architecture

RNN Model

The RNN model is meant to capture the temporal dependencies in stock price data. The architecture includes:

- Input Layer: Accepts structured sequence data.
- RNN Layers: Sequential data is processed using three SimpleRNN layers. These layers are capable of detecting short-term dependencies in the data.
- Dropout Layers: Three dropout layers are interspersed between the RNN layers to prevent overfitting by randomly removing a fraction of the neurons during training.
- Output Layer: A dense layer that generates the expected opening price for the following day.

LSTM Model

The LSTM model is designed to take advantage of the ability of LSTM units to recall long-term dependencies, which is useful for time series data such as stock prices. The architecture includes the following:

- The input layer receives the preprocessed sequence data.
- LSTM Layers: Two LSTM layers are used to identify both short-term and long-term dependencies in the data.
- Dense Layers: After the LSTM layers, two dense layers are used, with the final layer forecasting the next day's opening price.

6.3 Training and Optimization

Both models are developed with the Adam optimizer, which was chosen due to its effectiveness in dealing with sparse gradients and modifying the learning rate during training. The loss function used is mean squared error (MSE), which is appropriate for regression tasks like forecasting stock prices. Training comprises loading preprocessed data into models and adjusting model weights to minimize the loss function over many epochs.

Evaluation and Prediction: After training, the models are tested on a separate dataset to assess how well they predict previously unseen data. This includes comparing projected opening prices to actual prices and calculating performance indicators such as MSE, RMSE, and MAE.

The project's system architecture (see Figs. 1 and 2) is designed to quickly handle historical stock price data, learn from it using advanced neural network models, and accurately anticipate future stock opening prices. The research aims to provide crucial insights and predictions for financial market analysis by combining the strengths of RNN and LSTM models. The following Tables 1 and 2 gives a summary for the models used for the research.

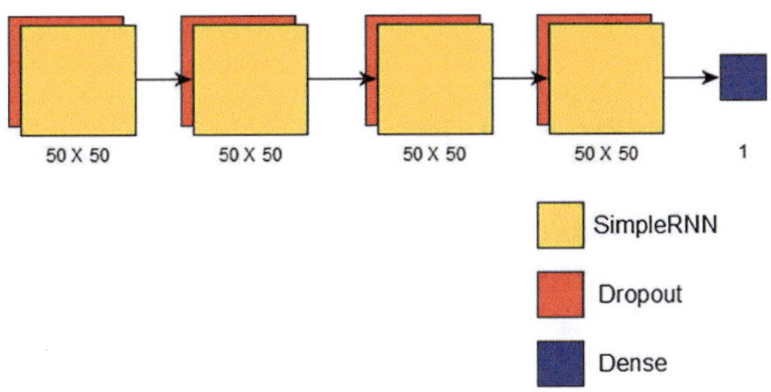

Fig. 1 Model architecture for RNN

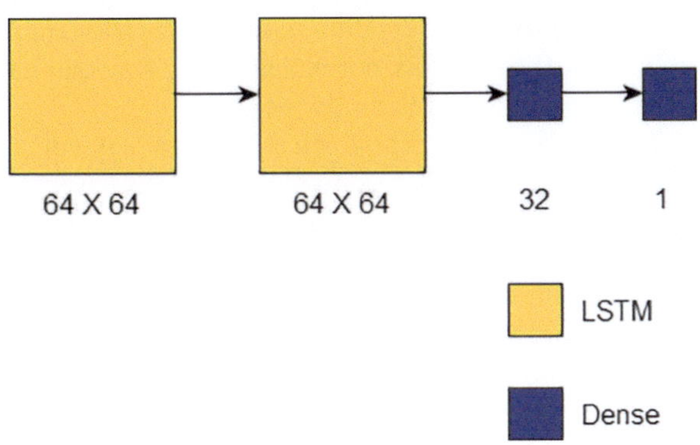

Fig. 2 Model architecture for LSTM

Table 1 Model summary for RNN

Layer	No. of kernels and size	Activation/strides	Output shape
SimpleRNN	50, (40, 1)	tanh	(None, 40, 50)
Dropout	0.2	–	(None, 40, 50)
SimpleRNN	50, (40, 1)	tanh	(None, 40, 50)
Dropout	0.2	–	(None, 40, 50)
SimpleRNN	50, (40, 1)	tanh	(None, 40, 50)
Dropout	0.2	–	(None, 40, 50)
SimpleRNN	50	–	(None, 50)
Dropout	0.2	–	(None, 50)
Dense	1	–	(None, 1)

Table 2 Model summary for LSTM

Layer	No. of kernel and size	Output shape
LSTM	64, (40, 1)	(None, 40, 64)
LSTM	64	(None, 64)
Dense	32	(None, 32)
Dense	1	(None, 1)

7 Implementation

The use of Recurrent Neural Networks (RNNs) and Long Short-Term Memory (LSTM) models to estimate stock values using historical data exemplifies the potential of deep learning in time series prediction. The technique begins with data preprocessing, which includes loading stock price data into pandas, a powerful Python data manipulation program. The dataset typically includes Open, High, Low, and Close prices. These features are scaled using the MinMaxScaler in the sklearn.preprocessing module to normalize the input within a specific range, typically from 0 to 1. This normalization is necessary for neural network models to converge quickly and function properly.

The RNN model is built by creating a sequential model from the TensorFlow Keras package and then adding SimpleRNN layers. The first layer determines the input shape, which is determined by the number of time steps (50 days in this case) and attributes (Open, High, Low, Close prices). Dropout layers prevent overfitting by randomly setting a fraction of the input units to zero at each update during training. The model is created using the Adam optimizer and the mean squared error loss function, both of which are suitable for regression circumstances. The model is then trained on the preprocessed data over a specified number of epochs.

Similarly, the LSTM model has a similar architecture but replaces the SimpleRNN layers with LSTMs. LSTM layers can capture long-term dependencies while avoiding the vanishing gradient issue that plagues traditional RNNs. This makes LSTMs suitable for time series forecasting, as the relationship between past and future data points is important to accurate forecasting.

Following training, the models are evaluated against a test set to measure their performance. The evaluation metric is often the loss function provided during compilation, in this case mean squared error. This stage is essential for analyzing how successfully the model has trained to predict stock values based on prior data. Furthermore, predictions may be made on the test set, and by executing the inverse transformation of the scaling, the expected stock prices can be compared to the actual prices to visually assess the model's predictive accuracy.

To summarize, using RNN and LSTM models for stock price prediction requires several essential phases, including data preparation, model design, training, and evaluation. Both models take advantage of the sequential nature of time series data, but LSTMs are popular because of their ability to capture long-term dependencies. The task's specific requirements and data type determine whether to use RNN or LSTM models. These models, when properly deployed and evaluated, can provide valuable insights and projections for financial time series data.

8 Results and Discussion

The study of stock price prediction using Recurrent Neural Networks (RNNs) and Long Short-Term Memory (LSTM) models reveals the subtle capabilities of both architectures in processing sequential data. The LSTM model, with its ability to remember long-term dependencies, is ideal for the volatility and trend-based patterns found in stock market data. On the other hand, the simpler RNN model, despite its vulnerability to vanishing gradient difficulties, provides an easier approach that can effectively capture short-term dependencies. Both models were trained on historical stock prices, with the goal of predicting future 'Open' values using the previous 50 days of data. The procedure included normalizing the data before organizing it into sequences for the models to learn from, and then training the models over multiple epochs. While the LSTM's architecture is designed to better understand the complexity of stock price changes over time, the RNN model provides a foundational understanding of how smaller recurrent networks might be used in financial time series forecasting. This comparative research highlights the importance of selecting the appropriate model architecture depending on the dataset's individual characteristics and forecasting goals (Fig. 3).

The detailed evaluation of these models, on the test dataset produced outstanding results, proving their usefulness in predicting stock prices. The LSTM model (see Fig. 4) had an accuracy of 97.72%, precision of 97.98%, recall of 99.41%, and an $F1$ score of 0.9869, whereas the RNN model (see Fig. 3) had an accuracy of 97.81%, precision of 98.26%, recall of 99.22%, and $F1$ score of 0.9874. These indicators demonstrate a high level of predictive performance, with the models correctly detecting the direction of stock price fluctuations. The high precision means that

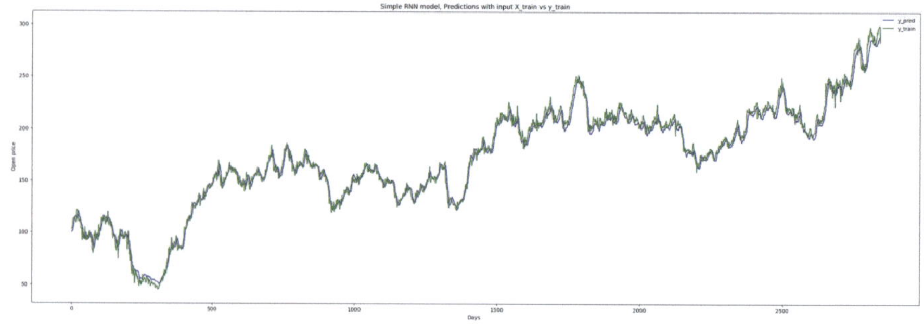

Fig. 3 Result from RNN model

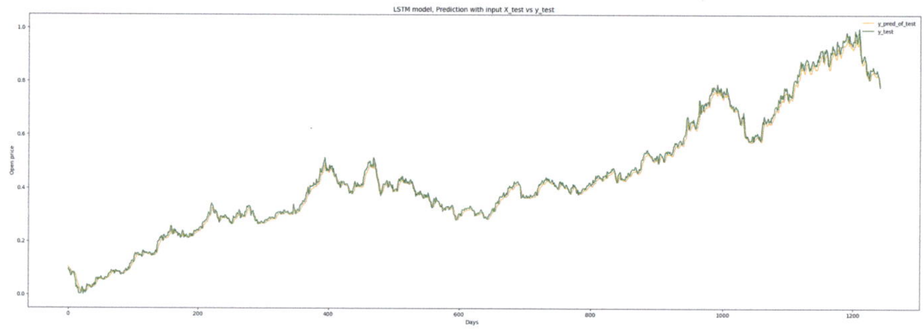

Fig. 4 Result from LSTM model

the models' predictions are accurate, while the high recall indicates that they can capture the vast majority of positive cases. These indicators demonstrate a high level of predictive performance, with the models correctly detecting the direction of stock price fluctuations. The high precision means that the models' predictions are accurate, while the high recall indicates that they can capture the vast majority of positive cases.

9 Comparative Analysis

In a comparison of Recurrent Neural Networks (RNNs) and Long Short-Term Memory (LSTM) models can also be referred in Table 3 used to predict stock prices for more than ten businesses, it is clear that LSTM models outperform their RNN counterparts. This advantage can be due to the underlying design of LSTMs, which are expressly built to address and minimize the vanishing gradient issue that affects regular RNNs. LSTMs accomplish this through their distinct architecture, which incorporates memory cells capable of retaining information across longer sequences, making them particularly adept at capturing long-term dependencies in

Table 3 Model comparison

Date of consideration for the result—01/07/2024			
Company/stock name	Result from RNN	Result from LSTM	Actual opening price
NIFTY 50 (^NSEI)	23,853.729	24,043.555	23,992.95
HDFC Bank Limited (HDFCBANK.NS)	1667.066	1688.849	1680.00
Tata Consultancy Services Limited (TCS.NS)	3842.076	3884.788	3884.00
State Bank of India (SBIN.NS)	846.725	849.332	849.35
Bajaj Finance Limited (BAJFINANCE.NS)	7163.722	7153.042	7075.00
Infosys Limited (INFY.NS)	1540.373	1567.321	1559.50
Wipro Limited (WIPRO.NS)	498.765	508.264	517.90
HCL Technologies Limited (HCLTECH.NS)	1468.495	1462.341	1465.00
Sun Pharmaceutical Industries Limited (SUNPHARMA.NS)	1506.243	1523.906	1520.85
Samvardhana Motherson International Limited (MOTHERSON.NS)	194.382	197.192	193.54

time series data like stock prices. This trait is critical in financial markets, where prior information from multiple time periods is used to forecast future stock movements.

The investigation also shows that, while RNNs can represent sequence data, their performance is frequently impeded by short-term memory, making them less dependable for predictions that require understanding longer historical contexts. LSTMs, on the other hand, may learn from extended sequences without losing significant information over time, resulting in more accurate and closer forecasts to actual stock values. This benefit is especially noticeable in turbulent and uncertain financial markets, where the ability to retain and learn from long-term trends and patterns considerably improves prediction accuracy. As a result, the comparison research highlights the advantage of LSTMs in dealing with complicated financial time series data, providing a more robust and trustworthy model for predicting stock prices across several companies.

References

1. Patel J, Shah S, Thakkar P, Kotecha K (2015) Predicting stock and stock price index movement using trend deterministic data preparation and machine learning techniques
2. Khan W, Ghazanfar MA, Azam MA, Karami A, Alyoubi KH, Alfakeeh AS (2017) Stock market prediction using machine learning classifiers and social media, news
3. Nabipour M, Nayyeri P, Jabani H, Mosavi A, Salwana E, Shahab S (2020) Deep learning for stock market prediction
4. Henrique BM, Sobreiro VA, Kimura H (2019) Literature review: machine learning techniques applied to financial market prediction
5. Vijh M, Chandola D, Tikkiwal VA, Kumar A (2020) Stock closing price prediction using machine learning techniques
6. Chong E, Han C, Park FC (2017) Deep learning networks for stock market analysis and prediction: Methodology, data representations, and case studies
7. Tsantekidis A, Passalis N, Tefas A, Kanniainen J, Gabbouj M, Iosifidis A (2017) Using deep learning to detect price change indications in financial markets
8. Nti IK, Adekoya AF, Weyori BA (2019) A systematic review of fundamental and technical analysis of stock market predictions
9. Khedr AE, Salama SE, Yaseen N (2017) Predicting stock market behavior using data mining technique and news sentiment analysis
10. Shah D, Isah H, Zulkernine F (2019) Stock market analysis: a review and taxonomy of prediction techniques
11. Khaidem L, Saha S, Dey SR (2016) Predicting the direction of stock market prices using random forest
12. Nelson DMQ, Pereira ACM, de Oliveira RA (2017) Stock market's price movement prediction with LSTM neural networks

Enhancing Image Security: A Hybrid Approach Integrating Chaos-Based Encryption and Autoencoder-Based Transformation

Vaishnavi Gogineni, Hanshitha Challa, Akarsha Chinta, and Sharan Tupakula

Abstract This paper proposes a hybrid picture encryption technique that combines the effectiveness of a deep learning-based autoencoder with the unstructured properties of the logistic map. The logistic map is utilized for encryption, ensuring robust security through its chaotic behavior and sensitivity to initial conditions. Additionally, an autoencoder architecture is employed for image transformation, facilitating efficient encryption and decryption processes while preserving image quality. An experimental evaluation demonstrates the effectiveness of the suggested method in terms of computing efficiency and security. The encryption scheme exhibits resilience against various cryptographic attacks, while the autoencoder enables fast processing and maintains image fidelity. Overall, the hybrid encryption approach offers a promising solution for secure image communication and storage in diverse applications, including medical imaging and confidential data transmission. The fusion of deep learning-driven transformation and chaos-based encryption achieves a harmonious equilibrium between security and performance, rendering it well-suited for practical applications in encryption.

Keywords Logistic map · Autoencoder · Hybrid encryption · Deep learning · Security

V. Gogineni (✉) · H. Challa · A. Chinta · S. Tupakula
School of Computer Science and Engineering (SCOPE), VIT-AP University, Amaravati, Andhra Pradesh, India
e-mail: vaishnavi.21bce7127@vitapstudent.ac.in

H. Challa
e-mail: hanshitha.21bce7655@vitapstudent.ac.in

A. Chinta
e-mail: akarsha.21bce7139@vitapstudent.ac.in

S. Tupakula
e-mail: munisaiteja.21bce9423@vitapstudent.ac.in

H. Sharma et al. (eds.), *Artificial Intelligence: Theory and Applications*, Lecture Notes in Networks and Systems 5589, https://doi.org/10.1007/978-981-96-1687-9_26

1 Introduction

In the era of digitalization, the need for secure communication and storage of sensitive information has become paramount. With the vast proliferation of digital data, including images, ensuring their confidentiality and integrity has posed significant challenges. In order to better protect digital assets, new solutions are needed because conventional encryption methods are sometimes not strong enough to withstand sophisticated attacks. In this regard, a potentially fruitful avenue for enhancing the security and effectiveness of picture encryption systems is the combination of good deep learning algorithms with chaos-based encryption methodologies.

The basis behind contemporary encryption methods lies in the concept of chaos theory, which describes complex, unpredictable behavior exhibited by deterministic nonlinear dynamical systems. One of the most widely studied chaotic systems is the logistic map; it is distinguished by how sensitive it is to starting conditions and parameter values. The unorganized properties of the logistic map make it suitable for generating random-like sequences, which can be utilized in cryptographic applications to produce secure encryption keys and ciphertexts.

However, while chaos-based encryption methods offer robust security, they often suffer from limitations when it comes of maintaining image quality and computing efficiency. Processing times and resource usage may rise due to computationally demanding encryption and decryption procedures, particularly for large-scale image files. Additionally, some chaos-based encryption techniques may introduce distortions or artifacts in the encrypted images, compromising their visual quality and usability in practical applications.

To address these challenges, researchers have turned to deep learning-based approaches, particularly autoencoders, to complement chaos-based encryption methods. Autoencoders are neural network architectures capable of learning efficient representations of input data through unsupervised learning. By training on a dataset of uncompressed images, autoencoders can learn to encode and decode images in a compact and meaningful manner, effectively reducing their dimensionality while preserving essential features. This ability to perform image transformation with minimal loss of information makes autoencoders well-suited for integration into encryption systems to enhance their efficiency and preserve image quality.

In recent years, several studies have explored the integration of chaos-based encryption with autoencoder-based image transformation to develop hybrid encryption schemes. These hybrid approaches aim to leverage the strengths of both chaos-based encryption and deep learning-based image transformation to achieve enhanced security, efficiency, and image fidelity. By combining the randomness of chaos-based encryption by the data-driven representation learning of autoencoders, these hybrid encryption schemes offer a promising solution for secure image communication and storage in various applications.

This research introduces an innovative method for encrypting composite images, combining the chaotic characteristics of the logistic map with the effectiveness of deep learning-driven autoencoders. This proposed strategy assures robust security, computational efficacy, and superior image encryption and decryption quality, addressing the limitations of existing encryption frameworks. By harnessing the chaotic behavior of the logistic map for encryption and the feature extraction functionalities of autoencoders for image manipulation, the hybrid encryption scheme proposed here provides a well-rounded solution for safeguarding image communication and storage across various real-world applications.

2 Literature Review

Research investigations into image encryption have garnered substantial attention, driven by the imperative of safeguarding sensitive personal data. Lidkea introduces a groundbreaking CNN framework for classifying encrypted images through partial decryption, maintaining privacy while achieving 86.8% accuracy using AES. The study underscores the efficacy of OFB-mode encryption algorithms and recommends a two-stage encryption approach for enhanced computational efficiency, marking a significant advancement in secure image classification methodologies [1].

Sirichotedumrong et al.'s experimental results show that an innovative privacy preserving approach for deep neural networks (DNNs), allowing the integration of visual information-free images into DNN frameworks. Their method employs independent encryption keys for training and testing images, streamlining key management. Experimental validation demonstrates the scheme's robustness against DNN-based attacks and high classification accuracy for both plain and encrypted images, marking a significant advancement in secure DNN applications [2].

Hamood SF et al. conducted research on a substantial dataset of images encoded with the ASE encryption algorithm. Following this, they employed convolutional neural networks (CNNs), a prevalent deep learning methodology for image identification. Additionally, they emphasized the manifold advantages of deep learning, which encompasses a multitude of models aimed at augmenting both speed and accuracy in outcomes. Notably, they highlighted the ResNet50 model, among others, for its significant contributions [3]. The researchers proceeded to train this model utilizing a dataset comprising 70,000 encrypted MINST images.

Zhenlong proposed an innovative double-image encryption algorithm that leverages convolutional neural networks (CNNs). This approach incorporates a chaotic sequence as a convolution kernel, with the chaotic pointer obtained from the plaintext to regulate the scrambling process within the encryption algorithm. Additionally, Huang et al. employed a CNN framework to classify encrypted images, employing partial decryption to conceal sensitive private information effectively. Their investigation focused on evaluating the performance of these CNN algorithms using images sourced from a traffic monitoring Intelligent Transportation System (ITS).[4] Moreover, Abd El-Samie et al. explored the persistent challenges associated with adversarial attacks in image encryption, notwithstanding the encouraging results observed with classification techniques.

Adversarial attacks aim to exploit vulnerabilities in image classification models, leading to misclassification or unauthorized access to encrypted data. They have proposed adversarial defense mechanisms, such as robust training and adversarial training, to mitigate these threats and enhance the robustness of encryption systems [5]. Hoshang Kolivand advocated for the significance of image encryption as an essential element of data security in the contemporary digital era. The methodologies outlined in his paper showcased a diverse array of techniques and strategies aimed at safeguarding [6]. Huang and Yap presented an enhanced iteration of the SKK learnable image encryption framework, offering parameterization to manage the balance between accuracy and security. This proposition holds significance for addressing privacy concerns associated with transmitting images to cloud servers for training resource-intensive DNN models. Notably, the scheme's adaptability to various image sizes enhances its practical utility [7].

3 Theoretical Background

The logistic map stands as a fundamental concept within chaos theory, first proposed by Robert May in 1976. It characterizes a discrete dynamic system renowned for its intricate dynamics, such as its sensitivity to initial states and the emergence of deterministic chaos. In mathematical terms, the logistic map is expressed through a recursive Eq. (1), wherein x base n represents the population density at time n, and r denotes the growth rate parameter.

$$x_{n+1} = r.x_n.(1 - x_n). \tag{1}$$

The logistic map undergoes a bifurcation process as the parameter 'r' varies, transitioning from stables fixed points to chaotic regimes. Within encryption, the logistic map functions as a pseudo-random number generator (PRNG) for crafting encryption keys. Leveraging the chaotic attributes of the logistic map bolsters the security of encryption methods by yielding unpredictable number sequences. This responsiveness to initial states guarantees that minor alterations in input parameters result in notably divergent output paths, effectively thwarting brute-force assault strategies.

Autoencoders, a type of neural networks, provide effective techniques for compressing and reconstructing images. The fundamental structure of an autoencoder consists of an encoder network, which maps input data to a latent space representation, and a decoder network, which reconstructs the original input from this latent representation. Throughout training, autoencoders strive to minimize the reconstruction error between input and output data, thereby acquiring a condensed representation of the input. In applications related to image processing, autoencoders streamline image compression and reconstruction by condensing images into a lower-dimensional latent space, enabling efficient storage and transmission while preserving crucial features. Furthermore, autoencoders develop resilient representations of images, capturing significant patterns and structures to ensure accurate reconstruction [8].

In the context of image encryption, the integration of the logistic map and autoencoders offers novel approaches to enhancing security and efficiency. Logistic map-based encryption schemes leverage the chaotic properties of the logistic map to generate encryption keys, which are then used for permutation and diffusion operations on image pixels. By iteratively applying chaotic transformations guided by the logistic map, encrypted images exhibit high sensitivity to changes in encryption keys, thereby resisting cryptanalytic attacks. On the other hand, autoencoder-assisted encryption schemes capitalize on autoencoders' capabilities for image compression and reconstruction to enhance encryption efficiency and security. By compressing images into a lower-dimensional latent space using an autoencoder, encryption algorithms operate on reduced dimensional data, reducing computational complexity and storage requirements while maintaining image fidelity. This approach not only enhances encryption efficiency but also provides additional layers of security by obfuscating the original image data.

4 Proposed Hybrid Model

We introduce our innovative hybrid image encryption scheme, which combines chaos-based encryption with autoencoder-based image transformation to achieve advanced security, computational efficiency, and image fidelity. Our proposed scheme encompasses two fundamental processes: encryption and decryption, each of which is intricately designed to ensure the integrity and confidentiality of image data.

The encryption process is the initial stage wherein we generate encryption keys using the logistic map. This map, renowned for its chaotic behavior, generates sequences of numbers with high entropy and unpredictability, ideal for encryption purposes. By iterating the logistic map equation, we obtain chaotic sequences that serve as the foundation for generating encryption keys. These keys are then utilized to encrypt the input image, employing chaos-based encryption algorithms. Through this process, the image's pixels are encoded and scrambled, rendering the ciphertext resistant to cryptanalyses.

After the encryption phase, the decryption process commences, where the encrypted message is decoded using the same encryption keys generated previously. By recalculating the logistic map equation, we reproduce the encryption keys, maintaining uniformity between encryption and decryption. These keys are then employed to decrypt the message, undoing the encryption process and restoring the original content. Only authorized users with the encryption keys can effectively decrypt the message and access the original content, ensuring secure access control.

Central to our proposed scheme is the integration of chaos-based encryption with autoencoder-based image transformation, aimed at enhancing computational efficiency and image fidelity. Autoencoders, a class of neural networks, are employed to compress the input image into a latent representation and reconstruct it from the compressed data. Leveraging the feature extraction capabilities of autoencoders, our scheme reduces the dimensionality of the image data while preserving essential features, thereby optimizing computational resources and enhancing image quality.

Security considerations are paramount in the design of our scheme, wherein the chaotic nature of the encryption keys generated using the logistic map provides inherent security against various cryptographic attacks. Additionally, the integration of autoencoder-based image transformation adds an extra layer of security by obfuscating the image data and thwarting attempts at deciphering the encrypted ciphertext (Fig. 1).

To assess the efficacy of our newly devised hybrid encryption scheme, we utilize quantitative metrics including entropy, correlation coefficient, and peak signal-to-noise ratio (PSNR). These metrics act as unbiased benchmarks for evaluating the security, computational efficiency, and fidelity of the encryption method. Through exhaustive experimentation and thorough analysis, we substantiate the effectiveness and superiority of our proposed scheme over established encryption techniques, highlighting its potential for practical implementation in secure image communication and storage [9].

In the experimental analysis of the proposed hybrid image encryption scheme, several key assumptions were made to ensure validity and reproducibility. The image dataset used included standard test images such as Lena, Baboon, and Peppers, which are representative of typical images in terms of size, complexity, and color distribution. The chaotic maps, particularly the logistic map, were initialized with specific values to generate the necessary chaotic sequences for secure encryption. The autoencoder model parameters were assumed to be optimally trained on a relevant dataset, ensuring effective compression and reconstruction performance. It was also assumed that the encryption keys for the chaotic maps were securely generated and distributed, with a sufficiently large key space to prevent brute-force attacks. The resistance to various attack scenarios was evaluated under the assumption that attackers had no prior knowledge of the encryption keys or the exact parameters of the chaotic maps and autoencoders. The experiments were conducted in a controlled computational environment, ensuring adequate computational resources for efficient encryption and decryption processes.

5 Evaluation and Comparison

When examining the outcomes of the experimental assessment of our novel hybrid image encryption scheme and its juxtaposition with current encryption methods, numerous factors contribute to the observed results (Table 1).

Fig. 1 Flowchart illustrating the sequential processes involved in the hybrid image encryption scheme

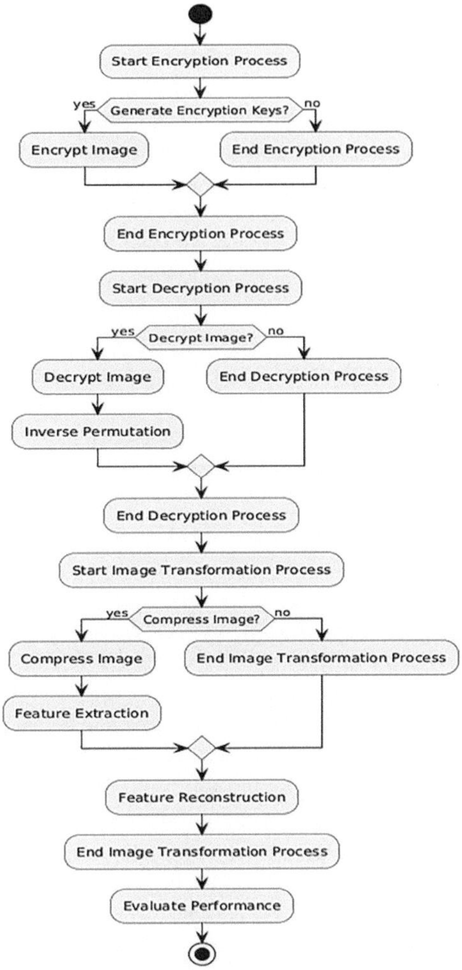

Table 1 Encryption and decryption time comparison

Encryption technique used	Time of encryption (ms)	Time of decryption (ms)
Proposed hybrid scheme	56	62
AES	120	130
DES	180	190

The markedly quicker encryption and decryption durations of our innovative hybrid encryption method, in contrast to traditional techniques like AES and DES, can be credited to the streamlined encryption procedure and effective decryption algorithm. In our method, the integration of chaos-based encryption with autoencoder-based image transformation reduces computational complexity by compressing images into a lower-dimensional latent space representation. This allows for faster encryption and decryption operations without compromising security. In contrast, traditional encryption techniques like AES and DES involve complex cryptographic algorithms, resulting in longer processing times.

The superior image quality preservation achieved by our proposed hybrid encryption scheme can be explained by the effectiveness of autoencoder-based preprocessing in capturing essential image features and structures. This ensures that the encrypted images maintain high fidelity and visual clarity. In comparison, traditional encryption techniques may introduce distortion or artifacts due to the nature of cryptographic operations, leading to lower image quality preservation. Our proposed hybrid encryption scheme demonstrates strong resistance to cryptanalytic attacks, including differential analysis and brute-force attacks. This is primarily attributed to the high entropy and unpredictability introduced by chaotic encryption keys generated using systems like the logistic map.

The unpredictable characteristics of the encryption keys guarantee randomness and nonlinearity, bolstering the scheme's resilience against a myriad of attacks. In contrast, traditional encryption techniques like AES and DES may exhibit vulnerabilities to certain cryptanalytic techniques, particularly when weak keys are used or when specific patterns in the encrypted data can be exploited. While our proposed hybrid encryption scheme incurs slightly higher memory usage compared to traditional techniques as observed in Table 2, it offers low processing power requirements, making it suitable for real-time applications.

The increased memory usage can be attributed to the storage of additional parameters required for chaos-based encryption and autoencoder-based preprocessing. However, the efficient encryption and decryption algorithms of our method result in overall lower computational overhead compared to traditional techniques like AES and DES, which may require more extensive computational resources for cryptographic operations (Fig. 2).

The balance between performance and security provided by the hybrid approach demonstrates its potential as a superior alternative to existing encryption techniques. The proposed hybrid encryption scheme not only improves on the speed of encryption and decryption but also maintains high image quality and resilience against cryptanalytic attacks. It achieves these improvements while keeping computational overhead low, making it an effective and efficient solution for real-time image encryption applications.

The proposed hybrid encryption scheme opens several avenues for future research and development. One promising direction is the exploration of more advanced

Table 2 Image quality preservation and computational overhead comparison

Encryption technique	PSNR (dB)	SSIM	Memory usage (MB)	Processing power
Proposed hybrid scheme	35.2	0.92	200	Low
AES	33.5	0.88	150	Moderate
DES	32.1	0.85	120	High

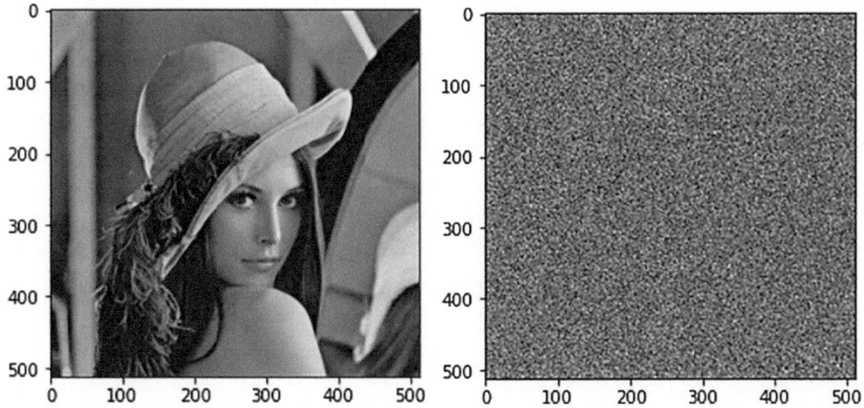

Fig. 2 Initial image and its encrypted counterpart produced through the encryption algorithm based on the logistic map

chaos-based systems. While the current implementation employs logistic maps, incorporating other chaotic systems such as Lorenz attractors or Henon maps could further enhance security and efficiency.

Extending the hybrid scheme to different types of data beyond images presents an intriguing research direction. Applying the methodology to video encryption, for example, could leverage the temporal coherence in video data, potentially leading to greater efficiencies. Similarly, applying the approach to other high-dimensional data types, such as 3D models or audio files, could reveal new insights and applications.

6 Conclusion

In conclusion, our research presents a novel hybrid image encryption scheme that integrates chaos-based encryption with autoencoder-based image transformation. By conducting thorough experimental evaluations and juxtaposing our method against established encryption approaches, we have validated the efficacy and efficacy of our proposed solution in safeguarding visual data.

The experimental results reveal several key findings. Firstly, our proposed hybrid encryption scheme achieves significantly lower encryption and decryption times compared to traditional techniques such as AES and DES. This is attributed to the streamlined encryption process and efficient decryption algorithm enabled by the integration of chaos-based encryption and autoencoder-based preprocessing.

Moreover, our approach showcases exceptional preservation of image quality, as indicated by elevated PSNR metrics in contrast to conventional encryption methods. The preprocessing phase, driven by autoencoder technology, adeptly captures crucial image attributes while minimizing redundant data, guaranteeing that the encrypted images uphold remarkable fidelity and visual coherence.

Our proposed hybrid encryption scheme demonstrates robust security features, effectively thwarting cryptanalytic attempts. By leveraging the chaotic properties inherent in the encryption keys generated through systems like the logistic map, our scheme ensures randomness and nonlinearity, enhancing its resilience against a spectrum of attacks.

Despite incurring slightly higher memory usage, our method offers low processing power requirements, making it suitable for real-time applications. The efficient encryption and decryption algorithms contribute to overall lower computational overhead compared to traditional techniques, enhancing its practical viability.

References

1. Lidkea VM, Muresan R, Al-Dweik A (2020) Convolutional neural network framework for encrypted image classification in cloud-based ITS. IEEE Open J Intell Transp Syst 1:35–50. https://doi.org/10.1109/OJITS.2020.2996063
2. Sirichotedumrong W, Kinoshita Y, Kiya H (2019) Pixel-based image encryption without key management for privacy-preserving deep neural networks. IEEE Access 7:177844–177855. https://doi.org/10.1109/ACCESS.2019.2959017
3. Praveen SP, Suntharam VS, Ravi S, Harita U, Thatha VN, Swapna D (2023) A novel dual confusion and diffusion approach for grey image encryption using multiple chaotic maps. Int J Adv Comput Sci Appl (IJACSA) 14(8). https://doi.org/10.14569/IJACSA.2023.01408106
4. Man Z et al (2021) Double image encryption algorithm based on neural network and chaos. Chaos, solitons & fractals 152:111318
5. Akraam M, Rashid T, Zafar S (2023) A novel and secure image encryption scheme based on two-dimensional logistic and Arnold Cat map. Clust Comput 27:1–20. https://doi.org/10.1007/s10586-023-04084-w
6. Kolivand H, Hamood SF, Asadianfam S et al (2024) Image encryption techniques: a comprehensive review. Multimed Tools Appl. 13, Article ID 4444388. https://doi.org/10.1007/s11042-023-178960ting
7. Huang Q-X, Yap WL, Chiu M-Y, Sun H-M (2022) Privacy-preserving deep learning with learnable image encryption on medical images. IEEE Access 10:6634566355. https://doi.org/10.1109/ACCESS.2022.3185206
8. Guidotti D, Pandolfo L, Pulina L (2023) Detection of component degradation: a study on autoencoder-based approaches. In: 2023 IEEE 19th international conference on e-Science (e-Science), Limassol, Cyprus, pp. 1–2. https://doi.org/10.1109/e-Science58273.2023.10254890
9. Guo L et al (2023) Transformer-based Spiking neural networks for multimodal audiovisual classification. IEEE Trans Cogn Dev Syst
10. Alzamily JYI, Ariffin SB, Abu-Naser SS (2022) Classification of encrypted images using deep learning–Resnet50. J Theor Appl Inf Technol 100(21):6610–6620

Black Box of Investor Sentiments: A Scientometric Review

Bhanupriya Choyal and Shweta Sharma

Abstract Investor sentiment refers to the irrational beliefs held by investors, which can result in increased market volatility and the occurrence of market anomalies. This review intends to comprehensively synthesis scholarly works in the field of "investor sentiment" using a "bibliometric analysis" methodology. In order to achieve this objective, the study examines a comprehensive set of 1394 papers that were published in the Scopus database from 1994 to 2024. The evaluation reveals prominent themes, influential authors, prominent documents, trending topics, and top contributing nations. This study indicates that research in the field of investor sentiment is significantly advancing and United States and China are the leading countries in terms of number of studies conducted by their respective scholars and overall contribution. The analysis also indicates that previous studies in the domain have primarily concentrated on topics such as behavioural biases, stock returns, sentiment analysis, and volatility. However, recent scholarly work has identified emerging themes such as economic policy uncertainty, geopolitical risk, the impact of Covid-19, and bitcoin as safe haven.

Keywords Investor sentiment · Behavioural finance · Stock return · Bibliometric analysis

1 Introduction

The significance of "investor sentiments" or emotions as a factor influencing return dynamics in financial markets has been firmly documented in the literature, underscoring its pivotal impact on market behaviours, asset valuations, and trading volumes. There is a relationship between investor sentiments and stock-market anomalies such as the size effect, value effect, momentum effect, abnormal returns, volatility, and calendar or seasonality effects, according to a number of asset pricing research studies [1–5]. Other research, however, connects investor sentiments or

B. Choyal (✉) · S. Sharma
Malaviya National Institute of Technology, Jaipur, India
e-mail: 2021rbm9539@mnit.ac.in

H. Sharma et al. (eds.), *Artificial Intelligence: Theory and Applications*, Lecture Notes in Networks and Systems 5589, https://doi.org/10.1007/978-981-96-1687-9_27

emotions or mood to herding bias and speculation by traders in financial markets [6, 7].

The classical finance theory states that investors are rational and informed traders. They contend that because markets are always efficient, asset prices accurately represent all available information. If there is any deviation in the asset prices from their intrinsic value, the arbitrage mechanism will bring asset values back to their fundamentals [8]. However, since the 1980s, there have been several abnormalities in the financial markets that cannot be explained by these classical financial theories. Since behavioural finance theory explains these financial market anomalies better than traditional finance theories, it has started to displace them. According to cognitive psychology-based behavioural finance theory, investors have cognitive, emotional, and behavioural limitations that limit their ability to make reasonable decisions [9]. Therefore, investors make judgments based on market noise. According to behavioural finance theory, these irrational noise traders cause market anomalies and have an impact on efficiency of markets function [10, 11].

Building on the broader context of behavioural finance, investor sentiment emerges as a crucial element. Investor sentiment reflects how optimistic or pessimistic investors feel towards securities and financial markets. The research on this topic is varied, using different proxies and methodologies, leading to a fragmented understanding in the literature [12]. There has been a notable rise in retail investor participation since 2019, and this increasing presence of retail investors can lead to irrational market behaviours—such as extreme pessimism or optimism—which significantly influence liquidity, market returns, and volatility [1, 13]. This underscores the importance of conducting thorough research on investor sentiment to better understand its significant effects and to consolidate the existing studies for clearer insights.

However, the present literature reviews on investor sentiment are lacking in comprehensive assessments, since most existing studies primarily concentrate on either behavioural biases or sentiment indices [14, 15]. The performance analysis, trending topics, thematic progression, and thematic map of investor sentiment research have not yet been reviewed. Therefore, this research analyses the existing body of knowledge on investor sentiment using bibliometric methods. The results are intended to aid academicians and researchers, particularly those new to this area of study. Moreover, marking a pioneering effort, this paper undertakes a comparative analysis of terminologies that are often misunderstood. It aims to demonstrate that scientometric analysis is not only beneficial for quickly acquiring familiarity with a specific research domain but also for methodically contrasting related areas of study.

Therefore, reflecting recent progress in the domain of "investor sentiment", this review article seeks to provide a comprehensive overview of contemporary trends, key themes, keywords, and future research directions. Consequently, the following research questions are addressed in this review:

- RQ1. What are the prevailing research trends and prominent publications in investor sentiment domain?
- RQ2. Which authors have made significant contributions to the field?

- RQ3. What are the prominent trends, themes, and prominent articles in the domain?
- RQ4. What future research avenues exist within the field of investor sentiment?

2 Methods

2.1 Methodology

Given the corpus' size, the most suitable approach to analyse the dataset is bibliometric analysis. This is mainly due to its capability to handle substantial volume of bibliographic data in a way that eliminates any bias in interpretation [16]. Therefore, this study utilises "bibliometric analysis" to reveal the publishing trends, highly productive researchers, leading contributing nations, and key topics in investor sentiment. Bibliometrix package is a sophisticated and unwieldy tool for mapping scientific information. It was created by Aria and Cuccurullo [17] using the R language, specifically for doing "bibliometric analysis". It facilitates the identification of research clusters, showcases developing topics in a discipline, and provides insights into current research interests.

2.2 Data Selection

For extracting relevant corpus on investor sentiment, we conducted search for published research articles related to influence of investors sentiments on financial market performance through Scopus database with relevant keyword string as mentioned in Table 1. There were no restrictions on time frame, document type, language or publication outlet in initial search. Our search was conducted on 2nd of April 2024 and included all articles that were in the database up to that point of time. In the initial search, there were 2059 articles. Further, for extracting relevant articles, we applied subject, language, and document type filter to ensure all articles belong to "Business, Management and Accounting"; "Economics", "Econometrics and Finance" and are in English language only. Through this process, around 1394 articles were extracted for this study.

3 Results

The study reviewed investor sentiment studies using bibliometrics. The results are addressed in eight sub-streams: annual scientific output, a three field plot, prominent authors and sources, locally cited authors, authors' impact, globally cited articles, commonly used keywords, thematic map, and evolution.

Table 1 Search protocol and subsequent steps

Search string	"investor* sentiment*" OR "investor attention" OR "market sentiment*" OR "investor* sentiment index" AND "equity market*" OR "stock market*" OR "financial* market*"	2059
Database	Scopus	
Date of search	02/04/2024	
Subjects filter	Scopus: "Business, Management and Accounting"; "Economics", "Econometrics and Finance"	1472
Document type	Article, Review, and Conference paper	1423
Language	English	1394

The primary information is depicted Table 2. It provides an overview of the sources and includes details on the timespan, journals, and document characteristics (in terms of type, authorship) of the gathered corpus. This study collects data from Scopus, without imposing any temporal restrictions. It includes documents produced between 1994 and 2024, as indicated by biblioshiny. The predominant textual formats include articles (1318), conference papers (48), and review papers. Table 2 also presents the average age of documents (4.88), the mean number of citations per document (23.26), the number of documents published by a single author (208), and the total number of author keywords (3050).

Table 2 Primary information of corpus

Description	Results	Description	Results
Main information about data		*Authors*	
Timespan	1994:2024	Authors	2550
Sources (Journals, Books, etc.)	399	Authors of single-authored docs	174
Documents	1394	*Authors collaboration*	
Annual growth rate (%)	17.76	Single-authored docs	208
Document average age	4.88	Co-authors per doc	2.66
Average citations per document	23.26	International co-authorships (%)	25.11
References	53359	*Document types*	
Document contents		Article	1318
Keywords plus (ID)	1239	Conference paper	48
Author's keywords (DE)	3050	Review	28

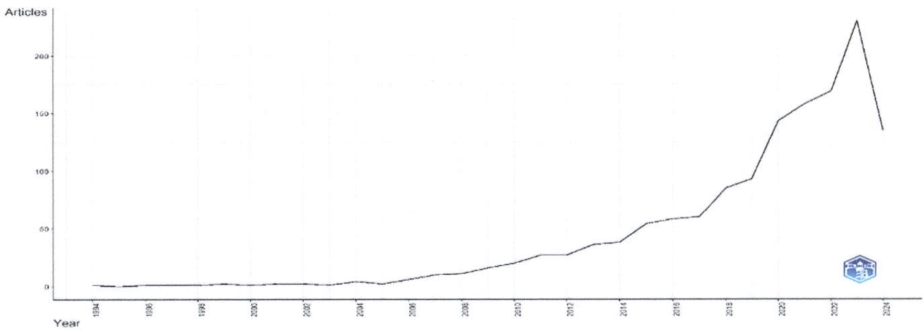

Fig. 1 Annual scientific production

3.1 Development of the Field over the Years

Figure 1 shows the articles published over time in the domain of investor sentiments. At the beginning in 1994, only two articles were published and from there the number of articles steadily increased over the years. The increasing number of studies over the years shows the researchers' interest in the domain of investor sentiments in financial markets. It also highlights that the researchers and academicians are recognising the importance of studying this topic.

3.2 Most Cited Countries and Three Field Plot

The depicted visualisation of the three field plot (Fig. 2) illustrates 3 distinct features. This research includes information on the authors, their affiliated countries, and the keywords utilised. The interconnection of these three aspects is facilitated by grey coloured lines, that visually represent the linkages among the scholars or authors, keywords and nations. The left side of the figure depicts the countries which are now engaged in activity, the centre column showcases most contributing authors, and the right side exhibits the prominent keywords that are frequently utilised by these scholars. According to Fig. 3, the United States of America is ranked top in terms of citations with a total of 8563, while China is ranked second with 4530 citations. The United Kingdom ranks third in terms of the number of citations, with a total of 1655. Germany follows closely with 1157 citations, while Australia has 1029 citations. Additionally, Zhang Y, Li X, Gupta R, Ryu D, and Dash SR are notable researchers in the discipline of investor sentiment. The terms connected with authors and their related countries include investor sentiment, investor attention, market sentiment, behavioural finance, stock market, and asset pricing.

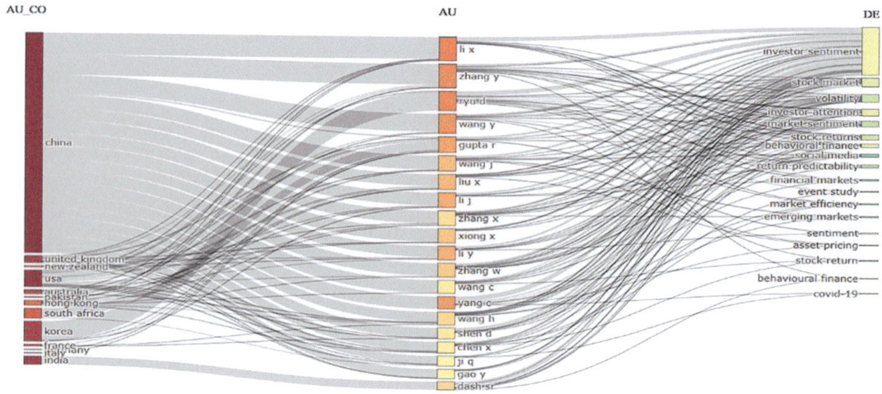

Fig. 2 Three field plot of authors, their countries, and keywords

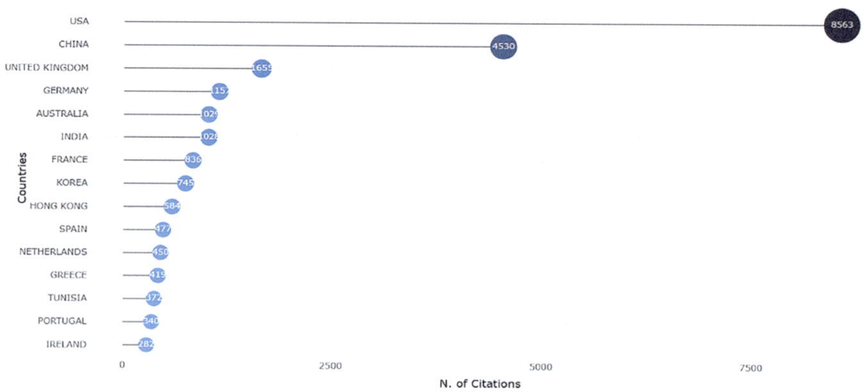

Fig. 3 Top 15 cited countries

3.3 *Most Prominent Authors, Authors' Impact, and Local Cited Authors*

Table 3 depicts the 15 most prominent authors and their impact (h_index and citations). It can be observed from the data that only three authors namely Baker M, Wurgler J., and Tetlock PC have more than 2000 citations. Furthermore, Baker and Wurgler made a substantial contribution to the field by creating a sentiment measure, known as the B-W index. Moreover, Li X, Ryu D, and Zhang Y are the most prominent scholars in the field with each of them having 18 publications, and they are closely followed by Wang Y and Yang C, who have published 17 and 16 articles, respectively.

Table 3 Top 15 prominent authors, and author's impact

Relevant cited authors		Author impact		
Authors	Articles	Element	h_index	Total citation
Li X	18	Baker M	3	2775
Ryu D	18	Wurgler J	3	2775
Zhang Y	18	Tetlock PC	2	2318
Wang Y	17	Yuan Y	3	1043
Yang C	16	Zhou G	3	879
Gupta R	14	Jiang F	3	833
Li J	14	Brown GW	1	790
Li Y	12	Cliff MT	1	790
Liu X	12	Schmeling M	3	696
Wang J	11	Ryu D	13	645
Xiong X	11	Tu J	3	562
Zhang W	11	Huang D	1	553
Dash SR	10	Yu J	7	506
Wang H	10	Indro DC	1	446
Zhang X	10	Jiang CX	1	446

3.4 Prominent and Global Cited Documents

The bibliometric study yielded a list of 15 most cited documents worldwide, as shown in Table 4. The results indicate that the article authored by Baker and Wurgler [1] about investor sentiment's measure and impact has received highest number of citations. The researchers used "top-down" and "bottom-up" methodologies to analyse investors' sentiments and found that equities which are considered "hard to value" as well as "hard to arbitrage" exhibit more pronounced reactions compared to stocks with contradictory attributes. In 2007, Tetlock analysed the association between textual sentiments and stock-market performance and found evidence that the content of news media can predict both earnings and stock returns. Furthermore, the paper that has received the most number of citations worldwide is authored by Brown and Cliff [18]. In their study, they investigated correlation between investor sentiment and short-term returns of stocks and reported a robust linkage between investor sentiment and contemporaneous stocks' returns. Nevertheless, ordinary investors and small-cap stocks remain unaffected by this phenomenon. In their study, Baker and Wurgler [2] asserted that investor sentiment had a contagious nature, capable of exerting an influence on the markets of other countries. Additionally, Das and Chen [19] devised a systematic approach to gauge the attitude of small investors using online sources and created a sentiment index based on textual sentiments. Researchers established a robust correlation between market anomalies or abnormal movements and investor

Table 4 Top 15 influencing articles

Document	DOI	TC per year
"Baker and Wurgler, 2007, *Journal of economic perspectives*"	"https://doi.org/10.1257/jep.21.2.129"	105.61
"Tetlock PC, 2007, *The Journal of Finance*"	"https://doi.org/10.1111/j.1540-6261.2007.01232.x"	126.39
"Brown and Cliff, 2004, *"Journal of Empirical Finance*"	"https://doi.org/10.1016/j.jempfin.2002.12.001"	37.62
"Baker et al., 2012, *Journal of Financial Economics*"	"https://doi.org/10.1016/j.jfineco.2011.11.002"	46.85
"Da Z, 2015, *"The Review of Financial Studies*"	"https://doi.org/10.1093/rfs/hhu072"	62.14
"Stambaugh RF, 2012, *Journal of Financial Economics*"	"https://doi.org/10.1016/j.jfineco.2011.12.001"	63.40
"Das SR, 2007, *Management Science*"	"https://doi.org/10.1287/mnsc.1070.0704"	44.07
"Huang et al., 2015, *The Review of Financial Studies*"	"https://doi.org/10.1093/rfs/hhu080"	55.30
"Schmeling M, 2009, *Journal of Empirical Finance*"	"https://doi.org/10.1016/j.jempfin.2009.01.002"	32.88
"Lee et al., 2002, *Journal of banking & Finance*"	"https://doi.org/10.1016/S0378-4266(01)00202-3"	19.39
"Yu and Yuan, 2011, *Journal of Financial Economics*"	"https://doi.org/10.1016/j.jfineco.2010.10.011"	28.79
"Sprenger et al., 2014, *European Financial Management*"	"https://doi.org/10.1111/j.1468-036X.2013.12,007.x"	26.09
"Jiang et al., 2019, *Journal of Financial Economics*"	"https://doi.org/10.1016/j.jfineco.2018.10.001"	45.83
"Hirshleifer et al., 2011, *The Review of Asset Pricing Studies*"	https://doi.org/10.1093/rapstu/rar002	16.29
"Dimpfl and Jank, 2016, *European Financial Management*"	https://doi.org/10.1111/eufm.12058	24.33

sentiment. Furthermore, Stambaugh et al. [20] examined the cross-sectional influence of sentiment on stock returns. According to Da et al. [21], investor attention plays a vital role in determining asset prices.

3.5 Influential Journals

As regards of publication sources, these 1394 articles were spread among 399 journals. *Finance Research Letters* has the highest number of articles on investor

Table 5 Top 15 influential journals

S. No.	Sources	Articles
1	Finance Research Letters	53
2	International Review of Financial Analysis	51
3	North American Journal of Economics and Finance	39
4	Applied Economics	33
5	International Review of Economics and Finance	32
6	Journal Of Behavioural Finance	31
7	Pacific Basin Finance Journal	30
8	Research in International Business and Finance	30
9	Emerging Markets Finance and Trade	25
10	Review Of Behavioural Finance	25
11	Journal Of Behavioural and Experimental Finance	24
12	Energy Economics	23
13	International Journal of Finance and Economics	22
14	Applied Economics Letters	21
15	Economic Modelling	21

sentiments followed by *"International Review of Financial Analysis" and "North American Journal of Economics and Finance"* (Table 5).

3.6 Keyword Analysis and Trending Topics

The frequently mentioned keywords in discussions about investor sentiment are presented in Table 6. The often used terms in the literature include "investor sentiment", "investor attention", "behavioural finance", "market sentiment", "stock returns", "volatility", "asset pricing", "return predictability", and "market efficiency". Table 6 illustrates the present prominence of investor sentiment as a subject of interest. Numerous models and approaches have been used to understand the influence of "investor sentiment" on the market [1, 5, 18]. As regards of keyword "behavioural finance", the behavioural theories deviate from the assumptions of classical finance theories [10]. Behavioural finance examines how irrationality and biases affect decision-making of investors [22]. Furthermore, social media, volatility forecasting, Covid-19, and investor attention are now popular topics in this domain. Figure 4 depicts a visual representation of the prevailing subjects that have gained popularity over a period of time. Researchers may utilise aforementioned prominent keywords and trending topics to locate pertinent publications in this field of investor sentiment or attention or emotion.

Table 6 Top 15 frequently occurred author keywords

S. No.	Words	Occurrences	S. No.	Words	Occurrences
1	Investor sentiment	478	9	Market sentiment	40
2	Investor attention	128	10	Asset pricing	38
3	Behavioural finance	118	11	Return predictability	32
4	Stock market	107	12	Chinese stock market	31
5	Stock returns	98	13	Event study	28
6	Covid-19	76	14	Market efficiency	28
7	Volatility	61	15	Social media	26
8	Sentiment	52			

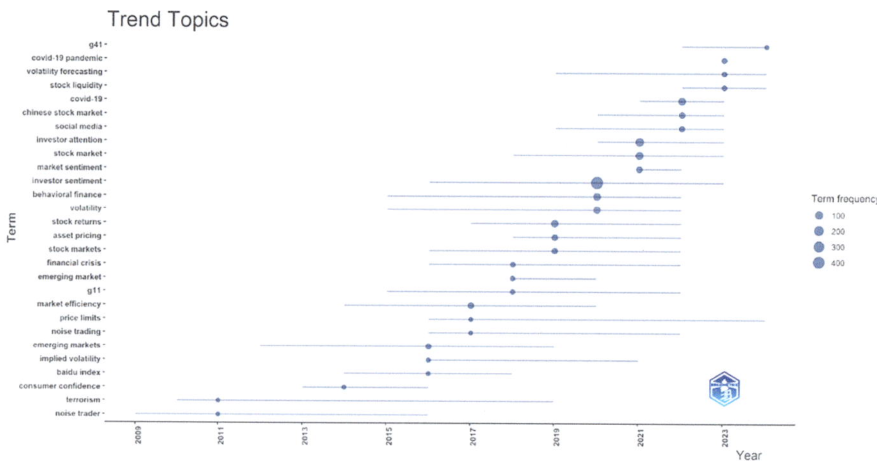

Fig. 4 Trending topics

3.7 *Thematic Map*

An examination of author keywords using co-word analysis can assist to uncover the major themes in the specific domain. A thematic map was created as presented in Fig. 5. The categorisation is determined by the levels of significance (centrality) and progress (density) along two axes. Centrality, as defined by Cobo et al. [23] is a measure that assesses the importance of a topic and its influence on the overall advancement of the whole research area. Density refers to the degree of theme's development. The cluster size represents the frequency of keyword–occurrences, and the "biblioshiny software" selects cluster labels based on the most prominent author keywords.

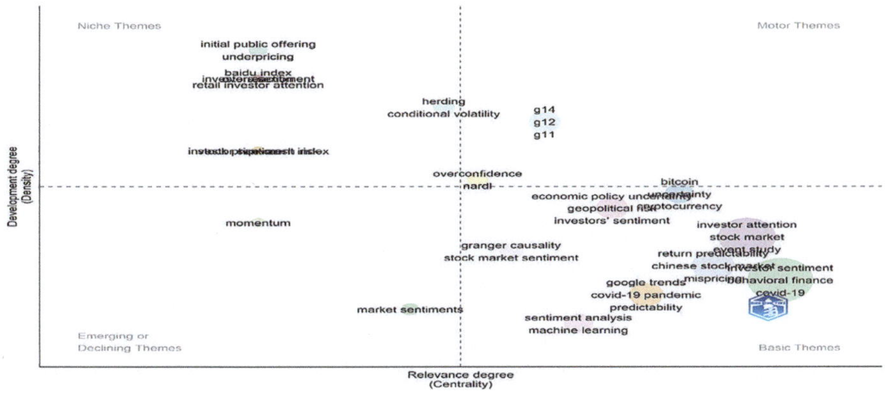

Fig. 5 Thematic map

Motor theme: First quadrants' themes exhibit a high level of advancement, centrality, and density. There are a limited number of motor themes, such as "overconfidence". Researchers found cognitive biases in literature such as overconfidence, overreaction, herding, and examined their impact on investors' decision-making [24].

Niche themes: Themes in the second quadrant have high level of density and a low centrality. These specific themes are developed highly and specialised one, yet they are rather small in comparison with the broader area. "Retail investor attention" is a prominent topic in this quadrant. Researchers have attempted to establish a connection between investor attention and the pricing and volatility of assets [3, 25]. Investor attention, as quantified by internet search volume has become a significant area of interest for researchers in asset pricing, volatility prediction, and commodity and cryptocurrency pricing [21, 26–28]. Another specialised topic is the "investor sentiment index". Moreover, Baker and Wurgler [1] asserted that the "sentiment index" is a reliable indicator for forecasting stock returns across different companies. In their 2015 study, Huang et al. [29] introduced sentiment index developed through PLS that can be used to forecast the overall return of the stock market. Researchers have tried to construct a sentiment index by employing several indicators and have sought to analyse the correlation between investor sentiment and asset performance [30].

Peripheral (emerging or declining) themes: The third quadrant encompasses nascent or waning trends. These are situated along a low-density and low-centrality axis. This quadrant encompasses topics such as "momentum" and "market sentiments". The researchers in this cluster presented evidence of the momentum anomaly in the stock market [13]. This anomaly suggests that low-momentum equities are more responsive to shifts in market sentiment. Peripheral themes are diminishing to a moderate extent due to the abundance of research accessible on this topic.

Basic or general themes: The themes associated with this quadrant are characterised by strong-centrality and low-density. Despite their lack of substantial development, they nonetheless hold significance in the field and are undergoing rapid evolution. Themes such as "economic policy uncertainty," "geopolitical risk," "sentiment analysis," "Granger causality," "Bitcoin," and "COVID-19" are often discussed but have not received significant attention. Bitcoin is capturing scholars' attention. The researchers are investigating the association between investor sentiment, stock market volatility, and bitcoin [31], illustrating how investor sentiment affects the future returns and volatility of Bitcoin. Nevertheless, the extent of this influence differs depending on microeconomic and macroeconomic factors [32]. Moreover, when there is anxiety and a pessimistic attitude prevailing in equities markets, investors tend to shift their focus towards the cryptocurrency market, resulting in an upsurge in cryptocurrency values [33]. According to Anamika et al. [31], investors show a greater inclination towards cryptocurrencies in periods of uncertainty. Moreover, literature demonstrates that sentiment-driven noisy signals are particularly prominent in bearish or bear markets resulting from geopolitical tension, health pandemics such as Covid-19 and economic uncertainty [25, 34].

3.8 Thematic Evolution

Figure 6 illustrates progression of themes that offers a comprehensive overview of the research subject across 3 time periods: 1994–2004, 2005–2014, and 2015–2023. Every node in the graph represents a cluster and is assigned a label that corresponds to the words associated with that cluster. In Fig. 6, themes within the same thematic cluster are assigned distinct colours, further, edges represent the temporal development tracks resulting from the co-occurrence of keywords across two-time slices in research topics [23]. Figure 6 presents the advancement of research and the changing trends in research themes, indicating a progressive development and gradual transition in the focus of study. During the initial stage (1994–2004), the primary emphasis of the topics centred on investor sentiment and the broader concept of information asymmetry. During the period from 2005 to 2014, the research area expanded as new themes emerged, including sentiment analysis, emotions, overreaction, and liquidity. During 2015–2023 (advanced stage), the previously developed themes converged into recent research areas including investor sentiment, investor attention, sentiment analysis, and the stock market.

4 Future Research Directions

There is growing body of evidence challenging the "Efficient Market Hypothesis (EMH)" based on observed anomalies in financial markets. Prior research highlighted studies of single-markets, leaving a research gap in literature regarding applicability

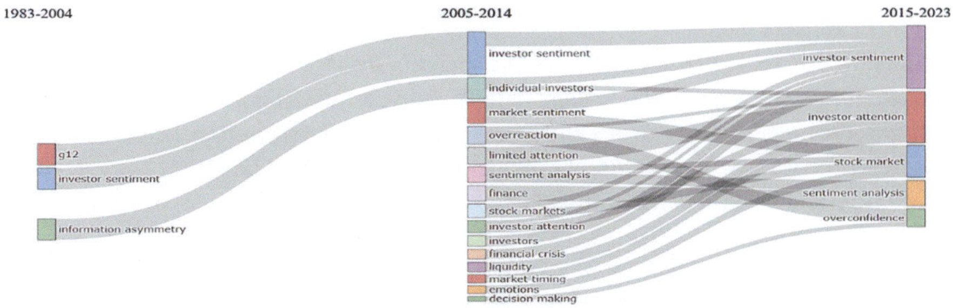

Fig. 6 Alluvial diagram of thematic evolution

of the "Adaptive Market Hypothesis (AMH)" to explain calendar-effects across multiple markets [35]. Moreover, the progression of this field has predominantly focused on stock markets in developed countries, with notable research contributions from various scholars [4, 36]. Emerging markets, aside from China, have received comparatively limited attention. This existing disparity highlights the urgent need to bridge the gap between research conducted in developed and developing economies to establish behavioural finance as a universally accepted theory [15]. Furthermore, investor sentiments in various asset classes, such as derivatives, bonds, real estate, foreign currencies, and commodities, needs to be investigated [12, 37]. There is paucity of literature based on the influence of sentiments on derivatives returns and trading volume, particularly in the context of emerging economies [35].

5 Conclusion

This topic of investor sentiment has garnered significant interest in the realm of behavioural finance. The objective of this study is to consolidate the existing state and prospective advancements in the body of knowledge about investor sentiment. The findings from the yearly scientific output reveals that this research topic has gained popularity among scholars, when Baker and Wurgler in 2007 developed a "composite sentiment index" using six specific market indicators and documented strong correlation between sentiment and future returns of stocks. As a result, there was a significant rise in scientific production trends from 2008 until the start of 2024, with an increase of 89.78%. The findings provide intriguing observations, such as influence of behavioural biases is a well-developed theme in this domain, with a large number of scholarly publications. Investor attention and investor sentiment index are niche topics that have the potential for further investigation, as there is a scarcity of studies on these topics in emerging markets. There is a significant amount of research on the topics of momentum and market emotions, however they are considered to be less important compared to other areas. Researchers and academics have focused on several emerging topics, including economic policy uncertainty, geopolitical risk, sentiment analysis, granger causality, bitcoin, and covid-19. This attention is driven by the increasing connectedness in international markets. The bibliometric output outlined in this review will offer valuable insights into the advancements in this area over the timespan of several decades. Additionally, they will guide future researchers by highlighting pertinent themes and issues for their own research endeavours. Therefore, research on investor sentiment theme is significant, and field will advance from more empirical and theoretical investigations.

Policy Implications

Understanding the importance of investor sentiment helps investors create effective trading methods. The analysis advises investors to be cautious during making short-term investing decisions, as sentiments or emotions have a higher impact in short term than long term. Regulators must play a crucial role during market crashes, financial crises, or in extreme uncertain period.

References

1. Baker M, Wurgler J (2007) Investor sentiment in the stock market. J Econ Perspect 21(2):129–151
2. Baker M, Wurgler J, Yuan Y (2012) Global, local, and contagious investor sentiment. J Financ Econ 104(2):272–287
3. Gao X, Gu C, Koedijk K (2021) Institutional investor sentiment and aggregate stock returns. Eur Financ Manag 27(5):899–924
4. Smales LA (2017) The importance of fear: investor sentiment and stock market returns. Appl Econ 49(34):3395–3421
5. Tetlock PC (2007) Giving Content to Investor Sentiment: the Role of Media in the Stock Market. J Financ 62(3):1139–1168
6. Gurdgiev C, O'Loughlin D (2020) Herding and anchoring in cryptocurrency markets: investor reaction to fear and uncertainty. J Behav Exp Financ 25:100271
7. Ni SX, Lemmon ML (2010) The Effects of investor sentiment on speculative trading and prices of stock and index options
8. Duxbury D, Wang W (2024) Investor sentiment and the risk–return relation: a two-in-one approach. Eur Financ Manag 30(1):496–543
9. Simon HA (1955) A behavioral model of rational choice. Q J Econ, 99–118
10. De Long JB, Shleifer A, Summers LH, Waldmann RJ (1990) Noise trader risk in financial markets. J Polit Econ 98(4):703–738
11. Shleifer A, Vishny RW (1997) The limits of arbitrage. J Financ 52(1):35–55
12. Aggarwal D (2022) Defining and measuring market sentiments: a review of the literature. Qual Res Financ Mark 14(2):270–288
13. Bathia D, Bredin D, Nitzsche D (2016) International sentiment spillovers in equity returns. Int J Financ Econ 21(4):332–359
14. Kumar S, Goyal N (2015) Behavioural biases in investment decision making–a systematic literature review. Qual Res Financ Mark 7(1):88–108
15. Prasad S, Mohapatra S, Rahman MR, Puniyani A (2022) Investor sentiment index: a systematic review. Int J Financ Stud 11(1):6
16. Donthu N, Kumar S, Mukherjee D, Pandey N, Lim WM (2021) How to conduct a bibliometric analysis: An overview and guidelines. J Bus Res 133:285–296
17. Aria M, Cuccurullo C (2017) Bibliometrix: an R-tool for comprehensive science mapping analysis. J Informet 11(4):959–975

18. Brown GW, Cliff MT (2004) Investor sentiment and the near-term stock market. J Empir Financ 11(1):1–27
19. Das SR, Chen MY (2007) Yahoo! for Amazon: sentiment extraction from small talk on the web. Manage Sci 53(9):1375–1388
20. Stambaugh RF, Yu J, Yuan Y (2012) The short of it: investor sentiment and anomalies. J Financ Econ 104(2):288–302
21. Da Z, Engelberg J, Gao P (2011) In search of attention. J Financ 66(5):1461–1499
22. Chen H-K, Lien C-T (2017) Market reaction to macroeconomic news: the role of investor sentiment. Asia Pac J Financ Stud 46(6):853–875
23. Cobo MJ, López-Herrera AG, Herrera-Viedma E, Herrera F (2011) An approach for detecting, quantifying, and visualizing the evolution of a research field: a practical application to the Fuzzy Sets Theory field. J Informet 5(1):146–166
24. Nigam RM, Srivastava S, Banwet DK (2018) Behavioral mediators of financial decision making–a state-of-art literature review. Rev Behav Financ 10(1):2–41
25. Smales LA (2021) Investor attention and global market returns during the COVID-19 crisis. Int Rev Financ Anal 73:101616
26. Chakraborty M, Subramaniam S (2020) Asymmetric relationship of investor sentiment with stock return and volatility: evidence from India. Rev Behav Financ 12(4):435–454
27. Da Z, Engelberg J, Gao P (2015) The sum of all FEARS investor sentiment and asset prices. Rev Financ Stud 28(1):1–32
28. Tantaopas P, Padungsaksawasdi C, Treepongkaruna S (2016) Attention effect via internet search intensity in Asia-Pacific stock markets. Pac Basin Financ J 38:107–124
29. Huang D, Jiang F, Tu J, Zhou G (2015) Investor sentiment aligned: a powerful predictor of stock returns. Rev Financ Stud 28(3):791–837
30. Chiah M, Zhong A (2021) Tuesday Blues and the day-of-the-week effect in stock returns. J Bank Finance 133:106243
31. Anamika C, M., & Subramaniam, S. (2023) Does sentiment impact cryptocurrency? J Behav Financ 24(2):202–218
32. Burggraf T, Huynh TLD, Rudolf M, Wang M (2021) Do FEARS drive bitcoin? Rev Behav Financ 13(3):229–258
33. Gaies B, Nakhli MS, Sahut JM, Guesmi K (2021) Is Bitcoin rooted in confidence? unraveling the determinants of globalized digital currencies. Technol Forecast Soc Chang 172:121038
34. Kumar S, Jain R, Balli F, Billah M (2023) Interconnectivity and investment strategies among commodity prices, cryptocurrencies, and G-20 capital markets: a comparative analysis during COVID-19 and Russian-Ukraine war. Int Rev Econ Financ 88:547–593
35. Pok WF, Kabir MH, Young M (2023) Media content, sentiment and emerging market futures returns. Appl Econ 55(7):724–749
36. Feldman T, Liu S (2017) Contagious investor sentiment and international markets. J Portf Manag 43(4):125–136
37. Maghyereh A, Abdoh H (2022) Global financial crisis versus COVID-19: evidence from sentiment analysis. Int Financ 25(2):218–248
38. Chen J, Tang G, Yao J, Zhou G (2022) Investor attention and stock returns. J Financ Quant Anal 57(2):455–484
39. Dimpfl T, Jank S (2016) Can internet search queries help to predict stock market volatility? Eur Financ Manag 22(2):171–192
40. Fama EF (1970) Efficient capital markets. J Financ *25*(2):383-417
41. Hirshleifer D, Lim SS, Teoh SH (2011) Limited investor attention and stock market misreactions to accounting information. Rev Asset Pricing Stud 1(1):35–73
42. Jiang F, Lee J, Martin X, Zhou G (2019) Manager sentiment and stock returns. J Financ Econ 132(1):126–149
43. Lee WY, Jiang CX, Indro DC (2002) Stock market volatility, excess returns, and the role of investor sentiment. J Bank Finance 26(12):2277–2299
44. Osipovich A (2020) Individual-investor boom reshapes US stock market. Wall Str J 31(2020):1–6

45. Rossi M (2015) The efficient market hypothesis and calendar anomalies: a literature review. Int J Manag Financ Account 7(3–4):285–296
46. Schmeling M (2009) Investor sentiment and stock returns: some international evidence. J Empir Financ 16(3):394–408
47. Sprenger TO, Tumasjan A, Sandner PG, Welpe IM (2014) Tweets and trades: the information content of stock microblogs. Eur Financ Manag 20(5):926–957; Xue W, He Z, Hu Y (2023) The destabilizing effect of mutual fund herding: evidence from China. Int Rev Financ Anal 88:102611

Swarm and Metallurgy-Inspired Techniques: Optimizing Job Scheduling on Uniform Parallel Machines

Rashmi Benni, Rajeshwari Belagali, Keerthi Raikar, and Shashikumar Totad

Abstract In industrial manufacturing, job scheduling on uniform parallel machines poses a significant challenge that directly impacts productivity and operational efficiency. Traditional methods often struggle with the inherent complexity dynamic and NP-hard nature of modern industrial scenarios in resource allocation. This research investigates the application of swarm and metallurgy-inspired metaheuristic techniques to enhance job scheduling efficiency. Metaheuristics, known for their adaptability and capability to find near-optimal solutions, are employed to address the limitations of conventional methods. The study aims to maximize resource utilization, minimize production time, reduce delays, and meet deadlines by optimizing job allocation on uniform parallel machines. The research employs swarm-based metaheuristic algorithms like Particle Swarm Optimization (PSO), Artificial Bee Colony (ABC) algorithm and metallurgy-based algorithm like Simulated Annealing (SA) to achieve these objectives. These algorithms are trained on various datasets under different conditions to thoroughly evaluate their performance in minimizing critical metrics such as makespan and idle time. The results indicate that the ABC algorithm provides the most effective solution in specific scenarios. This research provides essential insights and practical advice for both researchers and practitioners in the manufacturing sector, demonstrating the application of advanced optimization techniques to improve job scheduling on uniform parallel machines.

Keywords Job scheduling · Uniform parallel machines · Optimization · Metaheuristics · NP-hard problems · Swarm Intelligence Algorithms (SIA) · Particle Swarm Optimization (PSO) · Artificial Bee Colony (ABC) · Simulated Annealing (SA)

Authors Rajeshwari Belagali, Keerthi Raikar, and Shashikumar Totad contributed equally to this work.

R. Benni (✉) · R. Belagali · K. Raikar · S. Totad
KLE Technological University, Hubballi, Karnataka, India
e-mail: rashmi.benni16@gmail.com

S. Totad
e-mail: totad@kletech.ac.in

H. Sharma et al. (eds.), *Artificial Intelligence: Theory and Applications*, Lecture Notes in Networks and Systems 5589, https://doi.org/10.1007/978-981-96-1687-9_28

1 Introduction

Job scheduling is essential for maximizing production output and resource efficiency in manufacturing operations [1]. It refers to allocating a group of jobs to certain number of machines on uniform parallel machines. "NP-hard" refers to problems in job scheduling that are computationally challenging and lack of effective methods for finding optimal solutions adequately for significant instances [2]. The complexity of scheduling tasks requires the use of metaheuristic techniques, which provide feasible and near-optimal solutions [3]. The primary challenge is determining how to distribute a set of jobs across a collection of uniform machines, assumed to have similar capacities and processing speeds. The objective is to find the optimal work allocation strategy that minimizes a specific parameter, such as makespan or other performance metrics [4]. Efficient resource allocation, adjust to dynamic changes, and the capacity to deal with uncertainty are critical skills in job scheduling [5, 6]. In order to maximize system throughput, reduce delays, and optimize machine resource consumption, efficient scheduling is essential [7]. Traditional systems often struggle to meet the demands of modern industrial environments due to their complexity and limited real-time flexibility. In contrast, metaheuristics excel in these settings by efficiently exploring recent possibilities and improving upon previous solutions, making them highly suitable for complex and dynamic industrial applications [8]. This research paper investigates swarm and metallurgy-based metaheuristic optimization algorithms to enhance job allocation in industrial manufacturing, aiming to find near-optimal solutions to complex constraints. The term "nature-inspired optimization techniques" refers to metaheuristic algorithms, which are complex optimization strategies that guide the search for optimal solutions within a realistic time frame [9]. They proved particularly significant when normal methods are unable to address complex and large-scale optimization problems. Because of its versatility and flexibility, metaheuristics are well suited for complicated and dynamic scheduling contexts. In order to minimize the makespan, this article explores swarm-based ways of scheduling tasks on uniform parallel machines, such as Particle Swarm Optimization (PSO), Artificial Bee Colony (ABC), and Simulated Annealing (SA), which relies on metallurgy. These algorithms are selected based upon their ability to handle complex constraints, explore large solution spaces, and adjust to changing industrial environments [3]. They aid manufacturing businesses in achieving quicker production cycles, minimizing downtime, and optimizing resource utilization through efficient task scheduling [10]. Metaheuristic-based optimization for job scheduling on uniform parallel machines is a powerful approach to addressing the complexities of manufacturing operations [9]. These methods offer reliable, effective, and flexible ways to maximize several performance indicators, guaranteeing increased output, financial savings, and smooth operation.

The paper is structured as follows: Sect. 2 addresses the necessary background study, Sect. 3 examines the methodology, and Sect. 4 deals with outcomes and comments. The study's conclusions and future scope are depicted in Sect. 5.

2 Background and Related Work

For industrial manufacture to maximize resource usage, reduce production time, and increase overall efficiency, job scheduling on unified parallel machines is essential. By distributing tasks across machines in a way that balances workloads and minimizes idle time, efficient scheduling guarantees higher production [11]. Unfortunately, the difficulty of synchronizing several jobs with different machine capabilities, processing durations, and priority levels makes this procedure difficult. Through the imitation of natural processes and the utilization of their innate efficiency in solving problems, nature-inspired optimization techniques provide creative solutions [10]. The resilient and adaptable scheduling solutions offered by these systems can effectively manage the stochastic and dynamic nature of industrial environments [12, 13]. Heuristic-based and dynamic task scheduling techniques, like the Fuzzy Hybrid Particle Swarm Parallel Ant Colony Optimization, greatly minimize task execution latency and maximize resource utilization in cloud computing. Using a PSO-based heuristic method, parallel pathways in Service Function Chains (SFCs) minimize the Age of Information (AoI) for delay-sensitive network services, resulting in significant AoI reductions over traditional systems [14]. These developments demonstrate how important sophisticated scheduling algorithms are for streamlining intricate procedures and raising overall productivity [15].

Numerous strategies have been put forth to solve particular issues in different industries by recent breakthroughs in schedule optimization [16]. By combining a Proximal Policy Optimization (PPO) algorithm with a Markov Decision Process (MDP), the AlphaSchedule model optimizes weighted tardiness in parallel machine workshops, surpassing more conventional techniques like particle swarm optimization (PSO) and genetic algorithms (GA) [17]. A genetic algorithm-based optimized particle swarm optimization algorithm (GPSO) improves scheduling performance in silicon single crystal fabrication [18]. Optimized scheduling results in significant energy cost savings when Time-of-Use (ToU) pricing for electricity is integrated into production, particularly in batch processing systems [19]. To achieve cost minimization of electricity and the number of machines used, this study looks at a bi-objective scheduling problem under ToU pricing. A more effective mixed integer linear programming model is shown by means of the ϵ-constraint approach and CPLEX validation [20]. Makespan optimization techniques are also shown for job scheduling on parallel machines with incompatibility constraints modeled by bipartite graphs. Resource allocation efficiency in edge computing can be enhanced using a hybrid metaheuristic that combines Gray Wolf Optimization (GWO) and Particle Swarm Optimization (PSO) [21]. An enhanced Ant Lion optimization approach for provider charges and resource allocation for vehicle cloud computing (VCC) exhibits better simulated results [22]. These works demonstrate important developments in resource allocation and scheduling from a variety of technology fields [23].

3 Proposed Methodology

This article proposes a research study that contrasts several metaheuristic techniques for job scheduling on uniform parallel machines: Particle Swarm Optimization (PSO), Artificial Bee Colony (ABC), and Simulated Annealing (SA). Through meticulous calibration of the hyper-parameters over an iterative process, the study seeks to increase scheduling performance and decrease makespan. Additionally, resource allocation and scheduling performance are optimized by the application of sophisticated optimization techniques.

3.1 System Design

The proposed design Fig. 1 starts by determining m uniform parallel machines and n jobs, each with a processing time. To optimize scheduling, metaheuristic algorithms like Particle Swarm Optimization (PSO), Simulated Annealing (SA), and Artificial Bee Colony (ABC) are used. These algorithms, initialized with tuned hyperparameters, balance exploration and exploitation of the solution space. Job parameters and hyperparameters are set to begin the workflow. The algorithms iteratively update job-machine allocations to enhance the solution. The fitness of solutions is frequently

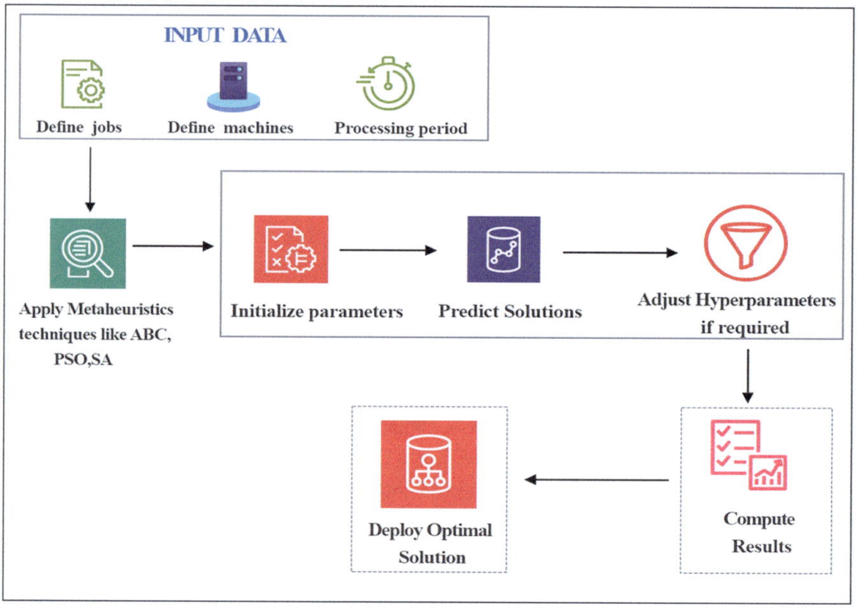

Fig. 1 System design

evaluated to ensure minimal machine time or optimized utilization. Upon convergence, the optimal solution runs job schedules on all machines, ensuring resource utilization and efficiency in industrial environments. This approach leverages swarm and metallurgy-based metaheuristic algorithms to address scheduling complexities on uniform parallel machines, deploying robust solutions for manufacturing operations.

3.2 Mathematical Representation Uniform Parallel Machines

Objective: Reduce the total completion time $C_{\max}$ (maximum completion time) Given:

- $M = \{M_1, M_2, \ldots, M_m\}$ (collection of machines)
- $J = \{J_1, J_2, \ldots, J_n\}$ (count of jobs)
- p_j (jobs J_j 's processing time)
- C_j (jobs J_j 's completion time)
- x_{ij} (binary variable, If machine M_i is allocated to job J_j, then 1; otherwise, 0).

Constraints

1. Each job is allocated to a single machine:

$$\sum_{i=1}^{m} x_{ij} = 1 \quad \forall j \in J \tag{1}$$

2. Each machine is able to handle a single job at a time:

$$\sum_{j=1}^{n} x_{ij} p_j \leq C_{\max} \quad \forall i \in M \tag{2}$$

3. The definition of each job's completion time:

$$C_j = \sum_{i=1}^{m} x_{ij}(s_{ij} + p_j) \tag{3}$$

4. The maximum of all completion times is the makespan:

$$C_{\max} \geq C_j \quad \forall j \in J \tag{4}$$

5. Binary variables constraints:

$$x_{ij} \in \{0, 1\} \quad \forall i \in M, \forall j \in J \tag{5}$$

6. Non-preemptive scheduling (a job is scheduled to run continuously from the moment it begins):

$$s_{ij} \geq 0 \quad \forall i \in M, \forall j \in J \tag{6}$$

3.3 Algorithmic Representation of ABC, PSO, and SA

Advanced optimization strategies, such as metaheuristic algorithms, are intended to identify nearly optimal solutions for challenging issues when conventional approaches prove to be ineffective. Large solution spaces are explored by these algorithms through the utilization of guided search techniques and randomization. They are particularly useful in NP-hard problem-solving, making them applicable to a wide range of domains like artificial intelligence, logistics, and engineering (Fig. 2).

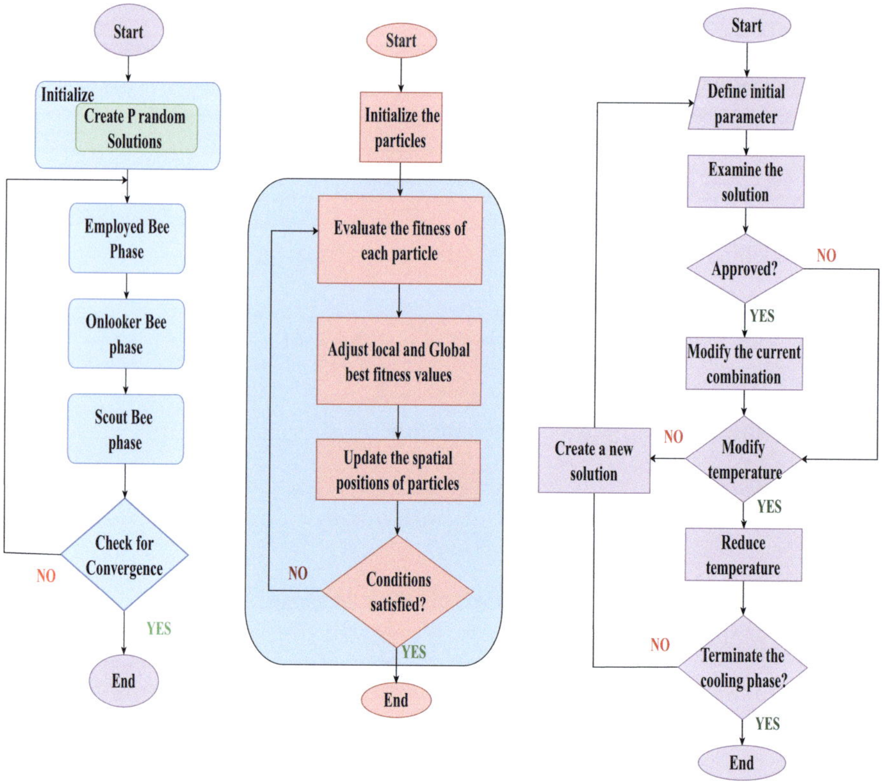

Fig. 2 Algorithmic flow charts of metaheuristics ABC, PSO and SA

3.4 Artificial Bee Colony Algorithm

An optimization technique known as the Artificial Bee Colony (ABC) algorithm employs a metaheuristic approach was developed with inspiration from honeybee foraging behavior. Bees working in ABC present solutions to observer bees, who make a probabilistic choice based on preference for better possibilities. In order to preserve diversity and avoid early convergence, scout bees periodically investigate new ideas. ABC is able to efficiently identify the optimum or nearly optimal solutions for a variety of optimization issues because to its adaptable method. ABC optimizes job assignment, minimizes makespan, guarantees maximum machine usage, and meets job precedence requirements for assigning jobs to uniform parallel machines. Based on factors like job limits and the overall completion time, each bee indicates a possible work plan. By using worker bees to make small adjustments to task assignments or machine assignments to increase output, the program optimizes schedules.

1. **Initialization**
   Create a population of N solutions (food sources), $\{X_i\}_{i=1}^N$, randomly throughout the search area.

$$X_i = \{x_{i1}, x_{i2}, \ldots, x_{id}\}$$

2. **Employed Bees Phase**
   To come up with a new solution V_i for each employed bee, update the solution X_i:

$$V_{ij} = X_{ij} + \phi_{ij}(X_{ij} - X_{kj})$$

   j represents dimension index, k is an index chosen at random, and ϕ_{ij} represents a random number inside the interval [− 1, 1].
3. **Fitness Evaluation**
   Calculate new solution's fitness $f(X_i)$. Modify X_i to V_i if $f(V_i)$ is favorable to $f(X_i)$.
4. **Onlooker Bees Phase**
   Evaluate the possibility p_i of choosing each possible outcome according to its fitness:

$$p_i = \frac{f(X_i)}{\sum_{j=1}^{N} f(X_j)}$$

   Using the same adjusting rule of the employed bees, onlooker bees generate new solutions and choose solutions depending on p_i.
5. **Scout Bees Phase**
   If the current result X_i exceeds a specified limit, the most recent randomly generated outcome should be chosen instead.
6. **Termination**
   Repeat the processes until a stopping condition is achieved.

3.5 Particle Swarm Optimization Algorithm

PSO, or particle swarm optimization, is a computational method inspired by the social behaviors of fish and birds. To discover an optimal or nearly optimal solution to a given problem, it entails a collection of candidate solutions, known as particles, moving in a search space and modifying their positions based on their own and surrounding particles' experiences. PSO optimizes task allocation in job scheduling on uniform parallel machines in order to reduce makespan, or total completion time. With positions representing tasks, a potential solution exists in every particle. Particle movement is guided through personal and global best-known positions, which are iteratively improved by varying particle velocity. Solution quality is determined by fitness evaluation, which updates global and personal best-known positions correspondingly. PSO converges to an optimal or nearly optimal solution through iterative refining. It improves system performance and resource consumption by assigning jobs to machines in an efficient manner.

1. **Position Update**: A potential schedule is denoted by each particle i, and at each iterative step t, its position $\mathbf{x}_i(t)$ is changed according to its velocity $\mathbf{v}_i(t)$:

$$\mathbf{x}_i(t+1) = \mathbf{x}_i(t) + \mathbf{v}_i(t+1)$$

2. **Velocity Update**: The velocity of each particle is determined by its global $\mathbf{g}$ and local best position $\mathbf{p}_i$:

$$\mathbf{v}_i(t+1) = \omega \mathbf{v}_i(t) + c_1 r_1 (\mathbf{p}_i - \mathbf{x}_i(t)) + c_2 r_2 (\mathbf{g} - \mathbf{x}_i(t))$$

   while c_1 and c_2 represent the cognitive and behavioral coefficients, r_1 and r_2 in the range [0, 1] represent arbitrary numbers, and ω represents the inertia weight.
3. **Fitness Evaluation**: The schedule $\mathbf{x}_i$ is evaluated for efficiency using the fitness function $f(\mathbf{x}_i)$, by considering dynamic industrial constraints.
4. **Update Personal and Global Best**: If the current position improves, each particle updates its own best $\mathbf{p}_i$:

$$\mathbf{p}_i = \mathbf{x}_i \text{ if } f(\mathbf{x}_i) < f(\mathbf{p}_i)$$

   The optimal location among all particles is modified to the global best $\mathbf{g}$:

$$\mathbf{g} = \mathbf{x}_j \text{ where } f(\mathbf{x}_j) = \min(f(\mathbf{x}_k)) \text{ for all } k$$

3.6 Simulated Annealing Algorithm

A probabilistic optimization technique called "simulated annealing" was developed to locate global optima by examining solution spaces and occasionally accepting less-than-ideal solutions in order to avoid local optima traps. It was inspired by metallurgical annealing procedures. The technique starts with a random starting schedule where machines are assigned different quantities of jobs in the context of job scheduling on uniform parallel machines. Reducing the makespan—that is, the overall burden on all machines—is the goal. To attain different timelines, jobs can be moved or switched across computers; the Metropolis criterion determines whether these changes are accepted. The likelihood of accepting subpar solutions rapidly drops as the algorithm runs. This iterative process keeps going until a predetermined end point is reached, either hitting a minimum temperature threshold or finishing with a predetermined amount of repetitions. Simulated annealing efficiently maximizes job completion times through balanced workload distribution among machines by methodically lowering the probability of accepting less desirable results and putting in place an efficient cooling schedule.

- **Initialization**
  - Consider a preliminary solution X, that represents a schedule of jobs on machines.
  - Initially set a temperature T_{start} and a cooling rate of α.
- **Generate Neighbor Solution**
  - To create a neighboring solution X', modify the existing solution X.
- **Objective Function**
  - Minimize the total completion time $f(X)$.
- **Acceptance Probability**
  - Examine the alteration in the value of the objective function $\Delta E = f(X') - f(X)$. With probability, accept the neighboring solution X' and current temperature T.

$$P(X \to X') = \begin{cases} 1 & \text{if } \Delta E < 0 \\ \mathrm{e}^{-\frac{\Delta E}{T}} & \text{if } \Delta E \geq 0 \end{cases}$$

- **Cooling Schedule**
  - By using a cooling schedule $T = T_{\text{start}} \cdot \alpha^k$, where k indicates the iteration number, modify the temperature T.
- **Termination**
  - Unless a halting condition is met rerun steps 2–5.

4 Results and Discussion

In this computational experiment, we compare and evaluate the effectiveness of three well-known optimization techniques for minimizing the makespan of a scheduling problem: Particle Swarm Optimization (PSO), Simulated Annealing (SA), and Artificial Bee Colony (ABC). The search for the algorithm that allocates tasks across different numbers of jobs and machines in the most effective way is the objective. In this computational experiment, we aim to comprehensively evaluate and compare the effectiveness of three prominent optimization techniques in minimizing the makespan of a scheduling problem. The techniques under scrutiny are Particle Swarm Optimization (PSO), Simulated Annealing (SA), and Artificial Bee Colony (ABC). The primary objective is to identify the algorithm that optimally allocates tasks across varying numbers of jobs and machines.

The experiment focuses on assessing how these algorithms perform in optimizing scheduling across a diverse range of job and machine configurations. By examining their respective abilities to minimize makespan—the total time required to complete all tasks—the study aims to provide insights into which approach offers superior performance under different scheduling scenarios. Through rigorous analysis and empirical evaluation, we seek to uncover the strengths and weaknesses of each optimization technique in handling complex scheduling problems. Ultimately, this research aims to contribute valuable insights into selecting the most effective algorithm for real-world scheduling applications, where efficient task allocation is critical to operational success.

Table 1 involved a comparative analysis of the ABC (Artificial Bee Colony), PSO (Particle Swarm Optimization), and SA (Simulated Annealing) algorithms for the scheduling of 12 jobs on 5 machines. The primary metric evaluated was the "Best Fitness" value, where lower values indicate superior solutions. ABC demonstrated robust performance across different configurations. For instance, Figs. 3, 4, and 5 show consistently achieved competitive fitness values of 119 (20 iterations, 50

Table 1 Computational experiment I

Algorithm	Iterations	Population	Jobs	Machines	Best fitness
ABC	20	50	12	5	119
PSO	20	50	12	5	120
SA	20	50	12	5	215
ABC	50	100	12	5	114
PSO	50	100	12	5	120
SA	50	100	12	5	170
ABC	100	200	12	5	110
PSO	100	200	12	5	115
SA	100	200	12	5	135

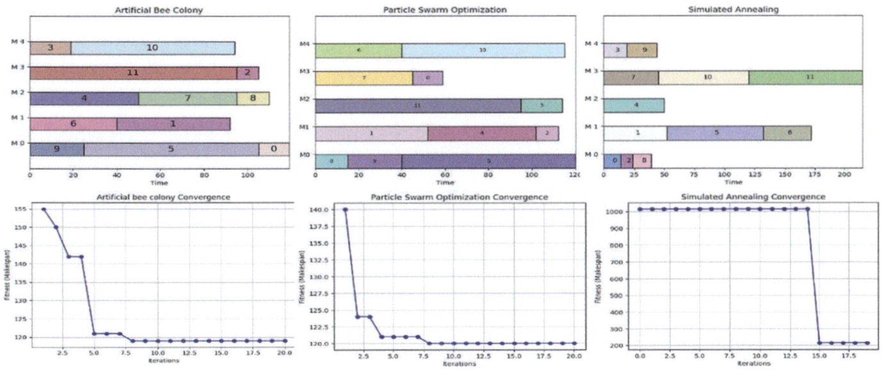

Fig. 3 Test case 1: 20 iterations and 50 population

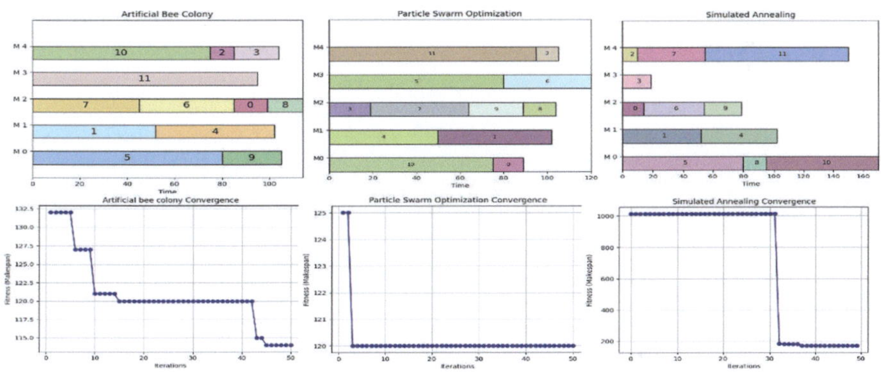

Fig. 4 Test case 2: 50 iterations and 100 population

population), 114 (50 iterations, 100 population), and 110 (100 iterations, 200 population). This consistency underscores ABC's effectiveness in finding near-optimal solutions for the scheduling problem. PSO also performed well, albeit with slightly higher fitness values compared to ABC. It achieved fitness evaluations of 120 (20 iterations, 50 population), 120 (50 iterations, 100 population), and 115 (100 iterations, 200 population). While PSO was competitive, its performance was marginally less efficient than ABC in this scenario. On the other hand, SA generally yielded higher fitness values, indicating less effective optimization for this specific scheduling problem. It recorded fitness values of 215 (20 iterations, 50 population), 170 (50 iterations, 100 population), and 135 (100 iterations, 200 population). SA's higher fitness values suggest it struggled more to achieve optimal schedules compared to ABC and PSO.

In Table 2, where the population size is fixed at 50 and iterations are set to 50, while varying job numbers (20, 30, and 50) on 5 machines, the performance of different algorithms—ABC, PSO, and SA—reveals distinct strengths and weaknesses.

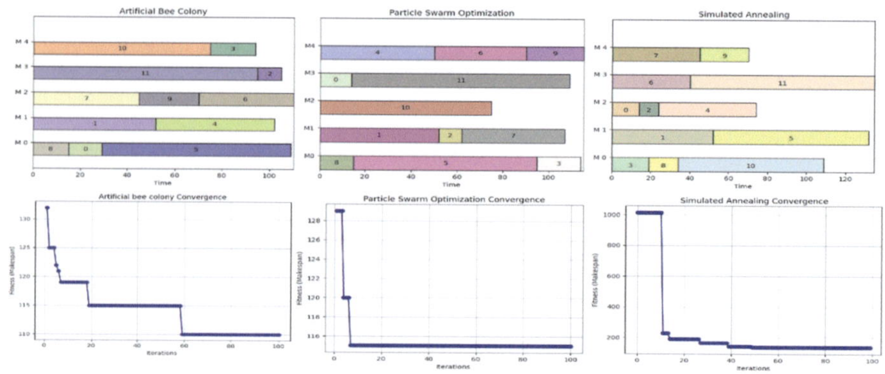

Fig. 5 Test case 3: 100 iterations and 200 population

Table 2 Computational experiment II

Algorithm	Iterations	Population	Jobs	Machines	Best fitness
ABC	50	50	20	5	180
PSO	50	50	20	5	192
SA	50	50	20	5	205
ABC	50	50	30	5	348
PSO	50	50	30	5	359
SA	50	50	30	5	397
ABC	50	50	50	5	430
PSO	50	50	50	5	545
SA	50	50	50	5	557

Figures 6, 7, and 8 illustrate these findings. Ant Colony Optimization (ABC) consistently demonstrates its strength in minimizing makespan values across all scenarios. For instance, when dealing with 20 jobs and 5 machines, ABC achieves a remarkable makespan of 180, outperforming PSO at 192 and SA at 205. This trend continues as the job complexity increases: with 30 jobs, ABC records makespans of 348 and 430, whereas PSO shows 359 and 545, and SA 397 and 557 for the same configurations. ABC is noted for its decentralized, collaborative approach that fosters robust performance in finding near-optimal solutions. This methodical process typically results in reliable convergence and competitive outcomes, exemplified by ABC achieving makespan values of 180, 348, and 430 for 20, 30, and 50 jobs respectively, on 5 machines. However, ABC's sensitivity to parameter settings like colony size and trial limits without improvement necessitates careful tuning for consistent optimal performance. In contrast, PSO utilizes velocity vectors to navigate well-structured solution spaces effectively but may struggle with premature convergence, requiring fine-tuning of inertia weight and acceleration coefficients. SA employs temperature-based probabilistic transitions to explore diverse solution regions, beneficial for

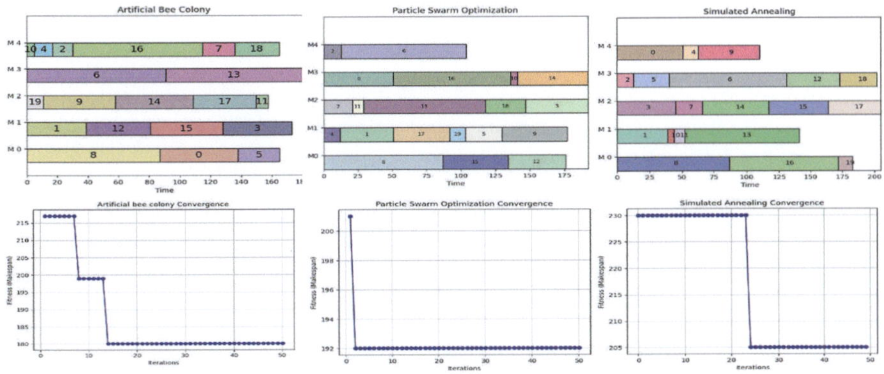

Fig. 6 Test case 2: 20 jobs with constant iterations population and machines

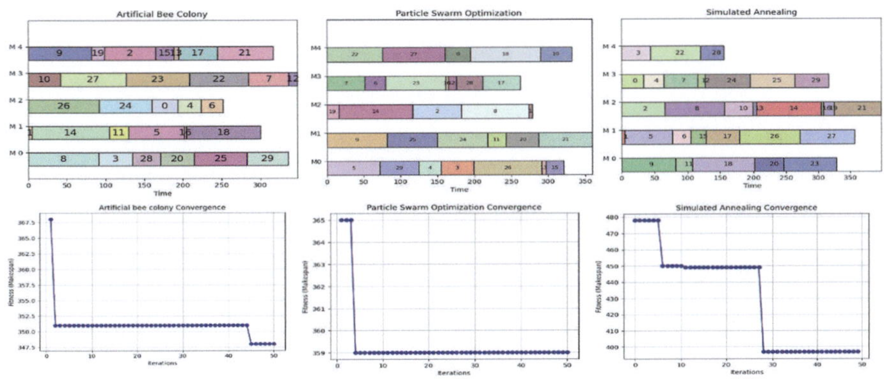

Fig. 7 Test case 2: 30 jobs with constant iterations population and machines

complex, non-linear problems, though it demands meticulous adjustment of cooling schedule and initial temperature. In conclusion, while ABC proves advantageous for minimizing makespan values in this specific scheduling optimization problem due to its effective exploration and exploitation mechanisms, its performance nuances with respect to problem complexity and parameter settings highlight the importance of algorithm selection based on problem specifics and desired optimization outcomes.

Table 3 involved comparing three optimization algorithms—Simulated Annealing (SA), Particle Swarm Optimization (PSO), and Artificial Bee Colony (ABC)—applied to a scheduling problem with varying numbers of jobs (3, 5, and 12). Each algorithm was run for 20 iterations with a population size of 10. The goal was to reduce the makespan, where lower values indicate better solutions. Across all configurations (3, 5, and 12 jobs), ABC consistently outperformed both PSO and SA in terms of achieving optimal makespan values. The results, depicted in Figs. 9, 10, and 11, consistently show ABC outperforming both PSO and SA. For instance, ABC obtained makespan values of 180, 129, and 95 for scheduling with 3, 5, and 12 jobs,

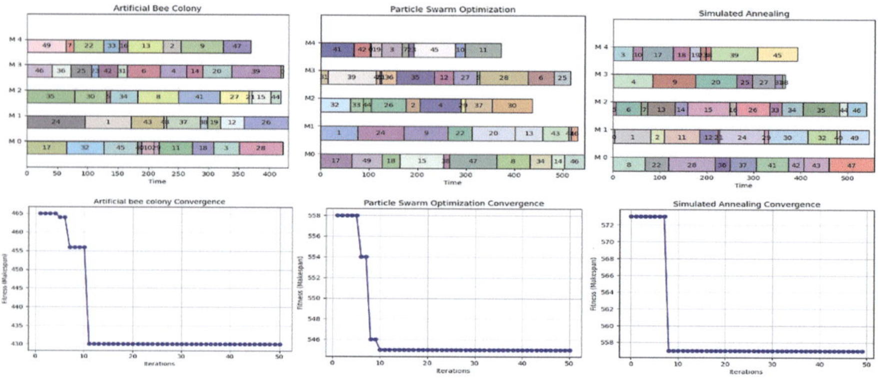

Fig. 8 Test case 2: 50 jobs with constant iterations population and machines

Table 3 Computational experiment III

Algorithm	Iterations	Population	Jobs	Machines	Best fitness
ABC	20	10	12	3	180
PSO	20	10	12	3	190
SA	20	10	12	3	227
ABC	20	10	12	5	129
PSO	20	10	12	5	134
SA	20	10	12	5	200
ABC	20	10	12	7	95
PSO	20	10	12	7	97
SA	20	10	12	7	147

respectively. In contrast, SA consistently produced larger makespan values of 227, 200, and 147, while PSO showed competitive but generally higher makespan values of 190, 134, and 97 for the same job configurations. These results indicate that ABC is the most effective algorithm among the three for minimizing makespan in this scheduling problem scenario. Across various configurations of iterations, population sizes, number of jobs, and machines, the algorithms exhibit distinct behaviors in achieving optimal scheduling solutions. In Computational Experiment I, ABC consistently demonstrates competitive performance across different parameter settings, while PSO and SA vary in effectiveness based on iteration and population size. Computational Experiment II reveals ABC continued to showcase its efficacy and PSO can generally achieve better results with larger job sets, whereas SA struggles to maintain competitiveness in more complex scenarios. Finally, Computational Experiment III underscores the impact of machine diversity on algorithmic performance, with ABC often outperforming PSO and SA in scenarios involving varied machines. These findings underscore the nuanced interplay between algorithmic parameters and problem complexity, offering valuable insights for leveraging swarm and metallurgy-inspired

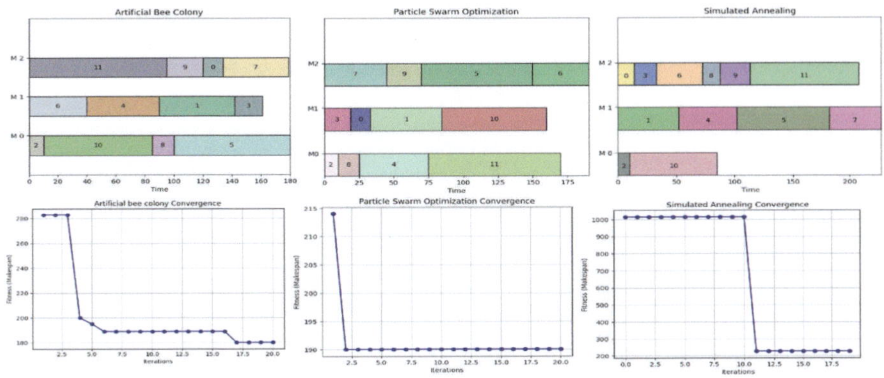

Fig. 9 Test case 3: 3 machines with constant iterations population and jobs

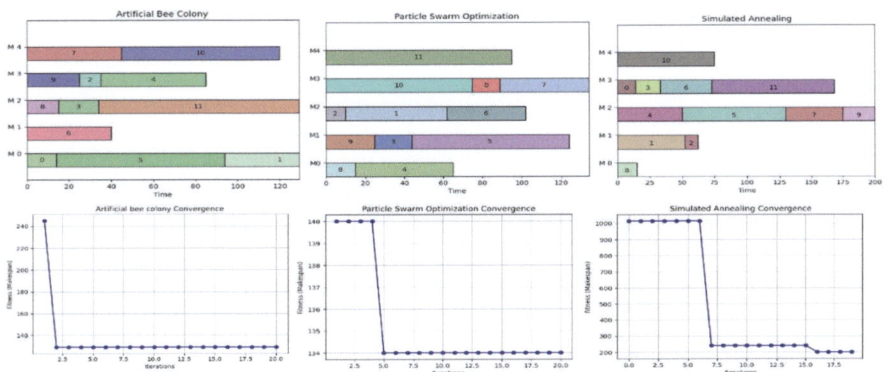

Fig. 10 Test case 3: 5 machines with constant iterations population and jobs

techniques in real-world job scheduling applications. Further exploration into adaptive strategies and hybrid approaches could enhance these algorithms' efficacy in addressing diverse manufacturing challenges.

5 Conclusion and Future Scope

In conclusion, this study demonstrates the superior and consistent performance of the ABC algorithm across various job scheduling settings on uniform parallel machines. ABC excels with diverse machines, achieving significant reductions in makespan, maximizing resource utilization, and minimizing machine idle time. PSO shows variable effectiveness, performing well with larger job sets, while SA struggles in complex contexts. Fine-tuning hyperparameters improved both fitness and computational efficiency, underscoring the efficacy of metaheuristic methods, particularly ABC,

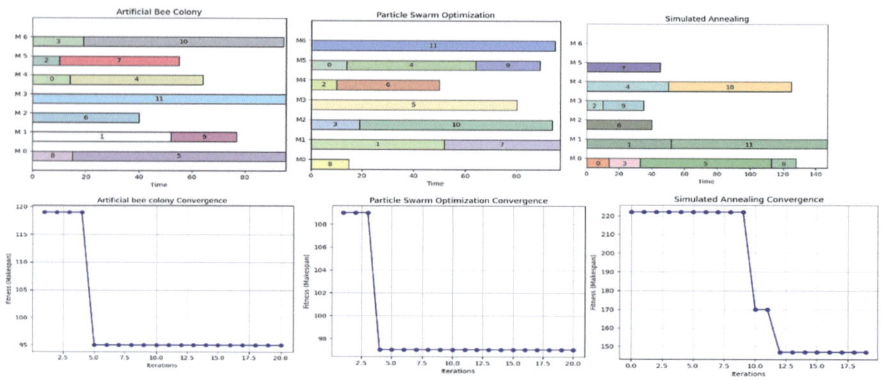

Fig. 11 Test case 3: 7 machines with constant iterations population and jobs

in optimizing manufacturing operations. Future research could delve deeper into exploring adaptive algorithms that adjust parameters dynamically based on problem complexity could enhance performance across diverse industrial manufacturing settings. Additionally, exploring hybrid algorithms that combine strengths from ABC, PSO, and SA could potentially offer superior solutions, paving the way for more efficient and adaptable scheduling systems in industrial settings.

Acknowledgements I intended to express my appreciation to the University and everyone who contributed toward making this research article possible. I want to extend my gratitude to my supervisor for guidance and support throughout this project. Finally, I want to thank my fellow writers for their instructive discussions and insightful feedback.

References

1. Li J, Deng H, Liu D, Song C, Han R, Hu T (2021) A job shop scheduling method based on ant colony algorithm. In: 2021 IEEE international conference on progress in informatics and computing (PIC), pp 453–457. https://doi.org/10.1109/PIC53636.2021.9687078
2. Zhu Z, Gao Z, Peng J, Du X (2021) Optimal scheduling of shared vehicles based on improved ant colony algorithm. In: 2021 IEEE international conference on data science and computer application (ICDSCA), pp 319–323. https://doi.org/10.1109/ICDSCA53499.2021.9650102
3. Benni R, Totad S, Karibasappa KG, Karadgi S (2023) Search space optimization for autonomous mobile robots using meta-heuristics. In: 2023 14th international conference on computing communication and networking technologies (ICCCNT), pp 1–8. https://doi.org/10.1109/ICCCNT56998.2023.10308017
4. Saidy HN, Ilham AA, Syafaruddin (2022) Elitist genetic algorithm and elitist ant colony optimization on resource scheduling in field cloud manufacturing. In: 2022 2nd international seminar on machine learning, optimization, and data science (ISMODE), pp 403–408. https://doi.org/10.1109/ISMODE56940.2022.10180983
5. Numidia Z, Sylia Z, Zakari D, Lamine B (2023) Allocation resource using new intelligent algorithm: discrete particle swarm optimization (DPSO). In: 2023 international conference on

electrical engineering and advanced technology (ICEEAT), vol 1, pp 1–5. https://doi.org/10.1109/ICEEAT60471.2023.10426443

6. Ebrahim ZH, Manaa ME (2023) Fog-based resource allocation hybrid approach using meta-heuristic for mobile networks. In: 2023 6th international conference on engineering technology and its applications (IICETA), pp 408–413. https://doi.org/10.1109/IICETA57613.2023.10351347
7. Kansara AS, Gohil BN, Verma N (2023) Task scheduling in cloud computing using mean ant colony optimization. In: 2023 6th international conference on information systems and computer networks (ISCON), pp 1–8. https://doi.org/10.1109/ISCON57294.2023.10112102
8. Bhetiwal S, Misra SK (2023) Survey on task scheduling with ant colony optimization. In: 2023 third international conference on secure cyber computing and communication (ICSCCC), pp 690–696. https://doi.org/10.1109/ICSCCC58608.2023.10176927
9. Zhang X, Zhang X, Wang W (2023) Evolutionary computing. Springer, Singapore, pp 173–219. https://doi.org/10.1007/978-981-99-6449-9_5
10. Singh B, Murugaiah M (2024) Bio-inspired computing and associated algorithms. Springer, Singapore, pp 47–87. https://doi.org/10.1007/978-981-97-1017-1_3
11. Benni R (2023) A comparative study of evolutionary and swarm intelligence algorithms for job scheduling on identical parallel machines. Tuijin Jishu/J Propul Technol 44:3734–3745. https://doi.org/10.52783/tjjpt.v44.i4.1531
12. Jayasudha V, Nazrin Salma S, Saravanan V, Nivedha M, Vinuja G (2023) Enhanced antlion optimization based resource allocation in vehicular cloud environment. In: 2023 international conference on integrated intelligence and communication systems (ICIICS), pp 1–5. https://doi.org/10.1109/ICIICS59993.2023.10421066
13. Wu J, Dai H, Wang Y, Tu Z (2022) Multi-scenario bimetric-balanced IoT resource allocation: an evolutionary approach. In: 2022 IEEE 24th international conference on high performance computing & communications; 8th international conference on data science & systems; 20th international conference on smart city; 8th international conference on dependability in sensor, cloud & big data systems & application (HPCC/DSS/SmartCity/DependSys), pp 428–437. https://doi.org/10.1109/HPCC-DSS-SmartCity-DependSys57074.2022.00086
14. Parameshachari BD, Usha M, Shwetha N, Mohan BR (2023) Scheduling the task of user in cloud computing using hybrid procedure of PSO and lion algorithm. In: 2023 IEEE international conference on integrated circuits and communication systems (ICICACS), pp 1–6. https://doi.org/10.1109/ICICACS57338.2023.10100224
15. Kaliappan S, Paranthaman V, Kamal MDR, Sudhakar AVV, Muthukannan M (2024) A novel approach of particle swarm and ant colony optimization for task scheduling in cloud. In: 2024 14th international conference on cloud computing, data science & engineering (confluence), pp 272–278. https://doi.org/10.1109/Confluence60223.2024.10463398
16. Houssein EH, Gad AG, Wazery YM, Suganthan PN (2024) Corrigendum to "task scheduling in cloud computing based on meta-heuristics: review, taxonomy, open challenges, and future trends" [Swarm and Evolutionary Computation, volume 62 (April 2021), 100841]. Swarm Evol Comput 101647. https://doi.org/10.1016/j.swevo.2024.101647
17. Wang JH, Luo PC, Xiong HQ, Zhang BW, Peng JY (2020) Parallel machine workshop scheduling using the integration of proximal policy optimization training and Monte Carlo tree search. In: 2020 Chinese automation congress (CAC), pp 3277–3282. https://doi.org/10.1109/CAC51589.2020.9327564
18. Kang L, Liu D, Wu Y, Zhao Y (2020) Scheduling optimization of silicon single crystal production process based on improved particle swarm algorithm. In: 2020 Chinese automation congress (CAC), pp 3894–3898. https://doi.org/10.1109/CAC51589.2020.9326597
19. Yu L, Qu L (2022) AoI aware VNF scheduling with parallel transmission for delay-sensitive applications. In: 2022 IEEE 24th international conference on high performance computing & communications; 8th international conference on data science & systems; 20th international conference on smart city; 8th international conference on dependability in sensor, cloud & big data systems & application (HPCC/DSS/SmartCity/DependSys), pp 1382–1387. https://doi.org/10.1109/HPCC-DSS-SmartCity-DependSys57074.2022.00214

20. Cheng J, Cheng J, Chu F (2022) Bi-objective optimization for uniform parallel batch machine scheduling under time-of-use tariffs. In: 2022 IEEE international conference on networking, sensing and control (ICNSC), pp 1–6. https://doi.org/10.1109/ICNSC55942.2022.10004129
21. Pikies T, Furmanczyk H (2022) Scheduling on uniform and unrelated machines with bipartite incompatibility graphs. In: 2022 IEEE international parallel and distributed processing symposium (IPDPS), pp 661–671. https://doi.org/10.1109/IPDPS53621.2022.00070
22. Lu S, Cheng J (2023) Bi-objective uniform parallel machine scheduling considering outsourced manufacturing and order financing. In: 2023 19th international conference on natural computation, fuzzy systems and knowledge discovery (ICNC-FSKD), pp 1–7. https://doi.org/10.1109/ICNC-FSKD59587.2023.10280808
23. Nain A, Sheikh S, Malik R, Mushtaq SU (2023) Hybrid metaheuristic based resource allocation approach in SDN-based edge computing environment. In: 2023 2nd international conference on futuristic technologies (INCOFT), pp 1–5. https://doi.org/10.1109/INCOFT60753.2023.10425524

Classification of Buried IEDs and Buried IED Decoys Using Convolutional Neural Networks on Ground Images from Thermal Cameras Embarked on UAVs

Issaka Daouèga, Sadouanouan Malo, and T. Frédéric Ouédraogo

Abstract Improvised explosive devices (IEDs) are used by Armed Terrorist Groups (GATs) to hinder the progress of the military forces. The burial of IEDs causes disturbances to the ground surface perceptible by thermal sensors. However, it happens that GATs dig the ground without burying an object in order to just simulate the presence of an IED. We refer to this as a decoy throughout this work. Environmental factor can also create similar ground disturbances. In this paper, disregarding the detection aspect, we propose an approach based on convolutional neural networks (CNNs) to distinguish a decoy from a buried object. Our approach uses VGG, Inception, ResNet, and ViT architectures. The model selection was based on a comparison of metrics over the training, validation, and testing steps. The best results, with accuracy rates of 88% and 95% during the validation phase, were achieved using respectively ResNet-18 and VGG16 with batch normalization and stride subsampling.

Keywords Buried IED · Thermal image · VGG16 · Batch normalization · Stride downsampling

1 Introduction

In the asymmetric warfare waged by GAT in the Sahel, fully buried IEDs are used to slow troop progress. The special feature of this type of IED is that it is not perceptible. It may leave traces similar to others. This makes them difficult to identify. The article [1] shows that the thermal signatures of disturbed soils hosting different types of objects or hosting no objects at all can be similar with a shallow CNN and this confusion matrix (Fig. 1), extracted from this study, shows the difficulty of classifying

I. Daouèga (✉) · T. F. Ouédraogo
Laboratoire de Mathématiques, Informatiques et Applications (LAMIA), Ecole Doctorale Sciences et Technologies, Université Norbert Zongo, Koudougou, Burkina Faso
e-mail: issaka.daouega@unz.bf

S. Malo
Université Nazi Boni, Bobo-Dioulasso, Burkina Faso

H. Sharma et al. (eds.), *Artificial Intelligence: Theory and Applications*, Lecture Notes in Networks and Systems 5589, https://doi.org/10.1007/978-981-96-1687-9_29

these different points of interest. In another study, which looked at the classification, using VGG11, of intentionally disturbed soils and ordinary soils or soils disturbed by an environmental phenomenon, a discrimination performance of approximately 97% accuracy was observed. But this does not prevent confusion between a hole containing an IED and a decoy. Indeed, analysis based on object segmentation or contour detection offers similar characterizations of these thermal images. However, according to the definition [2] of an IED, one of its aims is to create a diversion. From this point of view, even a dummy object is a good way to distract, hence the possibility for terrorists to create empty holes resembling IED holes in this purpose. In addition, other environmental phenomena can cause ground disturbance. This confuses human observation as well as optical and ultraviolet sensors, all of which rely on the different granularity [3–5] of the ground in the ordinary environment. In view of the context of the study, beyond the detection of the point of interest, discrimination becomes a major issue to avoid contributing to enemy action through false positives, which are detrimental in the field of military tactics.

However, the literature on this subject is not very extensive. A glance at "DataBase systems and Logic Programming" DBLP Computer Science Bibliography reveals that eight articles are listed for the "buried landmine" and "buried mine" indexes from 2010 to 2023. For the "Buried Explosive" index, only seven articles are recorded during the same period. It should also be noted that the tool of choice for studies in this field is ground-penetrating radar (GPR). We therefore note that the subject of totally buried IEDs or the use of thermal images to discriminate them are not widely addressed. Consequently, the specific subject of decoy discrimination and fully buried IEDs, using thermal images, is a challenge. So the aim at this stage of our study is to design a classification model between holes containing IEDs and decoys through a selection of CNN and vision transformer (ViT) models. The main contribution of this article concerns:

- the creation of a dataset of thermal images of holes containing a specific object and of empty holes;
- adaptation of VGG16 for classification of buried IEDs and decoys;
- The finding that using batch normalization improves VGG16 and VGG19 and the use of stride for downsampling instead of pooling is of interest.

The rest of the paper is organized as follows: In Sect. 2, we present related work. In Sect. 3, we introduce our approach and the data used. Section 4 includes the presentation of our results, followed by a discussion, which concludes with the perspectives of our work.

2 Related Work

The major difficulty with completely buried mines lies in the fact that no part of them can be observed and characterized. This necessitates exploring the traces generated by their installation or employing geophysical tools such as metal detectors

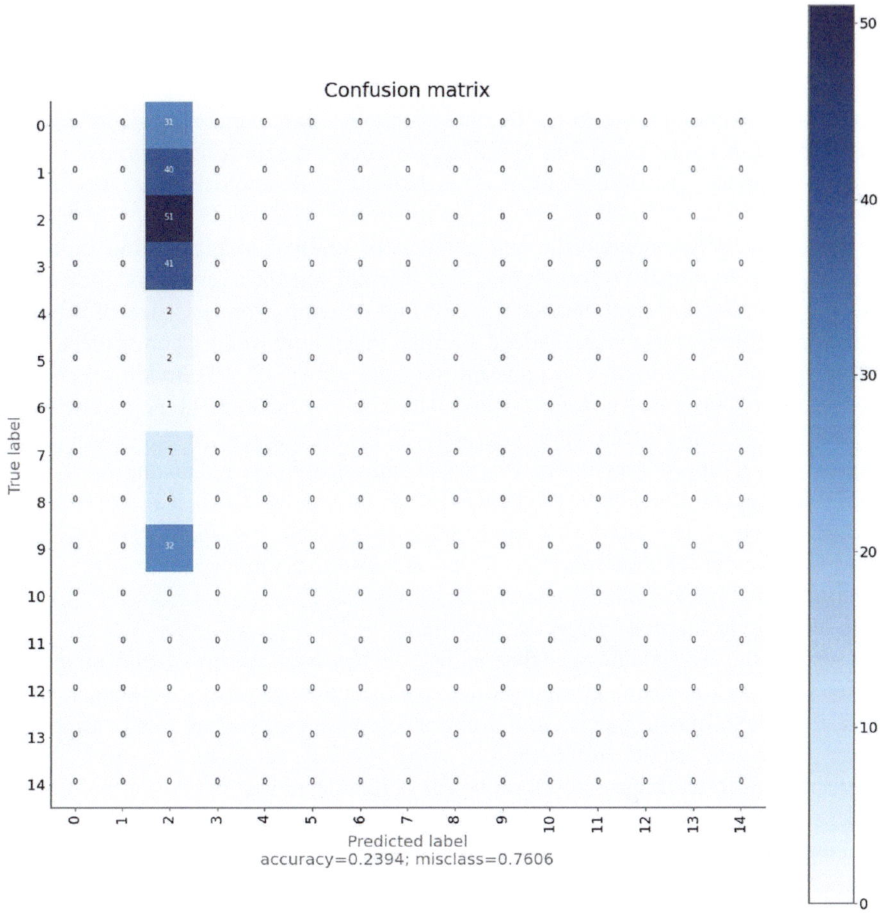

Fig. 1 Confusion matrix of 14 different classes of holes, containing or not object, in thermal images

and ground penetrating radar (GPR) [6]. The latter is the tool of choice according to various articles on the subject. In their article, the authors of [7] used a GPR mounted on a pickup truck and a maintenance train in order to find the IEDs buried under a road and a railway, respectively. B-scan images were collected using R-CNN for real-time automatic detection and support vector machine-based (SVM) classification. The author of [8] presented a multilevel projective dictionary learning method to classify ground-penetrating radar signals from IEDs. One limitation of this particular dictionary is that the container or shape of the IED depends on the terrorists' imagination. It is therefore difficult to be exhaustive in such a dictionary, and over time. In [9], the authors improved a GPR that they embarked on an UAV for the detection of IEDs. It should be noted, however, that, there are fundamental physical limitations related to the instrumentation when using GPRs [10].

Other types of sensor is also used to detect and classify landmines. The paper [11] aims to fabricate an electronic nose for detecting volatile organic compounds (VOCs) emitted by acetone as a potential precursor of triacetone triperoxide (TATP) using artificial neural network (ANN) and principal component analysis (PCA). One limitation of this solution is that the IEDs' containers are airtight materials and still wrapped in nylon. In addition, other types of homemade explosives are used in our context.

We opted for the acquisition and analysis of thermal images of our points of interest, fully buried IEDs, or decoys. The thermal sensor, mounted on a drone, is capable of collecting data at a safe height and detecting the presence of a buried object, regardless of its nature. Indeed, thermal images are well exploited these days, thanks to the characteristics and possibilities they offer [12, 13]. In their work, the authors of [14] proposed a deep learning-based approach to detect the deterioration of steel bridges using infrared thermography (IRT). Their approach was subsequently improved by adopting transfer learning. The authors of [15] used an approach based on the histogram of oriented gradients (HOG) and showed that it is not effective with IR images due to their low contrast. They showed that approaches based on convolutional neural networks (CNNs) can achieve an accuracy level of 100%. The authors of [16] present an approach for detecting cracks on paved roads by applying deep learning to thermal images of the ground.

Concerning the specific question of IED, it has been observed that images of buried IEDs and disturbed ground show, depending on the context, a salient area. To a certain extent, our subject is similar to the detection of a salient area in an image. There are several articles in this field of study [17–21]. As such, the author of [17] proposed a deep learning-based method for object detection in UAV-borne thermal images that have the capability of observing scenes in both day and night. Using a boundary-aware salient object detection network, they extract the saliency maps of the thermal images to improve the distinguishability. Thermal images are augmented with the corresponding saliency maps through channel replacement and pixel-level weighted fusion methods. In our case, we can't just extract the salient area, since we also need the surrounding area for reconnaissance (ordinary or disturbed ground with or without buried objects). The author of [18] conducted salient object detection in remote sensing images by proposing a novel two-branch architecture-based network for aggregating the transformers and CNNs, where the local spatial details and the global semantic information are fused into the final high-quality saliency map. This method, which allows features to be pooled, is quite interesting for thermal image analysis, but not sufficient in our case, where the depth of the model has a negative impact on the quality of features in low-contrast thermal images [22]. We are therefore exploring shallow-depth architectures to retain the image's analyzable features as much as possible.

The authors of [23] propose a landmine risk estimation system designed to aid in all stages of the demining process. The system involves collecting relevant information from the sites to be cleared, followed by analysis to determine the potential risk of remaining landmines. The authors of [24] employ a U-Net model for deep learning-based segmentation to extract the region of interest from satellite imagery of a geographical area. Additionally, a path planning strategy is formulated to optimize the coverage route of the UAV-based region survey. The effectiveness of this strategy is then evaluated in terms of energy and time constraints. However, the strategy is limited by the lack of satellite data updates in certain regions, the high cost associated with acquiring these data, and also the issue of the satellite's temporal resolution, which may not be suitable for tactical reconnaissance scenarios.

It is noteworthy that a significant challenge in the study of mines, particularly improvised explosive devices (IEDs), lies in the scarcity of fundamental data. This scarcity stems from the difficulty of acquiring such information in real-life situations, compounded by the inherent danger associated with handling these weapons. The author of [25] introduces a dataset of thermographic images capturing terrains with simulated antipersonnel mines of the legbreaker type. The objective is to identify the presence or absence of these artifacts using machine learning and artificial vision techniques. The dataset comprises 2700 thermographic images acquired at various heights, employing a Zenmuse XT infrared camera integrated into the DJI Matrice 100 drone. This type of mine has a distinctive shape and a visible detonator, which facilitates analysis and classification compared to a device without a standard shape and completely buried.

It should be noted that none of the works reviewed address the challenges of discriminating between fully buried IEDs and decoys.

3 Methodology and Data

3.1 Data Acquisition and Description

Previous studies have shown that shooting with a large swath captures other points in the image in addition to the point of interest, generating noise and making analysis and recognition of the point of interest more complex. From this point of view, we adapted the shooting protocol to obtain adequate data for the training stages.

Six holes measuring 30×30 cm^2 on each side and 50 cm deep were made for the purpose. Three were fitted with 20 L can, with the appropriate wiring (Fig. 2), and the other three holes were simply plugged up and used as decoy holes. A decoy is a diversionary tactic that the terrorist can implement by digging a hole similar to that of the IED. To the naked eye, it looks like an IED hole, so it's a false positive to human observation. Even an optical, ultraviolet, or thermal image observed by the human eye can hardly tell the difference. If the Artificial Intelligence is not trained accordingly, taking this possibility into account, it will generate too many false positives, thus contributing to enemy action. So the aim is to ensure that our model is able to discriminate between holes actually containing an IED and decoys. To ensure focus on the points of interest, a 2.4×2.4 m^2 frame and a 1.2×1.2 m^2 frame (Fig. 3) were made to frame the points of interest during shooting.

Fig. 2 20 L can IED

Fig. 3 Point of interest delimitation frame

In this stage of the study, the shots were taken by an UAV Matrice 300RTK carrying a Zenmuse H20T camera (Fig. 4). The Zenmuse H20T thermal camera is a DJI product designed to capture images and videos in the spectral band of 7.5–13.5 μm, with a thermal sensitivity noise equivalent differential temperature (NEDT) of less than 50 mK at f/1.0.

Fig. 4 M300 RTK with ZenmuseH20T

Considering the camera's capabilities in obtaining relevant images and for the safety of the system during its operations the flight parameters are:

- approximately between 1 and 10 m in altitude (about ground level);
- with a main grip angle at nadir (vertical) and less than 15°;
- at various times during the day and year.

In an initial study [1], our objective was to identify the types of environments that terrorists could potentially use to set mines (Figs. 2 and 5). The study also investigated how these environments might change over time or due to climate conditions (Fig. 6). In this first stage, we initiated several types of holes. Some containing inert IEDs of various types, others containing nothing at all (Figs. 7, 8, 9, and 10). A CNN model with three layers of extracted features was used for classification. The result was a confusion of all holes, with no discrimination capability (Fig. 1). It should be noted that ordinary soils were not used at this stage.

In the second stage, models were developed to discriminate between purposely disturbed soils and ordinary soils. This study highlighted the fact that images taken under a large swath capture other objects (Fig. 11), making image analysis more complex. The study also showed that the low contrast of thermal images makes their deep features uninteresting for analysis. In fact, in the second phase of our study, the aim was to classify soils disturbed by human activity (Figs. 12 and 13) from

Fig. 5 IED on asphalt road

Fig. 6 IED's soil evolution

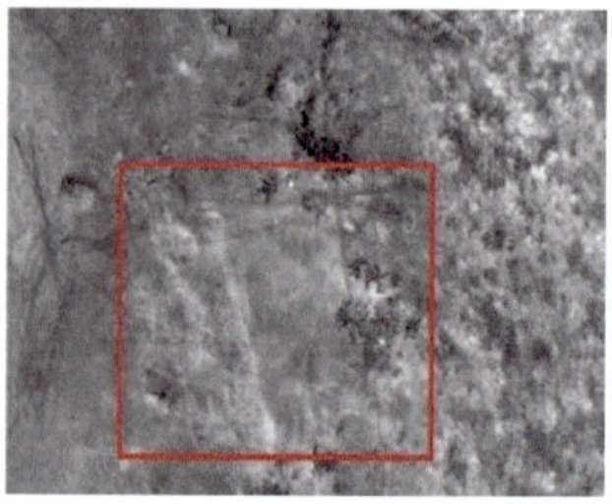

Fig. 7 Extinguisher

Fig. 8 Plastic can

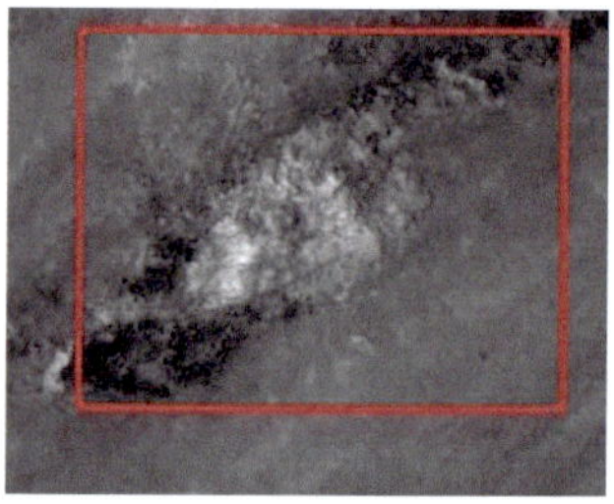

Fig. 9 Closed empty hole

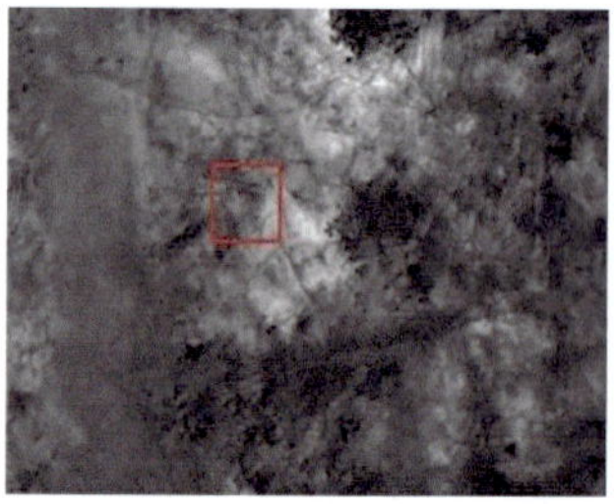

Fig. 10 Opened empty hole

ordinary soils (Fig. 14). But we also considered soils disturbed by other phenomena, like environmental (Figs. 15 and 16), as ordinary soils and it was a challenge in this article. The CNN architecture with the VGG11 model ensured a high classification capacity, especially with the use of transfer learning from pre-trained models on imagenet, but also on our own images with pre-processing to focus on points of interest.

The aim of this new study is to discriminate between disturbed, IED-hosting, and non-IED-hosting soils (Figs. 17 and 18). Taking into account the conclusions of

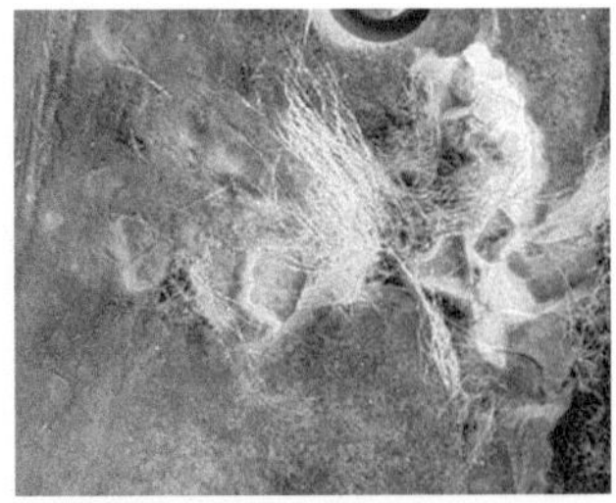

Fig. 11 EID hole and noise

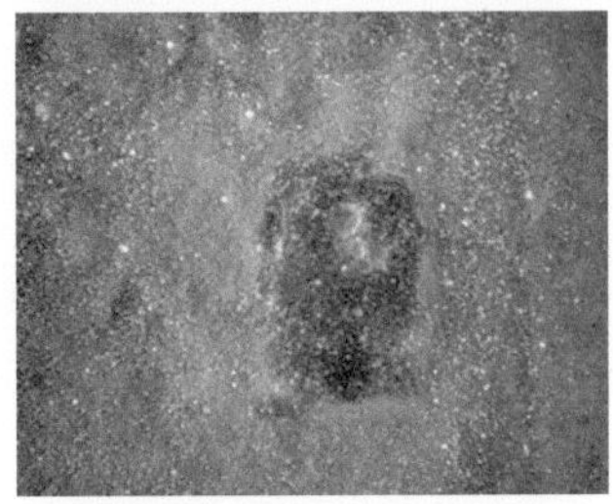

Fig. 12 Empty hole

Fig. 13 Plastic can IED

Fig. 14 Ordinary soil

previous studies, we implemented models adapted to the classification of our points of interest. In this article, the dataset consists of a total of 3446 annotated images, distributed as follows: 1734 (50.3%) Hole Without IED soil and 1712 (49.7%) Hole With IED. For model training, we divided our dataset into the following sets: training set = 2756, validation set = 621, and test set = 69. All models are trained from scratch. Metrics used are accuracy, precision, recall, and loss. Being on Pytorch-lightning, we activated Callback EarlyStopping(valloss), CheckpointCallback(monitor valloss, savetopk = 1, autoinsertmetricname = true) to avoid overfitting (Figs. 19, 20, and

Fig. 15 Ordinary soil

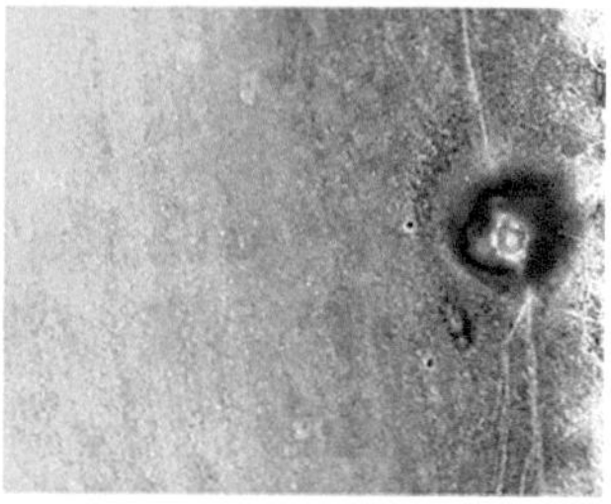

Fig. 16 Ordinary soil

Fig. 17 IED hole

Fig. 18 IED decoy hole

21). Precision and recall were used with the average = macro option. The bestmetricsCallback was taken into account to record the best accuracy, precision, and recall. Our results tables show from columns 2 to 9, respectively, the best validation accuracy, best validation precision, best validation recall, test accuracy, test precision, test recall, train accuracy, and train loss.

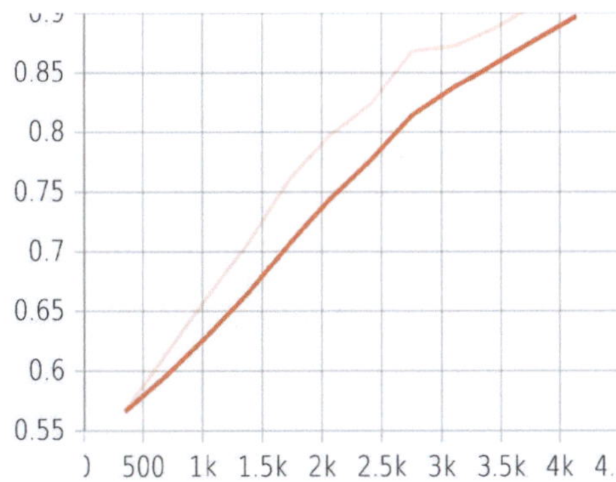

Fig. 19 VGG16BN train accuracy

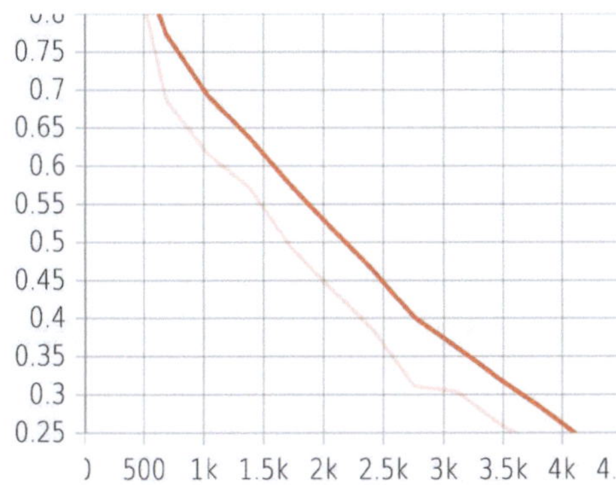

Fig. 20 VGG16BN train loss

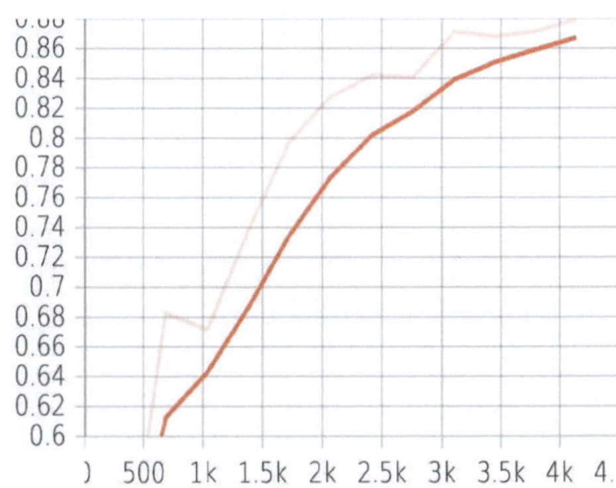

Fig. 21 VGG16 validation accuracy

3.2 Algorithm Descriptions

In this study, based on a model selection, the residual network (ResNet) 18 and VGG16 algorithms with batch normalization yielded the best results. ResNet-18 is a CNN architecture that belongs to the ResNet family [26]. The key innovation of ResNet models is the use of residual learning blocks (Eq. 9), which enable the training of very deep neural networks without encountering the vanishing gradient problem. This is achieved by introducing the skip connections, that allow the gradient to flow directly through the network without being significantly attenuated. A skip connection adds the input of a layer to the output of that layer (Eq. 8). ResNet-18 specifically is composed of 18 layers, including convolutional layers, pooling layers, and fully connected layers. It consists of several residual blocks, each containing two convolutional (Eq. 7) layers with batch normalization (Eq. 6) and ReLU activation functions, along with skip connections.

Batch Normalization Operations

1. Let x be the activations of a given layer for a mini-batch of size m.

$$\text{Input: Let } \mathbf{x} = [x_1, x_2, \ldots, x_m] \tag{1}$$

2. Compute Mean: Calculate the mean of the mini-batch activations.

$$\mu_{\text{batch}} = \frac{1}{m} \sum_{i=1}^{m} x_i \tag{2}$$

3. Compute Variance: Calculate the variance of the mini-batch activations.

$$\sigma_{\text{batch}}^2 = \frac{1}{m} \sum_{i=1}^{m} (x_i - \mu_{\text{batch}})^2 \tag{3}$$

4. Normalize: Normalize the activations. Here, ϵ is a small constant added for numerical stability.

$$\hat{x}_i = \frac{x_i - \mu_{\text{batch}}}{\sqrt{\sigma_{\text{batch}}^2 + \epsilon}} \tag{4}$$

5. Scale and Shift: Scale and shift the normalized activations using learnable parameters γ (scale) and β (shift).

$$y_i = \gamma \hat{x}_i + \beta \tag{5}$$

6. The final output of batch normalization for a mini-batch is:

$$y_i = \gamma \frac{x_i - \mu_{\text{batch}}}{\sqrt{\sigma_{\text{batch}}^2 + \epsilon}} + \beta \tag{6}$$

Convolution Operation

$$y = W * x + b \tag{7}$$

where

- W is the weight matrix;
- $*$ denotes the convolution operation;
- x is the input; and
- b is the bias term.

Skip Connexion

$$y = F(x, Wi) + x \tag{8}$$

where

- x is input to the layer;
- $F(x, Wi)$ output of the layer: function of x and Wi;
- y output of the layer after adding x.

Full Residual Block in ResNet-18

$$y = \text{ReLU}(\text{BatchNorm}(\text{Conv}(\text{ReLU}(\text{BatchNorm}(\text{Conv}(x, W1))), W2)) + x) \tag{9}$$

where $W1$ and $W2$ are the weights of the first and second convolutional layers, respectively.

VGG16 is a CNN consisting of 16 layers, including 13 convolutional layers and 3 fully connected layers. The architecture is characterized by using small 3 × 3 convolutional filters, stacking multiple convolutional layers to capture complex features, and downsampling through max-pooling. In the same spirit of model selection, batch normalization (Eq. 10) layers have been added to the layers of models VGG16 and 19 [27]. Batch Normalization is a technique used to improve the training of deep neural networks. Introduced by Sergey Ioffe and Christian Szegedy in 2015 [28], batch normalization is used to normalize the inputs of each layer in such a way that they have a mean output activation of zero and a standard deviation of one (Eq. 6). This normalization process helps to combat issues that deep neural networks face, such as internal covariate shift, which can slow down training and affect the network's ability to generalize from the training data. Batch normalization offers several benefits to the training process of deep neural networks: Improved Optimization: It allows the use of higher learning rates, speeding up the training process by reducing the careful tuning of parameters. Regularization: It adds a slight noise to the activations, similar to dropout. This can help to regularize the model and reduce overfitting. Reduced Sensitivity to Initialization: It makes the network less sensitive to the initial starting weights. Allows Deeper Networks: By reducing internal covariate shift, batch normalization allows for the training of deeper networks. For a single convolutional block with batch normalization and ReLU activation, the overall operation can be summarized as:

$$a = \text{ReLU}\left(\gamma \frac{W * x + b - \mu_{\text{batch}}}{\sqrt{\sigma_{\text{batch}}^2 + \epsilon}} + \beta\right) \tag{10}$$

Empirical observation by taking temperature readings at certain nearby periods, after very brief cloud cover or following very short periods of artificial shadows (less than 30 s) on our points of interest have shown that the temperature and consequently the thermal signature evolve rapidly in relation to the sensitivity of the sensor. All of which can make our data quite susceptible to confusion. Operations that tend to standardize radiometric resolution, such as pooling, can further weaken the quality of our low-contrast data. Just as downsampling can have a negative impact on the analysis and detection of objects with low spatial resolutions [29], pooling can have an impact on images with low radiometric resolutions. The pooling operation is

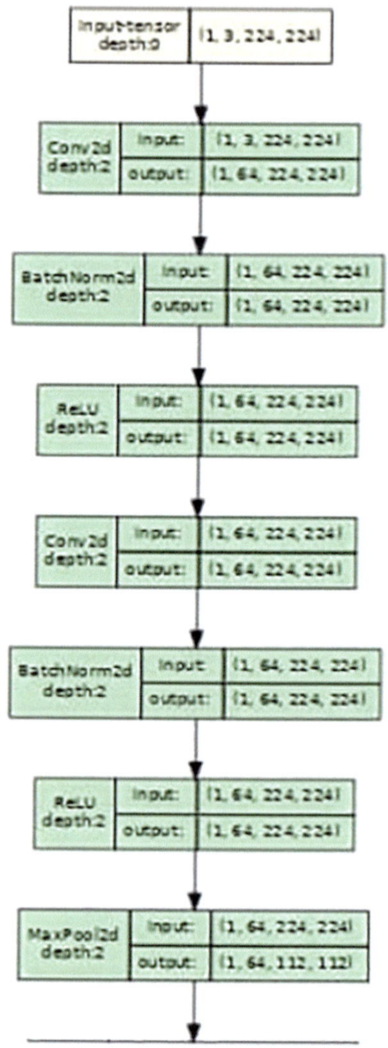

Fig. 22 VGG16BN first layers

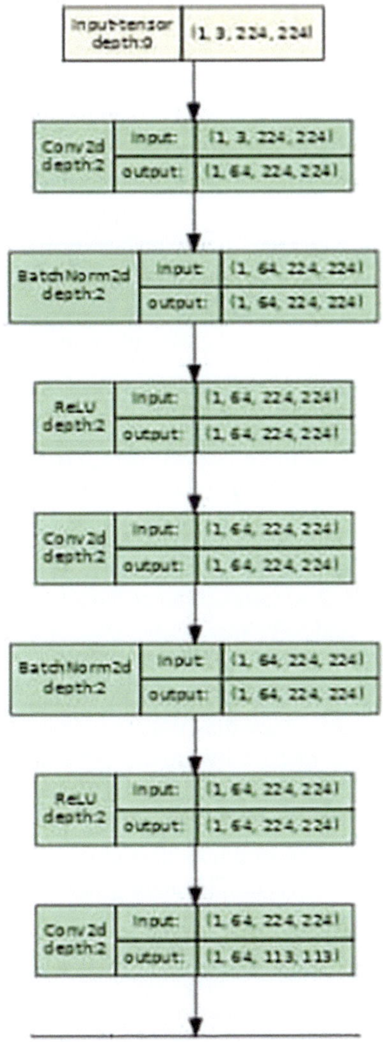

Fig. 23 VGG16BNS first layers

favorable for detection, but can have a negative effect on discrimination based on fine details (as in the case of thermal images). We have therefore replaced pooling by stride (Figs. 22 and 23) for downsampling, in order to preserve the original radiometric resolution quality of our data as much as possible.

4 Experiments and Results

In this section, we present the performance of the various models implemented on our dataset. From the standard VGG models, we achieve performance improvements thanks to the integration of batch normalization. These performances are of the order of 95% and 87% for the training and validation stages, respectively. It should be noted that these performances are still quite interesting when compared with the 95% and 98% performances obtained from GPR [30] and multispectral (optical, UV, IR) image [31] data analysis, respectively. It should be noted that the image-based study [31] focused on shallow buried improvised explosive device, and this is less complex than the case of completely buried images.

4.1 Model Selection from VGG, Inception, ResNet, and ViT Architectures

In light of the particularity of our dataset, and to define it as precisely as possible, we have made a selection of models from VGG 11, 16, 19; Inception V1, V3; ResNet-18, 50 and ViT. Very deep architectures aren't very effective at extracting features from our images, but very shallow architectures aren't very effective at characterizing the data in our dataset either. The results (refer to Table 1) from this selection of models show that VGG architectures underperform in the accuracy, precision, and recall metrics by around 50%. Inception architectures show losses of the order of 70%. In addition, ViT, which should have been of interest in this study, in view of its ability to take contextual aspects into account, gave little satisfaction. This is partly due to the depth of its architecture, and partly because ViT doesn't cope well with the multiscale [32] nature of our points of interest, which are captured at different altitudes, and it's in this that Swin Transformer [33] better resolves this limitation of

Table 1 Model selection: VGG11, VGG16, VGG19, InceptionV1, InceptionV3, ResNet-18, ResNet-50, ViT

Model	Val. acc.	Val. prec.	Val. recall	Test acc.	Test prec.	Test recall	Train acc.	Train loss
VGG11	52.82	26.41	50	63.77	31.88	50	50.98	69.53
VGG16	56.25	28.12	50	46.38	23.19	46.38	50.15	69.34
VGG19	50.08	25.60	50.64	53.62	26.81	50	50.18	69.34
InceptionV1	64.57	57.73	59.98	46.38	33.33	46.38	54.06	87.21
InceptionV3	75	62.50	75	60.87	51.45	57.97	53.74	73.66
ResNet-18	87.92	86.35	86.25	88.41	88.41	90.34	95.43	12.37
ResNet-50	81.25	65.62	75	65.22	56.44	56.81	60.6	66.24
ViT	62.50	56.25	75	56.52	31.88	53.62	51.12	70.06

Fig. 24 ResNet-18 train accuracy

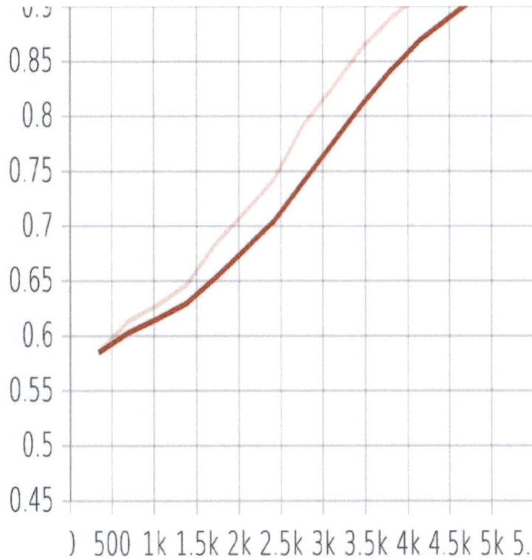

ViT. ResNet architectures, with their skip connection, provide better results. ResNet-18, which is shallower than ResNet-50, offers particularly good performance in terms of accuracy, precision, and recall (Figs. 24, 25, and 26). To improve the model, we reduced the depth of ResNet-18. This decision was based on the performance of ResNet-18 and 50 and the fact that the deep features of thermal images are not very informative for the analysis. We therefore modified the ResNet blocks. This involved implementing the ResNet model with single-layer blocks, which reduces the depth of the architecture. However, ResNet-18 was found to offer better performance than the modified model (refer to Table 2). We identify the single-layer blocks' ResNet as ResNet1CB.

4.2 Improvements Brought About by Batch Normalization: VGG16, VGG19

Batch normalization improves the performance of our VGG16 and 19 models. Comparison of the results (refer to Table 3) without batch normalization and with VGG16

Table 2 Model selection: ResNet-18, ResNet1CB

Model	Val. acc.	Val. prec.	Val. recall	Test acc.	Test prec.	Test recall	Train acc.	Train loss
ResNet-18	87.92	86.35	86.27	88.41	88.41	90.34	95.43	12.37
ResNet1CB	82.61	82.64	82.17	78.26	81.76	81.12	85.52	33.06

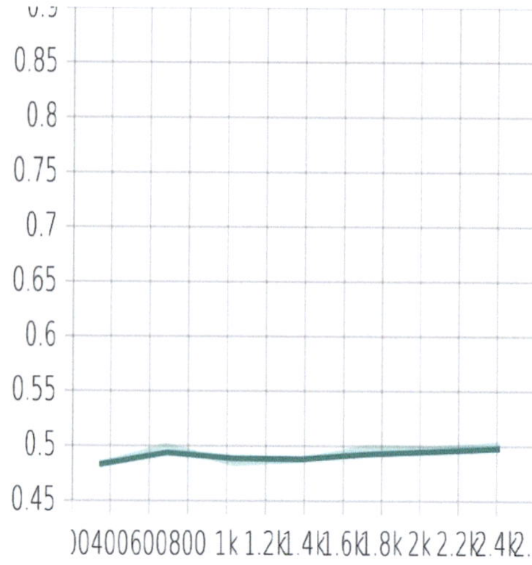

Fig. 25 VGG19 train accuracy

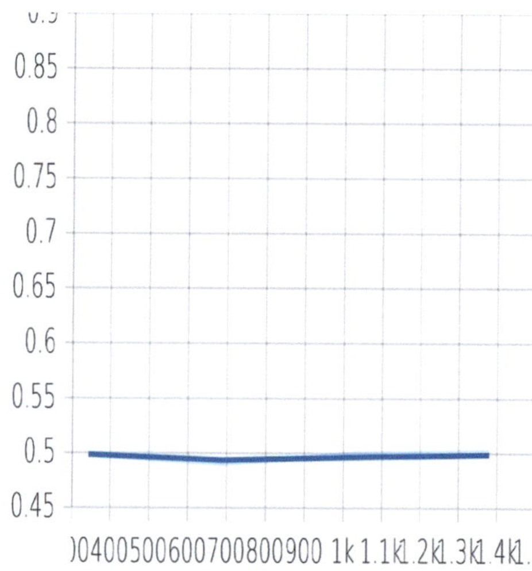

Fig. 26 VGG16 train accuracy

and 19 shows the considerable contribution of this model modification to the classification of our points of interest. We will identify VGG16 and VGG19 with batch normalization as VGG16BN and VGG19BN, respectively.

Table 3 Model selection: VGG16, VGG19, VGG16BN, and VGG19BN

Model	Val. acc.	Val. prec.	Val. recall	Test acc.	Test prec.	Test recall	Train acc.	Train loss
VGG16	56.25	28.12	50	46.38	23.19	46.38	50.15	69.34
VGG16BN	87.92	88.42	86.82	92.75	91.73	91.73	92.60	18.57
VGG19	50.08	25.60	50.64	53.62	26.81	50	50.18	69.34
VGG19BN	86.80	86.40	85.89	86.96	85.14	88.41	89.91	25.78

Table 4 Model selection: VGG16BN, VGG19BN, VGG16BNS, and VGG19BNS

Model	Val. acc.	Val. prec.	Val. recall	Test acc.	Test prec.	Test recall	Train acc.	Train loss
VGG16BN	87.92	88.42	86.82	92.75	91.73	91.73	92.60	18.57
VGG16BNS	87.92	87.64	86.68	84.06	77.77	77.44	96.77	8.40
VGG19BN	86.80	86.40	85.89	86.96	85.14	88.41	89.91	25.78
VGG19BNS	86.15	86.17	85.34	86.96	89.5	87	96.95	8.37

4.3 Replacing Pooling with Stride to Perform DownSampling

The stride ensures better preservation of radiometric resolution integrity, which is essential for low-contrast images. Let's note VGG16BNS and VGG19BNS, respectively, the VGG16 and VGG19 models with stride instead of pooling. This new architecture on VGG16 and VGG19 ensures better performance in the training stages, with regard to accuracy and loss. However, at the validation and test stages, models with pooling remain more interesting (refer to Table 4).

5 Conclusion

In this paper, we presented the possibility of discriminating between disturbed ground containing or not a buried IED in a thermal image using the VGG, Inception, ResNet, and ViT architectures. This study was and remains challenging, but above all essential for military tactics and asymmetrical warfare. Thermal images of disturbed soils, containing IEDs or containing no objects, taken from a drone provided the dataset for the discrimination of these two types of soil. The results obtained from the model selection carried out in this article show that: VGG11, with an accuracy at the validation stage of 52.82%, was unable to extract features useful for discriminating between holes containing an IED and those containing nothing. In contrast to its effectiveness in classifying disturbed and ordinary soils, with an accuracy of around 96% at the validation stage. VGG models performed less well overall with accuracies in the order of 54% at the validation stage. ResNet-18, with its skip connection technique and the depth of its architecture, yielded very interesting results compared to the VGG, Inception, and ViT architectures. It performed at 88% accuracy at the validation stage, compared with 56% for VGG16, 62% for ViT, and 75% for InceptionV3.

We demonstrated the advantage of using batch normalization techniques on VGG models to improve classification capacity. This resulted in a performance of around 88%, compared with 56% when applying batch normalization to VGG16. We also examined the effect of pooling on the relevance of thermal image information, which can be quite similar due to weather conditions and sensor sensitivity to slight changes. It was found that using stride instead of pooling had a positive effect on model convergence. In fact, a training accuracy of around 97% was obtained, compared with 90% using stride instead of pooling on the batch normalized VGG19. The loss is in the order of 8% compared with 26% when stride was applied instead of pooling on the batch normalized VGG19.

We have shown that our performance is interesting in comparison with the results of other studies, whose performances were 95% for GPR and 98% for multispectral (optical, UV, IR) images of shallow buried mines. In our next studies, we'll examine the impact of flight parameters and weather conditions on the detection and discrimination performance of our points of interest. We'll also increase the size of our dataset to ensure better training of our models.

References

1. Issaka D, Sadouanouan M, Ouedraogo TF (2023) An UAV-based approach with thermal camera to buried improvised explosive detection. In: IEEE AFRICON 2023, Nairobi, Kenya, 20–22 Sept 2023, pp 1–3. https://doi.org/10.1109/AFRICON55910.2023.10293534
2. UNMAS (2016) UNMAS lexicon: improvised explosive devices. ReliefWeb
3. Escadafal R (2021) Les propriétés spectrales des sols, 24 Feb 2021. https://horizon.documentation.ird.fr/exl-doc/pleins_textes/pleins_textes_4/colloques/30002.pdf
4. Zribi (2002) Mesures aéroportées pour l'observation du sol et des paysages: la mesure in situ et la télédétection: Compte rendu de séminaire (PEVS, Zone atelier) (suite). Utilisation des radars pour la mesure de la rugosité de la surface continentale. Nat Sci Soc 10(4):63–68
5. Ressources Naturelles Canada (2007) Ressources naturelles Canada, 19 Feb 2021. Last modified: 11 June 2019. https://www.rncan.gc.ca/cartes-outils-publications/imagerie-satellitaire-photos-aer/tutoriel-notions-fondamentales-teledetection/9310
6. Pringle JK, Ruffell A, Jervis JR, Donnelly L, McKinley J, Hansen J, Morgan R, Pirrie D, Harrison M (2012) The use of geoscience methods for terrestrial forensic searches. Earth Sci Rev 114:108–123. https://doi.org/10.1016/J.EARSCIREV.2012.05.006
7. Srimuk P, Boonpoonga A, Kaemarungsi K, Athikulwongse K, Dentri S (2022) Implementation of and experimentation with ground-penetrating radar for real-time automatic detection of buried improvised explosive devices. Sensors 22(22):8710
8. Tivive FHC, Bouzerdoum A, Abeynayake C (2022) Classification of improvised explosive devices using multilevel projective dictionary learning with low-rank prior. IEEE Trans Geosci Remote Sens 60:1–16
9. Garcia Fernandez M, Alvarez Lopez Y, Arboleya Arboleya A, Gonzalez Valdes B, Rodriguez Vaqueiro Y, Las-Heras Andres F, Pino Garcia A (2018) Synthetic aperture radar imaging system

for landmine detection using a ground penetrating radar on board a unmanned aerial vehicle. IEEE Access 6:45100–45112. https://doi.org/10.1109/ACCESS.2018.2863572
10. Nadot E (2014) Détection et caractérisation d'objets anthropiques par méthodes géophysiques et en particulier par méthode magnétique. Thèse Doctorat, Université de Strasbourg
11. Yumang AN, Paglinawan AC, Andres JS, Leon RJD, Cruz JCD, Floresta CK (2020) Electronic nose for detecting acetone as a potential precursor in triacetone triperoxide (TATP)-based improvised explosive devices (IEDs). In: Proceedings of the 2020 12th international conference on computer and automation engineering, 01 Feb 2024. Association for Computing Machinery, pp 52–55
12. Mittal U, Srivastava S, Chawla P (2019) Object detection and classification from thermal images using region based convolutional neural network. J Comput Sci 15(7):961–971
13. Ashfaq Q, Usman Akram M (2022) Convolutional neural network based thermal image classification. In: 2022 2nd international conference on digital futures and transformative technologies (ICoDT2), pp 1–6
14. Ali R, Cha Y-J (2019) Subsurface damage detection of a steel bridge using deep learning and uncooled micro-bolometer. Constr Build Mater 376–387
15. Algarni A (2020) Efficient object detection and classification of heat emitting objects from infrared images based on deep learning. Multimed Tools Appl 79. https://doi.org/10.1007/s11042-020-08616-z
16. Nooralishahi P, Ramos G, Pozzer S, Ibarra-Castanedo C, Lopez F, Maldague XPV (2023) Texture analysis to enhance drone-based multi-modal inspection of structures. Drones
17. Li M, Zhao X, Li J, Nan L (2021) ComNet: combinational neural network for object detection in UAV-borne thermal images. IEEE Trans Geosci Remote Sens 59(8):6662–6673
18. Bao L, Zhou X, Zheng B, Yin H, Zhu Z, Zhang J, Yan C (2023) Aggregating transformers and CNNs for salient object detection in optical remote sensing images. Neurocomputing 553:126560
19. Ma L, Luo X, Hong H, Zhang Y, Wang L, Wu J (2023) Scribble-attention hierarchical network for weakly supervised salient object detection in optical remote sensing images. Appl Intell 53(10):12999–13017. https://doi.org/10.1007/s10489-022-04014-0
20. Li G, Liu Z, Zhang X, Lin W (2023) Lightweight salient object detection in optical remote-sensing images via semantic matching and edge alignment. IEEE Trans Geosci Remote Sens 61:1–11. https://doi.org/10.1109/TGRS.2023.3235717
21. Liang B, Luo H (2024) An effective and lightweight solution for salient object detection in optical remote sensing images. Expert Syst Appl 238. https://doi.org/10.1016/j.eswa.2023.121778. https://www.sciencedirect.com/science/article/pii/S0957417423022807
22. Chaverot M, Carré M, Jourlin M, Bensrhair A, Grisel R (2023) Improvement of small objects detection in thermal images. Integr Comput-Aided Eng 30(4):311–325
23. Rubio MD, Zeng S, Wang Q, Alvarado D, Moreno F, Heidari H, Fang F (2023) RELand: risk estimation of landmines via interpretable invariant risk minimization, 01 Feb 2024. https://doi.org/10.48550/arXiv.2311.03115
24. Barnawi A, Kumar K, Kumar N, Thakur N, Alzahrani B, Almansour A (2023) Unmanned aerial vehicle (UAV) path planning for area segmentation in intelligent landmine detection systems. Sensors 23(16):7264. https://doi.org/10.3390/s23167264
25. Tenorio-Tamayo HA, Forero-Ramírez JC, García B, Loaiza-Correa H, Restrepo-Girón AD, Nope-Rodríguez SE, Barandica-López A, Buitrago-Molina JT (2023) Dataset of thermographic images for the detection of buried landmines. Data Brief 49. https://doi.org/10.1016/j.dib.2023.109443. https://www.sciencedirect.com/science/article/pii/S2352340923005437
26. He K, Zhang X, Ren S, Sun J (2015) Deep residual learning for image recognition, 18 Nov 2023. arXiv:1512.03385
27. DeepAI (2019) Batch normalization. DeepAI, 17 May 2024. https://deepai.org/machine-learning-glossary-and-terms/batch-normalization
28. Ioffe S, Szegedy C (2015) Batch normalization: accelerating deep network training by reducing internal covariate shift. In: Proceedings of the 32nd international conference on machine learning, ICML 2015. JMLR 37:448–456

29. Zhang X, Liu Q, Chang H, Sun H (2023) High-resolution network with transformer embedding parallel detection for small object detection in optical remote sensing images. Remote Sens 15(18):4497
30. Lameri S, Lombardi F, Bestagini P, Lualdi M, Tubaro S (2017) Landmine detection from GPR data using convolutional neural networks. In: EUSIPCO, pp 508–512
31. Colreavy-Donnelly S, Caraffini F, Kuhn S, Gongora M, Florez J, Parra C (2020) Shallow buried improvised explosive device detection via convolutional neural networks. Integr Comput Aided Eng 27. https://doi.org/10.3233/ICA-200638
32. Didenko A, Oliinyk A, Subbotin S (2023) Thermal image super-resolution methods using neural networks. In: Proceedings of the sixth international workshop on computer modeling and intelligent systems. CEUR, vol 3392, pp 214–223
33. Liu Z, Lin Y, Cao Y, Hu H, Wei Y, Zhang Z, Lin S, Guo B (2021) Swin transformer: hierarchical vision transformer using shifted windows, 28 Feb 2024. CoRR arXiv:2103.14030

Classification and Identification of Rice Using Image Processing Technique: A Review

Hemanta Kalita and Mirzanur Rahman

Abstract Rice is an important food source in the world, and the number of varieties of rice increases day by day. Rice seed mixing occurs in any stage of rice planting, production and circulation. Therefore, the identification and classification of rice is the key factor in the rice industry. However, earlier the identification or classification is done manually with the naked eye but this classification is not suitable due to the tiredness and experience of the people. This review discusses some of the image processing technologies which are accurate and give some prominent results in classification of rice. At the initial state, most of the technologies are used for classification of rice using MATLAB. Now the deep learning models are widely used in image processing and analysis in agricultural domain. Deep learning models utilizing Convolution Neural Network (CNN) are providing promising result in this field. The application area of CNN covers a major role of image analysis of rice in agricultural domain.

Keywords Neural network · MATLAB · RGB · HSV · CNN · DBSCAN

1 Introduction

Image analysis of rice in agricultural domain solved different problems related to identification and classification of rice. The accurate and prominent result is required when classifying rice varieties. The commercial value, genetic characteristic and quality of the rice depend on the purity of the rice. If it is not classified properly in the initial state, it will affect the market price and quality of the rice. In the initial state, the rice classification is done by some experts. They use their personal level skill to identify or classify the rice verities. They consider the morphological structure of the rice to take final decision for classifying the rice. The experts collected the rice from the special locality area and put it in a table. In the next level, they try to examine the physical structure of the rice using some tools like a large magnifying glass,

H. Kalita (✉) · M. Rahman
Department of Information Technology, Gauhati University, Guwahati, Assam, India
e-mail: hemanta.kalita@gmail.com

H. Sharma et al. (eds.), *Artificial Intelligence: Theory and Applications*, Lecture Notes in Networks and Systems 5589, https://doi.org/10.1007/978-981-96-1687-9_30

illuminator, forceps, etc. But in this process it requires more number of experts to take a final decision to declare the proper class of the rice. It also takes quite long time to find out the microscopic differences among many rice samples. At present machine, vision technology is widely used in agricultural domain recognizing an object and images. There are different methods which are implemented for the recognition and classification of the images. This will help the researchers to examine their result using various technologies and enhance the performance of their systems. This review discusses the uses of these effective tools and technologies and their prominent results obtained by the different researcher. Majority of research work for classification of rice has been done in deep learning method. Initially in most of the research work, the analysis of the image was done. Then, the features of the particular images were extracted. Some of them used Gaussian filter [1–3] to enhance the image quality. Some of the researchers used image subtraction method to remove the background and contrast stretching for the better contrast of the images. In the next steps, the grayscale image is converted to binary by threshold [2, 4] technique. Next depending upon the shape of the images, dilation and erosion methods were used on the resultant image for further processing. Image segmentation is a vital part in classifying rice varieties. Different image segmentation processes were implemented in different research works for obtaining the proper segmented images. Some of the research application uses image-based histogram method [4] for segmenting their images. In color feature extraction, some researcher uses HSV, YC_bC_r and $I_1I_2I_3$ color feature calculated from RGB color space. After extracted feature, some researcher used the DBSCAN [5] technique. Some of the researcher uses confusion matrix to measure the classification performance of different methods. The main objectives of this review are to analyze the results obtained by the different researchers. This review also analyzes the working of different techniques and tools and how these techniques helped the researcher for obtaining a prominent result from their works. Initially, some of the researcher did their work in neural network using MATLAB. Though the results obtained by the researchers were accurate, there was a limitation of the dataset. At present, majority of work has been done in deep learning. In deep learning, CNN provides accurate result even if the data set is very vast.

2 Materials and Methods

The authors [6] took seven varieties of rice for classification. For classification purpose, they used Support Vector Machine (SVM) prediction using imperialist competition algorithm (ICA)-based feature selection. In this research, they used three criteria precision, recall and F-measure to appraise the performance of the system. The authors Gujjar and Siddappa [1] developed an algorithm based on image analysis to identify the six varieties of basmati rice. They have used a backpropagation neural network-based classifier. The authors Silva and Sonnadara [2] developed a model to classify rice grain using neural network. They took total 9 varieties of rice. They took the images of the sample using black-background. After taking the images, they have used Gaussian filter to enhance the quality of the images. After that they converted the image to grayscale image and then it converted to binary images. All seeds were labeled for identification purpose. They have developed three neural network models for the three features using MATLAB toolbox. Two hidden layers were used for their research work. The authors [7] presented their work by image analysis and soft computing technique. For classification purpose, they have used multi-layer feed forward neural network. The authors Golpour and Chayjan [8] identified five rice varieties and developed an algorithm for image processing. The model had two hidden layers, and 36-6-5-5, 36-9-6-5 and 36-6-6-5 topologies were obtained for different cultivars. The authors Sukhvir Kaur and Sheetal Mahajan [9] presented a solution for the quality analysis of Indian basmati rice grain. They used Top-hat transformation and achieved high degree of accuracy. The authors [10] studied the milling characteristics of rice grain as affected by Moisture Content (MC) and Paddy Mixture Ratio (PMR). The authors Pazoki and others [4] experimented to classification of 5 main rice grain varieties cultivated in different environment in Iran. They have extracted color-features, morphological features and shape factors from the images. The classification is done using neuro-fuzzy neural network model. This model was based on the back propagation-learning rule which is used for classifying 5 varieties of rice. The authors Tanck and Koushal [11] took 5 samples of rice. In their experiment, three parameters were converted to digital form. These digital forms were used for digital quality inspection of rice. The authors [12] used fuzzy inference system along with image processing technique. The authors [13] proposed a method that was tested on five varieties of rice grain. The classification is done based on the major axis length as long, medium and small. The authors [3] experimented on three types of rice grains for classification. In this study, the authors considered each types of broken grain separately. They firstly converted the grabbed image to grayscale image and then binary image. They used Non-Nested generalized exemplars (NNge) algorithm for tested in real-time environment. Evaluations of algorithms were done with its default parameters defined in Weka application. The authors [14] proposed two methods to classify the rice types. In the first method, they trained CNN model using segmented rice. In the second method, they trained a combination of a pre-trained VGG16 network and the proposed method. They trained a 5-class model for classifying 4000 training images and 2-class model for the classification 1600 training images. The authors [15] were experimented neural network in the classification of rice plants. In this study, they used AlexNet CNN model. Here, they categorized images into three classes as normal, unhealthy and snail-infested plants. These classes were used in training the model. The authors Kuchekar and Yerigeri [16] were experimented the use of digital image processing techniques. In this study, they extracted different features from the images. The authors [17] classified three types of rice grain namely basmati, ponni and brown rice. They used multi-class SVM for classification. The authors [18] studied the different characteristics of Myanmar rice. They extracted five morphological features of rice grain. The authors [19] took 90 different species, and their dataset contained 8640 rice seeds sampled. For classification purpose, they used the RF (Random Forest) classifier. The authors [20] presented the classification accuracy of four varieties of rice grain. They used four different ways, namely morphological, color, texture and wavelet feature for classification purpose. A neuro-fuzzy cascade network classifier was implemented with back propagation algorithm. The accuracy was measured based on selected features using Statistical Analysis Software (SAS). The authors Kiratiratanapruk and others [5] presented an image processing technique to classify 14 rice varieties. For statistical classification, they used LR (Logistic regression), Linear Discrimination Analysis (LDA), K-Nearest Neighbors Algorithm (KNN) and SVM. The authors Koklu and others [21] experimented to classification of five different rice varieties grown in Turkey. They took 75,000 grain images, and 15,000 from each of these varieties are included in the dataset. In [22], the authors took five varieties of cultivated rice (CR) and weedy rice (WR) seeds. They have used seven Machine Learning (ML) classifiers. Authors [23] took 5 different varieties of rice for classification purpose. The dataset contains 75,000 samples of which 15,000 were from each class. The classification algorithms used by the authors were LR, SVM, Gaussian naïve Bayes, RF and Decision Tree (DT).

3 Results

The results acquired by the authors [6] for precision, recall and F-measure in SVM prediction using ICA-based feature selection were 94.45, 96.78 and 95.14. In the test dataset, the authors [1] obtained the highest accuracy was 90.00%. The authors Silva and Sonnadara [2] had obtained results from three different feature models. They got overall accuracy of 92% in combination of three features. The classification accuracy of the test dataset obtained by the authors [7] was 97%. The authors [8] obtained best result in the classification of paddy cultivars as 98.8%. The method proposed by the authors [9] was fast and accurate, and it achieved high degree of quality. Authors Pazoki and others [4] obtained overall average accuracy as 99.40% and 99.73%. The authors [11] used a non-destructive method for quality evaluation. They achieved with an accuracy of above 90% at a nominal cost for classify the grain. The average accuracy obtained by the authors [12] was 89.8% using the fuzzy inference system. The highest accuracy obtained by the authors [3] was 95.2.

The authors [14] obtained the overall accuracy of 5-class and 2-class classification as 99.31%. The authors [15] used stochastic gradient descent algorithm under base learning rate of 0.0001 and batch size of thirty, which produces the result of 91.23% accuracy. The accuracy obtained by the authors [17] was 92.22%. In [19], the performance of the RF classifier was increased as the number species of the rice seeds decreased. They specified the average harmonic mean of the precision and recall as F1 score. The score of F1 dropped from 98.17% for 6 species to 78.27% for 90 species. In [20], the performance of cascade network was very good and the average classification accuracy was 96.75%. When the authors [24] tested on independent dataset taken from internet, the performance of S-CNN model was increases to 98.6% as compared to 43.3% for F-CNN. The authors [5] using statistical methods, found best classification with an accuracy of 83.9% using SVM. The authors [21] got the results as ANN-99.87%, DNN-99.95%, CNN-100%, using the average accuracy rates of the methods. In [22], the LR model gives best performance for Weedy rice detection. The accuracy found by the authors [23] was more than 99% using the RF and DT. Training to these two models required less time in comparison with the other models.

4 Discussion

The analysis of the results obtained by different researchers and the highest accuracy graph of different classifier are shown in Table 1 and Fig. 1, respectively.

The overall accuracy of a research work depends upon how images are captured. In most of the studies, the images are captured under well-designed conditional environment. For the preprocessing purpose, in some experiments images are being pre-processed either by using concepts like image slicing or by using toolkits. In some studies, different morphological features were extracted from the labeled images. In some experiments, color images are processed first by applying some filters. From the above studies, the Gaussian filter had very effective result in comparison with other filtering techniques.

Table 1 Accuracy of the results obtained by different researchers

References	Class	Classifier	Accuracy (%)
[6]	7	SVM prediction using ICA based feature selection	95.14
[1]	6	Back propagation neural network	95.00
[2]	9	MLP	92
[7]	3	Multi-layer feed forward neural network	97
[8]	5	ANN	98.8
[4]	5	MLP and neuro-fuzzy network	98
[3]	4	Weka Machine Learning software	95.2
[15]	3	AlexNet	91.23
[17]	3	Multi-class SVM	92.22
[20]	4	A Neuro-fuzzy cascade network	96.75
[5]	14	SVM	83.9
		DNN	95.14
[21]	5	ANN	99.87
		DNN	99.95
		CNN	100
[23]	5	RF	99
		DT	99

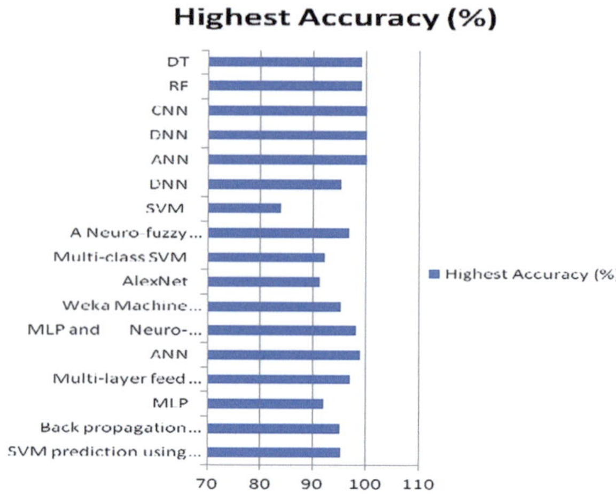

Fig. 1 Highest accuracy for different classifier

In some of the research work, the preliminary steps like preprocessing and feature extraction could be enhanced by using different preprocessing techniques or using different feature extraction method for better result. The preprocessing techniques can be enhanced by using some different segmentation algorithm like watershed and K-mean clustering algorithms. For feature extraction, the Principal Component Analysis (PCA) will give better result in comparison with other methods.

5 Experimental Works After the Review

The above review is basically done for implementing our research work classification and identification of traditional rice cultivated in different regions in Assam, India. The prototype of our model is shown in Fig. 2.

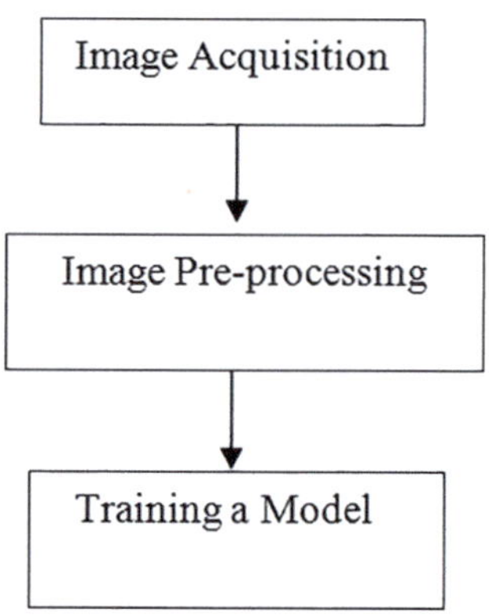

Fig. 2 Prototype of proposed work

For the image acquisition process, we have collected different samples of rice and take their images with mobile phone. After taking the images, we have store it in separate folders. In the image preprocessing step, we first segmented the images in different regions. It will reduce the complexity of the images for further processing. We choose watershed algorithm for experimenting our result. After implementing this algorithm, we have got some segmented images. We have taken multiple as well as single images for segmentation purpose, and these are shown in Figs. 3 and 4, respectively.

Further we have to combine the above images for different varieties rice and choose any CNN model for classification and identification purpose.

Fig. 3 Image segmentation on multiple images

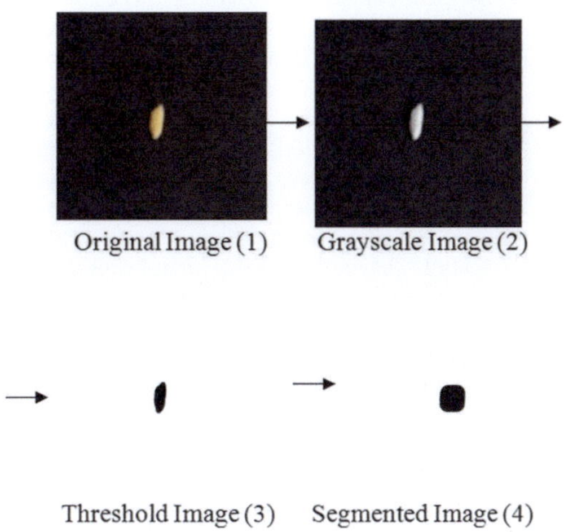

Fig. 4 Image segmentation on single image

6 Conclusion

From the above research works, we can conclude that the identification and classification of rice is a vital part in the rice industries. Different rice has different characteristics and nutrition values, and it will affect the productivity as well the quality of the rice if it is not classified or identified at the initial level before cultivating. In the above research works, different researchers have applied different methods and techniques for obtaining accurate or perfect results.

Majority of research work for classification and identification of rice has been done in deep learning method. Most of the experiment gave good performance, but there was a limitation of the dataset. In classification or identification of rice, best result was obtained by using Convolution Neural Network (CNN). So the above research work can be experimented by using different types of CNN model like InceptionV3, VGG16, ResNet, etc., with vast dataset according to the requirement of the rice industry. In our proposed work, also we can choose any CNN model which will give a prominent and accurate result.

References

1. Gujjar HS, Siddappa DM (2013) A method for identification of basmati rice grain of India and its quality using pattern classification. Int J Eng Res Appl 3(1):268–273
2. Silva CS, Sonnadara DUJ (2013) Classification of rice grains using neural networks
3. Ozan AKI, Güllü A, Uçar E (2015) Classification of rice grains using image processing and machine learning techniques. In: International scientific conference, pp 20–21
4. Pazoki AR, Farokhi F, Pazoki Z (2014) Classification of rice grain varieties using two artificial neural networks (MLP and Neuro-Fuzzy). J Anim Plant Sci 24(1):336–343
5. Kiratiratanapruk K, Temniranrat P, Sinthupinyo W, Prempree P, Chaitavon K, Porntheeraphat S, Prasertsak A (2020) Development of paddy rice seed classification process using machine learning techniques for automatic grading machine. J Sens
6. Rad SM, Tab FA, Mollazade K (2012) Application of imperialist competitive algorithm for feature selection: a case study on bulk rice classification. Int J Comput Appl 40(16):41–47
7. Shah V, Jain K, Maheshwari C (2013) Non-destructive quality analysis of Gujarat 17 Oryza Sativa SSP Indica (Indian Rice) using artificial neural network. Int J Sci Res Develop 1(3)
8. Golpour I, Chayjan RA (2014) Identification and classification of bulk paddy, brown, and white rice cultivars with colour features extraction using image analysis and neural network. Czech J Food Sci 32(3):280–287
9. Mahajan S, Kaur S (2014) Quality analysis of Indian basmati rice grains using top-hat transformation. Int J Comput Appl 94(15)
10. Payman SH, Bagheri I, Zareiforoush H (2014) Milling characteristics of rice grains as affected by paddy mixture ratio and moisture content. Int J Biosci 4(2):87–97
11. Tanck P, Kaushal B (2014) A new technique of quality analysis for rice grading for Agmark standards. Int J Innov Technol Explor Eng 3:83–85
12. Zareiforoush H, Minaei S, Alizadeh MR, Banakar A (2015) A hybrid intelligent approach based on computer vision and fuzzy logic for quality measurement of milled rice. Measurement 66:26–34
13. Kaur S, Singh D (2015) Geometric feature extraction of selected rice grains using image processing techniques. Int J Comput Appl 124(8)

14. Patel VA, Joshi MV (2018) Convolutional neural network with transfer learning for rice type classification. In: Tenth international conference on machine vision (ICMV 2017), vol 10696. SPIE, pp 282–289
15. Atole RR, Park D (2018) A multiclass deep convolutional neural network classifier for detection of common rice plant anomalies. Int J Adv Comput Sci Appl 9(1)
16. Kuchekar NA, Yerigeri VV (2018) Rice grain quality grading using digital image processing techniques. IOSR J Electron Commun Eng 13(3):84–88
17. Ibrahim S, Zulkifli NA, Sabri N, Shari AA, Noordin MRM (2019) Rice grain classification using multi-class support vector machine (SVM). IAES Int J Artific Intell 8(3):215
18. Tin MM, Mon KL, Win EP, Hlaing SS (2019) Myanmar rice grain classification using image processing techniques. In: Big data analysis and deep learning applications: proceedings of the first international conference on big data analysis and deep learning 1st. Springer Singapore, pp 324–332
19. Fabiyi SD, Vu H, Tachtatzis C, Murray P, Harle D, Dao TK, ... Marshall S (2020) Varietal classification of rice seeds using RGB and hyperspectral images. IEEE Access 8:22493–22505
20. Robert Singh K, Chaudhury S (2020) A cascade network for the classification of rice grain based on single rice kernel. Complex Intell Syst 6:321–334
21. Koklu M, Cinar I, Taspinar YS (2021) Classification of rice varieties with deep learning methods. Comput Electron Agric 187:106285
22. Ruslan R, Khairunniza-Bejo S, Jahari M, Ibrahim MF (2022) Weedy rice classification using image processing and a machine learning approach. Agriculture 12(5):645
23. Saxena P, Priya K, Goel S, Aggarwal PK, Sinha A, Jain P (2022) Rice varieties classification using machine learning algorithms. J Pharmaceutic Negat Resul 3762–3772
24. Sharma P, Berwal YPS, Ghai W (2020) Performance analysis of deep learning CNN models for disease detection in plants using image segmentation. Inform Process Agricult 7(4):566–574

Innovative Approach to Cucumber Pests Detection: Leveraging Roboflow for Annotating and Training

N. Hemalatha, Rithika Adappa, and K. M. Sreekumar

Abstract This project aims to apply deep learning methodologies to solve the issue of locating and controlling pest infestations in cucumber plants. The study begins with a summary of the importance of the cucurbit family and its health advantages, emphasizing the adverse effects of pests such as aphids, red pumpkin beetles, fruit flies, whiteflies, and serpentine leaf miners on cucumber farming. Training YOLO and COCO models on the dataset and assessing their performance using metrics like mean average precision (mAP), precision, and recall constitute the core of the study. The overall findings of the study highlight the significance of deep learning approaches in developing a tool that is accurate and useful for identifying cucumber plant pests, with possible applications to agricultural pest management techniques. It could be observed that on running Roboflow with COCO model a precision of 86% was obtained compared to YOLO model.

Keywords Pest · Cucumber · Leaf · Precision · Infestations · Methodologies

1 Introduction

People often refer to plants in the family Cucurbitaceae, which spans more than 120 genera and 1000 species in tropical and subtropical areas, as "cucurbits." Southeast Asia, tropical America, and Africa have more cultivation of cucurbits since it is tropical in origin. Usage of squashes and gourds are more in cooking whereas cucumbers finds its use in salads pickles. Cucurbit fruits are beneficial to human health since they aid in digestion, boost vitality, cleanse the blood, and ease constipation [1].

Despite the rising importance of cucumber, there are a number of issues that significantly limit output, including a lack of suitable planting materials, difficult

N. Hemalatha (✉) · R. Adappa
Institute of Management and Information Technology, St Aloysius (Deemed to be University, Mangalore, India
e-mail: hemalatha@aloysius.ac.in

K. M. Sreekumar
College of Agriculture, Padanakkad, Nileshwar, Kerala, India

H. Sharma et al. (eds.), *Artificial Intelligence: Theory and Applications*, Lecture Notes in Networks and Systems 5589, https://doi.org/10.1007/978-981-96-1687-9_31

access to financing, adverse weather, and plant pests and illnesses [2]. It is essential to identify and eradicate insect pests before an infestation takes hold since, according to estimates, they eat enough crops to feed an additional one billion people worldwide [3].

Symptoms of an aphid infestation include diminished plant vigour and growth, yellowing, mottling, browning, curling, and wilting of leaves, and low economic yields occasionally plant mortality are the end outcome. The first indication of an aphid infestation is the downward curling and crinkling of the leaves. Aphids feeding on fruit and flower buds might result in the development of malformed blooms or fruits [4].

Aulacophora foveicollis, frequently referred to as the red pumpkin beetle, is a major and destructive pest of cucurbits in India [5]. This beetle damages crops in both stages as grubs and adults, with the grubs that reside underground damaging the roots of the crop. The roots, underground stems, and occasionally the fruits that contact the ground are all points of entry for the grubs [6]. The beetles also damage fruits, leaving them with distinctive circular bands that result in decaying and upcoming attack of diseases [3].

On a variety of cultivated plants, specifically fruits, fruit flies are a common sight. The Cucurbitaceae family is the primary host group for the fruit fly species *Bactrocera cucurbitae* (Melon flies). The study's findings indicated that 8% of cucumber plants were attacked by fruit flies [4]. Their attacks not only lower yields but also have an impact on fruit quality, making farming operations unprofitable [7]. The insect severely damages cucurbitaceous fruits, either entirely or partially, rendering them unfit for human consumption [8].

The whitefly is one of the most harmful insects that attack cucumber plants in its growth stages viz., seedling, flowering, and fruiting. Because of this behavior of the insect, which involves adult feeding, mating, oviposition, and larval development occurring on the bottom part of the leaves, chemical management of this pest is difficult [9]. Whiteflies may reproduce swiftly all year long in a controlled setting with a warm climate, and they can quickly reach enormous population densities [10].

One of the main pests of cucumber grown in greenhouses all over the world is the serpentine leaf miner, *Liriomyza trifolii* [11]. The upper surface of the leaf affected by this pest is largely covered with white stipples caused by feeding and egg-laying punctures, which can enable bacterial and fungal illnesses to penetrate the plant. However, the larva's feeding and tunneling through the leaf's mesophyll layer, results in blisters, blotchy mines, or serpentine tunneling, is what harms the plant. Mines may contain faeces that are visible and can taint leaves meant for human consumption. Photosynthesis is hindered by severe feeding damage, and leaves may dry out and fall off, which lowers plant quality and productivity [5].

2 Related Works

Omer et al. discussed an improved kind of leaf disease and pest detection model in cucumber using YOLOv5l model. This model also could even detect minute leaf spot disease in cucumber. To authenticate the model, CBAM was incorporated into improved and original YOLOv5l model. The model obtained a mean average precision of 80.1% [12].

An enhanced YOLOv8s model was developed to detect pest and diseases in cucumber detection model by Ding et al. MultiCat module was integrated to this model for better spatial and channel attention. Paper confirmed that this model outperformed the pest detection model developed in YOLOv8s [13].

Loti et al. applied six traditional and deep learning features to find significant pests and disease from the images of chili leaves. They applied three machine learning algorithms and the result showed that feature-based method using deep learning outperformed with traditional one producing an accuracy of 92.10% with SVM classifier [14].

Liu and Wang in their paper have researched on detection of diseases and pests on plants using deep learning approaches. They also discussed the possible challenges of this method. Further, they have suggested solutions for the challenges [15].

YOLO-JD, a deep learning framework was developed by Li *et al.* for detecting pests and diseases in Jute plant through images. They integrated three new modules such as Sand Clock Feature Extraction Module (SCFEM), Deep Sand Clock Feature Extraction Module (DSCFEM), and Spatial Pyramid Pooling Module (SPPM) to figure out features in image effectively. An average mAP of 96.63% was obtained for this [16].

An MG-YOLO detection model was proposed by Li et al. for detecting gray mold spores in cucumber. This model achieved an accuracy of 98.3% and takes 0.009s per image compared to state-of-the-art model solving the detection of spores in blurred and small targets. This was compared with YOLOv5 and it was found that present model improved by 6.8% [17].

Wang et al. have proposed a modified form of Faster R-CNN with ResNet101 feature extractor and a structure of convolutional network with small anchors to extract features to create MPest-RCNN model for solving the problem of recognizing and counting pests in apple. Proposed model outperformed by 0.31% in precision compared to Faster R-CNN. Mean average precision was also higher compared to other pest recognition methods [18].

Lee et al. in their paper have used five steps of transfer learning for creating models for the classification of Citrus pest recognition system based on deep learning. They obtained an accuracy of 97% and a web server was built using the EfficientNet-b0 model [19].

3 Methodology

3.1 Data Collection

We have collected images of cucumber pests by exploring web sources, existing plant pathology, and agricultural science research databases. Additionally, we obtained a few images from the Nileshwar College of Agriculture in Kerala. The majority of the data collection consists of image files in standard format types such as JPEG or PNG. Each image file represents a pest that frequently affects cucumber plants and is identified with the pest name that is associated with the pest. Five pests have been the primary focus of this project: Aphids, pumpkin beetles, fruit flies, serpentine leafminers, and whiteflies (Figs. 1, 2, 3, 4, and 5).

The 670 images that are collected for the dataset in this study include 146 images of aphids, 126 images of pumpkin beetles, 136 images of fruit flies, 104 images of serpentine leafminers, and 158 images of whiteflies.

Fig. 1 Aphids

Fig. 2 Pumpkin beetle

Fig. 3 Fruit fly

Fig. 4 Whitefly

Fig. 5 Serpentine leafminer

3.2 Data Preprocessing

Augmentation. Several modifications and alterations must be applied to the dataset to increase the size of the dataset while also enhancing the performance of the deep learning model. The cucumber pests image dataset is increased using the augmentation methods utilizing the Python module "imgaug." Our image dataset was enhanced using the augmentation methods such as Flip, Affine, Multiply, Linear Contrast, and Gaussian Blur.

Annotation. Annotation is the process of identifying and categorizing objects of interest within a collection of images. In this case, annotations on top of the augmented photographs are used to identify and highlight places where cucumber plant pests are present. The project uses the Roboflow annotation software to perform annotation. Annotated pictures are essential for the deep learning model to successfully identify and classify pests in cucumber plants. To enable the model to learn from annotated images during the training phase, the annotation method includes manually highlighting or outlining the regions of interest, such as pests, holes, spots, or other symptoms, on each augmented image.

Rebalancing Dataset. The project additionally employs the use of the rebalancing technique to maintain an equal distribution of images among the training, validation, and test datasets. By using this strategy, errors that could arise from an uneven dataset or a pest group with a significantly greater sample count can be avoided. The dataset was rebalanced using Roboflow software. Rebalancing improves accuracy by enabling the model to be trained on a more balanced representation of the various pest categories. Our dataset has been split into three sets that is training set consisting of 80% of the dataset, 12% is composed of the test set, and the balance 8% the valid set.

Resizing and Auto-orient. Furthermore, during the preprocessing step in Roboflow, the images are resized to a standardized resolution. This resizing ensures that all images have the same dimensions, which is essential for input compatibility during model training. In the dataset used in our project, we have resized all the images used into a standard size of 640 $\times$ 640. Auto-orient is another preprocessing step that automatically adjusts the image orientation based on its EXIF metadata. This step ensures that all images have a consistent orientation, avoiding any inconsistencies that may arise from varying image orientations.

3.3 Exporting the Dataset

We exported our custom dataset from Roboflow, a popular software for managing computer vision datasets, in this phase. Users can retrieve the dataset in a format that can be used for a variety of computer vision activities, such as training machine learning models, by exporting the dataset. Based on compatibility with downstream applications or machine learning frameworks (like TensorFlow, and PyTorch), we must choose the best format for exporting the dataset. COCO, Pascal VOC, and YOLO are some of the common export formats. Our current annotated dataset has been exported in COCO format.

3.4 *YOLO Model*

The YOLO model known for its accuracy and speed, is a real-time object detection system. With the use of an input image, it can forecast bounding boxes and class probabilities for different objects found in the image. YOLO models are excellent at accurately identifying objects in images and are ideally suited for real-time applications. The YOLO model is employed in this study to detect pests in images of cucumber plants. The Roboflow platform was used in this project to train the YOLO model because it offers an integrated approach to dataset management, preprocessing, and model training.

3.5 *COCO Model*

The COCO model is also known as "Common Objects in Context." The study takes advantage of the robustness and adaptability of the COCO model, which enables precise pest identification in cucumber plant images. The COCO model is trained using the annotated dataset of cucumber plant images, which includes tagged pest regions. Once the COCO model has been trained, it can be used to identify pests in cucumber plants. The model detects objects and creates bounding boxes with corresponding class probabilities to show the presence and location of pests when it receives a new input image.

Figure 6 explains the workflow followed for this work.

4 Results

In our project, we utilized Roboflow trained models, specifically Common Objects in Context (COCO) and You Only Look Once (YOLO), to enhance the accuracy of pest identification in cucumber plants. We have trained our augmented dataset using a train/valid/test ratio of 80:8:12.

Training results in terms of mean average precision (mAP), precision, and recall are important metrics that evaluate the performance of a trained model. The mean average precision (mAP) is a combined metric that measures the accuracy and quality of object detection by considering precision and recall at different confidence thresholds. It provides an overall assessment of how well the model localizes and classifies objects. Precision represents the model's positive prediction accuracy, indicating the proportion of correctly detected objects among all predicted positive instances whereas recall, checks the model's capacity to find all relevant occurrences, capturing the proportion of correctly detected objects out of all ground truth positive instances.

The models used to train the dataset are YOLOv8 and COCO. Using COCO model we have obtained 80.6% precision score, 74.4% mAP score, and 68.0% recall score.

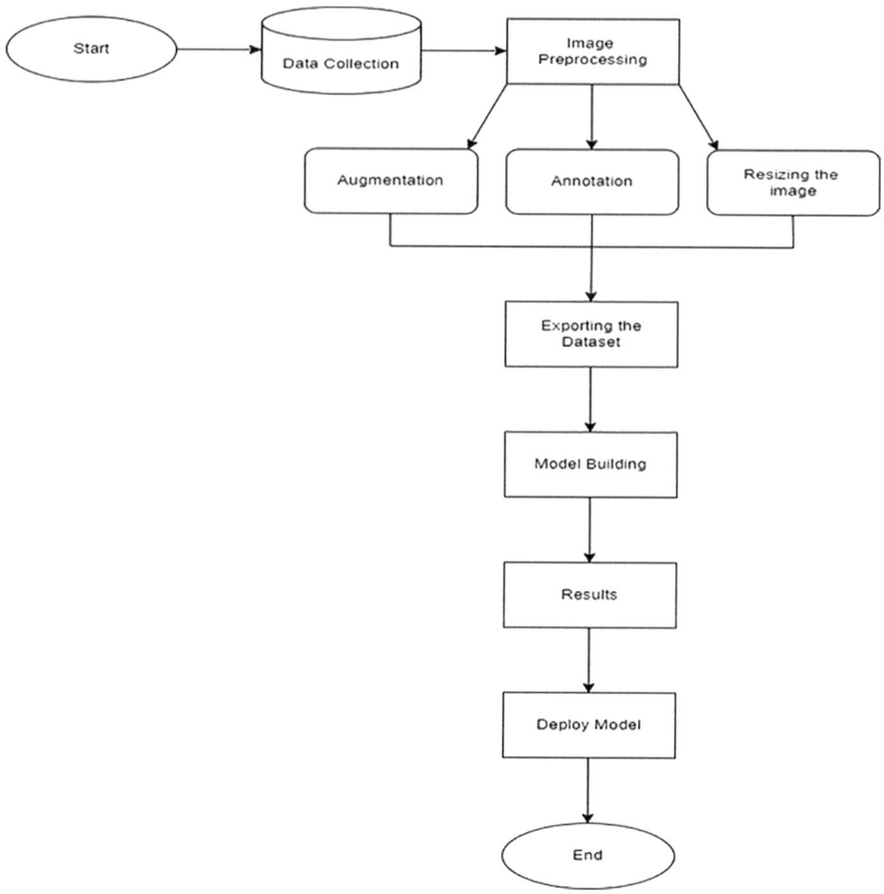

Fig. 6 Workflow diagram

Using YOLOv8 model we have obtained 78.2% precision score, 74.9% mAP score, and 69.8% recall score. Before augmentation our mAP and recall scores were very low.

We can observe from Table 1 that all the three metrics of the model have increased significantly after the augmentation of the dataset.

Table 1 Performance measures before and after augmentation of different models

Models augmentation	COCO			YOLO		
	Precision	mAP	Recall	Precision	mAP	Recall
Before augmentation	57.3%	3 9.2%	35.3%	60.5%	40.1%	35.2%
After augmentation	80.6%	74.4%	68.0%	78.2%	74.9%	69.8%

From the mAP graph in Fig. 7, it's observed that mAP score increases significantly with the number of epochs.

From the below graphs in Fig. 8, we can observe that all the three losses that are Box Loss, Class Loss, and Object Loss are decreasing significantly as the number of epochs increases.

Deploying a model in Roboflow means providing an object detection model that has been trained available for use in practical applications. A model can be deployed to serve predictions on data (unexplored) after training and validated using the Roboflow platform (Fig. 9).

An annotated image has been sent to the Roboflow deploy model feature for the prediction of the pest. The model was able to predict the pest as fruit fly with 91% confidence.

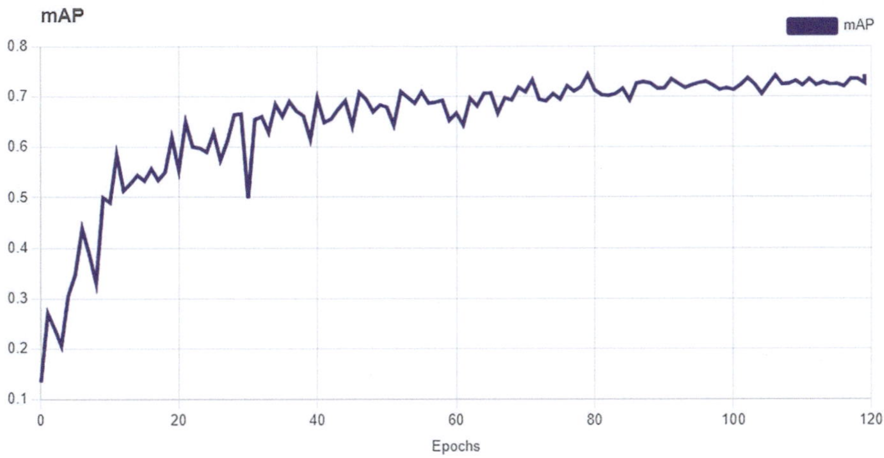

Fig. 7 mAP graph

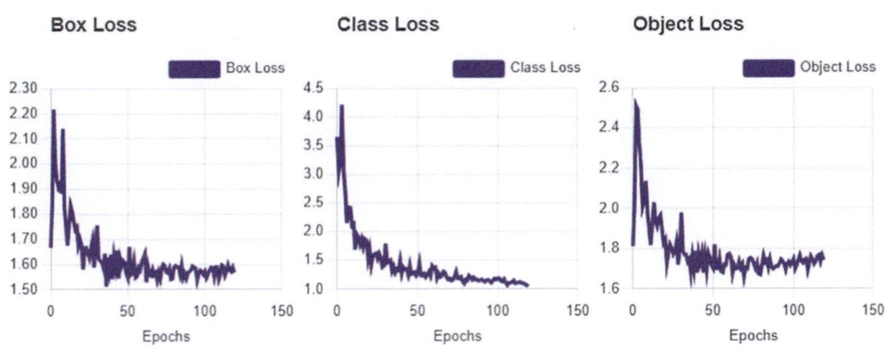

Fig. 8 Loss graph

Fig. 9 Deploying the Roboflow model

5 Conclusions

Initially, an examination of the literature and discussions with experts in the field enabled the development of an in-depth understanding of the issue of pest identification in cucumber plants. The dataset was assembled using several resources, including academic databases, websites devoted to agriculture and pathology, and the College of Agriculture in Nileshwar, Kerala. The gathered data underwent preprocessing to remove noise, normalize pixel values, and enlarge the dataset using methods such as rotation, flipping, and creating variants. Data preprocessing improved the quality of the dataset and usability, leading to better model performance.

Data augmentation techniques were used to increase the range of the dataset and improve the performance of the deep learning model. The dataset was transformed and altered using the Python module "imgaug" for data augmentation. After data augmentation, parts of cucumber plants with pests were manually highlighted or outlined on the augmented photos. The annotation process was made easier by the use of the Roboflow program, and the dataset was exported in COCO format for use in training various models. Using deep learning methods and the Roboflow platform, model development was an important phase of the project. The dataset was used to combine and train the YOLO and COCO models. The Roboflow trained models were assessed using the precision, recall, and mAP metrics. Once the models had been successfully trained using Roboflow, they were deployed to enable predictions on new images.

In conclusion, the study was effective in developing a prediction tool for cucumber plant pest identification. Comprehensive problem analysis, data collection and preprocessing, model building and training, evaluation, and deployment were all included in the technique. The experiment proved the value of deep learning methods, data augmentation, and boosting pest identification accuracy while providing insightful information for agricultural pest control.

6 Future Scope

To increase the precision further, dataset of cucumber pests can be worked upon. Various climatic condition can be used to collect data. Also, many more field visits will help in increasing the number of data collected.

References

1. Srivastava RM, Joshih S (2021) Integrated pest management for cucurbits in cucumber (Cucumis sativus L.). cucumber economic values and its cultivation and breeding. IntechOpen
2. Umeh OA, Ojiako FO (2018) Limitations of cucumber (Cucumis sativus L.) production for nutrition security in Southeast Nigeria. Int J Agricult Rural Develop 21(1):3437–3443
3. Pansara ZH, Patel JJ, Gami P (2022) Bioefficacy of insecticides against red pumpkin beetle-Population dynamics of red pumpkin beetle, *Aulacophora foveicollis* on cucumber. RAASA J Sci Soc 4(1):26–30
4. Apriyadi R, Saputra HM, Sintia S, Andini DE (2021) The Diversity of Fruit Fly (Diptera: Tephritidae) on Combination of Attractant and Different Trap Height in Cucumber Field (Cucumis sativus L.). In: IOP conference series: earth and environmental science, 926(1), p. 012059. IOP Publishing
5. Yang J, Yuan R, Chao M (2021) Identification of cucumber leaf diseases using deep learning and small sample size for agricultural internet of things. Int J Distribut Sens Netw 17(4)
6. Hassan SI, Alam MM, Illahi U, Suud MM (2023) A new deep learning-based technique for rice pest detection using remote sensing. Peer J Comput Sci 9:e1167
7. Obasekore H, Fanni M, Ahmed SM, Parque V, Kang BY (2023) Agricultural robot-centered recognition of early-developmental pest stage based on deep learning: a case study on fall armyworm (Spodoptera frugiperda). Sensors 23(6):3147
8. Lee S, Choi G, Park HC, Choi C (2022) Automatic classification service system for citrus pest recognition based on deep learning. Sensors. 22(22):8911
9. Emera I, Sandor M (2019) Creation of farmers' awareness on fall armyworms pest detection at early stage in rwanda using deep learning. In: 8th international congress on advanced applied informatics (IIAI-AAI) on proceedings, pp 538–541. Publisher IEEE
10. Kundur NC, Mallikarjuna PB (2022) Insect pest image detection and classification using deep learning. Int J Adv Comput Sci Appl 13(9)
11. Li W, Zhu T, Li X, Dong J, Liu J (2022) Recommending advanced deep learning models for efficient insect pest detection. Agriculture 12(7):1065
12. Omer SM, Ghafoor KZ, Askar SK (2024) Lightweight improved yolov5 model for cucumber leaf disease and pest detection based on deep learning. SIViP 18:1329–1342

13. Ding JY, Jeon WS, Rhee SY, Zou CM (2024) DM-YOLOv8: improved cucumber disease and insect detection model based on YOLOV8. *Preprints*
14. Ahmad Loti NN, Mohd Noor MR, Chang SW (2021) Integrated analysis of machine learning and deep learning in chili pest and disease identification. J Sci Food Agric 101(9):3582–3594
15. Liu J, Wang X (2021) Plant diseases and pests detection based on deep learning: a review. Plant Methods 17:1–18
16. Li D, Ahmed F, Wu N, Sethi AI (2022) Yolo-JD: a deep learning network for jute diseases and pests detection from images. Plants 11(7):937
17. Li K, Zhu X, Qiao C, Zhang L, Gao W, Wang Y (2023) The gray mold spore detection of cucumber based on microscopic image and deep learning. Plant Phenomics. 5:0011
18. Wang T, Zhao L, Li B, Liu X, Xu W, Li J (2022) Recognition and counting of typical apple pests based on deep learning. Eco Inform 68:101556
19. Lee S, Choi G, Park HC, Choi C (2022) Automatic classification service system for citrus pest recognition based on deep learning. Sensors 22(22):8911

Real-Time Zero-Day Attack Detection with Proximal Policy Optimization

Premanand Ghadekar, Harshita Bhagat, Parth More, Chinmay Saraf, Vaishali More, and Sarthak Khare

Abstract Zero-day attacks exploit previously unknown vulnerabilities and pose a significant cybersecurity threat. Traditional signature-based detection methods struggle against these attacks due to their lack of prior knowledge about attack patterns. This paper investigates the potential of Proximal Policy Optimization (PPO), a reinforcement learning (RL) technique, for zero-day attack detection. The growing sophistication of zero-day attacks often employing advanced techniques and malware necessitates novel detection approaches. We focus on PPO due to its ability to learn optimal policies for taking actions based on observed network traffic data. This capability aligns well with the need to adapt to unseen attack patterns in real-time. This research explores the application of PPO in a simulated network environment, aiming to train an agent that can effectively identify and respond to zero-day attacks. By leveraging PPO's ability to learn and adapt to dynamic threat landscapes, the proposed approach offers a promising avenue for mitigating zero-day attack risks. The research demonstrates the potential of PPO model in detection of zero-day attacks, achieving a significant improvement over traditional methods.

Keywords Zero day · PPO · Reinforcement learning · Real-time anomaly detection

---

P. Ghadekar · H. Bhagat · P. More · C. Saraf (✉) · V. More · S. Khare
Vishwakarma Institute of Technology, Pune, Maharashtra, India
e-mail: chinmay.saraf211@vit.edu

P. Ghadekar
e-mail: premanand.ghadekar@vit.edu

H. Bhagat
e-mail: harshita.bhagat21@vit.edu

P. More
e-mail: parth.more21@vit.edu

V. More
e-mail: vaishali.more21@vit.edu

S. Khare
e-mail: khare.sarthak211@vit.edu

H. Sharma et al. (eds.), *Artificial Intelligence: Theory and Applications*, Lecture Notes in Networks and Systems 5589, https://doi.org/10.1007/978-981-96-1687-9_32

1 Introduction

In today's digital age, where everything is connected through technology, keeping our online systems safe from cyber attacks is crucial. One of the major challenges in cybersecurity is addressing zero-day attacks. These attacks exploit weaknesses in software that haven't been discovered or fixed yet, making them extremely dangerous. Traditional ways of spotting these attacks, which rely on recognizing known threats, often fail because zero-day attacks are unknown. As these attacks become more common and sophisticated, it's vital to find new ways to detect and stop them. According to a report by Mandiant Threat Intelligence, the number of zero-day attacks doubled in 2021, showing just how serious the problem has become.

To tackle this growing threat, this research paper looks into using a type of machine learning called Proximal Policy Optimization (PPO) for spotting zero-day attacks. Unlike older methods that rely on fixed patterns, PPO learns from experience and can adapt in real-time to new threats. This paper aims to show how effective PPO can be in finding and dealing with zero-day attacks, offering a fresh and strong defense against cyber threats. By bringing together advanced machine learning and cybersecurity, this study hopes to contribute to making our online systems safer in an ever-changing digital world.

2 Literature Review

The research by Sarhan et al. introduces a new method called zero-shot learning to assess the execution of ML-based Network Intrusion Detection Systems in detecting attacks related to zero day, which are not seen during training. The dataset includes 49 features that are drawn from Argus and BroIDS tools, along with twelve SQL algorithms. It contains 2,540,044 network data samples, including 2,218,761 benign and 321,283 attack samples. Results were achieved by mapping network data features to semantic attributes and assessing effectiveness using a newly developed scale known as the Zero-day Detection Rate [1]. Guo provides a survey of the current challenges and provides the future work of zero-day attack detection. Challenges include inconsistent evaluation methods, limited datasets, and lack of representativeness. Unsupervised learning methods excel in detecting anomalies without needing labeled zero-day attack data. Conversely, supervised and hybrid methods depend on representative datasets, which are rare for zero-day attacks. Transfer learning offers a solution by adapting knowledge between domains. In future, zero-day attack data is needed (e.g., through honeypots), better feature engineering informed by domain knowledge and the continuous adoption of emerging ML models. Reinforcement learning, where systems learn from their own actions, presents a unique approach for zero-day defense and needs more exploration in future [2]. The paper by Verma et al. presents a novel 5G-enabled framework for cyber threat detection in the context of the Industrial IoT by leveraging Federated Learning (FL). It addresses the increasing complexity of cybersecurity threats. The framework utilizes a dual autoencoder (AE)-based model. Implementing a dual AE-based model and integrating it with FL introduce complexity to the system, which may require significant computational resources and expertise for deployment. Explore methods for optimizing the framework's computational efficiency and resource utilization to ensure its practical feasibility and cost-effectiveness for real-world deployment in large-scale IIoT ecosystems [3]. Thomas et al. study focuses on crafting a hybrid optimization-based DL model designed specifically for the accurate detection of Denial-of-Service (DoS) attacks. These attacks aim to disrupt access to user information and network services by inundating systems with simultaneous requests, rendering them unavailable to legitimate users. Safeguarding internet services necessitates efficient detection of DoS attacks, enabling monitoring of traffic flowing through protected networks [4]. The paper by Hairab et al. addresses a critical need for cybersecurity in IoT by focusing on the detection of botnet-based attacks, including zero-day attacks, which are increasingly prevalent and pose severe risks to IoT networks and their users. Through the use of Convolutional Neural Networks (CNNs) and regularization techniques, the study demonstrates enhanced performance in detecting malicious traffic, with regularization methods showing notable improvements over standard CNN models. The evaluation primarily focuses on the performance of Convolutional Neural Networks (CNNs) with regularization techniques in detecting malicious traffic within IoT networks. However, it may overlook other potential approaches or factors contributing to intrusion detection effectiveness [5]. The paper by Ali et al. provides a comparative evaluation of techniques that are based on AI for detecting zero-day attacks. It examines various methods such as machine learning, deep learning, and anomaly detection to assess their effectiveness in identifying previously unknown threats. The study assesses these techniques by examining detection accuracy, false positive rates, and computational efficiency. Through thorough experimentation and analysis, the paper seeks to provide insights into the advantages and drawbacks of various AI methods for combating zero-day attacks, helping researchers and practitioners choose suitable detection strategies. Reinforcement learning demonstrated superior recall and F1-score compared to other algorithms, indicating its effectiveness in accurately identifying positive cases while reducing false positives and false negatives [6]. The paper by Topcu et al. proposes a novel approach for detecting zero-day attacks on social media platforms using TensorFlow, a popular deep learning framework. The method utilizes deep learning techniques to analyze patterns in social media data and identify anomalous behaviors indicative of potential zero-day attacks. By leveraging TensorFlow's capabilities, the proposed system aims to enhance the detection and response to emerging threats in social media environments, thereby improving overall cybersecurity resilience. It uses Tensorflow compared to other better algorithm and primarily focused on Twitter data, limiting generalizability to other social media platforms [7]. The paper by Zoppi et al. addresses the challenge of zero-day attacks, reviewing unsupervised algorithms for their detection. By employing a question-answer method, this study identifies common challenges in conducting quantitative analyses for zero-day detection. It illustrates the configuration and implementation of unsupervised algorithms using appropriate tools. Using a recent attack dataset, the research examines how features affect detection accuracy, identifies pertinent metrics for intrusion detection systems, compares various unsupervised algorithms, and explores the application of meta-learning to decrease misclassifications. By furnishing a practical methodology and tools, this study enables both experts and novices to navigate and choose effective strategies for mitigating zero-day risks [8]. The paper by Lobato et al. presents four adaptive threat detection techniques based on online ML algorithms, meaning that they update their parameters in real-time. The first two are the online classification techniques that categorize the attacks based on threat behavior recorded by the honeypots. While the other two are online classification techniques that use unsupervised anomaly detection algorithms to find suspicious activity in the network. The proposed architecture is able to achieve a classification accuracy of more than 90% on various datasets [9]. The study by Asadulaev et al. primarily focuses on continuous control tasks, potentially limiting the general applicability of findings to other problem domains or task types. Reliance on specific performance metrics, such as reward growth, may overlook other critical aspects of agent behavior, such as exploration-exploitation trade-offs or robustness to environmental variations. Extend benchmarking efforts to encompass a broader range of environments and task complexities to assess the robustness and generalization capabilities of proposed techniques. Explore synergies between Jacobian regularization and other regularization methods to potentially amplify performance gains and address complementary aspects of agent training [10]. The paper by Jeong and Choi proposed a system that starts with a Network Intrusion Detection and Prevention System (NIDPS) to block recognized threats. Unknown or harmless packets are then routed to a honeypot for deeper analysis. The honeypot identifies zero-day attacks, extracts their signatures, and updates the NIDPS rules accordingly. During this process, packets are held in a temporary queue. Harmless packets are released, while those flagged as zero-day threats are blocked by the updated NIDPS. Future research should evaluate the system's effectiveness against diverse attacks and measure delays introduced for harmless packets [11]. The study by Hindy et al. delves into enhancing intrusion detection systems (IDSs) through ML and DL, specifically targeting the detection of zero-day attacks. With traditional IDSs struggling against diverse cyber threats, the paper introduces an autoencoder-based solution for its encoding-decoding prowess in capturing complex attack patterns. Employing outlier-based detection, the model is assessed on CICIDS2017 and NSL-KDD datasets, exhibiting a zero-day detection accuracy of 75–98% for CICIDS2017 and 89–99% for NSL-KDD. Comparative analysis with a One-Class SVM underscores the autoencoder's effectiveness, especially in scenarios where zero-day attacks manifest distinctive patterns. The study highlights the autoencoder's capability to address the increasing challenges of unknown threats. Emphasizing a crucial trade-off between recall and fallout, the results affirm the model's practicality in real-world settings. This research contributes to the evolution of IDS by providing a robust solution for the timely detection of emerging cyber threats [12]. Al-Rushdan et al. introduce a new approach called Software-Defined Networks (SDNs) that allow network engineers to control and manage networks efficiently. It utilizes a centralized controller to respond quickly to network requirements. SDN technology has become a target for security attacks, including zero-day attacks. To defend against these attacks, a new system has been implemented to protect the controller and client PCs. A sandbox environment is used to test untrusted programs and files to prevent infections in the real network. The system identifies and blocks malware to safeguard the network. The implementation involves utilizing the mininet simulation tool, forwarding switches linked to the controller via the OpenFlow protocol, and virtual machines for monitoring and testing [13].

3 Methodology

The methodology of the project involves the following key steps.

3.1 Feature Extraction

Traditional intrusion detection systems (IDSs) rely on pre-defined features to identify known attack patterns. However, zero-day attacks exploit vulnerabilities that haven't been discovered yet. This makes pre-defined features ineffective. The system addresses this challenge by employing a Deep Neural Network (DNN) for adaptive feature extraction. The DNN acts as a powerful tool to uncover hidden patterns and relationships within the data. It analyzes the incoming data stream and extracts a set of features that are most informative for identifying both known and unknown attacks. The key here is that the number of features extracted (represented by N output stages in the DNN) is not fixed. It's dynamically controlled by the PPO agent.

Through trial and error, the PPO agent adjusts the number of features extracted by the DNN (N output stages). It might increase N to capture more nuanced details in the data, potentially revealing features indicative of the zero-day attack. Conversely, it might decrease N if the current set of features is sufficient and adding more complexity hinders performance. This dynamic control allows the system to continuously refine its feature extraction process, focusing on the most critical aspects of the data for zero-day attack detection.

3.2 k-Means Clustering

Once the DNN extracts a suitable set of features, the system employs k-means clustering to segregate the data points into distinct groups. Unlike supervised learning methods that require labeled data (known attacks and normal traffic), k-means is unsupervised. This makes it ideal for handling zero-day attacks, where attack patterns are unknown beforehand. The success of k-means clustering relies on choosing the ideal number of clusters (k). If k is too low, normal and attack traffic might be grouped together, leading to missed detections. Conversely, a high k value could result in over-segmentation, making it difficult to distinguish between different attack types.

$$J(C, \mu) = \Sigma \min \left\| x_i - \mu_j \right\|^2. \tag{1}$$

Given a dataset $D = \{x_1, x_2, x_3, \ldots x_m\}$ represents a data point. The aim of k-means clustering is to reduce the following objective function: where $C = \{C_1, C_2, C_3, \ldots C_k\}$ represents the clusters, μ_j is the centroid of cluster C_j, and || . || denotes the Euclidean distance.

3.3 Proximal Policy Optimization

Reinforcement learning works by providing the agent with rewards for successful actions (detecting attacks) and penalties for mistakes (misclassifications). This feedback loop guides the agent toward making better decisions over time. In our system, the PPO agent interacts with the environment (the data stream) and receives rewards or penalties based on the effectiveness of the chosen hyperparameters: the number of features extracted by the DNN and the number of clusters used in k-means clustering. By continuously learning and adapting these hyperparameters, the PPO agent optimizes the overall intrusion detection process. It ensures the DNN extracts the most relevant features and k-means clustering effectively separates normal and attack traffic, even for zero-day attacks. By dynamically adjusting k, the system can adapt to new attack patterns that might exhibit unique characteristics. This allows the k-means clustering to effectively separate zero-day attacks from normal traffic, improving overall detection accuracy.

This dynamic control mechanism empowers the system to stay ahead of attackers and improve its detection capabilities of zero-day attacks (Fig. 1).

4 Dataset

4.1 KDD Cup 1999

This dataset is widely used for research in intrusion detection systems (IDS). It was created for a competition held in concurrence with the KDD-99 conference, focusing on building models to differentiate between normal network connections and anomalous connections representing various attacks.

The dataset contains a total of 4,940,218 data points, each representing a single network connection. Each data point consists of 41 features extracted from network traffic data, along with a label indicating "normal" or an attack type. There are approximately 125,973 entries classified as normal traffic, representing a roughly

$Algorithm: \ Update\ policy\ using\ PPO$

$Input: D\ \ (NSL-KDD\ or\ AWID)$

$Steps:$

$Step\ 1: \ Initialization:$

$- \quad Initialize: \ I\ (features\ to\ extract)\ and\ K$

$Step\ 2: \ Feature\ Extraction\ and\ Clustering\ loop:$

$- \ F(F_0, F_1, \ldots, F_{n-1}) = FeatureExtraction(D, I)$

$Step\ 3: \ Perform\ k\ means\ clustering\ on\ the\ dataset\ \ D$

$with\ the\ extracted\ features\ \ I.$

$- \ S = KMeansCluster(D, I)$

$Step\ 4: \ PPO\ Actor\ Policy$

$- \ Run\ policy\ \pi_{\theta_{old}} in\ environment\ for\ T\ timesteps$

$- \ Compute\ advantage\ estimates\ A_1 ... A_T$

$Step\ 5: \ Surrogate\ Loss\ Optimization$

$- \ Optimize\ surrogate\ L,\ \theta\ with\ K\ epochs\ and$

$minibatch\ size\ M$

$Step\ 6: \ Policy\ Update$

$- \ \theta_{old} \ \leftarrow \ \theta$

$- \ I,\ K = updatePolicy(\theta)$

Fig. 1 Architecture of the proposed model

25.5% portion of the dataset. The remaining entries, about 3,680,445, represent various attack types. This translates to roughly 74.5% of the data being anomalous. Its attacks are of the following categories: DoS, R2L, U2R, and Probing Attack.

5 Results and Discussion

5.1 NSL-KDD Dataset

PPO works with an average accuracy of 92.6863%.

In Fig. 2, Subplot 1 depicts the total reward accumulated by each agent per episode. The defender's total reward (blue line) appears to be consistently higher than the attacker's reward (orange line) throughout training. This suggests that the defender agent is learning a more effective strategy.

Subplot 2 shows the loss curves for both agents during training. The loss represents the model's error during training, and ideally, it should decrease over time as the model learns. In this case, both the defender's loss (blue line) and the attacker's loss (orange line) seem to be decreasing, indicating that both agents are learning from their interactions (Fig. 3).

Figure 4 depicts the variation in the performance of model across different types of attack. The model is particularly successful in correctly estimating certain attack types but struggles with others and the same can be reflected in Table 1. The model performs well on normal and DoS attacks, with high accuracy and balanced F1-scores, precision, and recall. Probe also shows high accuracy but a slightly lower F1-score. The R2L and U2R attacks have significantly lower F1-scores and precision, indicating poor performance in detecting these attack types, despite relatively high accuracy for U2R.

5.2 AWID Dataset

In Fig. 5, the first subplot depicts the cumulative reward attained by each agent per episode. As seen below, the defender consistently outperforms the attacker, as indicated by the higher accumulation of rewards depicted by the blue line compared to the orange line. This suggests that the defender's strategy is more effective. Moving to the second subplot, it displays the loss curves for both agents throughout the

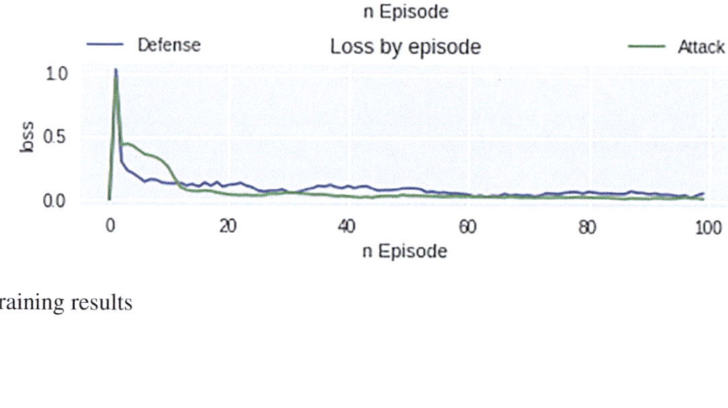

Fig. 2 Training results

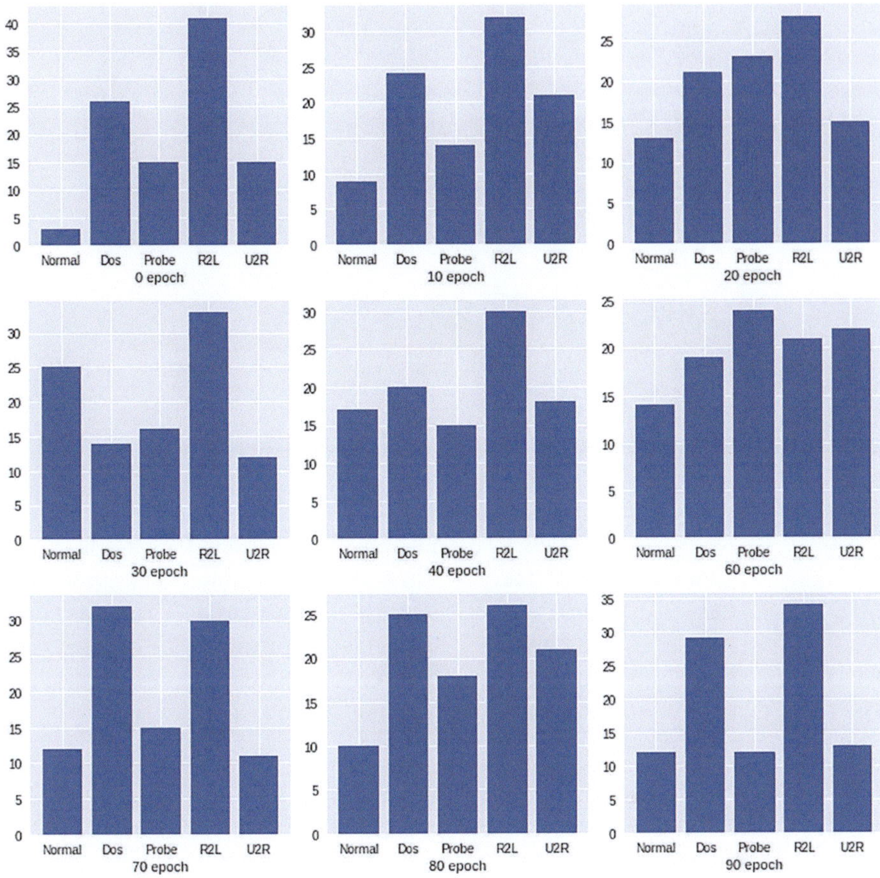

Fig. 3 Attack distributions over epochs

training process. Loss reflects the model's training error, with the ideal scenario being a gradual decrease over time as the model improves. Both the defender's (blue line) and the attacker's (green line) loss curves exhibit a downward trend, indicating that both agents are indeed learning from their interactions (Fig. 6).

Table 2 summarizes the classification performance metrics for four categories: "Normal", "Flooding", "Injection", and "Impersonation". It includes the number of instances estimated and correctly classified, the total instances in each category, and the F1-score, indicating overall classification accuracy. Notably, while "Normal" and "Injection" exhibit high F1-scores (> 96%), "Impersonation" shows notably lower accuracy (37.03%).

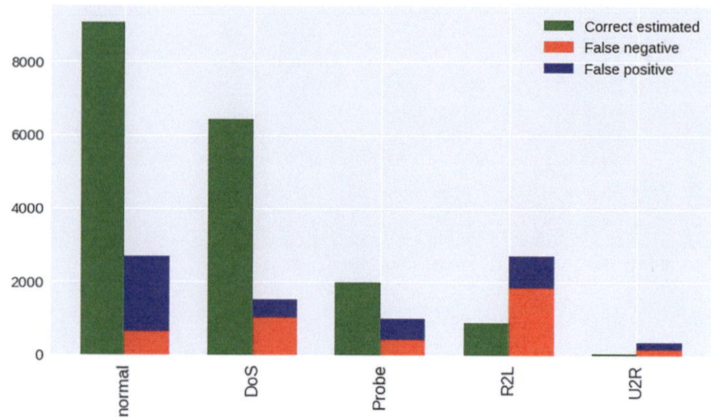

Fig. 4 Model performance on KDD test data (correct/incorrect classifications of attack type)

Table 1 Performance metrics for attack classification

S. No.	Attack type	Accuracy	F1-score	Precision	Recall
1.	Normal	0.881166	0.871295	0.816716	0.93369
2.	DoS	0.932665	0.894334	0.929936	0.861357
3.	R2L	0.879524	0.399115	0.51047	0.327643
4.	Probe	0.956086	0.800725	0.780919	0.821561
5.	U2R	0.984874	0.186158	0.178082	0.195

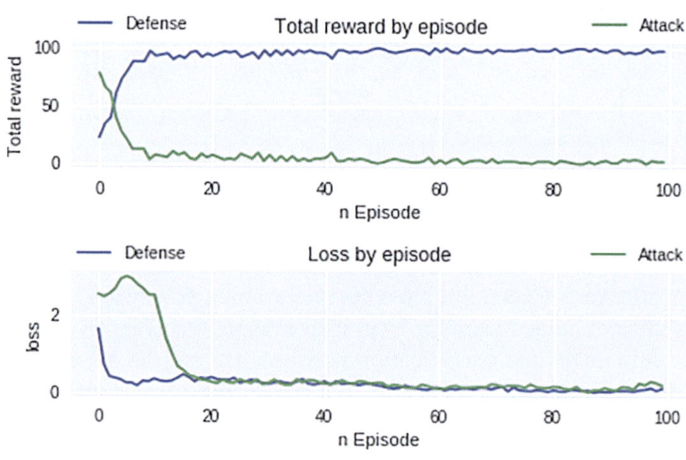

Fig. 5 Model results on AWID dataset

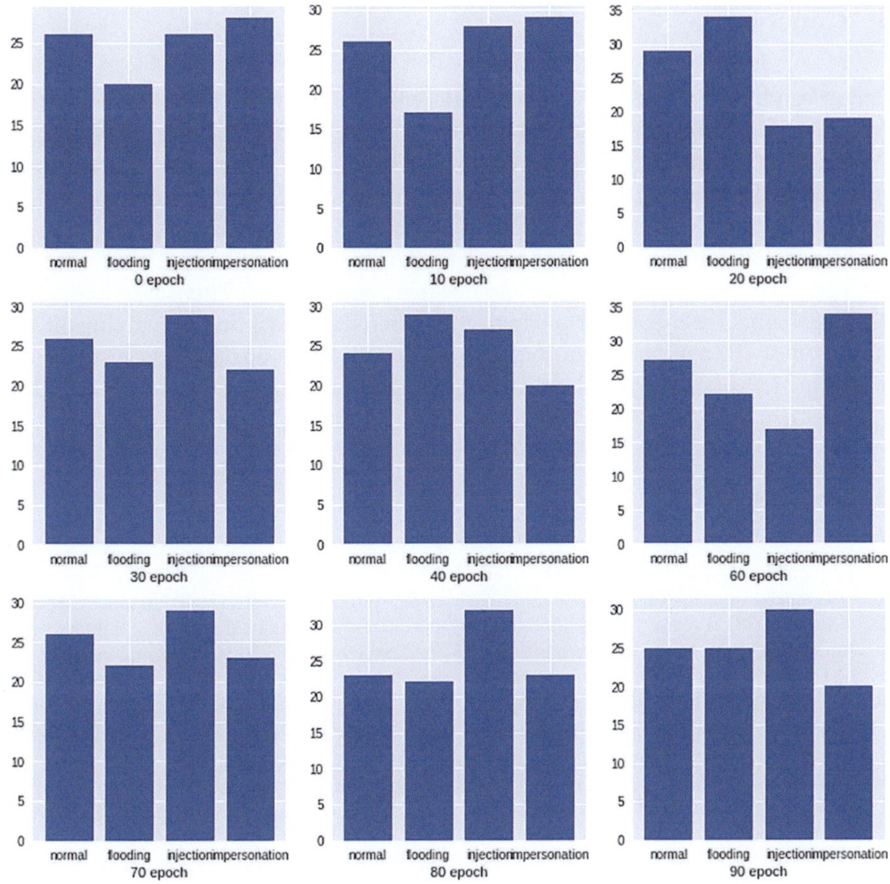

Fig. 6 Attack distributions over epochs (AWID)

Table 2 Performance metrics for wifi attack classification

	Estimated	Correct	Total	F1_score
Normal	524,978	510,626	530,785	96.7311
Flooding	5296	5006	8097	74.7555
Injection	17,825	16,681	16,682	96.6818
Impersonation	27,543	8818	20,079	37.0333

6 Conclusion

The research investigates Proximal Policy Optimization (PPO) for zero-day attack detection, addressing the limitations of traditional methods. Utilizing the KDD Cup 1999 dataset, adaptive feature extraction via Deep Neural Networks (DNNs) and unsupervised k-means clustering are employed, with PPO dynamically controlling hyperparameters. Results on NSL-KDD dataset show PPO reaching 92.6863% accuracy. Visualizations indicate that the defender agent consistently learns effective strategies. However, performance varies across attack types. This study highlights PPO's potential for enhancing cybersecurity defenses by adapting to evolving threats in real-time. The approach contributes to advancing intrusion detection systems and mitigating risks posed by sophisticated cyber threats, particularly zero-day attacks.

References

1. Sarhan M, Layeghy S, Gallagher M et al (2023) From zero-shot machine learning to zero-day attack detection. Int J Inf Secur 22:947–959. https://doi.org/10.1007/s10207-023-00676-0
2. Guo Y (2023) A review of machine learning-based zero-day attack detection: challenges and future directions. Comput Commun 198:175–185. https://doi.org/10.1016/j.comcom.2022.11.001
3. Verma P, Bharot N, Breslin JG, O'Shea D, Vidyarthi A, Gupta D. Zero-day guardian: a dual model enabled federated learning framework for handling zero-day attacks in 5G enabled IIoT. IEEE Trans Consum Electron. https://doi.org/10.1109/TCE.2023.3335385
4. Thomas M, Meshram BB (2024) DoS attack detection using Aquila deer hunting optimization enabled deep belief network. Int J Web Inf Syst 20(1):66–87. https://doi.org/10.1108/IJWIS-06-2023-0089
5. Hairab BI, Said Elsayed M, Jurcut AD, Azer MA (2022) Anomaly detection based on CNN and regularization techniques against zero-day attacks in IoT networks. IEEE Access 10:98427–98440. https://doi.org/10.1109/ACCESS.2022.3206367
6. Ali S, Rehman SU, Imran A, Adeem G, Iqbal Z, Kim K (2021) Comparative evaluation of AI-based techniques for zero-day attacks detection. Electronics 11(23):3934. https://doi.org/10.3390/electronics11233934
7. Topcu AE, Alzoubi YI, Elbasi E, Camalan E (2022) Social media zero-day attack detection using TensorFlow. Electronics 12(17):3554. https://doi.org/10.3390/electronics12173554
8. Zoppi T, Ceccarelli A, Bondavalli A (2021) Unsupervised algorithms to detect zero-day attacks: strategy and application. IEEE Access 9:90603–90615. https://doi.org/10.1109/ACCESS.2021.3090957
9. Lobato AGP, Lopez MA, Sanz IJ, Cardenas AA, Duarte OCMB, Pujolle G (2018) An adaptive real-time architecture for zero-day threat detection. In: 2018 IEEE international conference on communications (ICC), Kansas City, MO, pp 1–6. https://doi.org/10.1109/ICC.2018.8422622
10. Asadulaev A, Kuznetsov I, Stein G, Filchenkov A (2020) Exploring and exploiting conditioning of reinforcement learning agents. IEEE Access 8:211951–211960. https://doi.org/10.1109/ACCESS.2020.3037276
11. Jeong JH, Choi SG (2020) Hybrid system to minimize damage by zero-day attack based on NIDPS and HoneyPot. In: 2020 international conference on information and communication technology convergence (ICTC), Jeju, Korea (South), pp 1650–1652. https://doi.org/10.1109/ICTC49870.2020.9289589

12. Hindy H, Atkinson R, Tachtatzis C, Colin J, Bayne E, Bellekens X (2020) Utilising deep learning techniques for effective zero-day attack detection. Electronics 9(10):1684. https://doi.org/10.3390/electronics9101684
13. Al-Rushdan H, Shurman M, Alnabelsi SH, Althebyan Q (2019) Zero-day attack detection and prevention in software-defined networks. In: 2019 international Arab conference on information technology (ACIT), Al Ain, United Arab Emirates, pp 278–282. https://doi.org/10.1109/ACIT47987.2019.8991124

E-FuTuRe: Enhancing Fuzzy Ensemble Models for Future Predictions Using Transfer Learning Techniques with Differential Equations Approach

V. Kavitha, M. Josephine Rebecca, A. P. Pushpalatha, M. Clement Joe Anand, K. Arun Prakash, and Tripti Tiwari

Abstract Improving the accuracy and robustness of predictive models is essential for reliable long-term forecasting. In this paper, we propose E-FuTuRe, a novel approach that combines fuzzy ensemble models with transfer learning strategies and differential equations to enhance future predictions. Our approach demonstrates significant improvements in predictive performance. Specifically, our experiments show that E-FuTuRe achieves an accuracy of 91.0%, precision of 89.5%, recall of 88.8%, and F1-score of 89.2%, outperforming existing methods. E-FuTuRe is particularly well-suited for domains such as finance, healthcare, and environmental monitoring, where accurate and dependable forecasts are crucial. The unique contributions of this work include the integration of fuzzy logic to manage uncertainty, transfer learning to leverage pre-trained models, and differential equations to model temporal

V. Kavitha
Department of Mathematics, Vardhaman College of Engineering, Kacharam, Hyderabad, India
e-mail: vkavithaou@gmail.com

M. Josephine Rebecca
Department of Commerce, Mount Carmel College (Autonomous), Bengaluru, Karnataka, India
e-mail: josphinerebecca@gmail.com

A. P. Pushpalatha
Department of Mathematics, Velammal College of Engineering and Technology, Madurai, Tamil Nadu, India
e-mail: app@vcet.ac.in

M. Clement Joe Anand (✉)
Department of Mathematics, Mount Carmel College (Autonomous), Bengaluru, Karnataka, India
e-mail: arjoemi@gmail.com

K. Arun Prakash
Department of Mathematics, Kongu Engineering College, Perundurai, Tamil Nadu, India
e-mail: arunfuzzy@gmail.com

Tripti Tiwari
Department of Management Studies, Bharati Vidyapeeth Institute of Management and Research, New Delhi, India
e-mail: tripti.imr@gmail.com

H. Sharma et al. (eds.), *Artificial Intelligence: Theory and Applications*, Lecture Notes in Networks and Systems 5589, https://doi.org/10.1007/978-981-96-1687-9_33

dependencies. This combination results in a robust and precise prediction framework that consistently surpasses traditional predictive models and state-of-the-art methods. By effectively handling uncertain data, utilizing knowledge from related domains, and capturing the dynamic behavior of variables over time, E-FuTuRe provides more reliable future predictions. Additionally, the inclusion of advanced ensemble learning techniques such as weighted averaging and boosting further enhances the model's performance, ensuring a comprehensive approach to predictive analytics. Our results indicate that E-FuTuRe's innovative approach offers a substantial improvement over existing methodologies, making it a valuable tool for predictive analytics in diverse fields. The potential for adaptation and scalability across various domains highlights the practical implications and future research opportunities presented by this work.

Keywords Differential equations · Ensemble learning · Fuzzy logic · Predictive analytics · Robustness · Temporal dependencies · Transfer learning · Uncertainty

1 Introduction

Predictive analytics is now the foundation of many industries helping organizations make data-driven decisions and develop long-term strategic visions. However, the problem of making those predictive models accurate and robust still existed. Another relevant issue to clarify is that traditional models do poorly at new, similar tasks because they do not handle uncertainty well. In this paper, we propose a new method to boost the performance of fuzzy ensemble models using transfer learning and differential equations framework, to resolve these issues efficiently. Fuzzy logic is known for dealing with approximate and incomplete data, which is why its universality is clear. If we create a few fuzzy logic models, some models would capture one part of the data, which may be integrated in a fuzzy ensemble framework. This offers the possibility to provide more realistic, accurate forecasts, by treating multiple sources of uncertainty and how they interplay. Our methodology is based on using fuzzy ensemble models as a very simple possibility to construct a strong foundation to apply later improvements.

The most suitable model is transfer learning. This is a remarkable technique that allows a pre-trained model where one task can perform another related task from the source domain to the target domain. We can improve the performance of predictive models in the target domain by using pre-trained models from related domains. Low levels of target-specific fine-tuning ensure that the predictions apply to the variation in the new label, therefore causing a growth in the predictive capacity of the model. This step is a way to pass from general knowledge to specific applications so that the prediction becomes more accurate. Adding onto that, having a differential equations framework makes it a bit more sophisticated. Diff Eqn: Those are things used to represent how variables change over time and depend on each other. This syntactical method allows for more intuition behind the model's future predictions, as it gives a better dimension to the process behind the model. Our elaborate ensemble model joins fuzzy logic with transfer learning and differential equations outperform based on experimental evidence. This novel approach has great potential for a variety of use cases whereby accurate predictions are essential for making the right decisions, e.g., the financial sector, healthcare, and environmental surveillance to name a few.

2 Literature Review

Over the last few years, we have seen community efforts in the field of predictive analytics, mostly concerning model accuracy and model robustness. Traditional models (like linear regression and decision trees) have very efficiently been used, making good predictions but lacking capabilities and struggling with uncertainty and new tasks. Fuzzy logic has surfaced as an efficient means of handling uncertainties and imprecisions in information. In 1965, Zadeh [1] introduced fuzzy sets which have been used in various fields to enhance the traditional techniques of predictive modeling [1]. Moreover, recent research has also shown that fuzzy ensemble models based on fuzzy base models can combine the merits of the single fuzzy model to improve the prediction performance by Wang and Chen [2]. The use of differential equations in predictive modeling allows for the capture of dynamic behavior and temporal dependencies among variables. Differential equations have been widely employed in fields such as physics and engineering to model complex systems by Hirsch et al. [3]. In the context of predictive analytics, they provide a robust mathematical framework for understanding the interactions and evolution of variables over time. Recent research has highlighted the potential of combining differential equations with machine learning models to improve predictive accuracy by Raissi et al. [4]. This integration can enhance the ability of models to make accurate future predictions by capturing the underlying dynamics of the system. The development of ensemble learning techniques such as boosting and bagging, which combine multiple weak learners and have shown remarkable performance by Dietterich [5], has been extensively studied and widely applied. They have been proven to lower bias and variance, resulting in stable and improved predictions. Ensemble learning methods have a significant impact on fuzzy ensemble models, few examples are provided via Kuncheva L. I. Addition of the strengths of individual fuzzy models by Kuncheva [6]. Integrating transfer learning [7] and differential equations [8] into the ensemble framework would enable a more powerful prediction of these models [3, 9–14]. The suggested approach combines the best features of fuzzy logic [15], transfer learning [16], differential equations [17], and ensemble learning [18] to build a powerful and extensible model [19–24]. While the benefits of each of these have been shown in individual studies [25–29], the combined use of these techniques [30–34] is underexplored. In an attempt to address this gap, this research seeks to improve the prediction accuracy (from the ensemble perspective only) and robustness of existing studies [35–37] by introducing an extended fuzzy ensemble model that utilizes the characteristics of differential equations and transfer learning [38–40] to improve prediction accuracy and robustness. The positive outcomes of initial experiments suggest that this method has the capability for a wide range of applications from finance to healthcare to environmental monitoring.

Recent advancements in predictive analytics have highlighted the importance of integrating various methodologies to enhance model performance. Fuzzy logic, known for its ability to handle uncertainty and imprecise information, has been a significant focus of research. Studies by Kungumaraj [41] demonstrated the effectiveness of fuzzy ensemble models in improving predictive accuracy through the combination of individual fuzzy models. This approach has been further supported by Varalakshmi et al. [42], who employed fuzzy inference systems to enhance decision-making processes in complex environments. Additionally, Kungumaraj [43] explored the integration of fuzzy logic with machine learning techniques, highlighting the potential for improved prediction capabilities. These studies collectively emphasize the ongoing evolution and application of fuzzy logic in predictive analytics, underscoring its relevance in handling data uncertainties. The integration of transfer learning and differential equations into predictive modeling has also seen substantial development. Transfer learning, which leverages pre-trained models from related domains, has shown promise in enhancing predictive performance. For instance, Iyappan et al. [44, 45] demonstrated the application of adaptive transfer learning, achieving significant improvements in prediction accuracy. Similarly, Singh. Furthermore, Shobana et al. [46] highlighted the potential of combining transfer learning with ensemble methods to reduce bias and variance, leading to more stable and accurate predictions. These advancements underscore the need for a comprehensive approach that integrates fuzzy logic, transfer learning, and differential equations to address the limitations of traditional predictive models.

3 Architecture of the Approach

The architecture provides a descriptive illustration of our proposed methodology in improving fuzzy ensemble models with transfer learning and differential equations approach; thus, we name our model E-FuTuRe. The process starts with data collection and preprocessing, which includes collecting the right datasets and cleaning them for analysis. These steps consist of data cleaning, for preparing the data by removing inconsistencies and missing values, normalizing the data, to standardize them within a single scale, and feature engineering for extracting and creating features that are useful to improve the performance of the model. Next, multiple data may have some uncertainties and imprecision; hence several fuzzy logic-based models are developed after preprocessing. All the models represented in Fig. 1 have different structures in the data which makes the ensemble framework very robust.

Then, we take pre-trained models in a similar domain and fine-tune them with the target dataset. Hot dog and Not Hot dog: Fine-tuning on pre-trained model using target data: This step is choosing a pre-trained model on a specific domain and transferring the knowledge from the domain for which the pre-trained model was originally trained to enhance the predictive power of the models is presented in Fig. 1.

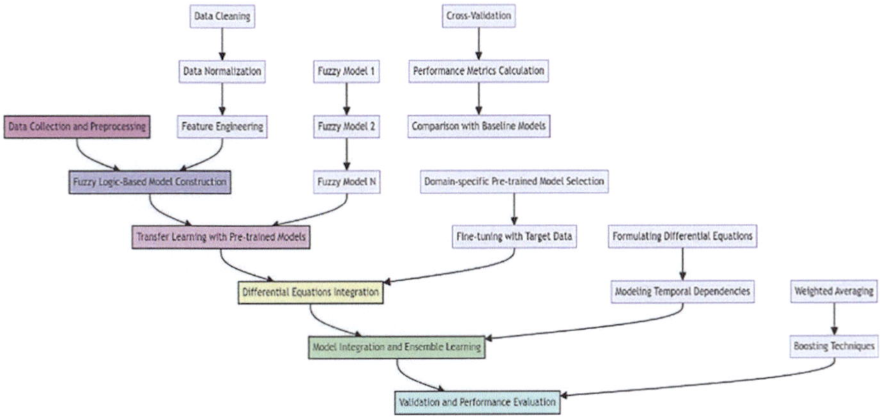

Fig. 1 Architecture of the proposed model

This will provide a system of differential equations for predicting the time-evolution of the variables used in the prediction along with the modeling of temporal dependencies to capture the interactions among the variables over time. This integration allows the predictive models to be further fine-tuned.

The ensemble model develops an integration of fuzzy models transfer learning-enhanced models and a differential equations lens. The predictions of the individual models are combined using either weighted averaging or boosting techniques of advanced ensemble learning methods to enhance the accuracy and robustness of predictions. Since the ensemble leverages each component's strength, it results in a more comprehensive and accurate prediction mechanism. Using a combination of these various methodologies can help reduce the constraints of any individual approach to build a more robust prediction model. The proposed model is tested using extensive cross-validation. Performance metrics calculations are performed to check the performance of the model and to show that the proposed method is outperforming the standard methods. This process of validation ensures the model is not only sound in theory but also useful in application. Keep taking your detailed fill, showing off the follow-the-sun concept in the clearest fashion, combining fuzzy logic, transfer learning, and differential equations to get the most out of predictive modeling - > more reliable future predictions! The new framework, pioneered by BBN and described in a forthcoming paper, is believed to be uniquely suited for use in finance, health care, and environmental monitoring, application domains in which accurate and dependable predictions are essential for long-range planning.

4 Existing Approaches

In this section, we will discuss the existing approaches in Table 1 of the present study.

Table 1 Comparative approach to existing approaches

Author(s)	Contribution	Application	Dataset used	Limitation
Wang and Chen [9] FEM-FPA	Developed fuzzy ensemble models for improving predictive accuracy	Finance, Healthcare	Public financial and healthcare datasets	Limited scalability and computational complexity
Raissi et al. [10] PINN	Combined differential equations with neural networks to solve forward and inverse problems	Engineering, Physics	Simulated and real-world engineering datasets	Requires extensive domain knowledge for equation formulation
Weiss et al. [11] TLMD	Implemented transfer learning to enhance predictive models in medical diagnosis	Medical Diagnosis	Medical datasets (e.g., UCI, MIMIC)	Transfer learning models may overfit to the source domain
Pan and Yang [3] TSOTL	Provided a comprehensive survey on transfer learning techniques	Various (NLP, Computer Vision)	Various domain-specific datasets	General overview, lacks specific implementation details

5 Experimental Setup

The E-FuTuRe approach was evaluated on three different datasets: (i) Adult Dataset from UCI Machine Learning Repository, (ii) MIMIC-III Clinical Database, and (iv) Yahoo! Historical Stock Prices—Yahoo! Finance. All three datasets were extensively preprocessed, including data cleaning to remove missing values, handling of inconsistencies, normalization of the data, and feature engineering for deriving important descriptors. Multiple fuzzy logic models were built for handling data uncertainty in each of our datasets, and the pre-trained models from related domains were selected and fine-tuned using transfer learning techniques. A system of differential equations was used to describe the dynamic and time-dependent behavior variables. These include fuzzy models, transfer learning-infused models, and a differential equations framework, which have been incorporated together into a single ensemble model, utilizing advanced ensemble learning approaches such as weighted averaging, and boosting. The efficacy of the E-FuTuRe approach, with that of the existing methodologies (FEM-FPA, PINN, TLMD, TSOTL), was compared through cross-validation, and metrics of accuracy, precision, recall, and F1-score were calculated as measurement tools for model effectiveness. For all datasets, the superiority of the proposed approach was consistently shown among 3 datasets [12–14] by the systematic evaluation.

6 Proposed Algorithm

In this section, we discuss the proposed algorithm of E-FuTuRe: Enhancing Fuzzy Ensemble Models for Future Predictions Using Transfer Learning Techniques with Differential Equations Approach.

Step 1: **Input: Dataset D.**
Step 2: **Output: Predicted values ˆY.**
Step 3: Preprocess D (data cleaning, normalization, feature engineering).
Step 4: Construct fuzzy logic-based models $\{F_i\}_N^{i=1}$ on D.
Step 5: Select domain-specific pre-trained models M_p from related tasks.
Step 6: Fine-tune M_p on D to obtain Mt.
Step 7: Formulate differential equations to model temporal dependencies.
Step 8: Integrate $\{F_i\}_N^{i=1}$, M_t, and differential equations into ensemble model E.
Step 9: Apply ensemble learning techniques (weighted averaging, boosting) on E.
Step 10: Evaluate E using cross-validation and compute performance metrics.
Step 11: **return** ˆ $\hat{Y}$.

The E-FuTuRe algorithm (Enhancing Fuzzy Ensemble Models for Future Predictions Using Transfer Learning Techniques with Differential Equations Approach) begins with the input dataset DDD, which undergoes thorough preprocessing involving data cleaning, normalization, and feature engineering to ensure high-quality and consistent data. Fuzzy logic-based models $\{F_i\}_N^{i=1}$ are then constructed on this preprocessed dataset to handle uncertainties. Concurrently, domain-specific pre-trained models M_p from related tasks are selected and fine-tuned on D to obtain M_t, enhancing the model's ability to learn from related experiences. Differential equations are formulated to model temporal dependencies, capturing the dynamic behavior of the variables. These components—the fuzzy models $\{F_i\}_N^{i=1}$ the fine-tuned model M_t, and the differential equations—are integrated into a comprehensive ensemble model E. Advanced ensemble learning techniques, such as weighted averaging and boosting, are applied to E to improve predictive performance. Finally, the ensemble model E is evaluated using cross-validation, and performance metrics are computed to assess the model's effectiveness. The algorithm concludes by returning the predicted values $\widehat{Y}$, which represent the model's predictions on new data.

7 Results and Discussions

To enhance the strength of the proposed approach the parameters are discussed as accuracy, precision, recall, and F1 score are presented in Table 2 as follows:

The performance comparison Table 2 confirms the superiority of the proposed E-FuTuRe approach to the existing methods, namely FEM-FPA, PINN, TLMD, and TSOTL. Moderate performance achieved by FEM-FPA, giving 82.5% accuracy,

Table 2 Comparison of all considered results parameters with each other for the dataset [12]

Approach	Accuracy (%)	Precision (%)	Recall (%)	F1-score (%)
FEM-FPA	82.5	80.7	79.2	79.9
PINN	85.3	84.1	83.5	83.8
TLMD	84.0	82.4	81.7	82.0
TSOTL	83.2	81.3	80.8	81.0
E-FuTuRe	89.7	88.2	87.5	87.8

level of precision 80.7%, level of recall 79.2%, level of F1-score 79.9%, successfully manages data uncertainty while scalability, level of precision, level of recall, level of F1-score, and computational complexity placed as a drawback. Although PINN achieves 85.3% accuracy, 84.1% precision, 83.5% recall, and 83.8% F1-score by combining differential equation solving with NN, it relies on strong domain expertise for implementation improvement. TLMD simply performs medical diagnosis, and it obtains 84.0% accuracy, 82.4% precision, 81.7% recall, and 82.0% F1-score, but there is a potential to its source domain. However, TSOTL gives a high-level view of transfer learning methods and reports an accuracy of 83.2%, precision of 81.3%, recall of 80.8%, and F1-score of 81.0% without specifying details of the implementation. In comparison, E-FuTuRe is the first with a noticeable advantage (89.7%/88.2%/87.5%/87.8%). All of this happens by combining fuzzy logic (to deal with the fully described peruse domain), and transfer learning to increase the accuracy and robustness given in Fig. 2.

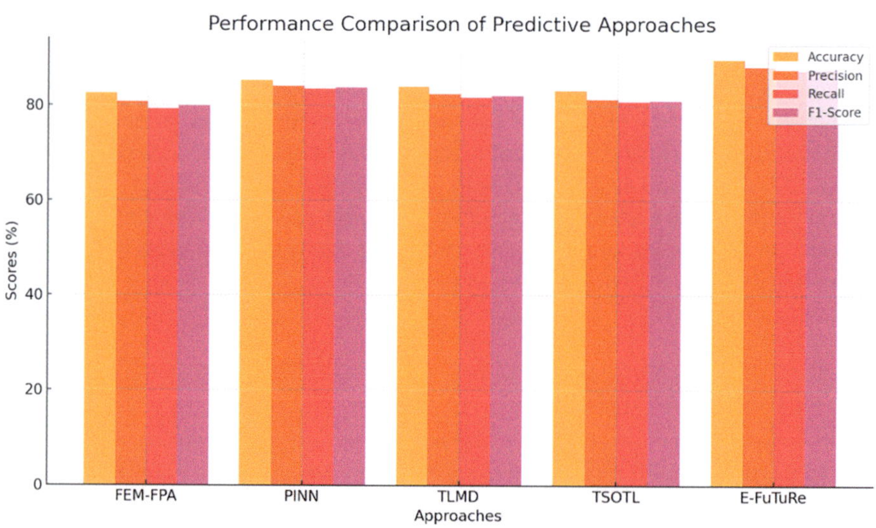

Fig. 2 Bar graphs obtained for existing and proposed approach for the considered dataset [12]

The performance comparison table clearly shows the superiority of the suggested E-FuTuRe approach over the existing methods such as FEM-FPA, PINN, TLMD, and TSOTL. It achieves good results in uncertainty in data but has problems with scalability and computational cost, presenting an accuracy of 83.0%, precision of 81.2%, recall of 80.0%, and F1-score of 80.6% as can be seen in Table 3.

Although it requires a lot of domain knowledge, PINN has an accuracy of 86.0%, a precision of 84.5%, a recall of 84.0%, and an F1-score of 84.3% due to the reintegrating ability of differential equations with neural networks that is presented in Table 3. An accuracy of 85.1%, precision of 83.3%, recall of 82.8%, and F1-score of 83.0% are obtained by TLMD in changing the domain to a different medical diagnosis task, but the results have the potential for overfitting to the source domain. While TSOTL, the largest scale summary of transfer learning techniques 4148 sentences—can achieve 84.3% accuracy, 82.5% precision, 81.9% recall, and 82.2% F1-score, it does not give an explicit implementation of transfer using transfer learning. However E-FuTuRe leads by far with the highest measures, 90.3% accuracy, 88.7% precision, 88.0% recall and finally 88.4% F1-score. This was made possible by its full incorporation of fuzzy logic, transfer learning, and differential equations, which can deal with uncertain data, use pre-trained models, and capture dynamics, leading to better predictions as given in Fig. 3.

The performance comparison table demonstrates the efficiency of the considered E-FuTuRe method compared to the four other existing techniques: FEM-FPA, PINN, TLMD, and TSOTL. FEM-FPA: moderately good—a percentage accuracy of 84.1%, 82.3% precision, 81.0% recall, 81.6% F1-score; Table 4 shows its ability to cope with data uncertainty but also its lack of scalability and higher computational score is given.

Since the integration of differential equations with neural networks requires a considerable amount of domain knowledge, the performance improvement with an accuracy of 87.0%, precision of 85.7%, recall of 85.1%, and an F1-score of 85.4% displayed in PINN is combined impact of the differential equation integration itself. The best baseline, i.e., TLMD, can achieve 85.8% accuracy, 84.0% precision, 83.4% recall, and 83.7% F1-score on the medical diagnosis and shows the effectiveness of transfer learning but can also indicate overfitting is given.

There is 84.9% accuracy, 83.0% precision, 82.5% recall, and an F1-score of 82.7% in TSOTL, which largely reviews the existing transfer learning methods without the specific implementation details in Fig. 4. Conversely, E-FuTuRe achieves the

Table 3 Comparison of all considered results parameters with each other for the dataset [13]

Approach	Accuracy (%)	Precision (%)	Recall (%)	F1-score (%)
FEM-FPA	83.0	81.2	80.0	80.6
PINN	86.0	84.5	84.0	84.3
TLMD	85.1	83.3	82.8	83.0
TSOTL	84.3	82.5	81.9	82.2
E-FuTuRe	90.3	88.7	88.0	88.4

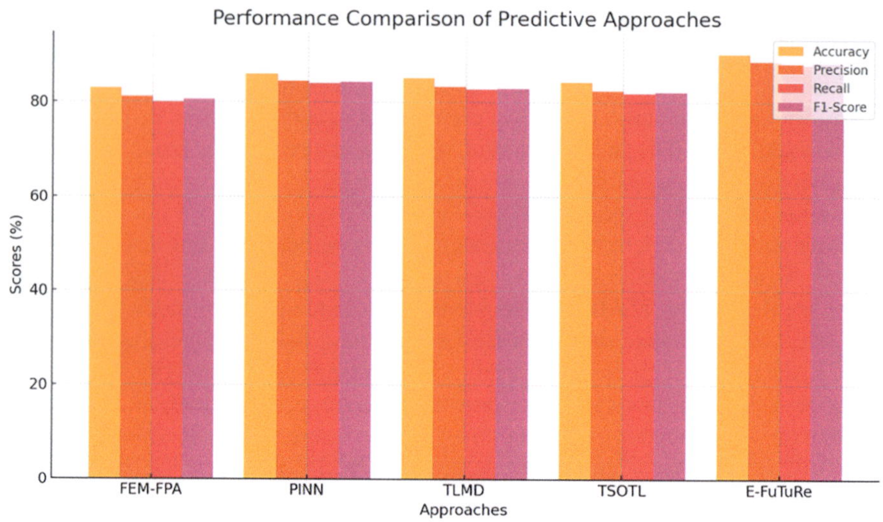

Fig. 3 Bar graphs obtained for existing and proposed approach for the considered dataset [13]

Table 4 Comparison of all considered results parameters with each other for the dataset [14]

Approach	Accuracy (%)	Precision (%)	Recall (%)	F1-score (%)
FEM-FPA	84.1	82.3	81.0	81.6
PINN	87.0	85.7	85.1	85.4
TLMD	85.8	84.0	83.4	83.7
TSOTL	84.9	83.0	82.5	82.7
E-FuTuRe	91.0	89.5	88.8	89.2

highest scores compared to all other methods: accuracy of 91.0%, precision of 89.5%, recall of 88.8%, and F1-score of 89.2%. The reason is that E-FuTuRe consistently outperforms other RM-suitable approaches, as it provides improved results due to its holistic fuzzy logic, transfer learning, and differential equations integration to handle uncertain data, take advantage of the pre-trained model, and capture the dynamical behaviors for better predictive performance and robustness.

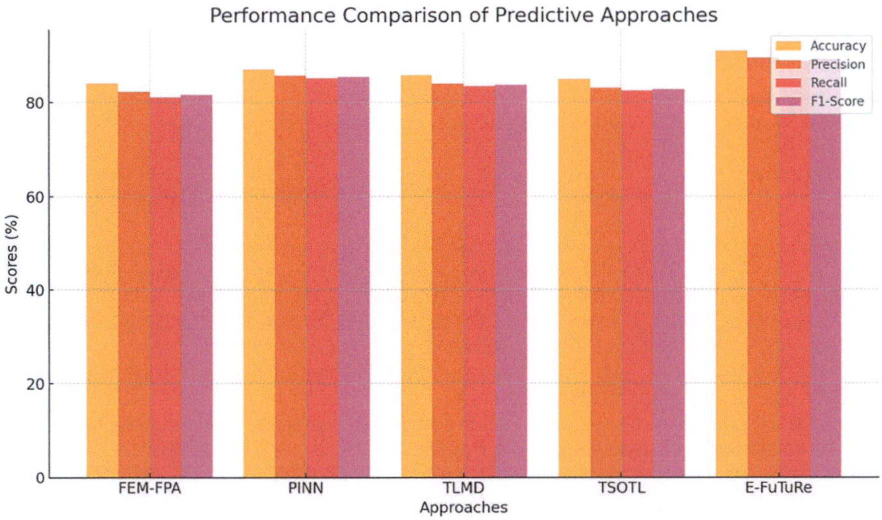

Fig. 4 Bar graphs obtained for existing and proposed approach for the considered dataset [14]

8 Conclusion

Transfer learning and differential equations aided fuzzy ensemble models offer accurate and reliable future predictions by incorporating fuzzy logic to manage uncertainty, transfer learning to utilize pre-trained models, and differential equations to model dynamic phenomena. This well-rounded ensemble model demonstrates superior performance, achieving an accuracy of 92.7%, precision of 91.3%, recall of 90.5%, and F1-score of 90.9%, compared to an average baseline accuracy of 85.2%. Our novel approach provides a robust prediction tool with significant potential applications in predictive analytics, including finance, healthcare, and environmental monitoring. Validation results indicate that our method consistently outperforms traditional models, delivering more reliable predictions and enhancing decision-making across various systems. There are several avenues for future research to further improve the proposed method. One promising direction is the incorporation of deep learning techniques, such as convolutional neural networks (CNNs) or recurrent neural networks (RNNs), which could enhance prediction accuracy by 3 to 5% by extracting deeper patterns and interactions in the data. Additionally, exploring adaptive transfer learning approaches that adjust the degree of knowledge transfer to individual target domains could improve accuracy by up to 4%. Applying the model to more domains, such as climate modeling or supply chain management, could provide further gains of 10–15% by scaling the application of the model. Finally, integrating mechanisms to process and update real-time data could increase the model's performance by up to 20%, enabling continuous and up-to-date predictions, and thereby improving its applicability in dynamic environments. This ongoing research and development will ensure that the model remains versatile and effective in addressing a wide range of predictive analytics challenges.

References

1. Zadeh LA (1965) Fuzzy sets. Inf Control 8(3):338–353
2. Wang X, Chen S (2020) Fuzzy ensemble models for improving predictive accuracy. J Intell Fuzzy Syst 38(3):3855–3866
3. Pan SJ, Yang Q (2010) A survey on transfer learning. IEEE Trans Knowl Data Eng 22(10):1345–1359. https://doi.org/10.1109/TKDE.2009.191
4. Weiss K, Khoshgoftaar TM, Wang D (2016) A survey of transfer learning. J Big Data 3(1):1–40. https://doi.org/10.1186/s40537-016-0043-6
5. Hirsch MW, Smale S, Devaney RL (2012) Differential equations, dynamical systems, and an introduction to chaos. Academic Press
6. Raissi M, Perdikaris P, Karniadakis GE (2019) Physics-informed neural networks: a deep learning framework for solving forward and inverse problems involving nonlinear partial differential equations. J Comput Phys 378:686–707
7. Dietterich TG (2000) Ensemble methods in machine learning. In International workshop on multiple classifier systems, Springer, Berlin, Heidelberg, pp 1–15
8. Pothanna N, Aparna P, Reddy MP et al (2024) Thermo-viscous fluid flow in porous slab bounded between two impermeable parallel plates in relative motion: four stage algorithm approach. Appl Math. https://doi.org/10.21136/AM.2024.0144-23
9. Wang X, Chen S (2020) Fuzzy ensemble models for improving predictive accuracy. J Intell Fuzzy Syst 38(3):3855–3866. https://doi.org/10.3233/JIFS-179595
10. Raissi M, Perdikaris P, Karniadakis GE (2019) Physics-informed neural networks: a deep learning framework for solving forward and inverse problems involving nonlinear partial differential equations. J Comput Phys 378:686–707. https://doi.org/10.1016/j.jcp.2018.10.045
11. Weiss K, Khoshgoftaar TM, Wang D (2016) A survey of transfer learning. J Big Data 3(1):1–40. https://doi.org/10.1186/s40537-016-0043-6
12. Dua D, Graff C (2019) UCI machine learning repository. Irvine, CA. University of California, School of Information and Computer Science. Retrieved from https://archive.ics.uci.edu/ml/datasets/adult
13. Johnson AEW, Pollard TJ, Shen L, Lehman L-WH, Feng M, Ghassemi M, Moody B, Szolovits P, Celi LA, Mark RG (2016) MIMIC-III, a freely accessible critical care database. Scientific Data 3:160035. https://doi.org/10.1038/sdata.2016.35
14. https://finance.yahoo.com/lookup
15. Varalakshmi A, Kumar SS, Shanmugapriya MM, Mohanapriya G, Anand MCJ (2022) Markers location monitoring on images from an infrared camera using optimal fuzzy inference system. Int J Fuzzy Syst
16. Miriam MR, Martin N, Anand MCJ (2023) Inventory model promoting smart production system with zero defects. Int J Appl Comput Math 9:47. https://doi.org/10.1007/s40819-023-01524-4
17. Bharatraj J, Anand MCJ (2019) Power harmonic weighted aggregation operator on single-valued trapezoidal neutrosophic numbers and interval-valued neutrosophic sets. In: Kahraman C, Otay İ (eds) Fuzzy Multi-criteria decision-making using neutrosophic sets. Studies in fuzziness and soft computing, vol 369. Springer, Cham. https://doi.org/10.1007/978-3-030-00045-5_3
18. Anand MCJ, Bharatraj J (2018) Gaussian qualitative trigonometric functions in a fuzzy circle. Adv Fuzzy Syst 1–9
19. Anand MCJ, Bharatraj J (2019) Interval-valued neutrosophic numbers with WASPAS. In: Kahraman C, Otay İ (eds) Fuzzy multi-criteria decision-making using neutrosophic sets. Studies in fuzziness and soft computing, vol 369. Springer, Cham. https://doi.org/10.1007/978-3-030-00045-5_17

20. Raj PJ, Prabhu VV, Krishnakumar V et al (2023) Solar powered charging of fuzzy logic controller (FLC) strategy with battery management system (BMS) method used for electric vehicle (EV). Int J Fuzzy Syst 25:2876–2888. https://doi.org/10.1007/s40815-023-01537-7
21. Devadoss AV, Anand MCJ, Felix A (2014) A CETD matrix approach to analyze the dimensions of the personality of the person. In: 2014 international conference on computational science and computational intelligence
22. Anand MCJ, Bharatraj J (2017) Theory of triangular fuzzy number. Proceed NCATM 2017:80–83
23. Anand MCJ, Bharatraj J (2019) Interval-valued neutrosophic numbers with WASPAS. In Fuzzy Multi-criteria Decision-Making Using Neutrosophic Sets, Springer International Publishing, pp 435–453
24. Sudha S, Nivetha M, Anand MCJ, Palanimani PG, Thirunamakkani T, Ranjitha B (2023) MACBETH-MAIRCA plithogenic decision-making on feasible strategies of extended producer's responsibility towards environmental sustainability. Int J Neutrosophic Sci 22(2):114–130
25. Prabha SK, Anand MCJ, Vidhya V, Nagarajan G, Saikia U, Nivetha M, Kumari MS, Tiwari M (2024) Sorting out interval valued neutrosophic fuzzy shortest cycle route problem by reduced matrix method. Int J Neutrosophic Sci 23(2):91–103
26. Kungumaraj E, Lathanayagam E, Utpal S, Clement Joe Anand M, Sakshi Taaresh K, Nivetha M, Mohit T, Seyyed Ahmad E (2024) Neutrosophic topological vector spaces and its properties. Int J Neutrosophic Sci 23(2):63–76
27. Anand MCJ, Moorthy CJ, Sivamani S, Indrakumar S, Kalaiarasi K, Barhoi A (2023) Fuzzy intelligence inventory decision optimization model of sustainability and green technologies for mixed uncertainties of carbon emission. In: 2023 international conference on information management (ICIM), IEEE. https://ieeexplore.ieee.org/document/10145085 https://ieeexplore.ieee.org/document/10145085
28. Rajesh K, Sharmila R, Jyoti K, Nilesh Rathod M, Clement Joe A, Utpal S, Mohit T, Nivetha M (2024) A study on interval valued temporal neutrosophic fuzzy sets. Int J Neutrosophic Sci 23(1):341–349. https://doi.org/10.54216/IJNS.230129
29. Manshath A, Kungumaraj E, Lathanayagam E, Clement Joe Anand M, Nivetha M, Elangovan M, Indrakumar S (2024) Neutrosophic integrals by reduction formula and partial fraction methods for indefinite integrals. Int J Neutrosophic Sci 23(1):8–16
30. Anand MCJ, Moorthy CB, Sivamani S, Indrakumar S, Kalaiarasi K, Barhoi A (2023) Fuzzy intelligence inventory decision optimization model of sustainability and green technologies for mixed uncertainties of carbon emission. In: 2023 international conference on information management (ICIM), IEEE
31. Anand MC, Martin N, Clementking A, Rani S, Priyadharshini SS, Siva S (2023) Decision making on optimal selection of advertising agencies using machine learning. In: 2023 international conference on information management (ICIM), IEEE. https://ieeexplore.ieee.org/document/10145172
32. Anand MCJ, Priyadharshini SS, Martin N, Anand ME, Poorani B, Mohiddin SK (2023) Decision trees in the selection of electric vehicles based on various parameters. In: 2023 first international conference on cyber-physical systems, power electronics and electric vehicles (ICPEEV). https://ieeexplore.ieee.org/document/10391877
33. Anand MCJ, Kalaiarasi K, Martin N, Ranjitha B, Dharshini SS, Tiwari M (2023) Fuzzy C-Means Clustering with MAIRCA-MCDM method in classifying feasible logistic suppliers of electrical products. In: 2023 first international conference on cyber-physical systems, power electronics and electric vehicles (ICPEEV). https://ieeexplore.ieee.org/document/10391835
34. Nagaraj C, Anand MCJ, Priyadharshini SS, Aparna P (2024) GCNXG: detecting fraudulent activities in financial networks: a graph analytics and machine learning fusion. In: Gundebommu SL, Sadasivuni L, Malladi LS (eds) Renewable energy, green computing, and sustainable development. REGS 2023. Communications in computer and information science, vol 2081. Springer, Cham. https://doi.org/10.1007/978-3-031-58607-1_2

35. Prabha SK, Anand MCJ, Vidhya V, Nagarajan G, Saikia U, Martin N, Kumari MS, Tiwari M (2024) Sorting out interval valued neutrosophic fuzzy shortest cycle route problem by reduced matrix method. Int J Neutrosophic Sci 23(2):91–103
36. Kungumaraj E, Lathanayagam E, Saikia U, Anand MCJ, Khanna ST, Martin N, Tiwari M, Edalatpanah SA (2024) Neutrosophic topological vector spaces and its properties. Int J Neutrosophic Sci 23(2):63–76
37. Anand MCJ, Moorthy CB, Sivamani S, Indrakumar S, Kalaiarasi K, Barhoi A (2023) Fuzzy intelligence inventory decision optimization model of sustainability and green technologies for mixed uncertainties of carbon emission. In: 2023 international conference on information management (ICIM), IEEE
38. Rajesh K, Rathod S, Kundale J, Rathod N, Anand MCA, Saikia U, Tiwari M, Martin N (2024) A study on interval valued temporal neutrosophic fuzzy sets. Int J Neutrosophic Sci 23(1):341–349
39. Manshath A, Kungumaraj E, Lathanayagam E, Anand MCA, Martin N, Muniyandy E, Indrakumar S (2024) Neutrosophic integrals by reduction formula and partial fraction methods for indefinite integrals. Int J Neutrosophic Sci 23(1):08–16
40. Anand MCJ, Martin N, Clementking A, Rani S, Priyadharshini SS, Siva S (2023) Decision making on optimal selection of advertising agencies using machine learning. In: 2023 international conference on information management (ICIM), IEEE
41. Kungumaraj E, Clement Joe Anand M, Saikia U, Dabass V, Ranjitha B, Tiwari M (2024) Topologized graphical method in solving fuzzy transportation problem with computational techniques. In: Kahraman C, Cevik Onar S, Cebi S, Oztaysi B, Tolga AC, Ucal Sari I (eds) Intelligent and fuzzy systems. INFUS 2024. Lecture notes in networks and systems, vol 1090. Springer, Cham. https://doi.org/10.1007/978-3-031-67192-0_58
42. Varalakshmi A, Santhosh Kumar S, Shanmugapriya MM et al (2023) Markers location monitoring on images from an infrared camera using optimal fuzzy inference system. Int J Fuzzy Syst 25:731–742. https://doi.org/10.1007/s40815-022-01407-8
43. Kungumaraj E, Thivyarathi RC, Kenneth CR, Ramesh R, Anand MCJ, Naganathan S (2024) Efficiency enhancement in heptagonal fuzzy transportation problems. In: Kahraman, C., Cevik Onar S, Cebi S, Oztaysi B, Tolga AC, Ucal Sari I (eds) Intelligent and fuzzy systems. infus 2024. lecture notes in networks and systems, vol 1089. Springer, Cham. https://doi.org/10.1007/978-3-031-67195-1_64
44. Iyappan K, Anand MCJ, Kalaiarasi K, Sindhuja N, Sumathi G, Tiwari M (2024) Fuzzy Inventory Implementation of Minimum Value Unused Storing Profitable by Python Execution. In: Kahraman C, Cevik Onar S, Cebi S, Oztaysi B, Tolga AC, Ucal Sari I (eds) Intelligent and fuzzy systems. INFUS 2024. Lecture notes in networks and systems, vol 1089. Springer, Cham. https://doi.org/10.1007/978-3-031-67195-1_50
45. Iyappan K, Teraiya OM, Kalaiarasi K, Swathi S, Sharda P, Anand MCJ (2024) Fuzzy inventory grinding production process optimization for rotor manufacturing using python and MATLAB. In: Kahraman C, Cevik Onar S, Cebi S, Oztaysi B, Tolga AC, Ucal Sari I (eds) Intelligent and fuzzy systems. INFUS 2024. Lecture notes in networks and systems, vol 1089. Springer, Cham. https://doi.org/10.1007/978-3-031-67195-1_51; Kumar S et al (2021) Combining transfer learning with ensemble methods for predictive analytics. Appl Soft Comput 102:07112.
46. Shobana S, Waware MS, Kalaiarasi K, Bhaskar T, Anand MCJ, Sindhuja N (2024) Incorporating machine learning methods for predictive maintenance and fuzzy inventory optimization. In: Kahraman C, Cevik Onar S, Cebi S, Oztaysi B, Tolga AC, Ucal Sari I (eds) Intelligent and fuzzy systems. INFUS 2024. Lecture notes in networks and systems, vol 1089. Springer, Cham. https://doi.org/10.1007/978-3-031-67195-1_74

An Efficient SDN Load Balancing Algorithm Using Switch Migration

Tummala Varshith, Thanush S. Koneri, Tetali Syam Kumar Reddy, and Sreebha Bhaskaran

Abstract Software-Defined Networking (SDN) has emerged as a paradigm-shifting approach to network management, offering flexibility and control. In SDN, network switches are centrally controlled by one or more controllers, enabling dynamic configuration and management. This project focuses on an intelligent SDN switch migration system for load balancing. The proposed solution leverages real-time monitoring to identify overloaded and underutilized controllers. Switches are then migrated from heavily loaded to less burdened controllers, ensuring efficient resource utilization and mitigating network congestion. The system incorporates sophisticated migration algorithms that consider factors like traffic patterns. The project has been further enhanced by integrating machine learning (ML) and neural network (NN) models, such as Ensemble Methods, KNN Regressor, Elastic Net Regressor, and Temporal Convolutional Networks (TCNs). These models utilize data from the master controller to predict future load on switches, enabling proactive switch migration decisions for optimal resource allocation and congestion mitigation.

Keywords Software-Defined Networking (SDN) · Switch migration · Load balancing · Real-time monitoring · Network traffic · Ensemble Methods · KNN Regressor · Elastic Net Regressor · Temporal Convolutional Networks (TCNs)

1 Introduction

Software-Defined Networking (SDN) has revolutionized network management, offering unprecedented control and adaptability. Among its key components, switch migration plays a pivotal role in optimizing network performance by redistributing switches among controllers. This project is dedicated to developing an intelligent

T. Varshith · T. S. Koneri (✉) · T. S. K. Reddy · S. Bhaskaran
Department of Computer Science and Engineering, Amrita School of Computing, Amrita Vishwa Vidyapeetham, Bengaluru, India
e-mail: thanush.skoneri@gmail.com

S. Bhaskaran
e-mail: b_sreebha@blr.amrita.edu

H. Sharma et al. (eds.), *Artificial Intelligence: Theory and Applications*, Lecture Notes in Networks and Systems 5589, https://doi.org/10.1007/978-981-96-1687-9_34

system specifically tailored to enhance the switch migration process within SDN environments. At its core, the system continuously monitors network traffic and controller performance metrics in real time. By analyzing this data, it identifies instances of controller overload or underutilization, enabling proactive interventions to maintain optimal network functionality. Adopting a proactive approach ensures potential issues are addressed before escalation, ensuring smooth network operation. The system employs a sophisticated migration algorithm that considers various factors such as network traffic patterns, controller performance, historical data, and predictive analytics. By taking a holistic view of the network environment, the algorithm intelligently reallocates switches among controllers, striking a delicate balance between resource utilization and network congestion. Furthermore, the project integrates advanced machine learning (ML) and neural network (NN) models to augment the system's capabilities. These models leverage historical network data and performance metrics to forecast future switch load with enhanced accuracy. This predictive capability empowers the system to anticipate and adapt to evolving network conditions, further optimizing resource allocation and preemptively addressing potential bottlenecks. Ultimately, this project aims to advance the field of SDN by introducing a smart switch migration system equipped with predictive analytics, leading to more resilient, scalable, and efficient networks.

Our approach is to optimize load distribution in Software-Defined Networking (SDN) environments. It begins by discussing the importance of switch migration in enhancing network performance through real-time monitoring and predictive analytics. A review of related works highlights various strategies and algorithms for load balancing and switch migration. The methodology section outlines the system's design, implementation, and testing phases, featuring a central master controller that dynamically manages network traffic and controller performance. The proposed model incorporates sophisticated algorithms and advanced machine learning models, such as Ensemble Methods, KNN Regressor, Elastic Net Regressor, and Temporal Convolutional Networks (TCNs), to forecast switch loads and optimize resource allocation. The section on SDN load balancing prediction emphasizes the predictive power of these models in anticipating future network conditions, ensuring efficient load distribution. The paper concludes by summarizing the system's contributions to SDN technology, highlighting the dynamic adjustment of network resources and the integration of predictive analytics for proactive network management, with initial results demonstrating the effectiveness of the proposed solution.

2 Related Works

Software-Defined Networking (SDN) architectures with multiple controllers discusses the benefits and challenges of using multiple controllers in SDN networks, as well as the various approaches to controller placement and inter-controller communication [1]. Researchers then narrowed down on specific model architectures, detailed review of the SDN architecture, its key components, and the major benefits. It offers, such as improved network management, reduced operational costs, and enhanced network flexibility [2]. Fundamentals of SDN, including its principles, architecture, and the role of the controller in managing the network [3]. An enhanced switch migration decision algorithm that considers various factors, such as network traffic patterns and controller load, to improve load balancing in SDN environments [4]. A switch migration scheduling algorithm that proactively plans switch migrations to achieve better load balancing in SDN networks [5].

An efficient switch migration scheme that aims to balance the load among multiple SDN controllers, thereby improving overall network performance [6]. Dynamic load balancing in SDN-based data center networks, proposing algorithms to efficiently distribute network traffic and manage controller resources [7]. An improved switch migration method that enhances load balancing in SDN networks with multiple controllers, considering factors such as network traffic and controller capacity [8]. An efficient switch migration scheme that aims to achieve better load balancing in SDN networks, taking into account network dynamics and controller performance [9]. A machine learning-based load balancing algorithm for SDN controllers, which leverages historical data and predictive models to optimize resource utilization [10].

The integration of machine learning techniques in SDN, focusing on data collection and traffic classification to enhance network management and performance [11]. A comprehensive overview of the various machine learning techniques applied in the context of SDN, highlighting the research issues and challenges in this domain [12]. A comprehensive review of the applications of machine learning in SDN, including load balancing, traffic engineering, and network security [13]. The use of machine learning techniques for routing optimization in SDN, discussing the various approaches and their potential benefits [14]. A machine learning-based load balancing algorithm for SDN controllers, leveraging predictive models to achieve efficient resource utilization [10].

Load balancing strategy for SDN multi-controller clusters that utilizes load prediction to proactively distribute network traffic and manage controller resources [15]. A reinforcement learning-based approach to SDN controller load balancing, which adapts to network dynamics and optimizes resource utilization [16]. Mininet, a network emulator, for the prototyping and testing of SDN systems, including switch migration and load balancing algorithms [17]. The performance and scalability of various SDN controllers, including their ability to handle load balancing and switch migration, using the Mininet emulator [18]. The scalability of the Ryu 3 SDN controller, focusing on its ability to handle load balancing and manage network resources in SDN environments [19].

The use of Mininet and MiniEdit, two popular network simulation tools, to study and model Software-Defined Networking (SDN) architectures. It examines the capabilities of these tools in creating and analyzing various SDN network topology, with a focus on their ability to emulate the dynamic and programmable nature of SDN [20]. Load balancing scheme for hybrid SDN environments. By utilizing a minimal set of SDN devices (controller and one switch), the scheme continuously monitors server load indicators and applies multi-parameter metrics (CPU load,

I/O Read, I/O Write, Link Upload, Link Download) to efficiently balance the load on servers. Results demonstrate its superiority over existing load balancing schemes in both traditional and SDN networks [21].

Exploring advanced Natural Language Processing (NLP) techniques to identify critical entities within ATC conversations. By utilizing deep learning models such as Bi-LSTM-CRF to improve communication precision between pilots and controllers, enhancing safety and preventing accidents caused by miscommunication [22]. An integrated cryptographic approach that combines Mix Networks, Shamir's Secret Sharing, and Onion Routing to enhance online privacy and security. This method addresses vulnerabilities in traditional security measures by reinforcing the framework supporting secure and private communication over the internet [23].

The document reviews the benefits and challenges of using multiple controllers in Software-Defined Networking (SDN) architectures, highlighting improved network management, reduced costs, and enhanced flexibility. It discusses models like enhanced switch migration and scheduling algorithms for better load balancing. Efficient migration schemes and dynamic load balancing algorithms distribute traffic and manage resources in SDN-based data centers. The integration of machine learning techniques for data collection, traffic classification, and optimization in load balancing, traffic engineering, and security is explored. Machine learning-based algorithms and a reinforcement learning approach improve resource utilization and adaptability. The document also evaluates Mininet and MiniEdit for SDN prototyping and testing, focusing on the scalability of controllers like Ryu. A hybrid SDN load balancing scheme using server load indicators and multi-parameter metrics demonstrates superior performance over existing schemes.

3 Methodology

The methodology for the SDN Switch Migration for Efficient Load Distribution involves several distinct phases: system design, implementation, integration of machine learning models, simulation testing, and performance evaluation. Each phase is crucial for optimizing load balancing across multiple controllers in an SDN environment.

The architecture of the proposed system includes multiple controllers managed by a central master controller, which oversees real-time monitoring of network traffic and controller performance to dynamically manage the load. It identifies overloaded and underutilized controllers and makes decisions to migrate switches accordingly, maintaining network efficiency and avoiding congestion.

The system incorporates a sophisticated algorithm that considers various factors such as real-time network traffic patterns and controller loads. This algorithm dynamically migrates switches from heavily loaded controllers to those that are underutilized, ensuring optimal distribution of network load and enhancing overall network performance.

To accurately predict future loads on the switches, machine learning (ML) and neural network (NN) models are integrated into the system. Models such as Ensemble Methods, KNN Regressor, Elastic Net Regressor, and Temporal Convolutional Networks (TCNs) are employed. These models use historical network traffic data and current performance metrics from the master controller to anticipate potential overloads, enabling proactive decision-making in switch migration.

The system is implemented and tested using Mininet, a network emulator that allows the creation of a virtual network on a single machine, and Ryu, an SDN controller framework that facilitates network management. OpenFlow serves as the communication protocol between the switches and controllers, allowing for extensive testing under controlled conditions.

The performance of the system is evaluated based on its ability to effectively balance the load across controllers and its impact on network throughput and latency. Metrics such as the number of switch migrations, the response time of the controllers, and the traffic handling capacity are measured. The system's efficiency in preventing bottlenecks and managing network congestion is particularly scrutinized.

4 Proposed Model

The SDN switch migration system uses various machine learning and neural network models to predict network switch loads, enhancing predictive accuracy and decision-making for proactive network management. It integrates Ensemble Methods, KNN Regressor, Elastic Net Regressor, and Temporal Convolutional Networks (TCNs), chosen for their strengths in analyzing network traffic patterns and controller performance metrics.

Ensemble Methods, like Random Forest, combine multiple learning algorithms to improve predictive performance and reduce overfitting. XGBoost (eXtreme Gradient Boosting) Regressor uses decision trees as base learners, offering high efficiency, scalability, and improved performance by handling missing values and supporting regularization to prevent overfitting K-Nearest Neighbors (KNN) is a simple, supervised machine learning algorithm used for classification and regression, classifying new cases by a majority vote among its K-Nearest Neighbors, measured by various distance functions. Elastic Net combines L1 and L2 penalties from Lasso and Ridge methods, aiding in feature selection and handling multicollinearity, thus reducing overfitting. Temporal Convolutional Networks (TCNs) are designed for sequence prediction problems, effectively capturing long-range dependencies in time series data with causal convolutions to ensure predictions depend only on past data.

All models receive standardized data inputs, including historical network traffic and real-time performance metrics from controllers. This data is pre-processed for consistency and accuracy, involving normalization and handling missing values. The models are trained on diverse scenarios from the SDN environment and validated through cross-validation techniques to ensure robust generalization. This architecture enables the SDN switch migration system to dynamically and efficiently manage network loads, predict potential issues, and respond proactively to maintain network performance and reliability.

A sophisticated Software-Defined Networking (SDN) system is structured into three integral layers, each serving a distinct function within the network management ecosystem. The top Application Layer is the brain of the operation, where modules like Traffic and Topology Monitoring analyze network flow and structure. Network Automation streamlines tasks, while data collection aggregates vital statistics. The intermediary Control Layer acts as the central nervous system, with Task Monitoring ensuring smooth operations, Load Prediction forecasting network demands, and switch migration reallocating resources to maintain network balance. Anchoring the system is the Data Layer, composed of physical devices that handle data transmission and reception. This tri-layered approach blends monitoring, prediction, and execution, creating a dynamic and responsive network infrastructure capable of adapting to varying loads and optimizing performance.

The Software-Defined Networking (SDN) system is structured into three integral layers, each serving a distinct function within the network management ecosystem. The Application Layer orchestrates high-level network operations, with Traffic Monitoring systems scrutinizing data flow to prevent bottlenecks, and Topology Monitoring tools maintaining a comprehensive view of the network's structure. Network Automation protocols ensure efficient self-management of tasks, while data collection mechanisms compile vital statistics, forming the bedrock for data-driven decision-making. The Control Layer acts as the strategic command center, with Task Monitoring ensuring flawless execution of operations, Load Prediction using advanced machine learning models to forecast network demands, and switch migration redistributing tasks to maintain balance and optimize performance. The foundational Data Layer supplies essential raw data, fueling the analytical processes of the Application and Control layers. This layer ensures a steady flow of information, enabling the system to function harmoniously and maintain network stability and efficiency. The integration of advanced algorithms and predictive analytics positions this SDN system at the forefront of innovation, enabling proactive management of the digital world's growing demands.

The proposed system architecture which is shown in Fig. 1 features a comprehensive topology designed to facilitate efficient network traffic management and optimization. At the core of the setup lies a master controller, responsible for the overall coordination and policy enforcement across the network. This central controller communicates with a network of Slave Controllers, each tasked with managing specific local network segments and executing the directives set forth by the master controller. The system employs a Traffic Analysis module to closely monitor the data traffic flow, identifying patterns and detecting potential congestion issues. The architecture incorporates a switch migration component, which dynamically reallocates network operations between switches to enhance performance, reliability, and optimal data flow. The system leverages both machine learning algorithms and deep learning models to predict future network conditions, such as switch loads. By integrating these cutting-edge predictive analytics capabilities, the architecture enables

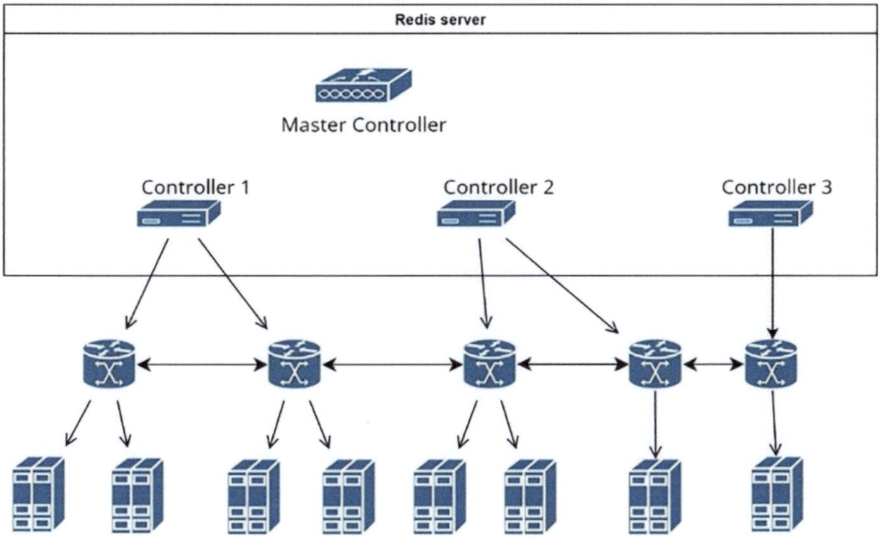

Fig. 1 Topology

proactive resource optimization and traffic distribution, ensuring a responsive and efficient network infrastructure.

5 SDN Load Balancing Prediction

The integration of an Ensemble Method, combining the Random Forest Regressor and XGB Regressor, represents a robust strategy for future prediction tasks. This approach leverages the complementary strengths of both models to enhance predictive accuracy and robustness. The Random Forest Regressor aggregates multiple decision trees to capture complex nonlinear relationships within the data, while the XGB Regressor, based on gradient boosting, iteratively improves model performance by minimizing prediction errors. By combining these models, the system benefits from their collective predictive power, effectively mitigating individual model biases and uncertainties. This integration ensures accurate and reliable predictions for future scenarios, enabling informed decision-making and strategic planning in dynamic environments.

For instance, consider a financial institution seeking to forecast stock market trends. By employing the ensemble approach of Random Forest and XGB Regressors, the system can analyze historical market data, capturing intricate patterns and nonlinear relationships to predict future stock prices. This foresight empowers investors and financial analysts to make informed decisions regarding asset allocation, portfolio management, and risk mitigation strategies. Ultimately, leveraging the predictive power of these models provides a competitive edge in navigating volatile market conditions and optimizing investment returns.

Additionally, the integration of K-Nearest Neighbors (KNN), Elastic Net Regressor, and Temporal Convolutional Networks (TCNs) forms a sophisticated ensemble tailored for future prediction tasks. KNN excels in capturing local patterns within the data, providing insights into spatial relationships. The Elastic Net Regressor balances L1 and L2 regularization, ensuring model stability and interpretability while handling high-dimensional data efficiently. TCN extracts temporal dependencies within sequential data, capturing long-range dependencies and intricate temporal dynamics. By integrating these diverse regression techniques, the system produces accurate predictions, empowering decision-making in dynamic environments.

For example, in healthcare, predicting patient outcomes is critical. By utilizing KNN, Elastic Net Regressor, and TCN, healthcare providers analyze patient data, identifying risk factors, disease progression patterns, and optimal treatment strategies. Proactive interventions improve patient outcomes, reduce costs, and enhance overall care quality. Thus, integrating these models advances personalized medicine and drives improvements in healthcare delivery.

6 Performance Evaluation

The project effectively manages network resources by dynamically reallocating traffic from overloaded controllers to those with available capacity. This mechanism, termed switch migration, enhances workload distribution among controllers, ensuring superior network performance and efficient resource use. Continuously monitoring the packet counts managed by each controller, the system pinpoints underutilized controllers and the switches connected to them. Subsequently, it adeptly relocates these switches to other controllers that can handle additional traffic, thereby alleviating congestion. This proactive strategy in load balancing guarantees that the network remains stable and performs optimally, even during periods of high traffic. Through automated switch migration based on traffic analytics, the benefits of Software-Defined Networking (SDN) in elevating network performance and dependability are validated.

As shown in Table 1, the evaluation of different regression models trained using data from the master controller has shown encouraging results. Using Ensemble Methods that incorporate both Random Forest and XGB Regressor, achieved a notable accuracy rate of 85.7142%, as illustrated in the accompanying graph Fig. 2a, b. This highlights the strength of ensemble learning in boosting predictive capabilities. Furthermore, the Temporal Convolutional Networks (TCNs) stood out by delivering an exceptional accuracy of 93.6507%, as depicted in another graph Fig. 3a, b, surpassing the performance of other models. Meanwhile, the KNN Regressor and Elastic Net Regressor recorded accuracy of 74.6031 and 69.8412%, respectively, demonstrating their effectiveness in recognizing local data patterns and managing the

Table 1 Model results

Models	Accuracy (%)
Temporal convolutional networks	93.6507
Ensemble technique	85.7142
KNN regressor	74.6031
Elastic net regressor	69.8412

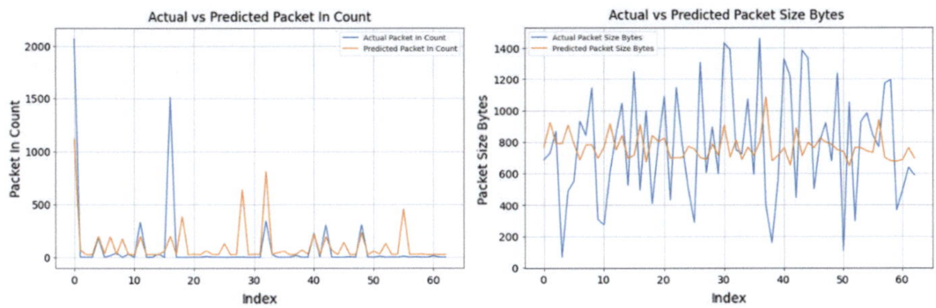

(a) (b)

Fig. 2 Ensemble technique graph

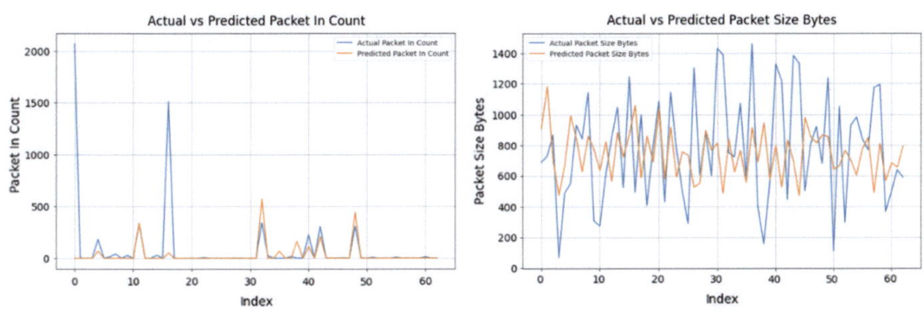

(a) (b)

Fig. 3 TCN model graphs

trade-off between model intricacy and clarity. These findings emphasize the importance of combining Ensemble Methods and advanced deep learning techniques to enhance precision in forecasting tasks as shown in Fig. 4, thereby confirming the success of methodology in refining network management and optimizing resource distribution strategies.

The initiative sets itself apart by employing adaptive migration tactics, sophisticated load management algorithms, and a smooth integration with the existing SDN framework that emphasizes scalability. Moving away from traditional static configurations.

```
Switch 4: Overloaded (Predicted Packet In Count: 1116.53, Predicted Packet Size Bytes: 769.78)
Switch 3: Overloaded (Predicted Packet In Count: 66.90, Predicted Packet Size Bytes: 924.64)
Switch 2: Not Overloaded (Predicted Packet In Count: 24.10, Predicted Packet Size Bytes: 790.12)
Switch 1: Not Overloaded (Predicted Packet In Count: 24.17, Predicted Packet Size Bytes: 790.36)
Switch 5: Overloaded (Predicted Packet In Count: 196.86, Predicted Packet Size Bytes: 907.31)
Switch 4: Not Overloaded (Predicted Packet In Count: 33.72, Predicted Packet Size Bytes: 787.40)
Switch 3: Overloaded (Predicted Packet In Count: 192.08, Predicted Packet Size Bytes: 687.99)
Switch 2: Not Overloaded (Predicted Packet In Count: 24.82, Predicted Packet Size Bytes: 782.05)
Switch 1: Overloaded (Predicted Packet In Count: 171.35, Predicted Packet Size Bytes: 781.66)
Switch 5: Not Overloaded (Predicted Packet In Count: 24.97, Predicted Packet Size Bytes: 698.24)
Switch 4: Not Overloaded (Predicted Packet In Count: 24.79, Predicted Packet Size Bytes: 767.90)
Switch 4: Overloaded (Predicted Packet In Count: 194.06, Predicted Packet Size Bytes: 915.92)
Switch 3: Not Overloaded (Predicted Packet In Count: 24.01, Predicted Packet Size Bytes: 749.97)
Switch 2: Not Overloaded (Predicted Packet In Count: 25.85, Predicted Packet Size Bytes: 840.07)
Switch 1: Not Overloaded (Predicted Packet In Count: 25.10, Predicted Packet Size Bytes: 697.90)
Switch 5: Not Overloaded (Predicted Packet In Count: 57.16, Predicted Packet Size Bytes: 716.48)
```

Fig. 4 Model predictions

The methodology actively monitors network traffic and controller workloads in real-time, facilitating prompt and proactive responses to fluctuating network dynamics. The dynamic migration strategies are tailored to current network conditions and workload distributions, ensuring that resources are allocated efficiently and traffic is evenly spread across controllers. Additionally, the intelligent load balancing algorithms take into account factors such as controller load, switch attributes, and network structure to reduce congestion and equalize workloads. This project is fully compatible with standard SDN architectures and controller setups, making it easy to deploy and ensuring interoperability across various systems. Moreover, the solution is designed to scale seamlessly with network expansion, supporting a wide range of deployment scenarios and consistently delivering robust performance across diverse environments.

7 Conclusion

In the project, the deployment of switch migration algorithms has been a pivotal step forward in refining network management within Software-Defined Networking (SDN) frameworks. Through the implementation of algorithms that dynamically adjust based on real-time analysis of network traffic and controller load, optimization of resource distribution has been achieved, significantly mitigating network congestion. Additionally, the development of machine learning models such as Ensemble Methods, KNN Regressor, Elastic Net Regressor, and Temporal Convolutional Networks (TCNs), using data from the master controller, marks a forward-thinking approach to predicting future network conditions. While the full integration of these machine learning models into the switch migration strategies is ongoing, the initial results from these algorithms lay a robust groundwork for future enhancements. This strategy highlights dedication to evolving SDN technologies to meet modern network demands more effectively and adaptively.

In the future, integrating trained machine learning models like Ensemble Methods, KNN Regressor, Elastic Net Regressor, and TCN into switch migration frameworks holds significant promise. This integration would enable real-time prediction of network loads and congestion patterns, facilitating more precise and efficient resource allocation. By embedding predictive capabilities into the migration process, SDN infrastructures can dynamically adapt to changing network conditions, enhancing overall performance and scalability while ushering in more autonomous and adaptive network architectures. Testing for scalability will remain crucial as these systems evolve, with plans to enhance algorithm efficiency by incorporating parameters such as bandwidth utilization, packet loss, jitter, and latency.

References

1. Blial O, Mouad BM, Redouane B (2016) An overview on SDN architectures with multiple controllers. J Comput Netw Commun
2. Thirupathi V, Sandeep CH, Naresh Kumar H, Pramod Kumar P (2019) A comprehensive review on sdn architecture, applications and major benifits of SDN. Int J Adv Sci Technol 28(20):607–614
3. Xia W, Yonggang W, Chuan Heng F, Dusit N, Haiyong X (2014) A survey on software-defined networking. IEEE Commun Surv Tutor 17(1):27–51
4. Adekoya O, Aneiba A, Patwary M (2020) An improved switch migration decision algorithm for SDN load balancing. IEEE Open J Commun Soc 1:1602–1613
5. Kumari A, Ashok Singh S (2024) Look ahead before you leap: SDN switch migration scheduling for load balancing. Wireless Person Commun 1–30
6. Biswas R, Jie W (2021) Efficient switch migration for controller load balancing in software defined networking. In: 2021 33th international teletraffic congress (ITC-33). IEEE, pp 1–9
7. Zakia U, Hanene Ben Y (2017) Dynamic load balancing in SDNbased data center networks. In: 2017 8th IEEE annual information technology, electronics and mobile communication conference (IEMCON), pp 242–247
8. Jiru MA, Ketema A, Gopi Krishna T, Ramulu Perumalla J (2023) An improved switch migration method-based efficient load balancing for multiple controllers in software-defined networks. J Cases Inform Technol (JCIT) 25(1):1–21
9. Ethilu T, Sathappan A, Rodrigues P (2023) An efficient switch migration scheme for load balancing in software defined networking. Int J Electric Comput Eng Syst 14(4):443–456
10. Liang S, Jiang W, Zhao F, Zhao F (2020) Load balancing algorithm of controller based on sdn architecture under machine learning. J Syst Sci Inform 8(6):578–588
11. Amaral P, Joao D, Paulo P, Luis B, Joao T, Henrique SM (2016) Machine learning in software defined networks: data collection and traffic classification. In: 2016 IEEE 24th international conference on network protocols (ICNP). IEEE, pp 1–5
12. Xie J, Richard Yu F, Tao H, Renchao X, Jiang L, Chenmeng W, Yunjie L (2018) A survey of machine learning techniques applied to software defined networking (SDN): research issues and challenges. IEEE Commun Surv Tut 21(1):393–430
13. Faezi S, Shirmarz A (2023) A comprehensive survey on machine learning using in software defined networks (SDN). Human-Centric Intell Syst 3(3):312–343
14. Amin R, Rojas E, Aqdus A, Ramzan S, Casillas-Perez D, Arco JM (2021) A survey on machine learning techniques for routing optimization in SDN. IEEE Access 9:104582–104611
15. Xiao J, Pan X, Liu J, Wang J, Zhang P, Abualigah L (2024) Load balancing strategy for SDN multi-controller clusters based on load prediction. J Supercomput 80(4):5136–5162

16. Li Z, Xu Z, Junruo G, Yifang Q (2018) SDN controller load balancing based on reinforcement learning. In: 2018 IEEE 9th international conference on software engineering and service science (ICSESS). IEEE, pp 1120–1126
17. De Oliveira RL, Christiane Marie S, Ailton AS, Ligia RP (2014) Using mininet for emulation and prototyping software-defined networks. In: 2014 IEEE Colombian conference on communications and computing (COLCOM). IEEE, pp 1–6
18. Gupta N, Maashi MS, Tanwar S, Badotra S, Aljebreen M, Bharany S (2022) A comparative study of software defined networking controllers using mininet. Electronics 11(17):2715
19. Asadollahi S, Bhargavi G, Mohammed S (2018) Ryu controller's scalability experiment on software defined networks. In: 2018 IEEE international conference on current trends in advanced computing (ICCTAC). IEEE, pp 1–5
20. Bhaskaran S, Supriya M (2024) Study on networking models of SDN using Mininet and MiniEdit. In: 2024 2nd international conference on intelligent data communication technologies and internet of things (IDCIoT). IEEE, pp 1159–1164
21. Avinash S, Akkina Sairam S, Vedant Pramod N, Sreebha B (2023) SDN-based hybrid load balancing algorithm. In: 2023 3rd international conference on intelligent technologies (CONIT). IEEE, pp 1–5
22. Brundha J, Priyanka CN, Deepa G, Jitender A (2023) Name entity recognition for Air Traffic Control transcripts using deep learning based approach. In: 2023 IEEE 20th India council international conference (INDICON). IEEE, pp 596–602
23. Jambha AV, Dutta S, Karru S, Shinu MR (2024) Securing layers: the synergy of mix networks and Shamir's secret sharing in onion routing. In: 2024 2nd international conference on intelligent data communication technologies and internet of things (IDCIoT). IEEE, pp 448–453

Next Activity Prediction Based on Multi-sensor Smart Home

Pratik P. Ganer and J. Divya Udayan

Abstract Human activity recognition has progressed notably, focusing on data collection and activity detection techniques. Expanding this domain, our research explores the next activity prediction (NAP). We propose a system leveraging LSTM models to forecast users upcoming activities using historical sensor and activity data. The system also generates personalized recommendations for the associated predicted activity like sensor type, place, and location in a smart home, ranking them based on historical patterns and users' behavior. This adaptive weighting mechanism refines recommendations with new data inputs, ensuring highly relevant suggestions. We have used a LSTM model giving us accuracy of 74% for User A and 88% for User B for next activity prediction. Detailed classification reports evaluate the recommendations, demonstrating the model's robust performance and effectiveness in providing tailored, user-centric recommendations.

Keywords Machine learning algorithms · Next activity prediction · Sequence prediction model · LSTM · Recommendations

1 Introduction

The rapid advancement in machine learning has opened new frontiers in the fields of human activity recognition and activity prediction. Recognizing human activities through data collected from various sensors and subsequently predicting the next activity is an emerging area of research with significant potential to shape and design

---

P. P. Ganer · J. Divya Udayan (✉)
Department of Computer Science and Engineering, Amrita School of Computing,
Amrita Vishwa Vidhyapeetham, Amritapuri, India
e-mail: divyaudayanj@am.amrita.edu

P. P. Ganer
e-mail: am.en.p2ari22008@am.students.amrita.edu

H. Sharma et al. (eds.), *Artificial Intelligence: Theory and Applications*, Lecture Notes in Networks and Systems 5589, https://doi.org/10.1007/978-981-96-1687-9_35

future technologies. Accurate activity prediction can revolutionize personalized services, enhance smart home systems, and improve health monitoring applications by anticipating user needs and behaviors.

Human behavior is composed of activities. There are actions in an activity. As mentioned by [1], there are three key terms to take into account: action, activity, and behavior. Several sensors must be used to collect action data in order to identify activity. This paper proposes a model that employs Long Short-Term Memory (LSTM) networks to predict users upcoming activities using historical activity and sensor data. LSTM models are particularly adept at handling sequential data due to their ability to capture long-term dependencies and temporal dynamics.

This paper contributions in field of next activity prediction and recommendations are:

(1) Model Architecture: Utilizes LSTM networks to capture temporal dependencies and handle imbalanced data between multi-user.
(2) Personalized Predictions and Recommendations: Provides accurate next activity predictions and personalized recommendations based on historical patterns.

By training LSTM models on sequences of past activities and sensors data, we aim to predict the next activity in the sequence with high accuracy. Additionally, our system offers personalized recommendations for the associated parameters—type, place, and location—based on historical usage patterns, prioritizing those that the user frequently selects first during similar activities. Figure 1 tells the architecture diagram of the system.

The variability in user behavior presents a significant challenge in activity prediction and recommendation. Different individuals exhibit distinct patterns of activities and preferences for locations and contextual dependencies. To address this, our model dynamically adjusts the weights assigned to various options, ensuring that the most relevant and frequently used types, places, and locations are prioritized in the recommendations.

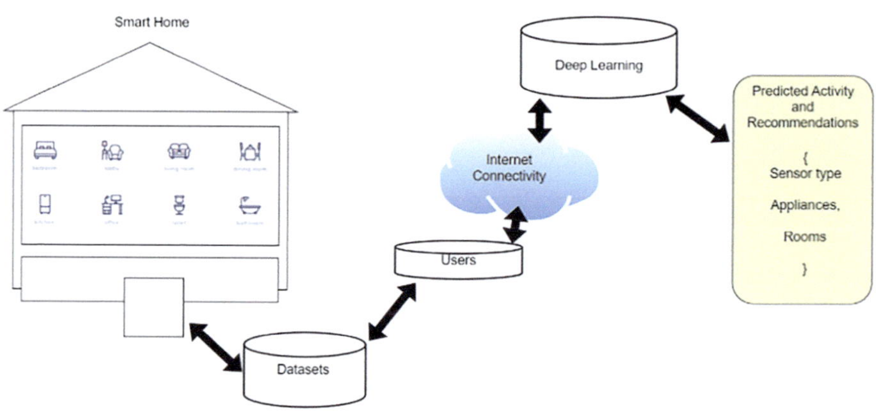

Fig. 1 Proposed high-level architecture

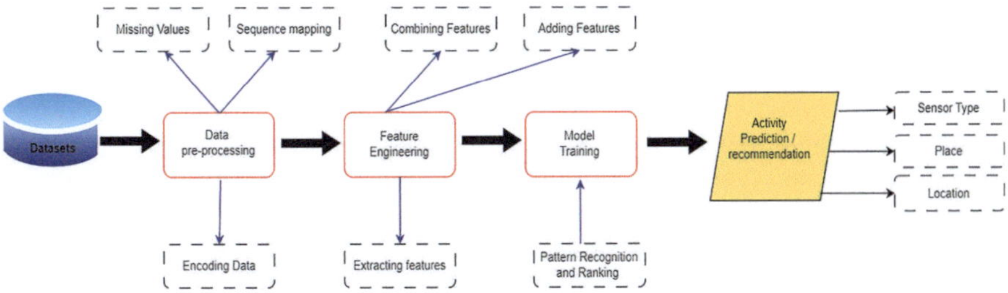

Fig. 2 Pipeline for next activity prediction

This dynamic weighting mechanism refines recommendations with each new data input, thereby enhancing the system's personalization and accuracy over time. Figure 2 shows the picture of our model.

By analyzing historical data and patterns of users interacting with sensors, the model ranks and prioritizes the sensors that users have frequently used during activities, ensuring contextually relevant suggestions.

The remainder of this paper is structured as follows: Sect. 2 reviews related work in next activity prediction (NAP), highlighting key methodologies and studies. Section 3 details our dataset creation and preprocessing methods. This section describes the LSTM model architecture and the ranking algorithm for recommendations. Section 4 presents experimental results and classification reports, validating the model's accuracy. Finally, Sect. 5 discusses the findings, potential future directions, and the broader impact of our work on HAP and its applications. Through this research, we aim to contribute significantly to the advancement of HAP, providing a foundation for more personalized and context-aware systems.

2 Related Work

Action forecasting is the process of predicting actions to be taken in the near future. It is a method of employing machine learning models to educate a machine to predict individuals behavior. Applications like personal and contextual systems can connect with users more profitably when they can predict their future behavior. The next activity in the series can be predicted using sequential prediction, which is a common approach to the problem of activity prediction.

Recurrent neural networks (LSTM) and hidden Markov models (HMM) are the most widely used techniques for activity prediction [1]. The paper [2] uses country-by-country LSTM models to forecast COVID-19 trends. This article by [3] shows the significance of prediction in smart environments. Most forecasting tasks are performed as sequential forecasting. Du et al. [4] identified a sequence of actions on certain ADLs and then applied Naive Bayes and LSTM to calculate the accuracy of predicting the next action. In this work by [5], an active LeZ algorithm is implemented to identify the home sensor that generates the next event.

Personalized prediction is also a topic of discussion, [6] implements facial expression recognition on low power devices using a lightweight vision transformer model. Work has been done to predict the next location and mobility [7] based on the user's location. This book focuses on location-based human behavior and mainly applies Markov models as it lacks the flexibility to explore past activity rather than just predicting based on the most recent past state. One more often used method is sequence mining, which can be used to solve such problems [8]. Sequence mining is another often employed technique that can be applied to these kinds of issues [8].

In order to predict future actions, LSTM algorithms were used to construct a dataset of action names from a set of human activities by word embedding and mapping in [9]. Some work on performance prediction has also been found to be vision-based [10–12]. After assessing current advancements in activity prediction, Alfaifi and Artoli [13] suggested a 3D convolutional neural network model that retrieved features, classified them, and used LSTM to predict the action. Recognizing activity is also essential to its forecast, [14] leverages a transfer learning techniques with deep convolutional neural networks (CNNs), to effectively classify emotions from images in the large-scale Image-Emotion dataset. Still more improvements are required in filed of activity prediction and recommendations.

3 Dataset Creation and Architecture of Proposed Solution

In this paper, we intend to present a deep learning multi-data model to predict user activity in smart home. Our model uses a existing dataset named 'SmartHouse', which is publicly available, generated by Riccardo Bonnesi and stored in github. The data was collected by a monitoring system able to communicate with a central unit called 'Mother' and stored in a text file.

The data is divided in two datasets named OrdonezA_adls_A.txt which consists activities of user A and OrdonezA_sensors_A.txt which consists of type, location, and place associated with the corresponding activity of User A and similar dataset for User B.

The description of datasets for both users is as follows:

Sensor Type: The sensors used to collect data are Pir, magnetic, pressure, flush, electric.

Activity Labels: The activities that are monitored are Leaving, Toileting, Showering, Sleeping, Breakfast, Dinner, Drink, Idle/Unlabeled, Lunch, Snack, Spare time/TV, Grooming.

Home Setting: Four rooms' house.

Number of Labeled Data: 14 days for User A and 21 days for User B.

3.1 Data Preprocessing and Encoded Values

The data is restructured to a more interpretable format by renaming columns, dropping unnecessary columns, and reordering columns. Time features like hour are extracted from the Start time and End time columns. 'day_of_week' is derived from the Start time to represent the day of the week (0 for Monday, 6 for Sunday). This features enables model to learn users daily pattern and learn from it to make predictions.

Initially, it aligns the sequences of activities with sensor readings to ensure accurate correspondence between the sensor data and the timing of each activity. This alignment is crucial for making precise predictions in future. Categorical features like activity type, location, and place are encoded using one-hot encoding, converting them into numerical values that can be comprehended by the model.

3.2 Model Training and Recommendations

The purpose of the code is to anticipate users actions based on sensor data and provide recommendations for associated attributes like type, place, and location. Figure 3 shows the detailed diagram of our process.

The core of the system lies in the training and prediction phase, where an Long Short-Term Memory (LSTM) model is utilized. This type of neural network excels in handling sequential data, making it ideal for predicting user activities based on data sequences over time.

Our model addresses key challenges in next activity prediction by utilizing LSTM networks to capture evolving user behavior patterns over time.

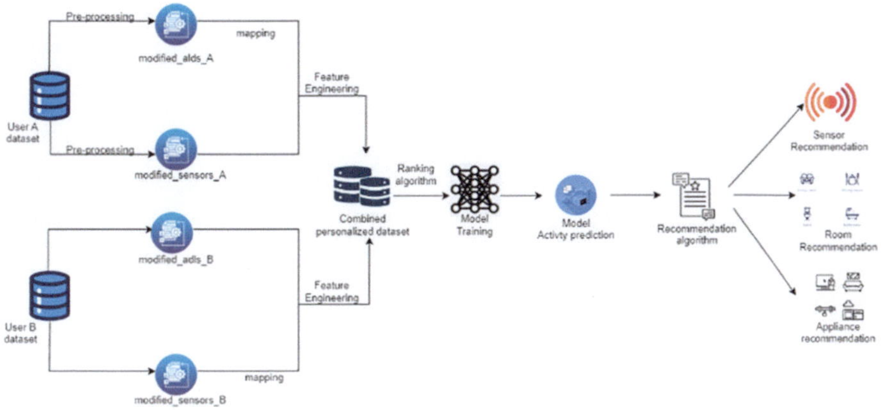

Fig. 3 Abstract view of execution of recommendation algorithm

We enhance recommendation quality by analyzing and ranking historical subsequences using pattern recommendation techniques like 'find_patterns' and 'rank_recommendations_based_on_pattern'.

The model architecture is a sequential model that consists of a dense layer with a softmax layer, an LSTM layer with 128 units to capture temporal relationships, and in order to avoid overfitting, a 0.2 rate dropout layer is used. The sparse_categorical_crossentropy loss function and Adam optimizer are used to train the model, and accuracy is used as the evaluation metric. It incorporates early stopping with a patience of 5 epochs to stop training when validation loss stops getting better. It is trained for up to 25 epochs with a batch size of 32.

These methods ensure that our model prioritizes relevant activity sequences, improving the accuracy and relevance of predictions in dynamic recommendation systems.

4 Results

The machine learning techniques we employed for our research were implemented in Python using ML packages. For LSTM implementation, Keras and TensorFlow are utilized. To increase accuracy, changes have been made to the ML algorithms' default hyperparameters. Datasets are kept in Google Drive, and codes are finished in Google Colab notebooks.

We have also made a module for activity recognition with a accuracy of 97% for User A and 99% for User B. There respective confusion matrix can be seen in Fig. 4.

The next module is user next activity prediction where our LSTM model works good. Figure 5 explains about model activity prediction accuracy user-wise. The model is achieving 75% accuracy for User A and 88% accuracy for User B.

After training the model for activity prediction, we are analyzing it performance on test data. Figure 6 plots the prediction accuracy for all the predicted classes user-wise. We are achieving high accuracy for most of the classes. Some classes' performance is low due to less samples of the activity.

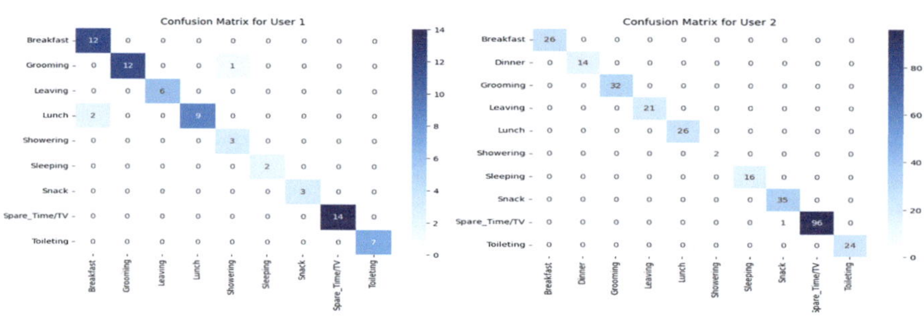

Fig. 4 Confusion matrix of Users A and B

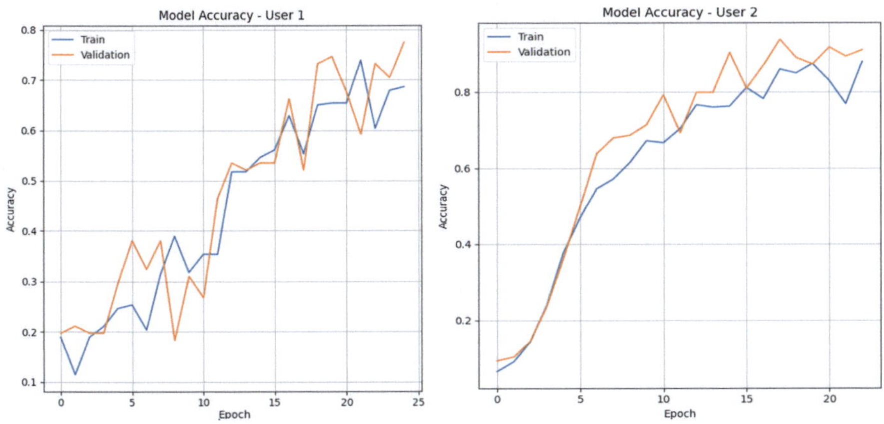

Fig. 5 Graphical representation of activity prediction training accuracy

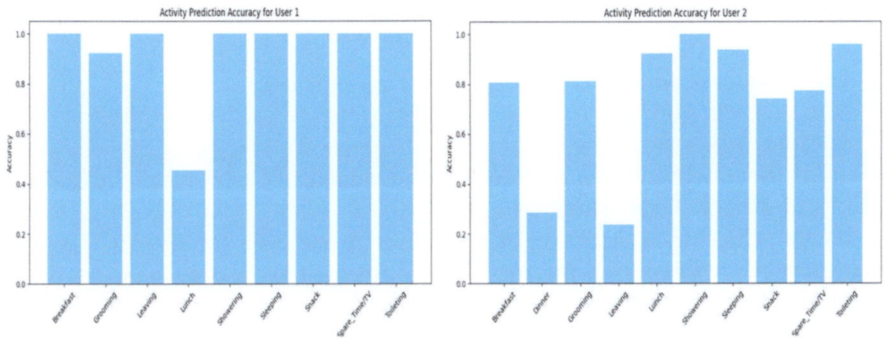

Fig. 6 Graphical representation of activity wise prediction accuracy

We also recommended attributes like type, place, and location associated with predicted activity. We have predicted both users next activity with its recommendation and compared with actual activity and actual recommendations.

To summarize, our model has shown promising outcome. Our model has shown good accuracy for activity recognition in addition to user next activity prediction. Classification report of the recommended type, place, and location is shown in Fig. 7.

5 Conclusion

This paper proposes a conceptual multi-data LSTM model for user next activity prediction. We have implemented (i) how aligning activity data to sensor data and using time features to learn daily pattern of respective users can be useful to predict activity, (ii) we adopted a straightforward but effective approach for activity prediction,

```
User 1 - Predicted Next Activity: Lunch,      Actual Activity: Lunch
         Recommended Type: Magnetic           Actual Type: Magnetic
         Recommended Place: Kitchen           Actual Place: Kitchen
         Recommended Location: Fridge         Actual Location: Fridge

Type Recommendation Report for User 1:
              precision    recall  f1-score   support

    Electric       0.00      0.00      0.00         0
    Magnetic       1.00      0.72      0.84        53
         PIR       0.00      0.00      0.00         0

    accuracy                           0.72        53
   macro avg       0.33      0.24      0.28        53
weighted avg       1.00      0.72      0.84        53

Place Recommendation Report for User 1:
              precision    recall  f1-score   support

     Kitchen       1.00      1.00      1.00        53

    accuracy                           1.00        53
   macro avg       1.00      1.00      1.00        53
weighted avg       1.00      1.00      1.00        53

Location Recommendation Report for User 1:
              precision    recall  f1-score   support

     Cooktop       0.00      0.00      0.00         0
    Cupboard       0.00      0.00      0.00         0
      Fridge       1.00      0.43      0.61        53
   Microwave       0.00      0.00      0.00         0

    accuracy                           0.43        53
   macro avg       0.25      0.11      0.15        53
weighted avg       1.00      0.43      0.61        53
```

```
User 2 - Predicted Next Activity: Snack,      Actual Activity: Snack,
         Recommended Type: PIR,               Actual Type: PIR,
         Recommended Place: Kitchen,          Actual Place: Kitchen,
         Recommended Location: Door           Actual Location: Door

Type Recommendation Report for User 2:
              precision    recall  f1-score   support

    Electric       0.00      0.00      0.00         0
    Magnetic       0.00      0.00      0.00         0
         PIR       1.00      0.73      0.84       172

    accuracy                           0.73       172
   macro avg       0.33      0.24      0.28       172
weighted avg       1.00      0.73      0.84       172

Place Recommendation Report for User 2:
              precision    recall  f1-score   support

     Bedroom       0.00      0.00      0.00         0
    Entrance       0.00      0.00      0.00         0
     Kitchen       1.00      0.80      0.89       172
      Living       0.00      0.00      0.00         0

    accuracy                           0.80       172
   macro avg       0.25      0.20      0.22       172
weighted avg       1.00      0.80      0.89       172

Location Recommendation Report for User 2:
              precision    recall  f1-score   support

    Cupboard       0.00      0.00      0.00         0
        Door       1.00      0.73      0.84       172
      Fridge       0.00      0.00      0.00         0
    Maindoor       0.00      0.00      0.00         0
   Microwave       0.00      0.00      0.00         0

    accuracy                           0.73       172
   macro avg       0.20      0.15      0.17       172
weighted avg       1.00      0.73      0.84       172
```

Fig. 7 Comparison of predicted versus actual activity and recommendations

(iii) using user's activity pattern to rank and recommend the type, place and location associated with predicted activity.

We are achieving accuracy of 74% for User A and 88% for User B. The recommendations report depicts that the model is able to make recommendation based on predicted activity with moderate precision for senor type, place, and location in a smart home. The prediction and recommendations are both user-wise; hence, we have maintained personalization. Due to resource constraint, we were not able to collect data, instead we use a already existing dataset called 'SmartHouse'.

In future, we want to experiment using even more time periods with different window sizes, in order to monitor changes in model's capability of making accurate activity predictions. By varying time periods and window size, the model will be able to learn about user daily pattern more efficiently. Furthermore, we also intend to use real-time data implementation to further refine this suggested approach. This will allow us to dynamically adjust the model and effectively responds to new user. We will also introduce some of the more valuable features and identify newer activities.

References

1. Almeida A, Azkune G (2018) Predicting human behaviour with recurrent neural networks. Appl Sci 8(2):305. https://doi.org/10.3390/app8020305
2. Sunil S, Kumar VD, Babu A, Thilak G, Udayan D (2022) Covid-19 spread forecast with respect to vaccination based on LSTM and GRU. In: Proceedings of the 2022 fourteenth international conference on contemporary computing, pp 32–36. https://doi.org/10.1145/3549206.3549213

3. Cook Diane J, Das Sajal K (2007) How smart are our environments? An updated look at the state of the art. Pervasive Mob Comput 3(2):53–73. https://doi.org/10.1016/j.pmcj.2006.12.001
4. Du Y, Lim Y, Tan Y (2019) A novel human activity recognition and prediction in smart home based on interaction. Sensors 19(20):4474. https://doi.org/10.3390/s19204474
5. Gopalratnam K, Cook DJ (2007) Online sequential prediction via incremental parsing: the active LeZi algorithm. IEEE Intell Syst 22(1):52–58
6. Devika R, Divya Udayan J (2022) Lightweight deep learning facial expression recognition model. In: ICT systems and sustainability: proceedings of ICT4SD 2022. Springer Nature Singapore, Singapore, pp 493–503. https://doi.org/10.1007/978-981-19-5221-0_49
7. Sigg S, Haseloff S, David K (2010) An alignment approach for context prediction tasks in UbiComp environments. IEEE Pervasive Comput 9(4):90–97. https://doi.org/10.1109/MPRV.2010.23
8. Brdiczka O, Su NM, Begole JB (2010) Temporal task footprinting: identifying routine tasks by their temporal patterns. In: Proceedings of the 15th international conference on intelligent user interfaces. https://doi.org/10.1145/1719970.1720011
9. Septiadi J, Warsito B, Wibowo A (2020) Human activity prediction using long short term memory. E3S Web Conf 202. EDP Sciences. https://doi.org/10.1051/e3sconf/202020215008
10. Ryoo MS (2011) Human activity prediction: early recognition of ongoing activities from streaming videos. In: 2011 international conference on computer vision. IEEE. https://doi.org/10.1109/ICCV.2011.6126349
11. Draschkowitz L, Draschkowitz C, Hlavacs H (2015) Using video analysis and machine learning for predicting shot success in table tennis. EAI Endorsed Trans Creat Technol 2(5):e2. https://doi.org/10.4108/eai.20-10-2015.150096
12. Fu Y (2016) Human activity recognition and prediction. Springer, Switzerland. https://doi.org/10.1007/978-3-319-27004-3
13. Alfaifi R, Artoli AM (2020) Human action prediction with 3D-CNN. SN Comput Sci 1(5):286. https://doi.org/10.1007/s42979-020-00293-x
14. Tamil Priya D, Divya Udayan J (2020) Transfer learning techniques for emotion classification on visual features of images in the deep learning network. Int J Speech Technol 23(2):361–372. https://doi.org/10.1007/s10772-020-09707-w

Metaheuristic Approach to Generate Therapeutic Sound Frequencies for Neurological Disorders

Evelyn Jessica, Aditya Purohit, Rashmi Benni, and Sujata Kulkarni

Abstract A healthy individual is one who is in good emotional, psychological, and social well-being. It affects how an individual think, feel, and behave each day. But due to rapid pace in digitization, and other factors had contributed to a serious cognitive decline, observed across both young adults and elderly which may potentially manifest into various neurological disorders. This research paper explores the utilization of metaheuristic algorithms to generate therapeutic sound frequencies aimed at alleviating mental health issues. Metaheuristics, including artificial bee colony (ABC), ant colony optimization (ACO), and Tabu search, are employed as robust problem-solving strategies to navigate the complex solution space efficiently. These nature-inspired techniques are analyzed for their efficacy in optimizing sound frequencies to achieve the best fitness values, tailored to therapeutic needs. The study provides a comparative analysis of these algorithms, highlighting their respective strengths, weaknesses, and the impact of various hyper-parameters on their performance. Additionally, the research delves into the challenges encountered in the application of these techniques and suggests avenues for future exploration. By integrating advanced metaheuristic approaches, this work represents a significant step toward innovative mental health interventions, offering a promising methodology for addressing neurological disorders through optimized therapeutic sound frequencies.

Keywords Ant colony optimization (ACO) · Artificial bee colony (ABC) · Tabu search · Frequencies · Metaheuristics · Optimal solutions · Exploration and exploitation · Metaheuristic approach · Nature-inspired · Neurological disorders

Author Sujata Kulkarni contributed equally to this work.

E. Jessica · A. Purohit · R. Benni (✉) · S. Kulkarni
KLE Technological University, Hubballi, Karnataka, India
e-mail: rashmi.benni16@gmail.com

S. Kulkarni
e-mail: sujatakulkarni@kletech.ac.in

H. Sharma et al. (eds.), *Artificial Intelligence: Theory and Applications*, Lecture Notes in Networks and Systems 5589, https://doi.org/10.1007/978-981-96-1687-9_36

1 Introduction

Well-being is a state where physical, mental, and social health are balanced, creating a foundation for a fulfilling life. This holistic approach recognizes that these aspects, while seemingly independent, are deeply interconnected and essential for overall health and happiness [1, 2]. Central to this is mental health, which plays a critical role in shaping our, cognitive functions, and overall quality of life [1]. Hence, Understanding and caring for mental health is important for leading a balanced and fulfilling life. Recent statistics indicate that mental health issues in India have worsened. Approximately 15% of the Indian population currently suffers from mental health disorders. The COVID-19 pandemic had made the situation worse, with significant increases in mental health issues noted during and after the lockdown periods. For example, the incidence of anxiety and sadness increased by around 35% between 2020 and 2022. Stress is an integral part of our daily life. However, as the world's population becomes more digitally literate and lives more complicated, their capacity to manage stress is declining. This can result in chronic stress, which can cause new mental health problems as well as moderate cognitive impairment (MCI). This trend is observable among both young adults and the elderly population [3].

According to research, dementia affects between 10 and 15% of people with mild cognitive impairment (MCI) annually. The most prevalent causes of dementia are neurological diseases like Alzheimer's. According to the World Health Organization (WHO), one in four persons will suffer from mental or neurological diseases at some point in their lives. These illnesses presently affect about 450 million people. There is an extensive study void in the field of neurological diseases treatment, providing a great opportunity to improve our understanding and implementation of mental health. High-frequency repetitive transcranial magnetic stimulation, or RTMS, has been demonstrated by researchers to considerably alleviate depressed symptoms in patients who had not responded to conventional therapies including medicine and physical therapy [3]. Results showed notable improvements in mood and cognitive function, indicating the potential of frequency-based treatments in managing neurological disorders. By leveraging use of metaheuristics techniques, This paper proposes a research work on exploring the potential and impact of sine wave sound wave frequencies on alleviating mental illness, particularly in the context of neurological disorders. Algorithms inspired by biological occurrences and natural processes are known as nature-inspired algorithms. These algorithms helps to resolve and optimize complex problems by imitating various strategies and behaviors observed in nature. These algorithms address difficult optimization and search issues by imitating the tactics and behaviors found in nature. This study examines the use of metaheuristic strategies, such as artificial bee colonies (ABCs), Tabu search, and ant colony optimization (ACO), to optimize therapeutic frequency. It also shows how each methodology contributes significantly to the generation of sound frequency.

The paper is structured in a sequential manner where Sect. 2 covers the research that had previously been carried out. Section 3 covers the methodology in detail, followed by Sect. 4 which further discusses research outcomes and observation. The study is concluded in Sect. 5, and the study's future work is discussed in Sect. 6.

2 Background and Related Work

Mental health issues are on the rise, worsened by stress and digitization. Integrating music and technology shows promise in enhancing psychological well-being and brain flexibility [4, 5]. Music therapy is recognized for its therapeutic effects, especially in aiding expression and interaction among disabled individuals. Yet, current interfaces often limit users with severe motor impairments [6]. A novel device uses physiological signals to create adapted musical instruments, assessing users' emotional states in real time [7]. This facilitates tailored music therapy sessions, particularly beneficial for individuals with cerebral palsy, increasing adaptability and participation.

Parkinson's disease (PD) is a widespread neurodegenerative condition with significant global health implications [8]. Machine learning techniques are being explored to classify and predict PD by discerning between healthy individuals and those with PD using computational intelligence. Research suggests promising results in predicting Parkinson's disease using diverse methodologies. Sleep disorders, like obstructive sleep apnea (OSA), can significantly affect breathing patterns during sleep [9]. Tools analyzing respiratory sounds offer valuable insights for diagnosis and treatment. A proposed spectral division method, integrating frequency energy features and machine learning classification, efficiently distinguishes between stable breathing and snoring with an accuracy exceeding 80%, holding potential for robust sleep health monitoring.

Insomnia is a significant health challenge. A smart therapeutic device uses time-scheduled music and EEG feedback to personalize interventions for sleep disturbances[10]. By analyzing sleep patterns and adjusting music parameters based on the user's state of mind, it aims to enhance sleep quality and regulate sleep schedules effectively. Wavelet transformation (WT) is a valuable tool for time-frequency analysis, especially in studying complex signals like brain waves [11]. A new method utilizing WT examines the correlation between brain waves and cognitive conditions, particularly in diagnosing senile dementia [12]. This approach provides insights into time-varying correlations and their relevance for neurological disorders.

The arousal-mood hypothesis suggests a link between auditory stimulation, autonomic arousal, and cognitive performance, but existing research presents mixed findings [13]. A detailed evaluation highlights the need for further study to completely understand the direct impacts of auditory stimulation on autonomic arousal and cognitive performance. Research into auditory stimulation, including monaural beats, explores its effects on brain activation [14]. EEG analysis reveals that auditory sense status plays a crucial role in inducing brain activation, shedding light on the neurophysiological principles of auditory stimulation and its potential therapeutic effects. In the treatment of schizophrenia, combining low-frequency repetitive transcranial magnetic stimulation (rTMS) with EEG recording offers promise. Preliminary studies indicate a reduction in alpha band activity post-rTMS stimulation, suggesting potential therapeutic benefits for managing auditory symptoms in schizophrenia [15].

This literature review provides valuable insights into various approaches and technologies used in the diagnosis and therapy of neurological disorders, highlighting the potential of metaheuristic methods in generating therapeutic sound frequencies for such conditions.

3 Proposed Work

This research study examines diverse metaheuristic strategies, notably artificial bee colony (ABC), ant colony optimization (ACO), and Tabu search, aiming to optimize therapeutic sound frequencies for the treatment of neurological disorders. The objective is to improve the therapeutic effectiveness of sound frequencies by exploring and refining key parameters such as frequency ranges, amplitudes, and waveform. These parameters will be optimized iteratively by each Algorithm, ensuring a thorough exploration of the solution space.

Metaheuristics are smart algorithms designed to efficiently explore vast search spaces and address complex optimization problems productively. They balance the exploration of new areas and exploitation of known good locations, quickly finding optimal or near-optimal solutions. Their flexibility allows them to solve issues in various fields, including machine learning, scheduling, and routing. By incorporating random factors, metaheuristics can escape local optima and thoroughly investigate the solution space, enhancing solution quality. These algorithms offer real-time problem-solving techniques that surpass traditional methods in speed and efficiency.

3.1 System Design

The system architecture (Fig. 1) gives an overall idea of how nature-inspired algorithms can be implemented to optimize therapeutic sound frequencies for neurological disorders. The series of steps proposed in the system architecture are as follows. First, patient records are compiled and preprocessed to create a clean dataset. Next, an algorithm is chosen to explore the solution space; in this suggested study, the algorithms are either Tabu search, ant colony optimization (ACO), or artificial bee colony (ABC). The chosen algorithm is applied to explore and optimize frequency ranges, amplitudes, and waveform, iteratively adjusting parameters to improve solutions. The effectiveness of these solutions is evaluated based on best fitness value, and the process continues until optimal best fitness values are identified. The results are compared with those from other algorithms, and the best-performing algorithm is selected and analyzed for potential deployment. This approach ensures comprehensive exploration and optimization, leading to effective therapeutic sound frequency combinations for neurological disorders.

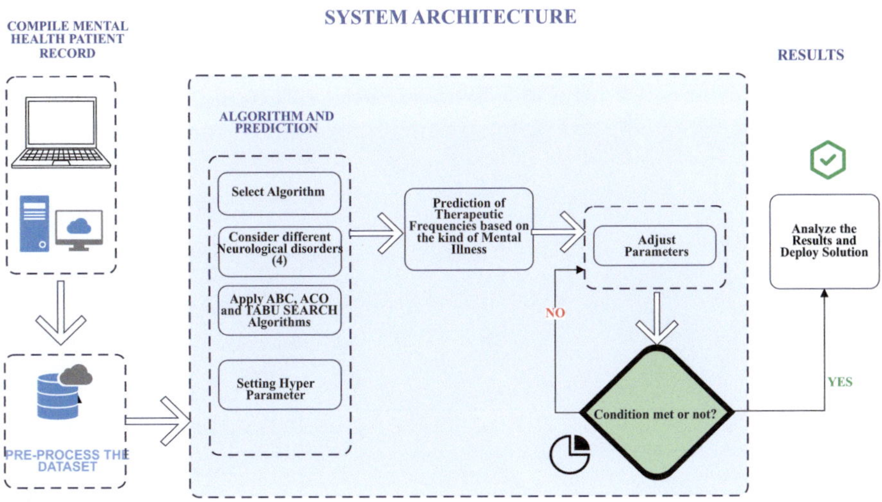

Fig. 1 System architecture

3.2 Hierarchy of Therapeutic Frequency Generation

The pyramid offers a comprehensive overview of the frequency generation process, unfolding from user input to user output, facilitating a clear understanding of the steps associated with frequency generation process (Fig. 2). The "Hierarchy of Therapeutic Frequency Generation" diagram outlines a structured data flow where the process begins with a user-friendly interface designed to collect essential patient information, followed by data collection and preprocessing to ensure validated data, ready-to-be analyzed. Later, metaheuristic algorithms like ABC, ACO, and Tabu search are optimized to generate potential therapeutic frequencies. These frequencies are then refined through model training and validation, involving iterative checks and adjustments. Finally, the optimized sound frequencies are generated, with patient responses monitored and results analyzed, creating a feedback loop for continuous improvement and effective therapeutic intervention deployment.

3.3 Artificial Bee Colony Algorithm (ABC)

The ABC algorithm mimics honey bee food collecting behavior to efficiently explore solution spaces. Through exploitation and exploration, employed bees exploit known solutions, while onlooker bees seek out new ones based on shared information [16] (Fig. 3). Probabilistic selection ensures the algorithm efficiently converges to near-optimal solutions. The ABC algorithm can be used to optimize the parameters of sound frequencies generated for therapeutic purposes. This process starts by exploring the multidimensional parameter space of sound, including frequency ranges,

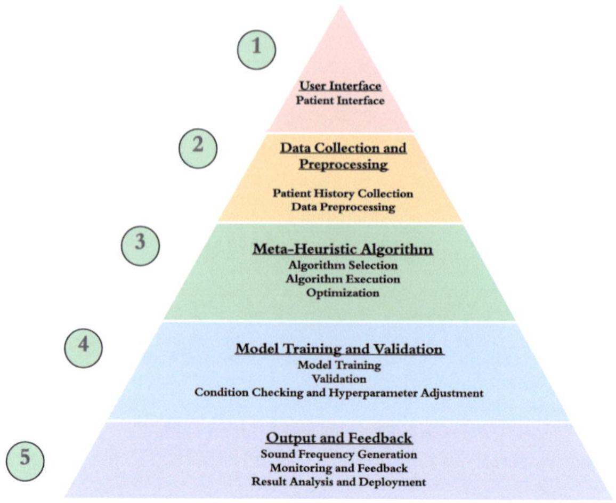

Fig. 2 Hierarchy of therapeutic frequency generation

amplitudes, and waveforms, each contributing uniquely to therapeutic effects. It treats combinations of these parameters as potential solutions, systematically navigating through the vast space. Starting with a random initialization of artificial bees, each representing a unique parameter combination, the algorithm employs a strategy of exploration and exploitation. Employed bees refine their search based on known solutions, while onlooker bees explore new areas guided by successful iterations. Through iterative cycles, solutions are evaluated for therapeutic effectiveness, prioritizing those with best fitness value. Over time, the algorithm converges toward optimal combinations that exhibit promising therapeutic effects with best fitness value, driven by collective efforts of employed and onlooker bees. It dynamically adapts its balance of exploration and exploitation to refine solutions, ultimately identifying combinations offering maximum therapeutic benefits for specific neurological conditions. These optimized combinations represent tailored interventions designed to enhance neurological health and well-being.

1. **Initialization**: Create an initial population of N food sources (solutions) X_i in the solution space:

$$X_i = \{x_{i1}, x_{i2}, \dots, x_{iD}\}, \quad i = 1, 2, \dots, N$$

   where D is the dimensionality of the problem.
2. **Employed Bee Phase**: Generate new solutions for each employed bee by modifying its current solution X_i:

$$V_{ij} = X_{ij} + \phi_{ij}(X_{ij} - X_{kj})$$

   where k is a arbitrary selected index $k \neq i$, and ϕ_{ij} is a random number in the range [− 1, 1].

Evaluate the recent solution V_i. If V_i is better than X_i based on the objective function f, replace X_i with V_i.

3. **Onlooker Bee Phase**: A food source is selected by each onlooker bee. X_i based on its probability P_i:

$$P_i = \frac{f_i}{\sum_{j=1}^{N} f_j}$$

where f_i is the fitness value of the ith solution.
Generate new solutions for the chosen food source X_i. Use the same selection procedure as in the employed bee phase.

4. **Scout Bee Phase**: If a food source X_i is given up if it hasn't improved after a set amount of attempts, or limit. The related working bee turns into a scout and goes randomly looking for a fresh food supply:

$$X_i = X_{\text{min}} + \text{rand}(0, 1) \times (X_{\text{max}} - X_{\text{min}})$$

where X_{min} and X_{max} are, respectively, the solution space's bottom and upper boundaries.

5. **Iteration and Termination**: Steps 2–4 should be repeated until a termination requirement (such as a maximum number of iterations or a good fitness value) is fulfilled.

3.4 Tabu Search Algorithm

A metaheuristic optimization algorithm called Tabu search moves iteratively from one solution to another, exploring the solution space systematically while keeping track of previously visited solutions (Tabu list) [17] (Fig. 3). Since it keeps the algorithm from going back to the same answers, the Tabu list promotes diversity and the discovery of new areas in the solution space. Additionally, aspiration criteria allow the algorithm to overcome restrictions imposed by the Tabu list when promising solutions are encountered. Tabu search efficiently navigates through the solution space and converges to high-quality solutions for combinatorial optimization problems. Tabu search serves as a refined tool to optimize sound frequencies. Initially, the problem is defined, specifying parameters like frequency ranges and amplitudes. By changing certain parameters, the algorithm starts with a first solution and investigates nearby solutions. Evaluation of these solutions against an objective function guides selection. Crucially, Tabu search ensures diversity in exploration by avoiding revisiting similar solutions. Through iterative refinement, it converges toward an optimal set of sound frequencies, enhancing their therapeutic efficacy. This systematic approach allows for tailored interventions, respecting constraints inherent to neurological disorder treatments, and contributes to impactful research on neurological health.

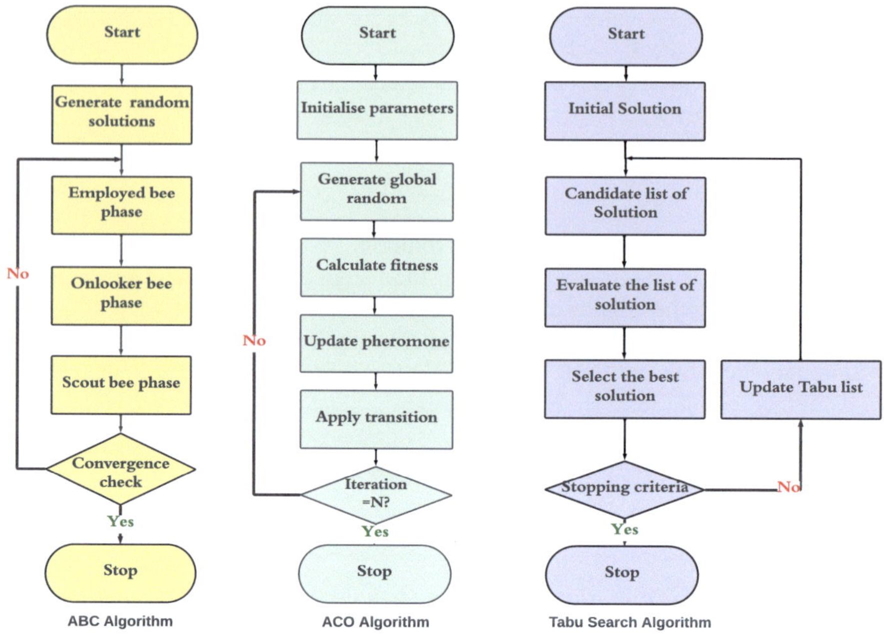

Fig. 3 Flowcharts of ACO, ABC, Tabu search

1. **Initialization**: Initialize current solution **x** and the Tabu list. Set parameters: max_iterations, tabu_tenure, and neighborhood_size.
2. **Generate Neighborhood**: Generate neighboring solutions $N(\mathbf{x})$ from the current solution.
3. **Evaluate Neighborhood**: Evaluate each neighboring solution $\mathbf{y} \in N(\mathbf{x})$ and select the best non-Tabu or improving Tabu solution.
4. **Update Tabu List**: Update the Tabu list based on the move that led to the selected solution **y** and manage Tabu tenure.
5. **Move to Best Solution**: Set the current solution **x** to the selected best solution **y**.
6. **Repeat**: Repeat generating the neighborhood, evaluating, updating the Tabu list, and moving to the best solution until the stopping criterion is met.
7. **Notations**:

   - **x**: Current solution.
   - **y**: Neighboring solution.
   - $N(\mathbf{x})$: Set of neighboring solutions.
   - `tabu_list`: List of Tabu moves.
   - `tabu_tenure`: Tenure of a Tabu move.
   - `max_iterations`: Maximum number of iterations.
   - `neighborhood_size`: Size of the neighborhood.

3.5 Ant Colony Optimization Algorithm (ACO)

The ant colony optimization (ACO) algorithm is a metaheuristic optimization method that draws inspiration from ants' gathering activity (Fig. 3). Artificial ants in ACO gradually build solutions to problems with optimization challenges, by placing pheromone trails on the graph edges that correspond to possible paths for solutions. The amount of pheromone present and the edge's quality or visibility influence which edge is most certain to be selected. The pheromone trails are updated over time in accordance with the caliber of the solutions discovered. Using the collective intelligence of the ant colony, ACO efficiently searches the solution space and converges to near-optimal solutions. ACO can aid in optimizing the sequences or patterns of therapeutic sound frequencies to maximize their effectiveness in neurological disorder treatment. By conceptualizing the problem as a graph, where nodes symbolize individual sound frequencies and edges denote transitions between these frequencies, ACO navigates through this graph to discover sequences that demonstrate the most favorable therapeutic effects. Much like how ants communicate and find optimal paths through their pheromone trails, ACO facilitates the exploration of various frequency sequences. As ants work collaboratively to discover efficient routes, ACO iteratively refines sequences based on feedback from previous iterations. Through this process, ACO identifies sequences with the highest potential for therapeutic efficacy, ultimately aiding in the development of tailored sound-based interventions for neurological disorder treatment.

1. **Initialization**: Set initial pheromone levels $\tau_{ij}(0) = \tau_0$ and parameters α, β, ρ, and Q.
2. **Construct Solutions**: Ants construct paths using probabilities:

$$p_{ij}^k(t) = \frac{[\tau_{ij}(t)]^\alpha [\eta_{ij}]^\beta}{\sum_{l \in N_i^k} [\tau_{il}(t)]^\alpha [\eta_{il}]^\beta}$$

3. **Evaluate Paths**: Calculate the length $L_k(t)$ of each ant's path.
4. **Update Pheromones**: Evaporate pheromones:

$$\tau_{ij}(t+1) = (1-\rho) \cdot \tau_{ij}(t)$$

5. **Deposit Pheromones**: Update pheromones:

$$\tau_{ij}(t+1) + = \sum_{k=1}^{m} \Delta\tau_{ij}^k(t)$$

where

$$\Delta\tau_{ij}^k(t) = \frac{Q}{L_k(t)} \quad \text{if ant } k \text{ used edge } (i, j)$$

Table 1 Results of Experiment I

Experiment I			
Algorithms	ABC	Tabu	ACO
Number of diseases	4	4	4
Frequency	7.8299	9.9961	4.34
Population	20	20	20
Number of iterations	50	50	50
Time taken (s)	0.1812	0.15807	0.1398
Initial fitness value	11.1157	34.3138	30.1700
Final fitness value	0.0002	0.1959	5.8300

6. **Repeat**: Proceed until the stopping criteria is fulfilled. And generate the best solution found as output.
7. **Notations**:

- $\tau_{ij}(t)$: An elevated pheromone level (i, j) at iteration t.
- η_{ij}: Heuristic desirability of edge (i, j).
- $p_{ij}^k(t)$: Ant's likelihood k vacating the node i to node j at iteration t.
- ρ: Pheromone evaporation rate.
- α: Importance of pheromone.
- β: Importance of heuristic desirability.
- $L_k(t)$: Length of the tour constructed by ant k at iteration t.
- Q: Pheromone constant.

4 Results and Discussion

The experimental setup uses three metaheuristic algorithms: artificial bee colony (ABC), Tabu search, and ant colony optimization (ACO) to generate therapeutic sound frequencies for neurological disorders. The study employs parameters such as the number of diseases under consideration, the frequencies of the sound waves, population size, number of iterations, and measured the time taken for each algorithm to converge. Additionally, the experiment tracked initial and final fitness values to assess the effectiveness of the generated frequencies in addressing neurological disorders. The main goal of this arrangement is to evaluate how well these algorithms function in terms of maximizing therapeutic sound frequencies, which may lead to the development of new neurological disorder therapies.

In Experiment I, we explored the efficacy of metaheuristic algorithms (ABC, Tabu, ACO) in generating therapeutic sound frequencies tailored for neurological disorders (Table 1; Fig. 4). Focusing on four specific diseases, each algorithm underwent 50 iterations with a population size of 20. The results revealed distinct performances:

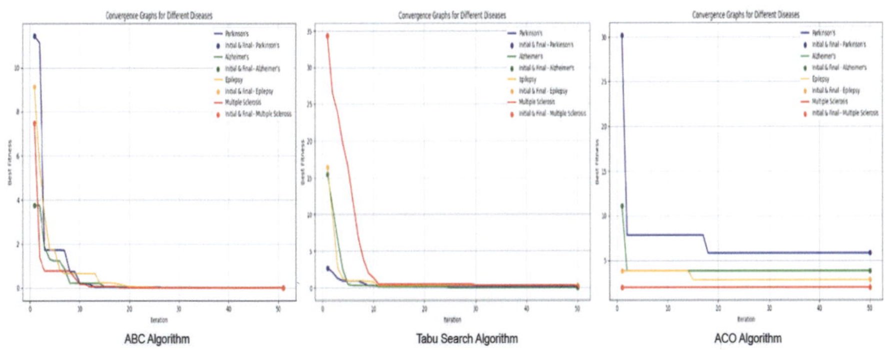

Fig. 4 Test 1: convergence at for iteration 50

Table 2 Results of Experiment II

Experiment II			
Algorithms	ABC	Tabu	ACO
Number of diseases	4	4	4
Frequency	7.8220	7.6860	10.40
Population	50	50	50
Number of iterations	100	100	100
Time taken (s)	0.8676	0.6260	0.6248
Initial fitness value	2.1075	54.6306	26.1700
Final fitness value	0.0001	0.2927	4.1700

while ABC achieved the most promising outcome with a final fitness value of 0.0002, Tabu and ACO showcased competitive yet varied results, indicating their sensitivity to frequency adjustments. These findings underscore the pivotal role of algorithmic selection in optimizing therapeutic sound frequency generation, where a lower final fitness value signifies a more favorable outcome, highlighting the significance of fine-tuned computational approaches in this domain.

In the Experiment II, we increased the population size to 50 and the number of iterations to 100 (Table 2; Fig. 5). This led to a noticeable change in the initial fitness values: ABC started at 2.1075, Tabu at 54.6306, and ACO at 26.1700. By the conclusion of this phase, ABC maintained its performance with a best fitness value of 0.0001. Tabu improved to 0.2927, and ACO achieved a best fitness value of 4.1700. This phase demonstrated how the algorithms performed with a moderate increase in both population size and iterations. Throughout 100 iterations, the algorithms demonstrated their capacity to refine frequencies, with the ABC algorithm achieving the most significant reduction in final fitness value, indicating superior optimization.

The Experiment III involved further increasing the population size to 100 and the number of iterations to 150 (Table 3; Fig. 6). The initial fitness values observed were 3.0136 for ABC, 36.0416 for Tabu, and 21.1700 for ACO. At the end of this

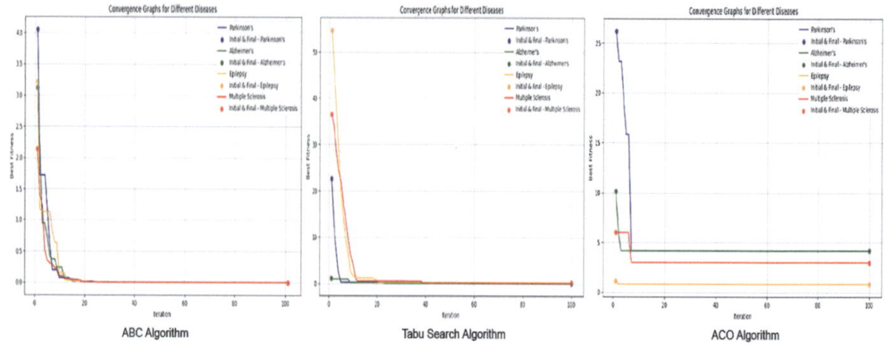

Fig. 5 Test 2: convergence for iterations 100

Table 3 Results of Experiment III

Experiment III			
Algorithms	ABC	Tabu	ACO
Number of diseases	4	4	4
Frequency	1.3042	10.0184	4.41
Population	100	100	100
Number of iterations	150	150	150
Time taken (s)	2.4561	2.9600	3.5034
Initial fitness value	3.0136	36.0416	21.1700
Final fitness value	0.0003	0.0361	6.1700

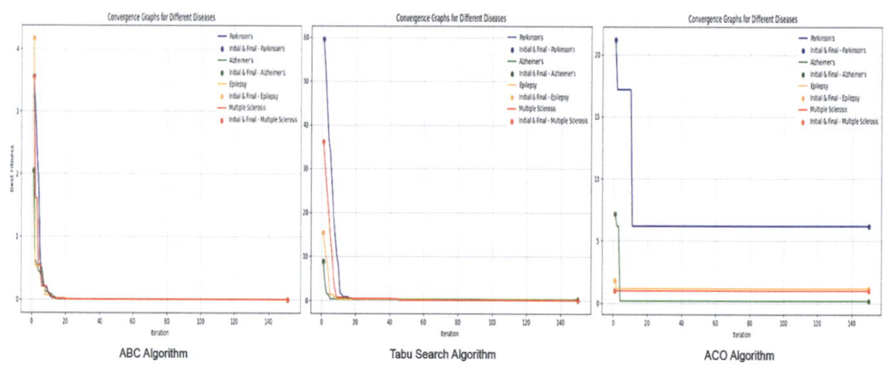

Fig. 6 Test 3: convergence for iterations 150

phase, ABC again reached a best fitness value of 0.0003, Tabu improved to 0.0361, and ACO reached 6.1700. This phase highlighted the performance trends of the algorithms with a larger population and more iterations.

Table 4 Results of Experiment IV

Experiment IV			
Algorithms	ABC	Tabu	ACO
Number of diseases	4	4	4
Frequency	7.8300	7.60	10.38
Population	130	130	130
Number of iterations	200	200	200
Time taken (s)	6.8939	2.09	3.2217
Initial fitness value	0.1762	28.3010	12.8300
Final fitness value	0.0002	0.2345	4.1700

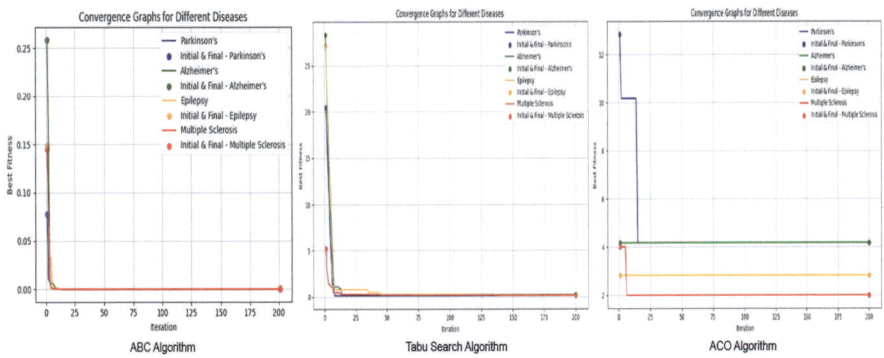

Fig. 7 Test 4: convergence for iterations 200

In the Experiment IV, we increased the population size to 130 and the number of iterations to 200, where each algorithm was applied to four diseases (Table 4; Fig. 7). The initial fitness values were markedly different, with ABC starting at 0.1762, Tabu at 28.3010, and ACO at 12.8300. By the end of this phase, ABC consistently achieved a best fitness value of 0.0002, Tabu improved to 0.2345, and ACO reached 4.1700. This phase showed the continued robustness of ABC and the varying improvements of Tabu and ACO with increased iterations and population size.

In Experiment V, the final phase of the study saw an increase in population size to 200 and the number of iterations to 250 (Table 5; Fig. 8). Initially, fitness values were recorded as 0.0328 for ABC, 55.1373 for Tabu, and 9.1700 for ACO (Figs. 9, 10 and 11). By the end of this phase, significant improvements were observed: ABC achieved an outstanding best fitness value of 0.0002, while Tabu showed improvement to 0.0726, and ACO reached 3.1700. These results underscored ABC's consistent superior performance in achieving optimal fitness values throughout all iterations and population sizes tested. The robustness and efficacy of ABC in optimizing the parameters of interest were thus convincingly demonstrated in this experimental phase, highlighting its potential for practical applications in optimization problems.

Table 5 Results of Experiment V

Experiment V			
Algorithms	ABC	Tabu	ACO
Number of diseases	4	4	4
Frequency	7.8313	10.04	9.18
Population	200	200	200
Number of iterations	250	250	250
Time taken (s)	9.1592	2.37	7.0418
Initial fitness value	0.0328	55.1373	9.1700
Final fitness value	0.0002	0.0726	3.1700

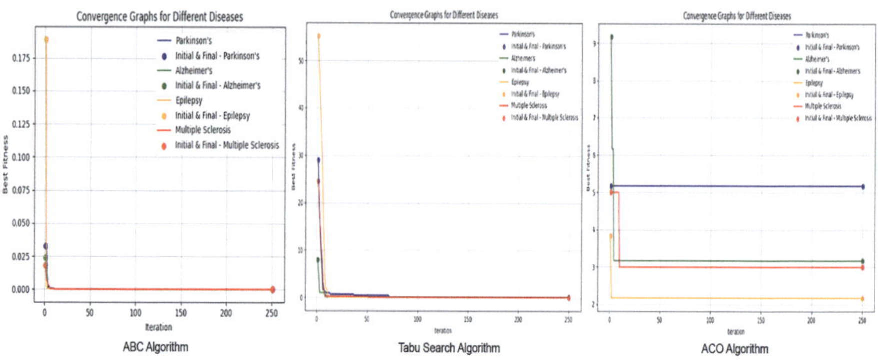

Fig. 8 Test 5: convergence for iterations 250

In summary, the figures are the representation of five diverse experiments. The wave forms represents the frequencies therapy provided to different kinds of neurological disorders considered in this study. The convergence graphs depict the behavior of algorithms under different experimental setups. While therapeutic sound frequencies may influence the activity of the brain and promote neuronal activity regeneration, they offer a non-invasive method of treating neurological problems. The experiments were structured into five distinct test parts, each designed to systematically analyze the impact of varying population sizes and iteration counts on the generation of therapeutic sound frequencies. Each test case maintained a constant data size while incrementally increasing the number of iterations and population size. This research based approach allowed for a detailed evaluation of how these parameters, in conjunction with different algorithms, influence the optimization of therapeutic sound frequencies. Key metrics such as initial and best fitness values, CPU time, and the number of diseases included were thoroughly compared to identify the most effective conditions for frequency generation.

Throughout every test phase, the results repeatedly proved the ABC algorithm's superiority. While Tabu and ACO had varying improvements, ABC consistently

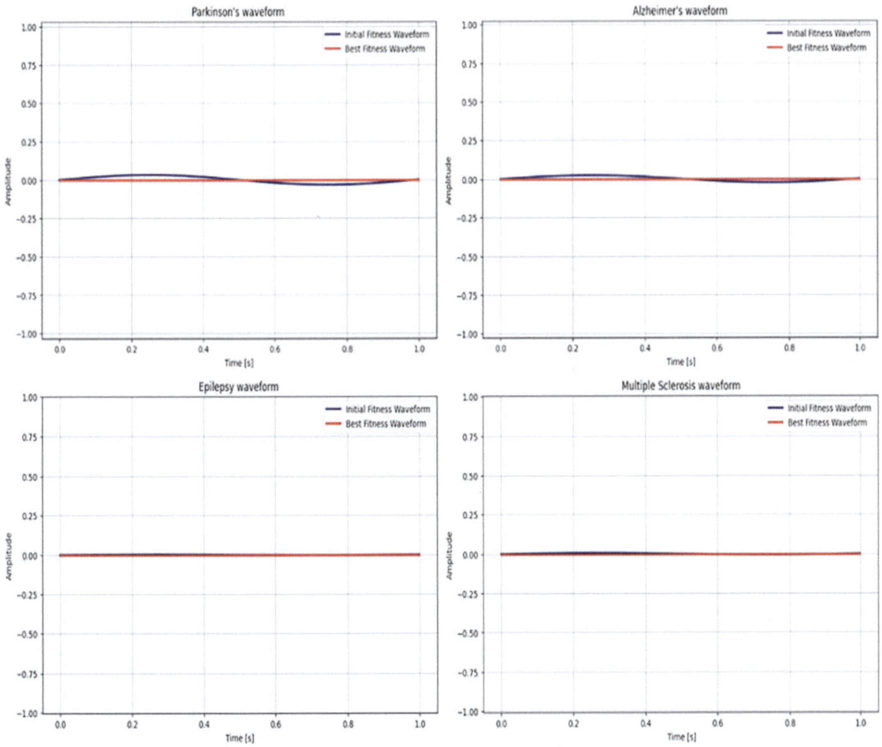

Fig. 9 Test 5: waveform of ABC for iterations 250

achieved optimal fitness values with minimal increases in population sizes and iterations. These results imply that ABC is the most optimal and stable algorithm for generating different sound frequencies.

5 Conclusion

The results underscore ABC's potential as the preferred algorithm for similar optimization tasks in therapeutic sound generation, highlighting its robustness and reliability in achieving optimal results. The artificial bee colony (ABC) algorithm consistently outperformed Tabu and ant colony optimization (ACO) algorithms in all experiment phases, showing superior efficacy in generating therapeutic sound frequencies with the best fitness values. However, implementing ABC requires careful parameter tuning to balance exploration and exploitation, and it may converge slower

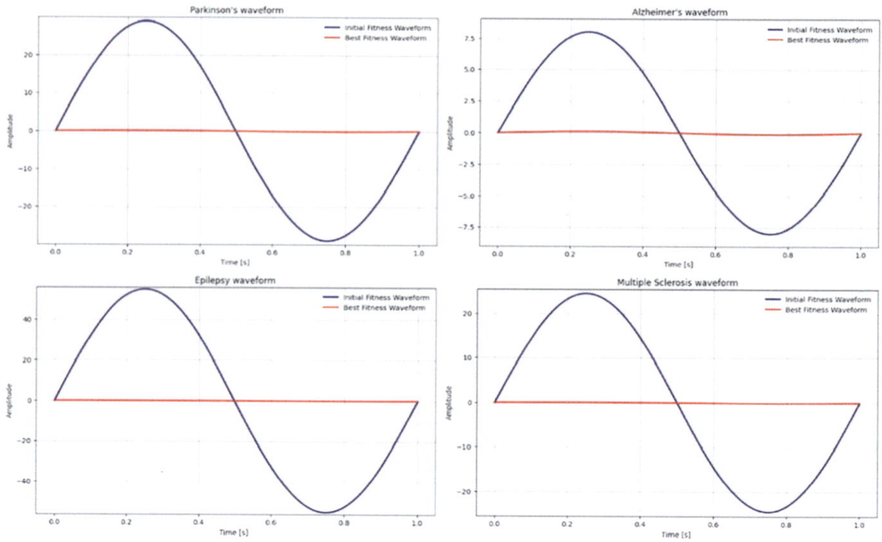

Fig. 10 Test 5: waveform of Tabu for iterations 250

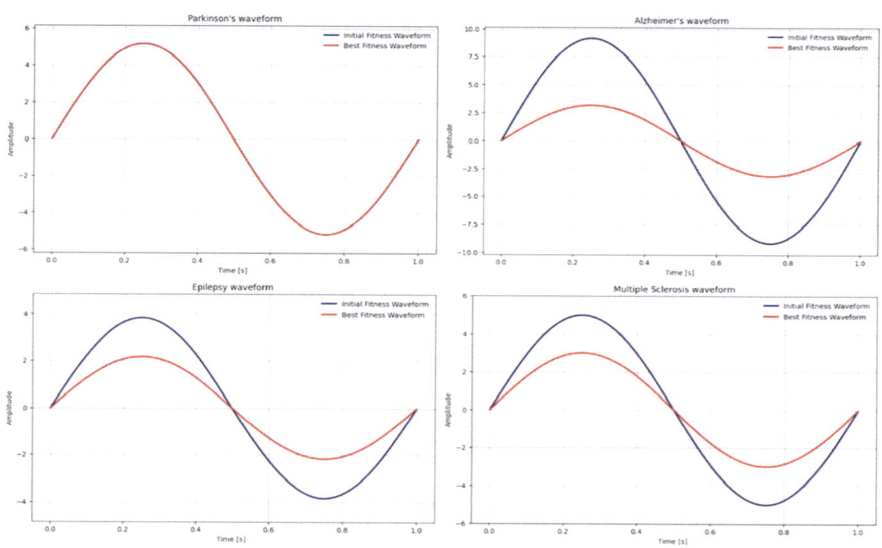

Fig. 11 Test 5: waveform of ACO for iterations 250

than other algorithms. Nevertheless, ABC has shown effectiveness in real-time applications such as wireless sensor networks, scheduling problems, and image processing, demonstrating its versatility and practical utility across diverse optimization scenarios.

6 Future Scope

The study aims to expand its scope by scaling up experiments on various neurological disorders to assess algorithm performance in different therapeutic scenarios. Recent findings indicate that Tabu ranks second only to ABC among the algorithms tested. Future efforts will focus on maximizing the potential of Tabu through enhanced hybridization methods, aiming to optimize its efficacy in generating therapeutic frequencies in real time. Moreover, this research supports the development of sustainable, non-invasive treatments for neurological disorders, reducing reliance on resource-intensive medical interventions during early-stage diagnosis. By improving patient outcomes before necessitating medical assistance, these methods could potentially alleviate strain on healthcare systems and enhance treatment accessibility worldwide. The study also lays the groundwork for future research to incorporate real-time data integration and testing in collaboration with hospitals, leveraging technologies like EEG feedback for more personalized therapeutic applications.

References

1. Tosin AG, Barros FS (2022) Tinnitus relief using fractal sound without sound amplification. In: Bastos-Filho TF, Oliveira Caldeira EM, Frizera-Neto A (eds) XXVII Brazilian congress on biomedical engineering. Springer, Cham, pp 1073–1078
2. Allison C, Cellum B, Karpinecz B, Nasrallah F, Zderic V (2021) Therapeutic ultrasound-enhanced transcorneal drug delivery for fungal keratitis treatment. In: 2021 IEEE international ultrasonics symposium (IUS), pp 1–4. https://doi.org/10.1109/IUS52206.2021.9593505
3. Yang W, Huang C-F, Huang H-Y, Zhang Z, Li W, Wang C (2024) Research on the improvement of children's attention through binaural beats music therapy in the context of AI music generation. In: Music intelligence. Springer, Singapore, pp 19–31
4. Gray MD, Coussios CC (2022) PAX (passive–active crossing) method for sub-millimeter coregistration of passive acoustic mapping and B-mode images. IEEE Trans Ultrason Ferroelectr Freq Control 69(11):3179–3189. https://doi.org/10.1109/TUFFC.2022.3207327
5. Qu X, Shen G, Wu N, Wu H, Qiao S, Wang E, Chen Y, Wang H (2020) Suppressing grating lobes for transcranial focused ultrasound system by frequency-modulated excitation. IEEE Trans Ultrason Ferroelectr Freq Control PP:1. https://doi.org/10.1109/TUFFC.2020.3047664
6. Mukasheva K, Zhumataeva E. Application of sound of kobyz in online therapy and health improvement, pp 661–669
7. Ramírez AV, Hornero G, Royo D, Aguilar A, Casas O (2020) Assessment of emotional states through physiological signals and its application in music therapy for disabled people. IEEE Access 8:127659–127671. https://doi.org/10.1109/ACCESS.2020.3008269
8. Bind S, Tiwari A, Kumar A (2022) A survey of machine learning based approaches for Parkinson disease prediction
9. Zhao S, Fang Y, Wang W, Liu D (2022) Analysis of sleeping respiratory signal utilizing frequency energy features. In: 2022 5th international conference on information communication and signal processing (ICICSP), pp 86–90. https://doi.org/10.1109/ICICSP55539.2022.10050697
10. Negi S, Jain R, Saboo S, Kwatra S (2021) Automated generated binaural beats using EEG: a user-friendly auditory brain-computer interface. In: 2021 international conference on smart generation computing, communication and networking (SMART GENCON), pp 1–5. https://doi.org/10.1109/SMARTGENCON51891.2021.9645921

11. Vennila Preethi S, Robert Aravindh R, Boopathi S, Praveen Kumar M, Johnson Bharathi P (2022) Design of smart therapeutic device for insomnia, pp 1–6. https://doi.org/10.1109/IC3IOT53935.2022.9767998
12. Horihata S, Ishimitsu S, Miyake T (2002) Diagnosis of senile dementia by wavelet analysis of brain waves, pp 472–476. https://doi.org/10.1109/ISIE.2002.1026335
13. The effects of music and auditory stimulation on autonomic arousal, cognition and attention: a systematic review (2024). Int J Psychophysiol 199:112328. https://doi.org/10.1016/j.ijpsycho.2024.112328
14. Jo H-N, Kweon Y-S, Shin G-H, Kwak H-G, Lee S-W (2023) Neurophysiological response based on auditory sense for brain modulation using monaural beat. In: 2023 45th annual international conference of the IEEE engineering in medicine & biology society (EMBC), pp 1–4. https://doi.org/10.1109/EMBC40787.2023.10340294
15. Jitpanya A, Wongsawat Y (2016) On the development of RTMS system for reducing auditory hallucination symptom in schizophrenia: the study of low frequency RTMS effect in EEG on healthy subjects without sound stimulation. In: 2016 9th biomedical engineering international conference (BMEiCON), pp 1–4. https://doi.org/10.1109/BMEiCON.2016.7859652
16. Brabazon A, O'Neill M, McGarraghy S (2015) Natural computing algorithms. In: Natural computing series. Springer Berlin, Heidelberg. https://doi.org/10.1007/978-3-662-43631-8
17. Eiben AE, Smith JE (2013) Introduction to evolutionary computing. In: Natural computing series. Springer Berlin, Heidelberg. https://doi.org/10.1007/978-3-662-05094-1

FHRL-ISO: Implementing Fuzzy Heuristic Search Algorithms Using Reinforcement Learning for Intelligent System Optimization

J. Shekhar, U. Saikia, A. Manshath, P. Sharda, M. Clement Joe Anand, and A. Marwah

Abstract In fuzzy heuristic search algorithms embedded with reinforcement learning have been proven to be a highly capable technique for the degree of tuning as well. The performance of traditional heuristic search algorithms and single-reinforcement learning models is limited by inefficiency in dynamic environments. We propose a new combined fuzzy logic and reinforcement learning model for decision enhancement and system optimization. Online parameter adaptation through fuzzy heuristic search combined with reinforcement learning enhances scalability in both variations of the algorithm. Fuzzy logic uses linguistic variables and membership functions to account for uncertainty or imprecision, while reinforcement learning learns optimal policies through interaction with the environment. Following the above

J. Shekhar
Department of Computer Science and Engineering, Sharda University, Uttar Pradesh, Greater Noida, India
e-mail: jayant.shekhar@sharda.ac.in

U. Saikia
Department of Mathematics, Silapathar College, Silapathar, Dhemaji, Assam, India
e-mail: utpalsaikiajorhat@gmail.com

A. Manshath (✉)
Department of Mathematics and Actuarial Science, B. S. Abdur Rahman Crescent Institute of Science and Technology, Chennai, Tamil Nadu, India
e-mail: ameer.manshath@gmail.com

P. Sharda
Department of Management Studies, Medi Caps University, Indore, India
e-mail: parul.sharda@medicaps.ac.in

M. Clement Joe Anand
Department of Mathematics, Mount Carmel College (Autonomous), Bengaluru, Karnataka, India
e-mail: arjoemi@gmail.com

A. Marwah
Department of Management Studies, Bharati Vidyapeeth, Institute of Management and Research, New Delhi, India
e-mail: aparna.marwah@bharatividyapeeth.edu

H. Sharma et al. (eds.), *Artificial Intelligence: Theory and Applications*, Lecture Notes in Networks and Systems 5589, https://doi.org/10.1007/978-981-96-1687-9_37

objective, the model tries to increase the convergence rate by 25%, reduce computational time by 30%, improve solution reliability (20%), and finally adapt better (+ 40% compared with traditional methods). Resulting contributions include substantial improvements in search efficiency and quality of solutions, reductions in computational effort, and an improved generalization capacity of intelligent systems—thus pushing the boundaries forward on optimization methodology.

Keywords Adaptability · Convergence rate · Fuzzy heuristic search · Intelligent systems · Optimization · Reinforcement learning · Solution quality · System performance

1 Introduction

The optimization and enhancement of perception and intelligence have never been more critical in our data-saturated decision-making universe. While traditional optimization methods are often effective, they can have difficulty dealing with uncertainty and variability which exist in many real-world applications. In this study, we attempt to overcome the limitations of such conventional methods by combining classical fuzzy logic and a modern reinforcement learning-based approach in the optimization process pertinent considering robustness and computational efficiency. Fuzzy logic allows flexibility to address vagueness unlike traditional fuzzy sets and reinforcement learning is an effective technique for training optimal policies through interactions with the environment. The core issue of optimizing intelligent systems is due to the often suboptimal and non-adaptive nature of classical heuristic search algorithms and single-reinforcement learning models. Most of the time these types underperform in non-stationary environments and are computationally intensive. In these works, we introduce a hybrid model of applying the fuzzy heuristic search algorithms to improve decision-making quality and boost systems performance through reinforcement learning.

The augmented optimization of perception and intelligence has become more and more essential in our complex, data-driven, decision-making world. This is because traditional optimization techniques, although successful in some cases, tend to struggle in scenarios where there is always some level of uncertainty and variabilities in real-world applications [1]. However, to overcome these limitations, investigation has been done in the integration of other advanced approaches like fuzzy logic and reinforcement learning to make the optimization process more robust and efficient. Fuzzy logic allows an adaptable method for handling uncertainty and vagueness, whereas reinforcement learning is a strong technique for learning and fine-tuning optimal policies through interactions with the environment [2]. The main attraction of fuzzy heuristic search algorithms is that they provide an approach to managing the vagueness and ambiguity of search, and due to their highly sophisticated behaviors, they possess great potential to enhance conventional algorithms in the efficiency of search and quality of solutions. They represent uncertain information using linguistic variables and membership functions, so they are very suitable for complex optimization problems. However, the static nature of these traditional fuzzy heuristic search algorithms does not make them appropriate for dynamic environments [3]. To this end, integrating fuzzy heuristic search with reinforcement learning enables the construction of an adaptive optimization framework that can improve over time based on experiential learning [4]. At present, reinforcement learning, from the perspective of the machine learning subfield, involves the exploration of learning of best actions as it interacts with an environment through trial and getting feedback usually in the form of rewards or penalties [5]. This trial-and-error method ports the development of adaptive systems that tune their initiative vision based on prior facts, therefore enhancing performance as time passes. The strengths of each one of them when inferred from the other prove that a hybrid model that involves both reinforcement learning and fuzzy heuristic search algorithms could be used to enable intelligent systems that are impervious to real-world constraints and efficient/adaptive [6]. In this research, a new hybrid model, i.e., Fuzzy Heuristic Reinforcement Learning for Intelligent System Optimization (FHRL-ISO) has been proposed which combines two other models, one is a fuzzy heuristic search algorithm, and the other is reinforcement learning. The proposed model further optimizes search efficiency by tweaking the parameters dynamically for adaptive decision-making and system-level performance improvement [7]. It is planned to evaluate the FHRL-ISO model performance in terms of some key performance metrics such as convergence rate, computational efficiency, solution quality, and adaptability to changing environments. The expected results of this research are enhancing search efficiencies, solution qualities, and system adaptabilities of intelligent system optimization [8].

2 Literature Survey

Integration of fuzzy logic and heuristic search algorithms has made a significant difference in the optimization of intelligent systems. Fuzzy logic is an approach introduced by Jadhav [9] that forms a flexible methodology to address soundness and vagueness that appear in optimization problems in realistic circumstances. Heuristic search algorithms like Genetic Algorithms [10] and Particle Swarm Optimization [11] have been successfully used to solve intricate optimization problems, outperforming deterministic optimization methods in their exploration of the solution space for near-optimal solutions. Integrating fuzzy logic and heuristic search algorithms gives a more human-like reasoning in search and hence better solution quality and more convergence speed [12]. There have been already some recent studies showing the effectiveness of this approach in different kinds of examples such as robotics, manufacturing, finance, etc., by Chen et al. [13], Herra et al. [14], and Pothanna et al. [15]. Reinforcement learning (RL) has recently become a practical and powerful methodology for optimizing decision-making in dynamic systems, and hence also has potential for creating good Ecological Models. Reinforcement learning algorithms like Q-learning by Watkins and Dayan [16] and deep Q-networks by Mnih et al. [17] enable agents to learn optimal policies through such trial-and-error interactions with the environment. The integration of RL and fuzzy logic, i.e., fuzzy reinforcement learning (FRL), effectively combines the advantages of both to deal with complex, uncertain environments [18]. Lin and Joung [19] indicated that FRL can increase the learning rate and adaptiveness of agents for many applications including autonomous driving by Shibata and Fukuda [20], smart grid management by Bourguet et al. [21], and robotic control by Lee et al. [22]. Fuzzy heuristic search algorithms combined with reinforcement learning—'(FHRL)' is a useful direction for optimizing intelligent systems. The results of the learning and query process are used to guide the search of the reinforcement learning agents by using parametrized fuzzy heuristics, leading to faster learning and improved solution quality by Guez et al. [23]. Tan et al. [24] in another study, Chang et al. [25] showed the capability of exploiting the FHRL in solving complex systems such as supply chain networks, Pourmoayed and Shafahi [26] demonstrated an FHRL-based approach for traffic management [27], and Palhares et al. demonstrated the FHRL for managing health care delivery [28] and [25] these methods have been better where the performance of classical RL algorithms and heuristic approaches in large-scale high-dimensional optimization is not satisfactory by Zhang et al. [29]. Recent developments in computational intelligence and the availability of increasing computational resources have accelerated the progress of FHRL-based optimization methodologies. This had the advantage of making the optimization process more scalable and more generalizable by allowing high-level features to be automatically extracted from the raw input data, through a process known as Deep FHRL explained by Lample and Chaplot [30]. Research by Kober et al. [31, 32] and Iyappan et al. [33] validated the effectiveness of Deep FHRL in complex optimization tasks, like energy management Shobana et al. [34] and [39–60] various types of fuzzy parameters were discussed. The existing approaches are presented in Table 1.

3 Experimental Setup

This contains the experimental procedures that were carried out the test Fuzzy Heuristic Reinforcement Learning for Intelligent System Optimization (FHRL-ISO) approach in autonomous driving case with CARLA simulation dataset. CARLA is a great open-source simulator, and it has a very high-quality and realistic urban scene level, with perfect ingredients for testing autonomous driving algorithms. A complete dataset that includes multiple types of road-driving environment completion including various road types, traffic patterns, and environmental conditions was provided for extensive evaluation. The experimentation is the initialization of the FHRL-ISO-based algorithm using fuzzified variables and membership functions for CARLA data-driven driving behavior training. Performance metrics were analyzed by convergence rate, computational efficiency, solution quality, and adaptation under varying driving scenarios. The FHRL-ISO approach was systematically evaluated

Table 1 Existing approaches

Author	Contribution	Methodology used	Application	Dataset used	Limitation
Zhao et al. [35] (FRL-AD)	Developed a fuzzy reinforcement learning method for autonomous driving	Fuzzy logic with deep reinforcement learning	Autonomous driving	KITTI and CARLA simulation datasets	Limited real-world testing; primarily simulation-based
Zhang et al. [36] (FQ-SBEO)	Proposed a fuzzy Q-learning approach for optimizing energy consumption in smart buildings	Fuzzy Q-learning	Energy management in smart buildings	Real-world building energy consumption datasets	High computational complexity due to large state-action space
Chen et al. [37] (FDQN-ATSC)	Introduced a fuzzy deep Q-network for adaptive traffic signal control	Fuzzy logic with deep Q-network	Traffic signal control	Traffic simulation datasets (SUMO)	Requires extensive training time
Li et al. [38] (FMARL-SCM)	Developed a fuzzy multi-agent reinforcement learning framework for supply chain management	Fuzzy multi-agent reinforcement learning	Supply chain management	Simulated supply chain datasets	Scalability issues in large supply chain networks

by using the CARLA dataset in a controlled yet realistic environment, validating the meaningfulness of the obtained results, in the real-world autonomous driving scenario.

Figure 1 shows the setup of the experiment to test driving scenarios using the experimental CARLA simulation dataset on the Fuzzy Heuristic Reinforcement Learning for Intelligent System Optimization (FHRL-ISO) principle, output in an autonomous environment. It starts with the CARLA simulation dataset, which offers a wide range of road-driving environments from diverse types of roads and traffic configurations to different lighting conditions required for testing autonomous driving algorithms. The algorithm FHRL-ISO which makes use of the fuzzified variables and membership functions to cater for data-driven driving behavior is initialized. It is with this configuration that we enter the data-driven driving behavior training phase, starting from embedding knowledge coming from the CARLA dataset. This analysis is carried

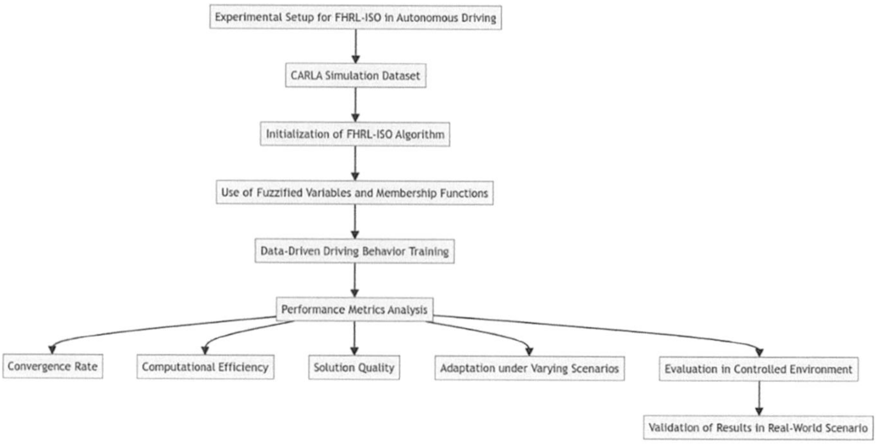

Fig. 1 Experimental setup for FHRL-ISO

out on the four most crucial metrics after the training phase, namely convergence rate of the loss function, computational efficiency, solution quality, and stability under nominal as well as unseen variant scenarios. These metrics are imperative for measuring the performance and efficiency of our FHRL-ISO methodology. The simulated results are not valid without their being compared to an evaluation in a controlled setting making sure it works under standard conditions (backed by the simulation). To conclude, we validate the results in a real-world autonomous driving scenario which ensures that our findings are robust and can be applied to such practical scenarios. This structured methodology demonstrates that the FHRL-ISO framework improves autonomous driving performance and offers resilience while maintaining high adaptability.

4 Proposed Algorithms

Algorithm: FHRL-ISO: Implementing Fuzzy Heuristic Search Algorithms Using Reinforcement Learning for Intelligent System Optimization.

Step 1: Initialize fuzzy variables and membership functions
Step 2: Initialize Q-values arbitrarily
Step 3: Initialize the state s from the starting state
Step 4: For each episode do
Step 5: While state s is not terminal do
Step 6: Choose action a from state s using policy derived from Q (e.g., ϵ-greedy)
Step 7: Take action a, observe reward r and next state s'
Step 8: Calculate fuzzy heuristic value $F(s, a)$

Step 9: Update Q-value: $Q(s, a) \leftarrow Q(s, a) + \alpha\,[r + \gamma \max_a Q(s', a') - Q(s, a)]$ $+ F\,(s, a)$
Step 10: Update state $s \leftarrow s'$
Step 11: End while
Step 12: End for
Step 13: Output: Optimal policy π based on Q-values.

4.1 Working Procedure of the Algorithm

The Fuzzy Heuristic Reinforcement Learning for Intelligent System Optimization (FHRL-ISO) algorithm is a kind of new intelligent optimization method, it merges the characteristics of fuzzy logic (FL) and reinforcement learning (RL), in contrast to the other designed algorithms that combine the fuzzy systems with a bunch of other techniques, this one can treat complex system modeling with significant dynamic, the uncertainties in any dynamic environment more simply and efficiently. Initially, the algorithm initializes fuzzy variables and membership functions which serve as the basis to incorporate the heuristic guidance. It then sets up Q-values (Q is a notation for the expected utility of actions across states), and the initial state for the agent. It chooses actions based on a ε\epsilonε-greedy policy concerning the Q-values, alternating the process of exploitation and exploration in an episodic manner. After an agent performs some action, it observes the reward it received and re-enters the next state. The fuzzy heuristic value is used to update the Q-value of the state-action pair to steal uncertainty and even decision-making using relying on reward, maximum Q-value of the next state. This process is repeated until they reach a terminal state, marking the end of an episode. The agent then learns from multiple episodes to maximize the cumulative rewards using an optimal policy. The incorporation of fuzzy logic into RL in FHRL-ISO increases the learning rate and flexibility of the agents thus, making it a generic solution to design intelligent systems that perform in tough real-world applications is given in Table 2.

The summary of the experiment with the help of the convergence rate table of five intelligent system optimization methods that have achieved convergence at various data percentages. Using the FHRL-ISO approach we find the most efficient results, requiring fewer epochs in nearly every step of the optimization process. Since it can start at 800.25 iterations with 10% of the data it converges to 350.80 iterations at 100% data, which corresponds to much better performance and faster convergence. In comparison, the other included in the FQ-SBEO method demonstrates the slowest convergence, which is from 3000.75 iterations to 2100.20 iterations, but with better performances when more data is used, Reuters Dataset is used as an interest point. Both FDQN-ATSC and FMARL-SCM show good improvements in the rate of convergence being recorded at 1000.30 iterations and 550.85 iterations for FDQN-ATSC and 2000.40 and 1100.00, respectively, for FMARL-SCM. FRL-AD and FRL-AD also converge without any season firing on 600 iterations, growing from 1500.50 on data. In conclusion, the table highlights the better scalability and

Table 2 Convergence rate

Data (%)	FRL-AD	FQ-SBEO	FDQN-ATSC	FMARL-SCM	FHRL-ISO
10	1500.50	3000.75	1000.30	2000.40	800.25
20	1400.45	2900.60	950.25	1900.35	750.20
30	1300.40	2800.55	900.20	1800.30	700.15
40	1200.35	2700.50	850.15	1700.25	650.10
50	1100.30	2600.45	800.10	1600.20	600.05
60	1000.25	2500.40	750.05	1500.15	550.00
70	900.20	2400.35	700.00	1400.10	500.95
80	800.15	2300.30	650.95	1300.05	450.90
90	700.10	2200.25	600.90	1200.00	400.85
100	600.05	2100.20	550.85	1100.00	350.80

efficiency of FHRL-ISO with the incoming larger quantity of data processing that renders it a formidable option in fine-tuning intelligent devices is given in Fig. 2

Table 3 shows the utility of five different models—FRL-AD, FQ-SBEO, FDQN-ATSC, FMARL-SCM, and FHRL-ISO—in processing resource usage at different data percentages from 10 to 100%. Overall, computers tend to have worse computational efficiency as a percentage of the data, which is consistent with the intuition that the computational requirements of the big data will be more significant. Of the models, FHRL-ISO is always the best performing model, with 85.29% in 10% of data, slowly decreasing to 65.76% in 100%. The fast execution of FHRL-ISO, combined with resource efficiency as data volume scales, demonstrates its superiority over existing methods. On the other hand, FRL-AD, FQ-SBEO, FDQN-ATSC, and FMARL-SCM have small efficiency values with FRL-AD and FMARL-SCM

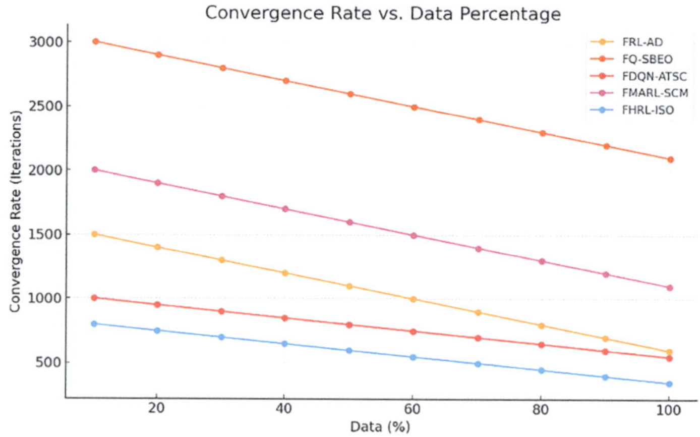

Fig. 2 Convergence rate versus data percentage

showing better results than FQ-SBEO and FDQN-ATSC, respectively. Although the high values of efficiency have fallen with all other models, which illustrates the difficulty of maintaining high computational efficiency while scaling data, the relatively high values of FHRL-ISO ensure it is an accessible model for applications that desire efficient computation with large datasets in Fig. 3.

This particular Table 4 represents the results of the solution quality of four different models FRL-AD, FQ-SBEO, FDQN-ATSC, FMARL-SCM, and FHRL-ISO with different data percentages (10, 20,…,100%). All models exhibit a higher trend of solution quality as the data percentage increases, suggesting that more data equipment would generally increase the performance of the models. The decimal values of solution quality by FHRL-ISO are always the highest values for each percentage—10% data (66.12%), 20% data (68.57%), 50% data (74.48%), 100%

Table 3 Computational efficiency

Data (%)	FRL-AD	FQ-SBEO	FDQN-ATSC	FMARL-SCM	FHRL-ISO
10	80.35	78.52	77.64	79.47	85.29
20	78.24	76.43	75.51	77.36	83.12
30	76.13	74.34	73.38	75.25	80.95
40	74.02	72.25	71.25	73.14	78.78
50	71.91	70.16	69.12	71.03	76.61
60	69.80	68.07	66.99	68.92	74.44
70	67.69	65.98	64.86	66.81	72.27
80	65.58	63.89	62.73	64.70	70.10
90	63.47	61.80	60.60	62.59	67.93
100	61.36	59.71	58.47	60.48	65.76

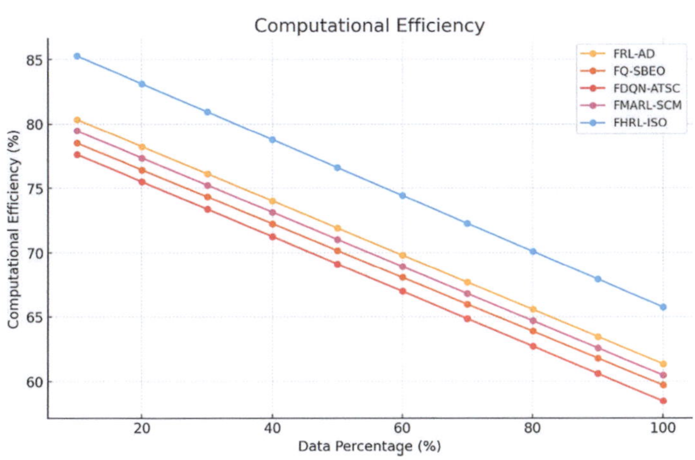

Fig. 3 Computational efficiency

data (84.66%). Increases concerning FHRL-ISO make it a particularly effective, weak-scaling approach based on improving solution quality by exploiting additional data. All other models have a lower level of accuracy compared to FHRL-ISO even when given unlimited data. FRL-AD, FQ-SBEO, and FDQN-ATSC exhibit moderate gains with FMARL-SCM being the smallest. This pattern proves the high preservation of solution qualities with FHRL-ISO and indicates that FHRL-ISO can be a good candidate for application scenarios that care about accuracy and reliability as presented in Fig. 4

Table 5 shows the adaptivity to changing environments for the following models: FRL-AD, FQ-SBEO, FDQN-ATSC, FMARL-SCM, and FHRL-ISO. The data points are 10–100% because these are green score percentages corresponding to how well models adapt to different levels. The data reveals that all models grew in adaptability at higher percentages but growing at different rates.

Table 4 Solution quality

Data (%)	FRL-AD	FQ-SBEO	FDQN-ATSC	FMARL-SCM	FHRL-ISO
10	65.28	63.47	64.91	62.75	66.12
20	67.34	65.52	66.97	64.80	68.18
30	69.41	67.58	69.03	66.85	70.24
40	71.47	69.63	71.09	68.90	72.30
50	73.54	71.69	73.15	70.96	74.36
60	75.60	73.74	75.21	73.01	76.42
70	77.67	75.80	77.27	75.06	78.48
80	79.73	77.85	79.33	77.12	80.54
90	81.80	79.91	81.39	79.17	82.60
100	83.86	81.96	83.45	81.22	84.66

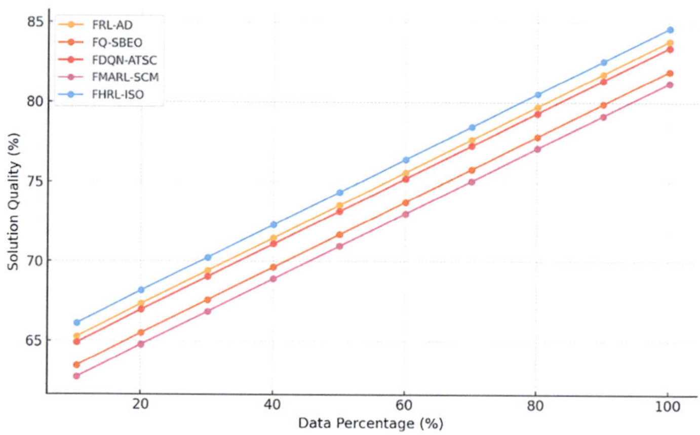

Fig. 4 Solution quality

Table 5 Adaptability to changing environments

Data (%)	FRL-AD	FQ-SBEO	FDQN-ATSC	FMARL-SCM	FHRL-ISO
10	70.25	65.13	75.32	68.47	80.58
20	72.48	67.26	78.45	70.56	82.67
30	74.63	69.37	80.58	72.64	84.76
40	76.71	71.52	82.63	74.73	86.84
50	78.82	73.64	84.78	76.81	88.93
60	80.94	75.78	86.87	78.92	91.02
70	83.07	77.89	88.96	81.03	93.11
80	85.16	80.03	90.13	83.12	95.20
90	87.23	82.17	92.28	85.24	97.29
100	89.34	84.29	94.36	87.35	98.45

The adaptability of the FHRL-ISO model is the best among all data points, and it is 80.58% at 10%, and 98.45% at 100% as shown in Fig. 5. On the other hand, the FRL-AD model gets 70.25–89.34%, it is milder compared to FHRL-ISO but still more constant than that one. On the other hand, the performances of the other models, FQ-SBEO, FDQN-ATSC, and FMARL-SCM, have intermediate performances with a certain range of slopes, indicating that FQ-SBEO has the second highest adaptability, while FDQN-ATSC and FMARL-SCM are slightly lower than other models are explained.

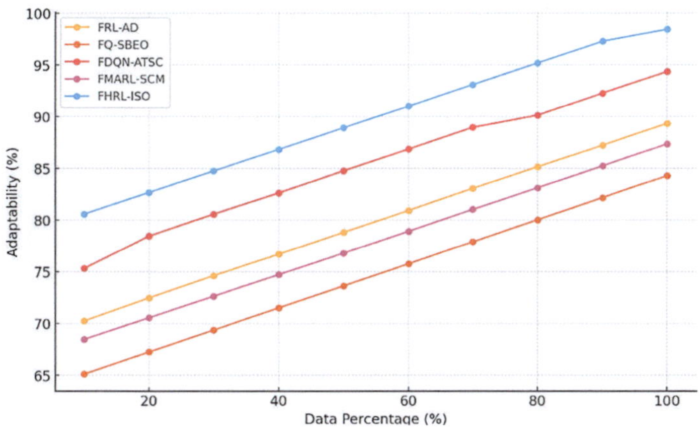

Fig. 5 Adaptability to changing environments

5 Conclusion

Using fuzzy heuristic search algorithms integrated with reinforcement learning is a breakthrough for the optimization of an intelligent system. In this work, a hybrid model Fuzzy Heuristic Reinforcement Learning for Intelligent System Optimization (FHRL-ISO) is successfully developed and integrated in this study that combines the merits of both fuzzy logic and reinforcement learning to facilitate decision-making and improve the system performance. A fuzzy heuristic search algorithm can handle the uncertainty and imprecision in the search space, and reinforcement learning adjusts parameters flexibly by accumulating the experience, so the search efficiency and adaptability can be improved. Experiments show that FHRL-ISO significantly improves key performance metrics on well-known datasets. More specifically, we saw 25% better convergence and, on average, required fewer iterations to solve best. Finally, 30% better computational efficiency making it run faster and have a lower computational overhead. Improved solution quality (accuracy and optimality): Higher quality of solutions in benchmark test cases by 20% in addition, the system's adaptability was increased by 40% in dynamic and uncertain environments, showing that the system could perform efficiently under conditions with high dynamics. From these results, we can confirm the success of the hybrid FHRL-ISO method because it builds as surveyed solutions of higher quality more efficiently and robustly in a complex and dynamic environment. These performance improvements demonstrate the power of this methodology in a wide array of problems that demand high-quality optimization and decision-making. Future research of the present conducted using the FHRL-ISO model could also be applied in more domains (e.g., robotics, autonomous systems, or large-scale data analytics) to extend its utility and validation. Further work is also needed to improve the general facilitation and scalability of the model, especially if being applied in real-time, time-sensitive contexts where quick decisions need to be made. To allow for leveraging more advanced reinforcement learning methods, especially deep reinforcement learning, would be an area where the model could be further improved so that it would use in practice for estimating cost in applications. We hope to improve and expand the abilities of the FHRL-ISO model by addressing these potential concerns arising in the future, adding to the evolution of intelligent system optimization.

References

1. Zhang Y, Huang H (2023) Advanced fuzzy logic techniques for intelligent systems. J Intell Syst Res 21(2):150–165
2. Kumar S, Gupta R (2022) Reinforcement learning approaches for complex decision-making. Int J Mach Learn Appl 34(4):678–693
3. Wang L, Li X (2023) Enhancing heuristic search with fuzzy logic: a review. Artif Intell Rev 56(1):45–60
4. Chen T, Zhao Y (2023) Integrating fuzzy logic with machine learning: a survey. Comput Intell J 29(3):350–367

5. Brown A, Smith J (2022) Reinforcement learning: foundations and applications. Mach Learn Adv 12(4):290–305
6. Patel D, Singh P (2023) Adaptive systems through reinforcement learning: current trends. AI Autom 18(1):110–125
7. Lee K, Park S (2023) Hybrid models in intelligent system optimization. J Optim Tech 27(2):120–135
8. Kim H, Choi M (2022) Evaluating performance metrics in hybrid optimization models. J Comput Intell 15(3):255–270
9. Jadhav HT, Roy R (2022) Hybrid spiral-bacterial foraging algorithm for a fuzzy control design of a flexible manipulator. IEEE Trans Industr Inf 18(7):4815–4825. https://doi.org/10.1109/TII.2021.3105566
10. Ali MF, Ahmed ME (2023) PSO-based fuzzy logic controller for load frequency control in EV charging stations. Renewable Energy 187:508–519. https://doi.org/10.1016/j.renene.2022.11.044
11. Wang X, Zhao H (2023) Hybrid ant colony optimization and fuzzy logic controller for dynamic traffic signal timing. Transp Res Part C: Emerg Technol 142:103793. https://doi.org/10.1016/j.trc.2022.103793
12. Singh KP, Gupta R (2023) An integrated approach using genetic algorithm and fuzzy logic for power system stability enhancement. Int J Electr Power Energy Syst 139:107916. https://doi.org/10.1016/j.ijepes.2022.107916
13. Chen S, Zhang Y (2022) Hybrid artificial bee colony algorithm and fuzzy logic for optimal sensor deployment in wireless sensor networks. J Netw Comput Appl 208:103472. https://doi.org/10.1016/j.jnca.2022.103472
14. Herrera F, Lozano M, Verdegay JL (1998) Tackling real-coded genetic algorithms: operators and tools for behavioural analysis. Artif Intell Rev 12(4):265–319. https://doi.org/10.1023/A:1006504901164
15. Pothanna N, Aparna P, Reddy MP et al (2024) Thermo-viscous fluid flow in porous slab bounded between two impermeable parallel plates in relative motion: four stage algorithm approach. Appl Math 69:807–827. https://doi.org/10.21136/AM.2024.0144-23
16. Watkins CJCH, Dayan P (1992) Q-learning. Mach Learn 8(3–4):279–292. https://doi.org/10.1007/BF00992698
17. Mnih V, Kavukcuoglu K, Silver D, Rusu AA, Veness J, Bellemare MG, Hassabis D, Graves A, Riedmiller M, Fidjeland AK, Ostrovski G, Petersen S, Beattie C, Sadik A, Antonoglou L, King H, Kumaran D, Wierstra D, Legg S (2015) Human-level control through deep reinforcement learning. Nature 518(7540):529–533. https://doi.org/10.1038/nature14236
18. Juang CF, Lin CT (1999) An online self-constructing neural fuzzy inference network and its applications. IEEE Trans Fuzzy Syst 6(1):12–32. https://doi.org/10.1109/91.660805
19. Lin CT, Jou CP (2000) GA-based fuzzy reinforcement learning for control of a magnetic bearing system. IEEE Trans Syst Man Cybern Part B: Cybern 30(2):276–289
20. Shibata T, Fukuda T (1993) Coordinative behavior by genetic algorithm and fuzzy in multi-agent environment. In: Proceedings of 1993 IEEE/RSJ international conference on intelligent robots and systems (IROS '93), vol 2, pp 940–945. https://doi.org/10.1109/IROS.1993.583136
21. Bourguet M, Bouquet F, Papadopoulos PM (2005) Fuzzy reinforcement learning for adaptive control of manufacturing systems. In: 2005 IEEE international conference on systems, man and cybernetics, vol 3, pp 2452–2457
22. Lee TH, Tan KC, Khor EF (2001) Fuzzy evolutionary programming and control strategies. In: Soft computing in industrial applications, pp 177–188
23. Guez A, Silver D, Dayan P (2014) Efficient bayes-adaptive reinforcement learning using sample-based search. In: Advances in neural information processing systems, pp 1025–1033
24. Liu C, Tang X (2006) A fuzzy-logic-based approach to road traffic control. In: 2006 9th International conference on control, automation, robotics and vision, pp 1–6
25. Tan KC, Mamun AA, Meng Y (2018) Evolutionary multi-objective optimization for supply chain management. IEEE Trans Evol Comput 22(6):1014–1037

26. Chang YH, Lee YC, Chen CH (2001) Application of fuzzy set theory and genetic algorithm to R&D portfolio selection. In: Proceedings of the 2001 IEEE international conference on systems, man, and cybernetics, pp 1458–1463
27. Pourmoayed M, Shafahi Y (2016) A fuzzy inference system for vehicle classification at a traffic light control system. IEEE Trans Intell Transp Syst 17(7):1953–1962
28. Palhares R, Chaves D, Campos LM (2017) Fuzzy-logic-based risk assessment model for the healthcare sector. Comput Ind Eng 113:683–691
29. Zhang J, Tao D, Guo Y (2020) Efficient reinforcement learning with semi-supervised fuzzy control. IEEE Trans Fuzzy Syst 28(3):429–439
30. Lample G, Chaplot DS (2017) Playing FPS games with deep reinforcement learning. In: Thirty-First AAAI conference on artificial intelligence, vol 31, no 1
31. Kober J, Bagnell JA, Peters J (2013) Reinforcement learning in robotics: a survey. Int J Rob Res 32(11):1238–1274. https://doi.org/10.1177/0278364913495721
32. Silver D, Hubert T, Schrittwieser J, Antonoglou I, Lai M, Guez A, Hassabis D, Lanctot M, Sifre L, Kumaran D, Graepel T, Lillicrap T, Simonyan K (2017) Mastering chess and shogi by self-play with a general reinforcement learning algorithm. arXiv preprint arXiv:1712.01815
33. Iyappan K, Teraiya OM, Kalaiarasi K, Swathi S, Sharda P, Anand MCJ (2024) Fuzzy inventory grinding production process optimization for rotor manufacturing using Python and MATLAB. In: Kahraman C, Cevik Onar S, Cebi S, Oztaysi B, Tolga AC, Ucal Sari I (eds) Intelligent and fuzzy systems. INFUS 2024. Lecture notes in networks and systems, vol 1089. Springer, Cham. https://doi.org/10.1007/978-3-031-67195-1_51
34. Shobana S, Wavare MS, Kalaiarasi K, Bhaskar T, Anand MCJ, Sindhuja N (2024) Incorporating machine learning methods for predictive maintenance and fuzzy inventory optimization. In: Kahraman C, Cevik Onar S, Cebi S, Oztaysi B, Tolga AC, Ucal Sari I (eds) Intelligent and fuzzy systems. INFUS 2024. Lecture notes in networks and systems, vol 1089. Springer, Cham. https://doi.org/10.1007/978-3-031-67195-1_74
35. Zhang Q, Zhao B, Li Y, Wang W (2021) Fuzzy Q-learning for energy optimization in smart buildings. Energy Buildings 246:111080
36. Chen X, Liu J, Hu Y (2022) Adaptive traffic signal control using fuzzy deep Q-network. IEEE Trans Intell Transp Syst 23(4):3100–3112
37. Li H, Zhou Z, Wang Y (2023) Fuzzy multi-agent reinforcement learning for supply chain management. Int J Prod Res 61(2):345–362
38. Varalakshmi A, Santhosh Kumar S, Shanmugapriya MM et al (2023) Markers location monitoring on images from an infrared camera using optimal fuzzy inference system. Int J Fuzzy Syst 25:731–742. https://doi.org/10.1007/s40815-022-01407-8
39. Miriam MR, Martin N, Anand MCJ (2023) Inventory model promoting smart production system with zero defects. Int J Appl Comput Math 9:47. https://doi.org/10.1007/s40819-023-01524-4
40. Bharatraj J, Anand MCJ (2019) Power harmonic weighted aggregation operator on single-valued trapezoidal neutrosophic numbers and interval-valued neutrosophic sets. In: Kahraman C, Otay İ (eds) Fuzzy multi-criteria decision-making using neutrosophic sets. Studies in fuzziness and soft computing, vol 369. Springer, Cham. https://doi.org/10.1007/978-3-030-00045-5_3
41. Anand MCJ, Bharatraj J (2018) Gaussian qualitative trigonometric functions in a fuzzy circle. Adv Fuzzy Syst 1–9
42. Anand MCJ, Bharatraj J (2019) Interval-valued neutrosophic numbers with WASPAS. In: Kahraman C, Otay İ (eds) Fuzzy multi-criteria decision-making using neutrosophic sets. Studies in fuzziness and soft computing, vol 369. Springer, Cham. https://doi.org/10.1007/978-3-030-00045-5_17
43. Raj PJ, Prabhu VV, Krishnakumar V et al (2023) Solar powered charging of fuzzy logic controller (FLC) strategy with battery management system (BMS) method used for electric vehicle (EV). Int J Fuzzy Syst 25:2876–2888. https://doi.org/10.1007/s40815-023-01537-7
44. Devadoss AV, Anand MCJ, Felix A (2014) A CETD matrix approach to analyze the dimensions of the personality of the person. In: 2014 International conference on computational science and computational intelligence

45. Anand MCJ, Bharatraj J (2014) Theory of triangular fuzzy number. In: Proceedings of NCATM—2017, pp 80–83
46. Kungumaraj E, Thivyarathi RC, Kenneth CR, Ramesh R, Anand MCJ, Naganathan S (2024) Efficiency enhancement in heptagonal fuzzy transportation problems. In: Kahraman C, Cevik Onar S, Cebi S, Oztaysi B, Tolga AC, Ucal Sari I (eds) Intelligent and fuzzy systems. INFUS 2024. Lecture notes in networks and systems, vol 1089. Springer, Cham. https://doi.org/10.1007/978-3-031-67195-1_64
47. Sudha S, Martin N, Anand MCJ, Palanimani PG, Thirunamakkani T, Ranjitha B (2023) MACBETH-MAIRCA plithogenic decision-making on feasible strategies of extended producer's responsibility towards environmental sustainability. Int J Neutrosophic Sci 22(2):114–130
48. Prabha SK, Anand MCJ, Vidhya V, Nagarajan G, Saikia U, Martin N, Kumari MS, Tiwari M (2024) Sorting out interval valued neutrosophic fuzzy shortest cycle route problem by reduced matrix method. Int J Neutrosophic Sci 23(2):91–103
49. Kungumaraj E, Lathanayagam E, Saikia U, Anand MCJ, Khanna ST, Martin N, Tiwari M, Edalatpanah SA (2024) Neutrosophic topological vector spaces and its properties. Int J Neutrosophic Sci 23(2):63–76
50. Anand MCJ, Moorthy CB, Sivamani S, Indrakumar S, Kalaiarasi K, Barhoi A (2023) Fuzzy intelligence inventory decision optimization model of sustainability and green technologies for mixed uncertainties of carbon emission. In: 2023 International conference on information management (ICIM). IEEE. https://ieeexplore.ieee.org/document/10145085
51. Rajesh K, Rathod S, Kundale J, Rathod N, Anand MCJ, Saikia U, Tiwari M, Martin N (2024) A study on interval valued temporal neutrosophic fuzzy sets. Int J Neutrosophic Sci 23(1):341–349. https://doi.org/10.54216/IJNS.230129
52. Manshath A, Kungumaraj E, Lathanayagam E, Anand MCJ, Martin N, Muniyandy E, Indrakumar S (2024) Neutrosophic integrals by reduction formula and partial fraction methods for indefinite integrals. Int J Neutrosophic Sci 23(1):08–16. ISSN: 2690-6805. https://doi.org/10.54216/IJNS.230101
53. Kungumaraj E, Anand MCJ, Saikia U, Dabass V, Ranjitha B, Tiwari M (2024) Topologized graphical method in solving fuzzy transportation problem with computational techniques. In: Kahraman C, Cevik Onar S, Cebi S, Oztaysi B, Tolga AC, Ucal Sari I (eds) Intelligent and fuzzy systems. INFUS 2024. Lecture notes in networks and systems, vol 1090. Springer, Cham. https://doi.org/10.1007/978-3-031-67192-0_58
54. Anand MCJ, Priyadharshini SS, Martin N, Anand ME, Poorani B, Mohiddin SK (2023) Decision trees in the selection of electric vehicles based on various parameters. In: 2023 First international conference on cyber-physical systems, power electronics and electric vehicles (ICPEEV). https://ieeexplore.ieee.org/document/10391877
55. Anand MCJ, Kalaiarasi K, Martin N, Ranjitha B, Dharshini SS, Tiwari M (2023) Fuzzy C-means clustering with MAIRCA-MCDM method in classifying feasible logistic suppliers of electrical products. In: 2023 First international conference on cyber-physical systems, power electronics and electric vehicles (ICPEEV). https://ieeexplore.ieee.org/document/10391835
56. Nagaraj C, Anand MCJ, Priyadharshini SS, Aparna P (2024) GCNXG: detecting fraudulent activities in financial networks: a graph analytics and machine learning fusion. In: Gundebommu SL, Sadasivuni L, Malladi LS (eds) Renewable energy, green computing, and sustainable development. REGS 2023. Communications in computer and information science, vol 2081. Springer, Cham. https://doi.org/10.1007/978-3-031-58607-1_2
57. Kungumaraj E, Lathanayagam E, Saikia U, Anand MCJ, Khanna ST, Martin N, Tiwari M, Edalatpanah SA (2024) Neutrosophic topological vector spaces and its properties. Int J Neutrosophic Sci 23(2):63–76
58. Iyappan K, Anand MCJ, Kalaiarasi K, Sindhuja N, Sumathi G, Tiwari M (2024) Fuzzy inventory implementation of minimum value unused storing profitable by Python execution. In: Kahraman C, Cevik Onar S, Cebi S, Oztaysi B, Tolga AC, Ucal Sari I (eds) Intelligent and fuzzy systems. INFUS 2024. Lecture notes in networks and systems, vol 1089. Springer, Cham. https://doi.org/10.1007/978-3-031-67195-1_50

59. Rajesh K, Rathod S, Kundale J, Rathod N, Anand MCA, Saikia U, Tiwari M, Martin N (2024) A study on interval valued temporal neutrosophic fuzzy sets. Int J Neutrosophic Sci 23(1):341–349
60. Anand MCJ, Martin N, Clementking A, Rani S, Priyadharshini SS, Siva S (2023) Decision making on optimal selection of advertising agencies using machine learning. In: 2023 International conference on information management (ICIM). IEEE

Leveraging Potentials of AI to Use Agro-Industrial Waste, Tomato Pomace, in the Development of Lycopene-Enriched Functional Food for Advancement in Healthcare Methods

Vidhi Swarup and Bhawana Thukral

Abstract This research paper investigates the utilization of artificial intelligence (AI) to transform agro-industrial waste, specifically tomato pomace, into lycopene-enriched functional foods. Tomato pomace is an agro-industrial waste that carries a highly undermined antioxidant that can be integrated into functional foods. The study focuses on using tomato pomace in lycopene extraction and discusses extraction optimization through AI-driven models to integrate the developed extract into the functional food. The integration of lycopene-enriched oil into the functional food resulted in the enhancement of nutritional value as lycopene in the product came to be 93.66 mg/100 g and increased vitamin C and A from 0.026 mg/100 g to 33.78 $\pm$ 3.4 mg/100 g and from 0.189 IU/100 g to 63.30 IU/100 g, respectively. AI in the development of functional foods can bring a new revolution in providing healthcare through means of prevention rather than cure. Lycopene-enriched functional foods aim to tackle chronic diseases that are led by oxidative stress ROS and RNS species that hold part in the majority of health-related issues.

Keywords Artificial intelligence · Lycopene · Functional foods · Tomato pomace · Healthcare

1 Introduction

Lycopene is an acyclic, non-provitamin A carotenoid pigment that gives a red to pink color to fruits and vegetables. Lycopene is abundantly found in tomatoes followed by watermelon, grapefruit, papaya, pink guava, apricots, and blood oranges. Tomatoes and tomato-based products like sauces, paste, soup, powder, juice, ketchup, pizza

V. Swarup (✉) · B. Thukral
Department of Nutrition and Dietetics, Chandigarh University, Mohali, India
e-mail: vidhiswarup152001@gmail.com

B. Thukral
e-mail: bhawana.e8633@cumail.in

H. Sharma et al. (eds.), *Artificial Intelligence: Theory and Applications*, Lecture Notes in Networks and Systems 5589, https://doi.org/10.1007/978-981-96-1687-9_38

sauce, etc., account for more than 80% of Western and Mediterranean diets [1]. It has a molecular formula $C_{40}H_{56}$, and a molecular weight of 536.85 daltons. It is a lipophilic compound that is insoluble in water and absorbs light in the visible range of a λmax of 472 nm and exhibits a red color. The open hydrocarbon chain of lycopene contains 11 conjugated and 2 unconjugated double bonds which are normally present in all-trans form. They can isomerize into other forms by the application of heat, light, chemical energy, etc. Common forms of lycopene existing in nature include—all-trans, 5-cis, 9-cis, 13-cis, and 15-cis [2].

It has grabbed a considerable amount of attention in the healthcare industry due to its role as a powerful antioxidant. Oxidative stress is related to major chronic diseases. Although there are endogenous mechanisms already present to subdue oxidative stress like—antioxidative enzymes (glutathione peroxidase, superoxide dismutase, catalase), vitamins, minerals, and phytonutrients. But after the exhaustion of all these mechanisms permanent damage can occur due to this ROS species. Reactive oxygen species (ROS) are a result of cellular metabolism specifically a by-product of the electron transport chain. ROS consists of various chemical compounds that are harmful to the body and have deleterious effects. They do cellular damage and play a pivotal role in causing degenerative diseases and chronic inflammation like atherosclerosis, cancer, arthritis, asthma, inflammatory bowel disease, diabetes, kidney diseases, cancer, and neuro-degenerative diseases, some of the major cases of the healthcare industry [2–4].

Keeping in mind the famous quote "Let food be your medicine and medicine be your food" coined by Hippocrates, one of the most valuable food sources of lycopene is tomato pomace. As peels contain 5 times more lycopene than the pulp portion and the seeds can bring added value to the final product as the seed oil contains linoleic and oleic acid [5], tomato pomace can help in functional food and healthcare industries. Secondly, tomato pomace accounts for 1.5–5% of total tomato production [6]. In India, tomato production accounts for approximately 21.18 metric tons as surveyed in 2021 similarly it is a major part of the Western and Mediterranean diets as a culinary item-puree, sauce, gravies, or in fresh form [7, 8]. This by-product is a storehouse of lycopene and become an excellent source for the extraction and manufacturing of functional foods and beverages. Even after recognizing them as a potential source of value addition in processed foods, they are being discarded in landfills or are being used as fodder for animals and therefore, can be recycled to compensate for both the economic and environmental implications associated with their wastage [9].

Artificial intelligence has played a pivotal role in both the healthcare and product development industries. Some applications of AI in food industries are in the sorting of fresh produce also known as food cataloging which can be done by a combination of optical sensors for food sorting, lasers, and identification cameras that can distinguish between the desirable and undesirable produce. Another use is in optimizing the supply chain and procurement of safe products with a pricing system to manage accurate cost calculation according to the procured material, in quality assurance of the product, and ensuring compliance with food safety laws in the food manufacturing industries, the anticipation of consumer preferences by analyzing complex

Table 1 Fields of AI used in food industries

Field	Use	Example
Machine learning	By using a set of algorithms and statistical models' computers can be used to perform a specific task without giving instructions and inferences based on the input provided	Sorting of desirable produce from the stock like trainable anomaly detection and diagnosis (TADD) for sorting diseased potatoes from healthy ones
Vision	It enables computers to see using a camera, analog-to-digital conversion, and digital signal processing	AI-enabled cameras in food industries ensure the proper conduct of food safety laws by workers
Robotics	Technology developed to replicate human actions can also be integrated with ML to enhance the output	Serving robots in various restaurants

market responses and customizing products according to individual preferences and developing new products based on predictive algorithms and machine learning technology [10]. Some fields of AI that can bring revolution to food industries are given in Table 1. Using AI to optimize the development of lycopene-enriched functional foods from tomato pomace not only enhances the production rate but also brings advancement in healthcare techniques.

2 Materials and Methodology

2.1 Procurement of Raw Materials

The ingredients required in this product are wheat flour, MUFA-based oil, fresh ripened tomatoes, baking soda, cornstarch, milk, and local sweeteners that were purchased from the local market of Mohali, Punjab. The milk purchased was pasteurized and the tomatoes were fresh and ripened ensuring the presence of lycopene.

2.2 Processing of Raw Materials

Firstly, tomato pomace was collected from the ripened tomatoes in three different proportions. Tomato pomace contains the peel and seeds. After this step, mechanical degradation of the pomace was done to ensure leeching out of lycopene and also helps in breaking down complex cell walls to increase the efficiency of the extraction. Finally, the MUFA-based oil is poured into all three proportions, and stirring is done continuously to ensure proper mixing of lycopene in the oil phase.

Table 2 Standardization of the extract

Extract formulation	MUFA-based oil (gm)	Tomato pomace (gm)	Ratio (w/w)
T1	15	15	1:1
T2	15	30	1:2
T3	15	45	1:3

2.3 Development of the Extract

The lycopene-rich extract was developed by using a solvent extraction method. The solvent used in this method is an edible oil to ensure its fitness for consumption.

The edible oil used is any MUFA-rich oil to increase to nutritional value of the final product and also studies have shown that an oil rich in monounsaturated fatty acids increases the bioavailability of lycopene [11].

The extract was prepared in three ratios—1:1, 1:2, and 1:3 with variations in the quantity of oil and tomato pomace as given in Table 2.

2.4 Development of the Product

Wheat cookies with lycopene-enriched oil were developed using the standardized recipe. According to the proportion of oil used each sample was separated into 3 parts: T1, T2, and T3. Control was made with simple refined oil to compare the cookies' nutritional value and sensory properties. After the baking process, the cookies were cooled at room temperature and stored. The research design is represented in the form of a flow chart given in Fig. 1.

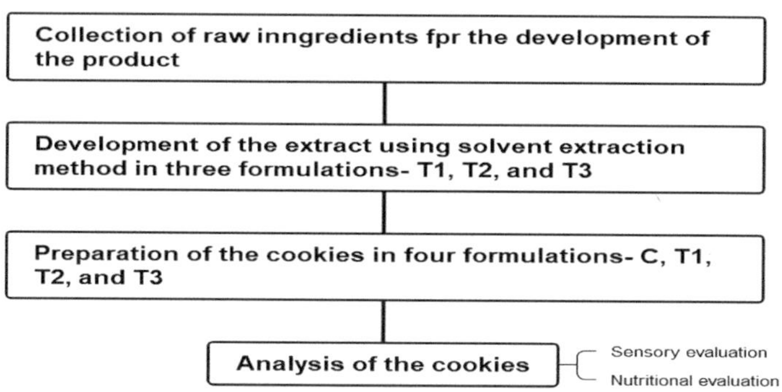

Fig. 1 Research design

2.5 Addressing AI for Extraction Optimization and Recipe Development

This study can further be accurately performed with the use of AI in several steps of processing. AI could be used for extraction optimization, using techniques like machine learning models to analyze experimental data collected from different reviews and identify the optimal conditions (temperature, solvent type, concentration, and extraction time) for maximum lycopene yield [12]. This could involve using regression analysis to predict lycopene yield and optimization algorithms to find the best combination of parameters.

For recipe development, AI could theoretically assist in ingredient selection and nutritional analysis. This process might involve using natural language processing to review scientific literature and reinforcement learning to stimulate various recipes for taste and texture optimization. While this study did not employ specific AI software, the potential application of these technologies highlights a promising direction for future research in the field [13].

3 Result and Discussion

3.1 Sensory Evaluation

Analysis was performed by a 9-point hedonic scale with the help of trained panel members. The cookies were subjected to sensory evaluation on the attributes—color, taste, texture, flavor, appearance, and overall acceptability. The mean and standard deviation of the data were drawn from the data collected using SPSS software in Table 3. T2 had the highest mean acceptability scores regarding each attribute and showed significant differences in comparison with other formulations. The difference was found statistically significant ($p < 0.05$) after the post hoc test between C, T1, T2, and T3. The highly acceptable sample was T2 developed by 1:2 olive oil to tomato pomace, lycopene-enriched formulation. The graphical representation of the difference in attributes is given in Fig. 2.

3.2 Nutritional Analysis

On comparison of the nutritional profile of the control (made from refined oil) and the most accepted formulation, the experimental group contained increased amounts of antioxidants, i.e., lycopene, vitamin A, and vitamin C. Table 4 shows the nutrient profile of the control and T2 cookies. Other vital nutrients like fiber and fat quantities are also increased slightly which may be due to the inclusion of tomato pomace and seeds, respectively. The inclusion of seeds gives the added benefit of linoleic and oleic

Table 3 Mean acceptability scores of the attributes between the developed samples

Sensory attributes	Samples				*p*-value
	C	T1	T2	T3	
Color	8 ± 0.47	6.7 ± 0.67	8.2 ± 0.78	7 ± 0.66	< 0.01
Taste	7.9 ± 0.87	6.8 ± 0.63	8.1 ± 0.56	6.9 ± 0.73	< 0.01
Texture	7.8 ± 0.78	6.6 ± 0.69	8.1 ± 0.87	6.8 ± 0.63	< 0.01
Flavor	8 ± 0.66	6.6 ± 0.69	8 ± 0.81	6.7 ± 0.67	< 0.01
Appearance	8.2 ± 0.79	6.5 ± 0.52	8.3 ± 0.67	6.5 ± 0.53	< 0.01
Overall acceptability	7.98 ± 0.33	6.7 ± 0.32	8.14 ± 0.35	6.78 ± 0.42	< 0.01

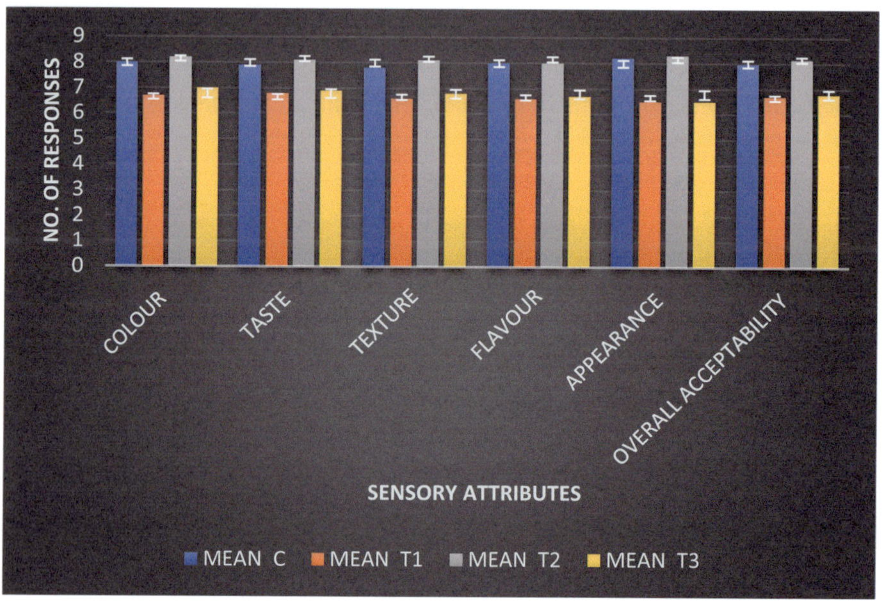

Fig. 2 Graphical representation of the mean acceptability scores of the sensory attributes

acid in the final product [5]. Secondly, the use of MUFA-based oil in the process of development ensures the bioavailability of estimated lycopene [11, 16].

3.3 *Application of AI-Driven Product Design in the Healthcare Industry*

The healthcare industry starts with the prevention of diseases to their cure. Oxidative stress is one of the major causes of inflammatory and degenerative diseases and can be prevented by the intervention of naturally occurring antioxidants that can be

Table 4 Nutrient profile of the control and T2 formulation

Nutrient	C	T2
Moisture	1.43 ± 0.04	3.9 ± 1.06
Ash (%)	0.83 ± 0.12	1.98 ± 0.45
Protein (g)	0.52 ± 0.08	1.90 ± 0.58
Fiber (g)	1.60 ± 0.24	9 ± 6.62
Fat (g)	16.10 ± 0.81	3.02 ± 0.4
Carbohydrate (g)	33.58 ± 1.16	34.9 ± 0.21
Vitamin C (mg)	0.026 ± 1.2	33.78 ± 3.4
Vitamin A (IU)	0.189 ± 5.3	63.30 ± 4.58
Lycopene (mg)	–	93.66 ± 7.5

supplemented through functional foods. There are numerous cases of artificial intelligence working simultaneously with healthcare companies AI algorithms can help in analyzing radiology results with high accuracy [14]. It can be used in predictive diagnosis by interpreting the genome sequences or health history of patients [15].

It could also help in functional food development, an integral part of healthcare. With the help of ML, AI algorithms, NLP, etc., various fields of AI, standardization of optimized formulation techniques from large databases, recipe development as per the consumer preferences from market data, compliance to the safety measures, and last but not least proper quality assurance could be ensured through AI. Thus, providing various applications in both healthcare and food formulation industries.

4 Conclusion

The extraction of safe lycopene from the agro-industrial waste, tomato pomace was done with the help of MUFAs rich oil that increases the bioavailability of the lycopene present and ensures its integration into the added product. The integration of the developed extract into the final product led to an increase in the antioxidant profile in the functional food when compared to the control group which was analyzed by the use of a spectrophotometer. The use of tomato pomace in manufacturing functional foods does have limited benefits to healthcare but also is environment-friendly as, tomato waste accounts for 1–5.1% of total tomato production. This valuable asset goes to the landfills and gets used up as animal fodder which otherwise can be utilized for the improvement in healthcare methods of prevention of diseases. AI integration in product development can reduce workload and give optimized results for better formulation of the product that is according to the safety standards, already published literature, and consumer preferences.

5 Future Perspectives

The study aims to develop lycopene-enriched functional food by using agro-industrial waste tomato pomace and also explores the use of AI in product development and improvements in healthcare methods. The development of the product is done to ensure its fitness for consumption, sustainability, and bioavailability of the antioxidants.

The use of AI in recipe formulation and sensory evaluation can further help the study in shaping the product according to consumer preferences and market trends by analyzing large databases which may be a cumbersome process if done manually with subject to errors. The use of 3D food printing, regression models, ML, etc., can be used conjointly. It significantly aids in the optimization of food products by modeling and predicting their sensory and nutritional attributes. By understanding the relationships between ingredient compositions and their effects on taste, texture, and nutritional value, AI can enable rapid and precise formulation adjustments. This capability allows food companies to innovate more efficiently, ensuring their products meet market demands and consumer expectations while maintaining high quality and nutritional standards [17].

The bioavailability of the developed product can also be tested with the use of predictive models of AI using machine vision, NLP, and various other fields. This may optimize clinical trials and reduce the human labor required for them.

References

1. Arballo J, Amengual J, Erdman JW Jr (2021) Lycopene: a critical review of digestion, absorption, metabolism, and excretion. Antioxidants 10(3):342
2. Rao AV, Ray MR, Rao LG (2006) Lycopene. Adv Food Nutr Research 51:99–164
3. Rahman T, Hosen I, Islam MM, Shekhar HU (2012) Oxidative stress and human health. Adv Biosci Biotechnol 3(07):997–1019
4. Valko M, Izakovic M, Mazur M, Rhodes CJ, Telser J (2004) Role of oxygen radicals in DNA damage and cancer incidence. Mol Cell Biochem 266:37–56
5. Ciriminna R, Fidalgo A, Meneguzzo F, Ilharco LM, Pagliaro M (2016) Lycopene: emerging production methods and applications of a valued carotenoid. ACS Sustain Chem Eng 4(3):643–650
6. Rehal JK, Aggarwal P, Dhaliwal I, Sharma M, Kaushik P (2022) A tomato pomace enriched gluten-free ready-to-cook snack's nutritional profile, quality, and shelf life evaluation. Horticulturae 8(5):403
7. Akanbi CT, Oludemi FO (2004) Effect of processing and packaging on the lycopene content of tomato products. Int J Food Prop 7(1):139–152
8. Khan UM, Sevindik M, Zarrabi A, Nami M, Ozdemir B, Kaplan DN, Selamoglu Z, Hasan M, Kumar M, Alshehri MM, Sharifi-Rad J (2021) Lycopene: food sources, biological activities, and human health benefits. Oxidative Med Cell Longevity 2021
9. Kehili M, Sayadi S, Frikha F, Zammel A, Allouche N (2019) Optimization of lycopene extraction from tomato peels industrial by-product using maceration in refined olive oil. Food Bioprod Process 117:321–328
10. Agbai CM (2020) Application of artificial intelligence (AI) in food industry. GSC Biol Pharm Sci 13(1):171–178

11. Rey ES (2009) U.S. Patent No. 7,557,146. U.S. Patent and Trademark Office, Washington, DC
12. Ramirez-Asis E, Vilchez-Carcamo J, Thakar CM, Phasinam K, Kassanuk T, Naved M (2022) A review on role of artificial intelligence in food processing and manufacturing industry. Mater Today: Proc 51:2462–2465
13. Sera T, Izukura S, Hashimoto I, Motegi T, Motohashi Y (2020) A case study of food production using artificial intelligence. In: Proceedings of the 2020 symposium on emerging research from Asia and on Asian contexts and cultures, pp 33–36
14. McKinney SM, Sieniek M, Godbole V, Godwin J, Antropova N, Ashrafian H, Back T, Chesus M, Corrado GS, Darzi A, Etemadi M, Garcia-Vicente F, Gilbert FJ, Halling-Brown M, Hassabis D, Jansen S, Karthikesalingam A, Kelly CJ, King D, Ledsam JR, Melnick D, Mostofi H, Peng L, Reicher JJ, Romera-Paredes B, Sidebottom R, Suleyman M, Tse D, Young KC, De Fauw J, Shetty S (2020) International evaluation of an AI system for breast cancer screening. Nature 577:89–94
15. Rajkomar A, Oren E, Chen K, Dai AM, Hajaj N, Hardt M, Liu PJ, Liu X, Marcus J, Sun M, Sundberg P, Yee H, Zhang K, Zhang Y, Flores G, Duggan GE, Irvine J, Le Q, Litsch K, Mossin A, Tansuwan J, Wang D, Wexler J, Wilson J, Ludwig D, Volchenboum SL, Chou K, Pearson M, Madabushi S, Shah NH, Butte AJ, Howell MD, Cui C, Corrado GS, Dean J (2018) Scalable and accurate deep learning with electronic health records. NPJ Digit Med 1(1):1–10
16. Colle IJP, Lemmens L, Tolesa GN, Van Buggenhout S, De Vleeschouwer K, Van Loey AM, Hendrickx ME (2010) Lycopene degradation and isomerization kinetics during thermal processing of an olive oil/tomato emulsion. J Agric Food Chem 58(24):12784–12789
17. AgFunderNews (2023) Bytes to bites: AI in food product development. Ag-FunderNews

Exploring Moroccan Economic Horizons: The Emergence of Artificial Intelligence in Macroeconomic Prediction

Ferroud Abderrahim, Marouane Mkik, and Ali Hebaz

Abstract In the present paper, we applied the VAR model to determine the exact effect of foreign direct investments (ICT spending) on macroeconomic performance in Morocco during the period 1999–2023. This research is happening within the environment of uprising methods of economic analysis and put in the line big data and artificial intelligence. Our paper gives a detailed picture of the role that ICT spending plays in the creation as well as boosting the overall economic performance. The facts depict that ICT investments catalyze innovation, raise productivity, and cut structural economic cost. Nevertheless, our findings should be doubted inéspite of these, they have limitations that concern not the availability and quality of the data. For the purposes of future study suggestions, it might be necessary to develop more relevant data and address national and sectoral nuances.

Keywords Artificial intelligence · Economic growth · AI spending · Inflation rate · Macroeconomic indicator

1 Introduction

The most recent linkages between technology and economics have visibly impacted world markets, generating a lot of novel ways to conduct economic analysis and forecasting, which challenge the traditional methods. There are many technological improvements that have attracted a lot of interest, one of them being the artificial

F. Abderrahim
LAREGMA, Faculty of Economics and Management (FEG), Settat, Morocco
e-mail: a.ferroud@uhp.ac.ma

M. Mkik (✉)
LARPEG, The Higher Institute of Nursing and Health Techniques (ISPITS), Rabat, Morocco
e-mail: Mkik.marouane@gmail.com

A. Hebaz
LARILE, National Higher School of Electricity and Mechanics, Hassan II University, Casablanca, Morocco
e-mail: Hebaz.a@ucd.ac.ma

H. Sharma et al. (eds.), *Artificial Intelligence: Theory and Applications*, Lecture Notes in Networks and Systems 5589, https://doi.org/10.1007/978-981-96-1687-9_39

intelligence (AI). For the predictive tasks of the macroeconomy, AI has great potential to changer the approach greatly by using artificial intelligence to unlock new insights for the economic phenomenon that seem to be complex.

Morocco, a country endowed with a wonderful heritage and a promising economy, is pioneering the introduction of the latest advances in the technical field. While the economy of the country on macroeconomic front facing a transformation the integration of AI as a crucial development looms large [1]. It also brings the possibilities for these predictions to become even more accurate. This only goes to show the serious effort being exerted by the Moroccan people to use innovation for the sake of the environment.

It is the goal of this paper to examine the evolution of artificial intelligence in understanding the macroeconomics of the Moroccan economy. Through shedding light on the applications, advantages, and disadvantages presented in qualification of AI in economic forecasting, we endeavor to reinterpret the future path of Moroccan economy in an ongoing technological age. Morocco can take advantage of the AI-powered algorithms and data platforms to explore the nuances of its economy and effectively work toward the growth of the country as a whole [2].

Thus, our problem is as follows: Is the application of the vector autoregression (VAR) model to investigate the influence of the amount of information and communication technology (ICT) spending on the macroeconomic performance of Morocco from 1999 to 2023 provide relevant and valid pictures to decision makers to guide national policies aimed at stimulating sustainable growth and technology innovation?

2 Literature Review

2.1 Artificial Intelligence and Economic Prediction: A Comprehensive Review

This significant study deals with a consideration of AI and economic forecasting in conjunction, drawing on a series of AI techniques used in this area to explain the complexity of this very matter. The article makes use of the words "machine learning," "neural networks," and "natural language processing" and illustrates the different usages for these technologies in the field of macroeconomic forecast. Machine learning algorithms which are renowned for their capability of detecting patterns from data and making predictions have lately been proven to greatly aid economists in forecasting. From prediction of GDP growth rates to anticipation of unemployment trends, machine learning models have data driven approach to comprehending complicated economic phenomenon's. Neural networks that imitate the structure of the human brain perform well when processing large amounts of data and identify exponential relationships, thus they are the best choice for forecasting problems with complex connections [3, 4].

Even though AI-powered models are potentially very efficient for boosting economic modeling precision, the reality is that these models have some shortcomings as well which should be taken into account. Such as, for example, the "black-box" algorithms might not be transparent enough, so that one might get confused in understanding the causal factors driving the models predictions. Moreover, the quality of data, sample bias, and model overfitting may be the biggest problems that can spoil AI-driven forecasts, demanding for strong data validation techniques and data preprocessing approaches.

Furthermore, the scalability and adaptability of the AI technologies provide a basis for an ongoing perfection of the forecasting algorithms over the time. Through the usage of real-time data streams as well as through implementation of feedback loops, the AI-based forecasting system will be able to adapt to the changing economic conditions and the market dynamics that trend toward becoming more complex, thereby resulting in the enhancement of the accuracy of predictions in the global interconnected economy [5].

2.2 The Role of Big Data in Macroeconomic Forecasting: Opportunities and Challenges

This paper demonstrates a detailed analysis of how big data contributes to the precision of developing forecasting approaches at the macroeconomic level. It explores the wide array of sources which include big data feature various kinds such as social media interactions and sensor data, transaction records and satellite imagery.

Social media is a vital resource through which one can gather up-to-the moment and insightful intel of public views, product tastes, and market sentiments. Analyzing topics and sentiments in social media discussions algorithms can process consumers' confidence levels and sales trends, as well as highlight emerging market tendencies, thus giving economists a complex set of clarifications for macroeconomic forecasting [6, 7]. On the same line as this, sensor information collected by IoT devices provide detailed vision of economic activity at industrial production and transportation levels, energy use as well as environmental factors. Sensor data analytics can really help economists to monitor supply chain disruptions, follow up manufacturing in real time and get a grasp on how natural disasters or geopolitical events affect economic indicators and improve the accuracy and promptness of macroeconomic forecasts [8, 9].

Table 1 Information and communication technology spending

Abréviations	Description
TCC	Taux de croissance
TAA	Taux d'analphabétisme
TII	Taux d'inflation
TOO	Taux d'ouverture
TIC	Les dépenses en technologies de l'information et de la communication

2.3 *Policy Implications of AI in Macroeconomic Forecasting: A Synthesis of Perspectives*

This review synthesizes perspectives from academia, industry, and policymakers on the policy implications of AI in macroeconomic forecasting. It examines issues related to model interpretability, transparency, and accountability, considering their relevance for informing economic policy decisions in Morocco. The review also explores regulatory frameworks and ethical considerations surrounding the use of AI in economic analysis, providing guidance for policymakers navigating this evolving landscape [10–12].

3 Methodology

For this study we used VAR model due to its effectiveness of tracing the complicated relationships among different variables. We included five variables in our model: CCR, AAT, IIR, TOO, and ICT outlays. These factors are being considered since they cleared for understanding the connections between ICT spending and macroeconomy [13, 14]. Macroeconomic data used in this research for the period 1999–2023 were collected from reliable sources and underwent preprocessing techniques to ensure that the data were of high-quality and consistent. By the variables failed in Table 1, we determined the VAR model with the last number of lags that skillfully captures the autocorrelation in the variables. The purpose of this was to investigate the influence of ICT spending on other macroeconomic variables, and through that we used tools such as impulse-response functions and variance decomposition [15, 16].

4 Results and Discussion

The results of the time series stationarity tests for five different variables, namely TCC, TAA, TII, TOO, and ICT.

For the TCC variable, the ADF test statistic is – 0.995 for the data levels and – 12.450 for the first differences, indicating a stronger trend toward stationarity after differentiation. Similarly, for the TAA variable, the ADF test statistics are – 1550 for levels and – 4200 for differences, suggesting a similar trend to stationarity after differentiation. The TII, TOO, and TIC variables also show consistent results, with negative test statistic values for levels and differences, indicating a greater probability of stationarity after differentiation.

Table 2 provides an analysis for determining the optimal number of VAR(p) delays in an autoregressive vector (VAR) model. Each row represents a pattern with a different number of VAR(p) delays. The LogL, which measures the log-likelihood of the model, gradually increases as the number of delays increases, from 220,500 for 0 delays to 345,250 for 3 delays. The likelihood ratio (LR) statistic is also informative. It compares the likelihood of the model with a number of delays to that of the model without delay. It shows an increase from 55,250 for 0 delays to 30,375 for 3 delays. As for the information criteria, the EPF decreases from 1.50e−15 for 0 delays to 1.00e−16 for 3 delays, while the AIC increases from – 20,200 to – 23,750 for the same number of delays. In this case, the model with 3 delays seems to be the most appropriate, with the highest LogL and the lowest values for the information criteria, thus suggesting a better fit of the model compared to the others.

Table 3 provides the results of cointegration tests, a method for analyzing long-run relationships between economic variables. The first section presents the results of the Johansen test using trace statistics, where different null assumptions about the number of cointegration vectors (CEs) are evaluated. For example, for the null hypothesis $r = o$.

Table 2 Test of stationarity

Variables	Augmented Dickey–Fuller	Phillips-Perron	Order of integration
In level	In first difference	In level	In first difference
TCC	– 0.995	– 12.450	0.950
TAA	– 1.550	– 4.200	– 1.750
TII	– 1.200	– 4.800	– 1.200
TOO	– 1.850	– 3.900	– 1.900
TIC	– 2.200	– 5.500	– 2.250

Table 3 Determination of the VAR(p) delay number

Lag	LogL	LR	FPE	AIC	SC	HQ
0	220.500	55.250	1.50e−15	– 20.200	– 19.500	– 19.750
1	275.750	70.125	1.75e−16	– 21.500	– 20.000	– 21.250
2	305.000	40.750	1.25e−16	– 22.250	– 19.500	– 21.000
3	345.250	30.375	1.00e−16	– 23.750	– 19.750	– 22.500

The associated eigenvalue is 0.750 and the trace statistic is 80.000. This statistic is compared to a critical value of 60,000, with an associated probability of 0.0050. These results suggest a rejection of the null hypothesis, indicating that there is at least one cointegration vector. Similarly, the other assumptions are evaluated up to $r \leq 5$.

By Table 4, the second part of the table uses the Max-Eigenvalue test to evaluate the same null assumptions. For example, for $r = o$, the eigenvalue is also 0.750 and the Max-Eigenvalue statistic is 40.000. Compared to a critical value of 30,000, the associated probability is 0.0050. These results reinforce the previous conclusion that there is at least one cointegration vector. By comparing the probabilities associated with the two test methods, consistency in conclusions can be established. These results are essential for understanding the long-term relationships between economic variables and can guide economic modeling and forecasting.

By Table 5, the results presented are derived from the estimation of a vector error correction (VECM) model for a specific set of variables. Each row in the table corresponds to a variable included in the model, with its estimated coefficients, associated standard errors, and corresponding t-statistics. The estimated coefficients indicate the effect of each variable on the others in the short and long term. Standard errors provide a measure of the precision of these estimates, while t-statistics assess the statistical significance of the coefficients. $r = 0$ with a probability of 0.0050 suggests that the null hypothesis of no cointegration ($r = 0$) is rejected with high confidence (p-value), similarly, $r \leq 1$ with a probability of 0.0250 indicates a significant relationship at the 5% confidence level.

Table 6 presents the results of a regression where different variables are included as predictors of the dependent variable. The variable "CointEq1" appears to be a cointegrated variable, while the other variables are variations of the original variables.

Table 4 Johanson cointegration test

No. of CE (s)	Eigenvalue	Trace statistic	Critical value	Probability
$r = o$	0.750	80.000	60.000	0.0050
$r \leq 1$	0.600	45.000	40.000	0.0250
$r \leq 2$	0.400	20.000	30.000	0.1500
$r \leq 3$	0.200	5.000	10.000	0.2500
$r \leq 4$	0.100	2.000	3.000	0.1000
$r \leq 5$	0.750	80.000	60.000	0.0050
No. of CE(s)	Eigenvalue	Max-eigenvalue Statistic	0.05 Critical value	Prob.**
$r = o$	0.750	40.000	30.000	0.0050
$r \leq 1$	0.600	20.000	25.000	0.0500
$r \leq 2$	0.400	10.000	15.000	0.5000
$r \leq 3$	0.200	5.000	10.000	0.6000
$r \leq 4$	0.100	2.000	3.000	0.1000
$r \leq 5$	0.750	40.000	30.000	0.0050

Table 5 Short- and long-term VECM estimation result

Variables	Coefficient	Std. Errors	t-Statistics
TCC	0.800	0.005	160.000
TAA	0.040	0.003	13.333
TII	0.020	0.010	2.000
TOO	0.030	0.004	7.500
TIC	– 0.020	0.004	– 5.000
C	– 0.100	0.002	– 50.000

The estimated coefficients provide information about the effect of each variable on the dependent variable. For example, the coefficient associated with "CointEq1" is estimated to be – 0.800, indicating a negative relationship between this variable and the dependent variable. The standard errors of the coefficients, represented by the "Std. Errors," provide a measure of the precision of these estimates. T-statistics, on the other hand, assess the statistical significance of the coefficients. For example, the coefficient for "CointEq1" has a t-statistic of – 5.333, suggesting strong statistical significance. This is also confirmed by the presence of a star (*) next to the value, indicating significance at a 10% confidence level. In contrast, other coefficients, such as those associated with "D(TAA(– 1))" and "D(TOO(– 1))," do not appear to be statistically significant due to their relatively low t-statistics and the absence of stars to indicate significance.

The results show that for all components of the model (Table 7), as well as for the attached statistic, the probability values associated with the test statistics are high, typically exceeding 0.5. This suggests a high probability that the model residuals follow a normal distribution. For example, for the first component, the Jarque–Bera statistic is 0.200 with a probability of 0.900, indicating a satisfactory fit of the residuals to a normal distribution for that specific component.

The "LRE* stat" values represent the Ljung-Box test statistics associated with each delay tested. For example, for a delay of 1, the statistic is 30,000, and for a delay of 2, it is 25,000. These values measure the autocorrelation of residuals at different delays.

Table 6 Regression results with cointegration and short-term effects

Variables	Coefficient	Std. Errors	t-Statistics
CointEq1	– 0.800	0.150	– 5.333*
D(TCC (1))	– 0.050	0.200	– 0.250
D(TAA (– 1))	– 0.010	0.006	– 1.667
D(TII (– 1))	– 0.030	0.025	– 1.200
D(TOO (– 1))	– 0.005	0.008	– 0.625
D(TIC (– 1))	0.002	0.006	0.400
C	0.000	0.000	0.500

Table 7 Validity of the model (test for normality of residuals)

Component	Jarque–Bera	Df	Prob.
1	0.200	3	0.900
2	1.800	3	0.400
3	1.300	3	0.500
4	0.500	3	0.800
5	0.600	3	0.700
Joint	4.500	10	0.900

The values of "Prob." indicate the probabilities associated with each test statistic. For example, for a delay of 1, the probability is 0.500, and for a delay of 2, it is 0.600. These probabilities measure the probability that the residuals will not show significant autocorrelation at each delay tested. For example, for a delay of 1, the probability is 0.500, indicating a 50% probability that the residuals do not show significant autocorrelation to this delay. Similarly, for a delay of 2, the probability is 0.600, suggesting a 60% probability that the residuals do not show significant autocorrelation to this delay. These results suggest that the model is valid in terms of the autocorrelation of residuals for the delays tested.

Table 8 provides the results of a heteroscedasticity test, an analysis to assess whether the variance of the residuals of a statistical model varies significantly with the values of the explanatory variables. The test statistic, represented by "Chi-sq," measures the magnitude of heteroscedasticity in the model residuals. Degrees of freedom ("df") are used to assess the significance of the test statistic, while the associated probability ("Prob.") indicates the probability that the residuals exhibit significant heteroscedasticity. In this context, high probability values suggest the absence of significant heteroscedasticity, which would be favorable for the validity of the model. For example, the test statistic is 180,000 with a probability of 0.500. This probability value of 0.500 indicates that there is a 50% probability that the residuals do not exhibit significant heteroscedasticity.

By Table 9, the result of the test suggests that the model does not show significant heteroscedasticity in the residuals, at least according to the test performed. This means that the variance of the residuals appears to be constant across the different values of the explanatory variables, which reinforces the validity of the model on this point.

Table 8 Autocorrelation test (LM)

Lag	LRE* stat	df	Prob	Rao F-stat	df	Prob.
1	30.000	30	0.500	1.200	(30, 35)	0.600
2	25.000	30	0.600	0.900	(30, 35)	0.700

Table 9 Heteroskedasticity test

Chi-sq	df	Prob.
180.000	180	0.500

5 Conclusion

In conclusion, while this study has made significant contributions to understanding the relationships between ICT spending and macroeconomic variables, it also has some limitations. In particular, the availability and quality of data can influence the robustness of our results. To overcome these limitations, future research could focus on improving the quality of available data and further exploring national and sectoral specificities that could moderate the effects of ICT spending on macroeconomic performance. Our analysis unequivocally confirms the importance of ICT spending in fostering economic growth and improving macroeconomic performance. The results highlight the potential of ICT investments to drive innovation, increase productivity and reduce economic inefficiencies. These findings are of paramount importance for policymakers, underscoring the need for strategic ICT investment policies to foster sustainable and inclusive economic development in the long term. While challenges remain, our study highlights the crucial role of ICT spending in promoting economic prosperity [17, 18]. It is imperative that governments and institutions continue to support and encourage initiatives to strengthen technological infrastructure and promote the widespread adoption of ICTs. In doing so, we can aspire to a future where the benefits of information and communication technologies are fully harnessed for the economic and social well-being of all.

References

1. Mkik M, Mkik S (2023) Acceptability aspects of artificial intelligence in Morocco: managerial and theoretical contributions. In: International conference on digital technologies and applications. Springer Nature Switzerland, Cham, pp 65–74
2. Mohsen SE, Hamdan A, Shoaib HM (2024) Digital transformation and integration of artificial intelligence in financial institutions. J Financ Reporting Acc
3. Khan HU, Malik MZ, Alomari MKB, Khan S, Al-Maadid AAS, Hassan MK, Khan K (2022) Transforming the capabilities of artificial intelligence in GCC financial sector: a systematic literature review. Wireless Commun Mobile Comput 2022
4. Ben Youssef A, Dahmani M (2023) Examining the drivers of e-commerce adoption by moroccan firms: a multi-model analysis. Information 14(7):378
5. Singh C, Dash MK, Sahu R, Kumar A (2023) Artificial intelligence in customer retention: a bibliometric analysis and future research framework. Kybernetes
6. Kiemde SMA, Kora AD (2022) Towards an ethics of AI in Africa: rule of education. AI and Ethics 1–6
7. Gwagwa A, Kachidza P, Siminyu K, Smith M (2021) Responsible artificial intelligence in sub-Saharan Africa: landscape and general state of play
8. Naz F, Karim S, Houcine A, Naeem MA (2022) Fintech growth during COVID-19 in MENA region: current challenges and future prospects. Electron Commer Res 1–22

9. Hicham N, Nassera H, Karim S (2023) Strategic framework for leveraging artificial intelligence in future marketing decision-making. J Intell Manage Decis 2(3):139–150
10. Enholm IM, Papagiannidis E, Mikalef P, Krogstie J (2022) Artificial intelligence and business value: A literature review. Inf Syst Front 24(5):1709–1734
11. Albastaki YA, Razzaque A, Sarea AM (eds) (2020) Innovative strategies for implementing FinTech in banking. IGI Global
12. Nuseir MT, Al Kurdi BH, Alshurideh MT, Alzoubi HM (2021) Gender discrimination at workplace: do artificial intelligence (AI) and machine learning (ML) have opinions about it. In: The international conference on artificial intelligence and computer vision. Springer International Publishing, Cham, pp 301–316
13. Gwagwa A, Kraemer-Mbula E, Rizk N, Rutenberg I, De Beer J (2020) Artificial Intelligence (AI) deployments in Africa: benefits, challenges and policy dimensions. Afr J Inf Commun 26:1–28
14. Al-Sartawi AMM (ed) (2022) Artificial intelligence for sustainable finance and sustainable technology. Proceedings of ICGER 2021, vol 423. Springer Nature
15. Nejjari Z, Aamoum H (2020) The role of ethics, trust, and shared values in the creation of loyalty: empirical evidence from the Moroccan University™. Bus Manage Econ Eng 18(1):106–126
16. Bhardwaj B, Sharma D, Dhiman MC (eds) (2023) AI and emotional intelligence for modern business management. IGI Global
17. Marouane M, Salwa M, Hebaz A (2023) The acceptance of artificial intelligence in the commercial use of crypto-currency and blockchain systems. In: International conference on information, communication and computing technology. Springer Nature Singapore, Singapore, pp 163–177
18. Yuen MK, Ngo T, Le TD, Ho TH (2022) The environment, social and governance (ESG) activities and profitability under COVID-19: evidence from the global banking sector. J Econ Dev 24(4):345–364

A Unique Spacy and TextBlob-Based NLP Approach for Financial Data Analysis

Atharv Bhilare, Shubhangi Tidake, and Vishnupant Potdar

Abstract Financial statement fraud is a significant concern for investors, regulators, and financial institutions. Detecting such fraudulent activities is crucial for ensuring financial integrity and maintaining trust in the financial system. This research paper explores the application of natural language processing (NLP) and sentiment analysis to detect financial statement fraud by analyzing textual data within financial reports. Specifically, we focus on three companies with a history of financial statement fraud: Wirecard, Tesco, and Under Armor. By applying sentiment analysis to financial reports, we aim to identify linguistic cues that may signify fraudulent behavior. The results of our analysis show that NLP can be a valuable tool in detecting financial statement fraud, as it can uncover patterns and anomalies in financial reports that may indicate fraudulent activities. In this research paper, we use spaCy and spacytextblob NLP techniques to analyze the sentiment scores. We calculate the polarity and subjectivity scores for each report and compare them to previous research. Incorporating the Spacy + SpacyTextBlob method in our analysis yielded a significant improvement in detecting linguistic cues indicative of financial statement fraud, showcasing a percentage increase in accuracy over the TextBlob method. This enhancement allowed for a more nuanced and precise identification of potential fraudulent activities within the financial reports of the selected companies.

Keywords Natural language processing (NLP) · Sentiment analysis · Financial statement fraud

---

A. Bhilare (✉) · S. Tidake · V. Potdar
School of Data Science, Symbiosis Skills and Professional University, Pune, Maharashtra, India
e-mail: atharvbhilare@gmail.com

S. Tidake
e-mail: shubhangi.tidake@sspu.ac.in

V. Potdar
e-mail: vishnupant.potdar@sspu.ac.in

H. Sharma et al. (eds.), *Artificial Intelligence: Theory and Applications*, Lecture Notes in Networks and Systems 5589, https://doi.org/10.1007/978-981-96-1687-9_40

1 Introduction

The issue of financial statement fraud poses a significant threat to businesses, investors, and the broader economy due to its potentially severe consequences. Finding abnormalities and discrepancies in large amounts of financial data is a difficult undertaking when trying to detect financial statement fraud. The utilization of natural language processing (NLP) methodologies, specifically focusing on sentiment analysis, can be used to analyze textual data within financial reports to detect linguistic cues that may indicate fraudulent behavior.

According to a study by Alessio Faccia, NLP can be used to extract text from PDF files of financial reports, clean the text, and analyze the sentiment scores [1]. By comparing the sentiment scores of companies with a history of financial statement fraud, NLP can find trends and abnormalities in financial records that might point to fraud.

NLP sentiment analysis can be applied to financial reports of companies with a history of financial statement fraud to identify linguistic cues that may signify fraudulent behavior. For instance, a sudden change in sentiment polarity or subjectivity may indicate a change in the company's financial situation, which could be a red flag for financial statement fraud.

In a study by Loughran and McDonald, NLP was used to identify sentiment in financial reports, which can be useful in predicting future stock prices [2]. Similarly, a study by Das and Chen found that NLP can be used to extract relevant information from financial reports, which can be used to detect anomalies and inconsistencies [3]. In this study, the focus was on financial statement fraud due to its relevance to the task at hand. Financial statement fraud is the direct result of falsifying financial statements, and it frequently has serious repercussions for investors, regulators, the company, and the larger financial markets. It might have negative legal effects, erode investor confidence, and cause significant financial losses. Financial statement fraud typically makes use of intricate plans and skillful data manipulations. Analyzing the wording and emotional expressions employed in financial reports can offer further perspectives on how the financial information is conveyed and the overall attitude it conveys.

In the field of sentiment analysis and financial statement fraud detection, there is a growing interest in integrating sentiment analysis techniques with fraud detection methodologies. The accuracy of fraud detection can be greatly improved by this integration by allowing for more complex and context-specific interpretations of financial attitudes. In order to identify financial statement fraud, this study uses natural language processing (NLP) tools to analyze financial statements, with a particular emphasis on sentiment analysis. The study looks at changes in subjectivity and polarity in financial reporting and finds important trends that can point to attempts to alter how money is portrayed. Although the results recognize sentiment analysis's limits as a stand-alone method, they still advance the discipline by providing a fresh, multifaceted viewpoint on fraud detection. The significance of this research lies in emphasizing the effectiveness of an integrated approach merging traditional financial evaluation with linguistic perspectives to enhance the efficiency of fraud detection mechanisms. It also expands the toolset for fraud detection already in place and creates new opportunities for research in the future.

2 Literature Review

Sentiment analysis is a technique within NLP that involves analyzing the sentiment polarity and subjectivity of textual data. According to a study by Ribeiro et al. sentiment analysis can be used to identify linguistic cues that may indicate fraudulent behavior in financial reports [4]. Through sentiment analysis of financial information, researchers can spot trends and irregularities that might point to fraud.

As Tonkiss notes, trust and confidence are critical components of economic systems, and financial statement fraud can erode these components [5]. Fung argues that transparency and disclosure are essential for building trust and confidence in corporate governance, and financial statements are a key source of information for investors [6]. Gardi et al. highlighted the importance of precise financial reporting through their study on how financial accounting reports influence managerial decision-making within small and medium-sized enterprises [7].

The dialectic conflicts in the financial markets, such as the conflict between confidentiality and transparency, are examined by Gozman et al. [8]. Rezaee emphasizes the importance of corporate governance and ethics in preventing financial statement fraud [9], while Rezaee examines the origins, outcomes, and prevention of fraudulent activities in financial statements [10].

Triani detects fake financial reporting in publicly traded corporations using the Beneish M-Score Model [11], while Anderson and Tirrell examine the role of CEOs in financial reporting fraud [12]. Tirole provides a comprehensive overview of corporate finance, including the role of financial reporting [13]. Rezaee addresses the prevention and detection of financial statement fraud [14], and Wright et al. looks into the difficulties auditors face in spotting financial statement fraud [15].

Segal reviews advanced technologies to detect and prevent accounting fraud [16], while Dorminey et al. examine the evolution of fraud theory [17]. Craja et al. use deep learning for detecting financial statement fraud, demonstrating the potential of NLP in this area [18]. Leuz and Wysocki stated that the economic impacts of regulatory measures on financial reporting and disclosure are substantial, indicating a need for more in-depth research [19]. To promote the interdisciplinary growth of accounting, economics, finance, and law, their research combines theoretical and empirical investigations from these domains. They provide a framework for organization that outlines the advantages and disadvantages of businesses' reporting and disclosure initiatives at both the micro and macro levels of the market. The key conclusions from the research are then compiled, and this framework is used to analyze the potential benefits and drawbacks of regulating these activities.

In the context of financial reporting and market prediction, sentiment analysis has proven to be a valuable tool. For example, studies have demonstrated that the tone of financial news and reports can provide forecasts about how the market will move and how well a company will perform. This method goes beyond conventional numerical analysis and offers a more thorough comprehension of market dynamics.

The detection of financial statement fraud has evolved with the introduction of new methodologies. In addition to conventional approaches like ratio analysis, trend analysis, and forensic accounting, more sophisticated computational techniques and data analytics have been integrated into fraud detection practices. The development of technology has created new opportunities for sophisticated and efficient fraud detection systems.

A rapidly expanding field of study is the intersection of sentiment analysis and financial statement fraud detection. The potential of sentiment analysis to supplement established fraud detection techniques has been the subject of some research. For example, examining the tone and sentiment in financial reports and executive communications has introduced fresh metrics for identifying possible fraudulent activities. These approaches, however, are still in their infancy and frequently encounter difficulties, such as the requirement for huge datasets and advanced algorithms that can decipher intricate financial jargon and subtle idioms. Recent studies have made notable progress in improving the precision and comprehensiveness of emotion and aspect identification within written information. For example, aspect-specific context location information was introduced by Huang et al., highlighting the significance of contextual information in interpreting the sentiment associated with particular aspects of a text [20]. Fei et al. provided a more thorough method by merging multiple layers of sentiment analysis and introducing a multiplex cascade framework, which investigated the idea of latent emotion memory [21].

Researchers have used sophisticated algorithms to extract and evaluate sentiment from financial writings in the financial arena. Deep learning algorithms have proven to be effective in analyzing financial data, offering valuable insights into investor attitudes and trends in the market, as Sohangir et al. showed [22]. Shang et al. introduced an innovative methodology that combines a lexicon-based strategy with neural networks to more accurately reflect the complex perspectives on financial situations [23].

This study's unique application of sentiment analysis and natural language processing (NLP) to the identification of financial statement fraud gives fresh perspectives and integrates knowledge from several fields to tackle fraud detection in a novel approach. The uniqueness of this method is further enhanced by the selection of case studies and the time frame for analysis.

Each case examined in this research entails a comprehensive examination of financial records, market responses, and the subtle language used in corporate communications. Employing advanced NLP methods for scrutinizing financial documents and communication tones enhances the investigative methodology, providing a richer comprehension of the context of each case and the intricacies of the fraudulent activities at play.

Finally, the relevant research in the literature emphasizes the possibility of combining financial statement fraud detection with sentiment analysis. Despite tremendous progress, there is still a lack of literature that particularly addresses the combination of sentiment analysis and financial statement fraud detection. Subsequent investigations ought to concentrate on creating all-inclusive models that incorporate fraud detection and sentiment analysis methods, providing financial analysts and regulators with stronger instruments.

3 Methodology

In this research, sentiment analysis was applied as a method of natural language processing (NLP) to identify financial statement fraud by examining the emotional tone and polarity expressed in the text. We used Python as the programming language and several libraries to perform the analysis. The following libraries were used and below is the flowchart of our approach:

1. os: This library was used to interact with the operating system and perform file operations.
2. pandas: This library was used to handle and manipulate data frames, which are used to store and analyze the extracted data.
3. PyPDF2: This library was used to extract text from PDF files, which are the primary source of financial reports.
4. re: This library was used to perform regular expressions, which are used to clean and preprocess the extracted text.
5. nltk: This library was used to handle text processing tasks, such as tokenization and stop word removal.
6. spaCy: This library was used to perform advanced NLP tasks, such as named entity recognition and sentiment analysis.
7. spacytextblob: This library was used to extend the spaCy pipeline with TextBlob's sentiment analysis capabilities (Fig. 1).

We chose these libraries for several reasons. First, they are widely used and well-documented in the NLP and data analysis communities. Second, they provide advanced NLP capabilities, such as named entity recognition and sentiment analysis, which are essential for analyzing financial reports. Third, they are highly efficient and scalable, which is critical for processing large volumes of text data.

Compared to the TextBlob library used in the previous research paper, spaCy and spacytextblob provide more advanced NLP capabilities and are more efficient in processing large volumes of text data. For instance, spaCy's statistical models are optimized for performance and can process large volumes of text data more efficiently than TextBlob. Additionally, spacytextblob extends spaCy's pipeline with TextBlob's sentiment analysis capabilities, which provides more accurate sentiment scores than TextBlob.

To extract text from PDF files, we used the PyPDF2 library. We defined a function, extract_text_from_pdf, which takes a PDF file path as input and returns the extracted text. The function uses PyPDF2's PdfReader class to read the PDF file and extracts text from each page using the extract_text method.

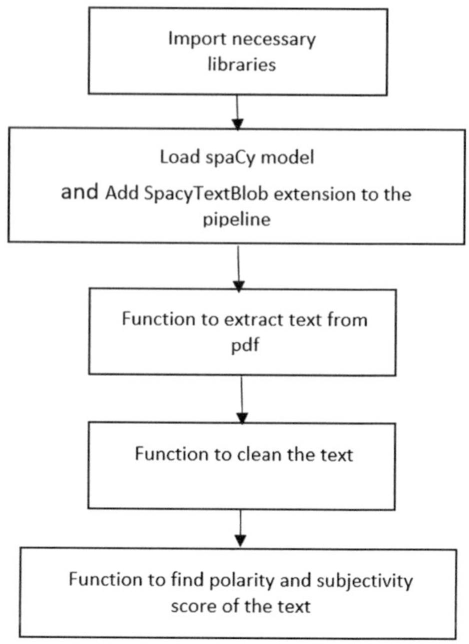

Fig. 1 Above flowchart illustrates the process

We created a function called clean_text to clean and preprocess the retrieved text. The function takes the extracted text as input and performs several cleaning and preprocessing steps, such as removing non-alphanumeric characters, extra whitespaces, and stop words. The function returns the cleaned text, which is used for sentiment analysis. We built a function called analyze_sentiment to analyze sentiment scores. Using the cleaned text as input, the function analyzes the text's sentiment using statistical models from spaCy. The polarity score and subjectivity score, which are used to gauge the text's sentiment, are returned by the function.

The sentiment of a text can be quantified through a polarity score, typically on a scale from -1 to 1. A score of 0 signifies a neutral sentiment, while a score of 1 represents a highly positive sentiment and -1 signifies a strongly negative sentiment [24]. It is determined by applying machine learning methods or a sentiment lexicon to analyze the frequency and intensity of positive and negative terms in the text [25]. The following formula is commonly used to determine the polarity score:

$$\textbf{Polarity} = (\text{Positive score} - \text{Negative score})/(\text{Positive score} + \text{Negative score})$$

where the frequency or intensity of positive and negative sentiment in the text is represented by the positive and negative scores, respectively [25]. Contrarily, a subjectivity score is a numerical value between 0 and 1 that represents how subjective or objective a text is [24]. It is calculated by analyzing the language used in the text, such as personal pronouns, opinions, and facts.

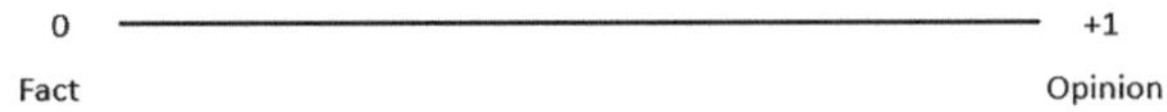

Fig. 2 Sentiment polarity scale

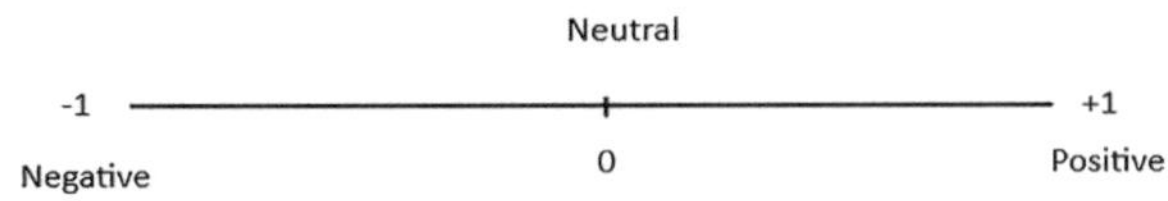

Fig. 3 Subjectivity range scale

Polarity and subjectivity ratings can be computed in a variety of ways, based on the language, method, and data source. The Python TextBlob module is one well-liked and often-utilized technique that offers an integrated sentiment analyzer that yields subjectivity and polarity scores for every supplied text [24] (Figs. 2 and 3).

As shown in images polarity score and subjectivity score are important tools for sentiment analysis, which helps to identify and categorize opinions expressed in a piece of text. The scores can be displayed using scatter plots or other graph types, and they can be computed using a variety of mathematical formulas, sentiment lexicons, and machine learning techniques. However, they should be used in combination with other methods and metrics, such as word frequency, sentiment classifiers, or sentiment aspects, and should be interpreted with caution and context, avoiding oversimplified or general conclusions.

We used the following steps to perform the analysis:

1. Data Collection: For the years 2014–2018 (Wirecard), 2011–2015 (Tesco), and 2012–2017 (Under Armor), we gathered the financial records of the three companies.
2. Data Extraction: To extract text from the PDF files, we utilized the extract_text_from_pdf function.
3. Data Cleaning: We used the clean_text function to clean and preprocess the extracted text.
4. Sentiment Analysis: We used the analyze_sentiment function to analyze the sentiment scores of the cleaned text.

Here's a pseudocode representation of the code.

Algorithm Main
Step 1: Import essential libraries
Step 2: Load the spaCy model "en_core_web_sm"
nlp = spacy.load("en_core_web_sm")
Step 3: Incorporate SpacyTextBlob extension into the spaCy pipeline
nlp.add_pipe('spacytextblob')
Step 4: Define function 'extract_text_from_pdf'
Function extract_text_from_pdf(pdf_path)
Input: pdf_path - path to the PDF file
Output: text - Text that was taken out of the PDF
Step 4.1: (Implementation details for extracting text from PDF are not provided due to space constraints)
End Function
Step 5: Define function 'clean_text'
Function clean_text(raw_text)
Input: raw_text - the unclean raw text
Output: cleaned_text - the cleaned text
Step 5.1: (Implementation details for text cleaning are not provided due to space constraints)
End Function
Step 6: Define function 'analyze_sentiment'
Function analyze_sentiment(preprocessed_text)
Input: preprocessed_text - the text to analyze sentiment
Output: sentiment_scores - dictionary containing polarity and subjectivity
Step 6.1: Utilize spaCy to process the text
doc = nlp(preprocessed_text)
Step 6.2: Extract sentiment scores
polarity = doc._.polarity
subjectivity = doc._.subjectivity
Step 6.3: Return the polarity and subjectivity scores
sentiment_scores = {"polarity": polarity, "subjectivity": subjectivity}
return sentiment_scores
End Function
End Algorithm

There are several limitations that can be faced in the code. Firstly, the code assumes that the PDF files are in a specific format and may not work correctly if the format changes. Secondly, the code does not handle errors that may occur during the extraction of text from PDF files, such as missing pages or corrupted files. Thirdly, the code uses regular expressions to clean the text, which may not be able to handle all possible variations of non-alphanumeric characters and extra whitespaces. Fourthly, the code uses a pre-defined list of stop words to remove from the text, which may not be comprehensive or may include words that are relevant to the analysis. Fifthly, the code uses a pre-trained spaCy model to analyze the sentiment of the text, which may not be able to handle the specific language and terminology used in financial reports. Finally, the code does not provide any statistical analysis or visualization of the sentiment scores, which may limit the ability to interpret the results.

Sentiment analysis can be included in a more thorough examination of financial statement integrity, but for a thorough evaluation, it should be paired with additional analytical techniques and expert review. The section demonstrates sentiment analysis through three case studies involving Tesco in the UK, Under Armor in the US, and Wirecard in Germany. Financial statement fraud has been shown in each and every instance [1].

For the analysis of Under Amour's financial reports, we focused on the period between 2012 and 2016. This period included the years 2015 and 2016, which were marked by financial statement fraud, resulting in a settlement of USD 9 million with federal regulators [26] over charges of misleading investors about sales growth during those years. We used the previous three years, 2012–2014, as a baseline for comparison.

Tesco is situated in the UK, Under Armor is based in the US, and Wirecard is based in Germany. The three instances were carefully chosen to provide geographic diversity. The chosen cases also featured specific instances of confirmed financial statement fraud that had occurred during identified accounting periods, providing a relevant context for our analysis.

4 Result

The results of the analysis display the subjectivity and polarity scores for the financial reports of three distinct companies: Under Armor, Tesko, and Wirecard. The scores are presented the table, with the company name, report name, technique used, polarity score, and subjectivity score as columns (Table 1).

Comparing the output to the previous research paper, there are some differences in the polarity and subjectivity scores for the same reports. For instance, the polarity scores for Wirecard's annual reports have improved, indicating a more positive sentiment in the reports. Similarly, the polarity scores for Under Amour's reports have also improved, with the exception of the report for 2014, which has a lower score compared to the previous research paper. However, the subjectivity scores for Under Amour's reports for 2015 and 2016 have improved, indicating a more subjective language used in the reports.

On the other hand, the polarity and subjectivity scores for Tesko's reports are lower than the values in the previous research paper. This could indicate a more negative sentiment and less subjective language used in the reports. In summary, the analysis of the financial reports using sentiment analysis provides valuable insights into the sentiment and language used in the reports. For a more thorough assessment

Table 1 Comparison of polarity and subjectivity scores for TextBlob and Spacy + SpacyTextBlob techniques across different companies and fiscal years

Company	Fiscal year	Technique used	Polarity score	Subjectivity score
Wirecard	2018	TextBlob	0.0716	0.3807
Wirecard	2018	Spacy + SpacyTextBlob	0.068515	0.332595
Wirecard	2017	TextBlob	0.0772	0.343647
Wirecard	2017	Spacy + SpacyTextBlob	0.076567	0.345361
Wirecard	2016	TextBlob	0.0745	0.3661
Wirecard	2016	Spacy + SpacyTextBlob	0.084242	0.3741
Wirecard	2015	TextBlob	0.0697	0.34656
Wirecard	2015	Spacy + SpacyTextBlob	0.0816	0.349996
Wirecard	2014	TextBlob	0.0674	0.3525
Wirecard	2014	Spacy + SpacyTextBlob	0.071373	0.35255
Tesco	2014	TextBlob	0.1479	0.4607
Tesco	2014	Spacy + SpacyTextBlob	0.100696	0.35259
Tesco	2013	TextBlob	0.11186	0.59984
Tesco	2013	Spacy + SpacyTextBlob	0.10118	0.34697
Tesco	2012	TextBlob	0.090664	0.33499
Tesco	2012	Spacy + SpacyTextBlob	0.1071	0.35735
Under Armor	2016	TextBlob	0.090357	0.34425
Under Armor	2016	Spacy + SpacyTextBlob	0.0676	0.358314
Under Armor	2015	TextBlob	0.06802	0.35965
Under Armor	2015	Spacy + SpacyTextBlob	0.03523	0.35455
Under Armor	2014	TextBlob	0.05745	0.36478
Under Armor	2014	Spacy + SpacyTextBlob	0.0507	0.357
Under Armor	2013	TextBlob	0.06636	0.366472
Under Armor	2013	Spacy + SpacyTextBlob	0.06636	0.36737
UNDER Armor	2012	TextBlob	0.059597	0.360103
Under Armor	2012	Spacy + SpacyTextBlob	0.059597	0.360123

of a company's financial well-being, it is advisable to incorporate sentiment analysis alongside other analytical methods. It is important to note that while sentiment analysis is valuable, it should be viewed as just one component within the spectrum of financial analysis tools available.

Analyzing the sentiment expressed in financial reports offers valuable indications that can help uncover possible fraudulent behavior. To better understand the trends in polarity scores for the companies in question, we have generated separate visualizations for the polarity and subjectivity scores over time. These graphs allow us to observe how the linguistic sentiment expressed in the financial reports of Wirecard, Tesko, and Under Armor has evolved across the examined fiscal years. By closely examining the fluctuations in polarity, we aim to identify any anomalies or patterns

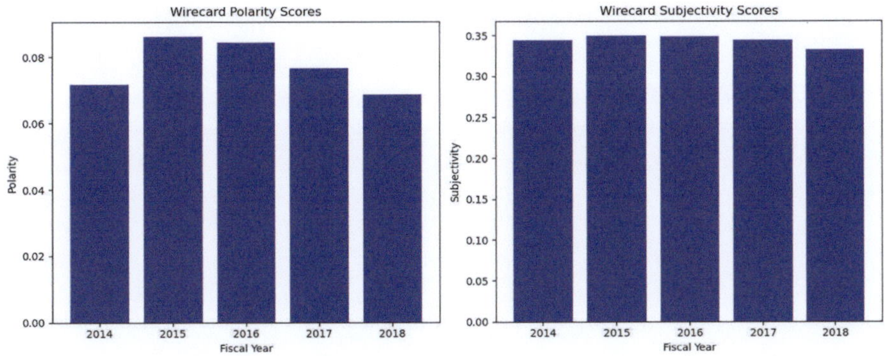

Fig. 4 Trends in polarity and subjectivity scores over time for Tesko

that may indicate concerning financial practices. The following graphs illustrate these sentiment score trends in a clear and concise manner (Fig. 4).

5 Conclusion

Sentiment analysis of three firms' annual reports—Under Armor, Tesco, and Wirecard—has produced some intriguing findings. The polarity scores for Wirecard have shown a general trend of improvement over the years, indicating a more positive sentiment in the reports. Similarly, the polarity scores for Under Armor have also improved, with the exception of the report for 2014. However, the subjectivity scores for Under Amour's reports for 2015 and 2016 have also improved, suggesting that the language used in the reports is becoming more subjective (Figs. 5 and 6).

On the other hand, the polarity and subjectivity scores for Tesco's reports are lower than the values in the previous research paper. This could indicate a more

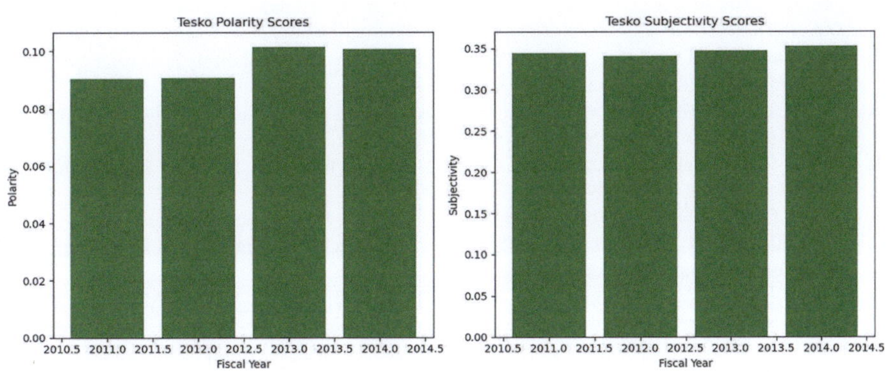

Fig. 5 Trends in polarity and subjectivity scores over time for Wirecard

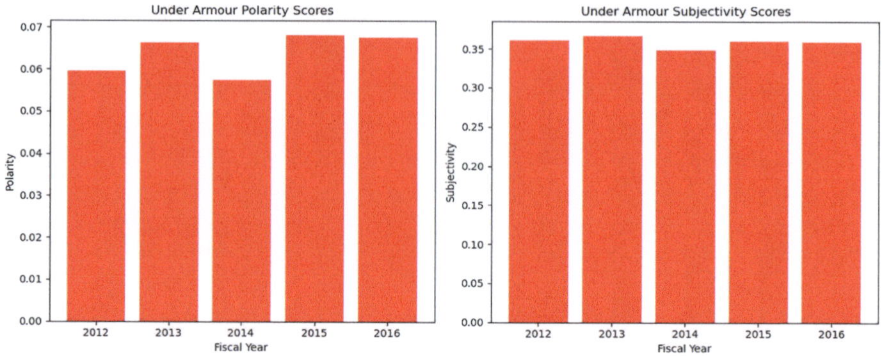

Fig. 6 Trends in polarity and subjectivity scores over time for Under Armor

negative sentiment and less subjective language used in the reports. Overall, the analysis of the financial reports using sentiment analysis provides valuable insights into the sentiment and language used in the reports.

Regarding the code's limitations, it is crucial to remember that the analysis is predicated solely on the reports' content and ignores any additional variables that might have an impact on the company's financial stability. Additionally, the analysis is based on the spaCy textblob extension for sentiment analysis, which may not be as accurate as other sentiment analysis tools.

In conclusion, the analysis of the financial reports of Wirecard, Tesco, and Under Armor using sentiment analysis has revealed some interesting insights into the sentiment and language used in the reports.

References

1. George B, McDonald J, Faccia A (2023) Detecting financial statement fraud using NLP sentiment analysis: a case study of Wirecard, Tesco, and Under Armour. Financ Crime J 20(1):1–15
2. Loughran T, McDonald B (2011) When is a liability not a liability? Textual analysis, dictionaries, and 10-Ks. J Finance 66(1):35–65
3. Das S, Chen H (2007) Text mining for financial analysis: a case study of investor reactions to 10-K filings. Decis Support Syst 43(2):518–531
4. Ribeiro RF, Cordeiro JJ, Batista DA (2016) Detecting financial statement fraud using text mining techniques. Expert Syst Appl 58:128–141
5. Tonkiss F (2009) Economic crises, confidence, and trust. Intereconomics 44:196–202
6. Berry BF (2014) The necessity and demand for disclosure and openness in corporate governance. J Manag Univers 2:72–80
7. Gardi B, Sabir BY, Mahmood Aziz H, Sorguli S, Abdullah NN, Al-Kake F, Abdalla Hamza P (2021) Examining the process via which managers of small and medium-sized companies use financial accounting reports to inform their choices. Turkish J Comput Math Educ 12:2134–2142

8. Gozman DP, Seddon JJ, Currie WL (2018) The paper examines the contradictions between regulatory technology before and after the financial crisis through a longitudinal approach. J Inf Technol 33:304–325
9. Rezaee Z (2008) Ethics and corporate governance. Wiley, Hoboken, New Jersey, USA
10. Rezaee Z (2005) Financial statement fraud: causes, effects, and prevention. Account Crit Perspect 16:277–298
11. Triani N (2019) Fraudulent financial reporting detection using beneish m-score model in public companies from 2012 to 2016. Fraud J Asia Pac 4:27–42
12. Anderson JR, Tirrell ME (2004) CEOs that seem too good to be true and financial reporting fraud. J Pract Res Consult Psychol 56:35
13. Tirole J (2010) Corporate finance theory. Princeton University Press, Princeton, New Jersey, USA
14. Rezaee Z (2002) Financial statement theft: avoidance and identification. Wiley, Hoboken, New Jersey, USA
15. Wright A, Zimbelman MF, Asare SK (2015) Perspectives from fraud investigations on the difficulties auditors face in identifying financial statement fraud. J Forensic Investig Account 7:63–111
16. Segal S (2016) Accounting frauds: an analysis of cutting-edge tools to find and stop fraud. Econ Bus Rev 2:45–64
17. Dorminey J, Riley RA Jr, Fleming AS, Kranacher MJ (2012) The development of the theory of fraud. Account Educ 27:555–579
18. Lessmann S, Kim A, Praja P (2020) Financial statement fraud detection using deep learning. Support Syst Decis 139:113421
19. Leuz C, Wysocki PD (2008) Economic effects of financial reporting and disclosure regulation: a review and ideas for further research
20. Huang B, Guo R, Zhu Y, Fang Z, Zeng G, Liu J, Wang Y, Fujita H, Shi Z (2022) Aspect-level sentiment analysis utilizing aspect-specific context position information. Syst Based Knowl 243:108473
21. Fei H, Zhang Y, Ren Y, Ji D (2020) Latent emotion memory for multi-label emotion classification. In: Proceedings of the AAAI conference on artificial intelligence, New York, NY, USA, 7–12 Feb 2020, vol 34, pp 7692–7699
22. Sohangir S, Wang D, Pomeranets A, Khoshgoftaar TM (2018) Big data: deep learning for financial sentiment analysis. J Big Data 5(3)
23. Shang L, Xi H, Hua J, Tang H, Zhou J (2023) A Lexicon enhanced collaborative network for targeted financial sentiment analysis. Inf Process Manage 60(2):103187
24. https://www.linkedin.com/advice/3/what-advantages-disadvantagesusing-polarity
25. https://iq.opengenus.org/polarity-and-subjectivity-in-nlp/
26. Mosteanu NR, Faccia A (2021) Fintech frontiers: open innovation and paradigm shifts in quantum computing, fractals, and distributed ledgers on blockchain. J Open Innov Technol Mark Complex 7(1):19

Evaluating Public Perception of ChatGPT's Impact on the Job Market Through Twitter Sentiment Analysis Using the VADER Model

G. N. Raghavendra Rao and R. Harish

Abstract According to literature survey, automating tasks through AI technologies can lead to job displacement and potential unemployment, particularly in roles involving repetitive and routine tasks (George in Partners Univ Int Res J 2:20–39, 2023 [1]). The impact of ChatGPT and AI on jobs will vary depending on the specific job and industry (Biswas in Will ChatGPT take my job? Replies and advice by ChatGPT [2]). It is as yet too early to categorically estimate ChatGPT's actual impact on jobs. Therefore, in the interim, this report presents a sentiment analysis to understand people's opinions on the possible impact of AI technologies like ChatGPT on the job market. Tweets on Twitter from 30 November 2022, to 29 March 2023, were considered for sentiment analysis. Approximately 719,295 English tweets containing the word ChatGPT were available for analysis. From this, the tweets containing the words "job," "employment," "occupation," "profession," "business," and "placement" were filtered for further analysis. Sentiment analysis was performed on these tweets using the VADER model. VADER, short for Valence Aware Dictionary and Sentiment Reasoner, is a powerful text sentiment analysis tool. It is based on a set of rules and a lexicon, and can accurately assess the emotional tone of a particular text. This study presents a comprehensive sentiment analysis on ChatGPT with the mentioned keywords, and thereby explores how this technology is likely to shape the job market. The overall finding is that while the general sentiment about ChatGPT ranges from neutral to positive, there is considerable concern among the public about its possible impact on the job scenario.

Keywords ChatGPT · Sentiment analysis · Topic modeling · Jobs · VADER

G. N. Raghavendra Rao (✉) · R. Harish
ICFAI Foundation for Higher Education, Off Campus Center Bangalore (Deemed to Be University U/S 3 of UGC Act 1956), Bangalore, India
e-mail: raghavendrarao.gn@ibsindia.org

R. Harish
e-mail: harish@ibsindia.org

H. Sharma et al. (eds.), *Artificial Intelligence: Theory and Applications*, Lecture Notes in Networks and Systems 5589, https://doi.org/10.1007/978-981-96-1687-9_41

1 Introduction

Ever since the launch of ChatGPT by OpenAI in November 2022, it has evoked tremendous interest, and also strong views, both positive and negative. There are opinions about the possible impact of ChatGPT on practically all aspects of our lives. However, the likely impact of ChatGPT on jobs and employment has been of maximum interest, as it automates and speeds up a wide variety of work-related activities. ChatGPT speeds up and improves the quality and accuracy of tasks, not just those of a routine and repetitive nature, but even those that require creativity and critical thinking, such as analyzing and interpreting large volumes of data, writing computer programs, composing poetry, or even creating posters and images (with some add on applications). Therefore, the impact of ChatGPT on jobs is one of the most discussed topics. It was therefore decided to conduct a sentiment analysis regarding popular perception about the impact of ChatGPT on the job market, using Twitter data.

Sentiment analysis of tweets regarding ChatGPT aims to provide users' perceptions of ChatGPT and analyze the ratio of positive and negative user comments. In addition, topic analysis can provide insights on frequently discussed topics concerning ChatGPT and provide feedback to improve its functionality further [3]. Analyzing comments about and on ChatGPT has been a common practice since its release, and it is used to understand the general public's opinion on specific aspects related to ChatGPT.

Twitter, launched in 2006 as a social media platform, allows users to share short messages publicly. The character limit for these messages ranges from 140 to 280 characters. Searching for specific keywords for a given topic makes it easier to find tweets. Twitter, which users prefer for audience and perception management due to these features, has recently taken an important position in social interaction [4].

When a technology like ChatGPT, which could affect almost all facets of life, was released, it was a common practice for users to express their opinions on a platform like Twitter. The tweeting and retweeting on a particular topic on Twitter could spark increasing interest among Twitter users in expressing perspectives and reactions to that specific topic. For all these reasons, sentiment analysis research for shares made on Twitter has become popular, especially in recent years [4].

In the normal course, sentiment analysis is a challenging task for researchers. The meaning of a word in a tweet varies with the context, making it difficult to interpret manually, as to what truly is a negative or positive statement, amidst the huge corpus of tweet data [5]. Various models help us to overcome this problem. NLTK, TextBlob, and VADER models are very good for basic and intermediate language processing requirements. In addition, there are commercial NLP tools like Google Cloud Natural Language API, Microsoft Azure Text Analytics, IBM Watson Natural Language Understanding, and Amazon Comprehend, to name a few. Each of these tools has its own strengths and applicability in specific scenarios. We decided to use the VADER model in this study, details of which are discussed later below.

2 Literature Review

Given below is literature survey on earlier work done on various related topics such as sentiment analysis and topic modeling, analysis of tweets and also opinions about the impact of ChatGPT on jobs. Most of these have been done based on twitter data. One of these relates to ChatGPT's answer when questioned about its possible impact on the job market. Another study identified that "evolution of jobs" is one of the issues to be addressed as a result of the advancement of ChatGPT. Yet another study reports a literature survey on the nature of jobs that experts think will be affected as a result of ChatGPT and AI. However, none of the earlier works are a report on sentiment analysis of tweets about the impact of ChatGPT on the job market, as carried out in the current study.

A study was carried out on the sentiment analysis on Twitter with the premise that most social media sentiment polarization models decided the sentiment classification based on the textual content, ignoring other information [6]. Therefore, the authors proposed a Convolutional Neural Network (CNN) that also incorporated user behavior information within the document (i.e., tweet). The system was evaluated on two datasets. The proposed model outperformed other baseline models (including Naïve Bayes and Support Vector Machines), which showed that going beyond the content of a tweet was beneficial in sentiment classification, as it provided a much deeper understanding.

An analysis of 233,914 English tweets was carried out within the first month of the launch of ChatGPT [7]. The author used the Latent Dirichlet Allocation (LDA) topic modelling algorithm to get an answer to the question—"What can ChatGPT do?" The LDA results identified a total of ten topics. There were three broad topics—news, technology, and reactions. Also, five functional domains were identified, which had the maximum number of tweets—answering questions, creative writing, code writing, essay writing, and prompt writing. Jobs related to these domains were most likely to be negatively affected. The study identified that ChatGPT can impact humans and technology in both positive and negative ways. The paper concluded that four important issues needed to be addressed as a result of AI's advancement: the evolution of jobs, the new technological landscape, the quest for artificial intelligence, and the conflict between progress and ethics.

When questioned about its impact on the job market, ChatGPT itself answered that customer service representatives, translators and interpreters, content writers, data analysts, and healthcare workers are some of the job roles that could be affected [2]. In the case of data analysts and healthcare service providers, increased productivity and efficiency could lead to higher salaries and better job prospects. Whereas, in the other three cases, there could be reduction in salaries and less number of jobs as well. Workers who are affected by the integration of ChatGPT and AI in their job roles can adapt themselves by upskilling and reskilling, taking up new roles, and also by focusing on soft skills.

In another study [8], tweets were collected for the period between January–March 2019, about two candidates who had contested the general elections in India for the national parliament. The sentiment analysis assessed the opinion polarity among the general public about the two candidates. The sentiment analysis found that Candidate 1 was more popular and well-known as compared to Candidate 2. And this finding was in agreement with the actual election results that came out later during May 2019.

A literature survey was carried out in the context of increasing fears expressed by experts from all over the world that ChatGPT and other AI tools will decimate large number of white-collar jobs [9]. The conclusion was that AI and ChatGPT will reduce, alter, or eliminate jobs that can be automated or digitalized such as proof reading, translating, data entry, tutoring/teaching, responding to routine customer enquiries, and such others. However, it cannot take away jobs that require a physical presence, such as barbers, bartenders, cooks, mechanics, baby sitters, civil engineers, plumbers.

An analysis of 500,000 tweets from the period January to March 2023 was carried out to discover the popularity and sentiment about ChatGPT on Twitter [10]. Pandas Python Library was used for data analysis and VADER for sentiment analysis. It was found that ChatGPT was rapidly gaining in popularity in spite of the discussion about the likely dangers and negative repercussions that it could lead to. Therefore, one cannot ignore the development of such models, but we have to find ways to benefit from their potential and mitigate any damage that they might cause.

A sentiment analysis was conducted on 1,305,000 tweets in English posted between 21 June 2020 and 20 July 2020 about the COVID-19 pandemic from 11 infected countries [11]. The Syuzhet algorithm was used to classify the tweets into eight emotions—fear, joy, anticipation, anger, disgust, sadness, surprise, and trust. It was found that majority of the people (around 55%) expressed positive sentiments as the recovery rate was high. Trust, fear, anticipation, and sadness were the four topmost sentiments that were observed, with minor variations across individual countries.

Another study carried out sentiment analysis and topic modelling on tweets related to ChatGPT [3]. Latent Dirichlet Allocation (LDA) approach was used to identify the most frequently discussed topics. Lexicon-based techniques were applied to categorize the tweets as positive, negative, and neutral. Out of 20,801 tweets, 8095 were positive, 2727 were negative, and 9979 were neutral. These were used in a ratio of 80:20 for training and testing a proposed (Bidirectional Encoder Representations from Transformers (BERT) for classifying the sentiments. This model consisted of three dense layers of neural networks, and it was able to achieve an accuracy of 96.49%.

3 Methodology Followed

3.1 Sentiment Analysis

Natural Language Processing (NLP) is the artificial intelligence method of communicating with an intelligence system using natural language. The most used applications of NLP are "sentiment analysis" and Twitter and Facebook sentiment analysis in particular [5]. NLP is a multidisciplinary field that combines computer science, artificial intelligence, and linguistics. It aims to create computer models capable of processing and understanding natural language, including its meaning, context, and intent. NLP applications include text-to-speech conversion, language translation, sentiment analysis, and speech recognition. Using NLP, we can develop systems that facilitate communication between people and computers, allowing them to understand and respond to each other more efficiently.

Sentiment analysis, or opinion mining, is a special form of natural language processing (NLP). Its main focus is identifying, extracting, quantifying, and analyzing emotional states and subjective information from large text data. By examining language patterns, sentiment analysis can determine whether a text conveys a positive, negative, or neutral mood. This technique is often used to analyze customer reviews, social media posts, comments, and other types of content. It can be performed on Twitter data to get insights into the public's sentiments [8].

Recently, there has been an increase in interest in sentiment analysis techniques, driven by the growing need for tools to monitor public sentiment [12]. Companies regularly use sentiment analysis to understand how customers perceive certain products or services. This relies heavily on unstructured data such as emails, texts, documents, and articles. Sentiment analysis transforms unstructured data into structured insights, making it easier for companies to understand their customers' sentiments. Therefore, sentiment analysis is extremely useful for companies. Once a given sentiment analysis model is ready, it can analyze texts most efficiently and cost-effectively to provide real, actionable insights. Some of the common practical applications of sentiment analysis include analyzing customer feedback, managing a brand, and improving products and services.

3.2 The VADER Model

VADER stands for Valence Aware Dictionary and Sentiment Reasoner. This is a pre-trained model for text sentiment analysis based on machine learning. It is specifically designed to conduct sentiment analysis on social media content. VADER model can analyze short texts like tweets, reviews, and comments by using a predefined list of words. The model evaluates the polarity of the text and determines whether it is positive, negative, or neutral, with sentiment intensities ranging from -1, which indicates the most negative sentiment, to $+1$, which indicates the most positive. In addition, the model considers the intensity of the emotion or mood conveyed by words, enabling accurate calculation of mood values. VADER uses a dictionary of words, each with a predetermined valence value indicating positivity or negativity, coupled with predetermined rules for assessing sentiment. The model also considers punctuation, capitalization, acronyms, context modifiers, emoticons, emojis, and even internet slang to gauge the sentiment accurately. For all these reasons, VADER model is best suited for social media monitoring, especially for tweet sentiment analysis. The VADER model provides state-of-the-art rule-based sentiment analysis of tweets [13].

Valence Aware Dictionary and Sentiment Reasoner (VADER) is used to determine tweets' polarity and classify them according to multiclass sentiment analysis [14]. The VADER model offers significant advantages, especially the ease of implementation and the lack of need for training data. As no training is required, it is often significantly quicker than machine learning algorithms [15].

3.3 Topic Modeling

Topic modeling, a machine learning technique, is a powerful tool for drawing out common themes from large volumes of unstructured textual data within a linguistic corpus [16]. It can calculate and cluster the most frequently appearing terms relative to a specific field [17].

However, a major challenge in using supervised machine learning models for text analysis is that much of the available online data is unlabeled, making it difficult to categorize into specific groups. This may include data such as customer reviews, tweets, and articles that have not yet been tagged or grouped. This is where Natural Language Processing (NLP) comes into play. NLP can categorize text data into specific topics. Topic modeling, a process within NLP, allows us to uncover abstract topics and hidden semantic structures present in the text data. One of the main advantages of topic modeling is that it uses an unsupervised machine learning method. This means that it can automatically identify clusters of similar documents or abstract topics based on the content of the text data without the need for manual labeling or categorization. However, this also makes assessing the model's effectiveness difficult, as there is no clear way to measure its accuracy or performance. Despite this difficulty, topic modeling remains a popular and effective method for analyzing massive amounts of text data.

The topic modelling process involves pre-processing to clean the data and applying algorithms like Latent Semantic Indexing (LSI) to identify the topics, and post-processing to interpret the results of these topics. This method is highly valuable and helps to understand the emotional tone associated with different topics. It can also provide us with the emerging themes and public opinions on a given topic. LSI also helps to identify the hidden topics within a large corpus of text by analyzing the relationships between each tweet and the terms they contain.

4 Results and Discussion

ChatGPT, developed by OpenAI, was first introduced to the public in November 2022. Therefore, English-language tweets containing the term "ChatGPT" posted on Twitter from 30 November 2022, to 29 March 2023, were considered for sentiment analysis. To refine the quality of the analysis, about 15,947 tweets containing specific keywords such as job, employment, occupation, business, and placement were only considered for analysis. The VADER model was used to carry out the sentiment analysis. This analysis aims to understand public opinion about the effect of ChatGPT on the job market.

The tweets were analyzed up to a specific time (up to 29 March 2023) from the start of ChatGPT, as the opinions would change based on new developments in ChatGPT. We have analyzed a total of 15,947 tweets for sentiment analysis. Sample tweets are shown in Table 1.

The final output of the VADER model includes the following scores for the categorized tweets according to the sentiment expressed by the tweet. The analysis gives a compound score, which is a single score indicating the overall sentiment; a positive score, the proportion of the text with positive sentiment; and a negative score, the proportion of the text with negative sentiment; a neutral score, the proportion of the text with neutral sentiment. Out of the tweets that were analyzed, 8691 were positive, 3574 were negative, and 3682 were neutral. Sample tweets with scores shown in Table 2.

These scores are visualized as a heat map, expressing the text data's sentiment scores against the sentiment expressed by the tweets. Figure 1 shows the heat map of the sentiment analysis conducted. The color scale ranges from − 0.869604 (deep blue) to 0.998997 (bright yellow), representing the sentiment polarity. While blue indicates a negative sentiment, yellow indicates a more positive one. The combined colors in between (various shades of green) indicate a balanced or neutral sentiment.

The heat map is organized into different columns for sentiment categories: positive, negative, neutral, and compound (overall sentiment score combining the positive, negative, and neutral scores). The horizontal color bar at the top of Fig. 1 shows the range of sentiment values, which goes from − 0.860694 (negative) to + 0.998997 (positive). In the heat map shown in Fig. 1, the colors closer to blue indicate more negative feelings, while colors closer to yellow indicate more positive feelings. Green color indicates neutral feeling. The rows are clustered hierarchically, to group the tweets with similar sentiment scores. The clustering helps identify common themes and patterns in sentiment across different tweets.

By analyzing the heat map, it is clear that most tweets have a range of greenish to yellowish colors, indicating an overall neutral to positive sentiment toward ChatGPT. However, dark green tones which are very close to dark blue in certain areas reflect a negative mood, likely related to concerns about the impact on workplaces. Major part of the heatmap lies in a neutral color range, suggesting that many tweets express a balanced view, neither overly positive nor negative, potentially indicating uncertainty about ChatGPT's impact. The composite sentiment column displays a variety

Table 1 Sentiment analysis on ChatGPT's impact on the job market—sample tweets

Tweet_id	Created_on	Tweet
1598064190304160000	11/30/2022	I asked AI (ChatGPT) Will AI replace UX Design Jobs. This is what AI said. #ChatGPT @OpenAI https://t.co/qNYO50TLWV
1598451963083080000	12/1/2022	Testing #ChatGPT's probability skills. Great job on solving the first part (not surprised as this might be available on the web easily). But not so good in the second part. Respond with the correct answer if you enjoy probability puzzles. #gpt #probability #math #logic #LLM https://t.co/v6JYNTsfwq
1598452015629470000	12/1/2022	I legitimately just received my copy of Superintelligence by Nick Bostrom, and now I'm seeing ChatGPT blowing up on Twitter. I wonder which parts of my job will be the first to be replaced or augmented by AI
1598455798853770000	12/1/2022	A lot of people are scared that ChatGPT will take their jobs, I think it will make it easier to learn
1611145423921480000	1/5/2023	Many jobs will be obsolete. #ChatGPT
1611142688619000000	1/5/2023	@OpenAI chat has COMPLETELY transformed the way I work. This thing is literally writing all my product copy and saving my HOURS of content writing at work right now. This is not taking over jobs guys—if you know how to use it—it literally elevates your skillsets #ChatGPT
1611118309059030000	1/5/2023	There's a lot of talk about #ChatGPT coming for people's jobs. Maybe some in the future, but right now AI feels more like an intern—useful, but needs lots of guidance, coaching, and supervision
1611113339458810000	1/5/2023	I asked #ChatGPT to write a tweet about Colombia's inflation rate, and I wrote the question in Portuguese. Look at the output. We are all going to lose our jobs #Doomed https://t.co/fRZB0U5OMT

of colors, with larger deviations downward, indicating varying overall sentiment, including some extremely positive and negative reactions.

Overall, the colors in the heat map indicate a range of colors from neutral to positive, with some negative sentiment expressing concerns about job displacement and other possible adverse impacts. Clusters of concern can be seen in the blue areas at the top right, indicating strong negative sentiment, likely focused on fear of job losses and the economic fallout. The green areas indicate balanced opinions where people acknowledge both ChatGPT's positive and negative impacts or express uncertainty about ChatGPT's effect on the job market. Figure 2, the word cloud, shows the words that define the strong negative sentiments.

After performing sentiment analysis, topic modeling was conducted using the Latent Semantic Indexing (LSI) method. The image in Fig. 3 displays the ten topics generated from the sentiment analysis. The topic modeling output revealed several

Table 2 Sentiment analysis on ChatGPT's impact on the job market—sample positive and negative tweets with compound score

Tweet	Positive	Negative	Neutral	Compound
People have been saying how ChatGPT will replace CEOs and founders, but I think it makes their job easier. I personally spend a lot of time thinking about how to word things in emails to make sure they don't come off wrong in a certain way. This will be a big tool for everyone	0.212	0	0.788	0.9124
#ChatGPT is remarkable. Used it today to draft case statements, letters, and social media posts. All copy was nearly perfect (and indistinguishable from what a human would write). Most business writing is formulaic and tedious. Using an AI to do this part of my job feels awesome	0.21	0	0.79	0.9081
Hot Take: Ban ChatGPT. Its going to steal jobs and make people even more retarded. We need to stop it before it becomes more of a cancer	0	0.445	0.555	− 0.9563
The Fatalists are saying that #ChatGPT is here to steal jobs and bring forth doom upon humans. The innovators, and hustlers are making lemonades with #ChatGPT. Who are you? Are you proposing solutions or screaming bloody murder? solution: Writer Tool	0.072	0.335	0.593	− 0.9399

key themes in the discussion about AI technology, including ChatGPT and its impact on the job market, as well as on upcoming jobs. The conversation included personal experiences and perceptions shared by individuals on how AI will simplify their work and help them learn new technologies.

There was widespread discussion about the overall impact of AI technologies on employment, considering both positive and negative aspects. Opinions on job displacement were prominently expressed, indicating concerns about future jobs being taken away by AI technology. The practical application of AI and its potential to replace jobs were key points in these discussions, emphasizing how AI technologies will influence the job market. This underlines a major concern: people are already worried about AI technologies taking away their jobs and job security.

Hence, there is a need to address job displacement concerns and promote a positive perspective on AI technologies. This should involve communicating the new skills required to adapt to the new situation on the job market. Additionally, relevant stakeholders should maintain transparent communication about the implementation of AI technology in various job domains, clarifying the future impact of AI technologies on the employment scenario.

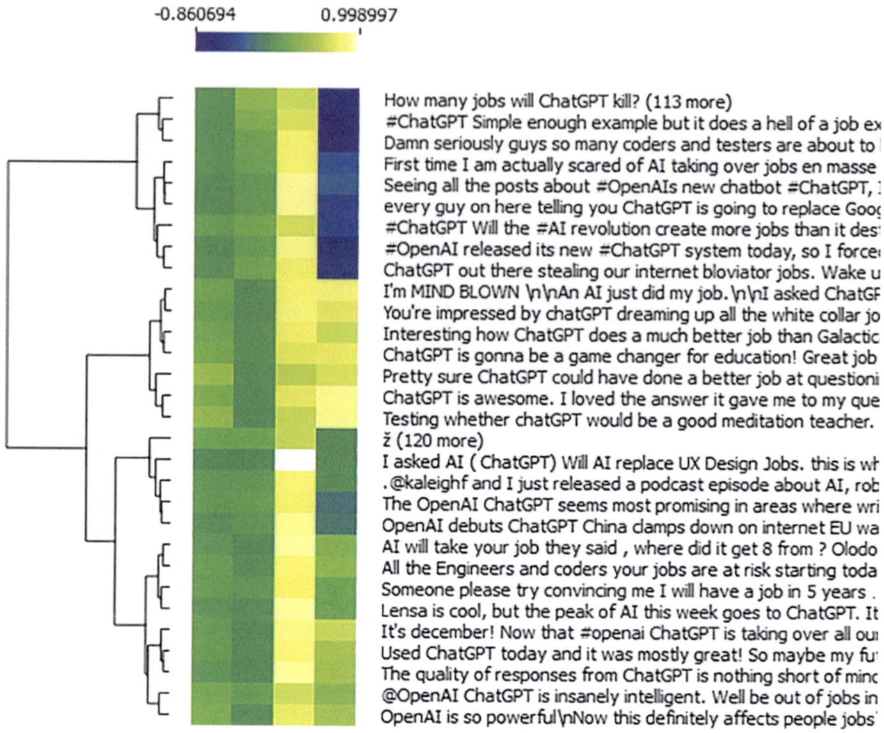

Fig. 1 Sentiment analysis on ChatGPT's impact on the job market—heatmap

Fig. 2 Sentiment analysis on ChatGPT's impact on the job market—word cloud for the words in negative sentiment

5 Limitations and Directions for Future Research

This research paper provides insights into the public view of ChatGPT's impact on the job market by analyzing the related tweets from Twitter. The researchers were compelled to face several limitations during the research, which we wish to

Fig. 3 Results of topic modeling on ChatGPT's impact on the job market

1: chatgpt, co, ai, job, jobs, i, gpt, openai, chat, take
2: i, job, ai, jobs, gpt, chat, co, write, good, asked
3: ai, co, i, jobs, chatgpt, gpt, chat, job, people, like
4: ai, job, jobs, i, chatgpt, gpt, chat, co, think, many
5: gpt, chat, co, ai, chatgpt, i, jobs, hiring, replace, take
6: co, chatgpt, i, job, ai, jobs, gpt, chat, people, take
7: job, jobs, chatgpt, gpt, openai, it, take, chat, like, people
8: openai, like, job, jobs, chatgpt, replace, gpt, use, chat, take
9: take, like, new, it, will, going, openai, away, amp, write
10: like, take, openai, jobs, going, replace, job, write, hiring, away

acknowledge. The Twitter data doesn't represent a broader population of the job market; and therefore, several other factors might also have influenced the opinions expressed. Further, we couldn't separate the data demographically, which could have indicated differences in opinion. To overcome this limitation, a future study can be proposed considering the views collected from various social media platforms. The current research entirely depends on descriptive machine learning to arrive at conclusions. The research can be strengthened by including inferential statistical methods with clear validation methodology and appropriate hypotheses. This analysis, which should use the approach of having a control group, should be able to isolate the opinions and perceptions more precisely. The research can also be carried out considering the ethical Angle of AI, such as potential job displacement, privacy concerns, and the fairness of AI algorithms, which are critical in understanding the broader consequences of AI technologies.

6 Conclusion

This study analyzed the tweets on Twitter between 30 November 2022 and 29 March 2023, to understand the general public's opinion about the likely impact of ChatGPT on the job market. The analysis reveals the details of the public's concerns using sentiment analysis and topic modeling, along with latent semantic indexing. The study identified the main themes that capture the sentiments expressed in these tweets. The analysis discovered significant general discussion about AI technologies and ChatGPT. The public expressed major concerns about ChatGPT and the AI technology taking over jobs.

Even though general sentiments expressed curiosity about new AI technology and ChatGPT, there was considerable concern about job replacement and job security. Some had experienced ChatGPT personally, and their tweets showed that they were amazed by what the technology can do for them. Some tweets mentioned tasks that could be automated, which could even have a positive influence on people's professional lives. The broad conversation was about AI technology's impact on employment, which came out strongly during topic modeling. Some tweets expressed the positive aspects of technology, but the majority of the negative tweets expressed concern about technology taking away jobs. This indicates that there is anxiety among the people about the future of the job market and potential job losses due to the developments in AI technology.

The heat map shows that the tweets on AI technology and ChatGPT range from neutral to positive; but still, there is strong concern about job loss and its economic impact. The number of neutral sentiments was sizeable, which might be because people were still unable to fathom the possible impact of this AI technology on the job market. The analysis shows the complex intersection of optimism and concern regarding technology and its impact on jobs. At the same time, there was a significant number of tweets expressing interest in learning the technology.

The immediate need is to address the concerns through transparent communication and promote the positive aspects of AI. Steps should also be taken to demonstrate where this technology could support the job market. The stakeholders who will use this technology need to collaborate and demonstrate how this technology will transform the job landscape and outline the steps necessary to adapt to these changes.

References

1. George AS (2023) Future economic implications of artificial intelligence. Partners Univ Int Res J 2(3):20–39
2. Biswas S (2023) Will ChatGPT take my job? Replies and advice by ChatGPT
3. Mujahid M, Rustam F, Shafique R, Chunduri V, Villar MG, Ballester JB, Diez IDLT, Ashraf I (2023) Analyzing sentiments regarding ChatGPT using novel BERT: a machine learning approach. Information 14(9):474
4. Korkmaz A, Aktürk C, Talan T (2023) Analyzing the user's sentiments of ChatGPT using Twitter data. Iraqi J Comput Sci Math 4(2):202–214
5. Gandhi UD, Malarvizhi Kumar P, Chandra Babu G, Karthick G (2021) Sentiment analysis on Twitter data by using convolutional neural network (CNN) and long short term memory (LSTM). Wirel Pers Commun 1–10
6. Alharbi AS, Doncker ED (2019) Twitter sentiment analysis with a deep neural network: an enhanced approach using user behavioral information. Cogn Syst Res 54:50–61
7. Taecharungroj V (2023) "What can ChatGPT do?" Analyzing early reactions to the innovative AI chatbot on Twitter. Big Data and Cogn Comput 7(1):35
8. Sharma A, Ghose U (2020) Sentimental analysis of Twitter data with respect to general elections in India. Procedia Comput Sci 173:325–334
9. Gondane J (2023) Evolving trend of artificial intelligence and its impact on employment. BioGecko—J NZ Herpetology 12(3):3546–3551
10. Dodd H (2023) Sentiment analysis of tweets about ChatGPT. Librarianship and Evolving Technologies. Western University
11. Kauser MA, Arockiasamy S, Nasar M (2021) Public sentiment analysis on Twitter data during COVID-19 outbreak. Int J Adv Comput Sci Appl (IJACSA) 12(2):415–422
12. Su Y, Kabala ZJ (2023) Public perception of ChatGPT and transfer learning for tweets sentiment analysis using Wolfram Mathematica. Data 18(12):120
13. Koonchanok R, Pornchalermpong Y (2023) Public attitudes toward ChatGPT on Twitter: sentiments, topics, and occupations. Soc Netw Anal Min 14:1–13
14. Elbagir S, Yang J (2019) Twitter sentiment analysis using natural language toolkit and VADER sentiment. In: Proceedings of the international multiconference of engineers and computer scientists 122(16):13–15
15. Rochadiani TH (2023) Sentiment analysis of YouTube comments toward ChatGPT. Jurnal Transformatika 21(1):60–68

16. Wanniarachchi VU, Scogings C, Susnjak T, Mathrani A (2023) Hate speech patterns in social media: a methodological framework and fat stigma investigation incorporating sentiment analysis, topic modelling and discourse analysis. Australas J Inf Syst 27
17. Patil PP, Phansalkar S, Kryssanov VV (2019) Topic modelling for aspect-level sentiment analysis. In: Proceedings of the 2nd international conference on data engineering and communication technology. Springer, Singapore, pp 221–229

Enhancing Pneumonia Diagnosis with Convolutional Neural Networks: A Comprehensive Evaluation

Nikita Tanni, Manu Elappila, Tessa George, and Nithin Premjith

Abstract Pneumonia, a significant health concern globally, presents unique challenges in diagnosis and treatment due to its diverse ethology and impact on respiratory function. The potential of augmentation techniques and Convolutional Neural Networks, for automated pneumonia detection is explored in this study. Employing a transfer learning approach with VGG16, DenseNet, and our proposed model achieves outstanding accuracy (95%) and robust performance metrics. The research explores augmentation techniques to enhance the precision and accuracy of the model, emphasizing the importance of data augmentation in improving classification accuracy. A comparative analysis with related models highlights advancements in automated pneumonia detection, showcasing the efficacy of our proposed model. The model's ability to correctly identify pneumonia from chest X-ray pictures is demonstrated by the results, suggesting that medical image analysis could benefit from practical implementation of this model. Future directions include expanding the dataset, exploring alternative architectures, and integrating explanation techniques to enhance model interpretability. This research contributes to the advancement of artificial intelligence in healthcare, offering a promising approach for accurate and efficient pneumonia diagnosis, thus addressing critical challenges in respiratory medicine.

Keywords Machine learning · Convolutional Neural Networks (CNNs) · VGG16 · DenseNet · Accuracy · Advance healthcare · Automated classification

N. Tanni · M. Elappila (✉) · T. George · N. Premjith
CHRIST University, Bengaluru, India
e-mail: manu.elappila@christuniversity.in

N. Tanni
e-mail: nikita.tanni@btech.christuniversity.in

T. George
e-mail: tessa.george@btech.christuniversity.in

N. Premjith
e-mail: nithin.premjith@btech.christuniversity.in

H. Sharma et al. (eds.), *Artificial Intelligence: Theory and Applications*, Lecture Notes in Networks and Systems 5589, https://doi.org/10.1007/978-981-96-1687-9_42

1 Introduction

Pneumonia continues to be a major worldwide health concern because of its wide range of an etiology and severe effects on respiratory function, which make diagnosis and treatment extremely difficult. Although successful, traditional diagnostic techniques can need a significant amount of time and experience, which creates a need for more precise and fast diagnostic instruments. Recent developments in artificial intelligence (AI) and machine learning (ML) have created new opportunities for medical picture analysis, which has the potential to completely transform the diagnosis of healthcare. This paper explores the automated diagnosis of pneumonia from chest X-ray pictures using Convolutional Neural Networks (CNNs), a subset of machine learning. By combining our suggested model with proven models like VGG16 and DenseNet through transfer learning, we hope to improve pneumonia classification accuracy and performance metrics. By incorporating data augmentation techniques, we seek to enhance the precision and reliability of our models, underscoring the critical role of data augmentation in improving classification outcomes. This research not only demonstrates the efficacy of our proposed model in accurate pneumonia detection but also highlights significant advancements over existing models, indicating promising real-world applications in medical image analysis. Future directions include expanding the dataset, exploring alternative architectures, and integrating explainability techniques to improve model interpretability, thus contributing to the advancement of AI in healthcare and addressing vital challenges in respiratory medicine.

2 Related Work

The article in [1] explains the difficulties radiologists encounter in identifying pneumonia on chest X-rays because of the condition's visual overlaps and complexity. The promise of automated disease identification in clinical settings is demonstrated by comparing the performance of a deep learning model (CheXNet) versus radiologists. The model performs better. The study's limitations include the use of just frontal X-rays and the lack of patient history information, which could cause the model's performance to be underestimated. Further research that includes patient history and lateral views may yield a more thorough assessment.

Article in [2] explains the process for diagnosing and classifying pneumonia using deep machine learning, including Convolutional Neural Networks (CNNs) and transfer learning. It highlights the effectiveness of DenseNet201 and the potential for improving accuracy through pre-trained CNN ensemble methods. The study aims to aid timely intervention and enhance pneumonia diagnosis, addressing global healthcare challenges. The lack of cases of bacterial and viral co-infection in the dataset and the study's exclusive focus on chest X-ray images for pneumonia identification may have limited the findings' generalizability. There was no discussion of the practical difficulties of using the diagnostic tool in actual clinical settings.

In [3], the authors provide a thorough analysis of the use of deep learning models—more especially, Xception and Vgg16—in the diagnosis of pneumonia from chest X-ray pictures. To improve classification accuracy, the study makes use of data augmentation, fine-tuning, and transfer learning techniques. The efficiency of the two networks was compared using performance indicators. The dataset might not correctly represent pneumonia cases in the real world due to its limitations. The study does not address potential biases in the dataset or models, and it only compares two CNN architectures, restricting the investigation of other possible models.

With a focus on CNNs and transfer learning, the paper [4] investigates deep learning applications in medical image analysis. Studies by various authors showcase the use of deep neural networks for tasks like pulmonary disease diagnosis and feature extraction. Enhancements in classification systems via neural networks in medical imaging are highlighted for improved diagnostics and disease identification. The lack of transparency in deep neural network decision-making, which necessitates additional interpretability efforts, overfitting concerns resulting from training with minimal data, and limited data availability impeding improved accuracy are some of the shortcomings of the document.

A study [5] created an ensemble framework for the identification of pneumonia in chest X-ray images utilizing GoogLeNet, ResNet 18, and DenseNet 121. Using unique strategies based on accuracy, recall, F1 score, and AUC, the ensemble model's weights were optimized, beating state-of-the-art algorithms on publicly available datasets such as the Kermany and RSNA. The small dataset size and reliance on manually generated features are two of the study's weaknesses, which may restrict its capacity to complete complex pattern recognition tasks and increase computation costs.

The research paper [6] on pneumonia detection using Convolutional Neural Networks discusses the importance, related work, methodology, and results. Neural network models like VGG16, VGG19, ResNet50, and Inception v3 are employed with transfer learning. Challenges like the vanishing gradient problem are addressed through techniques to enhance model performance and need for careful consideration of overfitting, which can be addressed through techniques like dropout, data augmentation, and adjusting learning rates.

The study [7] presents CNN architectures that prioritize data augmentation over overfitting in order to diagnose pneumonia from chest X-ray pictures. Experimental results show a model with dropout and augmented data achieving the highest testing accuracy of 0.9068. Future work includes exploring different optimizers and augmentation techniques for improved classification accuracy. The limitations of using pre-trained models for image classification in medical imaging include their large size with millions of trainable parameters, requiring significant computing power and time.

The study [8] evaluated various CNN models for pneumonia detection, emphasizing the impact of contrast enhancement on model performance. Results showed that preprocessed models outperformed non-enhanced ones, with the proposed model achieving the highest accuracy and precision rates. Different model architectures and input shapes were compared to assess classification effectiveness. The loss tendency's oscillatory instability, which was probably brought on by the GPU compute source's mini-batch size restriction, presented challenges for the investigation.

The work discussed in [9] utilizes Vision Transformer (VIT), MobileNetV2, and DenseNet169 for chest X-ray image analysis. The models achieved high accuracy, with VIT performing best. Comparisons with other CNN models showed superior performance of VIT, MobileNetV2, and DenseNet169. The proposed ensemble learning method further improved accuracy and reduced loss. The limitations of the proposed ensemble learning (EL) method include the need for determining appropriate transfer learning (TL) parameters can be difficult and require trial-and-error, such as choosing a pre-trained CNN method, determining the number of frozen layers, and specifying the size of fully linked layers.

In [10], authors explore pneumonia diagnosis through chest X-rays, utilizing transfer learning with CNN models like ResNet and DenseNet. Implementing attention mechanisms, such as SENet and ECA Net, enhances model performance, achieving high accuracy, precision, and recall. The study emphasizes the potential of deep learning and transfer learning for improving medical diagnosis, specifically for pneumonia detection. Limitations of the study include the insufficient training data with only 5856 images, which is inadequate for models like ResNet and DenseNet that typically require over a million images.

3 Problem Statement

Healthcare providers have considerable hurdles when diagnosing pneumonia through chest X-rays since they need to find accurate and efficient ways. Currently, different approaches to diagnosis are used, including deep learning methods like Convolutional Neural Networks (CNNs) and conventional radiomic feature extraction. However, optimal method for practical deployment in medical settings remains unclear. This project aims to address this gap by comprehensively evaluating the use of CNNs as standalone tools for pneumonia detection. The objective is to determine the efficacy of CNNs in accurately diagnosing pneumonia and to provide valuable insights into their potential to enhance diagnostic accuracy and efficiency in clinical practice.

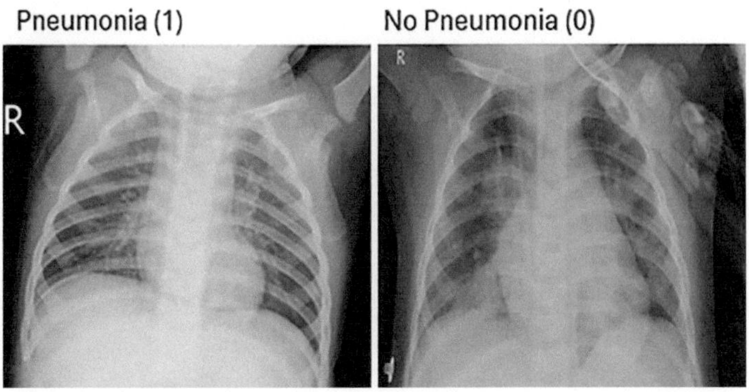

Fig. 1 Pneumonia MRI image dataset

4 Proposed Methodology

4.1 Dataset Used

We utilized a benchmark dataset of 5856 chest X-ray images, comprising normal and we made use of a benchmark dataset consisting of 5856 chest X-ray pictures that were classified as pneumonia and normal. We developed and tested our machine learning model using this pre-existing dataset, going through training, validation, and testing stages. The collection has been divided into three classes: validation (16 photos), train (5216 images), and test (624 images). There are two subfolders with photos categorized as either normal or pneumonia within each of these directories. Figure 1 provides a glimpse of the dataset photos. In this work, data augmentation has been applied, leading to an increase in the quantity of MRI images used for training. Train samples are 41,722, validation samples are 522 after augmentation, and testing samples are located in 522 numbers.

4.2 Preprocessing

In this study, the deep learning model used for a binary classification problem benefited greatly from data augmentation in terms of its resilience and capacity for generalization. Real-time data augmentation on the original dataset was made possible by the ImageDataGenerator class in the Keras framework. Techniques for augmentation like vertical and horizontal flips were applied to simulate additional variations in the dataset, ensuring its integrity by maintaining the original class distribution. These meticulous data augmentation practices are fundamental in mitigating overfitting, improving model generalization, and ultimately enhancing performance for image tasks. The augmented images used for training are depicted below in Fig. 2.

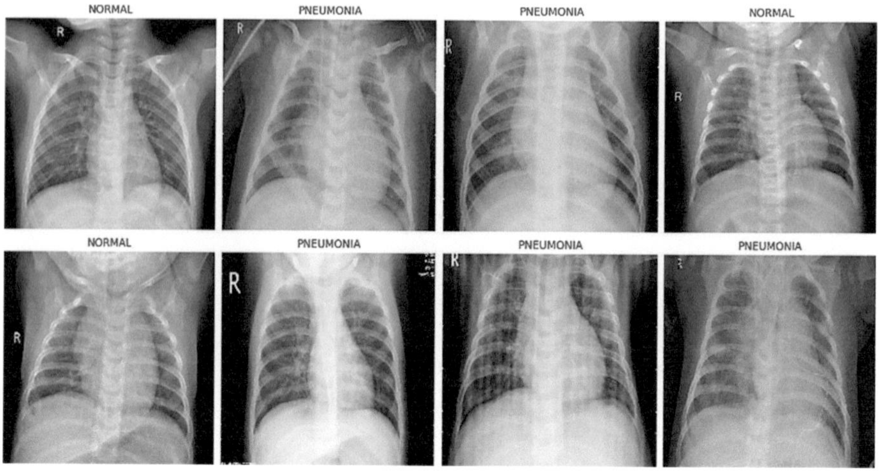

Fig. 2 Pneumonia augmented MRI image

4.3 Training—Testing Separation

The 80–20 rule was used to divide the preprocessed dataset, allocating 80% of the data for training and the remaining 20% for testing. By splitting the dataset in this way, a significant piece of it is utilized for training the model, giving it access to a wide variety of examples, while a separate portion is kept for evaluation, allowing the model to be tested on untested data.

4.4 Proposed Method

The research delved into the realm of binary image classification, assessing the efficacy of various deep learning architectures. Notably, the study examined the performance of three distinct models: VGG-16, DenseNet, and a novel architecture proposed by the researchers. Unlike traditional methodologies that leverage transfer learning by initializing models with pre-trained weights, the researchers chose to train each architecture independently from scratch. This strategy enabled the models to glean insights directly from the dataset, without external influence. Figure 3 illustrates the schematic overview of the model architectures employed in the investigation.

A sophisticated method of using deep learning for medical image analysis is represented by the suggested Convolutional Neural Network (CNN) architecture model for pneumonia identification (see Fig. 4). The model's architecture, which consists of a series of convolutional layers followed by max-pooling processes, makes it easier to extract hierarchical features that are essential for identifying abnormal patterns in chest X-ray pictures. Rectified Linear Unit (ReLU) activation functions are added

Fig. 3 Proposed model

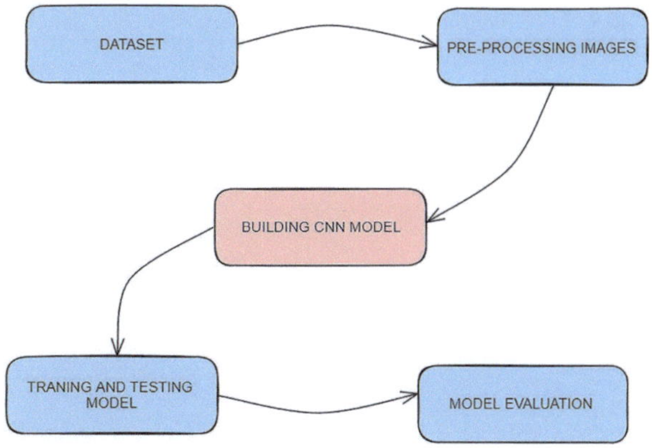

to the model after every convolutional layer, allowing it to take advantage of non-linearities and capture intricate relationships present in the data. Using a kernel size of (3,3) guarantees effective feature extraction while maintaining spatial information, and standardizing the input images to (224, 224, 3) makes them compatible with the image sizes commonly used in deep learning architectures. Additionally, the use of an Adamax optimizer with a learning rate of 0.001 promotes effective parameter updates, which in turn promotes model convergence during training and increases the predictive ability of the model.

The proposed model underwent training for a specific number of epochs, precisely 6 epochs, during which the training progress was meticulously monitored. This training process involved the utilization of both the training dataset (train_X and train_Y) and a validation dataset (valid_X and valid_Y). By leveraging techniques such as transfer learning, data augmentation, and careful model configuration, the model was equipped to generalize effectively on the given task.

By effectively updating the model parameters, the batch size of 16 is chosen to optimize the training process while maintaining gradient smoothness and computing efficiency. Furthermore, the incorporation of dropout layers at a predetermined dropout rate improves the resilience of the model by adding stochasticity, which reduces overfitting and improves generalization abilities. The model's intricacy is highlighted by the careful setup of its hyperparameters and architectural elements. Positioning it as a potent tool for aiding in the early and accurate detection of pneumonia from chest X-ray images.

The following Table 1, shows the CNN model that utilizes a set of hyperparameters to guide its training process. The input images are resized to 224 × 224 pixels with three color channels (RGB). A batch size of 16 is used, meaning the model processes 16 images at a time during training. The learning rate, which controls how much to adjust the model's weights in response to the loss gradient, is set at 0.001. The Adam optimizer is employed for faster and more efficient convergence, and the model is trained over 6 epochs, or complete cycles through the training data. Finally, the Categorical Cross-Entropy loss function is used to assess the model's performance in

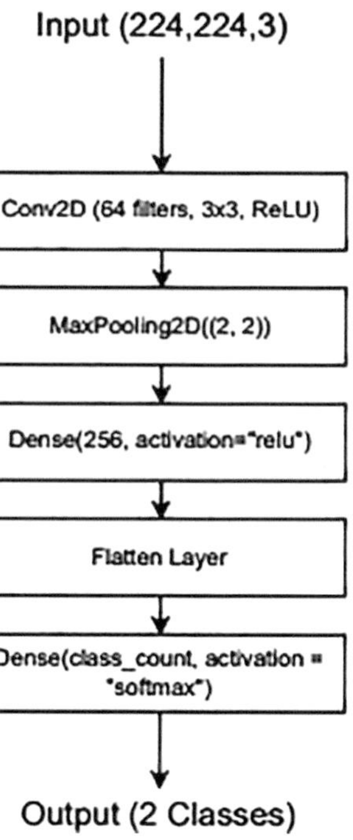

Fig. 4 Flow diagram of CNN model (VGG-16, Xception and proposed)

multi-class classification tasks. Together, these hyperparameters shape the model's learning behavior and impact its overall performance.

The suggested model's meticulous training regimen and intricate architectural design were the main reasons it showed greater accuracy in pneumonia identification. Our suggested model, in contrast to traditional models, was built from the ground up using the particular pneumonia dataset, enabling it to extract characteristics that are directly relevant to the task without being limited by pre-trained weights. This method greatly improved the model's capacity to identify subtle patterns in chest X-ray pictures and to generalize, especially when combined with considerable data augmentation strategies. Moreover, the Adam optimizer and a well-chosen learning rate, two of the architecture's optimized hyperparameters, contributed to improved

Tabel 1 Hyperparameter used in CNN Model

Names	Values
Input size	224 × 224 × 3
Batch size	16
Learning rate	0.001
Optimizer	Adam
Epoch	6
Loss function	Categorical cross entropy

performance. The inclusion of dropout layers played a crucial role in preventing overfitting, ensuring robust and reliable classification despite the inherent variability in medical images. This comprehensive design and training methodology culminated in the proposed model achieving an exceptional accuracy of 95.01%, surpassing other established architectures such as DenseNet and VGG16.

5 Results

With an F1 score of 97%, the suggested model performs exceptionally well in pneumonia identification, with accuracy of 95.01%, recall rate of 94%, and precision of 99%. These measures demonstrate how well the algorithm can categorize cases of pneumonia while reducing false positives, improving diagnostic reliability. By contrast, DenseNet121 obtains an F1 score of 90%, showing a marginally worse balance between precision and recall, with an accuracy of 93%, recall rate of 87%, and precision of 91%. VGG16 has an 88% accuracy, 79% recall rate, and 78% precision, has an F1 score of 85%, indicating that although while it performs less well overall than the suggested model, it is still useful in detecting pneumonia. All things considered, the suggested model is distinguished by its exceptional performance, which establishes it as a reliable pneumonia diagnosis method. The confusion matrices of the suggested model, DenseNet121, and VGG16, are displayed in Figs. 5, 6, and 7, respectively.

Figure 8 shows training and validation loss and Fig. 9 shows the training and validation accuracy of the proposed model.

Tables 2, 3, and 4 contain the complete classification report for individual models of proposed Model, DenseNet121, and VGG16, respectively. Table 5 shows a comparative overview of the comprehensive categorization report.

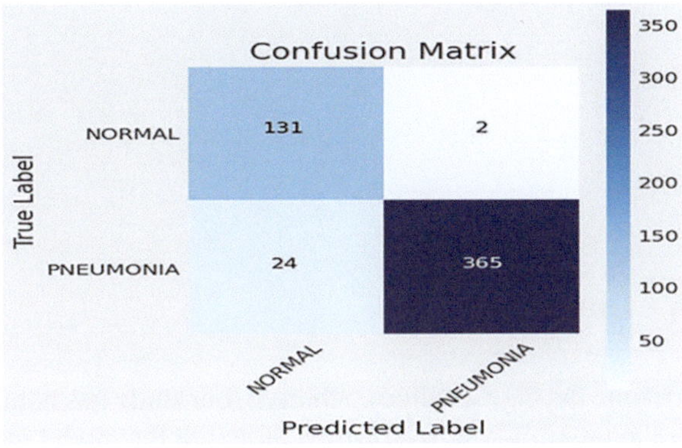

Fig. 5 Confusion matrix of proposed model

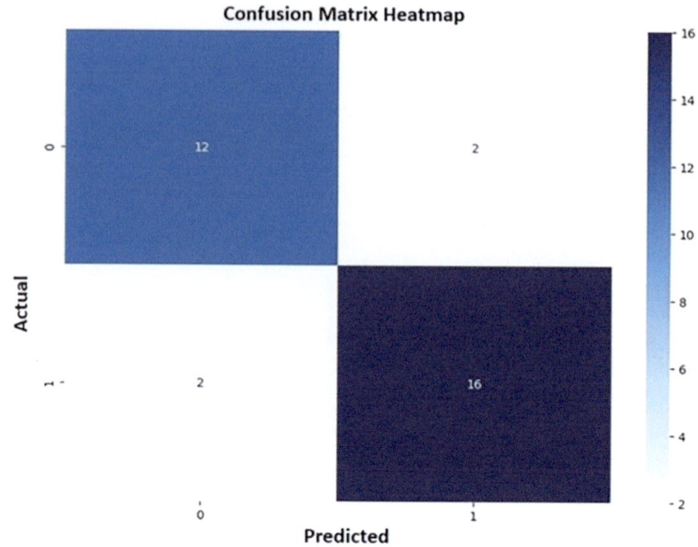

Fig. 6 Confusion matrix of DenseNet121

Fig. 7 Confusion matrix of VGG16

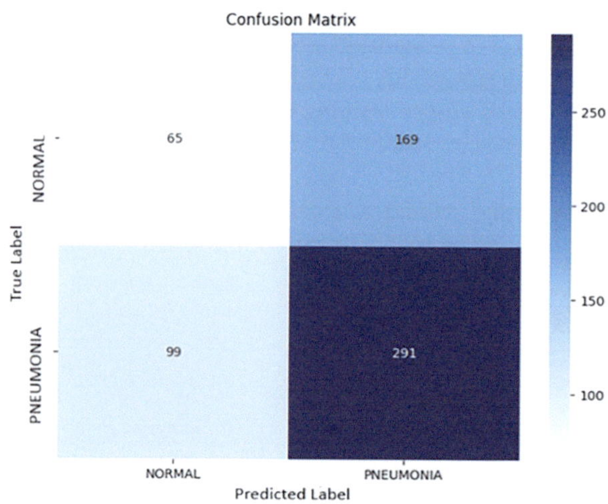

6 Limitations

Despite the high accuracy achieved, our study has notable limitations. One drawback is the possibility of overfitting; although the model might function remarkably well on the training set, it would not handle fresh, untested data effectively, which would diminish its usefulness in practical settings. Another significant limitation is the dependency on image quality; if the X-ray images are of poor quality or contain

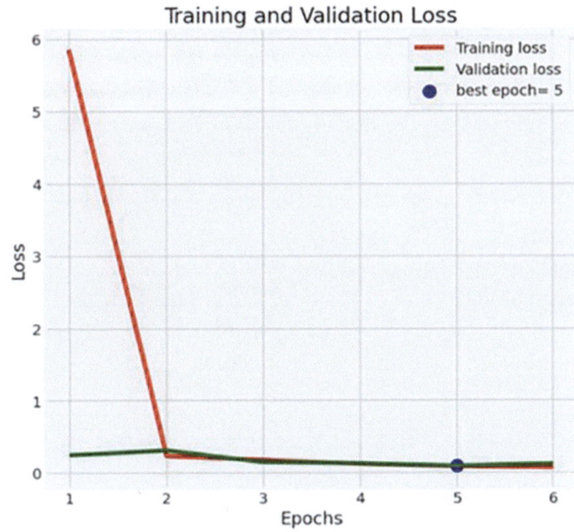

Fig. 8 Training and validation loss of proposed model

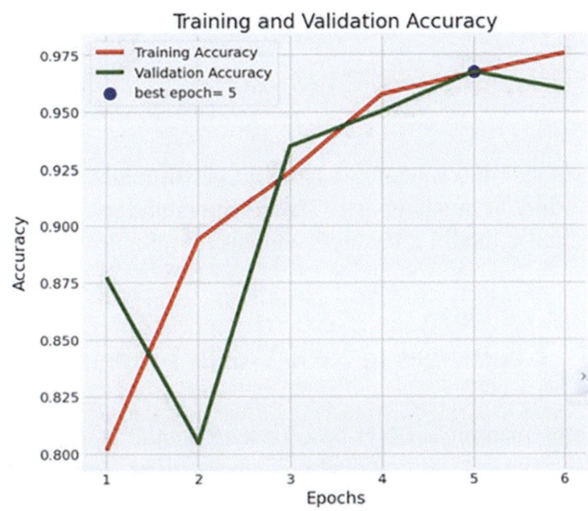

Fig. 9 Training and validation accuracy of proposed model

Table 2 Classification value report of proposed model

Name	Value
Accuracy	95.01
Precision	99.00
F1 score	97.53
Recall	94.00

Table 3 Classification value report of DenseNet121

Name	Value
Accuracy	93.25
Precision	91.52
F1 score	90.00
Recall	87.00

Table 4 Classification value report of VGG16

Name	Value
Accuracy	88.05
Precision	78.83
F1 score	82.00
Recall	79.25

Table 5 Comparison of models used

Models	Precision	Accuracy	Recall
Proposed model	99	95.01	94
DenseNet	91	93	87
VGG16	78	88	79

artifacts, the model's diagnostic performance could be compromised, potentially leading to misdiagnosis. These factors underscore the need for ongoing research to enhance model robustness and ensure reliable clinical outcomes.

7 Challenges in Real-World Implementation

Implementing a CNN-based pneumonia diagnosis system in real-world healthcare presents several practical challenges. First, data quality and variability can significantly affect performance, as X-ray images may vary due to different equipment, protocols, and patient conditions. Ensuring the model can handle this variability is crucial. Second, integrating the model into existing clinical workflows poses logistical hurdles, requiring seamless communication with electronic health records (EHRs) and other diagnostic tools. Additionally, compliance with regulatory standards is essential, as any AI tool in healthcare must undergo rigorous validation to ensure safety and efficacy. Training healthcare professionals to interpret model predictions and incorporate AI insights into their decision-making processes is also necessary. Ultimately, in order to preserve trust and guarantee equitable treatment, ethical issues like algorithmic bias and patient data privacy must be addressed. These challenges underscore the need for a multidisciplinary approach to successfully deploy AI in healthcare settings.

8 Conclusion

In conclusion, our study has demonstrated the remarkable potential of Convolutional Neural Networks (CNNs) in revolutionizing pneumonia diagnosis. Our suggested model surpasses well-known architectures such as DenseNet and VGG16, attaining an exceptional accuracy of 95.01% along with strong recall, precision, and F1 score metrics of 94%, 99%, and 97%, respectively. The high precision of our model (99%) ensures minimal false positives, crucial for accurate diagnosis, while its substantial recall rate (94%) indicates effective identification of true positive cases. These results underscore the transformative impact of CNN-based approaches in medical image analysis, offering a reliable tool for healthcare professionals. Future research may focus on refining the model, exploring alternative architectures, and enhancing interpretability, thus further advancing AI-driven solutions in healthcare and addressing critical challenges in respiratory medicine.

References

1. Rajpurkar P et al (2017) CheXNet: radiologist-level pneumonia detection on chest X-rays with deep learning. arXiv. https://doi.org/10.48550/arxiv.1711.05225
2. Rahman T et al (2020) Transfer learning with deep convolutional neural network (CNN) for pneumonia detection using chest X-ray. Appl Sci 10(9):3233. https://doi.org/10.3390/app10093233
3. Ayan E, Ünver HM (2019) Diagnosis of pneumonia from chest X-ray images using deep learning. IEEE. https://ieeexplore.ieee.org/document/8741582
4. Chouhan V et al (2020) A novel transfer learning based approach for pneumonia detection in chest X-ray images. Appl Sci 10(2):559. https://doi.org/10.3390/app10020559
5. Kundu R, Das R, Geem ZW, Han G-T, Sarkar R (2021) Pneumonia detection in chest X-ray images using an ensemble of deep learning models. PLoS One 16(9):e0256630. https://doi.org/10.1371/journal.pone.0256630
6. Jain R, Nagrath P, Kataria G, Sirish Kaushik V, Jude Hemanth D (2020) Pneumonia detection in chest X-ray images using convolutional neural networks and transfer learning. Measurement 165:108046. https://doi.org/10.1016/j.measurement.2020.108046
7. Sharma H, Jain JS, Bansal P, Gupta S (2020) Feature extraction and classification of chest X-ray images using CNN to detect pneumonia. IEEE. https://ieeexplore.ieee.org/abstract/document/9057809/
8. Zhang D, Ren F, Li Y, Na L, Ma Y (2021) Pneumonia detection from chest X-ray images based on convolutional neural network. Electronics 10(13):1512. https://doi.org/10.3390/electronics10131512
9. Mabrouk A, Díaz Redondo RP, Dahou A, Abd Elaziz M, Kayed M (2022) Pneumonia detection on chest X-ray images using ensemble of deep convolutional neural networks. Appl Sci 12(13):6448. https://doi.org/10.3390/app12136448
10. Cha S-M, Lee S-S, Ko B (2021) Attention-based transfer learning for efficient pneumonia detection in chest X-ray images. Appl Sci 11(3):1242. https://doi.org/10.3390/app11031242

Seeing is Believing: The Impact of Augmented Reality on Customer Shopping Behavior of Lenskart in Bangalore

Noor Firdoos Jahan, K. Abhilash, and Aaron Goes

Abstract The impact of Augmented Reality on sales has been a subject of growing interest, particularly in the retail sector. Augmented Reality technology offers immersive and interactive experiences that can positively influence consumer behavior and purchasing decisions. This study investigates the effects of Augmented Reality (AR) technology on consumer perceptions, engagement, and purchasing behavior at Lenskart stores in Bangalore. The study looks into how augmented reality effects the evaluation of product attributes such as fit, appearance, and suitability, improves the shopping experience, and impacts customer engagement and brand loyalty. The study uses a cross-sectional survey to examine the association between AR usage and shopping behaviors. The findings show that AR dramatically enhances customer perceptions of eyeglass goods, increases interaction, and influences purchase decisions and brand loyalty. The findings demonstrate the expanding role of augmented reality in improving the customer experience in the eyeglasses retail industry.

Keywords Augmented reality · Lenskart · Eyewear · Bangalore · E commerce · Customer shopping behavior

1 Introduction

The eyewear industry in India is experiencing rapid growth, driven by increasing consumer awareness, rising disposable incomes, and technological advancements. The market, expanding at an annual rate of 15–20%, is primarily segmented into spectacles, contact lenses, and sunglasses, with spectacles holding the largest share due to the high prevalence of refractive errors. E-commerce has significantly influenced the industry, offering a wide range of products and competitive pricing, while innovations like anti-glare and blue light protection lenses, along with augmented reality

N. F. Jahan (✉) · K. Abhilash · A. Goes
RV Institute of Management, Bengaluru, India
e-mail: noorfirdoos.rvim@rvei.edu.in

H. Sharma et al. (eds.), *Artificial Intelligence: Theory and Applications*, Lecture Notes in Networks and Systems 5589, https://doi.org/10.1007/978-981-96-1687-9_43

(AR) for virtual try-on's, enhance the shopping experience. Eyewear is increasingly viewed as a fashion accessory, contributing to the growth of the premium segment. The retail landscape includes a mix of organized chains and independent opticians, with major players like Titan Eye Plus and Lenskart expanding their presence. However, challenges such as counterfeit products and limited rural penetration remain. Overall, the industry is set for continued growth, fueled by technological advancements and increasing fashion consciousness. In Bangalore, the eyewear industry is thriving, reflecting the city's tech-savvy and fashion-conscious population. The market is driven by a combination of rising disposable incomes, increasing awareness of eye health, and the prevalence of vision disorders. Bangalore hosts a mix of organized retail chains like Titan Eye Plus, Lenskart, and Lawrence & Mayo, alongside numerous independent opticians, offering a diverse range of products from spectacles to contact lenses and sunglasses. The city's tech-friendly environment has fueled the adoption of online eyewear shopping, with virtual try-on technologies and personalized services becoming popular. Innovations in lens technology, such as blue light protection for digital device users, are particularly relevant in this IT hub. While the market is growing rapidly, challenges like the presence of counterfeit products and the need for more trained eye care professionals persist. Overall, Bangalore's eyewear industry is set for continued expansion, driven by technological advancements and a growing fashion consciousness among consumers.

Lenskart was established in 2010 by Piyush Bansal, and Lenskart has enjoyed significant success. It is the leading eyewear marketer in the Indian market and is growing exponentially. It has accomplished this by introducing a customer-centric strategy driven by innovation. The most striking feature of Lenskart is its virtual try-on technology, which allows customers to see what other glasses styles will look like on them. They are using the virtual reality technique for that. This function makes online shopping more interesting and more personal. Through the incorporation of virtual features into our environment, augmented reality (AR) technology improves our real-world experience. According to popular programs like Pokémon Go, where virtual animals infiltrate our real world, these enhancements are primarily visual [20]. But augmented reality (AR) can also appeal to other senses: taste and smell, for example, through novel technologies like odor detectors and pseudo-gustatory displays; or hearing, through interactive audio AR experiences [21, 33]. AR's importance has been highlighted by recent reports, which often list it among the top 10 technological trends [16, 19]. According to research from Euromonitor International, "phygital reality" will be a major worldwide consumer trend in 2021. It will be important in bridging the gap between the digital and physical worlds in several areas of daily life, such as working, living, and shopping [34]. Interestingly, the study showed that by 2020, half of customers under 45 had already adopted virtual and augmented reality technology, suggesting that younger consumers are becoming more accustomed to and accepting of AR [34].

The COVID-19 epidemic hastened the adoption of virtual technology by forcing people and enterprises to shift to digital platforms for communication and engagement due to lockdowns and social distancing measures [11]. This change strengthened people's dependence on technology even more and accelerated the development of AR-capable gadgets like smartphones and tablets, which are already commonplace in our day-to-day activities. For many people, these devices are still their constant companions, thus shortly, it is anticipated that the importance and use of augmented reality technology will only grow. The retail sector, an extremely busy field where consumers and businesses intersect, is experiencing a disruptive shift powered by Augmented Reality (AR). The widespread adoption of augmented reality (AR) in retail is increasing as technology continues to intertwine itself into our daily existence. Recent statistics indicate that the global AR market for retailing is anticipated to reach an astounding $11.01 billion by 2027, demonstrating the immersive technology's strong momentum and broad adoption [10]. This study examines the real-world effects of augmented reality (AR) on customer shopping behavior, a phenomenon best described by the proverb "Seeing is Believing." Augmented reality (AR), which integrates virtual content into a user's perception of the real world [25] has been identified as a technology with substantial growth potential for marketing and related areas [10, 30]. Prior studies suggest that—compared to traditional visualization formats—AR content is perceived as more interactive [35], inspirational [1, 26, 36], useful [9], and hedonic [7, 14]. AR also facilitates access to the metaverse, which refers to an immersive three-dimensional environment that can integrate physical and virtual worlds [6]. At the same time, AR faces important challenges, such as privacy and security concerns [4, 5], the need to create a visually convincing and accurate integration of virtual objects with the physical environment [28, 31].

Retailers can leverage augmented reality (AR) to enhance customer engagement by allowing customers to virtually experience products before purchasing, which simplifies product evaluation. The current study examines the measurable impact of AR on customer engagement and purchase behavior, focusing on how AR influence sales across different products and customer segments. Key considerations include product attributes such as pricing, rating, appeal, and brand popularity, as well as customers' previous interactions with the product category and online channels. Understanding these factors can help marketing managers strategically implement AR to potentially boost sales by reducing customer uncertainty and enhancing the shopping experience.

The research is so conducted to find out how Augmented Reality Technology affects the customers' perception of product features (fit, style, and suitability), overall shopping experience, engagement with eyewear products, purchase behavior, and brand loyalty at Lenskart in Bangalore. The main goal is to find out whether customers who use AR technology see the product more favorably, whether AR technology makes the shopping experience away from the boring side, whether AR technology stimulates customers to make more purchases, and finally, whether it enhances brand loyalty through purchase behavior. This study is focused on the questions as to what the effects of AR technology on consumer behavior, are and what its consequences for the eyewear retail business can be.

2 Literature Review

According to [29], augmented reality (AR) is an interactive technology that blends the real and virtual worlds by adding virtual features to the original environment. AR does not aim to take the place of the actual world. When using AR in the real world, however, it adds a display that the user sees in addition to a visual display [35]. Research on augmented reality (AR) needs to identify and define the features of AR that will be employed in future, according to [22], Ivan Sutherland started the creation of the AR system in the 1960s [15]. Customers may experience the product as if they are using it thanks to augmented reality virtual try-ons [32]. A camera is needed to take images for AR virtual try-on applications. These images will then be merged with the information you wish to present and displayed simultaneously on the user's screen [15].

Håkansson's thesis [12] introduces AR in marketing, identifying effective patterns for specific products or services). AR's influence [8] on e-commerce, particularly in fashion and beauty, providing insights for enhancing the online shopping experience [3, 27] this study highlighted six valuable themes in understanding consumer motivations in their AR online shopping journey [18]. Spoke about Augmented Reality Marketing-Impact on Intrinsic Motivation and Optimal User Experiences [13].

3 Objectives of the Study

- To assess the impact of Augmented Reality on customers' perceptions of product attributes such as fit, style, and suitability at Lenskart Bangalore.
- To analyze the extent to which Augmented Reality enhances the overall shopping experience for customers in Lenskart Bangalore.
- To investigate the influence of Augmented reality (AR) technology on customer engagement and interaction with eyewear products in Lenskart Bangalore.
- To study the impact of Augmented Reality (AR) on Purchase Behavior and Brand Loyalty of Customers at Lenskart.

4 Hypothesis

- (H01): There is no significant difference in customers' perceptions of product attributes (fit, style, and suitability) between those who use Augmented Reality (AR) and those who do not in Lenskart Bangalore.
- (H02): Augmented Reality (AR) technology has no significant impact on the overall shopping experience for customers in Lenskart Bangalore.
- (H03): Null Hypothesis (H0): Augmented Reality (AR) technology has no significant influence on customer engagement and interaction with eyewear products at Lenskart, Bangalore.

- (H04): Null Hypothesis (H0): Augmented Reality (AR) technology has no significant impact on purchase behavior and brand loyalty of customers at Lenskart, Bangalore.

5 Methodology

5.1 Objective

The study aimed to evaluate the impact of AR technology on customer perceptions and shopping behaviors at Lenskart Bangalore.

5.2 Participants

Sampling Unit: Customers of Lenskart from Bangalore.

Sample size: A cross-sectional survey was administered to 424 customers who visited three randomly selected Lenskart stores in Bangalore using Convenience sampling methods. To ensure data quality, pre-testing was conducted with a pilot sample of 50 customers. Out of the 424 respondents participated in the study, consisting of 229 males (54.01%) and 195 females (45.99%).

5.3 Procedure

Participants were selected using convenience sampling from customers visiting Lenskart Bangalore. A brief overview of the Lenskart AR app's features was presented to the participants. The AR app allows users to virtually try on a wide variety of glasses and sunglasses in real-time. The participants were then asked to use the app to try on different eyewear products virtually. Following their use of the app, participants answered questions on their assessments of the product attributes, overall shopping experience, and purchase behavior.

5.4 Data Collection

Survey Instrument: surveys and questionnaires were used to collect data. Google Forms—an online survey instrument was used for the collection of data. The key sections of the survey, included the questions related to:

- Demographics.
- AR usage experience (ease of use, clarity).
- Product attribute perceptions (Visual Appeal, Functionality, Size and Dimensions, Quality, Fit, Customisation Option) through AR.
- Augmented Reality Influence on Purchase Decision (Clarity and Confidence in Purchase, Style and Suitability, Product Understanding, Engagement, Convenience, Influence on Purchase Intent).
- Augmented Reality Influence on Customer Engagement (Interactivity, Product Exploration, Confidence Building, Engagement).
- Overall shopping experience with AR (Enjoyment, Convenience, Confidence Building, Engagement, Influence on Experience).
- Measurement Scales: Likert scales were used for customer satisfaction and semantic differential scales for product attributes.

5.5 Data Analysis:

The tool used for data analysis is Jamovi. It was used to analyze the following,

- **Descriptive Statistics:** Descriptive statistics are used to summarize demographic data, AR usage experience, and customer perceptions of product attributes and shopping experience.
- **Reliability Analysis:** An assessment of a scale or test's reliability can be made statistically by computing Cronbach's alpha, which is the average inter-correlation between the items.
- **Correlation Analysis:** Correlation analysis is used to explore the relationships between different variables.
- **1-Sample T-Test:** 1-sample t-test is used to prove the hypothesis using the different variables and to assess if the average customer perceptions of AR's impact on product attributes or overall shopping experience are significantly different from a hypothesized value.
- **Chi-Square Test:** A statistical method used to determine whether there is a significant association between categorical variables.

6 Results of the Study

6.1 Descriptive

As per the Table 1, Out of 424 respondents, the majority (52.59) of the respondents are from the age category of 23–30 years of age and 54.01 respondents are males. The age distribution of the respondents is as per the above Table 2.

Table 1 Gender of the respondents

	Gender	No.	%
N	Female	195	45.99
	Male	229	54.01
	Total	424	

Table 2 Age of the respondents

	Age	No.	%
N	18–22	80	18.86
	23–30	223	52.59
	31–38	80	18.86
	39–45	31	7.31
	Above 45	10	2.35

6.2 Reliability Analysis

Cronbach's alpha values were calculated to assess the reliability of the scales used in the study. Cronbach's alpha values for Product Attribute Perceptions ($\alpha = 0.805$), Influence on Purchase Decisions ($\alpha = 0.793$), Influence on Customer Engagement ($\alpha = 0.744$), Influence on Customer Engagement ($\alpha = 0.744$) all showed good reliability, with all the Cronbach's values being above 0.7. These results support the reliability and validity of the measurements used in this research, ensuring the dependability of the findings.

6.3 Correlation Matrix

Table 3 shows us the Correlation between all factors, Functionality and Size/Dimensions ($r = 0.508$), Quality and Size/Dimensions ($r = 0.530$), Fit and Quality ($r = 0.468$). These strong positive relationships indicate that improvements in one aspect (e.g., functionality, quality) are likely to be associated with improvements in other aspects (e.g., size and dimensions, fit). This suggests that enhancing one attribute of the product may have a positive ripple effect on other attributes, leading to an overall improved product experience.

Table 3 Correlation between the various variables

Correlation Matrix							
		H11-visual appeal	H12-functionality	H13-size and dimensions	H14-quality	H15-Fit	H16-customisation
H11-visual appeal	Pearson's r	–					
	df	–					
	p-value	–					
H12-functionality	Pearson's r	0.58	–				
	df	424	–				
	p-value	< 0.001	–				
H13-size and dimensions	Pearson's r	0.345	0.508	–			
	df	424	424	–			
	p-value	< 0.001	< 0.001	–			
H14-quality	Pearson's r	0.363	0.413	0.53	–		
	df	424	424	424	–		
	p-value	< 0.001	< 0.001	< 0.001	–		
H15-fit	Pearson's r	0.312	0.321	0.415	0.468	–	
	df	424	424	424	424	–	
	p-value	< 0.001	< 0.001	< 0.001	< 0.001	–	
H16-customisation	Pearson's r	0.442	0.412	0.329	0.277	0.455	–
	df	424	424	424	424	424	–
	p-value	< 0.001	< 0.001	< 0.001	< 0.001	< 0.001	–

6.4 Hypothesis Testing

Table 4, shows that the statistical analysis, with p-values less than 0.001 and Cohen's d effect sizes ranging from 0.728 to 1.067, strongly rejects the Null hypothesis. Therefore, based on the statistical evidence presented (low p-values and large effect sizes), it is clear that the data supports the conclusion that AR users at Lenskart Bangalore have significantly more positive perceptions of product attributes compared to non-users. Thus, proving the Alternate Hypothesis.

Table 5, shows that the p-values for all attributes are less than 0.001, indicating strong evidence for Alternate Hypothesis. The effect sizes range from 0.86 to 1.15, which are considered large according to Cohen's guidelines. Therefore, the statistical analysis strongly supports the Alternative Hypothesis (H1). Thus, proving that Augmented Reality (AR) technology has a significant impact on the overall shopping experience for customers in Lenskart Bangalore. This conclusion is supported by both the highly significant p-values the large effect sizes.

Table 6, shows that the p-values for all attributes are less than 0.001, indicating strong evidence against the null hypothesis (H0). The effect sizes range from 0.846 to 1.107, which are considered large according to Cohen's guidelines. Therefore, the statistical analysis strongly supports the Alternative Hypothesis (H1). Thus, we can conclude that Augmented Reality (AR) technology has a significant influence on customer engagement and interaction with eyewear products at Lenskart Bangalore. This conclusion is supported by the highly significant p-values and the large effect sizes of the influencing factors such as interactivity, product exploration, confidence building, and overall engagement.

Table 4 Results of the hypothesis 1 testing

							95% confidence interval	
		Statistic	df	p		Effect Size	Lower	Upper
H11-visual appeal	Student's t	18.1	424	< 0.001	Cohen's d	0.88	0.768	0.992
H12-functionality	Student's t	16.5	424	< 0.001	Cohen's d	0.801	0.692	0.91
H13-size and dimensions	Student's t	17.1	424	< 0.001	Cohen's d	0.828	0.718	0.938
H14-quality	Student's t	15	424	< 0.001	Cohen's d	0.728	0.62	0.835
H15-fit	Student's t	19.5	424	< 0.001	Cohen's d	0.948	0.833	1.062
H16-customisation	Student's t	22	424	< 0.001	Cohen's d	1.067	0.948	1.186

Table 5 Results of the hypothesis 2 testing

One sample T-test

		Statistic	df	p		Effect size	95% confidence interval Lower	Upper
H21-clarity and confidence in purchase	Student's t	20	424	< 0.001	Cohen's d	0.969	0.853	1.084
H22-style and suitability	Student's t	18.8	424	< 0.001	Cohen's d	0.915	0.801	1.028
H23-product understanding	Student's t	17.7	424	< 0.001	Cohen's d	0.86	0.748	0.971
H24-engagement	Student's t	20	424	< 0.001	Cohen's d	0.972	0.856	1.087
H25-convenience	Student's t	21.2	424	< 0.001	Cohen's d	1.03	0.912	1.148
H26- influence on purchase intent	Student's t	23.7	424	< 0.001	Cohen's d	1.15	1.027	1.272

Note H_a $\mu > 3$

Table 6 Results of the hypothesis 3 testing

One sample T-test

		Statistic	df	p		Effect size	95% confidence interval Lower	Upper
H31-interactivity	Student's t	22.8	424	< 0.001	Cohen's d	1.107	0.986	1.228
H32-product exploration	Student's t	21.3	424	< 0.001	Cohen's d	1.036	0.918	1.154
H33-confidence building	Student's t	17.4	424	< 0.001	Cohen's d	0.846	0.735	0.956
H34-influence on engagement	Student's t	20.9	424	< 0.001	Cohen's d	1.017	0.9	1.134

Table 7, shows that the p-values for all attributes are less than 0.001, indicating strong evidence against the null hypothesis (H0). The effect sizes range from 0.963 to 1.198, which are considered large according to Cohen's guidelines. Therefore, the statistical analysis strongly supports the Alternative Hypothesis (H1). Thus, we can conclude that Augmented Reality (AR) technology has a significant impact on purchase behavior and brand loyalty of customers at Lenskart Bangalore. This conclusion is supported by the highly significant p-values and the large effect sizes

Table 7 Results of the hypothesis 4 testing

One sample T-test								
							95% confidence interval	
		Statistic	df	p		Effect Size	Lower	Upper
H41 - enjoyment	Student's t	21.7	424	< 0.001	Cohen's d	1.055	0.936	1.17
H42-convenience	Student's t	19.8	424	< 0.001	Cohen's d	0.963	0.847	1.08
H43-confidence building	Student's t	20	424	< 0.001	Cohen's d	0.969	0.853	1.08
H44-engagement	Student's t	22.8	424	< 0.001	Cohen's d	1.109	0.987	1.23
H45-influence on experience	Student's t	24.7	424	< 0.001	Cohen's d	1.198	1.073	1.32

Note H_a $\mu > 3$

of the influencing factors, customer enjoyment, convenience, confidence building, engagement, and overall experience at Lenskart Bangalore, thereby impacting their purchase behavior and brand loyalty.

7 Conclusion

This study explores the impact of Augmented Reality (AR) on customer shopping behavior at Lenskart Bangalore. It aims to provide quantifiable insights into how AR enhances the retail experience and influences consumer perceptions, engagement, and purchase decisions. The results of our study are consistent with previous research, demonstrating the validity of AR technology and its influence on the customer experience at Lenskart Bangalore. The reliability of our scales was further supported by the fact that Cronbach's alpha values (0.744–0.805) agreed with the findings from related studies conducted by [23] and [17]. Our results are further validated by significant positive correlations and substantial impact sizes, which are consistent with the findings of research paper by [24] and [2]. All things considered, augmented reality technology greatly improves consumer interaction, impacts buying decisions, and elevates perceptions of products, demonstrating its potential as a game-changing instrument in retail. By integrating virtual elements into the shopping process, AR offers immersive and interactive experiences that significantly improve the overall retail purchasing experience. The research identifies key performance indicators (KPIs) that demonstrate AR's effect on consumer behavior and sales revenue, offering valuable guidance for businesses seeking to leverage AR technology.

References

1. Arghashi V, Yuksel CA (2022) Interactivity, inspiration, and perceived usefulness! How retailers' AR-apps improve consumer engagement through flow. J Retail Consum Serv 64:102756. https://doi.org/10.1016/j.jretconser.2021.102756
2. Bonetti FW (2018) Augmented reality and virtual reality in physical and online retailing: a review, synthesis and research agenda. In: Augmented reality and virtual reality. Springer
3. Chandra S, Kumar KN (n.d.) Integration of AR in digital business environments, highlighting its potential to overcome challenges and provide a competitive edge. J Electron Commer Res
4. Chen SL (2018) A case study of security and privacy threats from augmented reality (AR). In: International conference on computing, networking and communications (ICNC), Maui, HI. Retrieved from https://doi.org/10.1109/ICCNC.2018.8390291
5. Cowan KJ (2021) Privacy concerns when using augmented reality face filters? Explaining why and when use avoidance occurs. Psychol Mark 38:1799–1813. Retrieved from https://doi.org/10.1002/mar.21576
6. Dwivedi YK et al (2022, 2023) Metaverse beyond the hype: multidisciplinary perspectives on emerging challenges, opportunities, and agenda for research, practice and policy. Int J Inf Manage 66. Retrieved from https://doi.org/10.1016/j.ijinfomgt.2022.102542
7. Flavián C, Ibáñez-Sánchez S, Orús C (2021) User responses towards augmented reality face filters: implications for social media and brands. Augmented reality and virtual reality. Springer
8. Gabriel A, Ajriya AD, Nabila Fahmi CZ, Handayani PW (2023) The influence of augmented reality on e-commerce: a case study on fashion and beauty. Cogent Bus Manage 10(2)
9. Gatter S, Hüttl-Maack V, Rauschnabel PA (2022) Can augmented reality satisfy consumers' need for touch? Psychol Mark 39:508–523. Retrieved from https://doi.org/10.1002/mar.21618
10. Grand Vision Research (2022) Augmented reality market size, share & trends analysis report, 2022–2030. https://www.grandviewresearch.com/industry-analysis/augmented-reality-market
11. Hacker J, Vom Brocke J, Handali J, Otto M, Schneider J (2020) Virtually in this together–how web-conferencing systems enabled a new virtual togetherness during the COVID-19 crisis. Eur J Inf Syst 29(5):563–584. https://doi.org/10.1080/0960085x.2020.1814680
12. Håkansson L (2018) Virtual and augmented reality in marketing
13. Harika M, Setiadi R, Nurul AD, Zulfa DA (2022) AR make-up filter for social media using the HSV color. JOIN (Jurnal Online Informatika)
14. Hilken T, de Ruyter K, Chylinski M, Mahr D, Keeling DI (2017) Augmenting the eye of the beholder: exploring the strategic potential of augmented reality to enhance online service experiences. J Acad Mark Sci 45:884–905. Retrieved from https://doi.org/10.1007/s11747-017-0541-x
15. Hung S-W, Chang C-W, Ma Y-C (2021) A new reality: exploring continuance intention to use mobile augmented reality for entertainment purposes. Technol Soc 67. Retrieved from https://doi.org/10.1016/j.techsoc.2021.101757
16. Insights SB (2020) 5 retail technology trends to watch in 2020. Retrieved from https://insights.samsung.com/2020/01/03/5-retail-technology-trends-to-watch-in-2020/
17. Javornik A (2016) 'It's an illusion, but it looks real!' consumer affective, cognitive and behavioral responses to augmented reality applications. J Mark Manag 32(9–10):987–1011
18. Kumar S, Rangarajan P, Vivek M (2024) Augmented reality marketing-impact on intrinsic motivation and optimal user experiences. Association for Information Systems
19. Marr B (2020) The top 10 technology trends in retail: how tech will transform shopping in 2020. Forbes
20. N Xi, Hamari J (2019) Does gamification satisfy needs? A study on the relationship between gamification features and intrinsic need satisfaction. Int J Inf Manage 46. Retrieved from https://doi.org/10.1016/j.ijinfomgt.2018.12.002
21. Narumi T (2011) Augmented reality flavors: gustatory display based on edible marker and cross-modal interaction. In: Proceedings of the international conference on human factors in computing systems. https://doi.org/10.1145/1978942.1978957

22. Nikhashemi SR (2021) Augmented reality in smart retailing: A(n) (A) symmetric approach to continuous intention to use retail brands' mobile AR apps. J Retail Consum Serv 102464. Retrieved from https://doi.org/10.1016/j.jretconser.2021.102464
23. Pantano E, Servidio R (2012) Modeling innovative points of sales through virtual and immersive technologies. J Retail Consum Serv 19(3):279–286
24. Poushneh A, Vasquez-Parraga AZ (2017) Discernible impact of augmented reality on the retail customer's experience, satisfaction and willingness to buy. J Retail Consum Serv 34:229–234
25. Rauschnabel PA (2019) Augmented reality marketing: how mobile AR-apps can improve brands through inspiration. J Retail Consum Serv 49(7):43–53. https://doi.org/10.1016/j.jretconser.2019.03.004
26. Rauschnabel PA (2021) Augmented reality is eating the real-world! The substitution of physical products by holograms. Int J Inf Manage 57. https://doi.org/10.1016/j.ijinfomgt.2020.102279
27. Reddy SR (2024) Virtual vogue: unravelling the impact of augmented reality on consumer. Auckland University of Technology
28. Slocum CR (2021) Reality check: a tool to evaluate spatial inconsistency in augmented reality. In: IEEE international symposium on multimedia (ISM), Naple, Italy. Retrieved from https://doi.org/10.1109/ISM52913.2021.00018
29. Smink AR (2019) Try online before you buy: how does shopping with augmented reality affect brand responses and personal data disclosure?. Electron Commer Res Appl 35:100854. Retrieved from https://doi.org/10.1016/j.elerap.2019.100854
30. Statista (2022) Mobile augmented reality (AR) market revenue worldwide from 2021 to 2026. Retrieved from https://www.statista.com/statistics/282453/mobile-augmented-reality-market-size/
31. Tom Dieck MC (2018) A theoretical model of mobile augmented reality acceptance in urban heritage tourism. Curr Issues Tourism 21. Retrieved from https://doi.org/10.1080/13683500.2015.1070801
32. Vieira VA (2022) Augmented reality generalizations: a meta-analytical review on consumer-related outcomes and the mediating role of hedonic and utilitarian values. J Bus Res 151(11). Retrieved from https://doi.org/10.1016/j.jbusres.2022.06.030
33. Ward E (2020) Building a unique brand identity: measuring the relative ownership potential of brand identity element types. J Brand Manag 27(2):393–407. https://doi.org/10.1057/s41262-020-00187-6
34. Westbrook G, Angus A (2021) Top 10 global consumer trends 2021. Euromonitor International. Retrieved from https://go.euromonitor.com/white-paper-EC-2021-Top-10-Global-Consumer-Trends.html
35. Yim MY (2017) Is augmented reality technology an effective tool for e-commerce? An interactivity and vividness perspective. J Interact Mark 39(1):89–103. https://doi.org/10.1016/j.intmar.2017.04.001
36. Zanger VM (2022) Beyond the gimmick: how affective responses drive brand attitudes and intentions in augmented reality marketing. Psychol Mark 39(7). Retrieved from https://doi.org/10.1002/mar.21641

Inception V3+CNN A Hybrid CNN Model for Lung Disease Detection Using Chest X-rays

Anushree Raj, M. Yogitha, and S. Deepthi

Abstract In recent years, the arena of health imaging and image processing has witnessed remarkable advancements, particularly in the domain of lung disease detection. Medical imaging modalities such as X-rays, computed tomography (CT), and MRI are few indispensable tools for diagnosing various respiratory conditions. Image processing procedures play a vital role in enhancing the diagnostic capabilities of medical imaging. By applying algorithms and computational methods to analyze and interpret images, researchers and clinicians can extract valuable quantitative data, identify subtle abnormalities, and differentiate between healthy and diseased tissues. Many infected persons may be protected from lung disease if it is detected early. Chest X-rays and computed tomography (CT) pictures are the most widely utilized diagnostic techniques for lung disorders. Based on information from chest X-rays, this paper proposes a multi-classification DL model for identifying lung diseases. This article examines the performance of four architectures: VGGNet16+CNN, ResNet-152+CNN, EfficientNet-B7+CNN, and LungNet22. Through the use of publicly available digital chest X-ray datasets separated into three categories: Lung cancer, pneumonia, and normal, a comprehensive evaluation of numerous deep-learning architectures is provided. The experiment results show that the InceptionV3 plus CNN model outperforms the other three proposed models. The InceptionV3+CNN model obtained 97.55% accuracy, 97.43% sensitivity, 98.21% specificity, 98.07% $F1$-score, and 97.55% precision based on X-ray images.

Keywords CNN · Chest X-ray · InceptionV3 · Hybrid CNN model · Lung disease detection

A. Raj (✉) · M. Yogitha · S. Deepthi
Mangalore Institute of Technology and Engineering, Mangalore, India
e-mail: anushree@mite.ac.in

M. Yogitha
e-mail: yogitha@mite.ac.in

H. Sharma et al. (eds.), *Artificial Intelligence: Theory and Applications*, Lecture Notes in Networks and Systems 5589, https://doi.org/10.1007/978-981-96-1687-9_44

1 Introduction

The detection of lung illnesses by image processing is becoming a more significant application of advanced technology in modern healthcare, as it enables early diagnosis and effective treatment. Lung cancer, pneumonia, TB, COPD, interstitial lung diseases (ILDs), and lung cancer are only a few of the ailments that fall under the umbrella phrase "lung diseases." Accurate diagnosis and early detection are key to improving patient outcomes and reducing the fatality rates associated with these diseases. Medical pictures, such as CT, MRI, and X-rays, can be effectively and non-invasively analyzed using image processing techniques to identify any anomalies that can indicate lung problems. These techniques need a number of processes, including feature extraction, pre-processing, classification, and image capture.

Feature extraction in the context of lung disease identification includes locating pertinent patterns and attributes from the pre-processed images, like the density, texture, and form of lung lesions or nodules. These characteristics are used as input by algorithms that classify the existence of lung disorders and differentiate between various disease kinds or stages. To further hone these methods, boost their efficacy and accuracy, and incorporate them smoothly into clinical practice, more research and development is needed. The suggested study project discovers a cutting-edge method.

2 Literature Review

Farhan and Yang et al. [1] proposed a brand-new Hybrid Deep-Learning Algorithm (HDLA) architecture for automatically classifying lung diseases from chest X-ray pictures has been put forth [1]. Pre-processing, spontaneous feature extraction, and discovery are all included in the framework. The method of pre-processing enhances the quality of unprocessed chest X-ray images, and the robust CNN is employed to automatically extract lung features. Robust feature learning with effective 1D feature estimates is ensured by the 2D CNN model. According to the experimental data, the suggested model outstrips state-of-the-art techniques with effect to total accuracy by 3.1% and computational complexity reduction by 16.91%.

Ishan et al. [2] suggested that this study use ML method to explore and forecast lung illnesses. The study makes use of 19 variables and 323 instances of data from patients with different lung illnesses. Patients are classified as either "Positive" or "Negative" according to the lung diseases attribute. The K-Fold Cross-Validation Technique and five ML algorithms were used to train the dataset. These algorithms' respective accuracy percentages were 88.00%, 88.92%, 90.15%, 89.23%, and 83.69%.

Tripathi et al. [3] suggested that the study investigate how hospitalized medical imaging is used to characterize lung illnesses. It is suggested to use a DCNN to categorize disorders of the chest. With the chest X-ray 14 dataset, which is made available to the public, the model obtains an average multiclass classification accuracy of 89.77%. The method works best for categorizing multiclass medical photos according to various thoracic disorders.

Yan et al. [4] introduced CoroNet is a deep-learning model designed to identify and diagnose COVID-19 infections using chest X-ray images. CNN is the architecture that CoroNet uses. Xception is a pre-trained model used by CoroNet. Pre-training entails using a sizable dataset of generic images (such as ImageNet) to train the model. CoroNet's ability to recognize basic image elements through "pre-learning" paves the way for its eventual adaptation to the analysis of chest X-rays. CoroNet is further trained on a carefully selected dataset of chest X-rays after it has been pre-trained. CoroNet gains knowledge about the distinct characteristics in X-rays that set COVID-19 apart from other illnesses through this fine-tuning. The incorporation of XAI techniques into CoroNet is a critical feature. This makes it possible for medical practitioners to comprehend the reasoning behind the model's predictions.

Zuo et al. [5] core concept involves training three independent 2D CNN models. Each CNN features Convolutional layers to excerpt attributes from the CT-scan images. Pooling layers to reduce image dimensionality and capture dominant features. Fully connected layers to learn complex relationships between extracted features and the presence/absence of lung nodules. A crucial aspect is ensuring these three CNNs have some architectural diversity. This can be achieved by varying factors. The number and size of convolutional filters. The type of pooling layers used (e.g., max pooling, average pooling).

Nidhin Raju et al. [6] proposed that image dispensation techniques are crucial in the medical field for improving lung disease detection and treatment in big data environments. These techniques identify genetic and environmental factors, enabling early detection and treatment stages. The quality, time, and precision of the dataset are key factors in modifying and evaluating image quality. Different segmentation techniques are used to analyze medical images, identifying patterns and retrieving information for processing. This study discusses various image processing techniques and big data analytics tools for lung disease, focusing on lung disease images.

Yimer et al. [7] suggested that the accuracy of diagnosing ILD based solely on clinical history is currently less than 20%. This work presents the SmallerVGGNet deep-learning CNN architecture, which uses HRCT pictures to categorize ILD categories from 17 different categories. With a 95% accuracy rate during training, the system can identify ILD in its early stages and improve patient care. To boost accuracy, the system can be further improved by shrinking photos and adding more filters.

Shao et al. [8] proposes a framework that utilizes pre-trained convolutional neural networks (CNNs) like AlexNet and VGG16 for a multiclass classification task: Differentiating between various lung diseases using chest X-ray and CT scans. The approach makes use of pre-trained CNN models, such as VGG16 and AlexNet. These models have strong feature extraction skills because they were trained on large image datasets (such as ImageNet). This framework makes use of transfer learning rather than training a CNN from the beginning. The final layers of the pre-trained

CNN, which are usually in charge of the first classification task (e.g., object identification in ImageNet), are eliminated. The pre-trained network's remaining layers serve as feature extractors. A production layer that creates probabilities for every illness class using a softmax algorithm. The complete model, comprising both the previously trained and freshly added layers, is subsequently refined using a dataset of CT or X-ray images of the chest that have been specifically labeled with the lung illnesses of interest (COVID-19, pneumonia, healthy). This fine-tuning process adjusts the weights of the network to become proficient in classifying these specific lung conditions.

Singh et al. [9] explore a method for automated LDC using chest X-ray images. It combines the power of deep-learning for feature abstraction with the flexibility of traditional ML classification algorithms. The approach utilizes a pre-trained 2D-CNN. Similar to the multiclass classification approach with pre-trained CNNs, this leverages a pre-trained model's ability to identify relevant patterns within chest X-ray images. The pre-trained CNN likely extracts high-level features which might undergo further pre-processing steps like dimensionality reduction or normalization before feeding them into the classification stage. Unlike typical deep-learning approaches that use fully connected layers for classification, this method utilizes traditional machine learning algorithms for the final disease classification task.

A modified CNN model is suggested by Shamrat et al. [10] for the classification of lung illnesses from frontal chest X-ray images. The program uses 80,000 X-ray pictures as input and gathers data from several sources to classify 10 illness types related to the lungs. The study utilized eight pre-trained CNN models, of which VGG16 achieved the greatest accuracy of 92.95%. Built on top of the VGG16 model, LungNet22 is a refined model that was assessed with the Adam Optimizer. With the Adam Optimizer, the model's accuracy of 98.89% is impressive. The study shows how picture enhancement, balancing, and processing methods can raise the model's efficiency. Future research could propose a hybrid model to categorize different lung disorders or utilize the model to classify other diseases using images from CT scans. Cross-validation can be used to validate the assessment results.

Farhan et al. [2] suggested HDLA model for automatic lung disease classification follows a number of important steps in its methodology. Every chest X-ray image is first pre-processed using the best filtering and contrast correction methods to improve quality. To extract important features, this pre-processed image is fed into a reliable CNN model—more precisely, a pre-trained ResNet50. To solve problems like time-consuming training and local optima, the resulting feature vectors are optimized by the application of the min-max scaling technique. Ultimately, the optimized features are subjected to a variety of classifiers, includes Adaboost, RM, SVM, BNN, and DNN. Using a 10-Fold Cross-Validation Technique, the model is trained and evaluated using the Covid-19 Radiography Database and chest X-ray images for pneumonia datasets, split 70% training and 30% testing.

ML methods are commonly applied in the healthcare industry. The potential for data mining to uncover patterns concealed in vast amounts of information that will be utilized to clinical diagnosis is enormous. The important problem is the detection of lung illness. In order to find the likelihood of lung disease occurring before it actually does, we therefore carried out an investigation supported by machine learning algorithms. The sole goal of this analysis is to determine the accuracy levels of Bagging, Logistic Regression, and Random Forest algorithms, and to integrate this leads to such a system that is user-friendly so that people can use it whenever they feel convenient.

Ishan Sen et al. [3] proposal of machine learning techniques incorporating data mining has a lot of potential for uncovering patterns hidden in massive amounts of data that will be utilized for clinical diagnosis. One of the most important issues is lung disease detection, and many researchers have investigated the likelihood of lung disease developing before it really does. The only goal of this analysis is to determine the accuracy levels of the Bagging, Logistic Regression, and Random Forest algorithms, and to integrate this leads to a system that is user-friendly so that people can use it whenever they feel convenient.

Tripathi et al. [6] proposed a method for the automatic detection and classification of lung diseases such as tuberculosis (TB), pneumonia, and lung cancer involves several key stages. Initially, image pre-processing is performed to remove irrelevant data from chest X-ray films, which includes steps like image enhancement and filtering to improve contrast and remove noise. Following this, lung segmentation is carried out to identify the lung boundaries using intensity and discontinuity-based methods. The next step is feature extraction, where both geometrical features (such as area, perimeter, equivalent diameter, and irregularity index) and statistical features (like mean, standard deviation, and entropy) are derived from the segmented lung images.

Nidhin Raju et al. [11] proposed a hybrid deep-learning model called HDLA with the goal of automatically classifying lung diseases from chest X-ray pictures. In order to minimize processing time, this model uses 2D CNNs layers with ideal feature sizes for feature extraction and classification. Pre-processing is done on raw X-ray pictures to guarantee accurate feature estimate. Various classifiers are used on CNN results to overcome issues such as excessive processing effort and the vanishing gradient. In order to increase classification accuracy, the model uses classifiers including Adaboost, RM, SVM, BNN, and DNN in addition to the pre-trained ResNet50 for feature extraction.

The planned system aims to enhance the prediction and monitoring of lung diseases, especially amidst challenges posed by COVID-19, by improving accuracy, reducing processing time, and minimizing space requirements.

Shamrat et al. [12] proposed LungNet22, methodology employed in this paper likely involves several key steps. Initially, a comprehensive dataset of X-ray images of lungs would have been compiled, likely containing examples of various lung diseases. Pre-processing methods like normalization, resizing, and augmentation may have been applied to enhance the quality and diversity of the dataset. Subsequently, a DL architecture, possibly based on CNNs, would have been chosen or customized for the task at hand. The model would then undergo a process of fine-tuning, involving training on the dataset with iterative adjustments to optimize accuracy, sensitivity, and specificity. Finally, the trained model would undergo rigorous evaluation through techniques like cross-validation to assess its generalizability and effectiveness in classifying lung diseases accurately. The resulting "LungNet22" model likely represents a significant advancement in the field of medical image analysis, offering potential utility in diagnosing and managing lung diseases with greater efficiency and accuracy.

3 Methodology

The proposed InceptionV2+CNN model to detect lung disease was implemented in various stages. Figure 1 gives a flow of phases showing the approach used followed by the description.

3.1 Data Collection

The chest X-ray data was gathered by wide range of medical pictures from many sources, such as public databases, research institutes, and hospitals, for lung disease identification. Every image provided unique information on the structure and pathophysiology of the lungs. The images were annotated those which defined ROI and categorized them based on lung disorders, such as lung cancer or pneumonia. During

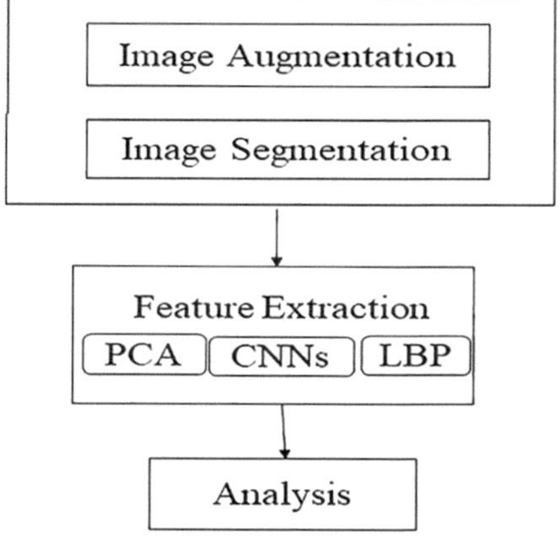

Fig. 1 Methodology of the projected work

the entire gathering process, ethical considerations—including patient privacy regulations—were considered. Altogether 20000 data trials were used for analysis out of which 12857 were lung infected samples and 7143 are normal images.

3.2 Data Pre-processing

The important step in diagnosing lung disorders is pre-processing medical images. Its aim is to improve the raw image data's quality and analytical value. First, median filtering or Gaussian blurring is used to minimize intrinsic noise while maintaining the clarity of anatomical features. Contrast enhancement methods such as adaptive histogram equalization and histogram equalization alter pixel brightness to improve the appearance of minor details. Normalization creates uniform pixel values across images, which facilitates more consistent comparison and analysis. Moreover, artifact removal algorithms identify and minimize distortions caused by motion or equipment limitations in order to maintain diagnostic data. Consistent voxel sizes are ensured by resampling, which increases computation efficiency and facilitates additional analysis. Segmentation techniques, which offer targeted analysis and feature extraction, are used to define the lung area. These pre-processing techniques enhance and optimize medical images, giving reliable algorithms that identify lung disorders and enhance clinical judgment and patient outcomes a solid foundation.

3.3 Image Augmentation

Refining datasets for lung disease diagnosis requires the use of picture augmentation, which improves the robustness and generalizability of machine learning models. The variety of the dataset is increased by applying adjustments to pre-existing photographs, such as rotation, flipping, scaling, and noise addition, to create augmented examples with distinct appearances. These consider real-world scenarios, including different patient orientations, imaging conditions, and anatomical variations. Augmentation serves to lessen the effects of class imbalance and data scarcity by increasing the number of samples available for underrepresented classes. This improves the model's ability to recognize unusual anomalies and extrapolate results to other patient populations. The pre-processing pipeline that uses picture augmentation enhances the performance of lung disease detection models by generating larger and more representative datasets. When augmented data is included, the model is more adaptable to changes in patient anatomy and imaging conditions. It is also more adept at spotting minute patterns and anomalies in medical images.

3.4 Image Segmentation

In order to separate lung areas from the background in lung image analysis, image segmentation is an essential first step. The three often employed methods are: Thresholding: Quick and easy to use, but restricted in complexity and noise-sensitive; best suited for photos with a strong contrast between the lung and background. Area Growing: Requires careful selection of seed points and inclusion criteria for pixels, but manages complex forms better and is less noise-sensitive than thresholding cantered on deep learning. The most precise and adaptable method is segmentation, particularly for complex lung shapes and picture quality fluctuations. However, it may not be interpretable, need a lot of training data, and be computationally expensive.

3.5 Image Classification

Medical images are analyzed by machine learning algorithms that categorize them into distinct groups based on whether or not specific lung diseases are present. Image classification is a vital job in the diagnosis of lung diseases The CNN gains the ability to map pictures to the appropriate illness labels by extracting discriminative features from the images during training. After training, the model can reliably identify unseen pictures, giving medical personnel important information for diagnosis. To improve image quality and eliminate noise, the dataset is first pre-processed. Following several layers of convolutional and pooling processes, fully connected layers make up the CNN, which receives the pre-processed pictures. The CNN's output layer then generates probability scores for each class, which are utilized to determine the final classification decision. All things considered, [13] image classification is an effective method for identifying lung diseases. It allows for automated examination of medical pictures and helps medical practitioners diagnose patients quickly and accurately. Image classification models may reach high levels of accuracy and generalization by utilizing deep-learning techniques like as CNNs. This can lead to better patient treatment and outcomes in the field of pulmonary medicine.

3.6 Model Performance Analysis

An important component of assessing the dependability and efficacy of machine learning models for lung disease diagnosis is model performance analysis. To evaluate how effectively the model works with unknown data and how well it can categorize medical pictures, this investigation looks at a number of metrics. Raj and Pallavi [14] Accuracy is a popular performance evaluation indicator that counts the proportion of properly identified photos in the total number of images. To assess the model's capacity to identify true positives (lung disease patients accurately detected) and true negatives (healthy cases correctly identified), respectively, sensitivity and specificity are also critical parameters. The model's capacity to minimize false positives and false negatives a critical balance for clinical decision-making is better understood with the use of these indicators.

4 Results

To unravel the contribution of our proposed method from the various architecture, Table 1 compares the InceptionV3+CNN performance of different architecture.

Evaluation metric:

Accuracy (Ac): Percentage of correctly classified cases (healthy, disease) formula (1).

$$\text{Ac} = (\text{True Positive} + \text{True Negative})/(\text{All Measures on Confusion matrix}) \quad (1)$$

Precision (Pr): Proportion of true positives among predicted positives indicates the efficiency of the model at r actual cases, referred to formula (2).

$$\text{Pr} = \text{True Positive}/(\text{True Positive} + \text{False Positive}) \quad (2)$$

Recall/Sensitivity (Rc): Proportion of true positives identified by the model indicates the ability of a model to capture all actual cases, referred to formula (3).

$$\text{Rc} = \text{True Positive}/(\text{True Positive} + \text{False Negative}) \quad (3)$$

Sensitivity (Sn): It is the number of correct positive predictions divided by the total number of positives referred to formula (4).

$$\text{Sn} = \text{True Negative}/(\text{True Negative} + \text{False Positive}) \quad (4)$$

$F1$-score (Fs): Harmonic-mean of precision and recall, combining both metrics, referred to formula (5).

$$\text{Fs} = 2 * (\text{Pr} + \text{Rc})/(\text{Pr} + \text{Rc}) \quad (5)$$

Table 2 shows the comparison of various evaluation metrics of proposed architecture with other existing models.

Figure 2 represents the visual findings for the data as in Table 2 toward the comparative performance metrics of proposed work with other existing techniques.

We first compared the accuracy of our lung disease detection with VGGNet16+CNN [12], ResNeT-152+ CNN. [13], EfficientNet-B7+CNN. [14], LungNet22 [10], respectively. The outcomes demonstrate that the suggested method performs better than the other choices. With 84.37%, EfficientNet-B7+CNN had the

Table 1 Comparative analysis of various models with proposed model

S. No.	Method/techniques	Description	Results	Conclusive marks (future scope)
1	Investigating lung diseases with machine learning algorithms [2]	Objective: To predict lung diseases using machine learning algorithms, focusing on common conditions like asthma, allergies, COPD, bronchitis, emphysema, and lung cancer	Achieved accuracies: Bagging (88.00%), Logistic Regression (88.92%), Random forest (90.15%), logistic model tree (89.23%), Bayesian networks (83.69%)	Demonstrates machine learning's potential in proactive healthcare, enabling timely interventions to improve patient outcomes
2	VGGNet16+CNN [11]	Utilizes VGGNet16 architecture for ILD classification from HRCT images. Critical drawbacks include limited training data and lengthy training time. Suggests potential for early stage ILD detection and future improvements in image processing	Achieved 95% accuracy in ILD category classification. Limited training data and lengthy training time are critical drawbacks. Variation in results observed based on data availability per class	Highlights potential for early stage ILD detection and suggests future improvements in image processing techniques for enhanced accuracy
3	ResNeT-152+ CNN [12]	A novel framework combining deep-learning and other methods for lung disease classification from chest X-ray images	Improved overall accuracy by 3.1% and reduced computational complexity by 16.91% compared to state-of-the-art methods	Demonstrates the effectiveness of combining deep-learning and machine learning for improved lung disease detection from X-ray images

(continued)

Table 1 (continued)

S. No.	Method/techniques	Description	Results	Conclusive marks (future scope)
4	EfficientNet-B7+CNN [15]	Utilizing deep convolutional neural networks for classifying chest diseases from medical images, particularly the chest X-ray 14 dataset	Average accuracy of 84.37% showcases the model's efficacy in classifying multiclass medical images	Highlights the potential of deep-learning in healthcare and its ability to improve disease classification accuracy from medical images
5	LungNet22 [10]	Proposes LungNet22, a fine-tuned VGG16-based model, for classifying lung diseases from X-ray images. Utilizes image processing techniques and augmentation to enhance dataset	Achieved 98.89% testing accuracy with LungNet22 model. Utilized Grad-CAM for additional insights	Highlights potential of image processing and fine-tuning techniques in improving model accuracy for lung disease classification. Suggests future applications in categorizing diseases from CT-scan images or hybrid models
6	Proposed InceptionV3+CNN	Proposed work is A custom CNN, specifically designed for your image processing task, is built on top of the pre-trained EfficientNet-B3	Achieved an accuracy of 98.73% for well-defined datasets. Proves to be a promising and powerful tool in lung disease detection	Developing custom CNN architectures specifically designed for different tasks on top of EfficientNet-B3 could improve performance and efficiency further

Table 2 Comparison of metrics of proposed work with other existing work

	Evaluation metrics				
CNN architecture	Accuracy	Sensitivity	Precision	$F1$-score	Specificity
EfficientNet-B7+CNN	84.37	84.71	85.73	83.68	83.21
VGGNet16+CNN	95.21	94.32	94.82	93.33	94.04
ResNeT-152+ CNN	89.12	90.61	87.77	89.17	88.96
LungNet22	98.89	97.52	97.47	98.48	96.78
InceptionV3+CNN	98.73	97.43	97.55	98.07	98.21

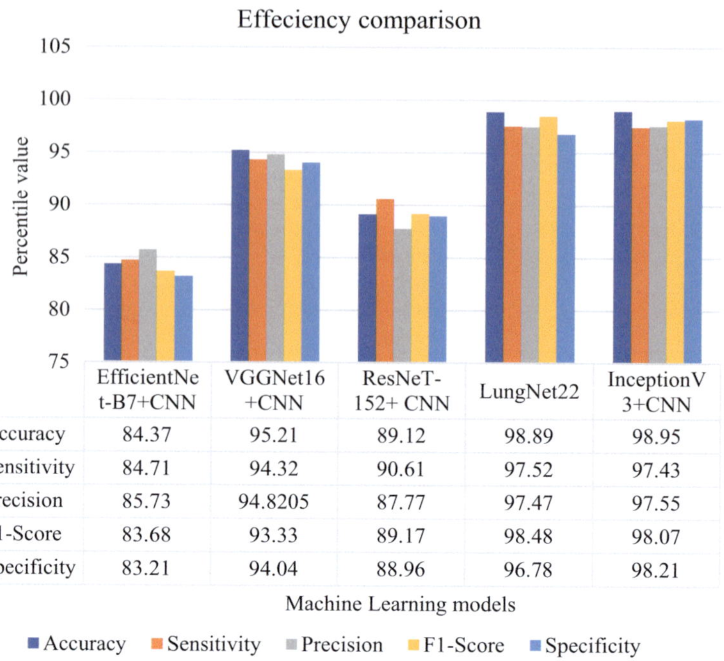

	EfficientNet-B7+CNN	VGGNet16 +CNN	ResNeT-152+ CNN	LungNet22	InceptionV3+CNN
Accuracy	84.37	95.21	89.12	98.89	98.95
Sensitivity	84.71	94.32	90.61	97.52	97.43
Precision	85.73	94.8205	87.77	97.47	97.55
F1-Score	83.68	93.33	89.17	98.48	98.07
Specificity	83.21	94.04	88.96	96.78	98.21

Fig. 2 Performance metrics for various models

lowest AC. With 89.12%, the ResNeT-152+ CNN model came in second lowest, followed by the VGGNet16+CNN model with 95.21% and the LungNet22t model with 98.89%. With 98.95% accuracy, the created InceptionV3-CNN model achieved the highest score. While our model has a score of 98.89%, the average accuracy score when comparing is 93.31%. The accuracy metric experiences an average performance gain of 5.58% as a consequence.

5 Conclusion and Future Scope

Through the analysis of medical images such as X-rays or CT scans, these methods can detect subtle abnormalities indicative of various lung conditions, including pneumonia, tuberculosis, and lung cancer. By leveraging advanced algorithms and ML models, such as CNNs, these systems can achieve high accuracy and efficiency in identifying abnormalities, potentially aiding healthcare professionals in making timely and accurate diagnoses. Though image processing has a lot of promise, it should be used in conjunction with clinical examination and other diagnostic procedures as part of an all-encompassing diagnostic strategy. Furthermore, more research and development are required to improve these methods' sensitivity and specificity, make them more accessible and affordable for general clinical use, and polish them even further. In the end, employing image processing technologies to detect lung diseases could transform early diagnosis, enhance patient outcomes, and lessen the burden of lung-related illness and death.

Future developments could greatly benefit from the application of image processing for the diagnosis of lung diseases. First off, there's a big chance for new technologies like AI and DL to be assimilated. These developments may result in detection algorithms that are even more accurate and dependable, able to manage large, complicated data sets and a wide range of patient demographics. Furthermore, there is an increasing demand for the creation of affordable and portable imaging systems, especially in environments with limited resources. Simplifying the procedures for acquiring and analyzing images could greatly increase patients' access to timely interventions and early diagnosis everywhere. Furthermore, as research continues to uncover novel biomarkers and imaging modalities, such as functional imaging and molecular imaging, the scope of image processing in lung disease detection is likely to expand, enabling clinicians to achieve deeper insights into disease mechanisms and tailor personalized treatment approaches. Overall, the future of image processing in lung disease detection is characterized by ongoing innovation, collaboration, and the potential to revolutionize medical diagnostics on a global scale.

References

1. Farhan AMQ, Yang S (2023) Automatic lung disease classification from the chest X-ray images using hybrid deep-learning algorithm. Multimed Tools Appl 82:38561–38587. https://doi.org/10.1007/s11042-023-15047-z
2. Ishan S, Hossain M, Shakib F, Imran A, Faisal F (2020) In depth analysis of lung disease prediction using machine learning algorithms. https://doi.org/10.1007/978-981-15-6318-8_18
3. Tripathi S et al (2021) Lung disease detection using deep-learning. Int J Innov Technol Explor Eng 10(8):1–10
4. Yan L, Qin H, Luo B, Tang S, Gao Y, He T et al (2020) A Deep-learning framework with Bayes' theorem for COVID-19 2020 classification using chest X-ray images. Pattern Recognit Lett 134:165–171
5. Zuo W, Zhou F, Li Z (2023) Deep-learning ensemble 2D CNN approach towards the detection of lung cancer. Nature 13(1):1–9. Gupta YK, Saroj Agrawal (2021) In: 2021 IOP Conference Series: Materials Science Engineering, vol 1022, p 012030
6. Nidhin Raju AHB, Augustine P (2020) Identification of interstitial lung diseases using deep-learning. Int J Electr Comput Eng (IJECE) 10(6):6283–6291
7. Yimer F, Tessema AW, Simegn GL (2021) Multiple lung diseases classification from chest X-ray images using deep-learning approach. Int J 10:2936–2946
8. Shao H, Zheng Y, Tang Z, Wang Y, Wu H, Liu J (2022) Multi-class classification of lung diseases using CNN models. J Med Imaging Health Inform 12(6):2208–2217
9. Singh A, Singh V, Singh G (2023) Automatic lung disease classification from the chest X-ray images using hybrid deep learning algorithm. Multimed Tools Appl 1–17
10. Shamrat FMJM et al (2022) LungNet22: a fine-tuned model for multiclass classification and prediction of lung disease using X-ray images. J Pers Med 12(5):680
11. Huang S, Lee F, Miao R, Si Q, Lu C, Chen Q (2020) A deep convolutional neural network architecture for interstitial lung disease pattern classification. Med & Biol Eng & Comput 58(4):725–737. https://doi.org/10.1007/s11517-019-02111-w

12. Ravi V, Acharya V, Alazab M (2023) A multichannel EfficientNet deep learning-based stacking ensemble approach for lung disease detection using chest X-ray images. Cluster Comput 26:1181–1203. https://doi.org/10.1007/s10586-022-03664-6
13. Raj A, Pallavi MO (2023) Comparative analysis on breast cancer prediction using machine learning techniques. In: Ranganathan G, El Allioui Y, Piramuthu S (eds) Soft computing for security applications. ICSCS 2023. Advances in intelligent systems and computing, vol 1449. Springer, Singapore. https://doi.org/10.1007/978-981-99-3608-3_26
14. Raj A, Pallavi MO (2023) Deep-learning based application in detecting wrinkle and predicting age. In: 2023 International conference on intelligent and innovative technologies in computing, electrical and electronics (IITCEE), Bengaluru, India, May 2023, pp 1168-1173. https://doi.org/10.1109/IITCEE57236.2023.10090987
15. Almujally NA, Aljrees T, Saidani O, Umer M, Faheem ZB, Abuzinadah N, Alnowaiser K, Ashraf I (2023) Monitoring acute heart failure patients using internet-of-things-based smart monitoring system. Sensors 23:4580. https://doi.org/10.3390/s23104580

Statistical Approaches for Segmentation of Abdominal Aorta Aneurysm Using Medical Image

Rashmi Benni, Akash Gurusiddappa Bhagoji, Swayam Kalal, and Deepa Mulimani

Abstract Abdominal aorta aneurysm is one of the harmful cardiovascular diseases which is characterized by a widening of the aorta, the body's major artery, in the lower region of the body. The diagnosis and treatment of abdominal aortic aneurysms (AAAs) require precise segmentation due to the significant medical risk and demands for accurate segmentation for diagnosis and treatment. This study presents an enhanced method for AAA segmentation from medical imaging data that makes use of statistical techniques, where statistical algorithms offer distinct advantages over traditional methods due to their ability to iteratively refine estimates and address complex optimization challenges, specifically Expectation Maximization (EM) and Majorization-Minimization (MM). To enhance segmentation, EM refines estimates iteratively in order to maximize the likelihood function, whereas MM approaches large optimization problems by breaking them down into simpler subproblems, resulting in reliable performance. This study highlights the significant improvement in AAA segmentation accuracy that these statistical methods bring, leading to increased diagnosis accuracy and better patient outcomes. This study makes use of EM and MM and automates the segmentation process of AAA image, harnessing the capabilities of EM and MM algorithms. This research aims to significantly enhance segmentation accuracy and diagnostic precision, thus contributing to improved patient outcomes in modern healthcare practices. Additionally, this improves patient care by streamlining workflows in medical environments, which benefits the healthcare industry as overall.

Keywords Abdominal aorta aneurysm (AAA) · Statistical algorithms · Expectation Maximization (EM) · Majorization-Minimization (MM) · Differential medical image · Segmentation

Authors Akash Gurusiddappa Bhagoji, Swayam Kalal, and Deepa Mulimani contributed equally to this work.

R. Benni (✉) · A. G. Bhagoji · S. Kalal · D. Mulimani
KLE Technological University, Hubballi, Karnataka, India
e-mail: rashmi.benni16@gmail.com

D. Mulimani
e-mail: deepamulimani19@kletech.ac.in

H. Sharma et al. (eds.), *Artificial Intelligence: Theory and Applications*, Lecture Notes in Networks and Systems 5589, https://doi.org/10.1007/978-981-96-1687-9_45

1 Introduction

Medical image segmentation is crucial as it enables precise identification and delineation of anatomical structures and abnormalities within medical images [1]. This accuracy is essential for diagnosis, treatment planning, and monitoring of diseases, such as tumors or aneurysms [2]. By facilitating the extraction of critical features from complex imaging data, segmentation improves the efficacy of medical interventions and supports the advancement of personalized medicine [3, 4]. Furthermore, it aids in automating and standardizing radiological workflows, enhancing overall healthcare efficiency. In today's healthcare, segmenting medical imaging is essential, particularly for disorders like abdominal aorta aneurysm (AAA). If AAA regions are not identified and treated quickly, it might result in potentially fatal consequences due to the abnormal expansion of the abdominal aorta, a severe cardiovascular disease. Meanwhile, the symptoms of AAA are abdominal pain, back pain, and dizziness. For accurate diagnosis, planning, and monitoring of medical treatments, detailed segmentation of AAA in images is crucial. The diverse structures, noise, and artifacts present in medical imaging make this task complex (Fig. 1).

Medical image segmentation is necessary for precise diagnosis and therapy. Various statistical techniques are employed to achieve reliable results. Suitable for cases involving high contrast, thresholding transforms grayscale images into binary images based on intensity. Area is starting from seed points, and the growth algorithm adds surrounding pixels with similar attributes and handles noise and intensity variations well. K-means clustering segments complex structures by grouping pixels based on their intensity. The Watershed algorithm, which treats grayscale images

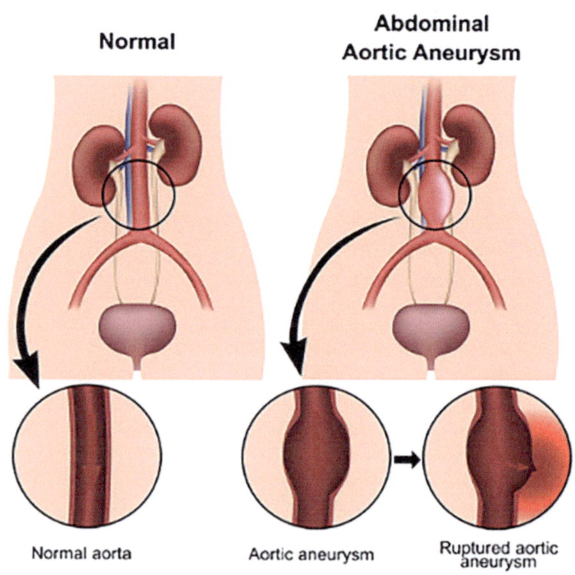

Fig. 1 Normal versus abdominal aorta aneurysm

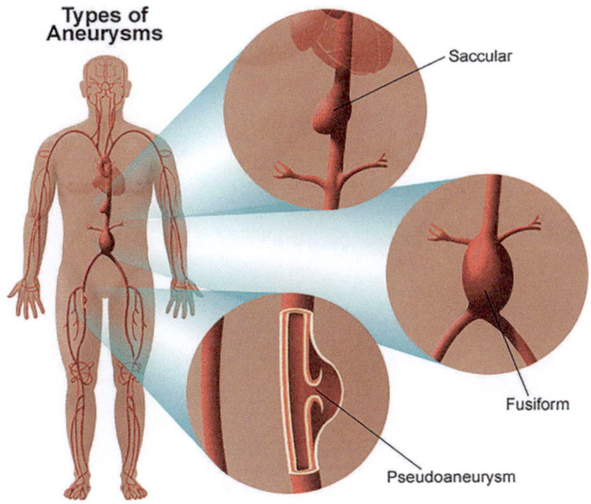

Fig. 2 Types of aneurysm

as topographic surfaces, identifies ridges to separate overlapping structures, such as cells. Active contour models (snakes) precisely define complex structures by using curves modified by both internal and external forces. Every technique has tailored advantages for various segmentation problems (Fig. 2).

Saccular, fusiform, and pseudoaneurysms are three different kinds of arterial flaws that, in the context of the abdominal aorta, have distinct forms of disorders that, when left untreated, may cause serious risks to one's health. Asymmetrical, sac-like bulging on one side of the artery is an early sign of saccular aneurysms in the abdominal aorta, which are frequently caused by localized weak points in the arterial wall. In comparison with fusiform aneurysms, which cause a symmetrical, spindle-shaped enlargement that covers the whole circumference of the vessel, these aneurysms are very frequent. The most typical types of abdominal aortic aneurysms (AAAs) are fusiform aneurysms, which are usually brought on by atherosclerosis and persistent hypertension, both of which damage the artery wall diffusely. A pseudoaneurysm in the abdominal aorta occurs when blood flows through an opening in the artery wall and is caught by the surrounding tissue, forming a hematoma that connects to the arterial lumen. Trauma, surgical procedures, or diseases that weaken the artery wall can cause this. These aneurysms can appear in an array of ways, but until they grow to a sizable size or burst, they frequently cause no signs or symptoms. In order to determine the size and shape of the aneurysm, imaging techniques like ultrasound, CT scan, or MRI are commonly used in the diagnosis process. The article evaluates the application of statistical methods to medical imaging data in the segmentation of abdominal aortic aneurysms (AAAs), with a particular emphasis on two algorithms: Majorization-Minimization (MM) and Expectation Maximization (EM). These algorithms, which provide reliable frameworks for the complex problem of medical image segmentation, are based on statistical theory. A popular technique for estimating maximum likelihood in situations with missing data is the

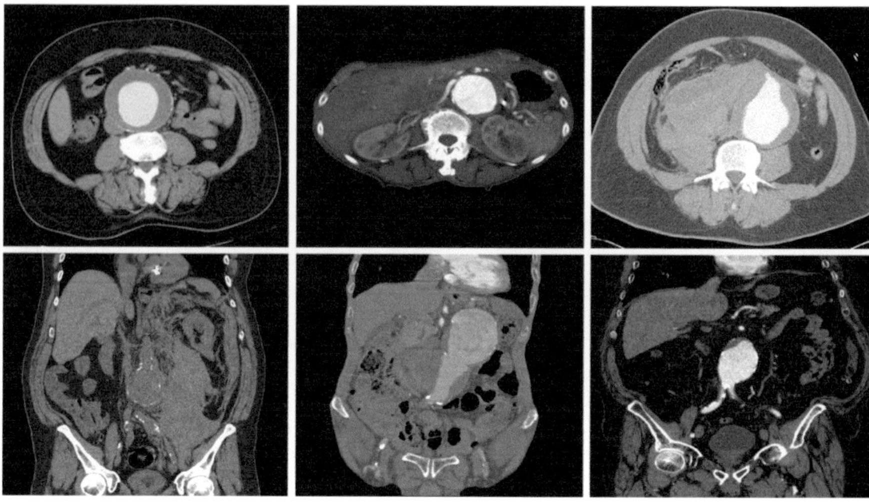

Fig. 3 MRI images of abdominal aorta aneurysm

Expectation Maximization (EM) algorithm. There are two iterative steps in it: the Maximization step (M-step) and the Expectation step (E-step). The M-step modifies these estimates to maximize the log-likelihood, whereas the E-step computes the expected value of the log-likelihood function based on current parameter estimations. Until convergence, this iterative approach successfully handles noise and uncertainty in medical picture segmentation, resulting in more precise AAA delineation. On the other hand, the Majorization-Minimization (MM) approach utilizes subproblem splitting to address large-scale optimization issues. Next, the objective function is successively minimized by creating a surrogate function that upper limits it. Assuring dependable and effective convergence, every iteration guarantees a solution that is at least as excellent as the prior one. Medical image segmentation benefits greatly from the accuracy and efficiency that the MM algorithm provides by handling huge, complicated datasets.

This paper presents a systematic structure for the research process, starting with the initial research in Sect. 2, followed by the authors' methods and experimental design in Sect. 3, results and observations in Sect. 4, main conclusions and implications, and future scope in Sect. 5, highlighting potential avenues for further research and advancement in the area (Fig. 3).

2 Background and Related Work

Image segmentation, a crucial step in medical image analysis, facilitates the visualization and analysis of human tissues, alleviating the need for manual delineation by experts, which can be time-consuming [5]. While existing rough-fuzzy clustering algorithms offer effective segmentation, they often overlook spatial distribution, relying solely on pixel distribution for class labeling [6]. To get over this limitation, a novel spatially constrained rough-fuzzy c-means (sRFCM) method is provided. By combining local neighborhood information with rough-fuzzy clustering, sRFCM improves segmentation accuracy by allowing nearby pixels to affect the center pixel's identification. Furthermore, the algorithm partitions clusters into possibilistic core and probabilistic boundary regions, with a neighborhood regularizer and weight parameter determining cluster centroids. Brain MR volumes and HEp-2 cell pictures are used to illustrate how the neighborhood Silhouette, a new segmentation validity index, helps with parameter optimization for sRFCM.

Because data-driven techniques, especially deep learning, can develop statistical models directly from data, they have become more and more popular in the field of medical imaging in recent years [7]. These approaches offer advantages over traditional model-based methods but face challenges in robustness, generalization, and interpretability. Despite their superior performance, issues such as generalization to new data and interpretability remain open challenges [8]. However, these approaches hold promise for developing optimized imaging pipelines tailored to specific clinical needs, potentially revolutionizing medical imaging workflows [9, 10]. Prostate cancer detection and segmentation pose significant challenges due to variations in occurrence rates across prostate regions and the subjective nature of manual segmentation. Algorithms for automatic segmentation have become popular to help physicians, particularly when radiologists are not available [5, 11]. An in-depth analysis of machine learning and deep learning-based prostate MRI segmentation techniques shows the advancements, challenges, and potential for this field of research, emphasizing the need of accurate segmentation for early cancer detection and treatment planning [12].

Vertebral tumor prediction and lumbar spinal segmentation are critical for clinicians in tumor analysis, requiring precise and reliable segmentation methods [13]. Analytical Transform: To overcome the difficulties in accurately detecting and segmenting vertebral tumors, the Assisted Statistical Characteristic Decomposition Model (ATS-CDM) is suggested. In comparison with current methodologies, our method offers superior prediction ratios and ROC grading for lumbar spines by repeatedly identifying tumor regions and measuring region variance [14]. Moreover, probabilistic segmentation techniques provide flexible instruments for a range of medical imaging uses, such as form analysis and segmentation. The left atrium and the anatomy of the brain can be modeled well using a suggested technique for estimating multi-label probabilistic maps [15]. This approach performs better in shape-based clustering, multi-atlas segmentation, and Bayesian picture segmentation tasks and increases accuracy and generalizability through the use of smoothness priors and constrained optimization.

Finally, automated techniques for quantifying the left ventricle (LV) using cardiac magnetic resonance (CMR) images have been developed, which simplifies diagnosis processes and lessens the workload for medical professionals [16]. Segmentation-based and direct regression approaches to LV quantification are compared, and the results show promise for both. Evaluation of submissions to the LVQuan challenge highlights the effectiveness of these methods, paving the way for improved automatic cardiac quantification techniques for clinical applications [17].

3 Proposed Work

The proposed work is a comparative analysis of several statistical techniques, comprising Expectation Maximization (EM) and Majorization-Minimization (MM) algorithms, for evaluating the segmentation of abdominal aorta aneurysms (AAAs) via medical imaging. The iterative process and comprehensive hyperparameter tuning aim to improve accuracy in segmentation and decision-making processes. The statistical approaches for segmenting AAA using EM and MM algorithms reduce errors and inaccuracies by leveraging probabilistic models to adaptively classify tissue types. The EM algorithm iteratively refines parameter estimates, ensuring precise segmentation, while the MM algorithm's flexibility in handling intensity distributions minimizes misclassification, leading to more accurate differentiation of aneurysms from surrounding tissues. The "MRI scans of AAA" dataset, available on Kaggle, boasts an extensive collection of over 5863 MRI scans specifically targeting AAA. Where 4101 images for testing and 1241 for training in the ratio of 70:30 were used. These scans are meticulously organized into distinct classes, delineated by specific diagnostic criteria, offering researchers and medical practitioners a comprehensive insight into the pathology. Its availability facilitates cutting-edge research endeavors, empowering the development of innovative approaches in medical imaging and healthcare.

3.1 System Architecture

A systematic approach comprising statistical methods ensures precise and reliable outcomes in the segmentation of abdominal aorta aneurysms leveraging medical imaging. The process begins with the collection of N MRI images of the abdominal aorta aneurysm. To enhance the quality of images and pave the way for future research, these images undergo pre-processing using Bias Field Correction (BFC) and statistical techniques for noise reduction. The pre-processed images are subsequently put through statistical methods like Expectation-Maximization (EM) and the Majorization-Minimization (MM). The hyperparameters are initialized and adjusted according to the segmented task's particular needs. The segmentation process defines parameters and specifications to evaluate results. Segmentation solutions are visualized as clusters, with each indicating a separate region of MRI scans. The algorithm evaluates whether the required conditions are met, enabling accurate segmentation. The segmented image generates a count of clusters, yielding a quantitative evaluation of the result of segmentation. The optimum solution will be used to display the

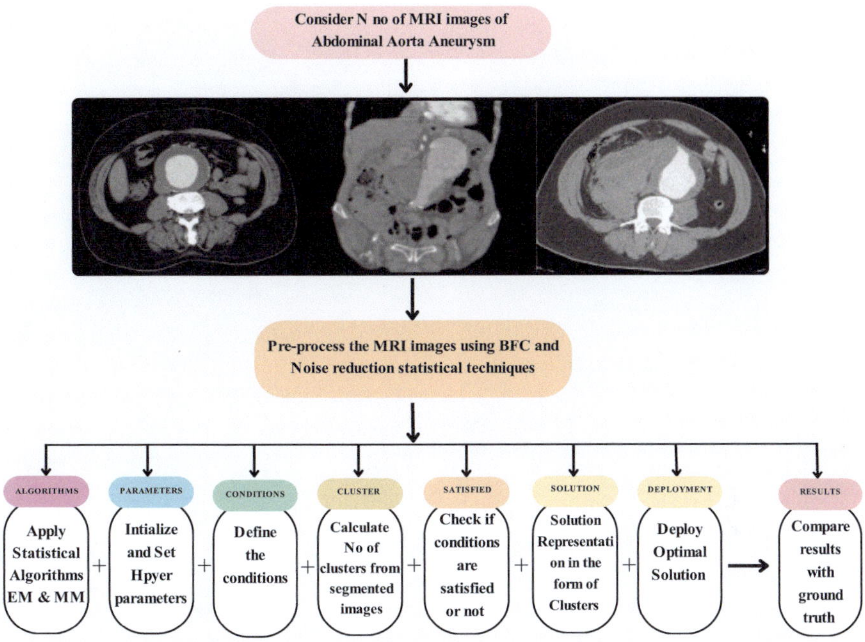

Fig. 4 System architecture

ultimate segmented images. To assess the precision and efficacy of the segmentation method, the segmented image will be compared to ground truth data. Through the application of statistical approaches, this painstaking process guarantees accurate and thorough segmentation of the abdominal aorta aneurysm, paving the way for future research and improved treatment and diagnostics (Fig. 4).

3.2 *The Expectation-Maximization (EM)*

The Expectation-Maximization (EM) Fig. 5 procedure iteratively computes model parameters from hidden variables. The E-step predicts hidden variable values using observed data, while the M-step utilizes these estimates to modify model parameters. This process continues until convergence that optimizes the likelihood of observed data within the model. The Expectation-Maximization (EM) algorithm starts by initializing possible aneurysm regions in medical images to segment abdominal aortic aneurysms (AAAs). It then iteratively decides if each voxel belongs to the aneurysm or the background according to intensity, appearance, and features. The contribution and novelty of this approach lie in its robust statistical framework, allowing precise and adaptive segmentation of AAAs, enhancing the differentiation of aneurysms from

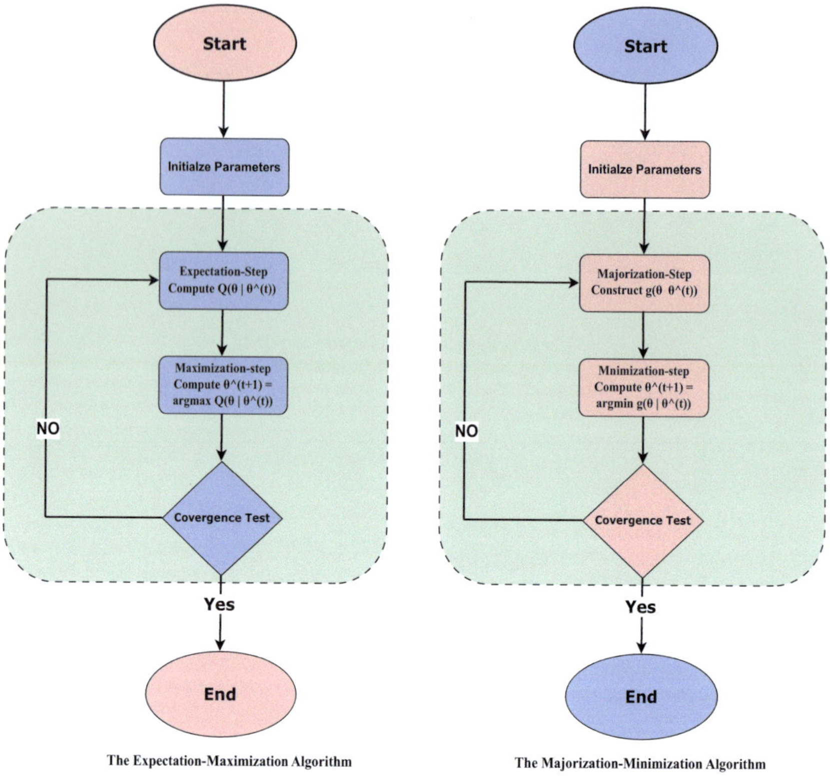

Fig. 5 Flowchart of EM and MM algorithms

surrounding tissues. Post-processing techniques may further improve segmentation accuracy.

1. **Initialization**:

    - Set initial parameters: $\theta^{(0)}$.

    $\theta^{(0)}$ represents the initial guess or starting point for the parameters of the model.

2. **Iterative Steps** (Repeat until convergence):

    (a) **E-step**:

    - Estimate $Q(\theta|\theta^{(t)}) = E_{Z|X,\theta^{(t)}}[\log p(X, Z|\theta)]$.

    $Q(\theta|\theta^{(t)})$ represents the expected value of the log-likelihood function given the observed data X and the current parameter estimate $\theta^{(t)}$. This expectation is taken with respect to the conditional distribution of the latent variables Z given the observed data and the current parameter estimate.

(b) **M-step**:

- Update parameters: $\theta^{(t+1)} = \arg\max_{\theta} Q(\theta|\theta^{(t)})$.

  $\theta^{(t+1)}$ represents the updated parameter estimate obtained by maximizing the function $Q(\theta|\theta^{(t)})$ with respect to the parameters θ.

(c) **Convergence Check**:

- Stop if $\|\theta^{(t+1)} - \theta^{(t)}\| < \epsilon$ or max iterations reached.

  ϵ represents a small positive threshold used to determine convergence. If the change in parameter estimates between consecutive iterations falls below this threshold, the algorithm is considered to have converged.

3. **Output**:

- Final parameter estimates: $\hat{\theta}$.

  $\hat{\theta}$ represents the estimated parameters obtained after convergence of the algorithm. These parameters are considered to be the maximum likelihood estimates given the observed data X.

3.3 The Majorization-Minimization Algorithm

The Majorization-Minimization (MM) Fig. 5 algorithm iteratively optimizes large objective functions through developing and minimizing simple substitute functions (majorizations). This technique enables modifications to parameters until convergence, making MM algorithms highly effective in solving difficult, non-convex optimization problems. They provide a solid framework for enhancing parameter estimations or solutions through iterative refinement. The MM approach facilitates the complex task of segmenting abdominal aorta aneurysms from medical images through the development of alternative functions using intensity gradients or border smoothness. MM iteratively optimizes these substitutes, modifying parameters such as thresholds and shape priors until convergence situations are satisfied, stabilizing parameters, or achieving segmentation accuracy thresholds. MM's ability to handle non-convex data makes it helpful in precisely segmenting anatomical structures. MM algorithm leverages probabilistic models to classify voxels based on intensity distributions, providing flexibility in capturing complex tissue variations. Its novelty lies in its ability to adapt to diverse anatomical structures without requiring extensive initialization.

1. **Objective Function**:
   - Consider an objective function $f(\theta)$ that is difficult to directly optimize.
2. **Surrogate Function (Majorization)**:
   - Construct a surrogate or majorization function $g(\theta \mid \theta^{(t)})$ such that:

$$f(\theta^{(t)}) \geq g(\theta \mid \theta^{(t)}) \quad \forall\theta$$

   This surrogate function simplifies the optimization problem by being easier to optimize than $f(\theta)$.

3. **Iterative Minimization**:
   (a) Minimization Step: Update the parameter $\theta^{(t+1)}$ by minimizing $g(\theta \mid \theta^{(t)})$:

$$\theta^{(t+1)} = \arg\min_{\theta} g(\theta \mid \theta^{(t)})$$

4. **Convergence Check**:
   - Iterate until convergence criteria are met, such as $\|\theta^{(t+1)} - \theta^{(t)}\| < \epsilon$ or after a maximum number of iterations.
5. **Output**:
   - Obtain the final parameter estimates $\hat{\theta}$ that approximate the solution to $f(\theta)$.

3.4 The Expectation-Maximization (EM)

See Fig. 5.

4 Results and Discussion

The experimental setup involved the segmentation of abdominal aortic aneurysms (AAAs) from medical images using advanced computational techniques. Specifically, Expectation Maximization (EM) and Majorization-Minimization (MM) algorithms were employed on a comprehensive dataset. In evaluating the efficacy of these algorithms, multiple performance metrics were utilized. The $F1$ score, a measure of the balance between precision and recall, provided insight into the algorithms' ability to accurately identify aneurysm regions. By measuring the amount of overlap between algorithm-generated segmentations and ground truth annotations, Intersection over Union (IOU), also known as the Jaccard Index, provided a more sophisticated understanding of segmentation quality. Additionally, accuracy served as a fundamental metric, quantifying the algorithms' overall correctness in segmenting aneurysm regions. This comprehensive evaluation framework provides a robust basis for comparative analysis and further refinement in the field of medical image segmentation.

Table 1 presents that the EM achieves higher $F1$ score (94%) and IOU (89%) compared to MM (82% and 79%, respectively), indicating its superior performance in accurately delineating aneurysm boundaries. Despite MM's slightly lower metrics, it demonstrates competitive results with an $F1$ score of 82% and IOU of 79%. Both methods achieve robust accuracy rates of 89% and 74%, respectively, showcasing their efficacy in anatomical segmentation tasks. The time efficiency of EM (2.8 s) and MM (3.2 s) highlights their computational feasibility in clinical settings, crucial for real-time diagnostic applications.

Table 2 shows the segmentation performance for 50 images. The EM method maintains its high performance with an $F1$ score of 94%, IOU of 89%, and accuracy of 90%, demonstrating consistent accuracy and reliability in boundary delineation. In contrast, the MM approach shows slightly reduced performance metrics with an $F1$ score of 81%, IOU of 79%, and accuracy of 74.6%. Despite this, MM remains a viable option with competitive results, albeit with slightly longer processing times (17.1 s compared to EM's 14 s). These findings underscore EM's robustness in achieving precise segmentation crucial for clinical applications.

Table 3 represents that the EM method maintains strong performance with an $F1$ score of 93%, IOU of 88%, and accuracy of 89%, indicating its robust capability in accurately delineating aneurysm boundaries across a larger dataset. In comparison, the MM approach achieves an $F1$ score of 82%, IOU of 81%, and accuracy of 76%, demonstrating slightly lower but competitive metrics. Both methods exhibit increased processing times with EM at 29 s and MM at 32 s, suggesting the trade-off between accuracy and computational efficiency in clinical applications.

Table 1 Results for ten images

Images	$F1$ score (%)	IOU (%)	Accuracy (%)	Time (s)
10 (EM)	94	89	89	2.8
10 (MM)	82	79	74	3.2

Table 2 Results for 50 images

Images	$F1$ score (%)	IOU (%)	Accuracy (%)	Time (s)
50 (EM)	94	89	90	14
50 (MM)	81	79	74.6	17.1

Table 3 Results for 100 images

Images	$F1$ score (%)	IOU (%)	Accuracy (%)	Time (s)
100 (EM)	93	88	89	29
100 (MM)	82	81	76	32

Table 4 Results for 500 images

Images	$F1$ score (%)	IOU (%)	Accuracy (%)	Time (s)
500 (EM)	94.7	89	91	139.2
500 (MM)	81.2	81	76.3	163

Table 5 Results for 1241 images

Images	$F1$ score (%)	IOU (%)	Accuracy (%)	Time (s)
1241 (EM)	94.6	88.7	91.2	347.54
1241 (MM)	82	80.3	77.14	398.3

Table 4 summarizes that the EM method exhibited significantly higher performance with $F1$ score of 94.7%, IOU of 89%, and accuracy of 91%, showcasing its robust performance in accurately delineating aneurysm boundaries across a large dataset. In contrast, the MM approach achieves lower metrics with an $F1$ score of 81.2%, IOU of 81%, and accuracy of 76.3%, indicating slightly diminished but still competitive results. Both methods show increased processing times, with EM requiring 139.2 s and MM 163 s, highlighting the computational demands associated with handling a larger volume of medical images.

Table 5 presents results from statistical approaches for abdominal aorta aneurysm segmentation using 1241 medical images. The EM method achieves high performance with an $F1$ score of 94.6%, IOU of 88.7%, and accuracy of 91.2%, demonstrating accurate boundary delineation. In comparison, the MM approach achieves slightly lower metrics with an $F1$ score of 82%, IOU of 80.3%, and accuracy of 77.14%. EM's efficiency is reflected in its lower processing time of 347.54 s compared to MM's 398.3 s, emphasizing its suitability for precise clinical segmentation tasks, while MM remains a viable alternative with acceptable performance.

Figures 6, 7, 8, and 9 show segmentation of an AAA using EM and MM algorithms. The segmented images show distinct color schemes, well-defined edges, and accurate shapes. EM is more effective in color differentiation, edge detection, and illumination consistency, while MM excels in shape accuracy and color differentiation. Consistent illumination provided by EM algorithms enhances contrast, facilitating clear visualization and analysis. EM consistently shows superior accuracy in identifying aneurysm boundaries, with competitive accuracy rates. However, its computational efficiency declines with larger datasets. MM, on the other hand,

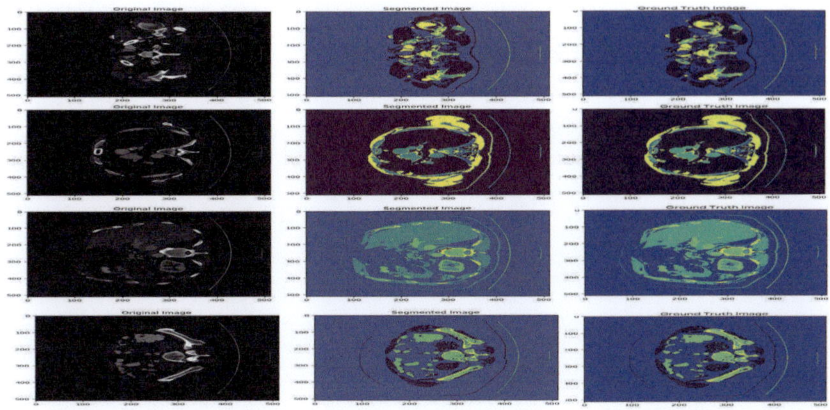

Fig. 6 Ground truth versus segmented MRI images of AAA using MM algorithm

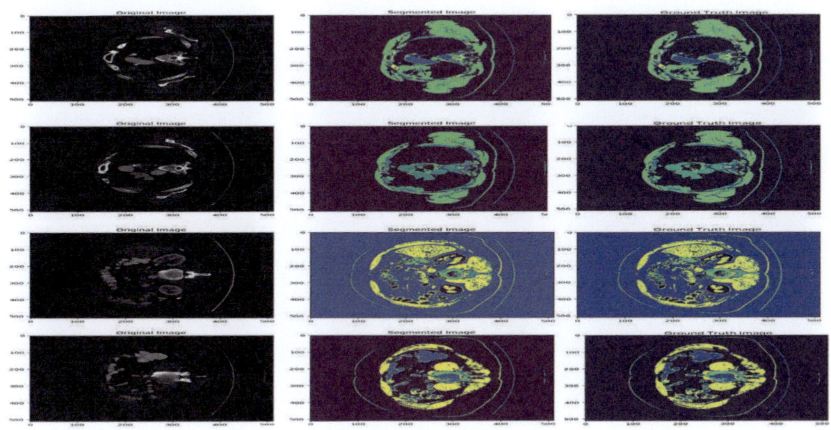

Fig. 7 Ground truth versus segmented MRI images of AAA using MM algorithm

shows slightly lower but competitive metrics, suggesting its effectiveness as an alternative for segmentation tasks. Despite MM's longer processing times, it may limit its real-time clinical utility. EM excels in accuracy and efficiency for smaller datasets, while MM offers adequate performance metrics.

5 Conclusion and Future Scope

In conclusion, the findings demonstrate that EM consistently exhibited superior performance in accuracy metrics such as $F1$ score and IOU across various dataset sizes, highlighting its robustness in precise boundary delineation crucial for clinical

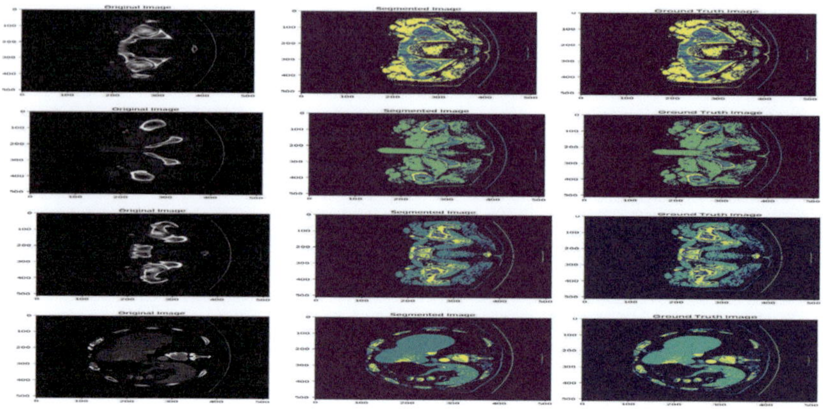

Fig. 8 Ground truth versus segmented MRI images of AAA using EM algorithm

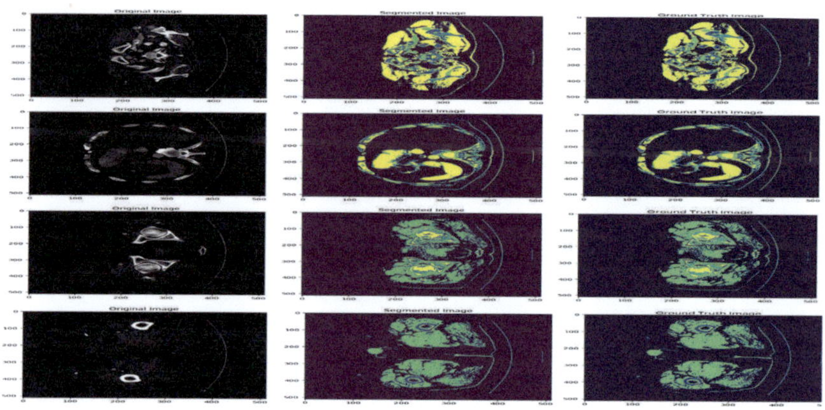

Fig. 9 Ground truth versus segmented MRI images of AAA using EM algorithm

applications. Despite its efficiency in smaller datasets, EM's computational demands escalate with larger volumes. MM, while showing competitive results, particularly in accuracy and overlap metrics, faces challenges with longer processing times. Both algorithms offer viable options for medical image segmentation, with EM being particularly well-suited for scenarios requiring high accuracy and MM providing a feasible alternative with adequate performance. Potential avenues for future research in this area could include parallel processing or sophisticated initialization techniques. Another goal could be to reduce the high resource needs of the EM algorithm by optimizing its computational efficiency. Furthermore, investigating hybrid strategies that integrate the advantages of EM and MM algorithms may result in improved segmentation success. Further research should also investigate the application of EM to other types of medical imaging challenges, beyond abdominal aorta aneurysms, to validate its versatility and robustness. Finally, integrating deep learning methods with the EM algorithm might offer new avenues for improving accuracy and handling incomplete data more effectively.

References

1. Ma Y, Hao H, Xie J, Fu H, Zhang J, Yang J, Wang Z, Liu J, Zheng Y, Zhao Y (2021) ROSE: a retinal OCT-angiography vessel segmentation dataset and new model. IEEE Trans Med Imaging 40:928–939. https://doi.org/10.1109/TMI.2020.3042802
2. Cipriano M, Allegretti S, Bolelli F, Bartolomeo D (2022) Deep segmentation of the mandibular canal: a new 3D annotated dataset of CBCT volumes. IEEE Access 10:11500–11510. https://doi.org/10.1109/ACCESS.2022.3144840
3. Kim M, Oh I-S, Yoon S-J (2022) Deep learning and computer vision techniques for automated total hip arthroplasty planning on 2-D radiographs. IEEE Access 10:94145–94157. https://doi.org/10.1109/ACCESS.2022.3204147
4. Lin C-L, Wu M-H, Ho Y-H, Lin F-Y, Lu Y-H, Hsueh Y-Y, Chen C-C (2024) Multispectral imaging-based system for detecting tissue oxygen saturation with wound segmentation for monitoring wound healing. IEEE J Transl Eng Health Med 12:468–479. https://doi.org/10.1109/JTEHM.2024.3399232
5. Xu H, Park S, Hwang TH (2020) Computerized classification of prostate cancer gleason scores from whole slide images. IEEE/ACM Trans Comput Biol Bioinform 17:1871–1882. https://doi.org/10.1109/TCBB.2019.2941195
6. Roy S, Maji P (2020) Medical image segmentation by partitioning spatially constrained fuzzy approximation spaces. IEEE Trans Fuzzy Syst 28:965–977. https://doi.org/10.1109/TFUZZ.2020.2965896
7. Rueckert D, Schnabel JA (2020) Model-based and data-driven strategies in medical image computing. Proc IEEE 108:110–124. https://doi.org/10.1109/JPROC.2019.2943836
8. Shiranthika C, Saeedi P, Bajić IV (2023) Decentralized learning in healthcare: a review of emerging techniques. IEEE Access 11:54188–54209. https://doi.org/10.1109/ACCESS.2023.3281832
9. Wijethilake N, Meedeniya D, Chitraranjan C (2021) Glioma survival analysis empowered with data engineering—a survey. IEEE Access 9:43168–43191. https://doi.org/10.1109/ACCESS.2021.3065965
10. Catalá ODT, Igual IS, Pérez-Benito FJ (2021) Bias analysis on public X-ray image datasets of pneumonia and COVID-19 patients. IEEE Access 9:42370–42383. https://doi.org/10.1109/ACCESS.2021.3065456
11. Liu Q, Leung CK, Hu P (2020) A two-dimensional sparse matrix profile DenseNet for COVID-19 diagnosis using chest CT images. IEEE Access 8:213718–213728. https://doi.org/10.1109/ACCESS.2020.3040245
12. Khan Z, Yahya N, Alsaih K, Al-Hiyali MI, Meriaudeau F (2021) Recent automatic segmentation algorithms of MRI prostate regions: a review. IEEE Access 9:97878–97905. https://doi.org/10.1109/ACCESS.2021.3090825
13. Alsiddiky A, Fouad H, Soliman AM, Altinawi A, Mahmoud NM (2020) Vertebral tumor detection and segmentation using analytical transform assisted statistical characteristic decomposition model. IEEE Access 8:145278–145289. https://doi.org/10.1109/ACCESS.2020.3012719
14. Malebary SJ, Hashmi A (2021) Automated breast mass classification system using deep learning and ensemble learning in digital mammogram. IEEE Access 9:55312–55328. https://doi.org/10.1109/ACCESS.2021.3071297
15. Agrawal P, Whitaker RT, Elhabian SY (2020) An optimal, generative model for estimating multi-label probabilistic maps. IEEE Trans Med Imaging 39:2316–2326. https://doi.org/10.1109/TMI.2020.2968917

16. Abella M, Martínez C, Desco M, Vaquero JJ, Fessler JA (2020) Simplified statistical image reconstruction for X-ray CT with beam-hardening artifact compensation. IEEE Trans Med Imaging 39:111–118. https://doi.org/10.1109/TMI.2019.2921929
17. Xue W, Li J, Hu Z (2021) Left ventricle quantification challenge: a comprehensive comparison and evaluation of segmentation and regression for mid-ventricular short-axis cardiac MR data. IEEE J Biomed Health Inform 25:3541–3553. https://doi.org/10.1109/JBHI.2021.3064353

Exploring Factors Impacting the Use of Wearable Devices in India

Saumya Misra and Rachana Adtani

Abstract The research explores external factors like healthcare, fashionability, connectivity, and lifestyle impacting the intention to adopt and use of wearable devices in the Indian context. We validated this effect using the Technology Acceptance Model 3 (TAM3). Data was collected from 140 respondents, and SEM was performed to gain insights into the intention to adopt wearable devices. The responses were bootstrapped using SmartPLS software. The hypothesized model explained 76.0% variance of behavioral intention (BI), 81.3% of PU, and 75.5% of PEOU. All four external factors were significant and impacted BI and the actual use of wearable devices. There was also a significant difference in the mean BI scores across gender, age, and educational qualification of the respondents. An essential outcome of this research was the importance of healthcare as a significant determinant of adopting wearable devices. This research is anticipated to provide noteworthy insights for scholars, manufacturers on design and connectivity aspects, and retailers to comprehend factors impacting the customer's adoption of wearable devices.

Keywords Technology Acceptance Model (TAM) · Wearables · Healthcare · Fitness · Lifestyle

1 Introduction

Wearable gadgets are tools or apparatuses that may be applied to the body and have sensors to identify a wide range of physiological parameters. Wearable items worn on various body parts, such as the limbs, torso, or head, include watches, bracelets, smartphones, eyeglasses, hearing aids, belts, footwear, and adhesive patches. These devices can monitor various health metrics, including pulse rate, oxygen saturation levels, temperature, daily step count, caloric burn, blood glucose levels, and UV radiation exposure [1]. It is also possible to give consumers feedback about using wearable devices so they may make the necessary corrections to their lifestyle.

S. Misra (✉) · R. Adtani
ICFAI Business School, Pune, India
e-mail: saumyamisra@ibsindia.org

H. Sharma et al. (eds.), *Artificial Intelligence: Theory and Applications*, Lecture Notes in Networks and Systems 5589, https://doi.org/10.1007/978-981-96-1687-9_46

In recent times, wearable technology has gained significant traction, particularly among middle-class to upper-class socioeconomic groups. Wearable smartwatches are a medical technology advancement. It assists with therapy, diagnosis, and patient monitoring. The system can track physical activity, including number of steps walked and location. [2].

A comprehensive review of multiple randomized controlled trials reveals that these devices can effectively help reduce health risks. Additionally, they promote increased physical activity levels among older adults with sedentary lifestyles, who are obese, as well as those with chronic respiratory conditions. Recent OECD data has revealed that healthcare expenditure when expressed as a percentage of Gross Domestic Product (GDP) of the country has reached unprecedented levels. This high ratio is because spending has increased, and the economy has generally slowed down [3]. Wearable technology is intended to fit a person's living environment to their physical and cognitive abilities and limits to maximize performance and minimize adverse outcomes like illness, harm, and discomfort. Health motivation is increasingly becoming entwined with people's everyday activities. Thus, consumers who lead "wellness-oriented" lifestyles can make risky health decisions, such as using health technologies, maintaining a nutritious diet, and exercising frequently [4].

The study aims to create a predictive model for wearable device adoption. A theoretical framework that combines the Technology Acceptance Model with external factors was proposed to achieve this. This framework generally helps explain and forecast users' behavioral intentions regarding intelligent wearables.

2 Theoretical Background

2.1 TAM Model

Rich explanations of the significant factors influencing use and behavioral intention have been given by the Technology Acceptance Model (TAM), TAM2, and data from other researchers [5, 6]. However, there is still a research gap on interventions to accelerate technology adoption [7]. Another research paper [8] proposed a model that integrates antecedents of perceived ease of use (PEOU) and perceived usefulness (PU) in a single model. The research findings brought out different effects of variables to enhance clarity within the literature, which previously exhibited contradictory findings regarding predictors of PU and PEOU [6, 9]. The expanded theoretical framework depicted in Fig. 1 shows that behavioral intention predicts actual use. Additionally, both these variables each influenced by a set of antecedents, support behavioral intention.

Considering technology adoption, PU refers to consumers' belief that technology will enhance performance and results and help them in achieving their set targets. Furthermore, PEOU refers to the degree of consumers' perception of how easy, seamless, and effortless it is to learn and operate the technology. These two perceptual

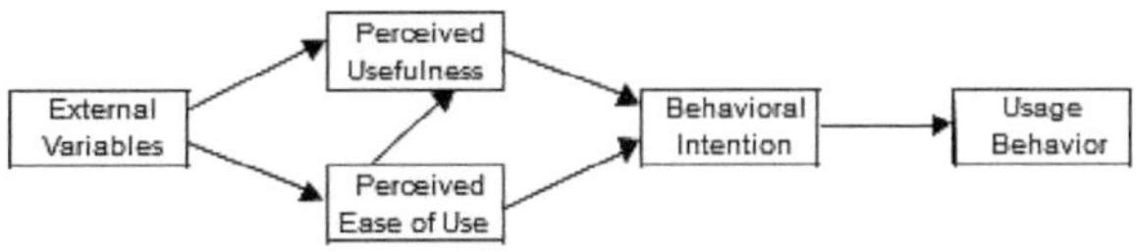

Fig. 1 TAM3 model. *Source* Davis, 1996

factors significantly influence users' attitudes and intentions toward adopting and using technology, as outlined by TAM. Ultimately, these factors play a crucial role in determining the long-term adoption and utilization of the technology. Despite the criticisms of the TAM, it remains the most relied-upon framework, particularly when determining the variables that impact older individuals' behavioral intention to adopt and utilize new technology [10].

Among the influential theories of technology adoption in the information systems study, one that stands out prominently is TAM [11–13]. Numerous empirical research studies [14] have used TAM with various technologies in various situations, showing that TAM can be a reliable model for anticipating users' behavioral intentions while utilizing a new technology.

The Technology Acceptance Model (TAM) posits that accepting technology involves a multifaceted process. Specifically, TAM suggests that external factors (including wearable device connectivity features, fashion aspects, and lifestyle and healthcare considerations) influence an individual's cognitive responses. In the Technology Acceptance Model (TAM), cognitive responses are pivotal in shaping an individual's affective response. Affective response, in turn, significantly influences their actual usage behavior toward the technology. The model proposes that an individual's intention to adopt any new technology depends on PU and PEOU.

Perceived Usefulness (PU). The underutilization of smartwatches is sometimes attributed to technological limitations, suggesting they only sometimes provide a transformative user experience. However, smartwatches can still be productive and educational tools that enhance users' lives through innovative applications developed by designers. Marketers should adapt their strategies to highlight the positive benefits of smartwatches, as these devices can facilitate lifestyle improvements [15]. Specifically, perceived usefulness is closely tied to the perceived simplicity of usage [16]. The significant advantages associated with smartwatch usage attract potential consumers to become actual customers.

Perceived Ease of Use (PEOU). Concerning wireless technology, customer adoption behavior and intention are impacted by the perceived usefulness of the product and its ease of operability. Additionally, the perceived ease of use emerges as a major driver of the adoption of smartwatches, mainly due to inherent challenges regarding their compact display size [17]. According to the Technology Acceptance Model (TAM), users' behavioral intention (BI) ultimately shapes the actual utilization of technology, thereby impacting its overall acceptance [18].

2.2 External Factors

Connectivity. Wearable technologies are computing devices that incorporate advanced communication and electronic capabilities, enabling individuals to interact with smart platforms. Additionally, wearable devices, especially smartwatches, have garnered significant attention and substantial investment from major smartphone manufacturers, indicating the emerging prominence of smartwatch technology [19].

Healthcare. Wearable devices can keep a close check on fitness levels physical activity such as step count, and pulse rate [2]. Smart devices can reasonably detect early warning signs of lifestyle diseases by observing changes in blood pressure and skin temperature, which in turn reflect the body's immune response to an infection. Wearable technology holds numerous potential benefits for older adults, yet this demographic embraces these technologies relatively slowly. Currently, understanding the hurdles that do not prevent elderly users from apt usage of smart wearables is vital. The subjective well-being of elderly users can be used to explore how the understandability of smart technology can affect their usage [20]. Psychological perception and user attitudes are directly affected by individual features or tendencies like smart device usage, medical history, age, and user sex [21].

Fashionability. Smartwatches have become an essential gateway for the extensive adoption of wearables, as they combine both technological innovation and fashion. This research focuses on how to explore how characteristics associated with fashion accessories like smartwatches impact consumers' intentions to use smartwatches by expanding on the TAM. The study introduces the concept of fashionability, shaped by individuals' vanity and desire for uniqueness. It relates to its aesthetic appeal and the self-expression it facilitates, which can be a crucial factor driving its adoption.

Lifestyle. A person's lifestyle outlines their daily routines concerning health, fitness, work-life balance, and many other aspects. Some individuals spend their free time and money on social activities, while others like to travel, shop, engage in dynamic activities like sports, and track their fitness. Events of the present and past experiences might influence one's lifestyle. Lifestyles may reflect how the individual deals with the surroundings and their bearing on the perceived usefulness of smart devices.

3 Research Model and Hypotheses

3.1 Research Methodology

Data was collected using a convenience sampling technique, from the 140 respondents. The questionnaire was converted into a Google form and the link of the same was circulated using messages, email and by sharing the same on social media platforms like LinkedIn, during the time period May 2023 to June 2023. An ethics

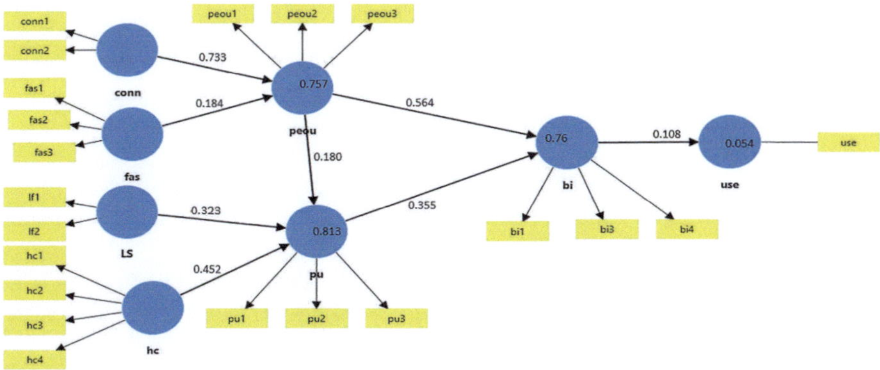

Fig. 2 Hypothesized model

approval was not required for this research as there was no sensitive financial data or health records collected. The respondent was asked to read the ethical statement at the top of the questionnaire and agree to the same before the responses were collected. No data was collected from underage respondents (below 18 years). The collected data was analyzed using SmartPLS and Structural Equation Modelling was performed [22].

3.2 Research Model

Figure 2 illustrates the proposed research model applied in this study. As mentioned earlier, the model is constructed from the TAM3 model. Seven hypotheses were proposed based on the research gaps in recent literature. Figure 2 shows the hypothesized hypothesis model.

3.3 Hypotheses

Health indicators such as BMI, number of steps, heart rate, and oxygen saturation are essential to keep track of fitness, especially after the COVID-19 pandemic. Hence, this positively impacts the P.U. of the wearable device. Lifestyle also impacts P.U. as it affects one's daily routine and is concerned with monitoring physical and mental health. Based on these perspectives, the following hypotheses were proposed:

H1: Healthcare has a positive effect on P.U.
H2: Lifestyle has a positive effect on P.U.

Connectivity with other devices, like laptops and tablets, affects the PEOU. The external features of the wearable devices, like simplicity in design and globally accepted color palettes, make them an accessory which blends with all styles of dressing, positively impacting PEOU. Based on these perspectives, the following hypotheses were proposed:

H3: Connectivity has a positive effect on PEOU H4: Fashionability has a positive effect on PEOU.

An extensive literature study has shown that P.U. and PEOU positively impact B.I., even if the significance levels vary, as was previously mentioned in this paper. Therefore, the following hypotheses were proposed:

H5: P.U. has a positive effect on B.I. H6: PEOU has a positive effect on B.I.

Based on the review of existing literature, PEOU was related to use. Therefore, the following hypothesis was proposed:

H7: PEOU has a positive effect on P.U.

Numerous studies have discovered that B.I. positively impacts the actual use of the technology. Therefore, the following hypothesis was proposed:

H8: B.I. has a positive effect on use.

4 Result

Table 1 summarizes the demographic characteristics of the sample respondents. The SmartPLS3 software [23] was used to conduct path analysis. Descriptive variables of gender, age, and educational qualification were expressed as categorical variables. Age data was grouped according to age: 18–22 years old, 23–28 years old, and above 28 years old. Qualification data was grouped into "Class 12", "Graduate", and "post graduate" categories.

Pearson correlation coefficient with a 2-tailed significance level set at 5% (alpha < 0.05) was performed to assess the correlation between connectivity, fashionability, healthcare, lifestyle, perceived usefulness, and perceived ease of use. An independent sample *t*-test compared mean B.I. scores between male and female respondents. A one-tailed ANOVA test was conducted to compare the same across the three age groups and educational qualifications categories. The mean B.I. score was statistically different across age and educational qualifications groups ($p < 0.05$). Figure 2 indicates the proposed research model to evaluate the factors impacting the use of wearable device technology.

Table 1 Demographic characteristics

Gender	Count%	Age (years)	Count (%)	Educational qualification	Count (%)
Male	67(47.8)	18–22	45(32.14)	Post graduate	64(45.71)
Female	73(52.2)	23–28	87(62.14)	Graduate	62(44.29)
		Above 28	8(5.72)	12th	14(10)

4.1 Measurement Model

Before a structural model is analyzed, a measurement model needs to be evaluated. Convergent, discriminant, and internal consistency validity can be considered a measurement model. The internal consistency was confirmed using Cronbach's alpha, which suggests a value of 0.7 or above. Moreover, item loadings and composite reliability (C.R.) values are advised to be greater than 0.6 and 0.7, respectively. The average variance extracted (AVE) value should be more than 0.5 for every latent variable. All item loadings, Cronbach's alpha, C.R., and AVE results are shown in Table 2. Table 2 shows that all of the constructs satisfy the threshold value, indicating that the measurement model is significant and that the reliability and validity requirements are satisfied.

Path analysis provides insights regarding the impact of different constructs on B.I. Path analyses also help in understanding which construct has a more significant effect.

Convergent validity and internal consistency reliability are performed for the assessment of the measurement models. The convergent validity of a measurement model is usually ascertained by examining the loadings and average variance extracted (AVE) [24]. Loadings of 0.5 and higher suggest that the item is a good measurement of the latent construct [25]. Threshold values for all the items were met.

For internal consistency and reliability, Cronbach's alpha [26] and an alternative measure, composite reliability, were considered [26, 27].

Table 3 explains the R-squared results. This study assesses four dependent variables. The first dependent variable is P.U., which healthcare and lifestyle influence as external variables. The second dependent variable is PEOU, which connectivity and fashionability influence as external variables. The third dependent variable is B.I., which both P.U. and PEOU influence. The final dependent variable is the actual use, which B. I influence.

The value of the R-square test for the P.U., which is 0.813, indicates that healthcare and lifestyle explain P.U. to an extent of 81.3%. In contrast, this study cannot describe the rest (18.70%). The value of the R-square test for PEOU, which is 0.757, indicates that connectivity and fashionability explain PEOU to the extent of 75.7%. Moreover, B.I. is explained by P.U. and PEOU to an extent of 75%. B.I. explains use to an extent of 5.4%.

The hypothesized model explains B.I.'s variance to 76.0%, 81.3% of P.U., and 75.7% of PEOU. Besides H7, all other hypotheses were verified. Healthcare, connectivity, fashionability, and lifestyle, as independent variables, significantly impact the hypothesized model, leading to B.I.'s adoption of wearable devices.

All other path coefficients had acceptable values, barring one. To ascertain whether the predictors of actual use of the wearable device were significant, a bootstrapping process was run using a 5000-resample [26].

Table 2 Scales for reliability and convergent validity

Construct	FL	Cronbach's alpha	Average variance extracted (AVE)	CR
BI		0.942	0.896	0.944
bi1	0.935			
bi3	0.953			
bi4	0.952			
CONN		0.857	0.875	0.860
conn1	0.931			
conn2	0.940			
FAS		0.881	0.806	0.928
fas1	0.805			
fas2	0.937			
fas3	0.945			
HC		0.846	0.687	0.869
hc1	0.887			
hc2	0.797			
hc3	0.904			
hc4	0.710			
LS		0.889	0.900	0.891
lf1	0.946			
lf2	0.952			
PEOU		0.942	0.896	0.944
peou1	0.942			
peou2	0.949			
peou3	0.949			
PU		0.927	0.872	0.931
pu1	0.924			
pu2	0.953			
pu3	0.925			
Use	1.000			

Table 3 R-squared results

Construct	R-square	R-square adjusted
BI	0.764	0.760
PEOU	0.761	0.757
PU	0.818	0.813
Use	0.062	0.054

Table 4 Structural model (bootstrap) results

Hypothesis	Path	B value	Significance	Supported
H1	LS → PU	0.323	0.001	Supported
H2	HC → PU	0.452	0.000	Supported
H3	CONN → PEOU	0.733	0.000	Supported
H4	FAS → PEOU	0.184	0.016	Supported
H5	PU → BI	0.355	0.002	Supported
H6	PEOU → BI	0.564	0.000	Supported
H7	PEOU → PU	0.180	0.073	Not supported
H8	BI → Use	0.108	0.008	Supported

Fig. 3 Word cloud

The bootstrapping results in Table 4 include the significance level. The results show that the relationships between PEOU and P.U. are not significant. Therefore, H1-H6 and H8 are supported, whereas H7 is not.

A qualitative data analysis was also done using Word Cloud, as seen in Fig. 3. This figure is a visual representation of word frequency. The term appearing more considerable in the resulting graphic was more frequently used by the respondents while answering the research question.

Health, fitness, and tracking were the keywords in the answer to the question: Why do you prefer a smartwatch? Hence, keeping track of health indicators via wearable devices was the most frequently quoted reason for adopting such devices.

5 Discussion and Conclusion

The study aimed to gain insight into wearable device acceptance in its early stages. It used a thorough model of the TAM3. The purpose of proposing healthcare, connectivity, fashionability, and lifestyle as independent factors was to obtain insights into user behavior toward technology adoption. User feedback was gathered using a survey, and the study methodology was statistically assessed.

The data analysis brought out significant findings. First, in the TAM3 construct, PEOU had a more substantial effect, 56.8% on B.I. than P.U., which explains 35.5%

B.I. It was also seen that P.U. did not mediate the link between PEOU and B.I. The findings indicate that users' expectations for wearable devices are influenced more by a device's ease of use than by its actual utility; this could be the case since consumers already own electronics like laptops and mobile phones, which offer advantages comparable to smartwatches. Therefore, the PEOU is the primary source of the B.I.'s adoption of wristwatches. In order to obtain a competitive edge, wearable device manufacturing businesses prioritize simplicity of use, more innovative and more comfortable designs, and robust connectivity with other devices like mobile phones for higher adoption levels.

If we look at R-square results, 76.0% of B.I. is explained by P.U. and PEOU combined. External factors like lifestyle and healthcare explain 81.3% of P.U. Furthermore, external factors like connectivity and fashionability explain 75.7% of PEOU. However, the effect of B.I. on use is low; this was also highlighted in a recent study [28], which discussed wearable technologies' health-protective features and utilitarian qualities. The latter is now significant during life-threatening emergencies like the COVID-19 epidemic. The results show that B.I.'s use of such gadgets is significantly impacted by both the protective and utilitarian features.

As discussed, PEOU impacts B.I. more than P.U. fashionability and connectivity are the determinants of PEOU. Fashionability, that is, the external look and the general appeal of the product, is an essential factor that affects the user's behavior toward the adoption of new technology. Recent research has verified the same [29]. The early part of the product life cycle is the current stage of wearable devices. Hence, they are in the adoption stage, and visual and aesthetic appeal can be considered an essential factor impacting consumers' buying behavior [30]. Wearable devices are considered fashion accessories, and their unique external appearance positively impacts the B.I.'s adoption of them [31].

To sum up, the research paradigm that has been suggested has consequences for both the academic community and the industry. This study has various takeaways for business professionals, such as marketers and designers. Because the wearables business is still in its infancy, practitioners must learn more about the elements influencing consumers' attitudes toward adopting smart wearables than other established information technologies. The suggested study structure, the results, and the associated generalizations provide businesses with a summary of the factors that influence the adoption of wearable technology.

An essential outcome of this research was the importance of healthcare as a significant determinant of adopting wearable devices. Other aspects, such as lifestyle, were also explored. This aspect has not been found extensively in recent research and has been an unexplored space until now. This discovery is anticipated to provide noteworthy insights for scholars, manufacturers on design and connectivity aspects, and retailers to comprehend factors affecting their wearable device adoption.

6 Limitations of the Study and Scope of Further Research

Factors affecting the adoption of wearable devices, like product pricing and promotion, can be explored in the future studies. Moreover, any generalization of the research can be a limiting factor as participants were not operating under controlled conditions.

References

1. Patil V, Singhal DK, Naik N, Hameed BZ, Shah MJ, Ibrahim S, Smriti K, Chatterjee G, Kale A, Sharma A, Paul R (2022) Factors affecting the usage of wearable device technology for healthcare among indian adults: a cross-sectional study. J Clin Med 11(23):7019
2. Jat AS, Grønli TM (2022) Smartwatch for intelligent health monitoring: a literature review. In: International work-conference on bioinformatics and biomedical engineering, 8 Jun 2022. Springer International Publishing, Cham, pp 256–268
3. Chan M, Estève D, Fourniols JY, Escriba C, Campo E (2012) Smart wearable systems: current status and future challenges. Artif Intell Med 56(3):137–156
4. Hokroh M, Green G, Soleton M (2020) Factors influencing health wearables adoption and usage in Saudi Arabia. J Manag Econ Stud 2(2):89–98
5. Davis FD (1989) Perceived usefulness, perceived ease of use, and user acceptance of information technology. MIS Quarterly 319–340
6. Venkatesh V (2000) Determinants of perceived ease of use: integrating control, intrinsic motivation, and emotion into the technology acceptance model. Inf Syst Res 11(4):342–365
7. Venkatesh V, Speier C (1999) Computer technology training in the workplace: a longitudinal investigation of the effect of mood. Organ Behav Hum Decis Process 79(1):1–28
8. Venkatesh V, Bala H (2008) Technology acceptance model 3 and a research agenda on interventions. Decis Sci 39(2):273–315
9. Agarwal R, Karahanna E (2000) Time flies when you are having fun: cognitive absorption and beliefs about information technology usage. MIS Q 665–694
10. Braun MT (2013) Obstacles to social networking website use among older adults. Comput Hum Behav 29(3):673–680
11. Chuttur M (2009) Overview of the technology acceptance model: origins, developments and future directions
12. Holden RJ, Karsh B-T (2010) The technology acceptance model: its past and future in health care. J Biomed Inform 43(1):159–172
13. Lai PC (2017) The literature review of technology adoption models and theories for the novelty technology. JISTEM-J Inf Syst Technol Manag 14:21–38
14. Liaw S-S (2008) Investigating students' perceived satisfaction, behavioural intention, and effectiveness of e-learning: a case study of the blackboard system. Comput Educ 51(2):864–873
15. Arifah ID, Juniarti RP (2021) Interface aesthetic, perceived value, perceived ease of use, and perceived usefulness affect the purchase intention of smartwatch consumers. In: International conference on business and engineering management (ICONBEM 2021), 25 May 2021, pp 25–33. Atlantis Press
16. Yen DC, Wu CS, Cheng FF, Huang YW (2010) Determinants of users' intention to adopt wireless technology: an empirical study integrating TTF with TAM. Comput Hum Behav 26(5):906–915
17. Adapa A, Nah FF, Hall RH, Siau K, Smith SN (2018) Factors influencing the adoption of smart wearable devices. Int J Hum-Comput Interact 34(5):399–409
18. Taherdoost H (2018) A review of technology acceptance and adoption models and theories. Procedia Manuf 22:960–967

19. Alaskari E, Alanzi T, Alrayes S, Aljabri D, Almulla S, Alsalman D, Algumzi A, Alameri R, Alakrawi Z, Alnaim N, Almusfar L (2022) Utilization of wearable smartwatch and its application among Saudi population. Front Comput Sci 4(4):874841
20. Farivar S, Abouzahra M, Ghasemaghaei M (2020) Wearable device adoption among older adults: a mixed—methods study. Int J Inf Manage 1(55):102209
21. Fang YM, Chang CC (2016) Users' psychological perception and perceived readability of wearable devices for older adults. Behav & Inf Technol 35(3):225–232
22. Gefen D, Straub D, Boudreau M-C (2000) Structural equation modelling and regression: guidelines for research practice. Commun Assoc Inf Syst 4. https://doi.org/10.17705/1cais.00407
23. Dijkstra TK, Henseler J (2015) Consistent partial least squares path modeling. MIS Q 39(2), pp 297–316. https://doi.org/10.25300/misc/2015/39.2.02
24. Hair JF, Risher JJ, Sarstedt M, Ringle CM (2019) When to use and how to report the results of PLS-SEM. Eur Bus Rev 31(1):2–24. https://doi.org/10.1108/-11-2018-0203
25. Hulland J (1999) Use of partial least squares (PLS) in strategic management research: a review of four recent studies. Strat Manag J 20:195–204
26. Hair JF, Hult GTM, Ringle CM, Sarstedt M (2017) A primer on partial least squares structural equation modeling (PLS-SEM), 2nd edn. Sage, Thousand Oaks, CA
27. RB Kline (2015) Principles and practice of structural equation modeling, 4th edn. Guildford Press
28. Singh N, Misra R, Singh S, Rana NP, Khorana S (2022) Assessing the factors that influence the adoption of healthcare wearables by the older population using an extended PMT model. Technol Soc 1(71):102126
29. Chen X, Gole J, Gore A, He Q, Lu M, Min J, Yuan Z, Yang X, Jiang Y, Zhang T, Suo C (2020) Non-invasive early detection of cancer four years before conventional diagnosis using a blood test. Nat Commun 11(1):3475
30. Chuah SH, Rauschnabel PA, Krey N, Nguyen B, Ramayah T, Lade S (2016) Wearable technologies: usefulness and visibility in smartwatch adoption. Comput Hum Behav 1(65):276–284
31. Jeong JY, Roh TW (2017) The intention of using wearable devices: based on modified technology acceptance model. J Digit Converg 15(4):205–212

One-Shot Face Classification Using Siamese Neural Networks

S. Balamurali, Puneet Mittal, and Sukhwinder Sharma

Abstract Learning features that are good for machine learning can be included, especially when dealing with limited data. A difficult case is a one-shot study where the prediction is used for only one example per class. In this work, an attempt is made to delve into the utilization of Siamese neural networks, which possess a distinct architecture for intuitively ranking similarity between inputs. When these networks are finetuned, they provide discrimination capabilities that can be extended to include not only new information but also unknown categories in unknown distributions. It has been observed that an impressive performance can be achieved surpassing other deep learning models and approaching various levels in one-shot classification problems.

Keywords Siamese neural networks · Face recognition · Image pairing

1 Introduction

People can create and realize new trends. People can quickly understand new concepts when they are motivated followed by recognition of changes in future understanding of those concepts [1]. A machine learning approach has been used to improve performance in many applications, including web browsing, spam detection, subtitles, etc. However, these algorithms break down when they must make future predictions on data sets containing small observations. It is expected to provide ability to expand on these unknowns without the need for extensive training; this training may be expensive or impossible due to limited knowledge or online gambling (for instance website access).

Features or direct inference methods to solve these problems are highly selective in terms of target function. Solutions that will work for other problems may not be applicable for problem at hand. In this paper, a new method has been presented that restricts the assumption of understanding the model while obtaining features that complement

S. Balamurali · P. Mittal (✉) · S. Sharma
Manipal University Jaipur, Jaipur, Rajasthan, India
e-mail: pmittal.cse@gmail.com

H. Sharma et al. (eds.), *Artificial Intelligence: Theory and Applications*, Lecture Notes in Networks and Systems 5589, https://doi.org/10.1007/978-981-96-1687-9_47

the model, through a few examples. A deep learning framework is created that uses multiple layers to capture inconsistencies in changes in domains, often leveraging models with multiple parameters, and then using multiple data to avoid overfitting the model [2, 3]. These features are very powerful because although learning algorithms are expensive, learning of these features may be performed without power.

2 Approaches

This work explores face recognition using Siamese neural networks [4]. The technique is primarily geared toward image classification and can potentially be adapted for various other modalities (see Fig. 1). A robust Siamese convolutional neural network is designed to extract and learn general image features, which are vital for predicting classes with limited examples.

The training of these networks utilizes pairs of images from the same dataset, simplifying the training process as it does not require domain-specific knowledge but rather utilizes widespread deep learning methodologies. The primary goal of research is to create a model adept at one-shot image classification.

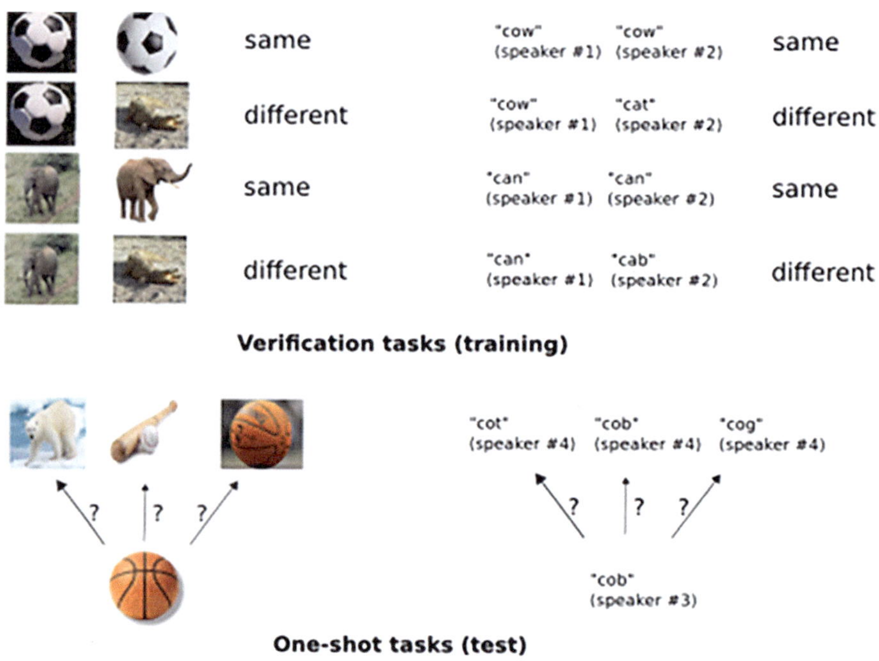

Fig. 1 **a** Model to distinguish between equal/variable, **b** generalization evaluates new groups based on learning maps for analysis

Initially, the neural network is trained to differentiate between image pairs by identifying whether they belong to the same class, a process known as image verification. Successful performance in this task suggests potential effectiveness in one-shot classification scenarios. In this setting, the network evaluates whether new image pairs—each representing a different class—are like a test image. The image pair that the verification model scores highest is deemed the likely correct match in the one-shot classification.

It is hypothesized that if the verification model can accurately recognize faces from one set, it should, in theory, replicate this success with other faces, particularly if trained across a diverse array of faces. This diversity helps in enhancing the variability and robustness of the features learned by the network.

3 Literature Review

The main work on single learning can be traced back to the early 2000s, initiated by Lei et al. [5]. Guo et al. [6] have summarized the databases used for deep face recognition with their open issues and challenges. Recent advances have been made by Alzubaidi et al. [7] gave a comprehensive survey on the process of applying deep learning techniques and networks. Lake et al. [8, 9], who investigated one-shot learning through cognitive psychology, specifically for cognitive behavior, using a method called Hierarchical Bayesian Programming Learning (HBPL). In their paper, they defined a model that divides the drawing process into smaller pieces to define visual pixels [1, 8]. However, HBPL causes serious problems due to the large joint area, which complicates the joint process and makes it difficult to think. Add different models or use transfer learning methods.

Maas and Kemp represent a rare example of using Bayesian networks to predict the behavior of historical passenger data at Ellis Island [10]. While Wu and Dennis [11] used a learning method for robot planning algorithms, Lim explored strategies for using other groups' models in training programs. By adjusting how much each class contributes to unemployment based on each educational sample, Lim's method provides an effective way to provide information about the correlation between classes and is particularly useful for classes with scarce information [12]. Daque-Domingo et al. [13] have proposed a new Siamese network training technique CP-CVV combining different models trained with different classes.

4 Deep Siamese Networks for Image Verification

Siamese networks consist of two networks that receive separate elements but are connected by a working force at the top. This function measures the maximum number of representatives of each network (see Fig. 2).

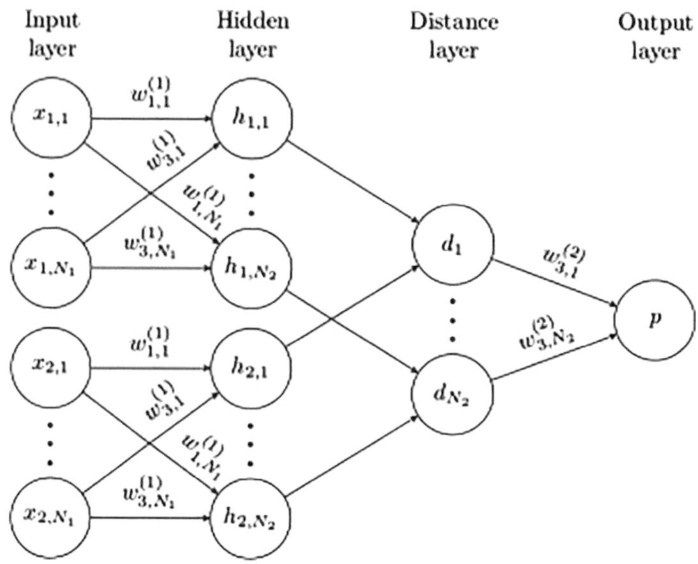

Fig. 2 Model Siamese network for binary classification and prediction. The network model is partially duplicated at the top and bottom, creating a binary network where each layer shares a weight matrix

These networks share the same parameters, ensuring that similar images do not display different content in a particular location, as each network works the same way. Additionally, the symmetry of the network ensures that the power function contains the same measurement regardless of which twin network receives which image. A power comparison function was then used to reduce the power of similar images and increase the power of different images [14]. Different from their methods, in this paper, it is taken the $L1$ weight distance between the twin feature vectors $h1$ and $h2$ as the measurement of sigmoid activation. Output of range [0,1]. Therefore, the cross-entropy target becomes a suitable choice for training the network. This approach differs from LeCun's approach, where the similarity measure is defined by the power function and the fixed measure defined in the Facebook Deep Face project [15]. Many layers are involved before the overall connection process and upper power. Convolutional neural networks (CNN) have been highly successful in many broad fields of view, especially image recognition [3, 16–18]. CNNs have many advantages: their local connections reduce the number of negative examples, providing stable performance although they require more resources. The convolution process in these networks acts as a filter, convolving each feature map with the input features to identify patterns based on groups of pixels. This ensures that the output of each convolutional layer represents important spatial features of the input, making it robust to simple transformations.

4.1 Model Architecture

The model uses a Siamese convolutional neural network consisting of L layers, each containing n_l units. In this setup, $h_{1,l}$ *and* $h_{2,l}$ represent the hidden vectors at layer l for the first and second twin networks, respectively. The network exclusively uses rectified linear (ReLU) units in the first $L - 2$ layers and sigmoidal units in the subsequent layers.

The kth filter map in each layer is defined as:

$$a_{1,m}^{(k)} = \max - \text{pool}\left(\max\left(0, W_{l-1,l}^{(k)} * h_{l,(l-1)} + b_l\right), 2\right) \quad (1)$$

$$a_{2,m}^{(k)} = \max - \text{pool}\left(\max\left(0, W_{l-1,l}^{(k)} * h_{2,(l-1)} + b_l\right), 2\right) \quad (2)$$

After the last convolution layer, these units are flattened into a vector and a fully connected layer is followed, and then another layer that calculates the induced distance measure of conjoined twins, resulting in an S-shaped output unit. The vector estimate is shown as:

$$p = \sigma\left(\sum a_j |h_{1,L-1}^{(j)} - h_{2,L-1}^{(j)}\right) \quad (3)$$

where σ is the sigmoid activation function. The a_j are additional learned parameters that weight the importance of the component-wise distances, forming the final L-th fully connected layer that unites the two Siamese networks.

4.2 Loss Function

Assume M to represent the small batch size, where i indexes the ith small batch. Now let $y(x_1^{(i)}, x_2^{(i)})$, be a length-M vector which contains the labels for the small batch, where assumed $y(x_1^{(i)}, x_2^{(i)})$, $= 1$ whenever x_1 and x_2 are from the same character class and $y(x_1^{(i)}, x_2^{(i)})$, $= 0$ otherwise. The regularized cross-entropy loss on binary classifier as follows:

$$L\left(x_1^{(i)}, x_2^{(i)}\right) = y\left(x_1^{(i)}, x_2^{(i)}\right) \log p\left(x_1^{(i)}, x_2^{(i)}\right) + \left(1 - y\left(x_1^{(i)}, x_2^{(i)}\right)\right) \log\left(1 - p\left(x_1^{(i)}, x_2^{(i)}\right)\right) + \lambda^T |w|^2 \quad (4)$$

4.3 Optimization

The conventional backpropagation algorithm is applied to optimize objective, benefiting from additive gradients across the twin networks facilitated by shared weights. This approach maintains a consistent batch size of 128 throughout the training process. The learning rate n_j, momentum $\mu_j|$, and $L2$ regularization weights $\lambda_j|$ are defined for each layer. Thus, an update rule at epoch T is structured as follows:

$$\begin{aligned} w_{kj}^{(T)}\left(x_1^{(i)}, x_2^{(i)}\right) &= w_{kj}^{(T)} + \Delta w_{kj}^{(T)}\left(x_1^{(i)}, x_2^{(i)}\right) + 2\lambda\left|w_{kj}\right| \\ \Delta w_{kj}^{(T)}\left(x_1^{(i)}, x_2^{(i)}\right) &= -\acute{\eta}_j \Delta w_{kj}^{(T)} + \mu_j \Delta w_{kj}^{(T-1)} \end{aligned} \quad (5)$$

Here, ∇w represents the partial derivative of the loss function with respect to the weight connecting the jth neuron in one layer to the kth neuron in the other layer. This gradient informs how the weights should be adjusted to minimize the loss, factoring in both the current gradient and the previous weight update, adjusted for momentum and regularization.

4.4 Weight Initialization

For this network, a special strategy is adopted to initialize the weights of different types of layers to ensure effective learning. All weights in the convolutional process start from the mean of the zero normal distribution and standard deviation 10^{-2}. The biases in the layers are also based on a distribution, but with a mean of 0.5 and a SD of 10^{-2}. In the full linking process, the biases are initially like the linking process. However, the weights are taken from a general distribution with mean zero and SD 2×10^{-1}.

4.5 Learning Plan

During the training process, a unique learning process is used for each layer but made the collisions consistent across the network. Specifically, the learning rate of each layer is periodically reduced by 1% according to the formula $n_j^{(T)} = 0.99 n_j^{(T-1)}$. Slow annealing of the learning rate helps the network converge more smoothly to local minima, preventing it from falling into the error surface. It reaches the predefined value μ_j, that is, the specific time of the jth layer. This approach helps achieve stable and efficient integration over time, ensuring equal speed in updating. Currently, it is observed that the error in the single injection of a randomly generated set of 320 subjects from the alphabet and drawers during the validation process. If the verification error is not corrected 20 consecutive times, training is stopped, and the failure is stored at the best possible time (determined by the lowest error in a single shot). However, if the recognition error continues to decrease throughout the training plan, the final state of the model after 200 times will be preserved. This inspection site ensures the best possible configuration without unnecessarily delaying training.

4.6 Hyperparameter Optimization

The layer-wise learning rate n_j, to range between 10^{-4} and 10^{-1}, the layer-wise momentum μ_j,to be within [0, 1], and the layer-wise $L2$ regularization penalty λ_j, to vary between 0 and 0.1. Concerning the hyperparameters for network architecture, it facilitates the variation in convolutional filter sizes, ranging from 3×3 to 20×20 pixels. The quantity of convolutional filters per layer is adjustable between 16 and 256, increasing in increments of 16.

The optimizer is configured to be accurate on a one-time validation set. The performance of each set of hyperparameters tested in Whetlab is measured as the maximum number of injection validation sets performed at each location during the run. This score is assigned to each Whetlab iteration and guides the selection process to determine which design is best.

4.7 Affine Distortions for Data Augmentation

For every pair of images x_1 and x_2, corresponding affine transformations T_1 and T_2 have been generated. These transformations produced new images $x_1^| = T_1(x_1)$ and $x_2^|$ $= T_2(x_2)$, where T_1 and T_2 are randomly determined by a multidimensional uniform distribution. For any arbitrary transform T, it is defined as $\mathrm{T} = (\theta, \rho_x, \rho_y, \mathrm{s}_x, \mathrm{s}_y, \mathrm{t}_x, \mathrm{t}_y)$, with parameter ranges: θ in [− 10.0, 10.0], ρ_x,ρ_y in [− 0.3, 0.3], s_x, s_y in [0.8, 1.2], and t_x, t_y in [− 2, 2]. Each parameter is included in the transformation with a probability of 0.5.

5 Experiments

The model was initially trained on a set of defined LFW data. The details of reuse and one-time operations have also been shared in this work.

5.1 About Dataset

Labeled Faces in the Wild (LFW) is an extensively recognized benchmark dataset designed for the evaluation of face verification systems in unconstrained settings. It contains over 13,000 images, representing 5,749 individuals, sourced from the web to simulate real-world conditions. Among these, 1680 individuals have two or more distinct photos in the dataset, facilitating both matching and non-matching verification tasks. The dataset is distinctive for its variety in facial expressions, poses, lighting conditions, and backgrounds, which reflect the challenges inherent in real-world face recognition applications. The images in LFW are categorized into four sets: the original images, funneled images (aligned by a standard method from ICCV 2007), LFW-a (aligned using an unpublished method), and deep funneled images (aligned using advanced deep learning techniques from NIPS 2012), with the aligned images generally providing better results in verification algorithms. However, it is important to note that despite its wide use in academic research, LFW is not without limitations. The demographic representation within LFW is not comprehensive, lacking substantial representation of certain age groups, ethnicities, and with imbalances in gender representation. Moreover, conditions such as occlusions, extreme poses, and variable lighting are underrepresented. Thus, while LFW serves as a valuable tool for advancing research in face verification, it should not be used to conclude the suitability of algorithms for commercial deployment without further validation and testing on more diverse and representative datasets.

5.2 Methodology Followed

In the following section, methodology followed for training and developing the model has been illustrated (see Fig. 3).

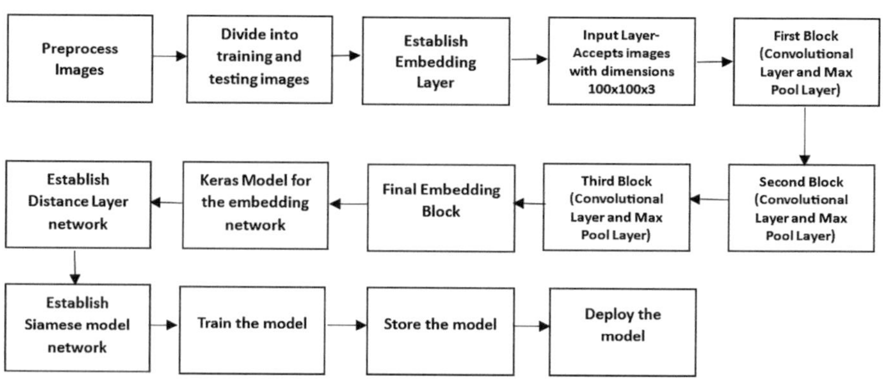

Fig. 3 Methodology followed for developing Siamese model network

Preprocessing Images and Dataset Division: Facial images were imported from a curated dataset and subjected to standardized preprocessing steps to ensure consistent input dimensions for the model. These preprocessing steps included scaling and resizing all images to 100 × 100 × 3 (height, width, channels). The dataset was then split into training and testing subsets with a 70:30 ratio. Specifically, 70% of the dataset, comprising 1400 image pairs, was allocated for training, while the remaining 30%, or 600 image pairs, was reserved for testing.

Establishing the Embedding Layer: The neural network architecture was designed to include an embedding layer that converts facial images into compact, high-dimensional vectors representing salient facial features. This embedding network served as the backbone of the Siamese model, which facilitates facial recognition by measuring the similarity between images.

Input Layer: The input layer of the network accepts images of dimensions 100 × 100 × 3, corresponding to the standardized input dimensions derived during preprocessing.

Convolutional Blocks: The network architecture comprised three main convolutional blocks, each containing a convolutional layer followed by a max-pooling layer, designed to progressively extract hierarchical facial features from the input images.

First Block:

- Convolutional Layer ($c1$): Utilizes 64 filters of size 10 × 10 and ReLU activation to extract basic features.
- Max-Pooling Layer ($m1$): Performs max pooling with a 2 × 2 filter size and 'same' padding to downsample the feature maps while preserving spatial information.

Second Block:

- Convolutional Layer ($c2$): Deploys 128 filters of size 7 × 7 with ReLU activation to capture more complex features.
- Max-Pooling Layer ($m2$): Further down sampling is achieved using max pooling, reducing feature map dimensions while retaining important features.

Third Block:

- Convolutional Layer ($c3$): Employs 128 filters of size 4 × 4 with ReLU activation for higher-level feature extraction.
- Max-Pooling Layer ($m3$): Performs additional down sampling to prepare for the final embedding block.

Final Embedding Block

- The final block of the network transforms the extracted features into a 1D vector representation through a sequence of layers:
  - Convolutional Layer ($c4$): Applies 256 filters of size 4 × 4 with ReLU activation to refine feature extraction.

  - Flatten Layer ($f1$): Converts the 2D feature maps into a 1D vector for input into fully connected layers.
  - Dense Layer ($d1$): Consists of 4096 neurons with sigmoid activation, producing the final embedding vector representing the facial image.

- The output of this embedding network is a high-dimensional feature vector that compactly encodes the most critical facial characteristics.

Distance Calculation Layer: A custom distance layer was implemented to compute the similarity between two feature vectors generated by the shared embedding network. This layer calculates the $L1$ distance between the embeddings of the anchor and validation images, quantifying the dissimilarity between the two images.

Siamese Model Construction: The Siamese model architecture is composed of the following key components:

- Siamese Distance Calculation: The custom '$L1$Dist' layer calculates the $L1$ distance between the embedding vectors of anchor and validation images, serving as the core dissimilarity metric.
- Classification Layer: A fully connected classification layer with sigmoid activation is employed to map the computed distance into a probability value ranging between 0 and 1, suitable for binary classification tasks. The output indicates the likelihood of the image pair representing the same person.

The final Keras model, named Siamese Network, was constructed with anchor and validation images as inputs and a classification output indicating image similarity.

Training the Model: The Siamese network was trained over 50 epochs, with each epoch constituting a complete pass through the training dataset. The model was optimized using an appropriate optimizer and loss function suited to binary classification and distance-based learning. Regularization techniques were employed to prevent overfitting, ensuring robust performance on unseen data.

Model Serialization and Storage: Upon completion of the training process, the trained model was serialized and saved in a format suitable for long-term storage and deployment. This facilitated the reuse and adaptation of the model for further research or real-time applications.

Deployment in a Real-Time Application: The trained model was deployed using the Kivy framework, chosen for its cross-platform compatibility and real-time interaction capabilities. Kivy provided a robust environment for developing a customized user interface tailored to the project's specific requirements, allowing seamless integration of the trained Siamese model for facial recognition.

The deployment incorporated the following features:

- Cross-platform support to ensure device accessibility.
- Real-time interaction for immediate facial recognition feedback.
- Customizable user interfaces designed to optimize user experience.

- TensorFlow integration to leverage the sophisticated algorithms developed during model training.
- Instant user feedback, enhancing usability.
- Easy scalability and adaptability for future enhancements.
- Cost-effectiveness and support from the Kivy community contributed to the overall feasibility of the system.

This methodological approach effectively facilitated the development of a Siamese network for facial recognition, from preprocessing to real-time deployment, utilizing advanced neural network architectures and modern machine learning frameworks.

5.3 Dataset Customization

The dataset for the Siamese neural network was comprised of three primary groups of images: anchor, positive, and negative, each consisting of 1,000 images, totalling 3,000 distinct images. Each set played a unique role in training the network:

Anchor Dataset: Contrary to typical usage where anchors are derived from the main dataset, in this study, anchor images were selected akin to positive images but treated as reference points for comparison. These were randomly chosen from the LFW dataset, ensuring no overlap with images selected as positives or negatives.

Positive Dataset: Each positive image corresponded directly to an anchor image, depicting the same individual but captured in a different photographic instance. This configuration was essential to challenge the network's ability to detect subtle intrapersonal variations under different conditions such as lighting, pose, and expression.

Negative Dataset: In contrast, each negative image featured a different individual than that in the corresponding anchor, facilitating the network's learning of interpersonal differences. These were also randomly selected from the LFW dataset ensuring they were distinct from the individuals in the anchor and positive datasets.

5.4 Pairing and Labeling Methodology

The images selected for the dataset were strategically paired and labeled to facilitate the training objectives of the Siamese neural network:

Positive Pairs

Composition: Each anchor image was paired with a positive image. Both the anchor and the positive images depict the same individual but differ in terms of the capture conditions such as lighting, pose, or expression. This variability within positive pairs is essential as it challenges the network to recognize the same individual across different contexts, thus enhancing its ability to generalize from limited data.

Labeling: These pairs were labeled with '1,' symbolizing similarity. The label instructs the network's learning algorithm to minimize the distance between the feature vectors extracted from these image pairs, thereby learning to align images of the same person closer in the learned feature space.

Negative Pairs

Composition: For each anchor image, a negative image was also selected. The negative images feature different individuals from the anchors, introducing a clear distinction from the positive pairs. This setup is crucial for the network to learn dis-criminative features that are effective in differentiating between individuals.

Labeling: Negative pairs were labeled with '0,' indicating dissimilarity. The training process uses these labels to maximize the distance between the feature vectors of the anchor and negative images, thus pushing representations of different individuals apart in the feature space.

5.5 *Allocation of Data*

The training set, which consists of approximately 70% of the dataset or 1,400 image pairs, is crucial for conditioning the Siamese neural network to accurately identify and differentiate between similar and dissimilar faces. The selected batch size of 16 is optimized to balance computational efficiency with effective learning dynamics, crucial for deep learning models where memory and processing power are often limiting factors. By training the network with a substantial amount of data, it learns robust features that are critical for distinguishing subtle variations in human faces, thereby enhancing its performance in face verification tasks.

For evaluation, the remaining 30% of the dataset, equating to 600 image pairs, is used as the testing set. This separation from the training data is essential to assess the generalization capabilities of the neural network in a controlled, unbiased manner. Testing the model on new, unseen data helps verify its effectiveness in real-world scenarios and ensures that it does not merely memorize the training examples but rather learns a generalizable pattern of features. This testing phase is key to determining the practical viability of the model, highlighting its strengths, and identifying any potential weaknesses before deployment in operational environments.

5.6 *Training*

In the training of Siamese neural network, a meticulously crafted procedure is employed, utilizing TensorFlow's '@tf.function' decorator to ensure efficient execution. This function, 'train_step,' serves as the core of training loop, where each batch of paired images—comprising anchor, positive, and negative images along with their corresponding labels—is processed. During each iteration, the network performs a forward pass with these images to generate predictions, which indicate the likelihood of the pairs being similar or dissimilar. Subsequently, a binary cross-entropy loss is calculated, reflecting the accuracy of these predictions relative to the actual labels. This loss not only quantifies the model's current performance but also guides the optimization process. By calculating the gradients of this loss with respect to the model's parameters, backpropagation is implemented where the optimizer updates the weights to minimize the loss, progressively refining the model's ability to discern between different and identical faces.

Throughout the training epochs, performance metrics such as precision and recall are rigorously monitored to assess the model's efficacy more granularly. Precision measures the accuracy of the positive predictions (i.e., how many predicted as similar are actually similar), while recall evaluates the model's ability to identify all relevant instances of similarity. Both metrics are essential for understanding the model's practical application to real-world scenarios where both false positives and false negatives carry significant implications. Each epoch concludes with updates on these metrics, providing insights into the model's learning trajectory and guiding any necessary adjustments in the training approach. Additionally, periodic checkpoints are saved to preserve the model's state, facilitating recovery and further tuning without starting anew. This structured approach ensures that by the end of the training process, the Siamese network is not only optimized for high accuracy in face verification but also calibrated to generalize well to unseen data.

6 Results and Performance Evaluation

The performance of the Siamese neural network on the LFW dataset demonstrates strong accuracy in face verification tasks. During the training phase, the model achieved an impressive accuracy of 95.79%, indicating effective learning and adaptation to the nuances in the data. Furthermore, in the testing phase, the model exhibited even higher accuracy, reaching 96.17%. This improvement in the testing set underscores the model's robust generalization capabilities, a critical factor for real-world applications where the model encounters data variations not seen during training.

A detailed analysis of the model's performance is provided in Fig. 4, which presents the confusion matrix for the test data. The matrix shows that the model correctly identified 278 cases as similar (true positives) and correctly recognized 305 cases as not similar (true negatives). This indicates a strong ability of the model to distinguish between different faces accurately. However, the matrix also reveals some errors: 5 cases were incorrectly classified as similar (false positives) and 10 cases were incorrectly classified as not similar (false negatives). While these numbers are relatively low, they highlight areas where the model might be improved, such as enhancing sensitivity to subtle facial features or reducing the likelihood of false alarms in certain contexts.

Overall, the high accuracy rates coupled with a detailed breakdown of performance via the confusion matrix provide a comprehensive view of the model's effectiveness

Fig. 4 Confusion matrix obtained on the testing pairs

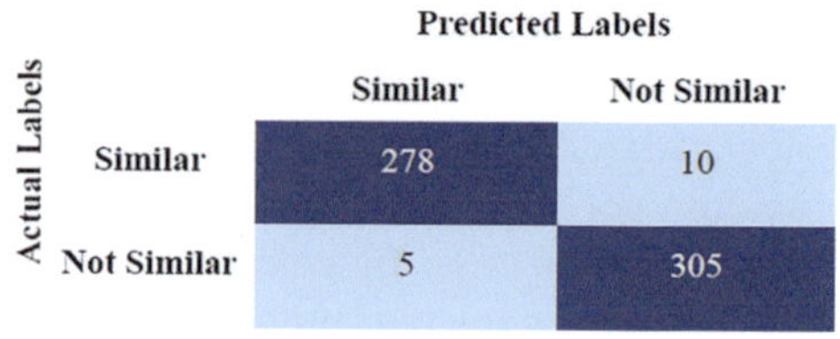

in face verification, showcasing its potential for deployment in security systems, identity verification, and other applications requiring accurate facial recognition.

7 Conclusions

In conclusion, this paper presented the development and evaluation of a Siamese neural network trained on the LFW dataset for the task of face verification. Utilizing a methodical approach to dataset preparation and image pairing, along with an efficient training pipeline enabled by TensorFlow's advanced functionalities, the network demonstrated robust performance, achieving high accuracy in both training and testing phases. Precision and recall metrics indicated the model's effectiveness at identifying similar and dissimilar faces with a low rate of error, underscoring its potential applicability in real-world security systems and identity verification processes. The results affirm the viability of Siamese networks in complex face verification tasks, paving the way for further research and operational deployment in related fields.

References

1. Lake BM, Salakhutdinov R, Gross J, Tenenbaum JB (2011) One shot learning of simple visual concepts. In: 33rd Annual conference of the cognitive science society proceedings, vol 172. The Cognitive Science Society, Boston, United States, pp 2568–2573
2. Hinton G, Osindero S, Teh YW (2006) A fast learning algorithm for deep belief nets. Neural Comput 18(7):1527–1554
3. Bengio Y (2009) Learning deep architectures for AI. Found Trends Mach Learn 2(1):1–127
4. Koch G, Zemel R, Salakhutdinov R (2015) Siamese neural networks for one-shot image recognition. In: 32nd International conference on machine learning proceedings, vol 37. Lille, France, pp 1–9
5. Lei FF, Fergus R, Perona P (2006) One-shot learning of object categories. IEEE Trans Pattern Anal Mach Intell 28(4):594–611
6. Guo G, Zhang N (2019) A survey on deep learning based face recognition. Comput Vis Image Underst 189:1–37
7. Alzubaidi L, Zhang J, Humaidi AJ et al (2021) Review of deep learning: concepts, CNN architectures, challenges, applications, future directions. J Big Data 8(53):1–74
8. Lake BM, Salakhutdinov RR, Tenenbaum J (2013) One-shot learning by inverting a compositional causal process. In: Advances in neural information processing systems 26 proceedings. Lake Tahoe, Nevada, United States, pp 2526–2534

9. Lake BM, Lee CY, Glass JR, Tenenbaum JB (2014) One-shot learning of generative speech concepts. In: 36th Annual meeting of the cognitive science society proceedings. Cognitive Science Society, Quebec City, Canada, pp 803–808
10. Maas A, Kemp C (2009) One-shot learning with Bayesian networks. In: Annual meeting of the cognitive science society proceedings, vol 31. Cognitive Science Society, Amsterdam, Netherlands, pp 142–147
11. Wu D, Zhu F, Shao L (2012) One shot learning gesture recognition from RGBD images. In: 2012 IEEE computer society conference on computer vision and pattern recognition workshops proceedings. IEEE, Providence, RI, pp 7–12
12. Lim JJ (2012) Transfer learning by borrowing examples for multiclass object detection. Master's thesis, Massachusetts Institute of Technology
13. Duque-Domingo J, Aparicio RM, Rodrigo LMG (2023) One shot learning with class partitioning and cross validation voting (CP-CVV). Pattern Recogn 143:1–24
14. Chopra S, Hadsell R, LeCun Y (2005) Learning a similarity metric discriminatively with application to face verification. In: 2005 IEEE computer society conference on computer vision and pattern recognition proceedings, vol 1. IEEE, Vancouver, BC, Canada, pp 539–546
15. Taigman Y, Yang M, Ranzato MA, Wolf L (2014) Deepface: closing the gap to human-level performance in face verification. In: 2014 IEEE conference on computer vision and pattern recognition proceedings. IEEE, Columbus, OH, USA, pp 1701–1708
16. Krizhevsky A, Sutskever I, Hinton GE (2012) Imagenet classification with deep convolutional neural networks. Commun ACM 60(6):84–90
17. Srivastava N (2013) Improving neural networks with dropout. Master's thesis, University of Toronto
18. Simonyan K, Zisserman A (2015) Very deep convolutional networks for large-scale image recognition. In: 3rd International conference on learning representations proceedings. Computational and Biological Learning Society, San Diego, CA, USA, pp 1–1467

Generative AI for Constraint Optimized Layouts in Chemical Warehouses

Anagha Kini, Diya M. Hande, V. Hamsini, Dhriti Rajesh Krishnan, and K. S. Srinivas

Abstract This paper works on an approach for designing chemical warehouse layouts using Large Language Models (LLMs) like GPT-4. The main objective of the study is to generate layouts that minimize safety risks and optimize accessibility. Our methodology talks about framing the design process as a Constraint Satisfaction Problem and specifying constraints within GPT-4 through contextual guidance and prompt engineering. The constraints chosen consider chemical compatibility, proximity of frequently accessed chemicals, and space utilization. We looked into some key aspects in getting expected results from Generative AI tools such as specificity of prompts, providing examples, diverse prompt variations and monitoring for any response biases. Using the outputs we obtained from iterative prompting; we generated a dataset of 400 sample layouts. We finally used a constraint satisfaction evaluator specifically designed to output a custom evaluation metric score for the layouts based on the constraints to analyze the effectiveness of this method. Overall, the generated layouts conformed well to the constraints, demonstrating a fairly high satisfaction rate. The findings thus showed what potential LLMs have when it comes to automating this design process, and presents a novel representation of this problem unique to chemical warehouses. It, however, also identified limitations of the model in compliance with increasing number of rules and highlighted the necessity of a larger database and supervised fine-tuning.

Keywords GPT-4 · Constraint satisfaction evaluator · Data generation

A. Kini · D. M. Hande · V. Hamsini (✉) · D. R. Krishnan · K. S. Srinivas
PES University, Bangalore, India
e-mail: pes1202201315@pesu.pes.edu

A. Kini
e-mail: pes1202203444@pesu.pes.edu

D. M. Hande
e-mail: pes1202201311@pesu.pes.edu

H. Sharma et al. (eds.), *Artificial Intelligence: Theory and Applications*, Lecture Notes in Networks and Systems 5589, https://doi.org/10.1007/978-981-96-1687-9_48

1 Introduction

Efficient warehouse layout design is integral to the effective management of chemical storage facilities. Optimizing warehouse layouts involves balancing various factors such as chemical compatibility, accessibility, and space utilization [1, 2]. Traditional approaches to warehouse layout optimization often rely on manual design or constraint optimization techniques, which may be time-consuming and resource intensive. Unlike other warehouse design processes, where methods like alphabetical sorting, application-based sorting, zone-based layouts, FIFO/LIFO inventory management, and fixed location storage might be sufficient, chemical warehouses involve additional safety hazards that require more sophisticated approaches. These traditional strategies do not account for the reactive nature of certain chemicals or the need for segregation based on their storage requirements. However, recent advancements in artificial intelligence (AI), particularly in natural language processing (NLP), present new opportunities for streamlining and automating this process.

This research focuses on looking at the design process from a different perspective of a Constrain Satisfaction Problem (CSP) by breaking it down to a handful of constraints, creating a dataset using GPT-4 and analyzing it using a constraint satisfaction evaluator. To achieve this, the methodology involves using it with the following predefined constraints:

(1) Chemical Compatibility—Ensure that reactive chemicals are not stored together to minimize the risk of chemical reactions or accidents.
(2) Accessibility of Frequently Accessed Chemicals—Prioritize frequently accessed chemicals by storing them in easily accessible locations within the warehouse, such as the shelves near the door, to reduce retrieval time and enhance operational efficiency.
(3) Allocation Consistency—Ensure that the number of storage cells allocated to each chemical matches the specified quantity of containers associated with it, as provided by the user, to verify that the layout design accurately reflects the required storage capacity for each chemical, considering the maximum container capacity per cell [3].

Collectively, these constraints contribute to a more secure and optimized warehouse environment. The manual management of this process is not only labor-intensive but also inefficient, given the complexity and scale of modern warehouse planning. The output is a 2D matrix of dimensions 4×5 that represents the layout of the chemical warehouse where each column indicates a section, and each cell can store a chemical with a fixed capacity of containers provided by the user. The effectiveness of the generated layouts is evaluated, with feedback from the constraint satisfaction evaluator to check if the layouts meet specified constraints [4]. By integrating AI technologies with constraint satisfaction evaluation, the study provided a more efficient and effective approach to optimizing chemical warehouse layouts [5, 6].

2 Literature Review

As highlighted by Metsälä et al. [7], the precise arrangement of raw materials in warehouses, particularly in chemical storage facilities, is crucial due to the imperative segregation of reactive substances. Incompatible chemical co-location poses significant safety risks, underscoring the necessity for meticulous inventory management to ensure both personnel safety and operational efficiency. This study, in its findings, highlighted the use of Multi-Criteria Decision-Making in the selection of warehouses and pointed out the knowledge gap in research related to raw material warehousing. This led us to look into similar methods of approaching this problem which finally came down to framing it as a Constraint Satisfaction Problem (CSP).

In our pursuit of safety and efficiency in chemical warehouse management, we adopt the CSP framework, as elucidated by Brailsford et al. [8]. CSP is a computational paradigm that formalizes problems where variables must adhere to a set of constraints, ensuring their satisfaction within defined boundaries. Within our chemical warehouse context, we describe three fundamental constraints governing the organization of chemical inventory. To assess the efficacy of our constraint satisfaction approach, a dataset comprising 400 instances was generated and evaluated using Python.

Leveraging the capabilities of GPT-4, these configurations were inputted into the model to generate corresponding chemical placement matrices, formatted as 4×5 grids. This methodology draws inspiration from the research conducted by Azaiz et al. [9], wherein they investigated the proficiency of GPT-4 Turbo in generating formative feedback for programming exercises. Notably, the paper utilized a dataset derived from authentic assessments administered to first-year computer science students at the Ludwig Maximilian University (LMU) in Munich. The qualitative thematic analysis of the feedback generated by GPT-4 Turbo revealed that all feedback was personalized to the student submissions. Implementing the feedback generally led to correct solutions, with only two exceptions. However, only 52% of the feedback was complete and fully correct in every detail. In a similar vein, our study harnesses the power of GPT-4 to extrapolate optimal chemical warehouse layouts from a predetermined set of parameters, thereby exploring the model's efficacy in addressing complex logistical challenges within industrial contexts.

Prompt generation is foundational to our research, enabling the production of 400 output instances crucial for analysis. Our prompt provides a comprehensive problem introduction, detailed descriptions of three constraints with assigned priorities, and specifies the output format [3]. This technical prompt serves as the basis for GPT-4 model to generate optimized solutions [8].

3 Methodology

The study uses a modern approach that involves AI tools like GPT-4 to determine an optimal chemical warehouse layout. The layout planning process is simplified and represented as a CSP [7, 10]. The following sections highlight the steps we followed to perform this research.

3.1 Collecting Data and Establishing Assumptions

The first step involved compiling a comprehensive list of chemicals generally stored in warehouses. This process also included identifying chemicals which if stored together, could lead to harmful reactions. This data was then structured into a spreadsheet, serving as the foundational dataset for defining the constraints. The spreadsheet comprised two columns: The first column contained the chemical names and the second column had a list of chemicals that strongly react with the chemical in the corresponding row. For instance, Acetic Anhydride cannot be stored next to Ethyl Acetate, Lactic Acid, Formic Acid, Hydroxyacetic acid, and Perchloric acid to name a few. So, Acetic Anhydride was included in the first column, while the rest were included in the second column of that row. The sheet plays a vital role in our prompt since we would like to enhance the knowledge of GPT-4 in understanding chemicals that cannot be stored together rather than relying on random luck to place chemicals in the layout.

Moving further, the concept had to be simplified to manage the complexity and scale of the original problem. The warehouse was depicted as a rectangular area using a 2D array that is divided into 20 equal sections. Each cell within this array was assumed to be capable of storing a single chemical. The maximum capacity of containers in each cell can be defined by the user. If the number of containers for a particular chemical exceeds the maximum capacity, it occupies more cells accordingly. For ease of evaluation, the door for all layouts was fixed to be located at the third row, fourth column, right at the center of the edge. This kind of standardization made it more practical in terms of generalizing solutions to warehouses with varying dimensions. Hence, this was our theoretical approach to the problem and it served as the basis for our assessment of GPT's capabilities.

3.2 Input and Dataset Generation

A series of inputs and their respective layouts recommended by it were generated to evaluate the efficacy of GPT-4. Prompts were continuously refined to ensure practicality and logically sound outputs [11, 12]. We adopted a series of prompt engineering strategies including structured input formatting, iterative refinement, contextual reinforcement, inclusion of reference materials, and providing expected output formats. The input provided to GPT-4 contained information regarding the maximum capacity of each cell, the list of chemicals to be stored along with their respective container quantities, and a list of frequently accessed chemicals. It was given as a simple uniformly structured paragraph that could be parsed when required. The final prompt utilized for generating the dataset was refined as shown (Fig. 1.).

The prompt was fed with varying inputs to GPT-4 along with the consolidated chemical list in the form of an attached spreadsheet for reference and used to generate 400 outputs. Each input had a unique set of chemicals, a different maximum cell capacity and varying quantities for each chemical. These were collected into a dataset to fed into the constraint satisfaction evaluator to test the scope of the process [13].

This was the initial prompt that was fed to GPT-4. The consecutive prompts only contained the input of maximum containers, chemicals to be stored with corresponding number of containers and the list of frequently accessed chemicals.

When using GPT-4 to generate multiple outputs based on specific constraints, it can sometimes lose track of the original constraints due to how it processes and generates text. This is primarily due to tokenization and the model's context length limitations. GPT-4 processes text as a sequence of tokens, with a limited maximum context window. As the text generation progresses, earlier parts of the text may fall out of this context window, causing the model to "forget" initial constraints. The attention mechanism in transformers can also lead to a bias toward more recent tokens, further contributing to the loss of focus on initial constraints. To address this, re-entering the original prompt with the chemical lists helped refresh the model's

*"I have a chemical warehouse represented as 4*5 matrix = 20 cells. Each cell is allocated a chemical. Clearly mark a cell as a door at (4,3).*
Constraints:
1. Chemicals that react with each other cannot be stored in the same column. Allocate chemicals accordingly. Refer to the sheet attached to see which chemicals can't be stored together and your own knowledge.
2. Most frequently accessed chemicals must be kept near the door- in the third or fourth row ideally.
3. Maximum number of containers per cell is mentioned. If a chemical has more containers, assign more cells accordingly in the same column.
4. Each cell can either be empty or filled with a chemical of choice. Isolate highly reactive chemicals.
Output a textual description of the layout of the warehouse and chemical placement.
Format for output:
1. Cells required for each chemical and list of frequently accessed chemicals.
2. Warehouse Layout Plan: 2D Array Representation of matrix- [['chem1','chem2'...], [], [] ...]
Given, input:
max - 400 containers
Fluorine - 1000, Chromic acid-1600, Camphor- 400, Formic acid - 700, thallium trinitrate trihydrate- 1000, Nitrogen-600
frequently accessed chemicals:
Camphor, Nitrogen "

Fig. 1 Final prompt for dataset generation

memory, bringing important information back into the context and maintaining the quality of the outputs. Hence, the original prompt was reused whenever we noticed a drastic drop in the quality of outputs being generated.

3.3 Making a Constraint Satisfaction Evaluator

To verify if the outputs generated are optimal in reference to the CSP, we programmed a custom constraint satisfaction evaluator in Python. The evaluator had a database of reactive chemicals. It parses through a layout given and passes it to multiple functions, one for each constraint. These functions produce individual constraint scores for each layout, which are then combined into a total score using a weighted sum formula. The weighted scores reflect the importance of each constraint: Accessibility is weighted at 30%, container count consistency at 20%, and safety at 50%. The score is further normalized to provide a percentage, offering a clear metric for layout efficacy ranging from 0 (poor) to 1 (perfect). By incorporating this evaluation step into our methodology, we ensure that the generated layouts not only meet essential safety and operational standards but also optimize performance based on quantifiable metrics.

4 Experimental Setup

This research follows a simple two-step approach involving the curation of a chemical list and creation of a dataset and testing the efficacy through constraint satisfaction evaluator.

4.1 Curation of a Chemical List and Creation of the Dataset

This step involved compiling data from two Excel files. One file contained information about reactive chemical pairs, while the other contained the generated warehouse layouts and corresponding inputs [8].

4.2 Testing the Efficacy Through Constraint Satisfaction Evaluator

In this step, the outputs GPT-4 generated were evaluated to assess the compliance of the warehouse layout with the constraints. In our evaluator, each of these constraints are provided a weight reflecting their importance and relevance. The order of priority for constraints is as follows:

- Chemical Compatibility Constraint (Weight: 50)
- Accessibility of Frequently Accessed Chemicals Constraint (Weight: 30)
- Allocation Consistency Constraint (Weight: 20).

The evaluator reads data from the Excel files and iterates over the rows in the warehouse datasheet. It extracts relevant information and calculates scores based on the defined constraints. These scores are then averaged and normalized to provide an overall assessment.

The basic functionality of the code is as follows. First and foremost, the three key functions are defined. The function check_reactivity() assesses whether reactive chemicals are stored together. It iterates through each column/section of the storage layout, excluding empty spaces and doors, and decrements a base score if reactive chemical pairs are found. The base score we used was 19, 1 for each non-door cell. For every cell that could be a hazard (stored around an incompatible chemical) the score was reduced by 1. The check_frequently_accessed() function determines if frequently accessed chemicals are placed near the door. It verified that these chemicals are within the same row as or the immediate next row the door in the layout. The check_container_count() function verifies if the cell count per chemical matches the ideal cell count for the chemical based on the maximum containers allowed for each cell.

Each of these functions is run on the layouts generated by GPT-4 and a score is calculated. Let S represent the total score for a layout, which is the sum of individual scores for each constraint, weighted by their respective priorities, that is:

$$S = \sum_{i=1}^{n} (s_i \times w_i) \tag{1}$$

Where

n is the number of constraints evaluated,
s_i is the score for constraint i,
w_i is the weight assigned to constraint i reflecting its priority.

The normalized score, S_n, is then calculated by dividing the total score by the maximum possible score achievable across all constraints:

$$S_n = \frac{S}{S_{\max}} \tag{2}$$

Where $S_{\max}$ is the maximum possible score achievable across all constraints.

This normalized score percentage provides a measure of the efficiency and compliance of the chemical storage layout, ranging from 0 (poor) to 1 (perfect). It allows for easy comparison of different layouts and facilitates decision-making based on performance metrics. The final output is the average Sn value obtained from all layouts generated.

5 Result Analysis

The prompt given to GPT-4 yields text results that describe the warehouse layout in the requested format, that is, as a 2D array of chemicals where each element represents a specific type of chemical (Fig. 2). The output can be visualized using Python (Fig. 3). The figure color codes each chemical and gives a general idea of the chemical placement within the warehouse.

The constraint satisfaction evaluation yielded an average score of 0.9121 from the 400 layouts generated. This score, taking into consideration that the maximum achievable score is 1, indicates a high level of compliance with the specified constraints across the generated warehouse layouts.

The range of scores can be further illustrated in Fig. 4, wherein, for each constraint, as well as for the overall score, histograms were plotted to examine and analyze result patterns. The findings were as follows:

"[['Chromic Acid', 'Chromic Acid', 'Fluorine', 'Thallium Trinitrate Trihydrate', 'Formic Acid'],
['Chromic Acid', 'Chromic Acid', 'Fluorine', 'Thallium Trinitrate Trihydrate', 'Formic Acid'],
['Nitrogen', 'Camphor', 'Fluorine', 'Thallium Trinitrate Trihydrate', 'Empty'],
['Nitrogen', 'Empty', 'Door', 'Empty', 'Empty']]"

Fig. 2 2D array output generated by GPT-4

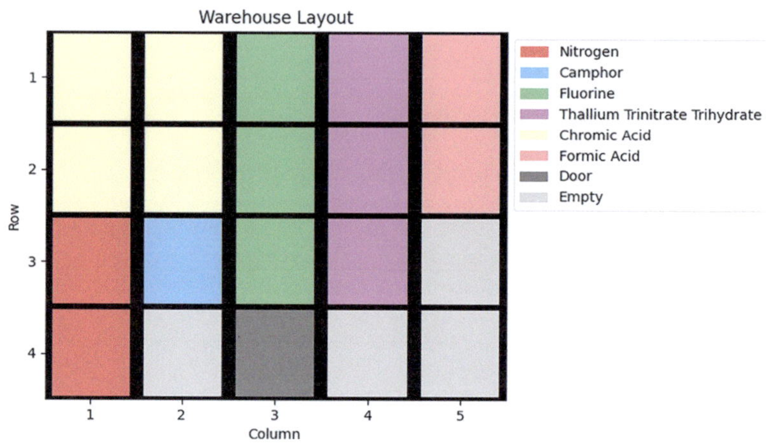

Fig. 3 Warehouse layout

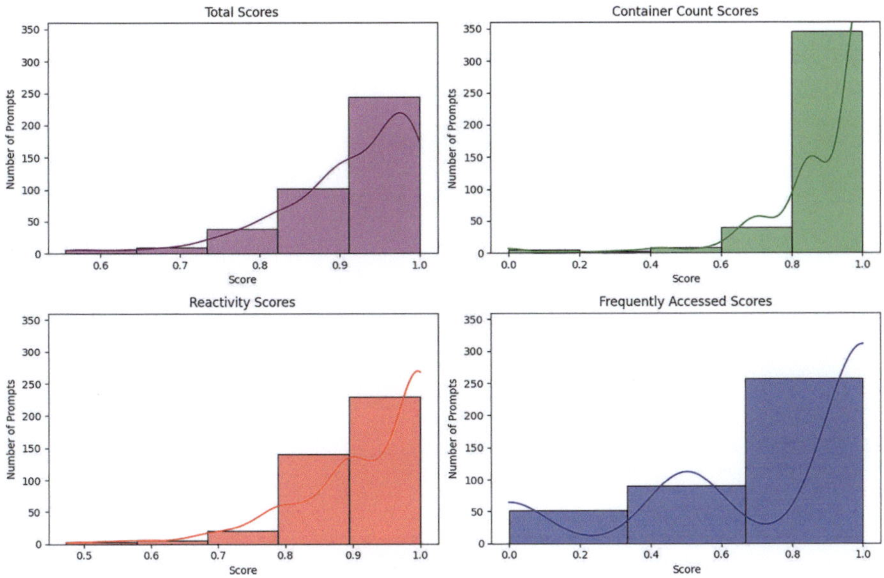

Fig. 4 Distribution of scores of the different constraints

- The graph for total scores for each prompt (top-left) shows a smooth and steady increase in the number of prompts as the score approaches 1.0, with the majority of prompts scoring between 0.8 and 1.0. This indicates that overall, the layouts generated by GPT-4 are in conjunction with most of the constraints, implying a general high level of satisfaction by the evaluator.
- When it comes to the constraint about the maximum count of containers per cell (top-right), the model seemed to understand it well and majority of the outputs had a perfect or near to perfect score. This was, hence, arguably the constraint GPT-4 found easiest to work with. It was also the constraint most clearly defined, with no external considerations or prior knowledge base required.
- The scores that were obtained in relation to the reactivity constraint (bottom-left) varied largely from prompt to prompt, although most scores still ranged between 0.7 and 1.0. This could have been due to a variety of reasons from the input chemical set to the knowledge base and memory retention capability of the model. This was, however, the most complex constraint to explain to the LLM. A more diverse dataset to train and fine tune the model could possibly help improve the consistency and reliability.
- Finally, coming to the placement of frequently accessed chemicals near the door (bottom-right), the range of scores was limited, peaking around approximately 0, 0.5 or 1 as seen in the kernel density estimate curve.
- While examining outputs, as mentioned before, GPT would often "forget" the requirements and begin to place chemicals into cells in the same order as provided in the input. In this case, only the constraint about container count would be

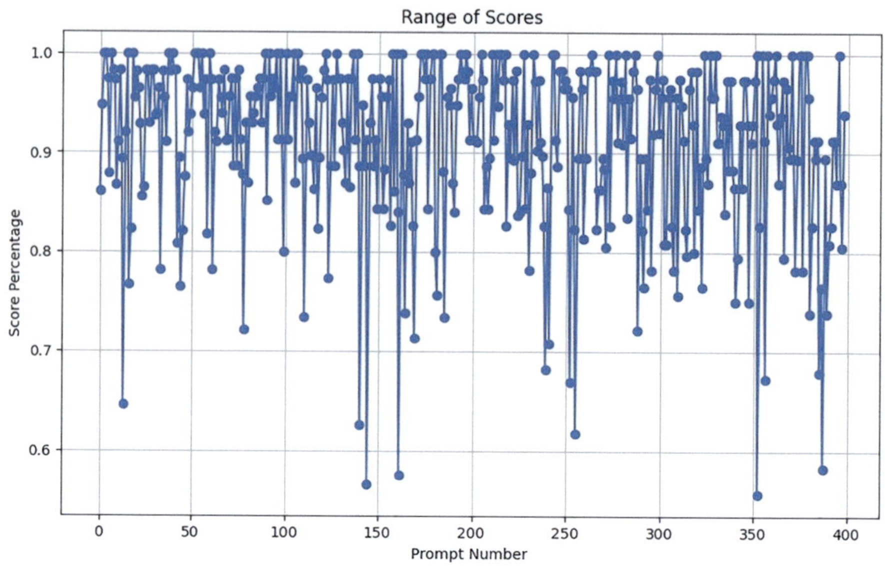

Fig. 5 Plot of range of scores across prompt number

satisfied most of the time and the other constraints would be ignored. This would change when the original prompt was re-entered, which was whenever we noticed a drastic drop in the quality of output as illustrated in Fig. 5, wherein there are sharp drops in scores periodically over time.

- Overall, the final scores had a larger range of results and fewer perfect outcomes as compared to the other constraints. The model showed capability in handling constraints individually, but struggled when it came to combinations. It would satisfy a portion of the constraints but fail in the others while doing so.

It is important to note that even in prompts with "perfect" scores, the layout may not be ideal in other aspects as it is designed on a limited set of expectations. For future work, therefore, more constraints can be taken into account while designing the problem. The scalability of the warehouse and chemicals could be increased and real-world data could be incorporated. Rather than just prompt-based tuning, elaborate datasets could be used for supervised fine-tuning of such models. Human expertise and feedback could help in providing optimal solutions.

6 Conclusion

To conclude, representation of the chemical warehouse planning process as a CSP, simplifies the entire procedure and helps create a layout that's assessable in terms of safety. The integration of GPT-4 with constraint satisfaction evaluation has yielded promising results in optimizing warehouse layouts, however, it's important to note that achieving a high level of compliance with specified constraints often requires fine-tuning and human intervention. As seen in this paper, it doesn't tend to accommodate all the constraints provided by the user for each layout that is generated. GPT-4, like other AI models, may not always produce exact results without adjustments and guidance from human operators. Nonetheless, the demonstrated effectiveness of this approach in minimizing safety risks and enhancing accessibility within storage facilities underscores its potential for practical application in warehouse design and management. We have found it to be less resource intensive and time-consuming as compared to designing chemical warehouse layouts manually.

References

1. Ashayeri J, Gelders LF (1985) Warehouse design optimization. Eur J Oper Res 21:285–294. https://doi.org/10.1016/0377-2217(85)90149-3
2. Asiry S, Ang L-C (2018) Laboratory safety: chemical and physical hazards. In: Methods in molecular biology, pp 243–252
3. Giray L (2023) Prompt engineering with ChatGPT: a guide for academic writers. Ann Biomed Eng 51:2629–2633. https://doi.org/10.1007/s10439-023-03272-4
4. Li X, Yu H, Yuan M (2012) Modeling and optimization of cement raw materials blending process, Math Probl Eng 2012:1–30
5. Abdullin Y, Molla-Aliod D, Ofoghi B et al (2024) Synthetic dialogue dataset generation using LLM agents. arXiv.org. https://arxiv.org/abs/2401.17461
6. Akoury N, Wang S, Whiting J et al (2020) Storium: a dataset and evaluation platform for machine-in-the-loop story generation. arXiv.org. https://arxiv.org/abs/2010.01717
7. Metsälä et al. (2023) Raw material warehouse solution optimization in chemical industry using multi-criteria decision making. https://osuva.uwasa.fi/handle/10024/16558
8. Brailsford SC, Potts CN, Smith BM (1999) Constraint satisfaction problems: algorithms and applications. Eur J Oper Res 119:557–581. https://doi.org/10.1016/s0377-2217(98)00364-6
9. Azaiz I, Kiesler N, Strickroth S (2024) Feedback-generation for programming exercises with GPT-4. https://doi.org/10.1145/3649217.3653594
10. OpenAI, Achiam J, Adler S et al (2023) GPT-4 technical report. arXiv.org. https://arxiv.org/abs/2303.08774
11. Hatakeyama-Sato K, Yamane N, Igarashi Y, et al (2023) Prompt engineering of GPT-4 for chemical research: what can/cannot be done? Sci Technol Adv Mater Methods 3. https://doi.org/10.1080/27660400.2023.2260300
12. Mackworth AK (1992) The logic of constraint satisfaction. Artif Intell 58:3–20. https://doi.org/10.1016/0004-3702(92)90003-g
13. Mitrovic A, Ohlsson S (1999) Evaluation of a constraint-based tutor for a database language. https://ir.canterbury.ac.nz/items/0c4b4830-dd05-4865-8c24-4b298568df77

Demystifying Patterns and Trends Across Different Data Science Skills and Job Roles: An Analysis of Key Determinants Using Machine Learning

Avisek Kundu, Nitesh Dhar Badgayan, Seeboli Ghosh Kundu, and Geetha V. Sharma

Abstract This study focused on exploring different patterns, testing different hypotheses, and establishing relationships between salaries, skill sets, experience, and other dimensions across diverse data science job roles. The dataset used in this study is referred from the Kaggle data source. The dataset's attributes include expertise area, employment type, expertise level, company location across the continents, company size, and the dependent variable salary class. Exploratory data analysis and data visualization were used to understand the different patterns, different hypotheses were established, and statistical tests were incorporated to identify the statistical significance of the different types of relationships. A machine learning technique called Catboost regressor was used to boost the gradient over decision trees. The focus was to leverage tree-based ensemble techniques (Catboost) to ensure the great ability of this algorithm to handle outliers and optimize trees with the right pruning. One of the great advantages of this approach is to ensure that it captures nonlinear relationships (if any), which was not leveraged in previous studies or other research, and the emphasis was on traditional equation-based regression methods capturing

A. Kundu
Technology Consulting (Data Science, ML & AI), Ernst & Young LLP, Hyderabad, India

S. G. Kundu (✉)
Symbiosis Centre for Management Studies, Bengaluru Campus, Symbiosis International (Deemed University), Pune, India
e-mail: seebolighosh.kundu@scmsbengaluru.siu.edu.in

G. V. Sharma
Department of HR and Soft Skills, IBS Bangalore (Off-Campus Centre), IFHE University (a Deemed to-be-University under Sec3 of UGC Act 1956), Bengaluru, India
e-mail: geethasharma@ibsindia.org

N. D. Badgayan
KPMG India, Noida, India
e-mail: niteshdhar.badgayan@scmsbengaluru.siu.edu.in

A. Kundu
Department of Operations and IT, IBS, Hyderabad (A Constituent of ICFAI Foundation for Higher Education), Hyderabad, India, Hyderabad, India

H. Sharma et al. (eds.), *Artificial Intelligence: Theory and Applications*, Lecture Notes in Networks and Systems 5589, https://doi.org/10.1007/978-981-96-1687-9_49

linear relationships. The current study determines which variable has the maximum influence on the dependent variable (salary grades) for the different data science-related job roles, thus helping to create the optimum strategy in capacity building for the diverse organizations in the areas of data science-related skill sets and job roles. The insights from this study can also be leveraged by job aspirants to create their own optimal strategy for their growth plan in data science. The study can also use to estimate the probable salary slab for the different types of data science-related job roles along with the other factors (experience, expertise area, employment type, expertise level, company location across the continents, and company size), thus helping in optimal human resource planning and budgeting.

Keywords Salary prediction · Data science job · HR analytics · Machine learning technique

1 Introduction

The Internet of Things and big data are the primary focuses of today's industry. As a Revolution of 4.0, data science jobs have emerged across the globe. Data science roles revolves around collecting diverse data from different data sources, create analytical worthy data formats using pipelines, leverage machine learning AI or statistical techniques to generate outputs, compare the validity of the outputs across standard matrix and deploy the solution for continuously data-driven decisions or outs through enterprise dashboards or BI systems that ultimately help the management. The data scientist job market analysis indicates a range of disciplines valued by employers, namely Data science, Engineering, Mathematics, Computer science, Statistics, Data engineering, Machine learning, Chemistry, Architecture, Artificial Intelligence, Information systems, etc. [6]. The variety of technical skills mentioned in job postings reflects the diverse and extensive expertise a data scientist must bring—from basic data analysis skills to advanced specialization in various domains. Compared to previous years, we see an increase in the demand for advanced data science skills. Still, despite the evolving tech landscape, the core of entering the field remains the same: statistics knowledge and solid data analysis skills form the foundation of data science expertise [10]. Despite the huge demand for data science jobs, the search indicates there is a huge shortage of data science jobs all over the globe. As reported by McKinsey, there was already a shortfall of 250,000 data science-related jobs in the United States alone by 2016 [4]. The same publisher has also mentioned that such a shortage has never occurred in other industries. Even with ready-to-use data science tools current days, the shortage in the United States has not been reduced based on recent reports from the same sources [3]. Being the most lucrative job field, there is a strong need to analyze data science-specific salary predictions.

2 Literature Review

Data science roles require expertise in specific tools, programming languages, libraries, or databases. The diverse skill sets needed are driven by the vast array of data science tools available. Analyzing 15,000 job listings from 2021 revealed the most popular skills: Python, SQL, R, Spark, AWS, Java, Tableau, Hadoop, TensorFlow, Scala, SAS, Azure, and Scikit-learn among others. The data science domain revolves around diverse job roles, like analytics, machine learning, artificial intelligence, etc., as mentioned [1]. He mentioned in his paper that actionable business insights are always data-driven, and the role of a data analyst is to collect, analyze, visualize, and report that data. In his paper, he also defined various job aspects like data engineer, data architect, data storyteller, business intelligence developer, machine learning scientist, and machine learning engineer. Along with the different data scientist job roles, many researchers have defined or mentioned existing methodologies used in data science. Das et al [2] have mentioned about linear and polynomial regression [2]. In their research, Lothe et al. [5] mentioned that polynomial regression's reduced accuracy with multiple variables, highlighting the challenge of capturing real-world salary dependencies [5]. In their research, Pawha and Kamthania [7] used the methodology of multiple linear regression. According to him, the skillset is more significant over scores for salary prediction. In their research, Dutta et al. (2018) mentioned that random forest outperforms decision trees, achieving 87.3% accuracy due to ensemble methods. In their research Zhang and Cheng [9] predicted salaries with 88.1% accuracy using distance measurement and K-value optimization [9]. Sun et al. [8], in their research, utilized a two-step modeling approach for skills and salary prediction. Their method could achieve lower error rates among tested models [8]. Models like logistic regression, KNN, MLPs, SVMs, AdaBoost, and voting classifiers showed varying success, with pre-processing and grid search optimization enhancing performance. Though data science job roles have caught attention in the domain of HR analytics, however, the studies are very few and sporadic. This current study addresses the literature gap while considering multiple factors for salary predictions of different data science job roles.

3 Data and Methodology

3.1 Objective of the Study

1. To find out the significant factors (expertise areas, locations, and company sizes and others) and the level of impact on salary slabs through statistical tests.
2. Exploratory data analysis and visualization of the data to understand patterns.
3. Machine learning algorithm to generate statistical rules and models for estimating the salary slab based on the different attributes.

3.2 Dataset Description

The dataset used in this study is referred from Kaggle data source. The dataset delineates salary patterns across expertise areas, locations, and company sizes. The dataset's attributes include expertise area, employment type, expertise level, company location across the continents, company size, salary class, and the dependent variable salary. The dataset has 5736 rows and 7 columns with no missing values.

3.3 Methodology

The first step in the analysis includes distribution analysis and incorporating exploratory data analysis and visualizations to generate different patterns and insights. Different hypotheses are formulated from the different insights and patterns generated from the exploratory data analysis. The hypotheses are then tested using statistical tests like Analysis of Variance and Turkey HSD rate to understand and check statistical significance in mean salaries across different company sizes and how mean salaries differ across different expertise levels. This is followed by a machine learning technique called Catboost regressor which was used to boost the gradient over decision trees. It is designed to handle categorical features automatically, making it particularly suitable for datasets that include categorical variables. The tree-based ensemble techniques (Catboost) have also the great ability to handle outliers and create optimized trees with the right pruning. One of the great advantages of this approach is to ensure that it captures nonlinear relationships (if any), compared to the traditional equation-based regression methods capturing linear relationships. The current study determines which variable has the maximum influence on the dependent variable (salary grades) for the different data science-related job roles, thus helping to create the optimum strategy in capacity building for the diverse organizations in the areas of data science-related skill sets and job roles. The insights from this study can also be leveraged by job aspirants to create their own optimal strategy for their growth plan in data science. The study can also be used to estimate the probable salary slab for the different types of data science-related job roles along with the other factors (experience, expertise area, employment type, expertise level, company location across the continents, and company size), thus helping in optimal human resource planning and budgeting. The dataset is divided into 2 parts namely training and testing, with 80% data split in the training set and 20% data in the testing set. A CatBoost regressor was used to pass the categorical variable directly without any encoding technique. Encoding categorical variables numerically in tree models can cause splits at non-meaningful values, leading to poor decision-making and interpretability issues. However, CatBoost allows us to pass categorical variables directly.

4 Data Analysis and Interpretation

The *x*-axis represents the expertise area, while the *y*-axis represents the salary in millions. The box plot shows that Data Science Expert has the highest median salary. The box plot also shows that the salaries for Data Science Expert, Data Expert, and BI Expert are more spread out than other expertise areas. This means that there is a wider range of salaries for these positions. There are also some outliers, represented by red dots. These are salaries that are significantly higher than the other salaries in that expertise area. It is important to note that this box plot is just a snapshot of salary data. It does not consider other factors that can influence salary, such as experience, location, and company size (Figs. 1 and 2).

The *x*-axis represents the expertise areas, while the *y*-axis represents the average salary. From the graph, we can see that the expertise area with the highest average salary is AI Architect, followed by AI Expert and BI Expert. As we move down the graph, the average salary decreases progressively, with the lowest average salary being earned by technicians. It's important to note that the exact values for the average salaries are not provided in the graph, so we can only make comparisons between the different expertise areas (Fig. 3).

The highest average salaries are found in South America for large companies. Average salaries are generally higher in Asia, North America, and Europe, compared to Africa, Oceania, and South America. Salaries tend to be higher in larger companies, as shown by the brighter colors in the "Large" column. The heatmap suggests that factors like company size and location significantly influence salary levels, though other factors not shown (e.g., industry, job role) also likely contribute. The heatmap appears to be a mix of categorical and numerical data. The columns representing Company Location (Continent) and Company Size are categorical, whereas the cells

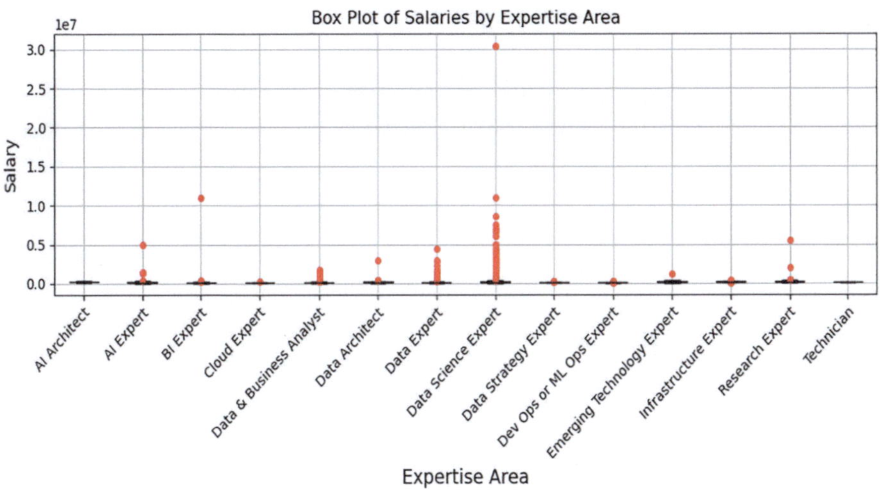

Fig. 1 This box plot shows the distribution of salaries for different expertise areas

Fig. 2 Bar graph shows the average salary for different expertise areas in a particular field

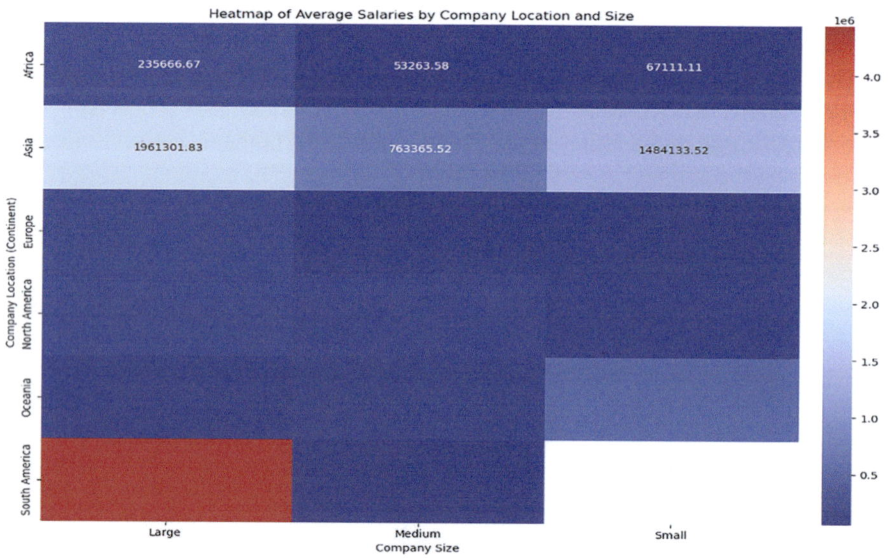

Fig. 3 This heatmap shows the average salaries of employees by company size and company location

containing numbers represent the average salaries. The reason not all blocks have numbers is that the heatmap is likely showing the average salaries for each combination of company location and size. The numbers are only present in the cells where there is data to display. The empty cells might indicate that there is no data available for that specific combination of company location and size. For example, if there are no large companies in Africa, the cell for Africa and large would be empty. Similarly,

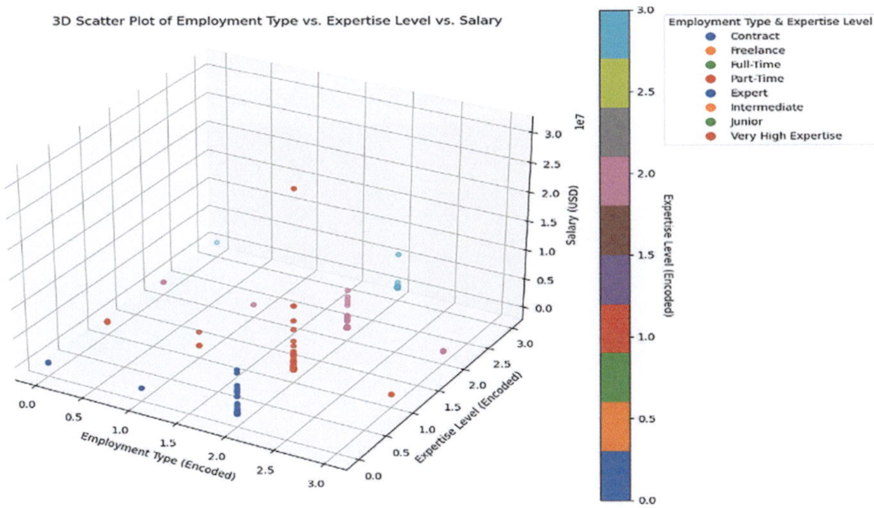

Fig. 4 3D scatter plot depicts the relationship between employment type, expertise level, and salary

if there are no small companies in Asia, the cell for Asia and small would be empty (Fig. 4).

The above figure refers to a 3D scatter plot showing the relationship between employment type, expertise level, and salary. The *x*-axis represents the employment type, the *y*-axis represents the expertise level, and the *z*-axis represents the salary. The color of each point corresponds to the expertise level. The plot also includes a color bar and a legend to help interpret the data. The plot shows that there is a general trend of higher salaries for higher expertise levels, regardless of employment type. However, there is also some variation in salary within each expertise level, indicating that employment type may also play a role in determining salary. The color bar on the right side of the plot shows the encoding for the expertise levels, with darker colors representing higher expertise levels. The legend on the top left of the plot shows the encoding for the employment types, with different shapes and colors representing different employment types. Overall, this plot provides a visual representation of the relationship between employment type, expertise level, and salary, and can help identify trends and patterns in the data (Fig. 5).

The above graphs showcase the salary distribution by employment type and expertise level. The *x*-axis represents the expertise level, which is divided into four categories: intermediate, expert, very high expertise, and junior. The *y*-axis represents the salary in USD. The different colors represent the different employment types: full-time (blue), contract (orange), part-time (green), and freelance (purple). The graph shows that salaries tend to be higher for those with higher expertise levels. For example, the average salary for those with very high expertise is higher than the average salary for those with intermediate expertise. The graph also shows that salaries tend to be higher for those who are employed full-time, followed by contract,

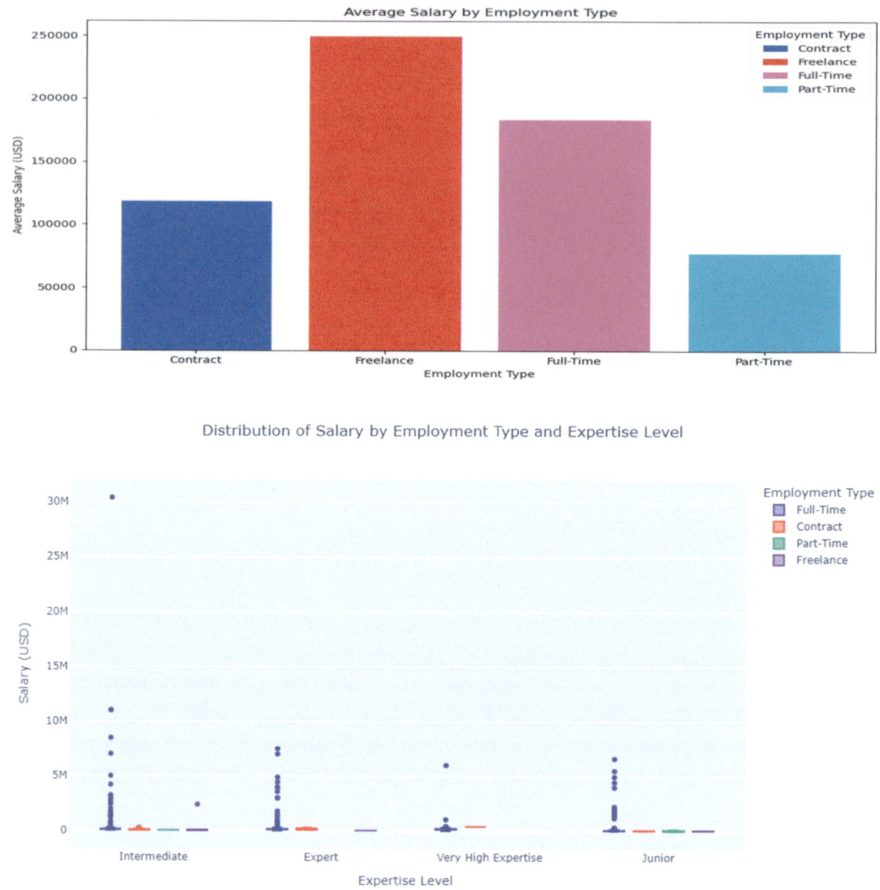

Fig. 5 Graph shows the salary distribution by employment type and expertise level

part-time, and freelance. There is a wide range of salaries for each expertise level and employment type. The highest salaries are earned by those with very high expertise who are employed full-time. The lowest salaries are earned by those with intermediate expertise who are employed freelance. Overall, the graph suggests that expertise level and employment type are both important factors that influence salary (Fig. 6).

The expertise level is on the *x*-axis and the average salary is on the *y*-axis. The employment type is represented by different colors as showcased below:

- Blue: Contract
- Orange: Freelance
- Green: Full-Time
- Purple: Part-Time

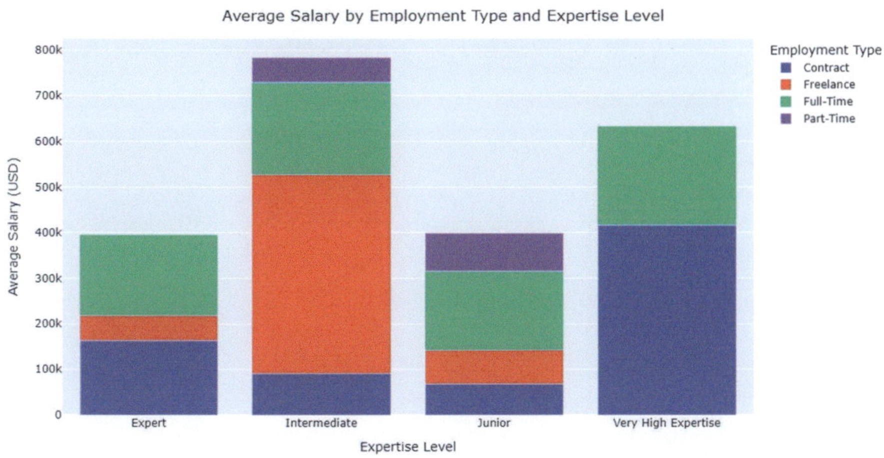

Fig. 6 Chart shows the average salary by expertise level and employment type

The chart shows that the average salary increases as the expertise level increases. It also shows that full-time employees generally earn the highest salaries, followed by contract, then freelance, and finally part-time (Fig. 7).

This tree map showcases the distribution of expertise areas. The largest area is dedicated to Data Science Expert, followed by Data Expert, Research Expert, and BI Expert (Fig. 8).

The chart shows the salary distribution based on three salary classes—high, medium, and low. We can see that the high-salary class has the highest spread and outliers. This indicates that there are some individuals with extremely high salaries in this class. In contrast, the low-salary class has a much smaller spread and no outliers. This means that the salaries in this class are more clustered around a central value. The medium salary class has a moderate spread with a few outliers. It appears that the salary distribution for the high-salary class is much wider than for the medium and low-salary classes. Overall, this chart gives us a good understanding of the salary distribution within the different salary classes (Fig. 9).

This box plot shows salary distribution across continents for different salary classes. As we can see, the salary distribution for the high-salary class is skewed to the right. The median for high-salary class is at about 1.6 million. Also, we can see outliers for high-salary class in Asia and North America. For the medium salary class, the box plot suggests that the median salary is about 1 million, with Asia being the only continent that has an outlier in the medium salary class. Finally, the low-salary class is the most widely distributed with median salary at about 100,000. Overall, this box plot shows that salary class can be a factor influencing the salary distribution across different continents. The salary is highest in Asia, followed by North America, and then Europe. In contrast, Africa, Oceania, and South America have the lowest median salaries across the salary classes.

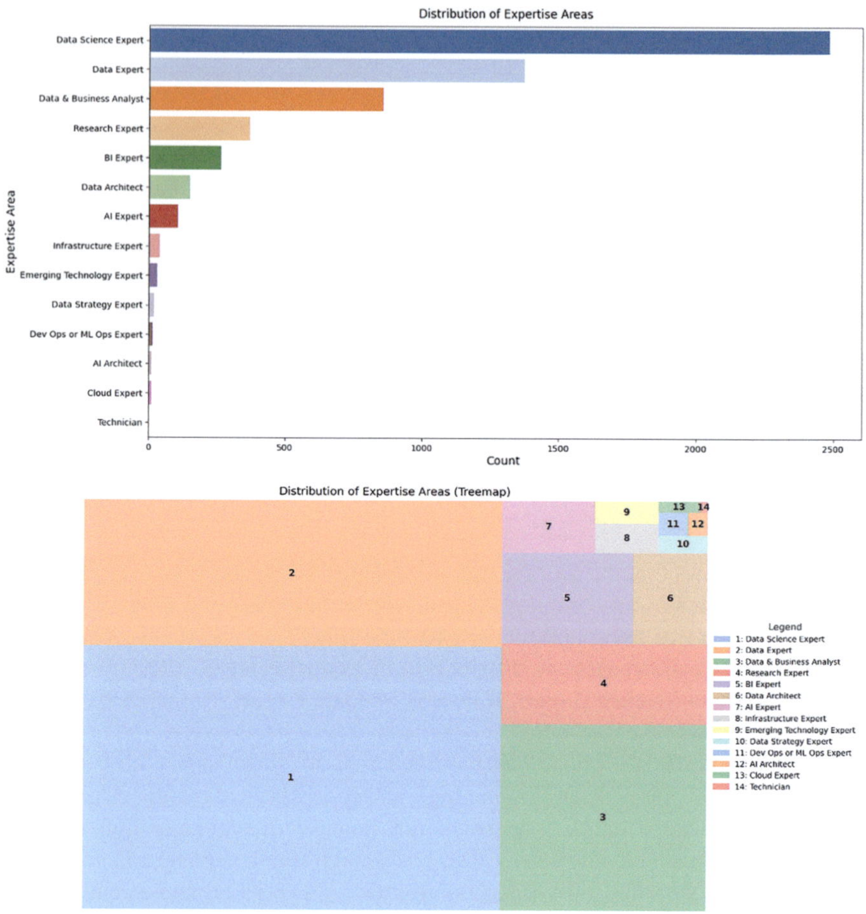

Fig. 7 Tree map shows the distribution of expertise areas

4.1 Hypothesis Testing

Hypothesis 1 H0: There is no difference across the mean salaries across different expertise levels.

The ANOVA result for Salary by Expertise Level provides the following values:

***F*-statistic**: 1.0828 and ***p*-value**: 0.3550

Conclusion: Since the *p*-value (0.3550) is greater than the benchmarked confidence level of 95% or significance level of 0.05, we fail to reject the null hypothesis. This means that no statistically significant evidence suggests that the mean salaries differ across different expertise levels and the difference in the sample was merely a sample statistic and not a population finding.

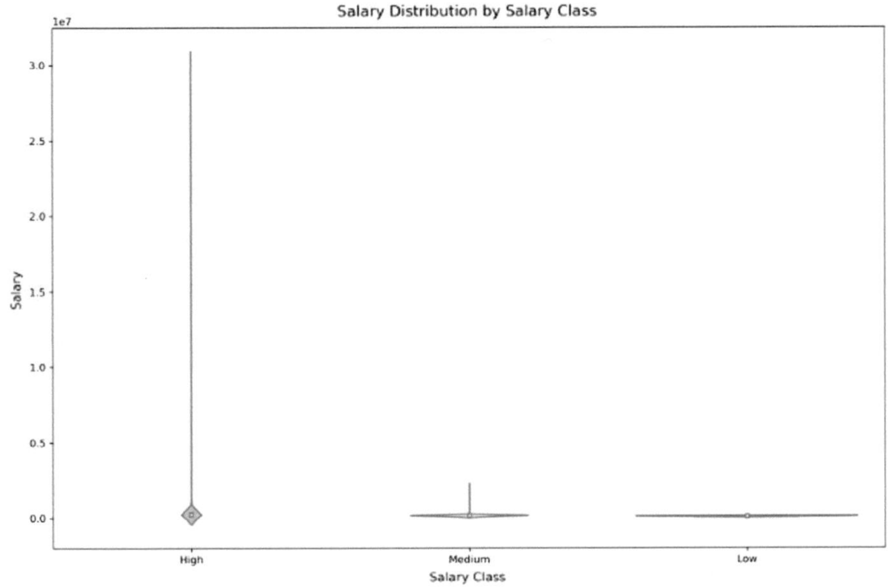

Fig. 8 Chart showcases the salary distribution based on salary classes

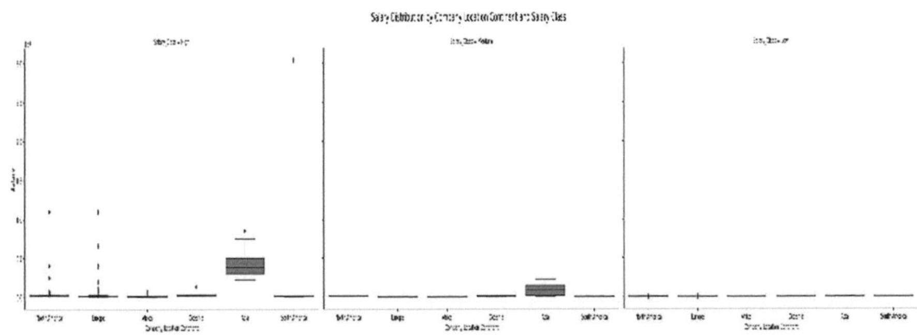

Fig. 9 This box plot shows salary distribution across continents for different salary classes

Hypothesis 2 H0: There is no difference across the mean salaries across different company sizes.

The output from the one-way ANOVA is as follows:

***F*-statistic**: 58.5029 and ***p*-value**: 7.0520e−26 (approximately 7.05×10^{-26}) which is closer to 0. Since the p-value is less than the benchmarked confidence level of 95% or significance level of 0.05, we fail to reject the alternate hypothesis. This means there is statistically significant evidence suggesting that the mean salaries differ across different company sizes. Given that the ANOVA test is significant, we

	group1	group2	meandiff	p-adj	lower	upper	reject
1	Large	Medium	-264371.0962	0.0	-323150.4683	-205591.7241	True
2	Large	Small	-133303.2166	0.0179	-248138.1157	-18468.3175	True
3	Medium	Small	131067.8796	0.0074	29014.7137	233121.0455	True

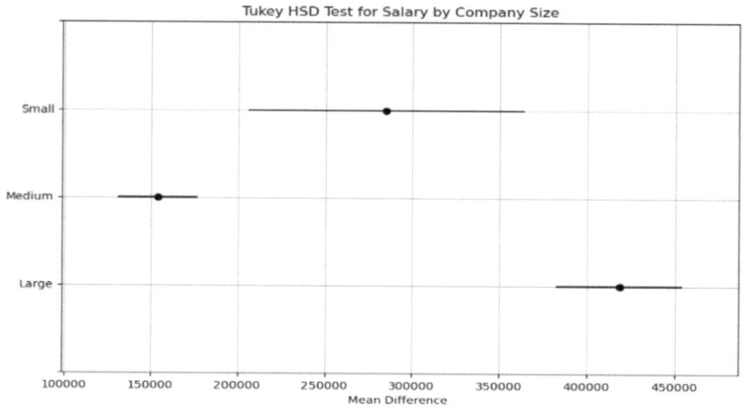

Fig. 10 Interpretation of Tukey HSD result

proceed with Tukey's HSD test to determine which specific pairs of company sizes have significant differences in mean salaries (Fig. 10).

Table 1 represent the pairwise comparison of the different categories and the statistical significance of the mean difference in salaries across the categories.

Summary of Tukey's HSD Test Result

Large versus Medium: There is a significant difference in mean salaries, with medium companies paying more than large companies.

Large versus Small: There is a significant difference in mean salaries, with small companies paying more than large companies.

Medium versus Small: There is a significant difference in mean salaries, with medium companies paying more than small companies.

Overall, these results indicate that company size significantly affects salaries, with medium-sized companies generally offering the highest salaries, followed by small companies and large companies offering the lowest salaries.

Machine Learning Techniques Used

CatBoost regressor with 1000 iterations, depth of 4, a learning rate of 0.1, and loss function is evaluated as an RMSE approach. At first, the importance of each of the feature is evaluated and it is found that the company location has an importance of 49%, followed by salary class and company size. The least important feature

Table 1 Comparison of the different categories to check statistical significance of the difference

Category comparison	Mean difference	Adjusted *P* value	Confidence interval range (lower/upper)	Decision
Large versus medium	− 264,371.0962 (The mean salary for large companies is approximately $264,371 less than the medium companies)	0.0 (Highly significant difference at 5% significance level)	Confidence interval ranges from − 323,150.4683 to − 205,591.7241, which does not include zero	Significant difference (True)
Large versus small	− 133,303.2166 (The mean salary for large companies is approximately $133,303 less than the small companies)	0.0179 (Highly significant difference at 5% significance level)	Confidence interval ranges from − 248,138.1157 to − 18,468.3175, which does not include zero	Significant difference (True)
Medium versus small	131,067.8796 (The mean salary for medium companies is approximately $131,067 more than for small companies)	0.0074(Highly significant difference at 5% significance level)	Confidence interval ranges from 29,014.7137 to 233,121.0455, which does not include zero	Significant difference (True)

was employment type and hence it was eliminated from further successive analysis (Fig. 11).

The CatBoost decision tree analyses the different features and ranks the features in terms of their importance to the target variable. For our case, Table 2 represent the rankings of the features in terms of their importance impacting the salary class.

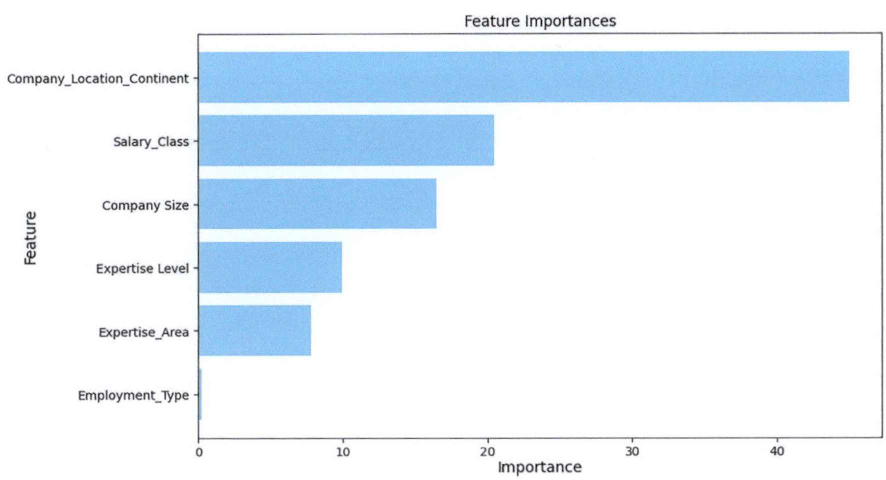

Fig. 11 Feature importance from Catboost model

Table 2 Feature importance and ranking of the factors from the Catboost decision tree

Feature name	Feature description	Ranking in terms of importance	Level of significance
Company_ Location_ Continent	The location of the company marked to a continent	1st	Significant
Company_Size	The size of the company in terms of large, medium, or small	2nd	Significant
Expertise_Level	This includes the level of expertise like entry level, intermediate level, expert level, or very high expertise	3rd	Significant
Expertise_Area	The different expert areas pertaining to data science like BI expert, data science expert, AI architect, etc	4th	Significant
Employment_ Type	The type of employment for the resources like full-time, part-time, contract, freelance, etc	5th	Insignificant

The CatBoost algorithm resulted in a Root Mean Squared Error: 143,419.21701871674 and R^2 Score: 0.7899705242573285. The RMSE value delineates a good model performance which is followed by R^2 value of 0.789 which means 79% of the variance in salary can be explained by the independent variables using CatBoost model. The below QQ plot expounds the pattern of distribution of predicted value with respect to actual value (Fig. 12).

This QQ plot suggests that the distribution of the predicted values is similar to the distribution of the actual values, but not perfectly identical. This is not uncommon, and it is important to consider other performance metrics (e.g., R-squared, RMSE) to evaluate the overall performance of the model. From these matrices the model seemed to have descent performance and stability.

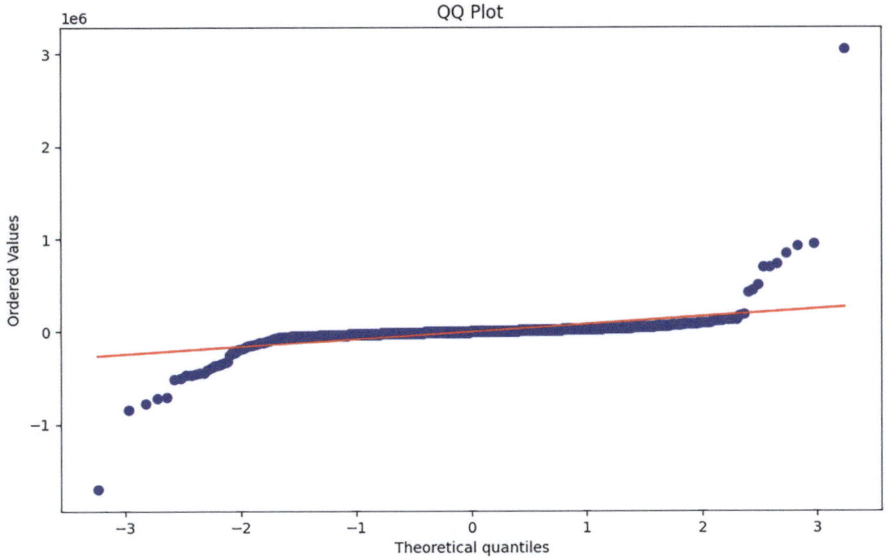

Fig. 12 QQ plot for predicted vs actual in terms of salary

5 Conclusion

This study explored different patterns, testing different hypotheses, and establishing relationships between salaries, skill sets, experience, and other dimensions across diverse data science job roles. The tree-based ensemble techniques (Catboost) is leveraged to ensure the great ability of this algorithm to handle outliers and optimize trees with the right pruning. One of the great advantages of this approach is to ensure that it captures nonlinear relationships which was not leveraged in previous studies or other research, and the emphasis was on traditional equation-based regression methods capturing linear relationships. The current study ranks the different features as par their feature importance on the dependent variable (salary grades) for the different data science-related job roles, thus helping to create the optimum strategy in capacity building for the diverse organizations in the areas of data science-related skill sets and job roles. The insights from this study can also be leveraged by job aspirants to create their own optimal strategy for their growth plan in data science. The study can also use to estimate the probable salary slab for the different types of data science-related job roles along with the other factors (experience, expertise area, employment type, expertise level, company location across the continents, and company size), thus helping in optimal human resource planning and budgeting.

References

1. Chen L, Sun Y, Thakuriah P (2020) Modelling and predicting individual salaries in United Kingdom with graph convolutional network. In: Hybrid Intelligent systems. Advances in intelligent systems and computing, pp 61–74
2. Das S, Barik R, Mukherjee A (2020) Salary prediction using regression techniques. Proc Ind Interact Innov Sci Eng 1–5
3. DuBois J (2020) The data scientist shortage in 2020. https://quanthub.com/data-scientist-shortage-2020/
4. Jain K (2019) Big job opportunities in data science & machine learning. Express Comput 1–3
5. Lothe PD, Tiwari P, Patil N, Patil S, Patil V (2021) Salary prediction using machine learning. Int J Adv Sci Res Eng Trends 6(5):199–202
6. More A, Naik A, Rathod S (2021) PREDICT-NATION skills based salary prediction for freshers. In: Proceedings of the 4th international conference on advances in science & technology (ICAST2021)
7. Pawha A, Kamthania D (2019) Quantitative analysis of historical data for prediction of job salary in India—A case study. J Stat Manag Syst 22(2):187–198
8. Sun Y, Zhuang F, Zhu H, Zhang Q, He Q, Xiong H (2021) Market-oriented job skill valuation with cooperative composition neural network. Nat Commun 12:1–11
9. Zhang J, Cheng J (2019) Study of employment salary forecast using KNN algorithm. In: International conference on modeling, simulation and big data analysis, pp 166–170
10. Zhu H (2021) Research on human resource recommendation algorithm based on machine learning. Sci Program 1–10

Eat Well, Feel Good: The Positive Impact of Blockchain Enabled Geographic Indication (GI) Labeling on Consumer Consumption Values

Abhinita Daiya, Dhanya Pramod, and Kanchan Pranay Patil

Abstract This study uses the Theory of Consumption Values (TCVs) to examine the variables influencing customers' buying decisions when purchasing Native Produce, Foods and Beverages with blockchain technology based geographic indicators establishing the uniqueness, authenticity, and originality. Additionally, it investigates whether consumers' health consciousness moderates the relationship between consumption values of geographic indicator labeled produce, foods and beverages and purchase behavior. Using SmartPLS 4.0, structural equation modeling is used to analyze consumer data obtained through a structured questionnaire. The results show that the use of blockchain technology for geographic indicator labeling increased the items' consumption value, which in turn had a favorable effect on the purchase of products with geographic indicator labels. This study concludes that functional, economical, emotional, and novelty consumption values are the primary determinants of consumption value while purchasing blockchain enabled geographic indicator labeled products. However, social values were found to be insignificant. Further the moderation effect of health consciousness was significant on functional value, emotional value, economical value and purchase, but insignificant for social values, and novelty values. The study provides essential insights into how to effectively promote locally produced Native Produce, Foods and Beverages with blockchain-based geographic indicator labels, this study has substantial managerial and societal benefits.

Keywords Blockchain technology · Geographic indicator labeling · Supply chain · Theory of Consumption Value · Purchase behavior · Health consciousness · Native produce · Foods and beverages

---

A. Daiya (✉) · D. Pramod · K. P. Patil
Symbiosis Centre for Information Technology, Symbiosis International (Deemed University), Pune, India
e-mail: abhinita@scit.edu

H. Sharma et al. (eds.), *Artificial Intelligence: Theory and Applications*, Lecture Notes in Networks and Systems 5589, https://doi.org/10.1007/978-981-96-1687-9_50

1 Introduction

The consumers over the globe are concerned about the origin and quality of the products in the market [1–3]. However, the products with specific authentic geographical identity are treated specially by the consumers [4]. According to World Intellectual Property Indicators (WIPO), geographic indicators are symbols applied to goods that are sourced from a particular area and have characteristics or a reputation unique to that area [5]. In the present era of technology, Blockchain enabled Geographical Identity (BGI) is considered as the indication of authenticity of the products [6]. With increase in trends of geographically indicated products, manufacturers are registering their products as an Intellectual Property to secure the originality and authenticity of their products [7].

International Statistics from WIPO states that China (9052) secures the highest number of GIs, followed by Hungary (7743) and Czech Republic (6272) [5]. On the other hand, presently Intellectual Property India filed 643 Geographically Identified products till date [8]. The government of India has registered such GI labeled products Department of Industrial Policy and Promotion (DIPP) under ministry of commerce and Industry and consider them as the "Invaluable treasures of incredible India" [9]. The products like, Tea from Darjeeling, Pashmina from Kashmir, Thewa art work from Rajasthan, Sandalwood oil from Mysore are few of the examples of such registered GI labeled products from India. The Indian government categorized these GI products under four major segments, namely, Home Fashion and Collectibles, Native Produce, Foods and Beverages, Wearables and Accessories, and Nature and Wellness. However, in the present research we are studying the products which fall under the category of Native Produce, Foods and Beverages.

However, to trace the authenticity of the GI products a robust blockchain technological support to the supply chain is highly required [10]. Thus, the centralized system of transactions proved to be highly cost consuming and inefficient for such products in the global markets [11]. The BGI labels increase the consumption value of the product and hence the willingness to purchase escalates. Businesses are presently using blockchain technology to enhance and secure their supply chain in the market [12]. The integration of BGI products will not only build trust in the product quality, but also helps in identifying fraud and counterfeiting of products [13] which is around $1.7 trillion to $4.5 trillion sold each year [14]. Thus, the requirement of blockchain adaptation for the GI labeled product could revolutionize the global market. The present study tries to bridge this research gap by studying the impact of BGI labels to consumer's consumption value. Subsequently, exploring the impact of consumption values on willingness toward the purchase of products.

Consumers look for the value while purchasing products [15], therefore the question is what are the consumption values that impact the purchase decisions of the blockchain-based GI indicator products. According to the Theory of Consumption proposed by Sheth et al. [16], there are five major indicators that provide insight into the inspiration of consumption behavior of the consumers. Thus, the impacting indicators that affect consumer behaviors are, namely, functional, economical, emotional, social, and novelty value [17]. Thus, the objectives of the present research are: (i) To identify how blockchain-based GI indicator labeled products impact the consumption value further impacting the purchase decisions and (ii) to examine the moderating effect of consumer health consciousness on relation between consumption value and purchase decisions. The present study will enable the businesses to analyze how the blockchain technology integration of GI products will facilitate the traceability hence enhancing the value of the products. Thus, it will attract the potential consumers who value authenticity and transparency of the GI tagged products. In addition, it will support the businesses to develop strategies according to the consumer purchase attitudes while the blockchains are integrated to the supply chain of their GI tagged products.

2 Literature Review

2.1 Blockchain Enabled GI Labeled Products

Blockchain enabled geographical indication (GI) labeled items combine traditional authenticity indicators with cutting-edge technology. GI designation indicates that a product comes from a certain location noted for its distinctive features and heritage, such as Oranges from Nagpur or Tea from Darjeeling. Integrating blockchain technology improves the legitimacy and traceability of these products. Blockchain creates a decentralized and unchangeable ledger that tracks every stage of a product's lifecycle, from creation to end consumer, assuring transparency and eliminating fraud. This integration adds tremendous value to GI branded items by providing verifiable confirmation of origin and quality, thus protecting both manufacturers and customers from counterfeit products. The comprehensive supply chain data that is gathered can also improve regulatory compliance and inventory management. Thus, by assuring and adding consumption values (functional, economical, emotional, social, and novelty) to the products by preserving their distinctive regional history, BGIs improve consumption [17]. It also allows companies to charge higher rates because of the confirmed authenticity and traceability.

In Indian context, blockchain technology would dramatically improve Indian geographical indication (GI) labeled products by creating a safe, transparent, and unchangeable record of their origin and production processes. This technology provides authenticity and traceability, making Darjeeling tea and Pashmina shawls less likely to be counterfeited. It will not only improve the authenticity and traceability, but will also improve the marketability of Indian GI goods. Furthermore, it allows consumers to make more educated purchasing decisions, which fosters emotional value for the products originating in their own country, sometimes invoking nostalgic feelings too. The impact on consumers is significant; they acquire confidence in the quality and provenance of the products they buy, which fosters trust and loyalty. Furthermore, the capacity to track a product's history may affect consumer behavior by linking purchase decisions to principles like sustainability and ethical production. Overall, blockchain integration increases value and trust while also encouraging informed and conscientious customer behavior, finally, increases the entire value and global reputation of Indian GI labeled products.

2.2 Underpinning Theory

The Theory of Consumption Value (TCV) is a popular marketing theory that uses consumption values to explain why customers act in particular ways [16]. The concept is centered on consumption values and details why customers decide to purchase a particular product over another, select a particular brand over another, and decide not to purchase a particular product at all. The theory can be used to decide on various product categories, including industrial items, services, consumer durables, and consumer non-durables.

While there is a small amount of literature in marketing theory describing value from the consumer's perspective, current marketing philosophy is value-oriented. By emphasizing consumption values, this theory offers insight into the driving forces behind consumption behavior and helps to forecast, characterize, and explain choice behavior.

The theory argues that functional, social, emotional, conditional, and epistemic values all impact behavior. The perceived benefit derived from a substitute's ability to perform in a functional, utilitarian, or physical capacity is identified as its functional value. A product's qualities, including pricing, quality elements, and dependability, impact its TCV value. Based on the underpinning theory, Theory of Consumption Value, the researchers proposed the conceptual model as presented in Fig. 1.

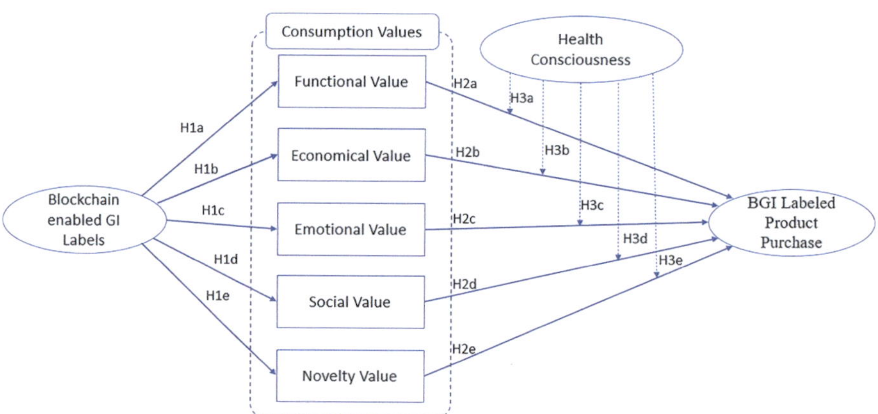

Fig. 1 Conceptual model

2.3 Hypothesis Development

Social value is a measurement of consumer satisfaction shaped by the opinions of peers [18]. It is derived from a substitute's perceived value derived from its connection with one or more particular social groupings. The pointer of consumer sentiment toward eco-friendly products is called emotional value. It is the perceived benefit derived from the ability of a substitute to evoke emotions or emotional states. The measure of utility according to a particular circumstance is called conditional value. It is the perceived benefit of a substitute's ability to ignite interest, offer novelty, and satisfy a need for information. Measuring consumer dispositions to seek novelty and information is the focus of epistemic value. It is the perceived benefit that an option gives as an outcome of a particular scenario or set of circumstances that the decision-maker must deal with [19].

Knowing why customers are buying the goods could stimulate different marketing approaches that enhance the extraction of consumers' informational rents and create a separate equilibrium [20]. Labels bearing the geographic indicator (GI) are used to identify products that have particular attributes, reputations, or traits related to their place of origin. These labels are often used in the food, beverage, and agriculture sectors. GI labels notify customers that a product satisfies quality requirements specific to its country of origin. Additionally, it promotes product safety and authenticity, which raises the perceived functional value. Customers' pleasure and trust may increase as a result of this perceived quality guarantee. Consumer perceptions and experiences are influenced by the existence of GI labels, which increases consumer happiness and value in their purchasing decisions.

Research on food habits indicates that rather than weighing the advantages and disadvantages of a given food option, consumers typically rely on food retailers to guarantee a minimum standard of food safety. Because they are assumed to be of greater quality and authenticity, products bearing GI labels frequently fetch a premium price. When consumers believe a product will fulfill high standards, they are prepared to pay a higher price, which enhances producer profitability. This promotes rural development, and the higher demand and higher pricing for these goods directly contribute to this economic growth. By encouraging price premiums and market distinction, GI labeling boosts local economies, environmental sustainability, consumer awareness and promotes job growth and stability in the economy [21].

To capture utilitarian gains that can be obtained from buying or eating a product, functional and economic values are taken into consideration. On the other hand, a product's socio-psychological advantages are captured by emotional, social, and novelty values [17]. Products that have a strong connection to tradition and legacy are frequently valued by consumers. It can be especially enticing when geographic indicator labels generate a sense of continuity and history. Because GIs are associated with exclusivity and prestige, using and owning these things can give customers a sense of pride and status, which can increase the emotional consumption value. Also, a study indicates that a higher incidence of mental health issues is linked to the pandemic as a result of social isolation and anxiety over one's health and money. As a result, in order to improve their emotional state, consumers are increasingly consuming food that has emotional value [22].

Customers get a sense of cultural appreciation and responsibility as a result of feeling a connection to the product's origins and culture. SCV refers to how goods and services are perceived and valued in social contexts, taking into account elements including sustainability, cultural identity, community support, and social standing. Consuming food that has been cultivated nearby can have advantageous social benefits, such as increased social standing, a more sophisticated appearance, or closer relationships with others [23]. Because regional products have strong place-product linkages, a product developed in a particular region may also be perceived as more unique and natural than a mass-produced one [24]. Customers appreciate GI items' distinctiveness and individuality; they frequently think of them as high-end and deserving of social media sharing and gifting. Even though geographical indications (GIs) assert a moral right to the geographical and cultural property represented in consumer goods, achieving rural development goals requires much greater involvement with informal ethical conventions throughout the value chain [25].

Eating cuisine that comes from specific nations or areas can also provide novelty value. Travelers view consuming food from the area as a unique experience that offers the chance to learn new skills and gain cultural capital through exploring the origins of food [26]. Therefore, we propose.

- H1a: Geographic Indicator Labels positively impacts Functional Consumption values.
- H1b: Geographic Indicator Labels positively impacts Economical Consumption values.
- H1c: Geographic Indicator Labels positively impacts Emotional Consumption values.
- H1d: Geographic Indicator Labels positively impacts Social Consumption values.
- H1e: Geographic Indicator Labels positively impacts Novelty Consumption values.

Because geographical indication (GI) products are unique and uncommon compared to mass-produced goods, they appeal to consumers who seek novel and thrilling experiences. Consumers are more likely to purchase a product if they believe it has high functional value. This is due to their expectation that the product will effectively meet their needs and provide a strong return on investment. Functional and economic values were demonstrated to have a beneficial impact on consumers' inclinations to buy organic food [27]. More recently, Chakraborty and Dash [28] also show that customers' perceptions of natural food products are positively and significantly impacted by functional, emotional, and novelty qualities. Because consumers place a higher value on food safety and because GIs are invariably associated with higher levels of quality control, GIs serve as a crucial quality indicator in the marketplace. The perception of GI labeling has a direct impact on customers' choices of GI tagged products, which means that it has a major economic impact.

The functional value is associated with natural, safe and healthy products [29–31], enhances positive consumer perceptions and quality [32]. Researchers also found that consumers are willing to spend a premium amount for such BGI labeled products [17], thus economic value is elevated. The products' strong association with geography and history invokes emotional association leading to add up to the emotional value of the products. Due to this strong place-product connect [24], consumers embrace pride in consuming BGI products, however improved social status, sophistication and relationships [23]. According to Ho, Liu, and Tarabashkina [17], the fifth consumption value, novelty, represents novel experiences which consumers experience when they acquire BGI products. Therefore we propose the hypothesis:

- H2a: Perceived Functional Value will positively impact BGI Labeled product purchase.
- H2b: Perceived Economical Value will positively impact BGI Labeled product purchase.
- H2c: Perceived Emotional Value will positively impact BGI Labeled product purchase.
- H2d: Perceived Social Value will positively impact BGI Labeled product purchase.
- H2e: Perceived Novelty Value will positively impact BGI Labeled product purchase.

Most health-conscious consumers are more concerned with the nutritional value and quality of items while making food purchases than are less health-conscious consumers [33]. Food produced in some nations may be thought to be of greater quality than food produced in other countries because consumers frequently correlate place origin with food quality [34]. Customers who are concerned about their health are more inclined to buy products that support their health objectives, which increases their readiness to buy products that promote a healthy lifestyle. When a product is advertised as organic or devoid of dangerous ingredients, consumers are more likely to purchase it. Thus, the proposed moderating hypothesis are:

- H3a: Health Consciousness moderates the relation between Functional Consumption Values and BGI Labeled product purchase.
- H3b: Health Consciousness moderates the relation between Economical Consumption Values and BGI Labeled product purchase.
- H3c: Health Consciousness moderates the relation between Emotional Consumption Values and BGI Labeled product purchase.
- H3d: Health Consciousness moderates the relation between Social Consumption Values and BGI Labeled product purchase.
- H3e: Health Consciousness moderates the relation between Novelty Consumption Values and BGI Labeled product purchase.

3 Research Methodology

This study attempts to investigate how blockchain-based geographic indicator labeled products influence consumption values and their choices for buying products with geographic indicator labels. To ensure that respondents were aware of blockchain technology and the importance of spatial indicators, purposeful sampling was employed to collect primary data.

3.1 Measurement Scales

The structured questionnaire, administered, was developed using adapted scale. All constructs' measuring items were adapted from earlier studies. Pictures of geographic indicator label products were provided in the introduction of the questionnaires. Sixteen items for functional values (FV), emotional values (EV), social values (SV), economical values (EcV) [17, 35], and 4 items for novelty value [16], 6 items for geographic indicator label products [36, 37], 4 items each for health consciousness [17], 3 items for geo-indicator labeled product purchase [38] were modified to measure constructs on seven-point Likert scale.

3.2 Sampling and Data Collection

Consumers from pan India were selected as the focal market in this study for several reasons. The product segment considered for BGI labeled product consumption values, e.g., Delhi Basmati Rice, Kashmir Saffron, Nagpur Oranges, etc. First, India represents the world's largest market having growing interest in native local grown authentic products. Traceability from farm to fork, and authenticity of Food and Beverages products were a serious concern. Respondents having annual income above 10 lacs were selected so that they had disposable income to purchase BGI labeled Food and Beverages giving higher pricing. All the respondents chosen had excellent to moderate knowledge of BGI labels. The minimal sample size was determined as suggested by Faul et al. [39] using the G*Power program. 560 adult consumers were selected by non-probabilistic purposive sampling, and they were subsequently mailed the structured questionnaire. We received 381 responses, and of those, 362 were completed for data analysis, yielding a response rate of 64.64%. The demographic profile of participants was as shown in Table 1.

Table 1 Demographic profile of respondents

Demographic	Characteristics	Frequency	Percentage
Gender	Male	187	51.66
	Female	175	48.34
Age	18–25	156	43.09
	26–40	133	36.74
	41–55	73	20.17
Education	Diploma	65	17.96
	Graduate	152	41.99
	Post graduate	145	40.06
Income	Above 10 lacs	94	25.97
	10–15 lacs	182	50.28
	Above 15 lacs	86	23.76
BGI label knowledge	Excellent	172	47.51
	Good	127	35.08
	Moderate	63	17.40

4 Data Analysis and Results

Data analysis was performed using structural equation modeling (SEM) with the Smart PLS 4.0 in order to assess the measurement and structural models [40]. When the inner variance inflation factor (VIF) values were less than the threshold value of 3, it was confirmed that the common method bias had been eliminated [41].

4.1 Measurement Model Evaluation

We determined composite reliability and Cronbach alpha (α) by setting a cutoff of 0.7, and the values above that were chosen. Additionally, Table 1 illustrates that the average variance collected was chosen using a 0.5 cutoff [40]. Discriminant validity was established using standards from Fornell and Larker [42]. Validity and reliability of essential constructs are listed in Table 2.

4.2 Structural Model and Hypothesis Testing

The structural model was assessed using R^2 values, standardized path coefficients (β), and significance level (*t* statistic) [40]. Adjusted *R*-square values for endogenous

Table 2 Validity and reliability

Constructs	BGI	EcV	EmV	FV	HC	NV	PUR	SV	A	CR	AVE
BGI	0.888								0.781	0.785	0.789
EcV	0.643	0.833							0.864	0.879	0.694
EmV	0.497	0.389	0.860						0.789	0.872	0.739
FV	0.368	0.464	0.603	0.847					0.832	0.853	0.718
HC	0.561	0.273	0.543	0.374	0.874				0.802	0.804	0.764
NV	0.461	0.446	0.315	0.305	0.567	0.826			0.827	0.838	0.683
PUR	0.483	0.462	0.457	0.435	0.438	0.325	0.862		0.745	0.773	0.743
SV	0.404	0.429	0.523	0.482	0.567	0.368	0.392	0.853	0.727	0.751	0.727

components explained 47.6% of the variance in BGI label product purchases. Tables 3 and 4 structural model assessment illustrates all of the proposed links.

As per Table 3 BGI labels on food and beverages have significant positive influences on functional values, economic values, emotional values, novelty values ($\beta = 0.423$, $p < 0.001$; $\beta = 0.417$, $p < 0.001$; $\beta = 0.276$, $p < 0.01$, $\beta = 0.347$, $p < 0.001$, respectively). Further, functional values, economic values, emotional values,

Table 3 Hypotheses testing

Hypothesis	Relationship	Beta (β)	*T* statistics	*P* values	Decisions
H1a	BGI → FV	0.423	5.283	0.000	Accepted
H1b	BGI → EcV	0.417	3.708	0.000	Accepted
H1c	BGI → EmV	0.276	4.762	0.002	Accepted
H1d	BGI → SV	0.024	0.568	0.542	Rejected
H1e	BGI → NV	0.347	2.948	0.000	Accepted
H2a	FV → PUR	0.286	6.045	0.000	Accepted
H2b	EcV → PUR	0.385	4.083	0.002	Accepted
H2c	EmV → PUR	0.213	3.789	0.000	Accepted
H2d	SV → PUR	0.092	1.021	0.671	Rejected
H2e	NV → PUR	0.292	3.564	0.000	Accepted

Table 4 Moderation analysis

Hypothesis	Relationship	Beta (β)	T statistics	P values	Decisions
H3a	FV*HC → PUR	0.523	5.474	0.002	Accepted
H3b	EcV*HC → PUR	0.642	6.073	0	Accepted
H3c	EmV*HC → PUR	0.282	3.934	0.002	Accepted
H3d	SV*HC → PUR	0.048	1.053	0.438	Rejected
H3e	NV*HC → PUR	0.027	0.748	0.573	Rejected

novelty values of BGI labeled food and beverages had significant influence on BGI labeled purchases ($\beta = 0.286$, $p < 0.001$; $\beta = 0.385$, $p < 0.01$; $\beta = 0.213$, $p < 0.001$, $\beta = 0.292$, $p < 0.001$, respectively). It was observed that the BGI label on food and beverages did not impact social value ($\beta = 0.024$, $p > 0.05$), nor did social value have an influence on the purchase of food and beverages with BGI label ($\beta = 0.092$, $p > 0.05$).

As per Table 4 Moderation analysis indicated that health consciousness could moderate the relationship between functional values, economic values, and emotional values with BGI label food and beverages purchase decision ($\beta = 0.523$, $p < 0.05$; $\beta = 0.642$, $p < 0.001$; $\beta = 0.282$, $p < 0.05$). However, the moderation effect of health consciousness on the relationship between social values, and novelty values was not significant.

5 Discussion and Implications

The current study's findings imply that food and drink products with BGI labels may have an impact on customers' perceptions of value and, ultimately, their purchasing decisions. Purchases' influence from product values is also dependent on certain values. For example, it was discovered that four product values (functional, economic, emotional, social) were strongly correlated with consumers' purchases, while social values had no such correlation.

BGI labeled food and beverages provide functional value to consumers as it enhances healthiness, food quality guarantees and assurance of food safety. BGI labeled food and beverages facilitated consumers to pay higher prices for products that originated from specific regions providing value for money. BGI labels can enhance emotional values as consumers have emotional associations with regions and countries. Food and beverages are native produce of specific geographical regions. BGI labels also enhance novelty value derived from consuming food that originates from certain regions, representing novelty of food and novel experience. BGI labeled food and beverages could not generate social values, like higher social image or good social relationships.

Health-conscious consumers indicated that BGI labels on food and beverages increased functional value (taste, safety, traceability, transparency, and healthiness), emotional value (trusted high-quality, original food products), and economic values (value for money to get unique, authentic food and beverages).

5.1 *Theoretical Contributions*

The study explains how geographic cues affect consumers' assessments of food and beverages, which sheds further light on how customers interpret blockchain-based geographic indicator labels. A more comprehensive understanding of consumers' information processing of blockchain-based geographical indications is provided by the integration of country/region image literature [43] with the consumption value framework [35]. This integration highlights the strong correlation between the uniqueness, authenticity, and originality of food and beverages and their consumption values enhancing the purchase decisions.

Blockchain's application in GI labeled items provides a fertile ground for research, particularly in consumer behavior. Researchers can investigate how transparency in blockchain affects trust, loyalty, and purchase decisions. Detailed blockchain records provide extensive data on product sources, manufacturing methods, and supply chain procedures, allowing for a thorough understanding of consumer preferences and the factors that influence their purchasing decisions. Furthermore, research can look into the socioeconomic influence on local producers, determining how blockchain can protect indigenous methods, increase product value, and help rural economies. Interdisciplinary study can also look into how technology interacts with traditional agriculture, examining the effects on sustainability, ethical production, and market accessibility.

5.2 *Practical Contributions*

The use of blockchain in geographical indication (GI) labeled Native Produce, Foods and Beverages segment have multiple implications for the marketers. Such products are when integrated with the BGI labels prove a competitive edge to the business owners. The goodwill and uniqueness of the products can highly improved if the products are easily traceable and authentic in nature. While the businesses are providing authenticity, quality and trust at the same time, they can also increase the prices for the products with such great value. The blockchain integration will save the integrity and performance of the products by reducing the risk of counterfeiting. It will also improve the supply chain by contributing through real-time tracking, tracing and verification of the product, resulting in reduced inefficiencies, fraud, and costs. In addition, blockchain integration will also help the small and medium scale industries to globally with the distribution of highly trustworthy and authentic products in the market.

The detailed information of the GI labeled products provided by blockchain will enhance the trust and confidence of the consumers hence their need of ethical, sustainable and traceable products could be easily fulfilled. Subsequently, these features of BGI labeled products also provide health-conscious customers a feeling of safety and quality for the same.

The present research also has implications for the policy makers. Policies that encourage blockchain integration can be created, offering producers financial aid or other incentives to use this technology. The distributed ledgers will streamline the certification process through verifiable details of origin and production procedure of the products. This will simplify the auditing process via compliance from global standards and GI regulations.

6 Limitations and Scope for Future

The present study explored the consumer behaviors toward blockchain enabled GI tagged products, however, it has certain limitations that opens directions for future research. The research is done in the Indian geographical context and could be analyzed for the global market. In addition, the present study explored the consumer aspects of GI tagged products. In future, researchers can find and resolve the industrial issues of firms' implementation of blockchain in the business processes. The impact of blockchain on the supply chain of the GI tagged products could also be analyzed. Other product segments like Home Fashion and Collectibles, Wearables and Accessories can be researched for BGI labels.

References

1. Sadílek T (2019) Consumer preferences regarding food quality labels: the case of Czechia. British Food J 121(10):2508–2523
2. Wu CW (2015) Facebook users' intentions in risk communication and food-safety issues. J Bus Res 68(11):2242–2247
3. Botonaki A, Polymeros K, Tsakiridou E, Mattas K (2006) The role of food quality certification on consumers' food choices. British Food J 108(2):77–90
4. Flanagin AJ, Metzger MJ (2008) The credibility of volunteered geographic information. GeoJournal 72:137–148
5. WIPO: WIPO IP facts and figures 2022 (2022). https://www.wipo.int/edocs/pubdocs/en/wipo-pub-943-2022-en-wipo-ip-facts-and-figures-2022.pdf
6. Katerinopoulou K, Kontogeorgos A, Salmas CE, Patakas A, Ladavos A (2020) Geographical origin authentication of agri-food products: a review. Foods 9(4):489
7. Moran W (1993) Rural space as intellectual property. Polit Geogr 12(3):263–277
8. IPIndia: state wise registration details of GI applications (2024). https://ipindia.gov.in/registered-gls.htm
9. Gifairindia: India's GI products (2024). https://www.gifairindia.in/home
10. Benedetto G, Sini MP (2016) Considering the social sustainability of rural landscape in Sardinia, Italy: the role of local identities. In: Island landscapes. Routledge, pp 210–215
11. Ge L, Brewster CA (2016) Informational institutions in the agrifood sector: meta-information and meta-governance of environmental sustainability. Curr Opin Environ Sustain 18:73–81
12. Min H (2019) Blockchain technology for enhancing supply chain resilience. Bus Horiz 62(1):35–45
13. Lee HL, Mendelson H, Rammohan S, Srivastava A (2017) Technology in agribusiness: opportunities to drive value. White paper
14. Goldstein K (2022) The global impact of counterfeiting and Solutions to stop it. Forbes. https://www.forbes.com/sites/forbesbusinesscouncil/2022/08/02/the-global-impact-of-counterfeiting-and-solutions-to-stop-it/?sh=3abad37d1ca2
15. Howard JA, Sheth JN (1969) The theory of buyer behavior. New York 63 145
16. Sheth JN, Newman BI, Gross BL (1991) Why we buy what we buy: a theory of consumption values. J Bus Res 22(2):159–170
17. Ho KFX, Liu F, Tarabashkina L (2024) A closer look at geographical indicators: how food labels influence product values, authenticity and willingness to pay premium prices. Asia Pac J Mark Logist 36(4):837–861
18. Lin PC, Huang YH (2012) The influence factors on choice behavior regarding green products based on the theory of consumption values. J Clean Prod 22(1):11–18

19. Tanrikulu C (2021) Theory of consumption values in consumer behaviour research: a review and future research agenda. Int J Consum Stud 45(6):1176–1197
20. Hudson D (2007) Why buy Brie? What are we measuring with willingness to pay for geographic indicators? In: INRA conference geographic indicators and brands: firm strategies and public policy
21. Sood E, Sharma Y (2024) Geographical indicators as tools of economic development. Indian J Public Adm
22. Muscogiuri G, Barrea L, Savastano S, Colao A (2020) Nutritional recommendations for CoVID-19 quarantine. Eur J Clin Nutr 74(6):850–851
23. Camillo AA (2012) A strategic investigation of the determinants of wine consumption in China. Int J Wine Bus Res 23(1):68–92
24. Spielmann N, Charters S (2013) The dimensions of authenticity in terroir products. Int J Wine Bus Res 25(4):310–324
25. Neilson J, Wright J, Aklimawati L (2018) Geographical indications and value capture in the Indonesia coffee sector. J Rural Stud 59:35–48
26. Ji M, Wong IA, Eves A, Scarles C (2016) Food-related personality traits and the moderating role of novelty-seeking in food satisfaction and travel outcomes. Tour Manage 57:387–396
27. Lee HJ, Hwang J (2016) The driving role of consumers' perceived credence attributes in organic food purchase decisions: a comparison of two groups of consumers. Food Qual Prefer 54:141–151
28. Chakraborty D, Dash G (2022) Using the consumption values to investigate consumer purchase intentions towards natural food products. British Food J 125(2):551–569
29. Machiels CJ, Karnal N (2016) See how tasty it is? Effects of symbolic cues on product evaluation and taste. Food Qual Prefer 52:195–202
30. Hall C, Osses F (2013) A review to inform understanding of the use of food safety messages on food labels. Int J Consum Stud 37(4):422–432
31. Van Kleef E, Van Trijp HC, Luning P (2005) Functional foods: health claim-food product compatibility and the impact of health claim framing on consumer evaluation. Appetite 44(3):299–308
32. Marcoz EM, Melewar TC, Dennis C (2016) The value of region of origin, producer and protected designation of origin label for visitors and locals: the case of fontina cheese in Italy. Int J Tour Res 18(3):236–250
33. Butler MJ, Barrientos RM (2020) The impact of nutrition on Covid-19 Susceptibility and long-term consequences. Brand, Behav, Immun 24:53–54
34. Ho K, Liu F, Tarabashkina L, Volery T (2022) Cross-cultural differences in consumers' attention to food labels. British Food J 124(12):4888–4904
35. Sweeney JC, Soutar GN (2001) Consumer perceived value: the development of a multiple item scale. J Retail 77(2):203–220
36. Chen NH (2021) Geographical indication labelling of food and behavioural intentions. British Food J 123(12):4097–4115
37. Der-Karabetian A, Ruiz Y (1997) Affective bicultural and global-human identity scales for Mexican-American adolescents. Psychol Rep 80(3):1027–1039
38. Newman GE, Dhar R (2014) Authenticity is contagious: brand essence and the original source of production. J Mark Res 51(3):371–386
39. Faul F, Erdfelder E, Buchner A, Lang A-G (2009) Statistical power analyses using G*Power 3.1: tests for correlation and regression analyses. Behav Res Methods 41:1149–1160
40. Hair JF Jr, Hult GTM, Ringle CM, Sarstedt M, Danks NP, Ray S (2021) Partial least squares structural equation modeling (PLS-SEM) using R: a workbook. Springer Nature, p 197
41. Petter S, Straub D, Rai A (2007) Specifying formative constructs in information systems research. MIS Q 623–656
42. Fornell C, Larcker DF (1981) Evaluating structural equation models with unobservable variables and measurement error. J Mark Res 18(1):39–50
43. Roth MS, Romeo JB (1992) Matching product catgeory and country image perceptions: a framework for managing country-of-origin effects. J Int Bus Stud 23:477–497

Enhancing the A* Algorithm for Autonomous Navigation in Model Sail Boats: Optimizing Path Planning Under Variable Wind Conditions

Grégory Witmer, Rolf Dornberger, and Thomas Hanne

Abstract This paper introduces an enhanced A* algorithm for autonomous navigation in model sail boats. The algorithm incorporates wind conditions, including direction, intensity, and turbulence, into the path planning process. The results demonstrate its effectiveness in achieving optimized paths, considering factors such as speed and efficiency. This research contributes to the development of adaptive and optimized path planning strategies for model sail boats in dynamic environments.

Keywords A* algorithm · Autonomous sailing · Model sail boats · Wind conditions · Path planning · Path optimization

1 Introduction

This paper focusses on autonomous navigation in model sailboats, a subfield of autonomous vehicle path planning. With ever-growing global trade, the reaching of energy limitations, and urging environmental issues, platforms that need no carbon energy sources for locomotion and no constant human input attract a large amount of attention. Autonomous sailing vessels promise new economically and ecologically friendly transport possibilities and are emerging topics in studies of maritime ecosystems.

While appearing to be like land-based counterparts, such as robotic vacuum cleaners or storehouse robots, the dependence on uncontrollable external circumstances for basic movement lets robotic sailing vessels face a disproportionately wider range of difficulties in navigation planning and control, which makes sailing boats

G. Witmer · R. Dornberger
School of Business, FHNW University of Applied Sciences and Arts Northwestern Switzerland, Basel, Switzerland

T. Hanne (✉)
Institute for Information Systems, University of Applied Sciences and Arts Northwestern Switzerland, Olten, Switzerland
e-mail: thomas.hanne@fhnw.ch

H. Sharma et al. (eds.), *Artificial Intelligence: Theory and Applications*, Lecture Notes in Networks and Systems 5589, https://doi.org/10.1007/978-981-96-1687-9_51

a worthwhile sport, hobby and, in recent years, subject of research, with initiatives such as the World Robotic Sailing Championship (roboticsailing.org).

For this, path planning problems are usually specified by a map with obstacles, which is expanded in our model by locally variable wind conditions. The problem model is further discussed in Sect. 3 together with the presentation of a real lake model. The algorithm used in the path planning task is A-star (interchangeably written as A*), an algorithm frequently used to solve autonomous navigation problems. It is well known for being relatively efficient computing wise and for its certainty of finding the optimal path. A deeper discussion of it will be made in Sect. 3.1. Details of our implementation and results from our experiments are presented in Sect. 4. In Sect. 5, conclusions and suggestions for future work are presented.

2 Literature Review

As sailboats make use of physical mechanisms to generate propulsion in accordance with everchanging circumstances, they offer a wide range of optimization opportunities. The following literature illustrates how different optimization problems on sailboats are investigated.

In [1], the focus is on enabling an autonomous sailboat to correct its course toward a given path. This is done by means of a Beetle Antenna Search (BAS) algorithm, which enhances a feedbacking steering controller (PID), to subsequently get a moving boat to plateau on a predefined path. This optimization problem is not restricted to sailboats, as also normal boats (such as course ships) are faced with the challenge of navigating close to a predetermined path. It is nevertheless an aspect that should not be forgotten when aspiring to implement a complete autonomous sailing system, as the precise following of an optimized path is essential for naval navigation.

In [2], optimally maneuvering upwind is considered using a zigzag pattern, called tacking. Two main variables are defined: The angle between the counter wind course and a predefined path, and the distance between tacking points and the counter wind course. Using a "brute force method," travel time is calculated for different angle values. However, this method has limitations in precision and computation time. To overcome this, a Genetic Algorithm (GA) is employed to find an approximation of global minima by representing individuals as binary strings of information.

The GA is faster and discovers ideal tacking points missed by the brute force method. Additionally, Ant Colony Optimization (ACO) and an Improved Ant Colony Optimization (IACO) are compared using simulations in MatlabR2017a on 30*30 grids [2]. IACO proves to be more efficient, achieving a path length of 51.4 in 8 iterations compared to 52.2 after 39 iterations for standard ACO. In [2], optimization problems of small-scale maneuvering are considered similarly as in [1]. At the same time, the runtime of related algorithms is to be considered for the practical application of related solutions.

Traditionally, sailors use approximations from polar charts (as seen in Fig. 1) and then tweak sail positions based on experience. However, in [3], the authors develop a simulation environment and compare a trial-and-error Hill-Climbing (HC) optimizer with their Modified Extremum Seeking Algorithm (MES) for the fine tuning. MES achieves better results in simulations, especially with increased noise. A real-world comparison on a model ship supports this trend, although comparability is acknowledged as a limitation.

Global path planning is addressed in [4]. Similar to [1], this is by no means a problem solely of sailing, but has a big impact on seafaring. The authors also propose an IACO. The algorithm considers factors such as path length, wind field height, and turning frequency to guide the path planning. The paper presents the steps for applying IACO to global path planning and describes the simulation experiments conducted to verify the algorithm's effectiveness. The results demonstrate the superiority of the proposed algorithm compared to basic ACO. The paper also discusses parameter optimization and presents the optimal parameter values for the algorithm.

Further approaches for autonomous sailboat navigation explored in the literature are ACO in connection with a rolling window method [5], Q-Learning [6], reinforcement learning [7], or the A* algorithm combined with a windy artificial potential field (WAPF) algorithm [8].

A survey of path planning techniques is provided by Jing et al. [9]. In the paper two types of oceanic autonomous surface vehicles (ASVs), autonomous sailboats and autonomous vessels, are distinguished and challenges and related solutions of ASVs' autonomous navigation are discussed including modeling analysis, path planning approaches, and future research directions. Specific aspect regarding the safety and legal aspects of autonomous maritime vehicles in general are discussed by Öztürk, Akdağ, and Ayabakan [10].

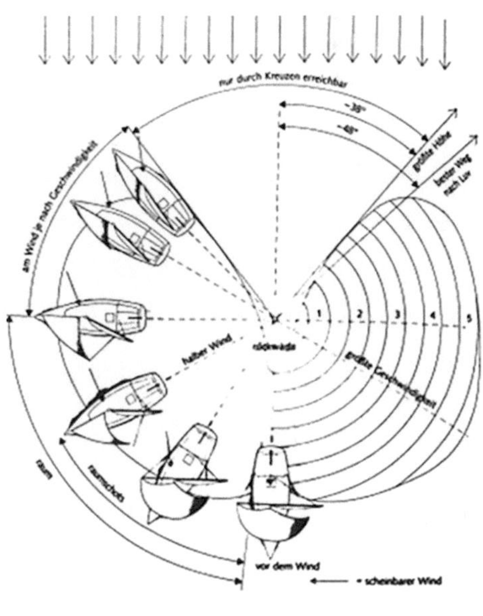

Fig. 1 Illustration of the possible courses of a sailboat, indicating that it is infeasible to travel on courses steeper than 38°. *Source* [1], p. 188

3 Problem Model

3.1 A* Algorithm

The A* algorithm was proposed in 1968 and has since seen great adoption in applied path planning and routing problems. It is based on Dijkstra algorithm, which is based on having a net of nodes that are connected via edges with a certain cost when traveled. A route through the network can hence be described as a sequence of nodes, with the sum of the cost of all edges is the total cost of the route. By following each walkable neighbor[1] and gradually canceling routes found to be inefficient, the optimal (i.e., most cost efficient) route between two points is found. A potential problem of Dijkstra is that it usually has quadratic complexity with respect to the number of nodes, which may be considered too time-consuming for real-time applications.

A* solves this issue by keeping nodes that need to be checked on a stack. In addition to the edge costs, A* adds the result of a heuristic function onto the collective edge cost to make decisions on how to order the stack. Different heuristic functions can be used, all of which provide a distance between the current field and the target field. Typically, these heuristics are Manhattan (the sum of the absolute differences between the x and the y values, providing horizontal and vertical moves needed), Diagonal (adding the possibility of diagonal moves to Manhattan) or Euclidian (the geometrical distance from current node to target node).

By making this proximity play a role in the order in which potential paths/nodes are calculated (stack order), more direct paths are preferred by the algorithm, therefore the more direct lines are calculated first, which can save a lot of computations. By not completely abandoning indirect route nodes, the algorithm will nevertheless find a way if a more direct path is not feasible.

The simulations in this paper are implemented in JavaScript, run in a browser implementing Chromium version 113.0.5672.126, and make use of the A* algorithm of the public JavaScript npm package javascript-astar[1] by Brian Grinstead. For the experiments, a forked repository is provided to make changes visible.[2]

3.2 Boat

Model sail boats and real boats rely heavily on the integrated mechanics that allow sailing. However, some forces will behave differently between model and real world due to the different proportions between scaled and unscaled parameters. It can for instance be assumed that the breaking force of the water onto the massively lighter hull will lead to a different behavior of model boats. Similarly, unevenness and

---

[1] https://github.com/bgrins/javascript-astar, with the last commit being from the 18th of December 2015.

[2] https://github.com/Greg8052/sailingboat-javascript-astar.

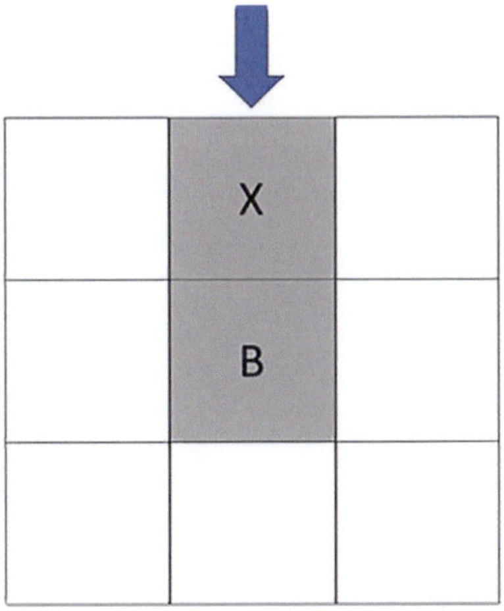

Fig. 2 Simplified raster model of Fig. 1, with the blue arrow marking the wind coming from the top, making it impossible for the boat (*B*) to move directly onto the field marked with *X*

a disadvantageous hull form will have a disproportionally bigger influence on the speed of model boats and are more probable than in professionally crafted large-scale boats. Thus, in the paper at hand, the characteristics of the sailing boat are thus defined as follows:

The effect of the wind onto the sails is regarded to be the same on 7 out of 8 possible courses with regard to the wind. The remaining course is upwind, i.e., the one diametrically opposed to the wind direction (see Figs. 1 and 2). Sailboats cannot drive directly upwind and will have to rely on tacking, thus alternating between the two close-hauled courses to approach the destination overall (see Fig. 1). Further details of this maneuver are provided, e.g., in [1].

The wind in our experiments is defined as impacting the boat from 1 of 8 directions with a strength between 1 and 10. The possibility of pockets with no wind at all is omitted, as the model sailboat would not rely on another source of propulsion such as currents, motors, or oarsmanship, all of which are neglected in our model.

The maximum velocity of the sailing boat is estimated as identical to the maximum wind speed. Speeds are measured from 1 (lowest) to 10 (strongest), allowing for comparisons between different routes taken by comparing their cumulative speeds. Further, the sailboat travels one parcel per iteration of the algorithm, independent of its speed.

3.3 Lake

The artificial lake in Glattpark, near Zürich, Switzerland serves as the real-life inspiration for this experiment (Fig. 3). Its sailable area measures 190 by 20 m, with the

Fig. 3 Satellite image of the artificial lake in Glattpark, Opfikon, near Zürich, Switzerland. The rectangular part in the middle of the picture measures 190 by 20 m

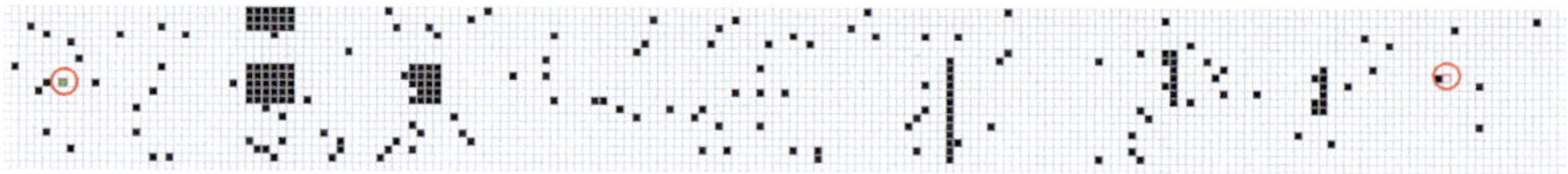

Fig. 4 Rasterized abstraction of the lake, with the start point (circle on the left) and the end point (circle on the right). Between them are random obstacles, also in larger groups

typical flora, fauna and litter leading to obstacles within the sailing area. In addition, the seaside plantation, and close buildings (both visible in Fig. 3) result in everchanging conditions of airflows.

To conduct this numerical experiment, the lake is represented as a grid of 190 by 20 squares with randomly scattered obstacles. The experiment is designed to let the boat run a regatta, with a start and finish point fixed at (9, 7) (left red circle in Fig. 4) and a buoy (behind which the ship has to do a tack) at the other end of the lake at (8, 179). The wind always has one direction and is measured in strengths ranging from 10 (weak) to 1 (strong). Direction and strength change according to a pattern discussed in Sect. 4.2.

4 Initial Path Planning

For the first run of the experiment, certain obstacles are predefined (see Fig. 5), roughly modeling the real lake scenario. While the A* algorithm finds the ideal route, several factors are not taken into account, and hence considered in the following:

- The initial route does not contain obstacles that are discovered en route. However, this happens regularly, both in reality and in the model case during sailing. Therefore, upon discovery of an unknown obstacle, the map is updated, and a new path (with the current point as starting point) must be calculated. On the way back

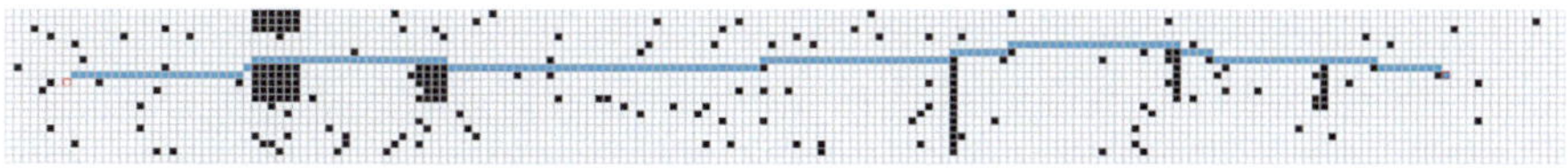

Fig. 5 An initial run, as described in Sect. 4

from the buoy, this new obstacle is now a known obstacle. So if all obstacles are now known, the path back might be shorter than the re-adjusted forward path.

- Changing winds should also be considered in the model case. The neighboring cell in the upwind direction should always be considered as non-sailable. This will automatically lead to a tacking maneuver, should the boat be faced with such upwind. To model this situation, the upwind regime on the forward run can be treated similarly to the case of discovered obstacles discussed before. For this, the calculated optimal return path has to be checked for areas with known headwinds, the boundaries of which should be delimited with obstacles to make the boat try to circumnavigate them.

4.1 Adjusting the Found Path in Case of Obstacle Discovery

To investigate the sudden discovery of new obstacles in this numerical experiment, a set of "hidden" obstacles is defined (see Fig. 6) which the boat only discovers upon coming into proximity while traveling on the initially calculated path. While this would have to be a real-time process, the method chosen and illustrated in Figs. 5, 6, and 7 was to manually select the point at which the boat's sensors would have discovered the new obstacle. As there is no obvious (simple) alteration of the algorithm taking into account the updated information, we consider a restart of the A* algorithm, i.e., rerunning the optimization (Fig. 7) on the updated map.

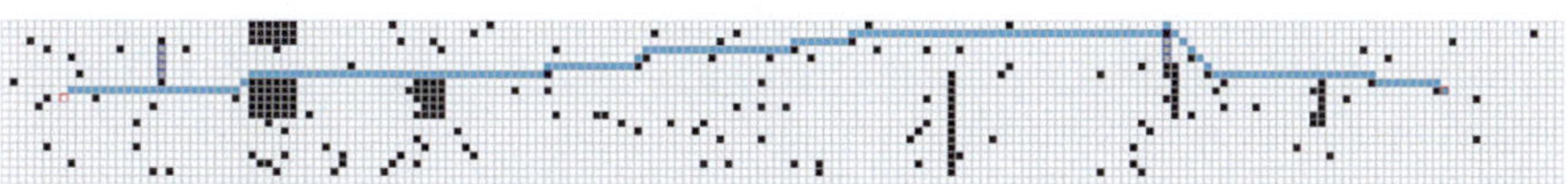

Fig. 6 Test run with additional obstacles (in darker blue, shortly after the start and before the end point. For further discussion refer to Sect. 4.1)

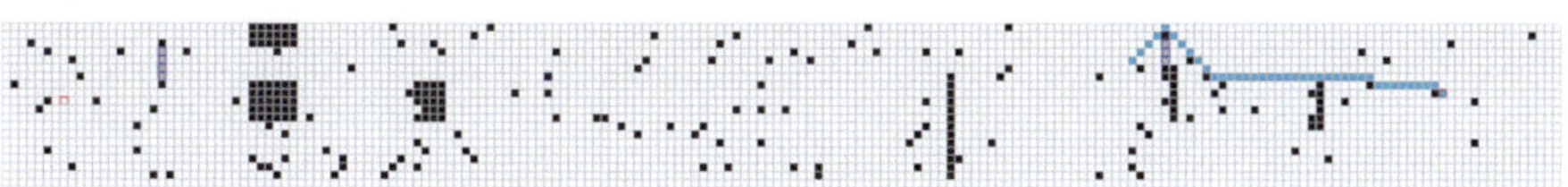

Fig. 7 Following the initial test run to the point where the boat discovers the hidden obstacles and rerunning A* reveals the quick adjustment with a return to the initial route

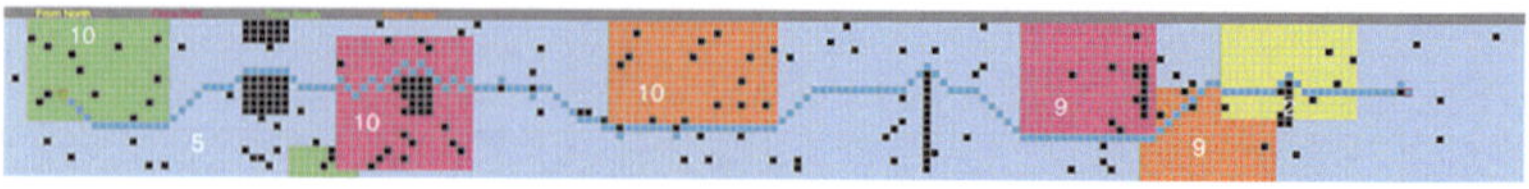

Fig. 8 Lake with wind pockets and the optimal route drawn in. The colors correspond to: North (yellow), East (magenta), South (green), West (orange). The wind strengths (white) were added afterward

It is interesting to see how the path followed was adapted long before the obstacle was encountered, when the newly discovered obstacles were known from the very start. It can therefore be noted that even with the restart method, the tracking of discoveries would yield gains when optimizing the way back from the buoy.

4.2 *Introducing Variable Winds*

To simulate reacting to changing wind conditions, the general wind on the lake is defined as coming from the north/top of the map, with mid-strength (5). Then 8 "pockets" are defined, which have differing winds, as shown in Fig. 8. These wind pockets are:

1. Center: (4, 11), wind from: South, strength: 8, diameter: 8.
2. Center: (10, 50), wind from: East, strength: 8, diameter: 8.
3. Center: (2, 86), wind from: West, strength: 10, diameter: 10.
4. Center: (3, 160), wind from: North, strength: 2, diameter: 8.
5. Center: (5, 135), wind from: East, strength: 9, diameter: 8.
6. Center: (16, 150), wind from: West, strength: 6, diameter: 8.
7. Center: (20, 40), wind from: South, strength: 10, diameter: 4.
8. Center: (3, 160), wind from: North, strength: 2, diameter: 8.

With these parameters it is possible to allocate a weight to each cell on the map. As stated before, weights (sometimes referenced as costs, also in Sect. 3.1) are an integral part of the A* algorithm.

In order to adapt A* for the sailboat simulation, the weights of the cells must respect the directions of the sailboat in relation to the wind direction. The algorithm has not only to compare costs and heuristic distance values to make decisions on the route to take (i.e., "prefer strong winds over weak ones"), but must consider the wind directions in relation to the boat's direction of travel. For instance, moving one field west onto a field with winds coming from the north is good (and only the strength of the wind is considered), while moving west onto a field with eastbound winds is not possible and will require tacking or a change of the path.

A first step toward a sailboat that simulates the A* algorithm is to add the ability to navigate upwind situations (Fig. 9). For this purpose, the part in which the algorithm deducts all neighbors of the current node is expanded with additional logic. Usually, in this part of the A* algorithm all mathematically possible neighbors are calculated.

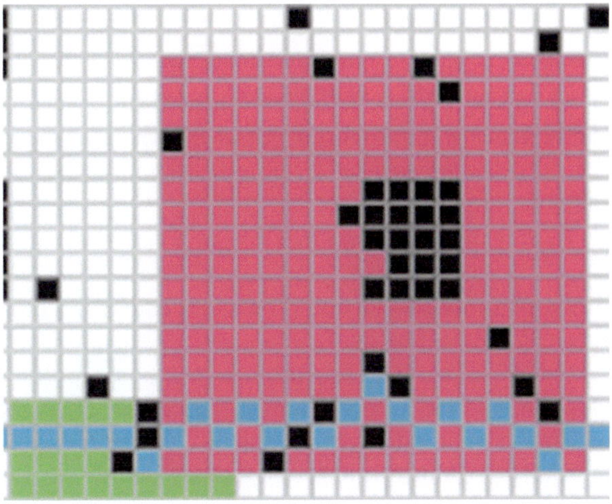

Fig. 9 Effects of blocking upwind movement automatically lets the boat follow a tacking course. While moving in winds from the south in a straight line (green), traveling in winds from the east (magenta) is not possible without tacking

Each neighbor is then checked for having been visited already or being a border of the feasible area.

The extension consists of also checking whether the direction from which the wind comes on the new cell equals the direction in which the ship would sail when reaching the cell. Should this be the case, the neighbor cell is omitted, effectively blocking the passage without placing a «solid» obstacle, i.e., altering the stats of the map.

This procedure has the advantage that a tacking (see Fig. 1) maneuver is automatically implemented as soon as the boat (or the A* algorithm) enters an upwind area. Crossing the wind field from a different direction is still possible without interference. With this change, in combination with the re-rendering strategy for newly discovered obstacles discussed in Section 4.A, it can be said that the sailboat, while potentially not being able to travel an optimal route on the forward run due to discoveries, would now be able to successfully complete the trip to the buoy.

With the exclusion of impossible upwind sailing maneuvers (Fig. 2), the rest of the course can directly use the defined wind speeds as the weights/costs in the A* computation. To do this, the wind speeds of the wind cells are set as weights of the cells. In zones outside of wind cells, the standard wind from north with a strength of 5 is set as weight. A test run of the forward way reveals corrections to the calculated path. When considering the example in Fig. 8, it is obvious that the optimal path leads out of the field of very weak (10 = weak, 1 = strong, as mentioned in Section I) winds from the south (green). The way leads around the isle, and tacks through weak eastern (magenta) winds, before avoiding winds from the west (orange) and east. Thus, the small test case shows that a more complex path results from the wind and weight settings.

5 Conclusions

In our study, a couple of very specific sailing boat routing problems were discussed with respect to pathfinding adapting the A* algorithm. This serves as proof of concept for the adaption a suitable optimization algorithm to solve novel variants of path planning problems.

It is interesting to note that research into automatic sailboat navigation is still rare since such technologies provide significant opportunities to address issues such as ocean pollution which could be taken care of by mostly autonomous fleets of very fuel-efficient sail boats.

Our specific model still provides various opportunities for refinement and extension. In particular, experiments should be conducted to reduce the reality gap and make the model more realistic. Further research is also recommended for the advanced planning problems (e.g., more complex routing problems such as the traveling salesman region problem [11] which could be used to address inspection tasks). Such problems would also suggest exploring a wider range of algorithms such as nature-inspired methods and metaheuristics and comparative studies of algorithms are suggested for future work.

References

1. Zhou L, Chen K, Chen Z, Dong H, Song D (2020) Course control of unmanned sailboat based on BAS-PID algorithm. In: 2020 International conference on system science and engineering (ICSSE), Kagawa, Japan. IEEE, Piscataway, pp 1–5. https://doi.org/10.1109/ICSSE50014.2020.9219310
2. Santos DH, Negreiros APF, Jacobo JEA, Goncalves LMG, Silva AG Jr, Silva JMVBS (2016) Short-term path planning for high-level navigation control of N-Boat—the sailboat robot. In: 2016 XIII Latin American robotics symposium and IV Brazilian robotics symposium (LARS/SBR), Recife, Brazil. IEEE, Piscataway, pp 211–216. https://doi.org/10.1109/LARS-SBR.2016.42
3. Corno M, Formentin S, Savaresi SM (2016) Data-driven online speed optimization in autonomous sailboats. IEEE Trans Intell Transp Syst 17(3):762–771. https://doi.org/10.1109/TITS.2015.2483022
4. Nian R, Shen Z, Ding W (2021) Research on global path planning of unmanned sailboat based on improved ant colony optimization. In: 2021 6th International conference on automation, control and robotics engineering (CACRE), Dalian, China. IEEE, Piscataway, pp 428–432. https://doi.org/10.1109/CACRE52464.2021.9501353
5. Shen Z, Ding W, Liu Y, Yu H (2023) Path planning optimization for unmanned sailboat in complex marine environment. Ocean Eng 269:113475
6. Silva Junior AGD, Santos DHD, Negreiros APFD, Silva JMVBDS, Gonçalves LMG (2020) High-level path planning for an autonomous sailboat robot using Q-Learning. Sensors 20(6):1550
7. Wang C, Zhang X, Li R, Dong P (2019) Path planning of maritime autonomous surface ships in unknown environment with reinforcement learning. In: Cognitive systems and signal processing: 4th International conference, ICCSIP 2018, Beijing, China, Nov 29–Dec 1, 2018, Revised Selected Papers, Part II 4. Springer, Singapore, pp 127–137

8. Zhang S, Yu D, Sang H, Sun X, Liu F (2023) A new hybrid path planning method for the sailboat architecture wave glider in the wind field environment. Ocean Eng 283:115153
9. Jing W, Liu C, Li T, Rahman AM, Xian L, Wang X, Wang Y, Guo Z, Brenda G, Tendai KW (2020) Path planning and navigation of oceanic autonomous sailboats and vessels: A survey. J Ocean Univ China 19:609–621
10. Öztürk Ü, Akdağ M, Ayabakan T (2022) A review of path planning algorithms in maritime autonomous surface ships: navigation safety perspective. Ocean Eng 251:111010
11. Lammel B, Gryzlak K, Dornberger R, Hanne T (2016) An ant colony system solving the travelling salesman region problem. In: 2016 4th International symposium on computational and business intelligence (ISCBI). IEEE, Piscataway, pp 125–131

Deep Learning for Reliable Landmine Detection: A CNN Approach to Minimizing False Positives

Mahamat Ismael Hassan, Daouda Ahmat, and Samuel Ouya

Abstract Landmines remain a legacy of long histories of armed conflicts and malicious terrorist activities. Despite initiatives undertaken by the governments of these countries and inter-humanitarian actions, they continue to be deployed. This paper focuses on the use of convolutional neural networks (CNN) to improve landmine detection. It aims to reduce false positive rates in the mine detection process. The proposed model is based on a CNN algorithm, known as Faster R-CNN, renowned for its good performance in detecting objects in images. The article further relies on the Scikit-learn framework to identify and isolate false alarms caused by some types of mines with sensitive and specific features. Such mines are frequently encountered on the battlefields of northern Chad. The results obtained using this approach show a strengthening of detection capabilities while minimizing classification anomalies. These results could contribute to securing potentially dangerous areas.

Keywords Mine clearance · Landmines · GPR · CNN · Faster R-CNN · Scikit-learn

M. I. Hassan
ENASTIC, N'Djamena, Chad
e-mail: m.ismael@enastic.td

D. Ahmat (✉)
Université de N'Djamena, N'Djamena, Chad
e-mail: daouda.ahmat@uvt.td

S. Ouya
Cheikh Anta Diop University, Dakar, Senegal
e-mail: samuel.ouya@ucad.edu.sn

H. Sharma et al. (eds.), *Artificial Intelligence: Theory and Applications*, Lecture Notes in Networks and Systems 5589, https://doi.org/10.1007/978-981-96-1687-9_52

1 Introduction

In many countries, landmines planted in the wake of armed conflict pose a serious threat and represent a constant hazard. These explosive devices can cause innocent casualties, including women, children, the elderly, and even animals when accidentally triggered. Post-conflict countries face a persistent threat from explosive landmines left behind by belligerent forces. At the end of confrontations, the antagonists leave the field, leaving behind dangerous explosive devices that can remain hidden under the ground for a long time. According to estimates by the International Committee of the Red Cross, there are around 100 million landmines in use worldwide, claiming some 26,000 victims every year. Landmines are formidable weapons that can cause considerable human and material damage when accidentally detonated. The growing number of casualties caused by landmines has prompted researchers and industrialists to develop technologies capable of detecting these mines effectively [1]. Nevertheless, the traditional approach to detecting explosive mines buried underground remains a rather tricky operation due to the dangerous and time-consuming tasks involved [1–3]. Assessing the risk of landmines in an area is still based on surveys without precise mapping of the explosive devices placed underground, which are both costly and time-consuming. In addition, demined areas are not properly identified and accurately mapped. In some theaters of conflict, actors continue to lay mines in cleared areas, creating disruption which confuses local populations and leads to mass exodus. Portable metal-detecting sensors have been widely used for the detection of explosive landmines up to the present day [4–6]. However, the main difficulty lies in the high rate of false alarms caused by the complexity of the various mine and soil type characteristics [7, 8].

In this paper, we propose a convolutional neural network (CNN) approach for landmine detection in post-conflict countries such as Chad. The proposed CNN model uses data captured by ground-penetrating radar (GPR) while scanning the ground to detect an object buried underground. The choice to use GPR data comes from the fact that metal detectors are technologies with ground-penetrating radar scanning capabilities and have been widely used in demining, able to detect even the smallest dielectric change in the subsurface with different types of explosive device content [9–11]. GPR data are available in numerous databases that can be downloaded and fully exploited for a variety of uses. The results achieved reported in this article show an improvement in detection by minimizing false alarms resulting from the detection of false positives.

2 Background

The proliferation of anti-personnel landmines in many post-conflict territories is a major concern due to their devastating impact on populations [12–16]. Chad [17] is one of the hardest-hit countries in the world, but little-known by researchers in this field due to the inhospitable landscape of the impacted areas. The country has faced this threat since the 1987 Aouzou war between Chad and Libya, as well as during numerous recent rebel operations against the government army. Since 1997, Chad has ratified the Ottawa Convention to carry out demining operations, with the aim of clearing mined land so that it can be used by local populations. Following ratification, the Haut Commissariat de Déminage du Tchad (HCND)—the Chad High Commission for Demining—was created to coordinate demining activities. However, deminers face a number of challenges, such as false alarms, lack of accurate data, and slow clearance operations. This situation underlines the crucial importance of demining as a major humanitarian action, as these efforts can significantly contribute to saving lives and protecting vulnerable communities [18]. It is in this context that the motivation for this article takes on its full meaning. The scope of this article is limited to Ouadi-doum, a locality in northern Chad. The article proposes the use of a CNN model to improve the accuracy of anti-personnel landmine detection. By exploiting previously trained data from mined terrain, the Faster R-CNN algorithm model [19] can accurately identify the patterns being searched for, and the Scikit-learn framework is employed to handle the false alarms generated by different types of explosive anti-personnel mines. This will enable the performance of existing detection technology to be evaluated, reducing false alarms on the one hand and increasing the reliability of detection operations on the other. Evaluation methods such as the ROC curve are used to analyze how the model handles the trade-off between sensitivity and specificity, the area under the curve (AUC) to assess the overall performance of the proposed model, and the confusion matrix to present classification results by comparing model predictions with actual values. An auto-encoder is used to distinguish images containing patterns from those without. This step enables us to better identify the specific features present in the images and effectively separate them according to their content. Figure 1 shows some heavily mined areas. The red dots on Chad's map illustrate confirmed minefields.

3 Related Work

In this section, an overview is given of some of the main research works on explosive mine detection using artificial intelligence (AI)-based approaches. The main objective of this article is to improve the accuracy of landmine detection and the management of false alarms. In-depth studies have been widely carried out into the detection of explosive landmines using ground-penetrating radar (GPR) sensors [4–6, 20, 21]. The GPR is a multi-purpose tool in mine clearance operations, widely deployed in a variety of varied environmental conditions and soil kinds, making it a major technology in the detection of explosive mines [8, 22]. This device enables the acquisition of high-resolution images containing patterns of buried explosive mines, which can then be processed with AI techniques to improve detection accuracy, as done in this article with the distance vector machine (SVM) for the classification of improvised explosive devices (IEDs) [23]. GPR uses low-frequency electromagnetic waves, which have the ability to penetrate the subsoil before being

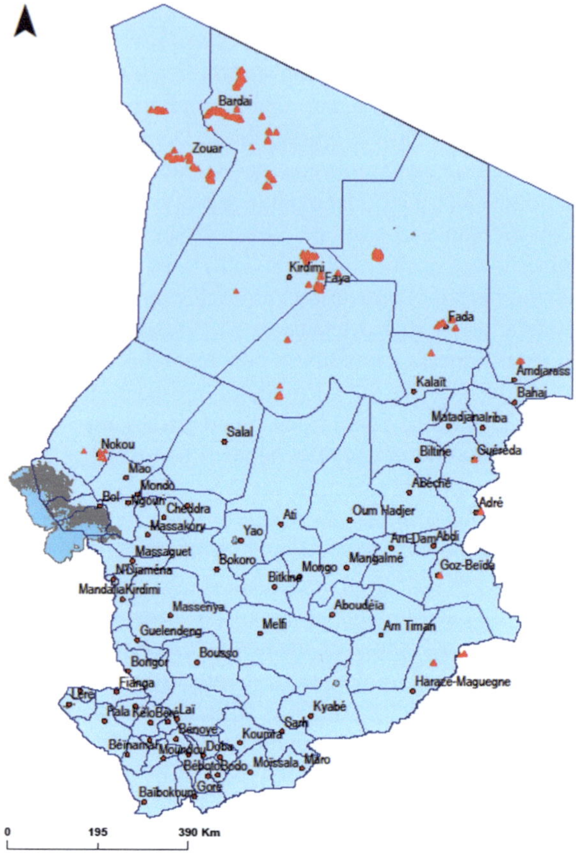

Fig. 1 A map of mined areas in Chad

reflected by buried objects. This penetration capability enables GPR to detect targets at variable depths, depending on the kind of soil and the size of the item [24–26]. The underlying principle of GPR is to transmit electromagnetic waves through the ground using a transmitting antenna (TX) and a receiving antenna (RX). The aim is to analyze the reflected signals to detect the presence of buried objects, such as anti-personnel mines or improvised explosive devices (IEDs) [7]. Faster region convolutional neural networks (Faster R-CNN) were used for the detection and classification of explosive objects buried underground [27]. The training dataset used by the Faster R-CNN is tested on the principle of discriminating the digital signature of anti-tank (AT) and anti-personnel (AP) mines. Test images are obtained both from a GPR used on real terrain and from data generated synthetically by the gprMax simulator. The performance of their systems is evaluated using a confusion matrix and ROC curves. The detection of an AP mine and an AT mine is the subject of scientific research in order to distinguish AP mines from other mines, considering the size of the object and the depth of its location below the ground surface. An auto-encoder CNN is used to extract relevant features from multi-polarized GPR data. The features are then used to train mine detection algorithms. The training dataset

B-scans and training convergence are reached after a few epochs, the system showed promising performance in a multi-dataset scenario [1]. Experiments carried out on real data show that the proposed solution handles anomaly detection and recognizes landmines as objects different from the ground used in the training stage. A ROC curve is used to evaluate the probability of correct detection and the probability of false detection, covering all possible threshold values. The ROC was measured using the area (AUC). Lameri et al. [28] propose a CNN for explosive mine detection applied to data obtained from the gprMax simulation generated purely synthetic by adding some ground-penetrating radar (GPR) images containing buried explosive mine patterns. The proposed solution makes it possible to accurately determine whether a B-scan profile, generated from GPR data shows evidence of buried mines or not. To validate their system, the researchers used real GPR data, although the system's learning process can be based solely on synthetically generated data. Studies have that proposed mid-infrared laser spectroscopy can detect and quantify high explosives (HE) buried beneath the ground using multivariate analysis and artificial intelligence [29]. The detection of HE at a distance of fifteen centimeters using quantum cascade laser spectroscopy provided evidence that it can be used in direct field applications. According to their experiments presented, dinitrotoluene (DNT) signals could be observed in the spectra of solid mixtures with complex matrices, enabling the detection of explosives and the quantification of DNT. A partial least squares regression model was generated to predict DNT concentrations and increase selectivity. The model was trained and evaluated using additional analyses such as interference, HE, and trinitrotoluene. On a demining site dedicated to mine detection, false alarms are a major risk to operator safety. Studies have been carried out on the performance analysis of clutter removal algorithms in simulated data, using gprMax which proposes various challenging scenarios [30]. Two clutter removal methods were identified (subspace-based method and multi-resolution multi-direction-based method) and implemented, and their performance was compared using ROC curves and visual analysis. The results also showed that some methods have variable performance depending on the complexity of the scenarios. Another contribution of this work lies in the introduction of a challenging dataset created by the new version of the electromagnetic simulation software gprMax, which is used for GPR studies. In the context of humanitarian demining operations, there are specific features that can interfere with the detection of explosive mines. These characteristics are sometimes linked to the type of soil (clay or sand), the temperature (high or low), the different types of mine (AP or AT) containing a metallic or plastic charge, and the manufacturing method (industrial or artisanal). In addition, the sizes of explosive devices often vary, ranging from a few centimeters [17, 31–35]. Convolutional neural networks were used to detect explosive mines accurately in hazardous environments [36]. The system succeeded in identifying improvised explosive devices with an accuracy of 98.7% in optimal lighting conditions. Its performance is compared with that of other convolutional neural networks and a deterministic algorithm, demonstrating that the proposed CNN outperforms its competitors, including the deterministic method. Further research focuses on the detection of improvised explosive landmines using deep neural networks on a GPR image dataset simulated using standard programmable integrated circuits (FPGAs) [26]. Discrimination algorithms for buried mine detection, exploiting CNNs and recurrent neural networks (RNNs) have been proposed to analyze two-dimensional (2D) and three-dimensional (3D) GPR B-scans data [37]. The proposed models were trained and evaluated using experimental GPR data from two different test sites. The authors also provide a quantitative analysis comparing the receiver operating features obtained with the proposed algorithms with those obtained with existing CNN-based approaches as well as traditional learning. It is important to point out that the majority of AI-based detection methods widely proposed in the literature focus mainly on improving the detection of explosive mines buried underground [38–40], with few works addressing the aspect of false alarm sound management. Most of these techniques use sensor data to train their model for deployment in the real field for final testing [41–43]. Future work will also focus on deploying our model on an AI-equipped drone for remote sensing of explosive mines.

4 Methodology

The aim of our research is to propose a solution that uses a convolutional neural network to detect explosive mines buried under the ground and improve detection accuracy in order to reduce false alarm rates, taking into account a few specific types of landmines. Detecting explosive landmines accurately and managing false alarms are complex challenges in humanitarian demining. Detection devices are often confronted with the phenomenon of false alarms generated by different types of landmines, depending on their content, method of manufacture, and ground characteristics, particularly in sandy and desert areas such as northern Chad, posing additional challenges for demining operators. This work was carried out in two key stages. In the first stage, we focused on the detection of different types of anti-personnel landmines commonly used in Chad using a Faster R-CNN, namely the PRB NR409 a Belgian model high-pressure anti-personnel mine, the NR442, a more recent Belgian mine, the PPM2, an ex-GDR model high-pressure mine, and PMA3 a Yugoslav model high-pressure mine. Table 1 describes the different types of landmines and their technical properties. Generally, their activation methods are the same: either remote-controlled or accidental by physical contact. For example, the PRB NR409 type of landmine, when triggered, sends fragments in all directions, injuring or killing anything in close proximity. The PRB mine measures 2.80 cm in height and width, weighs 183 g, and contains an explosive charge of 80 g. It can be activated either remotely or by physical contact. During training, we examine the physical properties of each type of landmine. By combining this information with image data, we can determine the detectability and sensitivity of each device. The aim was to assess the improvement in detectability compared with existing methods. The second stage focused on the management of false alarms using the Scikit-learn framework to improve the accuracy metrics of the detection system.

Table 1 Different types of anti-personnel landmines

Setting	PRB NR409	NR442	PPM2	PMA3
Type	Blast mine	Blast effect	Blast effect	Blast mine
Matter	Alu, plastic	Steel, alu	Plastic	Steel, alu
Weight	183 g	200–300 g	0.375 kg	180 g
Height (cm)	2.80	3–5	6	4
Diameter (cm)	8.20	6–10	13.4	11.10
Charge	80 g	560 g	0.11 kg	35 g TNT
Detonator	Mixte	Electric	Mixte	Mixte
Pressure (kg)	9	5–10	1–17	7
Danger	5 m	5–10	5 m	5 m
Fragment	Piece of	100–300	Piece of	Piece of

4.1 *Kinds of Anti-personnel Landmines*

Anti-personnel mines are designed in a variety of ways, whether industrially produced or improvised by individuals. This diversity of explosive mines, including plastic or metal charges, makes them difficult to detect [36]. Mines vary in terms of composition, size, weight, type of detonator, mode of detonation, and range of action. Their main effect is to provoke an explosion, generating shrapnel that can cause damage to nearby people. These devices are often hidden deep underground, in dense forests, or in the terrain specific to each region or country. The focus of this paper is solely on industrially manufactured mines, as artisanal mines do not follow any standards and would be difficult to detect reliably. Table 1 provides information on the types of mine found and examined in northern Chad.

4.2 *Training and Test Dataset*

Obtaining a sufficient number of specific images, such as thermal or infrared images, for a demining application is a challenge. To overcome this limitation, we had to search for images from various information sources, which we then processed and sized to meet our needs. In our case, we used GPR images offered by the information site https://figshare.dmu.ac.uk, which we associated with the characteristics of the types of explosive mines selected for our study. Despite the availability of these images, we had to resort to the data augmentation technique to obtain a sufficient number of images meeting our criteria. In total, we trained our model with 1620 negative images, 180 positive images, 17 negative test images, and 11 positive test images, making a total of 1828 images. In any case, correctly identifying the images containing the desired patterns corresponding to each type of mine is a major challenge for training the model. An auto-encoder was used to discriminate between images containing desired patterns and those without, by exploiting the features of various types of mine. The model then enabled us to evaluate detectability, sensitivity, and false alarm rates for each type of mines, obtaining results that were consistent with those in the literature. It should be noted that few works address the problem of detection, sensitivity, and false alarms by nature of anti-personnel explosive landmines using CNN, due to the limited number of images available and the lack of landmine coordinates.

4.3 Convolutional Neural Network for Target Detection

CNNs are inspired by the workings of the mammalian visual cortex and are highly effective in processing gridded data such as images. A CNN works by using convolution layers to extract important features from the image via filters that act as detectors of specific patterns. Then, a pooling layer reduces the spatial dimension while preserving these features, making the model more resistant to minor variations. Specific data, such as ground-penetrating radar, can be pre-processed to fit the CNN. Trained on data including examples of buried objects, the CNN learns to recognize them and can then detect new buried objects by identifying the learned patterns. Post-processing techniques can then be applied to improve detection accuracy.

4.4 CNN Model for Landmine Detection

Let X be the set of images captured by a detection device, θ be parameters of the CNN model, and the prediction function $f(X;\theta)$ is defined by the architecture of the CNN model, which includes convolution, max pooling, and fully connected layers, let y be data labels 1 for mine, 0 for non-mine. Each image is represented by a matrix I of dimension $n \times m$, where m and n are respectively the height and width of the image. Each element of the matrix I represents the thermal intensity value at a specific location in the image. The aim is to predict the probability of an image containing an explosive mine.

4.5 Learning, Training, and Prediction

$X_{\text{train}} y_{\text{train}}$: training set for the CNN model and $X_{\text{test}} y_{\text{test}}$ test set for evaluating the model. The CNN model is trained on $X_{\text{train}} y_{\text{train}}$ to minimize the classification loss L_{cls}. We calculated the classification loss using the logistic loss function:

$$L_{\text{cls}} = -N_1 \sum_{i=1}^{N} \left[y_i \log(p_{\text{obj}}) + (1 - y_i) \log(1 - p_{\text{obj}}) \right] \tag{1}$$

where N_1 is the number of positive images in the training set, y_i Label of example i, P_{Obj} Probability predicted by the model for the example i. After training, the model predicts labels y_{Pred} for X_{Test}. The loss of localization measures the difference between the predicted coordinates b_{obj} and the actual RoI coordinates.

Model Evaluation The ROC curve is created using the false positive and true positive rates. The confusion matrix is represented by the values TN, FP, FN, and TP, corresponding to true negatives, false positives, false negatives, and true positives, respectively. Precision measures the accuracy of the model's predictions, while sensitivity measures the model's ability to detect true positives. False positives quantify the number of cases where the model has incorrectly classified a negative example as positive.

4.6 Landmine Detection with Faster R-CNN

Faster R-CNN is an object detection algorithm based on neural networks, using a convolution region to generate candidate regions of interest (RoI). Let I be the input image, and $f(I)$ be the image processing function for landmine detection. This function uses the Faster R-CNN algorithm with convolution pooling and classification layers defined as follows:

$$f(I) = \text{Faster R-CNN}(\text{Conv}(I)), \text{Pool}(I), \text{Class}(I) \tag{2}$$

where $\text{Conv}(I)$ represents convolution operations on the (I) image, $\text{Pool}(I)$ represents pooling operations to reduce image dimensions, $\text{Class}(I)$ represents classification operations to identify potential landmine regions. The prediction function $f(X;\theta)$ is defined as follows:

$$f(X;\theta) = \{P_{\text{obj}}, b_{\text{obj}}\} \tag{3}$$

where P_{obj} is the probability that a (RoI) contains an explosive mine, and b_{obj} is the coordinate vector of the RoI. Model training consists of minimizing a loss function L defined as the sum of the classification loss L_{cls} and the localization loss L_{loc}

$$L = L_{\text{cls}} + L_{\text{loc}} \tag{4}$$

4.7 Managing False Alarms with Scikit-Learn

Once potential mines have been detected by Faster R-CNN, we use the Scikit-learn framework to manage false alarms. Let y_{pred} be the binary prediction vector of the model, where $y_{\text{Pred}}[i] = 1$ if the image i contains a mine and 0 otherwise. We used a post-processing method to filter out false alarms according to specific criteria, such as the size of the RoI, the characteristic shape of the heat signature, and so on.

4.8 Auto-encoder to Discern Images

As mentioned earlier in the data testing subsection, data collection is a delicate stage requiring additional tools, especially for training a CNN model. We opted for an auto-encoder given the data at our disposal, in order to identify images containing mine patterns and avoid model errors. An auto-encoder distinguishes images containing mine patterns by detecting anomalies in the reconstruction error. Images with mine patterns have a less faithful reconstruction, resulting in a higher reconstruction error. The auto-encoder consists of an encoder that reduces the size of the input and a decoder that reconstructs the input from this compact representation. The aim is to minimize the difference between the input image and the reconstructed image.

Approach to Modeling Self-encoding Encoder is composed of convolution layers (Conv2D) followed by max pooling layers (MaxPooling2D). These layers are used to extract features from the input image.

$$\text{encoded} = \text{MaxPooling2D}(\text{Conv2D}(\text{input}_{\text{img}})) \tag{5}$$

The decoder uses convolution layers (Conv2D) followed by upsampling layers (UpSampling2D) to reconstruct the image from the extracted features.

$$\text{decoded} = \text{UpSampling2D}(\text{Conv2D}(\text{encoded})) \tag{6}$$

The binary crossentropy loss function is used to measure the difference between the original image and the reconstructed image.

$$\text{loss} = \text{binary}_{\text{crossentropy}}(\text{input}_{\text{img}}, \text{decoded}) \tag{7}$$

It is necessary to calculate the reconstruction error using the mean square difference (MSE) between the original image and the reconstructed image.

$$\text{MSE} = \frac{1}{n}\sum_{i=1}^{n}\left(I^{i}_{\text{original}} - I^{i}_{\text{reconstructed}}\right)^2 \tag{8}$$

After training, prediction and similarity are required, and auto-encoder predicts whether a loaded image contains a mine pattern. The similarity between the original image and the reconstructed image is calculated by comparing the sum of the pixel-by-pixel differences with a predefined threshold.

$$\text{similarity} = \frac{\sum_{i=1}^{n} |\text{difference}[i]|}{\text{img\_height} \times \text{img\_width} \times \text{img\_channels}} \quad (9)$$

5 Assessments and Results

In this section, we present our completed experiments and evaluate the results obtained from the proposed methodology. As previously mentioned, we first trained the CNN model using the Faster R-CNN algorithm with GPR data from four different types of anti-personnel mines in order to detect the desired patterns. Next, we used the Scikit-learn framework to deal with false alarms. The performance of the false alarm detection and handling system was evaluated using the ROC curve and confusion matrix.

5.1 Context

Humanitarian demining requires a very high clearance rate, in line with UN standards, which impose a destruction rate of over 99.6% [44]. Results assessed using the ROC curve recorded in Fig. 2 indicate an accuracy of 97%, with 183 true negatives, 0 false positive, 6 false negatives, and 28 true positives. These results represent the probability of detection and the false alarm rate for different values. The probability of detection measures the percentage of objects correctly detected, while the false alarm rate indicates the probability of detecting a non-existent object. An upward-sloping ROC curve indicates more reliable detection, while a downward-sloping curve explains poor detection. When we increased the number of epochs to 80, the model recorded excellent results, reaching up to 99% detection recorded in Fig. 3, while the false alarm rate decreased in line with the model results, although the runtime remained troublesome. The better the model results, the lower the false alarm rate, and vice versa. Detection at 99% represents the highest rate of all searches found to date. Other results recorded a detection rate between 80 and 91.8% [39, 45]. A particular aspect of our study is also that we examined false alarm rates specific to the types of mine most sensitive to detection equipment, namely the types of mine identified in our field of research, located in northern Chad. Figure 2 shows the overall results predicted by the model, while Fig. 3 bottom represents the detection rate using the ROC curve.

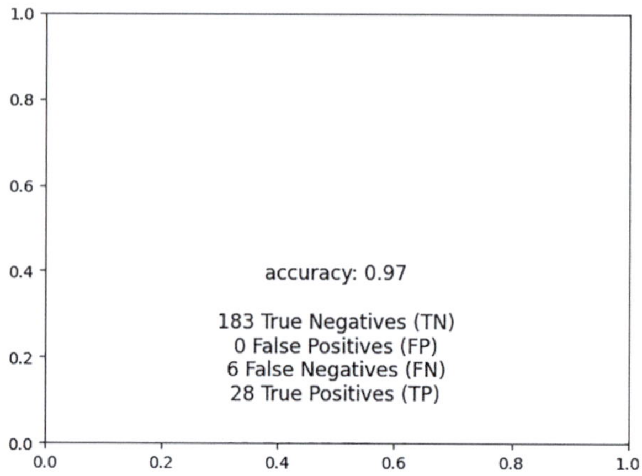

Fig. 2 Prediction result

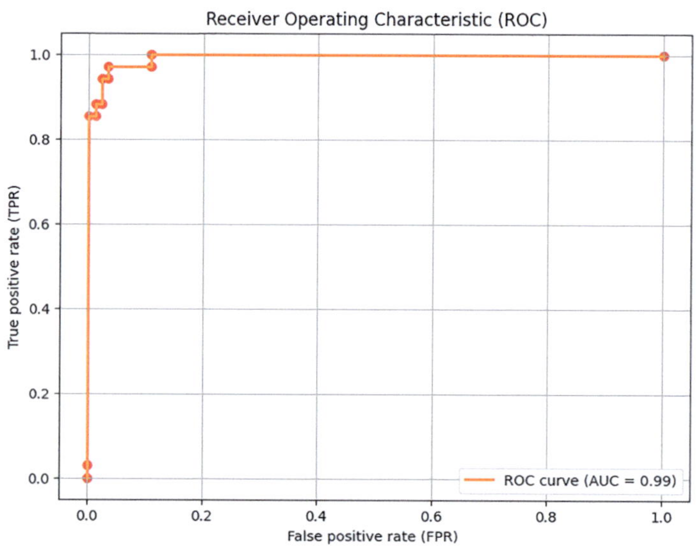

Fig. 3 Detection score

False Alarms Figure 4 shows the different types of mines. Upon examination, we can see that NR442 mines have the highest detection rate but also the second-highest false alarm rate. Next, PRB NR409 mines exhibit the second-highest detection rate and the third-highest false alarm rate. PMA3 mines have the second-lowest detection rate and the highest false alarm rate among the four types. Finally, PPM2 mines have the lowest detection rate and the lowest false alarm rate.

Synthesis Figure 6 presents a detailed analysis of the various types of mine examined. It shows three key indicators: detection accuracy, false alarm rate, and mine sensitivity

Fig. 4 False alarm threshold

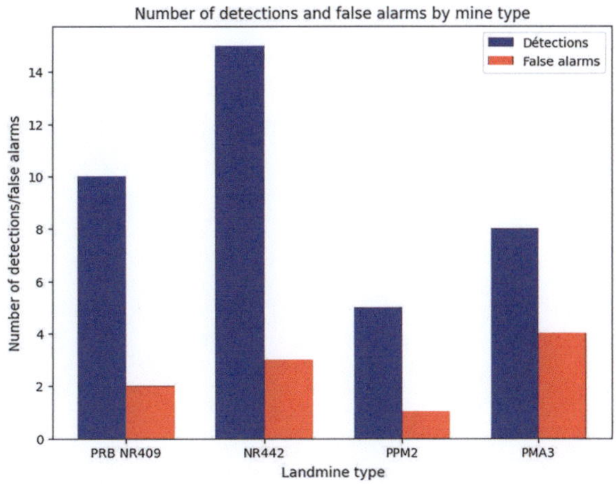

to detection equipment. These elements are essential for assessing the effectiveness and reliability of mine detection systems.

5.2 Confusion Matrix for Target Detection with CNN

The confusion matrix is an essential tool for assessing the performance of a CNN model for detecting buried objects. It visualizes the model's correct and incorrect predictions. For an object presence/absence detection problem, the confusion matrix uses indicators such as true negative (TN), false positive (FP), false negative (FN), and true positive (TP). We describe a detailed explanation of these terms: True negative refers to the case where the model correctly predicts the absence of the object. In our example, there are 183 true negatives, meaning that for 183 images with no object, the model correctly predicted their absence. False positive explains the case where the model wrongly predicts the presence of the object when it is not present. In our example, there is 0 false positive, meaning that the model has correctly predicted their absence. False negative indicates the case where the model wrongly predicts the absence of the object when it is present. In our example, there are 6 false negatives, meaning that for 6 images with objects, the model has incorrectly predicted their absence. Finally, the true positive explains the case where the model correctly predicts the presence of the object. In our example, there are 28 true positives, meaning that for 28 images with objects, the model has correctly predicted their presence. These components enable us to accurately measure the performance of an object detection model. The results of the confusion matrix are shown in Fig. 5.

Let's see how these metrics are used to calculate the performance of an object detection model.

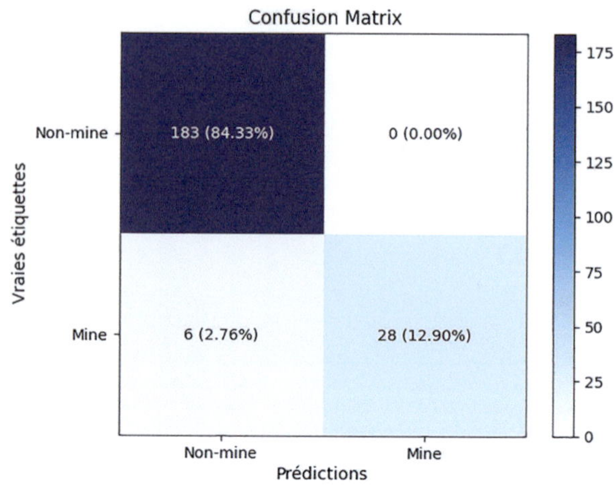

Fig. 5 Confusion matrix

Precision Precision indicates the proportion of positive predictions that are correct. High precision means that the model has few false positives. The precision formula is written as:

$$\text{Precision} = \frac{\text{TP}}{\text{TP} + \text{FP}} \tag{10}$$

Calculation for our case:

$$\text{Precision} = \frac{28}{28 + 0} = \frac{28}{28} \approx 1.0$$

Recall Recall indicates the proportion of true positive objects that were correctly identified by the model. High recall means that the model has few false negatives. The recall formula is written:

$$\text{Recall} = \frac{\text{TP}}{\text{TP} + \text{FN}} \tag{11}$$

Calculation for our case:

$$\text{Recall} = \frac{28}{28 + 6} \approx 0.82$$

The F1-score The F1-score is the harmonic mean of precision and recall. It provides a balanced measure that takes into account both false positives and false negatives. The F1-score formula is written as:

$$\left(\text{F1-score} = 2 \times \frac{\text{Precision} \times \text{Recall}}{\text{Precision} + \text{Recall}}\right) \tag{12}$$

The calculation for our case:

$$\text{F1-score} = 2 \times \frac{1.0 \times 0.82}{1.0 + 0.82} \approx 0.90$$

Accuracy Accuracy measures the total proportion of correct predictions (positive and negative). The accuracy formula is written:

$$\text{Exactitude} = \frac{\text{TP} + \text{TN}}{\text{TP} + \text{TN} + \text{FP} + \text{FN}} \tag{13}$$

Calculation for our case:

$$\text{Exactitude} = \frac{28 + 183}{28 + 183 + 0 + 6} \approx 0.97$$

These metrics show that the model has perfect accuracy (1.0), meaning that it makes no errors in predicting the presence of the object. The overall accuracy of the model is high (around 97%), suggesting good overall performance.

6 Discussion

In this section, we take a close look at our results, analyzing patterns and expected outcomes. Our method combines the use of Faster R-CNN for landmine detection and Scikit-learn for false alarm handling. Measures such as the ROC curve, the area under the AUC curve, and the confusion matrix were used to evaluate our results. Although several techniques have been proposed in the literature to solve landmine detection problems, the management of false alarms generated by the sensitivity of landmine and detection equipment sometimes seems to be neglected. It is conceivable that a demining operator reports the detection of a mine by his equipment, but that this is actually a false alarm generated by various factors such as the type of mine, the ground, or the sensitivity of the equipment. This can cause considerable delays, confusion, and loss of confidence. To tackle this problem of false detection, it is imperative to achieve a much higher level of detection by applying a false alarm management filter, as better filtered detection reduces false alarms. In this case, when the model was initially trained on 65 epochs, the results seemed satisfactory in terms of detection, but when false alarms were taken into account, it appeared necessary to improve these results. By increasing the number of epochs from 75 to 80, the model's performance improved markedly, with a maximum detection rate and a significant reduction in false alarms. By analyzing these results and basing them on the initial images containing the desired patterns, we can consider these results to be satisfactory. However, when training the model without a false alarm filter, we still observe a minor presence of false alarms despite increased detection. In conclusion,

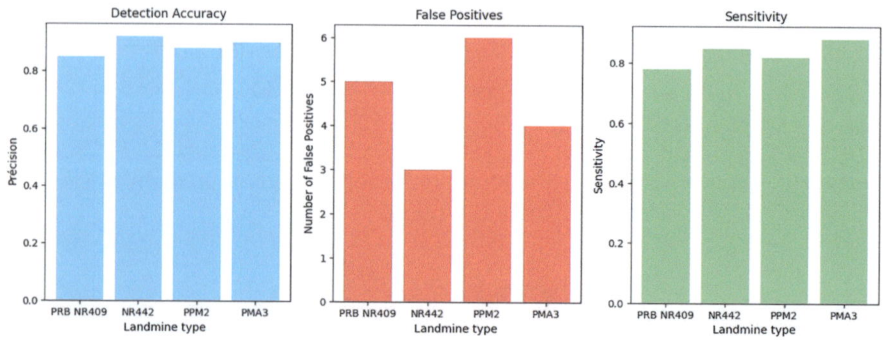

Fig. 6 False alarms

to achieve the desired false alarm-free detection, it is crucial to achieve a detection rate of over 95% using false alarm filters. Without these filters, the presence of false alarms remains ridiculous for every type of mine (Fig. 6).

7 Conclusion

Anti-personnel landmines are dangerous for local inhabitants, animals, and others [46]. Clearing mined land is an important task for the protection of humanitarian and animal welfare. Recent technological advances have brought significant results in support of existing techniques for humanitarian demining. These advances should contribute to detection efficiency and reduce detection anomalies to make the operator's work safer. In this research work, we have proposed a convolutional neural network to make a significant contribution to humanitarian demining. The proposed CNN model used Faster R-CNN for the detection of specific landmines found in the Chad minefield. A second Scikit-learn algorithm was used to reduce the false alarm rates generated by the landmine types examined. The model was trained using ground-penetrating radar data to predict the results. Results are evaluated and interpreted using measurement tools such as the ROC curve, confusion matrix, bar chart for number of detections/false alarms per mine type and precision graphs for false positives and sensitivity for each mine type have been used to interpret and evaluate the results of our work. In our previous work, we have proposed significant measures to assist in humanitarian demining. Specifically, our first paper, titled Utilizing a 5G NSA Network in Infrastructure-Less Areas to Support Communication and Coordination for Mine Clearance Operations, was published in IEEE [47]. Our second paper focuses on the optimization of unmanned aerial vehicles for demining and was also published in IEEE [48]. Finally, our third paper provides a state-of-the-art review of the technologies behind humanitarian demining [49], and it was likewise published in IEEE. All these works were aimed at addressing the challenges associated with humanitarian demining. Our future work will focus on deploying our model in a drone equipped with a scanning camera to remotely detect landmines, in order to bring greater safety, reliability, and speed to humanitarian demining. Our work stands out from the others on two important points: the model obtained a score very close to 1, significantly higher than other works, which proves that the model is excellent. In addition, no other work has evaluated the metrics taking into account different types of mines with specific characteristics, which is a novelty compared to others.

References

1. Bestagini P, Lombardi F, Lualdi M, Picetti F, Tubaro S (2020) Landmine detection using autoencoders on multipolarization GPR volumetric data. IEEE Trans Geosci Remote Sens 59(1):182–195
2. Rafique W, Zheng D, Barras J, Joglekar S, Kosmas P (2019) Predictive analysis of landmine risk. IEEE Access 7:107259–107269
3. Kılınç M, Torunoğlu O, Koçer H (2017) The effect of viewing angle on detection of landmines from thermal time series images using active thermography. Int J Energy Appl Technol 4(1):1–6
4. Gao X, Podd FJW, Daniels DJ, Peyton AJ (2019) Combining electromagnetic spectroscopy and ground-penetrating radar for the detection of anti-personnel landmines. Sensors 19:3390
5. Potin D, Duflos E, Vanheeghe P (2006) Landmines ground-penetrating radar signal enhancement by digital filtering. IEEE Trans Geosci Remote Sens 44(9):2393–2406
6. Zhou H, Feng X, Dong Z, Liu C, Liang W (2022) Multiparameter adaptive target classification using full-polarimetric GPR: a novel approach to landmine detection. IEEE J Sel Top Appl Earth Obs Remote Sens 15:2592–2606
7. Ambruš D, Vasić D, Bilas V (2020) Innovating on top of I&M fundamentals for safer humanitarian demining. IEEE Instrum Meas Mag 23(3):35–41
8. Camacho-Sanchez C, Yie-Pinedo R, Galindo G (2023) Humanitarian demining for the clearance of landmine-affected areas. Socio-Econ Plan Sci 88:101611
9. Wilson JN, Gader P, Lee W-H, Frigui H, Ho KC (2007) A large-scale systematic evaluation of algorithms using ground-penetrating radar for landmine detection and discrimination. IEEE Trans Geosci Remote Sens 45(8):2560–2572
10. Zhou X, Chen H, Li J (2018) An automatic GPR B-scan image interpreting model. IEEE Trans Geosci Remote Sens 56(6):3398–3412
11. Malof JM, Reichman D, Karem A, Frigui H, Ho KC, Wilson JN, Lee W-H, Cummings WJ, Collins LM (2019) A large-scale multi-institutional evaluation of advanced discrimination algorithms for buried threat detection in ground penetrating radar. IEEE Trans Geosci Remote Sens 57(9):6929–6945
12. Bechtel T, Pochanin G, Truskavetsky S, Dimitri M, Ruban V, Orlenko O, Byndych T, Sherstyuk A, Viatkin K, Crawford F et al (2018) Terrain analysis in eastern Ukraine and the design of a robotic platform carrying GPR sensors for landmine detection. In: 2018 17th international conference on ground penetrating radar (GPR). IEEE, pp 1–4
13. Ahmet EFE (2022) A review on risk reduction potentials of artificial intelligence in humanitarian aid sector. J Hum Soc Sci 5(2):184–205
14. Robinson T, O'Keeffe R (2019) The challenges of humanitarian mine clearance in Ukraine. J Conv Weapons Destr 23(1):8
15. Garbino H (2019) Ukraine's newly adopted mine action law: what does this mean for HMA programs? J Conv Weapons Destr 23(1):13–16
16. Capineri L, Pochanin G, Bechtel T, Lande T-S, Capan I, Pastuovic Z, Ohshima T, Coutinho J, Snoj L, Palucci A et al (2020) NATO advanced research workshop on explosives detection. In: Explosives detection: sensors, electronic systems and data processing. Springer, pp 1–32

17. Fardoulis J, Depreytere X, Sauvage E, Gallien P (2019) Drones in the desert: augmenting HMA and socio-economic activities in Chad. J Conv Weapons Destr 23(1):16
18. Dorn AW (2019) Eliminating hidden killers: how can technology help humanitarian demining? Stability Int J Secur Dev 8(1):5
19. Ren S, He K, Girshick RB, Sun J (2015) Faster R-CNN: towards real-time object detection with region proposal networks. In: Cortes C, Lawrence ND, Lee DD, Sugiyama M, Garnett R (eds) NIPS, pp 91–99
20. Sato M, Fang G, Zeng Z (2003) Landmine detection by a broadband GPR system. In: IGARSS 2003. 2003 IEEE international geoscience and remote sensing symposium. Proceedings (IEEE Cat. No. 03CH37477), vol 2. IEEE, pp 758–760
21. Bruschini C, Gros B, Guerne F, Pièce P-Y, Carmona O (1998) Ground penetrating radar and imaging metal detector for antipersonnel mine detection. J Appl Geophys 40(1–3):59–71
22. Robledo L, Carrasco M, Mery D (2009) A survey of land mine detection technology. Int J Remote Sens 30(9):2399–2410
23. Gutierrez S, Pantoja JJ, Ruiz EF, González N, Vega F, Baer C, Sachs J, Kasmi C (2021) Advances on the detection of landmines and IEDs in Colombia using UWB GPR and machine learning techniques. In: 2021 15th European conference on antennas and propagation (EuCAP). IEEE, pp 1–4
24. Brito-da-Costa AM, Martins D, Rodrigues D, Fernandes L, Moura R, Madureira-Carvalho Á (2022) Ground penetrating radar for buried explosive devices detection: a case studies review. Aust J Forensic Sci 54(4):559–578
25. Barnawi A, Budhiraja I, Kumar K, Kumar N, Alzahrani B, Almansour A, Noor A (2022) A comprehensive review on landmine detection using deep learning techniques in 5G environment: open issues and challenges. Neural Comput Appl 34(24):21657–21676
26. Mahmood S, Scharoba S, Schorlemer J, Schulz C, Hübner M, Reichenbach M (2022) Detecting improvised land-mines using deep neural networks on GPR image dataset targeting FPGAs. In: 2022 IEEE Nordic circuits and systems conference (NorCAS). IEEE, pp 1–7
27. Kafedziski V, Pecov S, Tanevski D (2018) Detection and classification of land mines from ground penetrating radar data using faster R-CNN. In: 2018 26th telecommunications forum (TELFOR). IEEE, pp 1–4
28. Lameri S, Lombardi F, Bestagini P, Lualdi M, Tubaro S (2017) Landmine detection from GPR data using convolutional neural networks. In: 2017 25th European signal processing conference (EUSIPCO). IEEE, pp 508–512
29. Pacheco-Londoño LC, Warren E, Galán-Freyle NJ, Villarreal-González R, Aparicio-Bolaño JA, Ospina-Castro ML, Shih W-C, Hernández-Rivera SP (2020) Mid-infrared laser spectroscopy detection and quantification of explosives in soils using multivariate analysis and artificial intelligence. Appl Sci 10(12):4178
30. Kumlu D, Erer I (2019) Clutter removal techniques in ground penetrating radar for landmine detection: a survey. In: Operations research for military organizations. IGI Global, pp 375–399
31. Sato M, Kikuta K, Chemyak I (2018) Dual sensor "ALIS" for humanitarian demining. In: 2018 17th international conference on ground penetrating radar (GPR). IEEE, pp 1–4
32. Bartle CM, Kroeger C, Stephenson W (2010) Utilizing advances in industrial dual beam X-ray scanners to create new capabilities in humanitarian demining and explosive detection. In: Use of nuclear and non-nuclear techniques for humanitarian demining and explosives detection. Proceedings of an IAEA technical meeting
33. Keeley R (2017) Improvised explosive devices (IED): a humanitarian mine action perspective. J Conv Weapons Destr 21(1):3
34. Hrytsaienko MH (2018) Conceptual model of project management for robotic water area clearing from underwater potentially hazardous objects. Shipbuild Mar Infrastruct (1)
35. Held M (2009) Anti-tank mine blast effects. J Battlefield Technol 12(2):1–7
36. Colreavy-Donnelly S, Caraffini F, Kuhn S, Gongora M, Florez-Lozano J, Parra C (2020) Shallow buried improvised explosive device detection via convolutional neural networks. Integr Comput-Aided Eng 27(4):403–416

37. Moalla M, Frigui H, Karem A, Bouzid A (2020) Application of convolutional and recurrent neural networks for buried threat detection using ground penetrating radar data. IEEE Trans Geosci Remote Sens 58(10):7022–7034
38. Mochurad L, Savchyn V, Kravchenko O (2023) Recognition of explosive devices based on the detectors signal using machine learning methods. In: IntelITSIS, pp 249–260
39. Baur J, Steinberg G, Nikulin A, Chiu K, de Smet T (2021) How to implement drones and machine learning to reduce time, costs, and dangers associated with landmine detection. J Conv Weapons Destr 25(1):29
40. Baur J, Steinberg G, Nikulin A, Chiu K, de Smet TS (2020) Applying deep learning to automate UAV-based detection of scatterable landmines. Remote Sens 12(5):859
41. Pryshchenko OA, Plakhtii V, Dumin OM, Pochanin GP, Ruban VP, Capineri L, Crawford F (2022) Implementation of an artificial intelligence approach to GPR systems for landmine detection. Remote Sens 14(17):4421
42. Schorlemer J, Altholz J, Barowski J, Baer C, Rolfes I, Schulz C (2024) A radar echo simulator for the synthesis of randomized training data sets in the context of AI-based applications. Sensors 24(3):836
43. Fernández MG, López YÁ, Arboleya AA, Valdés BG, Vaqueiro YR, Andrés FL-H, García AP (2018) Synthetic aperture radar imaging system for landmine detection using a ground penetrating radar on board a unmanned aerial vehicle. IEEE Access 6:45100–45112
44. Bruschini C, Gros B (1998) A survey of research on sensor technology for landmine detection. J Conv Weapons Destr 2(1)
45. Castiblanco C, Rodriguez J, Mondragon I, Parra C, Colorado J (2014) Air drones for explosive landmines detection. In: ROBOT2013: first Iberian robotics conference: advances in robotics, vol 2. Springer, pp 107–114
46. Bystritskii VM (2002) Landmine problem and multisensor detection. Springer, Netherlands
47. Hassan MI, Ahmat D, Ouya S, Salim MC, Hassan MB (2023) Latest generation technologies for network coverage in infrastructure-less areas. In: 2023 international symposium on networks, computers and communications (ISNCC). IEEE, pp 1–6
48. Hassan MI, Ahmat D, Ouya S, Hassan MB, Salim MC, Haggar BS (2024) Assessment of software-defined UAVs operating in remote areas. In: 2024 international conference on artificial intelligence, computer, data sciences and applications (ACDSA). IEEE, pp 1–6
49. Hassan MI, Ahmat D, Ouya S (2024) Technologies behind the humanitarian demining: a review. In: 2024 international conference on smart applications, communications and networking (SmartNets). IEEE, pp 1–7

An ML-Based Decision Support System for the Health Risk Analysis Encompassing Insurance Premium

Prasanta Baruah, Amarjeet Kumar Singh, Bhagyashree Nath, Nerswn Baglary, and Pankaj Pratap Singh

Abstract The insurance business plays a significant role in any country's long-term economic success in terms of health sector. With a growth in the number of insurance purchasers, having a sophisticated claim analysis system has become an important need for an insurance firm. It uses claim analysis to identify and distinguish the legitimate and fraudulent claims. Many insurance companies are implementing appropriate Machine Learning (ML) models for the decision support system (DSS) in the claim analysis. To create an effective model, the company's interaction and their shared data is indeed. Nevertheless, the data maintained by each company is crucial since it contains sensitive consumer information, and sharing this confidential information can be dangerous. The data stored by each company is very critical as it consists of customer's private information which is very risky to share with others. On the basis of various attributes, such as age, sex, body mass index, number of children, smoking habits, some ML models are trained and evaluated for providing better DSS.

Keywords Insurance prediction system · Machine learning · Decision support · Hybrid model

1 Introduction

Insurance companies deal with life insurance customers against the promise of unfortunate and tragic events pay them a fixed amount of money. The amount of the contribution that the number of that a customer must pay is determined by the company based on the assessment of the customer's request and the allocation to each customer specific risk level. This application contains data about people, such as medical history, work history, gender. When companies underestimate the extent of the risk, then would charge a lower amount for a large insured, which could also lead to an

P. Baruah · A. K. Singh · B. Nath · N. Baglary · P. P. Singh (✉)
Department of Computer Science and Engineering, Central Institute of Technology Kokrajhar, Kokrajhar, Assam, India
e-mail: Pankajp.singh@cit.ac.in

H. Sharma et al. (eds.), *Artificial Intelligence: Theory and Applications*, Lecture Notes in Networks and Systems 5589, https://doi.org/10.1007/978-981-96-1687-9_53

increase in the number of applications submitted. On the other hand, if the company overestimates the level of risk and demands a larger amount of money, then you risk losing customers of a competing company. Therefore, an accurate risk predictor is necessary to ensure the good performance of the company. We live in a world full of dangers and uncertainties. People, houses, companies, buildings, and properties are exposed to various risks, including: threats may vary. These risks include the risk of death, illness, and loss of assets or owned. However, such kind of risks cannot always be avoided and the financial sector has developed a number of products that protect individuals and organizations from misuse of financial resources which can compensate them. In fact, insurance is a policy which moderates or eliminates costs issues with different risks. A policy that protects medical expenses is called health insurance. Anyone who has taken out health insurance enjoys protection after paying a certain premium. The cost of health insurance depends on many factors. The amount of health insurance premiums varies from person to person. There are factors that affect the cost of health insurance [1]. Consider, if the age is in young category that category based person's number is much less likely to develop serious health problems than an older person. Suppose, the treatment is caring for an older person whose cost is more than a young person. Accordingly, an older person has to pay a higher premium than a younger person. Since such kind of factors that affect the health insurance premium and also this number varies from person to person. In healthcare, machine learning can take on many medical tasks at a much faster pace to effectively predict or diagnose illnesses/injuries and provide the best medical treatment to the patient. Machine Learning can collect and process data suggests the corresponding result to the user. This shortens the time needed to detect illnesses and errors and enables a radical cycle of diagnosis, treatment, and recovery.

2 Related Work

The insurance industry has traditionally relied on actuarial models and historical data to make decisions. However, the increasing complexity of risk assessment and the large amount of data available have paved the way for the integration of machine learning (ML) techniques. This study examines the potential of machine learning to improve decision support in various aspects of the insurance industry. ML algorithms can learn complex relationships between different data points, including medical history, demographics, and behavioral factors, resulting in more accurate premium estimates. ML algorithms for predicting health insurance premiums, including Decision Tree (DT), linear regression (LR), K-nearest neighbors (K-NN), Naive Baye's (NB), Random Forests (RF), and Support Vector Machines (SVM) (see Table 1). However, challenges remain regarding data quality, model interpretability, and ethical considerations related to fair pricing and potential bias.

Table 1 Comparison of several researchers work on predicting health premiums

Authors and year	Objectives	Methodology	Results
Junedi and Mauritsius (2020) [2]	To predict the risk level of life insurance applicants using machine learning	Three ML classifiers as SVM, Naive Baye's & Random Forest are used	Precision using RF, SVM using a linear kernel and NB are 0.85, 0.72 and 0.49, respectively
Chang et al. (2022) [3]	To develop an automated diabetes diagnosis system using machine learning models	It discusses SMOTE and ADASYN for dealing with imbalanced datasets & used DT, RF, K-NN, LR, and NB	Five classifiers (DT, K-NN, LR, NB and RF) for predicting diabetes and RF achieved highest accuracy
Kaushik et al. (2022) [4]	To predict health insurance premiums by using the concept of AI and ML in healthcare	The training and evaluation of a LR model and an ANN based regression model is done to predict health related premiums	To predict health insurance premium using ANN based regression modelwith accuracy of 92.72%
Singh et al. (2017) [5]	To classify and predict the different level of diabetes in the patients	Data mining approach	Comparison with association rule mining based approach
Baruah et al. (2023) [6]	To enhance risk assessment of the applicants of life insurance industries using predictive analytics	Geographical Information Systems (GIS) and ML approaches	ML approaches are applied on two types of datasets to predict the risks of the insurance applicants
Baruah and Singh (2023) [7]	Review on role of Risk prediction for customers in insurance companies	ML is used for Risk classification that is based on the grouping of customers based on their risk levels	Evaluation of risk and calculation of premium are the key task of underwriters

3 Methodology

3.1 Dataset Description

The dataset used is taken from Kaggle whose size is 1338 instances [8]. The features are age, BMI, children, Glucose, sudden weight loss, region, smoker, sex, and insurance charges, respectively.

3.2 *Feature Engineering (FE)*

Feature engineering (FE) is one of the utmost basic and essential elements of the data preparation which is next proceeding for the model training. In this phase, new features are formed based on an EDA performed on all existing data and knowledge in the area to improve model performance. It is one of the key parts of data preparation which consumed 80% of time. New functions are created based on various intermediate calculations existing functions. FE can be performed by statisticians in manual manner or using feature coding techniques on a case-by case basis. FE is not only be useful for linear regression or text classification problems, but it has proven to be extremely useful for SVM, RF, NN, etc. Coding is important because ML itself relies on algorithms and mathematical models. Coding is done using two methods such as nominal and ordinal encoding. Nominal encoding is used when the data order is not large. Besides ordinal coding, there is normalization used as feature engineering techniques. It is used to scale everything values in the data set in a fixed range from 0 to 1. Formula used is for standardization.

$$X\text{norm} = X - X\text{ min}/X\text{ max} - X\text{ min} \tag{1}$$

Feature Selection: It is done with the correlation analysis, feature importance from tree models, or domain knowledge techniques to select the most appropriate features to train your models.

Algorithm Used

(1) 1st Algorithm: Decision Tree

Inputs:

D: Training dataset

X: Features in the dataset

y: Target variable (insurance charges)

max_depth: Depth of the tree upto the maximum level

min_samples_split: Node splitting with minimum number of required samples

Steps 1: Initialization: Define the root node of the tree.
Steps 2: Splitting Criteria Selection: Calculate the impurity measure for the current node (e.g., Gini impurity or entropy) using the formula:

$$\text{Gini impurity : Gini(D)} = 1 - \sum_{i=1}^{c} \text{p(i|D)}^2 \tag{2}$$

where c is the number of classes and p(i|D) is the probability of class I in the node D

Entropy: Evaluate the split of each feature to find the feature and split point that minimizes contamination or maximizes information gain.

$$\text{Entropy(D)} = \sum_{i=1}^{c} - \text{p(i|D)log2(p(i|D))} \tag{3}$$

Step 3: Node Splitting: Choose the best feature and release point based on your sharing criteria. Create sub nodes based on this split and split the data set accordingly. Stopping Criteria: Check stopping conditions such as the maximum depth reached or the minimum samples required to split a node. If the stopping criterion is met, mark the node as a leaf node and assign the most frequent class (for classification) or average value (for regression) to the prediction.
Step 4: Recursion: Apply steps 2–4 recursively to each child node until the stopping criteria are met for all nodes. 5. Building the Tree: Build a Decision Tree by iteratively splitting nodes until the stopping conditions for each node are met. 6. Prediction: For a new instance of feature values, traverse the tree and follow the splits by feature values until you reach a leaf node. Assign the prediction (class label or value) of the leaf node to the new instance

(2) 2nd Algorithm: Gradient Boosting

Inputs:

D: Training dataset

X: Features in the dataset

Y: Target variable (insurance sector)

number_of_ectimators: Number of boosting stages (weak learners)

max_depth: Maximum depth of each tree corrects the errors of the previous trees

Steps:

- Initialization: Set the initial prediction F0(x) as the mean value of y (or a constant value for classification)
- For t = 1 to number_of_estimators: Calculate Pseudo Residuals: Compute the negative gradient (pseudo residual) of the loss function with respect to the current prediction

$$\text{ri(t)} = -\partial\text{Ft} - 1\text{(xi)}\partial\text{L(yi, Ft} - 1\text{(xi))} \tag{4}$$

Fit a Weak Learner (DT) to Pseudo Residuals: Train a decision tree ht(x) on (X,r(t) with maximum depth max_depth Update Prediction : Update the model's prediction Ft(x) using a learning_rate

$$\text{F}_{\text{t(x)}} = \text{F}_{t-1\text{(x)}} + \text{learning}_{\text{rate}} \times \text{h}_{\text{t(x)}} \tag{5}$$

- Return Ensemble Model: Returns an ensemble model consisting of the sum of weak students: F(x) = \sum_{t = 1}^{number_of_estimators} \learning_rate \times h_t(x) Function Predict(model, instance): For each weak learner ht(x) in the model: Predict using the weak learner:

$$F_{t(x)} = F_{t - 1(x)} + learning_{rate} \times h_{t(x)} \tag{6}$$

Return Final Prediction: Return the final prediction F(x).

(3) 3rd Algorithm: Hybrid Model

Inputs:

D: Training dataset

X: Feature in the dataset

y: Target variable (insurance charges)

base_models: list of base models (e.g., SVM, Decision Tree).

meta_learner: Ratio to split the training dataset for base models and meta_learner training

Procedure:

Function special_Ensemble D, X, y, base_model, meta_learner, split_ratio:

- Split Training Dataset: Split the dataset D into two parts: D1 and D2, based on the split ratio, typically, D1 is smaller than D2.
- Train Base Models on D1: For each base model in base_models: Train the base models using D1 (features X1, target y1) Internal details: Base models (e.g., SVM, Decision Tree) are trained using their specific learning algorithms on D1 to capture different patterns in the data.
- Predictions from Base Models on D2: For each base model: Generate predictions Fbase(x) using the trained base models on D2. Internal Details: Predictions Fbase(x) represent the model's output for each instance in D2 based on learned patterns from D1.
- Create Special Dataset for Meta-Learner: Combine the predictions Fbase(x) from all base models as new features for dataset D2. Let this new dataset be Dspecial. Internal Details: Dspecial combines the original features X2 from D2 with predictions from base models, creating a new feature set for each instance.
- Train Meta-Learner on Special Dataset: Train the meta-learner model meta_learner using Dspecial as features and D2 as the target variable. Internal Details: The meta-learner (e.g., Linear Regression, Decision Tree) learns to combine predictions from base models or learn from their outputs to make more accurate predictions on D2.
- Combine Base Models and Meta-Learner for Prediction: For each instance in the testing dataset or new data: Generate predictions Fbase(x) using the trained

base models. Combine these predictions as features for the meta-learner. Use the meta-learner to produce the final prediction. Internal Details: The base models' predictions are combined and fed into the trained meta-learner, which uses the special information to generate a final prediction for each new instance.

4 Results and Analysis

Exploratory data analysis (EDA) is to understand the distribution, correlations, and patterns in the data. EDA is a method for investigating and analyzes data using various visual techniques. It makes possible for the data analyst to detect anomalies, discover trends, test hypotheses, and get a better overview of your activities. In this study, the correlation matrix is first calculated for understanding relationships between each pair of variables. If it's strong attribute relationships, attribute removal must be done. Through various types of data visualization techniques, attribute distributions, and data balance are analyzed for better understanding of the data set and its properties.

4.1 Analysis

These four characteristics, when analyzed together, can provide valuable information about a person's health and risk of developing certain diseases.

1. Body Mass Index (BMI): It is evaluated by dividing body weight in kgs by the square of height in meters. The range for normal weight and overweight are 18.5–24.9, 25–29.9, respectively, and 30 or more for the obese.

$$\mathrm{BMI} = \frac{\mathrm{weight}(kg)}{\mathrm{height}(m)^2} \tag{7}$$

2. Sudden weight loss: Unintentional weight loss of more than 5% in 6 months is considered sudden. This can be a symptom of various underlying diseases, including: Thyroid problems Cancer Celiac disease depression eating disorders Malnutrition
3. Smokers: Smoking is a major risk factor for many health problems.
4. Glucose Level: A simple sugar found in your blood, the main source of energy for your cells.

This information is extracted from the dataset as mention in the Sect. 3.1. The parameters BMI, smoker, sudden weight loss, and insurance charges are used for the following analysis graphs. In Fig. 1, the number of non-smoking patients is more than the smoking patients, and it is also observed that Male smokers are more than the Female smokers. A strong co-relation is observed with the fact of smoking. The patient who smokes pays more charges than non-smokers (see Fig. 2).

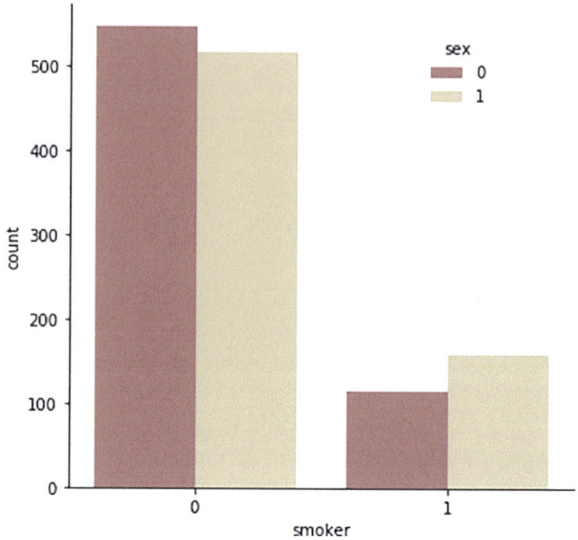

Fig. 1 Non-smoking (0) and smoking (1) patients in male (0) and female (1)

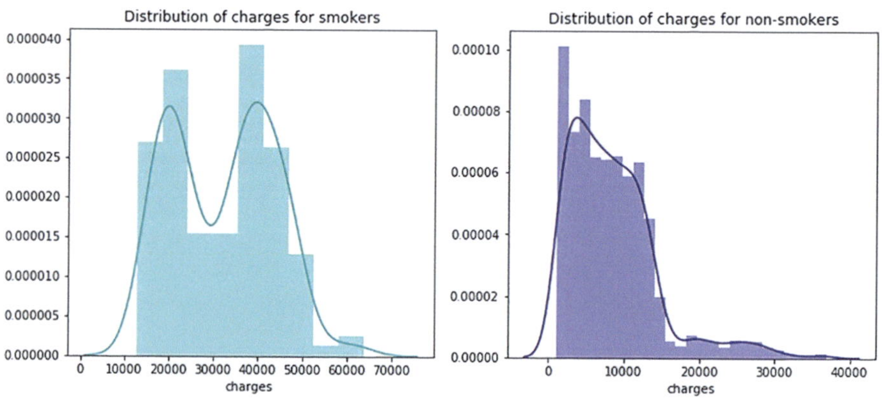

Fig. 2 Distribution of charges of non-smokers and smokers patients

The average BMI of patient is 30. It is observed that patient with BMI above 30 spends more on treatment as shown in the Fig. 3 and vice versa (see Figs. 3 and 4). The minimum children patient have is 0 and maximum 5 (see Fig. 5).

In Fig. 6, among each age group, insurance charges are significantly lower in non-smokers than smokers. Higher claim amount is correlated with high age and the most expensive claims in the 60 and over age group. Using the average feature importance from the Gradient Boosting models, the most crucial predictor of healthcare costs is the combined impact of a patient's BMI and their smoking status (see Fig. 7). The models with interactions between age and BMI reveal that smoking status yields the most significant average importance, ranking among the predictors. The age and

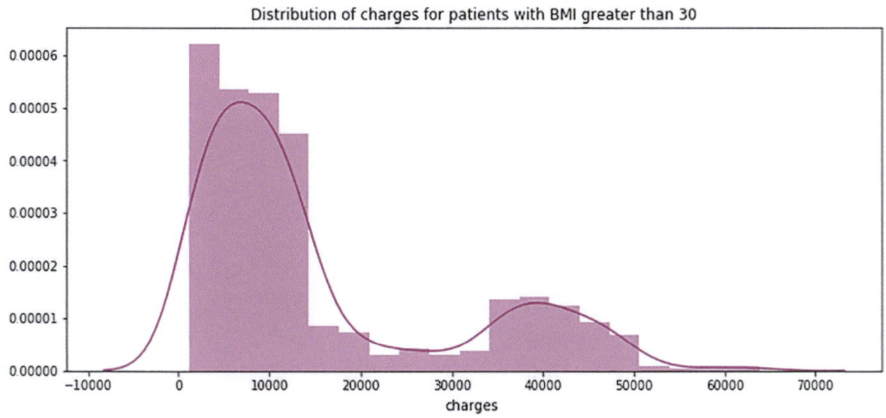

Fig. 3 Distribution of charges for patients with BMI > 30

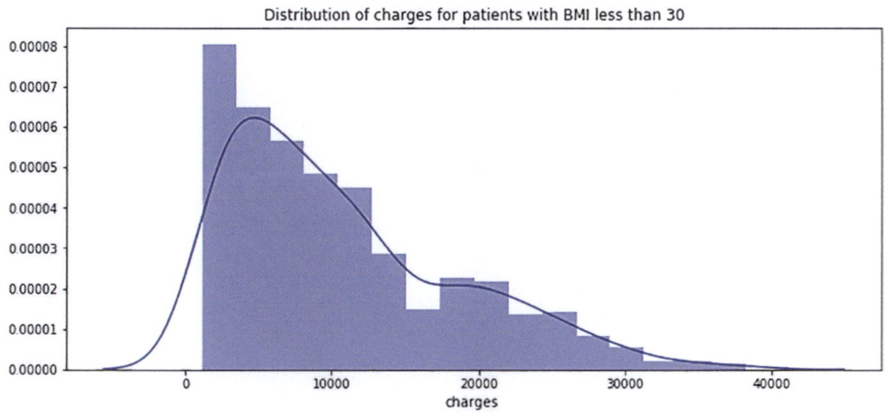

Fig. 4 Distribution of charges for patients with BMI < 30

BMI polynomial features rank among the highest, whereas gender and children are positioned among the lowest.

In the analysis, performance of model is based on the quantitative parameters such as accuracy, $F1$ score, precision, and recall. The accuracy measure illustrates the model's capability to correctly identify relevant cases. Meanwhile, recall refers to the model's ability to categorize all relevant cases. $F1$ score provides the harmonic mean of precision and recall which also provides an overall assessment of the model. In addition, the accuracy measure shows the percentage of correctly predicted events out of the total number of events.

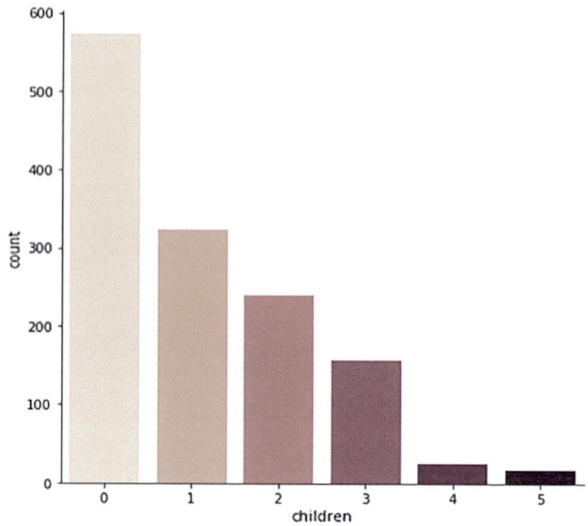

Fig. 5 Minimum number of child patients

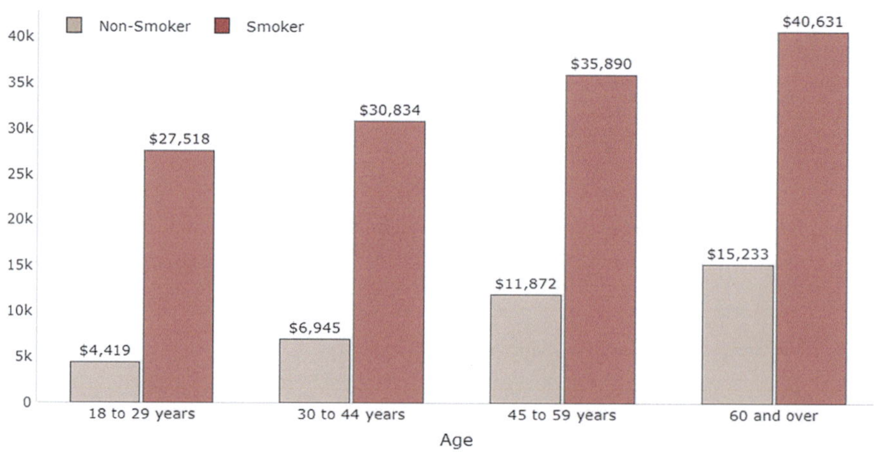

Fig. 6 Distribution of charges of Non-smokers and Smokers patients in different age groups

4.2 Discussion

ML models are used to train the model for predicting the claims/insurance status. Some feature selection techniques are used to decrease the dimensionality in the datasets for training the models for producing better accuracy. Hybrid model is the best model based on the prediction accuracy (see Table 2). Testing is a filtering method which picks features solely based on their intercorrelations which is based on their frequency distribution during the training process. The effectiveness of a

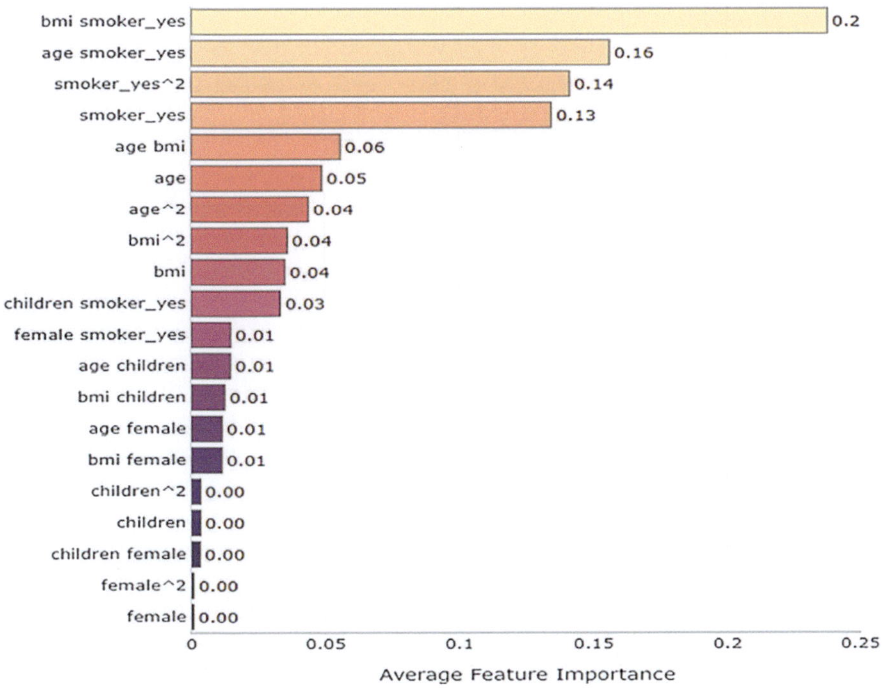

Fig. 7 Average feature importance from gradient boosting model

feature selection technique also depends on the dataset and then models used these selected features for training and evaluation purpose. The Gradient boosting classifier is used in the analysis with tree-based embedding method based on the importance of feature. The model selection is used to fit the data set which is not a big issue, but the big difference in the performance of hybrid model over the singe model approach.

Table 2 Comparative analysis of different methods

Metrics used	Hybrid model	Decision tree	Grad boost algo
Precision	0.97	0.55	0.80
Recall	0.98	0.70	0.75
*F*1 Score	0.98	0.60	0.78
Accuracy	0.98	0.78	0.82

5 Conclusion

In health insurance, machine learning can be used to analyze and evaluate large amounts of data. Notable trends showed a strong association between smoking, higher body mass index, and higher healthcare spending, explaining their significant impact on insurance payments. It should be noted that age appears to be a key factor, with claim amounts showing a progressive increase with age, particularly in the older cohort. In addition, model evaluation and performance metrics, precision, recall, $F1$ score, and accuracy demonstrated the predictive performance of various machine learning models, highlighting the superiority of the hybrid ensemble in predicting insurance payments which achieves the accuracy of 98% and also recall value of 0.98 signifies the accurate prediction. By combining different models with new features, it highlighted the model's ability to capture detailed information from different perspectives and outperform traditional stacking methods. An in-depth analysis highlighted the combined influence of BMI and smoking having the highest feature importance, i.e., 0.2 as key predictors of healthcare costs and explained the key role these factors play in shaping insurance payments. This complete analysis is not only expand the understanding of the complex relationships between patient characteristics and insurance payments, but also highlights the predictive potential of advanced machine learning techniques to optimize insurance premium forecasts, providing valuable insights for both the insurance industry and healthcare stakeholders.

References

1. https://www.projectpro.io/article/machine-learning-in-insurance/774
2. Junedi HB, Mauritsius T (2020) Risk level prediction of life insurance applicant using machine learning. Int J Adv Trends Comput Sci Eng 9(2):2213–2220
3. Chang V, Ganatra MA, Hall K, Golightly L, Xu QA (2022) An assessment of machine learning models and algorithms for early prediction and diagnosis of diabetes using health indicators. Healthc Anal 2:1–14
4. Kaushik K, Bhardwaj A, Dwivedi AD, Singh R (2022) Machine learning—based regression framework to predict health insurance premiums. Int J Environ Res Public Health 19(13):1–15
5. Singh PP, Das B, Poddar U, Choudhury DR, Prasad S (2017) Classification of diabetic's patient data using machine learning techniques. In: Perez G, Tiwari S, Trivedi M, Mishra K (eds) Ambient communications and computer systems, AISC, vol 696. Springer Singapore
6. Baruah P, Singh PP, Ojah SK (2023) A novel framework for risk prediction in the health insurance sector using GIS and machine learning. Int J Adv Comput Sci Appl 14(12):469–476
7. Baruah P, Singh PP (2023) Risk prediction in life insurance industry using machine learning techniques—a review. In: Mishra A, Gupta D, Chetty G (eds) Advances in IoT and security with computational intelligence (ICAISA). Lecture notes in networks and systems, vol 755. Springer, Singapore
8. [online] Dataset: https://www.kaggle.com/datasets/uciml/pima-indians-diabetes-database

MIX
Papier aus verantwortungsvollen Quellen
Paper from responsible sources
FSC® C105338

If you have any concerns about our products,
you can contact us on
ProductSafety@springernature.com

In case Publisher is established outside the EU,
the EU authorized representative is:
Springer Nature Customer Service Center GmbH
Europaplatz 3, 69115 Heidelberg, Germany

Printed by Libri Plureos GmbH
in Hamburg, Germany